Dermatophytes and Dermatophytoses

Jean-Philippe Bouchara • Pietro Nenoff •
Aditya K. Gupta • Vishnu Chaturvedi
Editors

Dermatophytes and Dermatophytoses

Springer

Editors
Jean-Philippe Bouchara
Univ Angers, Université de Bretagne
Occidentale, CHU, SFR ICAT
Groupe d'Etude des Interactions Hôte-
Pathogène (GEIHP, EA 3142)
Angers, France

Pietro Nenoff
Laboratory for Medical Microbiology
Rötha, Sachsen, Germany

Aditya K. Gupta
Mediprobe Research Inc.
London, ON, Canada

Vishnu Chaturvedi
New York State Department of Health
Wadsworth Center
Mycology Laboratory
Albany, NY, USA

ISBN 978-3-030-67423-6 ISBN 978-3-030-67421-2 (eBook)
https://doi.org/10.1007/978-3-030-67421-2

This Springer imprint is published by the registered company Springer Nature Switzerland AG.
The registered company address is: Gewerbestrasse 11, 6330 Cham, Switzerland

Preface

The study of dermatophytes and dermatophytoses is continually evolving. This edited volume provides an up-to-date snapshot of the basic and translational research and contemporary approaches in use in the dermatology clinics and diagnostic laboratories. Both clinical researchers and physicians will find valuable information about the molecular, genetic, and epidemiological aspects. The diagnosis and treatment of dermatophytoses are discussed in detail, which is increasingly challenging due to resistance to current antifungal drugs. The spectrum of pathogens causing dermatophytoses has expanded in recent years. The reasons for this include the direct migration from Africa and Asia to Europe and North America. In addition, the increased travel and trade due to globalization have brought old and new dermatophytes to new geographic locations. Molecular methods for direct detection of fungi or fungal DNA are increasingly part of diagnosing dermatophytoses alongside conventional techniques. Similarly, many dermatophyte species require exact identification by DNA sequencing or mass spectrometry.

The future of dermatophytosis treatment will probably involve innovative treatment regimens and novel pharmaceuticals to fight the acquired mutations in fungal pathogens. The increase in mixed infections and geographic distribution also poses barriers to successful treatment, demonstrating the necessity for more effective diagnostic methods. Similar concerns exist about the increase in the incidence of chronic recurrent and treatment-recalcitrant dermatophytoses. As a result of the "globalization of people and fungi," the first isolates of terbinafine-resistant dermatophyte are now known from India, the Middle East, Iran, Germany, Finland, and the USA. In vitro resistance testing of dermatophytes, including mutation analysis of squalene epoxidase, has experienced an unexpected boom in the last few years. It is likely to become a mainstay of clinical management of dermatophytoses.

The Part I "Current Taxonomy of Dermatophytes" is a formal introduction to fungal pathogens classified as "dermatophytes." The focus is on a current understanding of their evolution with human and animal hosts, changing epidemiology, recommended identification methods, and a list of valid species in genera *Trichophyton*, *Epidermophyton*, *Nannizzia*, *Microsporum*, *Paraphyton*, *Lophophyton*, and *Arthroderma*.

The Part II "Clinical Aspects" includes five contributions. "Dermatophytes and Dermatophytic Infections Worldwide" is a sweeping overview of dermatophytes and dermatophytic infections worldwide. Many maps are provided to highlight geographic distribution and to postulate future trends. "Onychomycosis in the Twenty-First Century: An Update on Epidemiology and Diagnosis" focuses on the global prevalence of fungal nail infections (onychomycosis), epidemiological factors, and contemporary laboratory diagnosis methods. "Invasive Dermatophytoses: Clinical Presentations, Diagnosis, and Treatment" introduces invasive or deep dermatophytoses, relatively rare but serious conditions, which afflict humans with underlying immune dysfunctions. An excellent overview is presented about the combined medical and surgical management of deep dermatophytoses. "Unusual Dermatophytosis Presentations and New Emerging Dermatophytes Species" describes newly emerging species of dermatophytes and unusual presentations of dermatophytosis. Chapter "Non-dermatophyte Dermatoses Mimicking Dermatophytoses in Animals" includes dermatopytoses in animals, dermatoses that mimic dermatophytoses, and the fungal species involved. The scope is inclusive of pets, domestic animals, and wildlife with an enhanced focus on differential diagnosis.

The Part III focuses on the pathogenesis of dermatophytosis. The three contributions, "Experimental Models of Dermatophytosis", "Transcriptome of Host–Dermatophyte Interactions Using Infection Models" and "Genetic Predictors of Susceptibility to Dermatophytosis" highlight contemporary experimental approaches to study pathogenic potential, host responses, and antifungal efficacy. "Experimental Models of Dermatophytosis" provides a comprehensive overview of in vitro, in vivo, and ex vivo models to study dermatophytoses and evaluation of new antifungal agents. The complexity of infectious propagules, pathogen recognition, immune responses, histopathology, and antifungal effects are explained with exquisite illustrations. "Transcriptome of Host–Dermatophyte Interactions Using Infection Models" introduces experimental approaches to understand host-pathogen interplay in dermatophytoses. Novel gene expression studies in experimental animal models are applied to gain insights into pathogenic factors and host defense mechanisms in dermatophytoses. "Genetic Predictors of Susceptibility to Dermatophytosis" covers genetic factors that predispose humans to infections with dermatophytes. A critical analysis of existing data is presented to summarize putative underlying host mechanisms that determine susceptibility to dermatophytosis.

The Part IV highlights epidemiology of dermatophytes and dermatophytosis. "The *Trichophyton rubrum* Complex" focuses on the *Trichophyton rubrum* complex, which includes agents of nail, skin, and hair infections. An elaborate polyphasic scheme with morphological, physiological, and DNA methods is highlighted as "gold standard" for *T. rubrum* complex. "*Microsporum ferrugineum*: The Renaissance of a Forgotten Dermatophyte" details unusual findings of *Microsporum ferrugineum* in Germany, an anthropophilic dermatophyte not seen in the country for nearly 50 years. Clinical workup, enhanced laboratory diagnostic tools, and treatment approaches are detailed. "*Trichophyton mentagrophytes* ITS Genotype VII from Thailand" highlights emerging European cases of *T. mentagrophytes* ITS-genotype VII, which causes inflammatory and abscessing dermatophytoses. The unusual geographic emergence is linked with human-to-human transmission. "Trends in Epidemiology of Dermatophytes in Iran" is a comprehensive review of epidemiology of dermatophytes in Iran in the last 70 years. It details changes in species spectrum, antifungal treatment, emergence of resistance, and improved laboratory diagnosis. "Onychomycosis in Adults: A Clinical Perspective from Mexico" summarizes onychomycosis with emphasis on South America and Latin America. "*Tinea capitis* in School Children: Current Status" is about *Tinea capitis* with emphasis on epidemiology, clinical presentations and complications, treatment, and community management.

Part V "Laboratory and Molecular Diagnosis of Dermatophytosis" includes three contributions. "The Potential of Molecular Diagnostics in Routine Dermatology" widens the traditional mycological examination of clinical specimens to molecular diagnostics. In-house PCR and hybridization tests are compared to commercial kits, and challenges and strategies for their implementation. "Laboratory Diagnosis of Dermatophytosis" summarizes the traditional mycological examination of clinical specimens to confirm dermatophytoses. "MALDI-TOF-Based Identification of Dermatophytes" focuses on the application of mass spectrometry for the identification of dermatophytes. It presents a useful comparison of the commercial systems, their databases, and relative performances.

Parts VI and VII describe antifungal resistance of dermatophytes and treatment of dermatophytoses. "Antifungal Susceptibility Testing of Dermatophytes" deals with antifungal susceptibility testing of dermatophytes. The description includes broth microdilution and other testing formats, endpoint reading, clinical breakpoints, and the diagnosis of terbinafine resistance. "Terbinafine and Itraconazole Resistance in Dermatophytes" widens the perspective on antifungal susceptibility testing with a focus on terbinafine and itraconazole resistance. "New Antifungal Agents and New Formulations Against Dermatophytes" introduces new antifungal drugs and new formulations of existing drugs for the treatment of onychomycosis. The value of clinical management, combined with patient education, is emphasized. "Are Natural Products an Alternative Therapy for Dermatophytosis?" involves the potential of natural products used in traditional methods for dermatophytoses treatment. As scientific evidence accumulates for their efficacy, it is expected that these plant-based products and their derivatives would be valuable sources of new drugs to treat dermatophytoses.

Angers, France — Jean-Philippe Bouchara
Rötha, Sachsen, Germany — Pietro Nenoff
London, ON, Canada — Aditya K. Gupta
Albany, NY, USA — Vishnu Chaturvedi

Contents

Part I

Current Taxonomy of Dermatophytes

Introduction to Dermatophytes

Sybren de Hoog, Chao Tang, Xue Kong, Hazal Kandemir, and Yanping Jiang

Abstract

Dermatophytes (family *Arthrodermataceae*) are a group of fungi in the order *Onygenales* which have evolved relatively recently, almost exclusively with mammal hosts. The ecological subdivision in geophilic, zoophilic, and anthropophilic species broadly reflects their course of evolution. *Homo sapiens* is phylogenetically the most recent host. Evolution in anthropophilic dermatophytes by definition has proceeded for less than a million years and is still ongoing. Species have not yet diverged much in diagnostic markers and are, therefore, difficult to distinguish. Dermatophytes are classically identified by a combination of clinical features, colony characteristics, and microscopy. While this is effective in dermatological routine with large numbers of strains with limited diversity from a single host, *Homo sapiens*, this approach is inadequate to grasp is the large diversity of *Arthrodermataceae* from other sources. Recent precise multilocus studies combined with physiological and ecological data have been able to resolve some of the groups that were distinguished in classical dermatomycology but were problematic with the molecular approach. With the recent emergence of dermatophyte antifungal drug resistance and increased virulence, it is recommended to perform molecular identification and antifungal susceptibility testing routinely in cases of chronic, recurrent, and/or recalcitrant and atypical presentations.

S. de Hoog (✉) · C. Tang · X. Kong · H. Kandemir
Center of Expertise in Mycology of Radboud University Medical Center, Canisius Wilhelmina Hospital, Nijmegen, The Netherlands
e-mail: Sybren.deHoog@radboudumc.nl

Y. Jiang
Center of Expertise in Mycology of Radboud University Medical Center, Canisius Wilhelmina Hospital, Nijmegen, The Netherlands

Department of Dermatology, The Affiliated Hospital, Guizhou Medical University, Guiyang, China

J.-P. Bouchara et al. (eds.), *Dermatophytes and Dermatophytoses*,
https://doi.org/10.1007/978-3-030-67421-2_1

Dermatophytes (family *Arthrodermataceae*) are a group of fungi in the order *Onygenales* which have evolved relatively recently, almost exclusively with mammal hosts. The ecological subdivision in geophilic, zoophilic, and anthropophilic species broadly reflects their course of evolution. Geophilic species are found in soil around burrows of terrestrial mammals. Most species, particularly of the large genus *Arthroderma*, are sexually active, producing gymnothecial fruitbodies after mating [1–4]. Conidia of geophilic and zoophilic species are often large, multi-celled and have thick, rough walls; additional microconidia are found distributed in many species of the *Arthrodermataceae*. In the course of evolution, carriage by the host became more important, many species residing asymptomatically in the fur of the mammal (or rarely bird) and traveling back to the environment to complete the sexual part of their life cycle. Having a double life-cycle, the zoophiles have been characterized as environmental pathogens [5].

Homo sapiens is phylogenetically the most recent host. After initial eradication of classical agents associated with poor hygienic conditions, new pathogens are emerging, such as the ones associated with pets [6]. Globally, the prevalence of dermatophytosis is rising, and it is estimated that over 20–25% of the global populations are affected [7]. Infections are more prevalent in tropical and subtropical countries like India, where temperature and humidity are high most of the year [8].

Humans are exceptional among mammals in lacking a fur, which more rapidly leads to infection rather than colonization. This makes the anthropophilic species fundamentally different from both other ecological categories. Transmission is now between hosts, the intermittent environmental phase being limited to small cells known as arthroconidia which serve directly as infectious propagules; they were referred to as zoophilic pathogens [5]. The invading thallus can be regarded as the main form of presentation of the fungus. In some species, conidia are absent or produced only reluctantly, and none of the anthropophilic *Trichophyton* species is known to have a sexual phase. A similar loss of conidiation is noted in zoophilic *Microsporum* species, where clonal offshoots on humans, known as *Microsporum audouinii* and *Microsporum ferrugineum*, show a lack of sporulation. Mating being absent, the *MAT* location genes can be unequally distributed in human-dependent species, which is a further factor in their loss. For example, strains of *Trichophyton rubrum* are all of the – (*MAT1–1*) type [9], with only a few isolates, referred to as *Trichophyton megninii* being + (*MAT1–2*) [10, 11], but as yet strains of the opposite mating type have not been observed to mate successfully.

Evolution in anthropophilic dermatophytes by definition has proceeded for less than a million years and is still ongoing. Species have not yet diverged much in diagnostic markers and are therefore difficult to distinguish. The process starts with small differences in ecologically relevant parameters by which lineages drift away from each other. In addition, geographic isolation may play a role. Differentiation is enhanced by the lack of sexuality, whereby mutations acquired in local epidemics are maintained. Su et al. [12] noted three lineages in the *Trichophyton rubrum* complex, which on average are associated with different clinical pictures and/or have geographic differences. *Trichophyton violaceum* and *Trichophyton soudanense*, the latter being prevalent in arid parts of Africa, are more frequently involved in *Tinea capitis* than *T. rubrum*, which generally causes mild skin infections. The slight, gradational differences between these groups indicate a recent evolutionary process, which has sympatric elements (i.e., in the same area but specializing on different habitats) and is partly allopatric, i.e., taking place in different parts of the world. The absence of sexuality enhances rapid genetic drift.

There is a remarkable diversity of strictly anthropophilic species in the genus *Trichophyton*, comprising the *T. rubrum* group (with *T. violaceum* and *T. soudanense*), *Trichophyton interdigitale* (with the recent emerging species *T. indotineae* in India), *Trichophyton concentricum*, *Trichophyton tonsurans*, and *Trichophyton schoenleinii*. None of the other mammals is known to have given rise to such repeated specialization. Several of the human-associated fungi have a closely similar counterpart on mammal, mostly domesticated hosts which are the probable origins of the anthropophilic lineages: *T. schoenleinii* via *Trichophyton sarkisovii* from camels [13], *T. interdigitale* via *Trichophyton mentagrophytes* from rabbits [14], *T. tonsurans* via *Trichophyton equinum* from horses [15]. In all these cases the human colonizing agents are only doubtfully distinct from their zoophilic ancestors. Kandemir et al. [15] described the process of evolution whereby the two mating types drift apart, with the human lineage differing in being autotrophic for nicotinic acid, which is naturally available on horse skin. Also in *Trichophyton benhamiae*, a zoophilic species primarily found on guinea pigs, a diversification process of the two mating types can be observed, with differential infective potential on humans [16, 17].

The majority of the prevalently zoophilic species can remain asymptomatic in the fur of the host: in addition to *T. benhamiae* [18], *Trichophyton erinacei* occurs on hedgehogs [19], *Trichophyton quinckeanum* on mice [20], *Trichophyton bullosum* on horse [21], *Trichophyton verrucosum* on cattle [22], *Arthroderma redellii* on bats [23], *Arthroderma eboreum* on badgers [24], *Microsporum canis* on cats and dogs [25]. Occasional human infection by such species is often highly inflammatory (e.g., [26, 27]). The furs of these animals provide entirely different habitats in which the species may specialize, or eventually be transmitted to hosts with similar fur types, but most zoophilic species have insufficiently been studied to explain the mechanism of host jumps. Geophilic members of the family *Arthrodermataceae* are found in the vicinity of sources of keratin, for example, around burrows of terrestrial animals such as rodents or badgers. A wide diversity of species exists. Most of these are known to have sexual forms with ornamented gymnothecia that contain deliquescent asci with one-celled ascospores. The main genus is *Arthroderma*, currently containing 21 recognized species [28]. Species are much wider apart from each other than the anthropophilic *Trichophyton* species, and mostly are easily recognizable by molecular and phenotypic parameters. Phylogenetic distances are considerable to the ancestral genera *Ctenomyces* and *Guarromyces*. Within the family, several clusters of species show strong phylogenetic support which coincides with morphological parameters, which led de Hoog et al. [28] to propose a novel generic classification for the *Arthrodermataceae*. This system comprises the genera *Trichophyton* (anthropophilic or zoophilic mostly on mammals living in close vicinity of humans; with microconidia; macroconidia, if produced, smooth- and thin-walled), *Epidermophyton* (anthropophilic, with smooth-walled macroconidia only), *Nannizzia* (ecologically undefined, mostly abundant sporulation with elongate, rather thick-walled and slightly ornamented macroconidia), *Microsporum* (on pet animals, if sporulating, producing large, thick-walled, ornamented macroconidia), *Lophophyton* (originally described from fowl, but also reported from diverse mammals; very large, fusiform, thick-walled, multi-celled macroconidia), *Paraphyton* (ecologically undefined, morphologically similar to *Lophophyton* but forming a separate cluster in ribosomal phylogeny), and *Arthroderma* (soil-inhabiting species, mostly easily sporulating with short fusiform, thick-walled macroconidia, and sexually active). For an overview of accepted species, refer to de Hoog et al. [29] (Fig. 1).

Dermatophytes are classically identified by a combination of clinical features, colony characteristics, and microscopy. While this is effective in dermatological routine with large numbers of strains with limited diversity from a single host, *Homo sapiens*, this approach is inadequate to grasp in the large diversity of *Arthrodermataceae* from other sources. Also occasional infections by rare species are unlikely to be recognized in most laboratories. In addition, while fresh cultures may show optimal characteristics, subcultures tend to lose sporulation, or, conversely, non-sporulating species like *T. verrucosum* may produce microconidia over time. Molecular methods have therefore become indispensable for confident identification, and to produce reliable results with geophilic and the majority of zoophilic species. However, the recent evolution of anthropophilic dermatophytes is reflected in limited molecular variation between species. Gräser et al. [30, 31] applied the rDNA Internal Transcribed Spacer (ITS) region and were able to resolve a large number of species. The main structure of their phylogenetic tree and separation of species have been confirmed by the use of other generic regions such as beta-tubulin [32, 33], translation elongation factor 1-alpha [34], and high-mobility group mating type [35]. Most of the derived anthropophilic species remain, however, difficult to distinguish with any of these markers. This became clear with the identification of outbreaks of virulent, multi-resistant genotypes [36]. Particularly in India, the prevalence of severe dermatophytosis is emerging rapidly and has become a challenge for dermatologists [37]. Corticosteroids and oral antifungal agents, which are frequently applied over the counter, may have been instrumental in this epidemic [38–40]. The creams which are applied without medical prescription suppress inflammation and improve the symptoms of *Tinea* initially, but irrational use of drug combinations enhance acquired resistance [38]. The current epidemic causes significant distress to the patients socially, emotionally, and financially [37]. The agents of the main Indian elevation [41] were variously identified as *T. interdigitale* [42] or *T. mentagrophytes* [43]. This goes beyond simple naming, since *T. interdigitale* is known to be an anthropophilic species while *T. mentagrophytes* is classically treated as a zoophilic species; the identification as either one of these species thus implies a different cause of the outbreak. The use of the recently proposed separate name, *T. indotineae*, even though based on fragmentary data [44], may therefore be advisable.

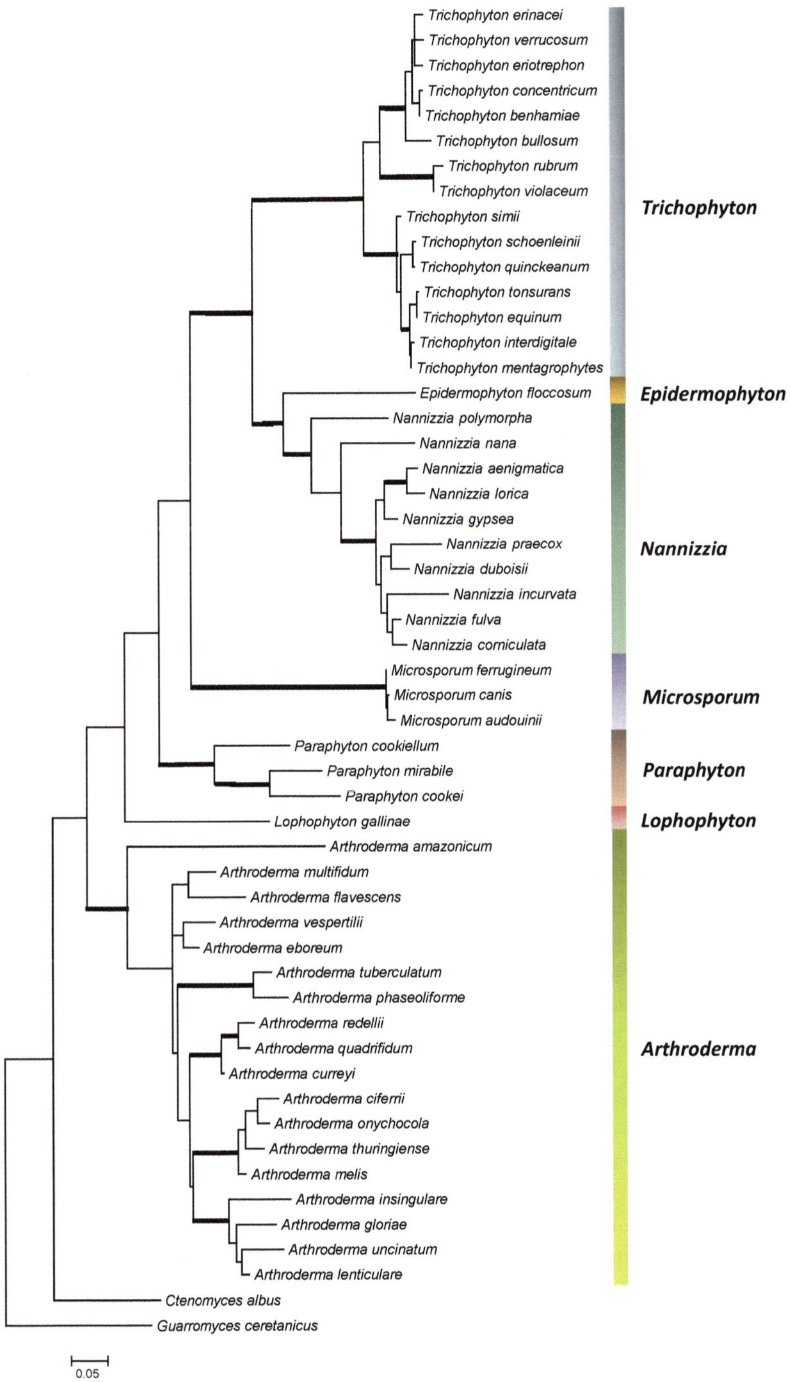

Fig. 1 Maximum likelihood tree (RAxML v8.2.12) of rDNA ITS sequences of the family *Arthrodermataceae*, using GTRCAT as model, with 1000 bootstrap replications. *Guarromyces ceretanicus* was selected as outgroup

Recent, precise multilocus studies combined with physiological and ecological data have been able to resolve some of the groups that were distinguished in classical dermatomycology but were problematic with the molecular approach, as mentioned above [12, 15]. For taxonomic entangling species in the middle of their evolutionary process, a combination of molecular, phenotypic, and clinical data is required. As a consequence of this dynamic course of adaptation, we also have to accept the fact that taxonomic borderlines may not be sharp, and that numerous intermediate genotypes may be encountered. As diagnostic parameters with a high degree of predictivity, ITS sequencing as yet is recommended. In some cases, e.g., in the *T. rubrum* group, this broadly matches with clinical, classical, and cultural features that have been applied to distinguish species.

Members of the *T. rubrum, T. interdigitale/mentagrophytes, T. tonsurans*, and *M. canis* complexes have become the major species of urban populations globally [6]. During recent years, a sufficient choice of antifungal drugs was available on the market to control dermatophytes which generally respond well to topical antifungal treatment. Therapy failure is particularly common with onychomycosis and skin lesions with folliculitis, where prolonged systemic therapy is required [45]. Standard guidelines for antifungal susceptibility are available from Clinical Laboratory Standards Institute (CLSI), European Committee on Antimicrobial Susceptibility Testing (EUCAST), and British Society of Antimicrobial Chemotherapy (BSAC) [46]. A large study from Iran using the CLSI method in 1502 direct microscopy and culture-proven cases showed that terbinafine was the most potent antifungal against all isolates, with an MIC range of 0.002–0.25μg/mL, followed by itraconazole (0.004–0.5μg/mL), griseofulvin (0.125–8μg/mL), and fluconazole (4–128μg/mL) [47].

A significant increase in the number of dermatophytosis has been noted in recent years in India. Numerous cases with chronic recalcitrant disease, atypical presentations, frequent relapses, and treatment failures have been reported [40, 48, 49], sometimes in epidemic proportions [39]. As explanation of this phenomenon, two hypotheses have been put forward: this may have been caused by zoonotic emergence of dermatophytes from pets or pest animals, or by irrational use of corticosteroid-containing antifungal combinations sold over the counter, which has enhanced acquired multi-resistance to common antifungals [40, 46]. Emergence of resistance is also known in other geographical locations, such as Denmark, Switzerland, and China [50–53], Belgium [54], Germany [55], Japan [51], Iran [56], Finland [57], Switzerland [58] and Bahrain [59], which concerned Leu393, Phe397, Phe415, and His440 substitutions in SQLE (squalene epoxidase) of *T. interdigitale, T. mentagrophytes*, and *T. rubrum* clinical isolates [51]. Earlier studies from North India showed the occurrence of non-responders to griseofulvin therapy among *Tinea capitis* patients [60]. *Trichophyton rubrum* is known to develop resistance to azoles, amorolfine, and terbinafine after prolonged exposure to sub-inhibitory concentrations of these drugs leading to treatment failures and consequently contributing to persistence and chronicity of the infections. Azole resistance in dermatophytes has been reported as high as 19% worldwide [46, 61]. After the recent emergence of dermatophyte antifungal drug resistance and increased virulence, it is recommended to perform molecular identification and antifungal susceptibility testing routinely in cases of chronic, recurrent, and/or recalcitrant and atypical presentations.

References

1. Stockdale PM, Mackenzie DWR, Austwick PKC. *Arthroderma simii* sp. nov., the perfect state of *Trichophyton simii* (Pinoy) comb. nov. Sabouraudia. 1965;4:112–23.
2. Currah RS, Abbott SP, Sigler L. *Arthroderma silverae* sp. nov. and *Chrysosporium vallenarense*, keratinophilic fungi from arctic and montane habitats. Mycol Res. 1996;100:195–8.
3. Choi JS, Gräser Y, Walther G, Peano A, Symoens F, de Hoog GS. *Microsporum mirabile* and its teleomorph *Arthroderma mirabile*, a new dermatophyte species in the *M. cookei* clade. Med Mycol. 2012;50:161–9.
4. Symoens F, Jousson O, Packeu A, Fratti M, Staib P, Mignon B, et al. The dermatophyte species *Arthroderma benhamiae*: intraspecies variability and mating behavior. J Med Microbiol. 2013;62:377–85.
5. de Hoog GS, Ahmed SA, Danesi P, Guillot J, Gräser Y. Distribution of pathogens and outbreak fungi in the fungal kingdom. In: Seyedmousavi S, de Hoog GS, Guillot J, Verweij PE, editors. Emerging and epizootic fungal infections in animals. Berlin: Springer; 2018. p. 3–16.
6. Zhan P, Liu W. The changing face of dermatophytic infections worldwide. Mycopathologia. 2017;182:77–86.
7. Havlickova B, Czaika VA, Friedrich M. Epidemiological trends in skin mycoses worldwide. Mycoses. 2008;51:2–15.
8. Lakshmanan A, Ganeshkumar P, Mohan SR, Hemamalini M, Madhavan R. Epidemiological and clinical pattern of dermatomycoses in rural India. Indian J Med Microbiol. 2015;33:134–6.
9. Li W, Metin B, White TC, Heitman J. Organization and evolutionary trajectory of the mating type (MAT) locus in dermatophyte and dimorphic fungal pathogens. Eukaryot Cell. 2010;9:46–58.
10. Gräser Y, de Hoog S, Summerbell RC. Dermatophytes: recognizing species of clonal fungi. Med Mycol. 2006;44:199–209.
11. Metin B, Heitman J. She loves me, she loves me not: on the dualistic asexual/sexual nature of dermatophyte fungi. Mycopathologia. 2020;185:87–101.
12. Su H, Packeu A, Ahmed SA, Al-Hatmi AMS, Blechert O, Ilkit M, et al. Species distinction in the *Trichophyton rubrum* complex. J Clin Microbiol. 2019;57:e00352–19.
13. Probst S, de Hoog GS, Gräser Y. Development of DNA markers to explore host shifts in dermatophytes. Stud Mycol. 2002;47:57–74.
14. Van Rooij P, Detandt M, Nolard N. *Trichophyton mentagrophytes* of rabbit origin causing family incidence of kerion: an environmental study. Mycoses. 2006;49:426–30.
15. Kandemir H, Dukik K, Hagen F, Ilkit M, de Hoog GS. A polyphasic approach for the differentiation *Trichophyton tonsurans* and *T. equinum*. Mycopathologia. 2019;185:113–22.
16. Symoens F, Jousson O, Packeu A, Fratti M, Staib P, Mignon B, et al. The dermatophyte species *Arthroderma benhamiae*: intraspecies variability and mating behavior. J Med Microbiol. 2011;62:377–85.
17. Hiruma J, Kano R, Harada K, Monod M, Hiruma M, Hasegawa A, et al. Occurrence of *Arthroderma benhamiae* genotype in Japan. Mycopathologia. 2015;179:219–23.
18. Bartosch T, Frank A, Günther C, Uhrlaß S, Heydel T, Nenoff P, et al. *Trichophyton benhamiae* and *T. mentagrophytes* target Guinea pigs in a mixed small animal stock. Med Mycol Case Rep. 2019;23:37–42.
19. Abarca ML, Castellá G, Martorell J, Cabañes FJ. *Trichophyton erinacei* in pet hedgehogs in Spain: occurrence and revision of its taxonomic status. Med Mycol. 2017;55:164–72.

20. Beguin H, Pyck N, Hendrickx M, Planard C, Stubbe D, Detandt M. The taxonomic status of *Trichophyton quinckeanum* and *T. interdigitale* revisited: a multigene phylogenetic approach. Med Mycol. 2012;50:871–82.
21. Lyskova P, Hubka V, Petricakova A, Dobias R, Cmokova A, Kolarik M. Equine dermatophytosis due to *Trichophyton bullosum*, a poorly known zoophilic dermatophyte masquerading as *T. verrucosum*. Mycopathologia. 2015;180:407–19.
22. Ming PX, Ti YLX, Bulmer GS. Outbreak of *Trichophyton verrucosum* in China transmitted from cows to humans. Mycopathologia. 2006;161:225–8.
23. Lorch JM, Minnis AM, Meteyer CU, Redell JA, White JP, Kaarakka HM, et al. The fungus *Trichophyton redellii* sp. nov. causes skin infections that resemble white-nose syndrome of hibernating bats. J Wildl Dis. 2015;51:36–47.
24. Campbell CK, Borman AM, Linton CJ, Bridge PD, Johnson EM. *Arthroderma olidum*, sp. nov. a new addition to the *Trichophyton terrestre* complex. Med Mycol. 2006;44:451–9.
25. Cafarchia C, Romito D, Capelli G, Guillot J, Otranto D. Isolation of *Microsporum canis* from the hair coat of pet dogs and cats belonging to owners diagnosed with *M. canis* tinea corporis. Vet Dermatol. 2006;17:327–31.
26. Degreef H. Clinical forms of dermatophytosis (ringworm infection). Mycopathologia. 2008;166:257–65.
27. Braun SA, Jahn K, Westermann A, Bruch-Gerharz D, Reifenberger J. Tinea barbae profunda by *Arthroderma benhamiae*. A diagnostic challenge. Hautarzt. 2013;64:720–2.
28. de Hoog GS, Dukik K, Monod M, Packeu A, Stubbe D, Hendrickx M, et al. Towards a novel multilocus phylogenetic taxonomy for the dermatophytes. Mycopathologia. 2017;182:5–31.
29. de Hoog GS, Guarro J, Gené J, Ahmed SA, Al-Hatmi AMS, Figueras MJ, et al. Atlas of clinical fungi. 4th ed. Utrecht/Reus: Westerdijk Institute; 2019.
30. Gräser Y, Kuijpers AFA, Presber W, de Hoog GS. Molecular taxonomy of *Trichophyton mentagrophytes* and *T. tonsurans*. Med Mycol. 1999;37:315–30.
31. Gräser Y, Kuijpers AFA, Presber W, de Hoog GS. Molecular taxonomy of the *Trichophyton rubrum* complex. J Clin Microbiol. 2000;38:3329–36.
32. Rezaei-Matehkolaei A, Mirhendi H, Makimura K, de Hoog GS, Satoh K, Najafzadeh MJ, et al. Nucleotide sequence analysis of beta tubulin gene in a wide range of dermatophytes. Med Mycol. 2014;52:674–88.
33. Pchelin IM, Zlatogursky VV, Rudneva MV, Chilina GA, Rezaei-Matehkolaei A, Lavnikevich DM, et al. Reconstruction of phylogenetic relationships in dermatomycete genus *Trichophyton* Malmsten, 1848 based on ribosomal internal transcribed spacer region, partial 28S rRNA and beta-tubulin genes sequences. Mycoses. 2016;59:566–75.
34. Mirhendi H, Makimura K, de Hoog GS, Rezaei-Matehkolaei A, Najafzadeh MJ, Umeda Y, et al. Translation elongation factor 1-α gene as a potential taxonomic and identification marker in dermatophytes. Med Mycol. 2015;53:21524.
35. Kosanke S, Hamann L, Kupsch C, Moreno Garcia S, Chopra A, Gräser Y. Unequal distribution of the mating type (MAT) locus idiomorphs in dermatophyte species. Fungal Genet Biol. 2018;118:45–53.
36. Luchsinger I, Bosshard PP, Kasper RS, Reinhardt D, Lautenschlager S. Tinea genitalis: a new entity of sexually transmitted infection? Case series and review of the literature. Sex Transm Infect. 2015;91:493–6.
37. Pathania S, Rudramurthy SM, Narang T, Saikia UN, Dogra S. A prospective study of the epidemiological and clinical patterns of recurrent dermatophytosis at a tertiary care hospital in India. Indian J Dermatol Venereol Leprol. 2018;84:678–84.
38. Bishnoi A, Vinay K, Dogra S. Emergence of recalcitrant dermatophytosis in India. Lancet Infect Dis. 2018;18:250–1.
39. Singh A, Masih A, Khurana A, Singh PK, Gupta M, Hagen F, et al. High terbinafine resistance in *Trichophyton interdigitale* isolates in Delhi, India harbouring mutations in the squalene epoxidase gene. Mycoses. 2018;61:477–84.

40. Singh A, Masih A, Monroy-Nieto J, Singh PK, Bowers J, Travis J, et al. A unique multidrug-resistant clonal *Trichophyton* population distinct from *Trichophyton mentagrophytes/ Trichophyton interdigitale* complex causing an ongoing alarming dermatophytosis outbreak in India: genomic insights and resistance profile. Fungal Genet Biol. 2019;133:103266.
41. Nenoff P, Verma SB, Vasani R, Burmester A, Hipler UC, Wittig F, et al. The current Indian epidemic of superficial dermatophytosis due to *Trichophyton mentagrophytes*—a molecular study. Mycoses. 2019;62:336–56.
42. Chowdhary A, Singh A, Singh PK, Khurana A, Meis JF. Perspectives on misidentification of *Trichophyton interdigitale/Trichophyton mentagrophytes* using internal transcribed spacer region sequencing: urgent need to update the sequence database. Mycoses. 2019;62:11–5.
43. Nenoff P, Verma SB, Uhrlaß S, Burmester A, Gräser Y. A clarion call for preventing taxonomical errors of dermatophytes using the example of the novel *Trichophyton mentagrophytes* genotype VIII uniformly isolated in the Indian epidemic of superficial dermatophytosis. Mycoses. 2019;62:6–10.
44. Kano R, Kimura U, Kakurai M, Hiruma J, Kamata H, Suga Y, Harada K. *Trichophyton indotineae* sp. nov.: a new highly terbinafine-resistant anthropophilic dermatophyte species. Mycopathologia. 2020;185:947–58.
45. Thatai P, Sapra B. Critical review on retrospective and prospective changes in antifungal susceptibility testing for dermatophytes. Mycoses. 2016;59:615–27.
46. Dogra S, Shaw D, Rudramurthy SM. Antifungal drug susceptibility testing of dermatophytes: laboratory findings to clinical implications. Indian Dermatol Online J. 2019;10:225–33.
47. Ansari S, Hedayati MT, Zomorodian K, Pakshir K, Badali H, Rafieie A, et al. Molecular characterization and *in vitro* antifungal susceptibility of 316 clinical isolates of dermatophytes in Iran. Mycopathologia. 2016;2181:89–95.
48. Sahoo AK, Mahajan R. Management of tinea corporis, tinea cruris, and tinea pedis: a comprehensive review. Indian Dermatol Online J. 2016;7:77–86.
49. Poojary S, Miskeen A, Bagadia J, Jaiswal S, Uppuluri P. A study of *in vitro* antifungal susceptibility patterns of dermatophytic fungi at a tertiary care center in Western India. Indian J Dermatol. 2019;64:277–84.
50. Saunte DML, Hare RK, Jørgensen KM, Jørgensen R, Deleuran M, Zachariae CO, et al. Emerging terbinafine resistance in *Trichophyton*: clinical characteristics, squalene epoxidase gene mutations, and a reliable EUCAST method for detection. Antimicrob Agents Chemother. 2019;63:e01126–19.
51. Yamada T, Maeda M, Alshahni MM, Tanaka R, Yaguchi T, Bontems O, et al. Terbinafine resistance of *Trichophyton* clinical isolates caused by specific point mutations in the squalene epoxidase gene. Antimicrob Agents Chemother. 2017;61:e00115–7.
52. Kano R, Hsiao YH, Han HS, Chen C, Hasegawa A, Kamata H. Resistance mechanism in a terbinafine-resistant strain of *Microsporum canis*. Mycopathologia. 2018;183:623–7.
53. Taghipour S, Pchelin IM, Zarei Mahmoudabadi A, Ansari S, Katiraee F, Rafiei A, et al. *Trichophyton mentagrophytes* and *T. interdigitale* genotypes are associated with particular geographic areas and clinical manifestations. Mycoses. 2019;62:1084–91.
54. Sacheli R, Harag S, Dehavay F et al. Belgian national survey on tinea capitis: Epidemiological considerations and highlight of terbinafine-resistant *T. mentagrophytes* with a mutation on SQLE gene. J Fungi 2020;6:E195.
55. Nenoff P, Verma SB, Ebert A et al. Spread of terbinafine-resistant *Trichophyton mentagrophytes* type VIII (India) in Germany – the tip of the iceberg? J Fungi 2020;6:E207.
56. Taghipour S, Shamsizadeh F, Pchelin I et al. Emergence of terbinafine resistant *Trichophyton mentagrophytes* in Iran, harboring mutations in the squalene epoxidase (SQLE) gene. Infect Drug Resist. 2020;13:845–50.
57. Järv, H, Uhrlaß, S, Simkin, et al. Terbinafine resistant *Trichophyton mentagrophytes* genotype VIII, Indian type, isolated in Finland. J Fungi. 2019;95:P39.

58. Hsieh A, Quenan S, Riat A et al. A new mutation in the SQLE gene of *Trichophyton mentagrophytes* associated to terbinafine resistance in a couple with disseminated tinea corporis. J Mycol Med 2019;29:352–5.
59. Süß A, Uhrlaß S, Ludes A et al. Ausgeprðgte Tinea corporis durch ein Terbinafin-resistentes *Trichophyton-mentagrophytes*-Isolat vom indischen Genotyp bei einem Sðugling aus Bahrain in Deutschland. Hautarzt 2019;70:888–96.
60. Singal A, Rawat S, Bhattacharya SN, Mohanty S, Baruah MC. Clinico-mycological profile of tinea capitis in North India and response to griseofulvin. J Dermatol. 2001;28:22–6.
61. Ghannoum M. Azole resistance in dermatophytes: prevalence and mechanism of action. J Am Podiatr Med Assoc. 2016;106:79–86.

Part II

Clinical Aspects

Dermatophytes and Dermatophytic Infections Worldwide

Ping Zhan, Guanzhao Liang, and Weida Liu

Abstract

Dermatophytoses are a common public health problem with high prevalence. The etiological agents of dermatophytoses, termed dermatophytes, change with geography and socioeconomic status. *Trichophyton rubrum* is the predominant species for skin and nail infections followed by *Trichophyton mentagrophytes*/*Trichophyton interdigitale* complex. However, we notice a *T. rubrum* to *T. mentagrophytes* shift in some parts of Iran, India, and Australia. The causative agents of *Tinea capitis* changed most greatly with obvious geography co-relation and time, with *Microsporum canis*, *Trichophyton tonsurans*, and *Trichophyton violaceum* present as the predominant dermatophytes in different regions. In this chapter, we provide a brief introduction of dermatophytes history and review their present epidemiology worldwide by continents and related diseases. The aim is to get a comprehensive understanding of prevailing geographic distribution of these pathogens, and prospect the trends of future changes in their occurrence and prevalence.

Ping Zhan and Guanzhao Liang contributed equally to this work.

P. Zhan
Institute of Clinical Medicine and Dermatology Department, Jiangxi Provincial People's Hospital, Nanchang, Jiangxi, China

Center of Expertise in Mycology of Radboud University Medical Center, Canisius Wilhelmina Hospital, Nijmegen, The Netherlands

G. Liang · W. Liu (✉)
Department of Mycology, Institute of Dermatology, Chinese Academy of Medical Sciences and Peking Union Medical College, Nanjing, Jiangsu, China

J.-P. Bouchara et al. (eds.), *Dermatophytes and Dermatophytoses*,
https://doi.org/10.1007/978-3-030-67421-2_2

1 Introduction

Dermatophytes are a highly specialized group of filamentous fungi, which are highly adapted to keratinized tissues of humans and animals. Dermatophytes are true fungal pathogens, usually colonizing and invading immunocompetent individuals. Their distribution and evolution are distinctly influenced by geography and socioeconomic status, e.g., host age, housing and sanitary conditions, population density, water supply, health service availability, and the use of antifungal agents [1–6].

Dermatophytosis remains a very important public health problem globally, especially in developing countries. Till today, it has been recognized that dermatophytosis are the most common fungal infections worldwide, affecting 20–25% of the world population [6]. Due to the development of new antifungal drugs and improved socioeconomic status, the prevalence of dermatophytosis decreased prominently in most areas and the spectrum of dermatophytes changed greatly. We found an anthropophilic to zoophilic shift within *Tinea capitis* during the past 60 years in mainland China [1]. Also, India has a *Trichophyton rubrum* to *Trichophyton mentagrophytes* shift for superficial fungal infections [7, 8].

In this chapter, we give a brief introduction of dermatophytes history and review their present epidemiology worldwide. The aim is to get a comprehensive understanding of these organisms and prospect the trends of their changes in future.

2 Natural and Taxonomy History of Dermatophytes

The first report of dermatophytosis was recorded in Romans' word around 30 AD and described as suppurative infections on scalp, again suggesting these organisms have lived with human beings for a long history [9]. From that faraway era and down through the Middle Ages, these kind of dermatophytosis were described as *Tinea* (*Tinea* refers to a cloth moth that destroys garment with circular holes) and ringworm [10]. Although doctors have been aware of superficial fungal infections for centuries, the causal agents and treatment of these infections remained unknown until the mid-1800s [9].

Trichophyton schoenleinii was firstly discovered in the crust of *Tinea favosa* around the 1830s and this episode was usually regarded as the start of medical mycology [9, 10]. In 1837, Robert Remark, a Polish-German physician, noted hyphae in the crusts of favus and named the etiological agent as *Achorion schoenleinii* (*T. schoenleinii*) in honor of Johann Lucas Schoenlein in whose laboratory he was working that time [10]. In 1841, David Gruby described several different types of dermatophytosis in his publication, including *Tinea favosa*, and the ectothrix and endothrix trichophytosis and microsporiosis, and reported the contagiousness of *Tinea favosa* [10].

Paul Grawitz and Emile Duclaux independently isolated the pure culture of dermatophytes in 1886 [10]. Henceforth, *Microsporum audouinii*, *T. mentagrophytes*, *Trichophyton tonsurans*, and *Epidermophyton floccosum* were successively either reported on the basis of the appearance of clinical materials or described under the microscope [2, 10, 11]. Around the 1900s, Raymond Sabouraud improved a newly developed isolation medium for fungal culture (Sabouraud Dextrose Agar, SDA), identification and preservation of dermatophytes, and established the taxonomic criteria for these fungi. He integrated the mycological and clinical aspects into a comprehensive concept, and classified the dermatophytes into four genera, *Achorion*, *Microsporum*, *Trichophyton*, and *Epidermophyton*. Emmons deleted the genus *Achorion* in 1934 and the other three genera were reserved till today [9].

Earlier, the genera and species of dermatophytes were described on the basis of morphology in pure culture, clinical reaction of the host, and the species of animals involved. This complicated nomenclature afterward led to confusion. In 1935, Dodge introduced 118 species of dermatophytes in his book *Medical Mycology* which reflected the complicated dermatophyte differentiation around the first half of the twentieth century [2, 9].

Two major contributors to dermatophyte classification were Chester W. Emmons and Libero Ajello, whose work influenced the training of numerous medical mycologists, and organized a more logical and usable dermatophytes classification system [10, 12]. After the 1950s, the dermatophytes were classified on the basis of spore morphology, physiological characteristics, and the sexual reproduction. However, there are still problems due to high diversity of filamentous fungi, which lead to one fungus being assigned multiple names. Since the 1980s, the application of molecular methods made the taxonomy of dermatophytes more rational and simplified [13]. International co-workers contributed greatly to the nomenclature working. In 2011, the Amsterdam Declaration on Fungal Nomenclature promoted the "One Fungus One Name" concept, which was adopted quickly worldwide as a comprehensive naming system, although this new system still keeps some limits [14].

3 Dermatophyte Evolution in the Past Century

The spectrum of dermatophytes has changed in the past 60 years due in parts to improved diagnosis, treatment, and better public health awareness [2, 4]. Before the 1950s, *Tinea pedis* and onychomycosis were uncommon, and *T. mentagrophytes*/*Trichophyton interdigitale* were the main agents for *Tinea corporis*. Meanwhile, *Tinea capitis* was a severe public problem threatening school-age children [4, 15–17]. The anthropophilic agents *M. audouinii*, *Microsporum ferrugineum*, and *T. schoenleinii* were the predominant agents of *Tinea capitis* in British islands, Northern and Western Europe, and America [2, 5, 15–18]. Few original reports could be reviewed before the 1950s for Africa, Asia, and Middle East. Before the 1960s, the anthropophilic dermatophytes *E. floccosum*, *M. ferrugineum*, *Trichophyton violaceum*, and *T. schoenleinii* were frequent in these areas as the main causes of scalp ringworm, yet their respective frequency varied by regions [1, 3, 15].

Molecular analysis showed that *T. rubrum-T. violaceum* complex possibly originated from West Africa, moving to east Asia at some timepoint. Finally, these fungi were brought to Europe and North America in the later part of the nineteenth century and become widespread around the whole world with globalization [19, 20]. Interestingly, in its original homeland west Africa, *T. rubrum* does not cause *Tinea pedis* due to no footwear worn by the local people [21]. After the Second World War, *T. rubrum* increased quickly worldwide, replacing the *T. mentagrophytes* as the first important pathogen for superficial fungal infections in most areas [2]. Based on the above information, we would propose possible evolution pathway for dermatophytes, which might survive within human environment for thousands of years and become pathogenic with human civilization. Among them, *T. rubrum* is the typical example which obtained specific virulence and became adapt for human body after the nineteenth century. Inevitably, the spectrum of dermatophytes is not static due to more frequent global immigration, convenient tourism, and changing socioeconomic status.

The first broad-spectrum antifungal drug for dermatophytosis was griseofulvin, discovered in 1939 from *Penicillium*, and successfully applied for treating experimental ringworm in guinea pigs by Gentles in 1958 [9, 11, 22]. The epidemiology of dermatophytosis changed greatly after the Second World War, perhaps because of the widespread use of oral griseofulvin especially for pandemic favus, and improvement in hygiene due to rising living standards. *Tinea capitis* declined significantly in many areas while *Tinea pedis* and onychomycosis increased globally thereafter [1, 2, 18]. *Trichophyton rubrum* became the most common cause of superficial fungal diseases worldwide more recently, except for infections on scalp [2, 4, 15]. Since the 1950s, the frequency of *M. audouinii* and *T. schoenleinii* in British Island and North and Western Europe declined along with an increased frequency of *Microsporum canis* and *T. mentagrophytes* as the most important agents isolated from scalp [3, 5, 15, 17]. In North and South America, *T. tonsurans* took the place of *M. audouinii* as the most common agent for *Tinea capitis* followed by *M. canis* [16, 23]. Etiology of dermatophytosis in Australia was much similar to that of America except that *M. audouinii* was rarely reported in Australia [23, 24]. It is thought that its spread was prevented by *M. canis* [2, 23, 24].

Chinese nationwide campaign of "eliminating *Tinea capitis*" exhibits a typical example of the change in occurrence and distribution. Sixty years ago, most provinces in China were hyper-endemic for favus, with *T. schoenleinii* as the predominant pathogen [1, 2]. In the middle of the 1960s, a nationwide campaign was sponsored by the government aiming to eliminate favus. Griseofulvin and topical antifungal agents were prescribed for free for all involved patients, as well as health interventions. Favus largely disappeared across the country, and the prevalence of *Tinea capitis* decreased to 7.2 per 100,000 in the year 2000. With such a sharp decline of *T. schoenleinii*, infections due to *T. mentagrophytes* and *T. violaceum* occurred more frequently afterward. According to Wu et al. [25], in 1986 *T. mentagrophytes* and *T. violaceum* accounted for 52.3% and 17.4% of all cases of *Tinea capitis*, respectively, but from 1986 to 2006, *M. canis* gradually increased and it has now become the predominant organism in areas with high economic growth. Among the superficial fungal infections, *Candida* species showed a notable increase, while *T. rubrum* and *T. mentagrophytes* were the most common dermatophytes for non-scalp infections [25].

4 General Epidemiology of Dermatophyte Worldwide Today

Population mobility, the advent of new antifungal agents, popularization of leather shoes and sneakers, and improvement of human hygiene are all responsible for a major shift in the dermatophyte spectrum noticeable since the Second World War.

New antifungal drugs were great advancement for the treatment of dermatophytosis, but their use also led to evolution in the etiological agents. External formulations of antifungals such as nystatin, miconazole, clotrimazole, and econazole were introduced successively since 1944 [9], and ketoconazole became available in 1981, as the first azole derivative with systemic antifungal action. The three other most important systemic antifungal drugs, itraconazole, fluconazole, and terbinafine available in the early 1990s, became the most preferred drugs for treatment of dermatophytosis [9, 26, 27].

In the last 20 years, the spectrum of dermatophytes appears to change at a slower speed. *Trichophyton rubrum* remains as the predominant species for *Tinea* on skin and nails. *Trichophyton mentagrophytes* ranks the second, and even the main etiologic agent in some regions including southwestern Iran, Croatia, and Australia [2, 24, 28, 29]. Some anthropophilic species are quite rare beyond Africa, e.g., *Trichophyton soudanense, M. audouinii*, and *M. ferrugineum*, which can be imported into other countries, and lead to small outbreaks traced to migration from Africa. Today, *E. floccosum* and *T. schoenleinii* are mainly reported from Asia and Africa, the former even becoming the common dermatophytes in Iran [28, 30, 31].

5 Dermatophytes and Their Distribution by Continents

In the newly proposed taxonomy, there are more than 50 species of dermatophytes distributed into nine genera [32]. Considering the new taxonomy yet to be used in clinical practice, in this chapter, we apply the traditional classification system with three genus, *Trichophyton*, *Microsporum*, and *Epidermophyton*, for the review of literature but use the proposed new taxonomy name in the manuscript. Table 1 lists the most common clinical dermatophytes causing human disease. Within *T. mentagrophytes* complex, there are still many problems to differentiate the most common species, *T. mentagrophytes* and *T. interdigitale*, in clinical laboratories [33]; therefore, we summarize these sibling species as *T. mentagrophytes* in this review.

Table 1 Dermatophytes and their clinical infections

Species	Host preference	Major distribution	Major infection type
E. floccosum	Anthropophilic	Africa and Asia	Skin infections
M. audouinii	Anthropophilic	Africa, sporadic in other areas	Scalp infections
M. canis	Zoophilic	Worldwide	Scalp infections
M. ferrugineum	Anthropophilic	Africa and Asia	Scalp infections
N. gypsea	Geophilic	Worldwide	Scalp infections
T. mentagrophytes	Anthropophilic/zoophilic	Worldwide	All kinds of infections
T. rubrum	Anthropophilic	Worldwide	Skin and nail infections
T. schoenleinii	Anthropophilic	Africa, sporadic in Asia	Scalp infections
T. soudanense	Anthropophilic	Africa	Scalp infections
T. tonsurans	Anthropophilic	America, Europe, UK	Scalp infections, small outbreaks and asymptomatic carriers
T. verrucosum	Zoophilic	Worldwide	Skin and scalp infections
T. violaceum	Anthropophilic	Africa, Europe and Asia	Scalp infections and asymptomatic carriers

E Epidermophyton, M Microsporum, N Nannizzia, T Trichophyton

Concerning their natural habitat and host preference, dermatophytes can be divided into anthropophilic (on human), zoophilic (on animal), and geophilic (in soil) species. Generally, anthropophilic dermatophytes are geographically limited except for *T. rubrum*, which is most frequently isolated in developing countries and usually associated with low socioeconomic status, while zoophilic and geophilic species, which are related to pet feeding and farming work, increase dramatically nowadays, and show a global distribution [2]. *Microsporum canis* and *T. mentagrophytes* are the major pathogenic species belonging to this category [2]. However, there is an increase in anthropophilic infections, which used to be rare in Europe and America until recently; this increase is most likely related to tourism and migration [2, 34].

Generally, the pathogenic agents of skin and nail infections are less diverse than those responsible for scalp infections worldwide, which show high regional variations [2, 3, 35, 36]. Many reports about superficial fungal infections originate from Europe while very few published data are available for North America and are especially rare for Australia (see Tables 2, 3, 4 and 5). However, we can speculate that superficial fungal diseases are more prevalent in Africa and Asia due to the ecological diversity and population density in tropical/subtropical areas and low socioeconomic status in these areas. Therefore, we discuss the global distribution of dermatophytes by continents and cite reports with relatively big data after the year 2000.

Table 2 Major surveys on dermatophytoses in Europe

Areas	Period	Disease (cases)	Predominant species[a]	Ref
Croatia, Split	1996–2002	Skin infections (465) Scalp infections (268)	Tr 30%, Tm 28%, Mc 25% Mc 74%, Tm 7%, Tt 5%	[113]
Croatia, Zagreb	1999–2008	Skin and nail infections (9304) Scalp infections (1767)	Tr 12%, Tm 68%, Mc 17% Mc 92%, Tm 6%	[29]
Denmark	2003	Skin and nail infections (5329) Scalp infections (185)	Tr 80%, Tm 15%, Mc 2% Mc 57%, Tv 35%, Ma 3%	[114]
France, Grenoble	2001–2011	Skin and nail infections (1285) Scalp infections (63)	Tr 82%, Tm 3%, Mc 1% Tt 24%, Tm 10%, Tv 1%, Tve 1%	[115]
Greece, Crete	2010–2015	All infections (294)	Mc 36%, Tr 35%, Tm 23%	[116]
Southwestern Greece	2000–2008	Skin and nail infections (261) Scalp infections (28)	Tr 47%, Mc 43%, Tm 10% Mc 75%, Tr 14%, Tm 11%	[117]
Italy, Rome	2002–2004	All infections (252)	Mc 28%, Tm 22%, Tr 18% Ma 10%, Tt 6%	[118]
Poland, Krakow	1995–2010	Skin and nail infections (1559) Scalp infections (39)	Tr 55%, Tm 38%, Ef 2%, Mc 2% Tr 39%, Tm 26%, Mc 15%, Ng 11%	[119]
Sweden, Stockholm	2005–2009	Skin and nail infections (8182) Scalp infections (680)	Tr 90%, Tm 8%, Tv 1% Tv 64%, Tso 18%, Ma 8%, Tt 6%	[120]
Switzerland, Lausanne	1993–2000	Skin and nail infections (3806) Scalp infections (387)	Tr 68%, Tm 26%, Mc 3% Mc 26%, Tv 16%, Tm 14%	[121]
Southwest UK[b]	2005	All infections (1056)	Tr 70%, Tm 22%, Tt 5%	[122]

Ef E. floccosum, Ma M. audouinii, Mc M. canis, Ng N. gypsea, Tm T. mentagrophytes/T. interdigitale, Tr T. rubrum, Tso T. soudanense, Tt T. tonsurans, Tv T. violaceum, Tve T. verrucosum

[a]In this column, the numbers indicate the species percentage among all positive dermatophyte isolates (excluding non-dermatophyte infections); only the most common species were listed

[b]Only the data of 2005 were retrieved and the disease could not be differentiated as no data were available in this report

5.1 Dermatophytes in Europe

Tinea pedis and onychomycosis are the most common dermatophytosis in Europe with noticeable increase since the Second World War [5]. *Trichophyton rubrum* is the main causative agent in *Tinea unguium*, *Tinea cruris*, *Tinea corporis*, *Tinea pedis*, and *Tinea manuum*, followed by *T. mentagrophytes* (Fig. 1) [2, 5, 9, 19].

Microsporum canis remains the principal agent for *Tinea capitis* in most European countries, i.e., Central Europe, Southern Europe, and Mediterranean regions, including Austria, Spain, Italy, Greece, Hungary, Germany, and Poland [2, 15]. *Trichophyton tonsurans* is the most common agent in the United Kingdom

[3, 37]. Interestingly, some tourist hotspots, such as Paris and Milan, witnessed a rapid rise in the incidence of imported anthropophilic infections [34, 38]. Factually, an increased frequency of *Tinea capitis* due to *T. tonsurans*, *T. violaceum*, and *M. audouinii* is observed in Europe with African immigrants together with a decreased frequency of zoophilic *M. canis* [17, 18, 34, 38, 39]. For example, *T. violaceum*–related *Tinea capitis* was rarely reported before 2000 in Switzerland [40]. However, this dermatophyte shows persistent increased occurrence [40, 41]. Additionally, the reappearance of *M. audouinii* was reported from Switzerland, Belgium, and Germany [2, 41, 42]. *Trichophyton violaceum* and *T. soudanense* increased to be the predominant agent of *Tinea capitis* in some Italian cities [34, 43].

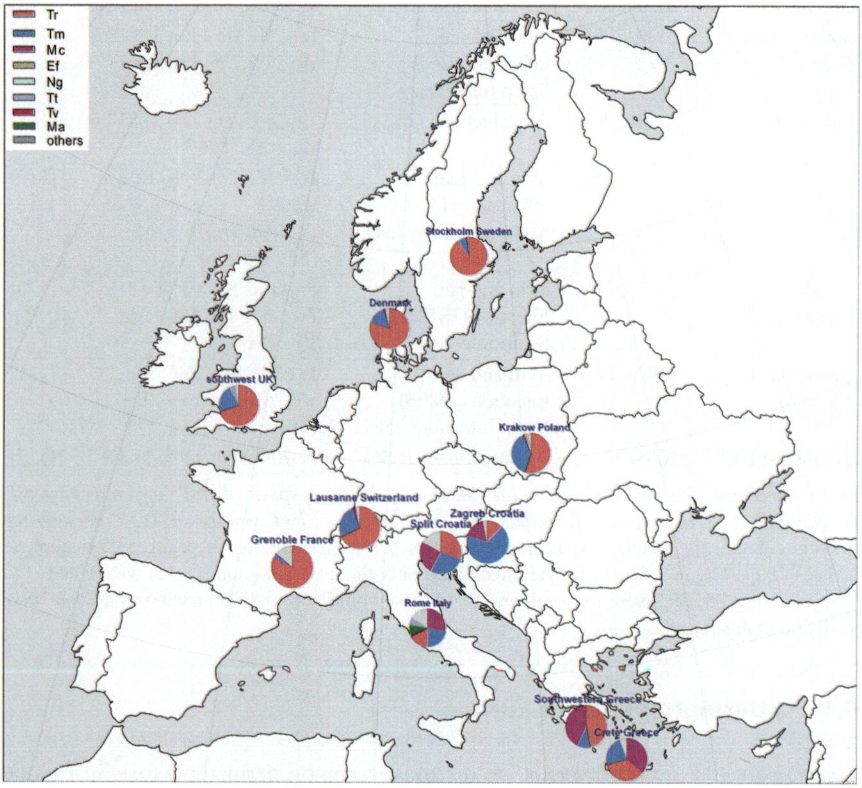

Fig. 1 Dermatophyte distribution in Europe isolated from skin and nail infections since 2000. Only major surveys with detailed and large data of dermatophyte spectrum were retrieved. The relative frequencies refer to the percentages among all dermatophyte isolates, excluding non-dermatophytes. *Ef E. floccosum*, *Ma M. audouinii*, *Mc M. canis*, *Ng N. gypsea*, *Tm T. mentagrophytes/T. interdigitale*, *Tr T. rubrum*, *Tt T. tonsurans*, *Tv T. violaceum*

5.2 Dermatophytes in Asia

Dermatophytosis remain a severe public health problem with high prevalence in Asian countries, as most of these regions have high population density, less ideal hygienic environment, and warm, rainy climate that supports dermatophytes survival and proliferation [1, 44].

Trichophyton rubrum is the principal agent for glabrous and nail infection in Asia (Fig. 2), with a percentage ranging from 23% to 83% (Table 3). In a report by Wu et al. [25] covering a period of 30 years with more than 30,000 cases in China, *T. rubrum* showed a little decrease from 75% of the dermatophytes in 1986 to 57% in 2006, while *T. mentagrophytes* slightly increased from 18% to 24%. Meanwhile, *Candida albicans* showed significant increase and peaked around 2006 (26.9%). In Cai's report from South China with 475 cases of skin and nail infections, *T. rubrum* and *T. mentagrophytes* complex took a percentage of 81% and 15%, respectively [45].

Dermatophytes distribution in Iran shows quite a different pattern than China. In Tehran and Khuzestan, *T. mentagrophytes* complex and *E. floccosum* replace

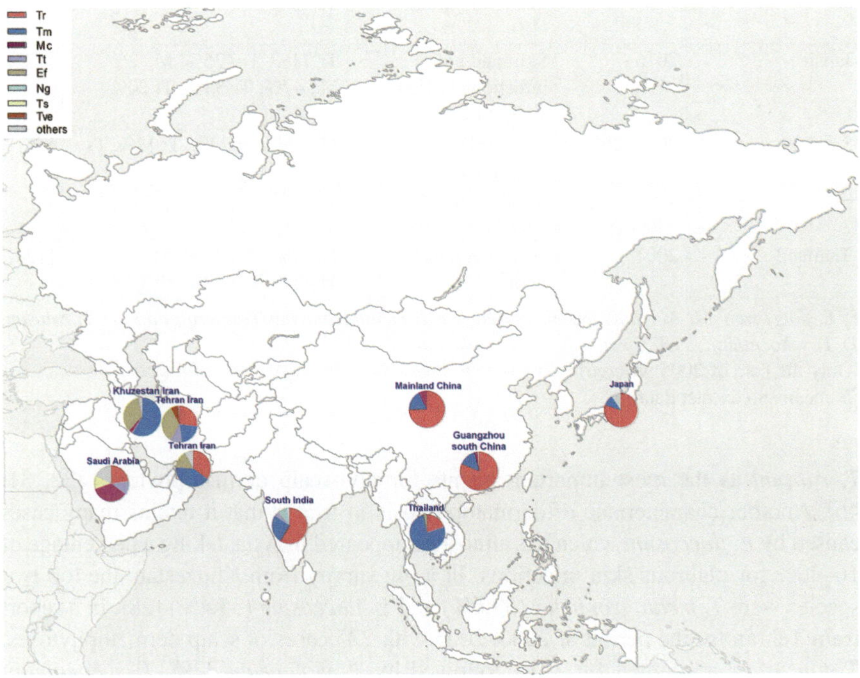

Fig. 2 Dermatophyte distribution in Asia isolated from skin and nail infections since 2000. Only major surveys with detailed and large data of dermatophyte spectrum were retrieved. The relative frequencies refer to the percentages among all dermatophyte isolates, excluding non-dermatophytes. *Ef E. floccosum, Mc M. canis, Ng N. gypsea, Tm T. mentagrophytes/T. interdigitale, Tr T. rubrum, Ts T. schoenleinii, Tt T. tonsurans, Tve T. verrucosum*

Table 3 Major surveys on dermatophytoses in Asia

Area	Period	Disease (cases)	Predominant species	Ref
Mainland China[a]	2006	All infections (13,209)	Tr 75%, Tm 18%, Mc 8%	[26]
South China, Guangzhou	2004–2014	Skin and nail infections (479) Scalp infections (109)	Tr 81%, Tm 15%, Mc 3% Mc 54%, Tm 20%, Tv 18%	[45]
South India	2012	All infections (150)	Tr 59%, Tm 25%, Ng 8%, Tt 5%	[48]
Iran, Khuzestan	2013–2014	Skin and nail infections (911) Scalp infections (120)	Tm 59%, Ef 35%, Mc 3% Tm 83%, Tv 2%, Mc 13%	[28]
Iran, Tehran	2000–2005	Skin and nail infections (4321) Scalp infections (225)	Tr 27%, Tm 21%, Ef 34%, Tt 10%, Tve 7% Tv 32%, Tt 37%, Mc 12%, Ts 12%	[31]
Iran, Tehran	2008–2010	Skin and nail infections (762) Scalp infections (15)	Tm 41%, Tr 35%, Ef 16% Mc 33%, Tt 20%, Tm 20%, Ts 13%	[46]
Japan	2016	Skin and nail infections (1253) Scalp infection (15)	Tr 71%, Tm 25%, Mc 2% Mc 47%, Tr 33%, Tt 20%	[123]
Saudi Arabia	2003–2005	Skin and nail infections (45) Scalp infections (26)	Mc 35%, Tr 23%, Tt 12%, Ts 12% Tm 58%, Mc 29%, Tr 7%	[47]
Thailand	2001	Nail infections[b] Foot[b]	Tr 77%, Tm 20%, Ef 3% Tr 41%, Tm 47%, Ef 12%	[124]

Ef E. floccosum, Mc M. canis, Ng N. gypsea, Tm T. mentagrophytes/T. interdigitale, Tr T. rubrum, Ts T. schoenleinii, Tt T. tonsurans, Tv T. violaceum, Tve T. verrucosum

[a]Only the data of 2006 was retrieved

[b]N means no extract data

T. rubrum as the most important agents for non-scalp dermatophytoses [28, 31, 46]. Another characteristic of dermatophytosis in Iran is that it retains many cases caused by *E. floccosum* which has almost disappeared in Asia, taking a percentage of 16–40% for glabrous skin infections. In a big survey from Khuzestan, the top two species were *T. mentagrophytes* (58.6%) and *E. floccosum* (35.4%) [28]. In a report from Tehran for the period of 2000–2005 with 225 cases of scalp dermatophytoses, *T. tonsurans* accounted for 37%, followed by *T. violaceum* 32%, *T. schoenleinii* 12%, and *M. canis* 12% [31]. But in another report from Khuzestan, southwest Iran, with 120 cases of *Tinea capitis* from 2013 to 2014, *T. mentagrophytes* was the most common species (83%) [28]. Interestingly, Saudi Arabia, neighboring Iran, has *T. mentagrophytes* as the most common agent for *Tinea capitis* coincidently [47]. The prevalence of *T. mentagrophytes* in India is similar to that reported in

Iran, and a switch from *T. rubrum* to *T. mentagrophytes* has been observed around 2018 [7, 8, 48]. In a multicenter study covering West, North, East, and South India, 92.62% dermatophytes were identified as *T. mentagrophytes* [7, 8]. The sequences revealed a solely occurring *T. mentagrophytes* "Indian ITS genotype" that might be disseminated due to the widespread abuse of topical clobetasol and other steroid molecules mixed with antifungal and antibacterial agents [7].

The scalp infection shows diversity in Asia with great geography correlation. *Microsporum canis* remains as the leading agent of scalp infections in most areas, while some anthropophilic dermatophytes are endemic. *Trichophyton violaceum*, *M. ferrugineum*, and *T. schoenleinii* are more prevalent in South Xinjiang province, West China [1]. Two other provinces of Central China, Jiangxi and Anhui, have *T. violaceum* as the predominant agent [1, 36]. There is an increasing trend of *T. violaceum* after 2000 in South China and *T. tonsurans* is rare in China today [1, 36, 44].

5.3 Dermatophytes in North and South America

The epidemiology of dermatophytes in America is similar to that seen in Europe, except that *T. tonsurans* is more common in Europe (Fig. 3).

There is limited data from North America describing dermatophytes spectrum after 2000. In an epidemiologic surveillance from 1999 to 2002 including 15,381 superficial specimens, dermatophytes remained the most commonly isolated fungal organisms. *Trichophyton rubrum* was the most prevalent pathogen (32–82.9%) for superficial fungal infections, and increased incidence of this species was observed for glabrous skin infections (onychomycosis, *Tinea corporis* and *Tinea cruris*, *Tinea manuum*, and *Tinea pedis*) [49].

At the beginning of the past century, *M. audouinii* and *M. canis* were the most common cause of *Tinea capitis* in the United States [4, 50]. The introduction of griseofulvin for treatment of *Tinea capitis* in the 1950s, together with a vigilant school surveillance program, led to a marked decline of superficial fungal infections in the United States [9]. Currently, *T. tonsurans* is the primary etiological agent of *Tinea capitis* instead of *M. audouinii* and *M. canis* [4, 35, 50]. In nationwide surveys, *T. tonsurans* was found to be most prevalent in the United States [51–53]. However, some rare species such as *M. audouinii*, *T. violaceum* and *T. soudanense* were also noticed in recent years due to migration from Africa [52].

The spectrum of dermatophytes in South America is similar, with a few exceptions. In Brazil and Venezuela, *T. rubrum, T. mentagrophytes* and *T. tonsurans* were the major pathogens [54–58]. For instance, in Mato Grosso (central-west Brazil), *T. rubrum* (33.3%), *T. mentagrophytes* (15.2%), and *T. verrucosum* (11.1%) together with *T. tonsurans* (13.1%) were found as the major causative species for *Tinea* in military [58].

Fig. 3 Dermatophyte distribution in America isolated from skin and nail infections since 2000. Only major surveys with detailed and large data of dermatophyte spectrum were retrieved. The relative frequencies refer to the percentages among all dermatophyte isolates, excluding non-dermatophytes. *Ef E. floccosum, Ma M. audouinii, Mc M. canis, Ng N. gypsea, Tm T. mentagrophytes/T. interdigitale, Tr T. rubrum, Tt T. tonsurans, Tve T. verrucosum*

Table 4 Major surveys on dermatophytoses in America

Area	Period	Disease (cases)	Predominant species	Ref
North America				
Dominican Republic	2010	Scalp infections (118)	Tt 61%, Ma 24%, Mc 12%	[125]
Mexico	1996–2006	Skin and nail infections (1789) Scalp infections (83)	Tr 83%, Tm 6%, Tt 5%, Mc 4% Tt 70%, Mc 30%	[126]
USA[a]	2002	*Tinea pedis* (15,381) Scalp infections (775)	Tr 83%, Tm 10% Tt 96%	[49]
South America				
Brazil, Mato Grosso	2009	All infections (99)	Tr 33%, Tt 13%, Tve 11%, Ng 8%	[54]
Northeast Brazil	2013–2014	Skin and nail infections (46) Scalp infections (13)	Tr 46%, Tt 20%, Mc 6.5% Ng 6.5% Tt 46%, Mc 38%, Tr 15%	[55]
Brazil, São Paulo	2005–2011	All infections (2626)	Tr 96% Tm 2%, Ng 1%	[56]
Venezuela, Caracas	2001–2014	All infections (1098)	Tr 70%, Tm 15%, Mc 9.4%, Ef 4%	[57]

Ef E. floccosum, Ma M. audouinii, Mc M. canis, Ng N. gypsea, Tm T. mentagrophytes/T. interdigitale, Tr T. rubrum, Tt T. tonsurans, Tve T. verrucosum

[a]For this report, we only use the data of the year 2002 and *Tinea pedis* for skin infections as there are no precise data for other clinical forms of dermatophytosis in this report

5.4 Dermatophytes in Africa

There are limited data of dermatophytes and dermatophytosis from Africa. Most extensive epidemiological data came from children surveys in Africa [59].

Tinea capitis has a high prevalence throughout African continent, infecting about 10–70% children of school-age children [59–63]. For adults, *Tinea corporis* seems to be the most common clinical type of dermatophytosis in Africa [59, 60]. In a study from Libya, *Tinea corporis* accounted for 45.9% of the superficial mycoses, and 85% of the cases occurred in children under 15 years of age [64].

Anthrophilic dermatophytes are predominant for *Tinea capitis* in Africa and *T. violaceum*, *T. soudanense* and *M. audouinii* are the most frequently isolated pathogens [59]. However, the distribution of the etiologic agents showed a high diversity. In most countries of west Africa such as Mali, Senegal, Ivory Coast, Gabon, and Nigeria, *T. soudanense* and *M. audouinii* are the main cause for *Tinea capitis*, while *T. violaceum* is the prevalent species in the remaining areas, followed by *M. canis* (North and East Africa) and *M. audouinii* (Central Africa) [59, 60, 65]. However, in Libya, *T. schoenleinii* is the most common agent, followed by *T. violaceum* and *M. audouinii.*

An extensive study for *Tinea corporis* in Dar-es-Salam, Tanzania, reported *T. rubrum* (27%), *T. mentagrophytes* (23%), *M. canis* (19%), and *T. violaceum* (18%) in a survey of nearly 500 infections [66]. In Morocco, among 4940 cases of onychomycosis, *T. rubrum* was the overwhelming agent (84%) [67], while in Sfax, Tunisia, *T. rubrum* took a percentage of 87% among etiological agents of glabrous skin infections, followed by *T. mentagrophytes* (8%) (Fig. 4) [65].

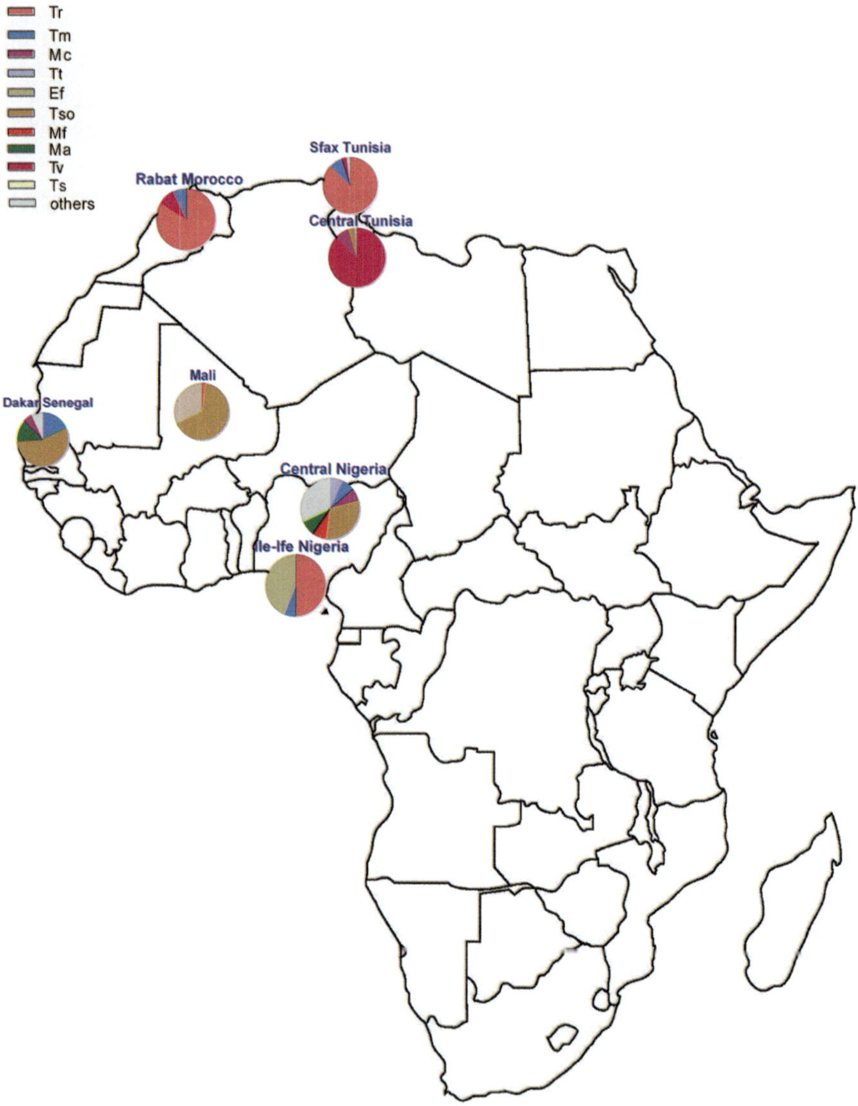

Fig. 4 Dermatophyte distribution in Africa isolated from skin and nail infections since 2000. Only major surveys with detailed and large data of dermatophyte spectrum were retrieved. The relative frequencies refer to the percentages among all dermatophyte isolates, excluding non-dermatophytes. *Ef E. floccosum, Ma M. audouinii, Mc M. canis, Mf M. ferrugineum, Tm T. mentagrophytes/T. interdigitale, Tr T. rubrum, Ts T. schoenleinii, Tso T. soudanense, Tt T. tonsurans, Tv T. violaceum*

Table 5 Major surveys on dermatophytoses in Africa

Area	Period	Disease (cases)	Predominant species	Ref
Mali	2001	Scalp infections (515)	Tso 66%, Ma 32%, Tr 2%	[62]
Morocco, Rabat	1982–2003	Nail infections (4940)	Tr 84%, Tv 9%, Tm 7%	[67]
Nigeria, Ile-Ife	2011	Skin and nail infections (17) Scalp infections (96)	Ef 44%, Tr 50%, Tm 6% Ma 46%, Tr 26%, Tm 18%, Ts 10%	[127]
Central Nigeria	2004	Scalp infections (245)	Tso 31%, Mf 8%, Ma 8%, Tm 7%, Tt 7%, Mc 7%	[128]
Senegal, Dakar	2008–2013	Scalp infections (566)	Tso 56%, Tr 18%, Ma 13%, Mc 6%	[57]
Central Tunisia	1986–2007	Nail infections (2887)	Tr 87%, Tv 8%, Tm 4%	[129]
Tunisia, Sfax	1998–2007	Skin and nail infections (12,873) Scalp infections (1350)	Tr 87%, Tm 8%, Tv 3% Tv 58%, Mc 30%, Tm 3%	[61]

Ef E. floccosum, Ma M. audouinii, Mc M. canis, Mf M. ferrugineum, Tm T. mentagrophytes/T. interdigitale, Tr T. rubrum, Ts T. schoenleinii, Tso T. soudanense, Tt T. tonsurans, Tv T. violaceum

6 Dermatophytoses and Related Diseases

Dermatophytosis are clinically classified according to the infectious sites as *Tinea pedis* (foot), *Tinea manuum* (palm), *Tinea corporis* (body), *Tinea unguium* (nail), *Tinea barbae* (moustache areas of the face), *Tinea capitis* (scalp), and invasive dermatophytoses. Considering the similar fungal spectrum of *Tinea pedis, Tinea manuum, Tinea barbae, Tinea corporis,* and *Tinea unguium,* we can simplify all these diseases into three classes [2]. The first class comprises all infections on glabrous skin and nails, with *T. rubrum* and *T. mentagrophytes* as the predominant pathogens, and the second class is the hair infection, with *M. canis*, *T. tonsurans* and *T. violaceum* as the major pathogens, but with variations according to geographic locations and studied populations [2]. The last group invades beyond the epidermis, which is termed as invasive dermatophytoses.

6.1 Tinea Capitis

Tinea capitis refers to the dermatophytoses of scalp hair follicles caused by species within *Microsporum* and *Trichophyton* genera.

In the United States, a 3% prevalence rate is reported for the scalp infection [68]. Yet the incidence of *Tinea capitis* is unknown in most countries, even though it remains an important public health problem, especially in prepubescent children.

The agents of *Tinea capitis* showed much geographical difference and time variation. Nowadays, *M. canis* is the leading agent of *Tinea capitis* in most parts of Europe and Asia [2]. *Trichophyton tonsurans* presents as the predominant pathogen of *Tinea capitis* in North/South America and the United Kingdom [15, 29, 30], while *T. violaceum*, *T. schoenleinii*, *M. ferrugineum*, and *M. audouinii* were still epidemic in Africa, eastern Europe, and Asia [15, 31–35].

Microsporum species invade the hair with an ectothrix pattern, in which arthrospores are formed on the outside of the hair shaft. Invasion of the hair by *Trichophyton* species usually presents as endothrix type, in which spores develop into the hair. In *T. schoenleinii* infections, the pathogen invades the hair medulla, but then regresses and leaves tunnels containing air within the hair shaft, named as the favic pattern [4, 69]. *Microsporum* species rarely infect adults and *Trichophyton* species infection can persist after puberty.

6.2 Dermatophytosis of Glabrous Skin and Nail

Most cases of *Tinea unguium*, *Tinea cruris*, *Tinea corporis*, *Tinea pedis*, and *Tinea manuum* are caused by *T. rubrum*, which is the foremost dermatophyte all over the world. *Tinea pedis* and *Tinea unguium* did not become common until the recent century, as a result of changes of modern lifestyle, e.g., popular of leather shoes and sneakers, the use of common bathing and sport facilities. Factually, this species is relatively new to human beings compared to the other main dermatophytes. Microsatellite analysis suggested that it originated from Africa, followed by the emergence of a new genotype in Asia with subsequent spread of this genotype over Europe and the United States by population expansion and migration after the Second World War [20, 21].

Species within *T. mentagrophytes* complex are the other important dermatophytes for skin and nail infections. Even in Australia, parts of Iran, and India, these fungi have already become the principal agents of dermatophytosis [2, 19, 24]. Other species, including *T. violaceum, T. tonsurans, E. floccosum, Nannizzia gypsea, T. soudanense, M. audouinii, M. ferrugineum*, and *T. verrucosum*, are also common dermatophytes causing superficial infections [2].

6.3 Deep Dermatophyte Infections

Dermatophytes rarely cause infections beyond the epidermis, and their exact incidence and frequency remain to be investigated. There are four different clinical types: (1) Majocchi's granuloma (MG), which is a dermatophytic infection accompanied with perifollicular granulomatous inflammation; (2) deeper dermatophytosis (deep dermal invasion due to dermatophytes, which spread not only to the perifollicular but also to the interfollicular area); (3) disseminated dermatophytosis (dermatophytic infection having extracutaneous spreading to other organs); and (4) mycetoma and pseudomycetoma (dermatophytes affecting the scalp and presenting as tumor-like growth with or without sinus tracts) [70–91].

6.4 Majocchi's Granuloma

Majocchi's granuloma (MG) was named after an Italian dermatologist, Prof. Domenico Majocchi (1849–1929), who first described a case of dermal granuloma caused by *T. tonsurans* in 1883 [70]. Later on, the description of MG was expanded to describe all dermatophytes, even extended to use for non-dermatophyte (*Aspergillus* species, *Phoma* species, *Malbranchea* species) perifollicular and subcutaneous infections. Currently, Durdu et al. [74] summarized the literature about diagnostic methods and criteria, and reinforce this definition of MG, that is, exclusively dermis infections due to dermatophytes. Thereafter, MG should only refer to dermatophyte-related infections, whereas perifollicular granulomatous disorders by non-dermatophyte species are suggested to be called "MG-like non-dermatophytic infections".

Fungal culture from dermis tissue and histopathology are the main diagnostic methods for MG. The former establishes the etiological agent, and the later presents perifollicular granulomatous inflammation as a confirmatory feature. Identification of the causative agent may be done from cultures by morphology and molecular methods.

Durdu et al. [74], Boral et al. [85], and Ilkit et al. [86] reviewed the literature including total 115 cases of MG and MG-like non-dermatophytic infections, in which dermatophytes are the dominant etiologic agents. Top three fungal pathogens were *T. rubrum* (65.7%), *T. mentagrophytes* (10.2%), and *T. tonsurans* (7.4%), followed by *M. canis* (3.7%), *N. gypsea* (2.8%), *T. violaceum* (1.9%), *T. interdigitale* (1.9%), *M. ferrugineum* (0.9%), *T. verrucosum* (0.9%), *M. audouinii* (0.9%), and *E. floccosum* (0.9%). The non-dermatophytic molds infection took up 2.8% cases, with *Phoma*, *Aspergillus* and *Malbranchea* species involved. By analyzing the epidemiology of MG, it was found that cases reported kept rising over the recent 5 years.

Pre-existent dermatophytosis is the major risk factor of MG and chronic dermatophytosis; for instance, *Tinea unguium* and *Tinea corporis* promote MG occurrence [75–79]. Other behaviors promoting this kind of infection include shaving the hairs, sexual contact, and specific occupation, which might damage the skin barrier [76, 79, 80]. MG can also be acquired by contact with animals, for example, Guinea pigs, indicating the clinicians should pay attention to the zoophilic characteristics of dermatophytes, as pet associated fungal infections have been increasing in the past decades [81–83].

The patient's immune status is closely associated with the type of invasion. In fact, among the mentioned diseases, only MG and pseudomycetoma have been observed in immunocompetent individuals [84]. MG can be detected in both immunocompromised (37.4%) and immunocompetent (62.6%) patients; however, almost all patients with deeper and disseminated dermatophytosis tend to have immunocompromised status [74, 85, 86]. Immunosuppressed conditions have been reported in patients with MG including malnutrition, lymphoma, leukemia, Cushing's syndrome, AIDS, long-term use of steroids, systemic chemotherapy, and immunosuppressants [74, 85–89].

In immunocompetent patients, the facial region is the most commonly infected area (37.5%), while the lower extremities (66.7%) are involved in immunosuppressed patients [85]. Other involved sites include groin and gluteal region (4.1%), trunk (4.1%), ear (2.7%), vulva (1.3%), and scrotum (1.3%) [82, 86, 90, 91]. Various types of lesions appear; however, the most common lesion in patients with MG is nodules (63.5%), followed by plaques (43.5%), papules (24.3%), ulcers (3.5%), and abscess (2.6%). About one third (29.6%) of patients have pustules on these lesions. The number of reported MG cases has increased significantly in the recent 9 years ($n = 40$), compared to that between 1883 and 2011 ($n = 79$). Furthermore, the frequency of facial infection was also obviously higher (36.4%) than that formerly reported (6.3%) [92].

6.5 Severe Dermatophytosis

We recognize three types of deep dermatophytic infections as severe invasive dermatophytosis. Different from MG, these invasive dermatophytoses feature the extension of the infection beyond the perifollicular area. The lesions appear as infiltrated papules, nodules, plaques, abscess, and necrosis, associated with itching, pain, discharge, and disfigurement. The pathogenic dermatophytes invade the dermis, interfollicular area, and even spread to lymph nodes, or organs either by contiguity (e.g., bone) or by hematogenous dissemination (e.g., central nervous system) [93–95].

Invasive dermatophytoses have been reported in immunocompromised patients with defective cellular immunity, most of which have identifiable innate or acquired immunodeficiency [93, 96]. The underlying risk factors responsible for invasive dermatophytosis include:

1. Solid organ transplantation (SOT). From the year 1987, about 32 cases of non-MG invasive dermatophytosis in SOT recipients have been published, and *T. rubrum* is the species most frequently involved. Rouzaud et al. [96] reported 12 cases of severe dermatophytosis in SOT recipients and found *Trichophyton rubrum* in 92% (11/12) patients, among which three had extensive infections.
2. HIV infection. Ten cases of invasive dermatophytosis in HIV-infected patients have been reported between 1999 and 2016. *Trichophyton rubrum*, *M. canis*, *N. gypsea*, *T. mentagrophytes*, *T. tonsurans*, and *Lophophyton gallinae* (formerly *Microsporum gallinae*) were the etiological agents [93, 97–104].
3. Genetic immune disorder. CARD9 deficiency predisposes to severe infections, including invasive dermatophytosis [105]. Twenty-three patients with deep dermatophytosis were reported to have CARD9 deficiency [89, 105–108]. The patients came from China, the United States, Tunisia, Algeria, Morocco, Egypt,

and Brazil. Clinical lesions started in childhood with recurrent and persistent superficial lesions. The isolated species were mainly *T. violaceum*, *T. rubrum*, and *M. ferrugineum*. Extensive dermatophytosis can occur in patients with chronic mucocutaneous candidiasis associated with STAT1 mutations [109], and in patients presenting as prurigo-induced pseudoperforation caused by *T. interdigitale* associated with STAT3 mutations [110]. Some patients have an underlying disorder of keratinization such as the KID (keratitis, ichthyosis, deafness) syndrome [111, 112]. Although some cases were reported without mycological identification or apparent risk factors, we can propose that underlying immunodeficiency might be persistent.

4. Secondary immunodeficiencies. Systemic corticosteroid treatment, immunosuppressive drugs (i.e., azathioprine, cyclosporine, cyclophosphamide, methotrexate, infliximab, and tacrolimus), hematological malignancy, autoimmune hepatitis, Cushing disease, congenital adrenal hyperplasia, atopy, and diabetes mellitus are reported as underlying risk factors for invasive dermatophytosis. In these patients, *T. rubrum* remains the predominant species; *Microsporum canis*, *T. mentagrophytes*, *T. violaceum*, *T. verrucosum*, and *E. floccosum* were sometimes involved, and *T. violaceum* is more frequently isolated from Africa [93, 96].

7 Summary and Conclusions

We systematically reviewed the evolution and epidemiological status of dermatophytosis worldwide with three axes: by time, by continents, and by related diseases. The aim is to better understand the dermatophytic epidemiological profile and its trends. In the past century, dermatophytes have experienced great evolution and their spectrum has remarkedly changed over the world with differences depending on the geographic area and socioeconomic development [2, 5]. The distribution and pattern of dermatophytosis will keep being dynamic in future. Nowadays *T. rubrum* is the predominant species for skin and nail infections, followed by *T. mentagrophytes/T. interdigitale* complex [2, 5]. Other species—*M. canis*, *T. tonsurans*, *T. violaceum*, *E. floccosum*, *M. audouninii*, etc.—are popular depending on geographic locations and populations [2, 5]. We notice a shift in the major pathogenic species from *T. rubrum* to *T. mentagrophytes* in some countries, especially in India, Iran, and Australia [6–8, 24].

Microsporum canis is the leading agent of *Tinea capitis* in most parts of Europe and Asia; however, *T. tonsurans* is the predominant species in North and South Americas and the United Kingdom. Anthropophilic *T. soudanense*, *M. audouinii*, and *T. violaceum* are highly epidemic in Africa with high diversity and imported infections due to these dermatophyte species were continuously reported in North America and Europe. Additionally, in some areas of Asia, including Iran, and south and western China, *T. violaceum* and *T. schoenleinii* related *Tinea capitis* are frequently diagnosed.

Due to the use of immunosuppressive therapy, SOT, HIV infection, etc., severe cases of dermatophytoses are increasingly reported. Although usually strictly superficial and limited, dermatophytosis have been playing a negative role in reducing the quality of life worldwide. Most severe, disseminated, chronic, and recalcitrant dermatophytic infections can lead to a poor outcome, even death. Considering the imbalance of regional medical conditions, we speculate that prevalence of these infections may be underestimated and wide epidemiological investigations are still needed.

Acknowledgments This work was funded by the National Natural Science Foundation of China (grant No. 81960367 and No. 81972949).

References

1. Zhan P, Li D, Wang C, Sun J, Geng C, Xiong Z, et al. Epidemiological changes in tinea capitis over the sixty years of economic growth in China. Med Mycol. 2015;53(7):691–8.
2. Zhan P, Liu W. The changing face of dermatophytic infections worldwide. Mycopathologia. 2017;182(1–2):77–86.
3. Fuller LC. Changing face of tinea capitis in Europe. Curr Opin Infect Dis. 2009;22:115–8.
4. Elewski BE. Tinea capitis: a current perspective. J Am Acad Dermatol. 2000;42(1 Pt 1):1–20.
5. Hayette M-P, Sacheli R. Dermatophytosis, trends in epidemiology and diagnostic approach. Curr Fungal Infect Rep. 2015;9(3):164–79.
6. Havlickova B, Czaika VA, Friedrich M. Epidemiological trends in skin mycoses worldwide. Mycoses. 2008;51(Suppl 4):2–15.
7. Nenoff P, Verma SB, Vasani R, Burmester A, Hipler UC, Wittig F, et al. The current Indian epidemic of superficial dermatophytosis due to *Trichophyton mentagrophytes*—a molecular study. Mycoses. 2019;62(4):336–56.
8. Thakur R, Kalsi AS. Outbreaks and epidemics of superficial dermatophytosis due to *Trichophyton mentagrophytes* complex and *Microsporum canis*: global and Indian scenario. Clin Cosmet Investig Dermatol. 2019;12:887–93.
9. Negroni R. Historical aspects of dermatomycoses. Clin Dermatol. 2010;28(2):125–32.
10. Espinel-Ingroff A. Medical mycology in the United States: a historical analysis (1894–1996). 1st ed. Dordrecht: Springer; 2003. p. 113.
11. Ajello L. Natural history of the dermatophytes and related fungi. Mycopathol Mycol Appl. 1974;53(1):93–110.
12. Espinel-Ingroff A. History of medical mycology in the United States. Clin Microbiol Rev. 1996;9:235–72.
13. Cafarchia CIR, Latrofa MS, Gräser Y, Otranto D. Molecular epidemiology, phylogeny and evolution of dermatophytes. Infect Genet Evol. 2013;20:336–51.
14. de Hoog GS, Haase G, Chaturvedi V, Walsh TJ, Meyer W, Lackner M. Taxonomy of medically important fungi in the molecular era. Lancet Infect Dis. 2013;13(5):385–6.
15. del Boz-González J. Tinea capitis: trends in Spain. Actas Dermosifiliogr. 2012;103(4):288–93.
16. Philpot CM. Geographical distribution of the dermatophytes: a review. J Hyg (Lond). 1978;80 (2):301–13.
17. Ginter-Hanselmayer G, Weger W, Ilkit M, Smolle J. Epidemiology of tinea capitis in Europe: current state and changing patterns. Mycoses. 2007;50(Suppl 2):6–13.
18. Ameen M. Epidemiology of superficial fungal infections. Clin Dermatol. 2010;28 (2):197–201.
19. Ohst T, de Hoog S, Presber W, Stavrakieva V, Gräser Y. Origins of microsatellite diversity in the *Trichophyton rubrum-T. violaceum* clade (dermatophytes). J Clin Microbiol. 2004;42 (10):4444–8.

20. Gräser Y, Kuijpers AF, Presber W, de Hoog GS. Molecular taxonomy of the *Trichophyton rubrum* complex. J Clin Microbiol. 2000;38(9):3329–36.
21. Dismukes WE, Pappas PG, Sobel JD. Clinical mycology. Oxford: Oxford University Press; 2003.
22. Gentles JC. Experimental ringworm in Guinea pigs: oral treatment with griseofulvin. Nature. 1958;182(4633):476–7.
23. Dolenc-Voljč M. Dermatophyte infections in humans: current trends and future prospects. In: Razzagghi-Abyaneh M, Shams-Ghahfarokhi M, Rai M, editors. Medical mycology: current trends and future prospects. Boca Raton, FL: CRC; 2016. p. 3–27.
24. Hainsworth S, Hamblin JF, Vanniasinkam T. Isolation of dermatophytes (and other fungi) from human nail and skin dust produced by podiatric medical treatments in Australia. J Am Podiatr Med Assoc. 2015;105(2):111–20.
25. Wu SX, Guo NR, Li XF, Liao WQ, Chen M, Zhang QQ, et al. Human pathogenic fungi in China—emerging trends from ongoing national survey for 1986, 1996, and 2006. Mycopathologia. 2011;171:387–93.
26. Nenoff P, Krüger C, Ginter-Hanselmayer G, Tietz HJ. Mycology - an update. Part 1: Dermatomycoses: causative agents, epidemiology and pathogenesis. J Dtsch Dermatol Ges. 2014;12(3):188–209; quiz 210, 188–211; quiz 212.
27. Seyedmousavi S, Rafati H, Ilkit M, Tolooe A, Hedayati MT, Verweij P. Systemic antifungal agents: current status and projected future developments. Methods Mol Biol. 2017;1508:107–39.
28. Rezaei-Matehkolaei A, Rafiei A, Makimura K, Gräser Y, Gharghani M, Sadeghi-Nejad B. Epidemiological aspects of dermatophytosis in Khuzestan, southwestern Iran, an update. Mycopathologia. 2016;181(7–8):547–53.
29. Miklić P, Skerlev M, Budimcić D, Lipozencić J. The frequency of superficial mycoses according to agents isolated during a ten-year period (1999–2008) in Zagreb area. Croatia Acta Dermatovenerol Croat. 2010;18:92–8.
30. Aghamirian MR, Ghiasian SA. Dermatophytes as a cause of epizoonoses in dairy cattle and humans in Iran: epidemiological and clinical aspects. Mycoses. 2011;54(4):e52–6.
31. Bassiri-Jahromi S, Khaksari AA. Epidemiological survey of dermatophytosis in Tehran, Iran, from 2000 to 2005. Indian J Dermatol Venereol Leprol. 2009;75(2):142–7.
32. de Hoog GS, Dukik K, Monod M, Packeu A, Stubbe D, Hendrickx M, et al. Toward a novel multilocus phylogenetic taxonomy for the dermatophytes. Mycopathologia. 2017;182 (1–2):5–31.
33. Frias-De-Leon MG, Martinez-Herrera E, Atoche-Dieguez CE, Cespon JLG, Uribe B, Arenas R, et al. Molecular identification of isolates of the *Trichophyton mentagrophytes* complex. Int J Med Sci. 2020;17(1):45–52.
34. Mapelli ET, Cerri A, Bombonato C, Menni S. Tinea capitis in the paediatric population in Milan, Italy: the emergence of *Trichophyton violaceum*. Mycopathologia. 2013;176 (3–4):243–6.
35. Mirmirani P, Tucker LY. Epidemiologic trends in pediatric tinea capitis: a population-based study from Kaiser Permanente northern California. J Am Acad Dermatol. 2013;69(6):916–21.
36. Zhan P, Geng C, Li Z, Jin Y, Jiang Q, Tao L, et al. Evolution of tinea capitis in the Nanchang area, southern China: a 50-year survey (1965–2014). Mycoses. 2015;58(5):261–6.
37. Alexander CL, Brown L, Shankland GS. Epidemiology of fungal scalp infections in the west of Scotland 2000–2006. Scott Med J. 2009;54(2):13–6.
38. Gits-Muselli M, Benderdouche M, Hamane S, Mingui A, Feuilhade de Chauvin M, Guigue N, et al. Continuous increase of *Trichophyton tonsurans* as a cause of tinea capitis in the urban area of Paris, France: a 5-year-long study. Med Mycol. 2017;55(5):476–84.
39. Kolivras A, Lateur N, De Maubeuge J, Scheers C, Wiame L, Song M. Tinea capitis in Brussels: epidemiology and new management strategy. Dermatology. 2003;206(4):384–7.
40. Gaviria Morales E, Iorizzo M, Martinetti Lucchini G, Mainetti C. *Trichophyton violaceum*: an emerging pathogen in southern Switzerland. Dermatology. 2019;235(5):434–9.

41. Kieliger S, Glatz M, Cozzio A, Bosshard PP. Tinea capitis and tinea faciei in the Zurich area - an 8-year survey of trends in the epidemiology and treatment patterns. J Eur Acad Dermatol Venereol. 2015;29(8):1524–9.
42. Zink A, Papanagiotou V, Todorova A, Seidl HP, Niedermeier A, Ring J, et al. Outbreak of *Microsporum audouinii* in Munich—the return of infectious fungi in Germany. Mycoses. 2014;57(12):765–70.
43. Farina C, Fazii P, Imberti G, Lombardi G, Passera M, Andreoni S, et al. *Trichophyton violaceum* and *T. soudanense*: re-emerging pathogens in Italy, 2005–2013. New Microbiol. 2015;38:409–15.
44. Zhu M, Li L, Wang J, Zhang C, Kang K, Zhang Q. Tinea capitis in southeastern China: a 16-year survey. Mycopathologia. 2010;169:235–9.
45. Cai W, Lu C, Li X, Zhang ZP, Xi L, et al. Epidemiology of superficial fungal infections in Guangdong, southern China: a retrospective study from 2004 to 2014. Mycopathologia. 2016;181:387–95.
46. Rezaei-Matehkolaei A, Makimura K, de Hoog S, Shidfar MR, Zaini F, Eshraghian M, et al. Molecular epidemiology of dermatophytosis in Tehran, Iran, a clinical and microbial survey. Med Mycol. 2013;51:203–7.
47. Abanmi A, Bakheshwain S, El Khizzi N, Zouman AR, Hantirah S, Al Harthi F, et al. Characteristics of superficial fungal infections in the Riyadh region of Saudi Arabia. Int J Dermatol. 2008;47:229–35.
48. Hanumanthappa H, Sarojini K, Shilpashree P, Muddapur SB. Clinicomycological study of 150 cases of dermatophytosis in a tertiary care hospital in South India. Ind J Dermatol. 2012;57 (4):322–3.
49. Foster KW, Ghannoum MA, Elewski BE. Epidemiologic surveillance of cutaneous fungal infection in the United States from 1999 to 2002. J Am Acad Dermatol. 2004;50(5):748–52.
50. Kemna ME, Elewski BE. A U.S. epidemiologic survey of superficial fungal diseases. J Am Acad Dermatol. 1996;35:539–42.
51. Leung AKC, Hon KL, Leong KF, Barankin B, Lam JM. Tinea capitis: an updated review. Recent Pat Inflamm Allergy Drug Discov. 2020;14(1):58–68.
52. Grigoryan KV, Tollefson MM, Olson MA, Newman CC. Pediatric tinea capitis caused by *Trichophyton violaceum* and *Trichophyton soudanense* in Rochester, Minnesota, United States. Int J Dermatol. 2019;58(8):912–5.
53. Panackal AA, Halpern EF, Watson AJ. Cutaneous fungal infections in the United States: analysis of the National Ambulatory Medical Care Survey (NAMCS) and National Hospital Ambulatory Medical Care Survey (NHAMCS), 1995–2004. Int J Dermatol. 2009;48 (7):704–12.
54. Lopez-Martinez R, Manzano-Gayosso P, Hernández-Hernández F, Bazán-Mora E, Méndez-Tovar LJ. Dynamics of dermatophytosis frequency in Mexico: an analysis of 2084 cases. Med Mycol. 2010;48:476–9.
55. Silva-Rocha WP, de Azevedo M, Chaves GM. Epidemiology and fungal species distribution of superficial mycoses in Northeast Brazil. J Mycol Méd. 2017;27(1):57–64.
56. Di Chiacchio N, Madeira CL, Humaire CR, Silva CS, Fernandes LH, Dos Reis AL. Superficial mycoses at the hospital do Servidor Publico municipal de Sao Paulo between 2005 and 2011. An Bras Dermatol. 2014;89(1):67–71.
57. Capote AM, Ferrara G, Panizo MM, Garcia N, Alarcon V, Reviakina V, et al. Superficial mycoses: casuistry of the mycology Department of the Instituto Nacional de Higiene “Rafael Rangel”, Caracas, Venezuela (2001–2014). Invest Clin. 2016;57(1):47–58.
58. Leite DP Jr, Amadio JV, Simões Sde A, de Araujo SM, da Silva NM, Anzai MC, et al. Dermatophytosis in military in the central-west region of Brazil: literature review. Mycopathologia. 2014;177:65–74.
59. Coulibaly O, L'Ollivier C, Piarroux R, Ranque S. Epidemiology of human dermatophytoses in Africa. Med Mycol. 2018;56(2):145–61.

60. Nweze EI, Eke IE. Dermatophytes and dermatophytosis in the eastern and southern parts of Africa. Med Mycol. 2018;56(1):13–28.
61. Ndiaye M, Diongue K, Seck MC, Badiane AS, Diallo MA, Deme AB, et al. Epidemiological profile of tinea capitis in Dakar (Senegal). A 6-year retrospective study (2008–2013). J Mycol Méd. 2015;25(2):169–76.
62. Maïga II, Dicko DS, Guindo M, Diawara-Konaré H, Rocherau A, Kéita S. Epidemiology of tinea capitis in school circle in Bamako. J Mycol Méd. 2001;11:143–8.
63. Oladele RO, Denning DW. Burden of serious fungal infection in Nigeria. West Afr J Med. 2014;33(2):107–14.
64. Ellabib MS, Khalifa ZM. Dermatophytes and other fungi associated with skin mycoses in Tripoli, Libya. Ann Saudi Med. 2001;21:193–5.
65. Neji S, Makni F, Cheikhrouhou F, Sellami A, Sellami H, Marreckchi S, et al. Epidemiology of dermatophytoses in Sfax, Tunisia. Mycoses. 2009;52:534–8.
66. Komba EV, Mgonda YM. The spectrum of dermatological disorders among primary school children in Dar Es Salaam. BMC Public Health. 2010;10:765.
67. Boukachabine K, Agoumi A. Onychomycosis in Morocco: experience of the parasitology and medical mycology laboratory from Rabat children hospital (1982–2003). Ann Biol Clin (Paris). 2005;63(6):639–42.
68. Al Sogair S, Hay RJ. Fungal infection in children: tinea capitis. Clin Dermatol. 2000;18 (6):679–85.
69. Hay RJ. Dermatophytosis (ringworm) and other superficial mycoses. In: Bennett JE, Dolin R, Blaser MJ, editors. Mandell, Douglas, and Bennett's principles and practice of infectious diseases, vol. 2; 2015. p. 2985–94.e1.
70. Majocchi D. A new trichophytic granuloma: clinical and mycologic studies. Bull R Acad Med Roma. 1883;9:220–3.
71. Smith KJ, Welsh M, Skelton H. *Trichophyton rubrum* showing deep dermal invasion directly from the epidermis in immunosuppressed patients. Br J Dermatol. 2001;145(2):344–8.
72. Marconi VC, Kradin R, Marty FM, Hospenthal DR, Kotton CN. Disseminated dermatophytosis in a patient with hereditary hemochromatosis and hepatic cirrhosis: case report and review of the literature. Med Mycol. 2010;48(3):518–27.
73. Isa-Isa R, Arenas R, Isa M. Inflammatory tinea capitis: kerion, dermatophytic granuloma, and mycetoma. Clin Dermatol. 2010;28(2):133–6.
74. Durdu M, Kandemir H, Ilkit M, de Hoog GS. Changing concepts and current definition of Majocchi's granuloma. Mycopathologia. 2020;185(1):187–92.
75. Kranseler J, Gupta D. Erythematous annular plaques on a 9-year-old's arms and chest. Clin Pediatr (Phila). 2018;57(4):488–90.
76. Rallis E, Katoulis A, Rigopoulos D. Pubic Majocchi's granuloma unresponsive to itraconazole successfully treated with oral terbinafine. Skin Appendage Disord. 2016;1(3):111–3.
77. Su HA, Sun PL, Sung WW, Cheng SY, Chang HC, Yang JH, et al. Deep dermatophytosis caused by zoophilic strain of *Trichophyton interdigitale* with successful treatment of itraconazole. Mycopathologia. 2017;182(7–8):715–20.
78. Kim JE, Won CH, Chang S, Lee MW, Choi JH, Moon KC. Majocchi's granuloma mimicking Kaposi sarcoma in a heart transplant patient. J Dermatol. 2011;38(9):927–9.
79. Gallo JG, Woods M, Graham RM, Jennison AV. A severe transmissible Majocchi's granuloma in an immunocompetent returned traveler. Med Mycol Case Rep. 2017;18:5–7.
80. Parmar NV, Asir GJ, Rudramurthy SM. Atypical presentation of Majocchi's granuloma in an immunocompetent host. Am J Trop Med Hyg. 2017;96(1):1–2.
81. Liu C, Landeck L, Cai SQ, Zheng M. Majocchi's granuloma over the face. Indian J Dermatol Venereol Leprol. 2012;78(1):113–4.
82. Wargo JJ, Berg A, Gudjonsson JE. Ongoing pleuritic chest pain and a Guinea pig: missed pulmonary embolism and Majocchi's granuloma. Am J Med. 2016;129(9):e165–7.

83. Mazur M, Lodyga M, Łańczak A, Adamski Z. Majocchi's granuloma (granuloma trichophyticum) in a Guinea pig owner: a case report and literature review. J Mycol Méd. 2018;28(3):523–6.
84. Kramer SC, Ryan M, Bourbeau P, Tyler WB, Elston DM. Fontana-positive grains in mycetoma caused by *Microsporum canis*. Pediatr Dermatol. 2006;23(5):473–5.
85. Boral H, Durdu M, Ilkit M. Majocchi's granuloma: current perspectives. Infect Drug Resist. 2018;11:751–60.
86. Ilkit M, Durdu M, Karakas M. Majocchi's granuloma: a symptom complex caused by fungal pathogens. Med Mycol. 2012;50(5):449–57.
87. Radentz WH, Yanase DJ. Papular lesions in an immunocompromised patient. *Trichophyton* rubrum granulomas (Majocchi's granuloma). Arch Dermatol. 1993;129(9):1189–90, 1192–3.
88. Kim ST, Baek JW, Kim TK, Lee JW, Roh HJ, Jeon YS, et al. Majocchi's granuloma in a woman with iatrogenic Cushing's syndrome. J Dermatol. 2008;35(12):789–91.
89. Li FQ, Lv S, Xia JX. Majocchi's granuloma after topical corticosteroids therapy. Case Rep Dermatol Med. 2014;2014:507176.
90. Raducan A, Constantin MG. To be or not to be Tinea—misdiagnosis and inappropriate treatment. DermatoVenerol (Buc). 2017;62(3):29–32.
91. Lyra MR, Muniz Alvarez B, Lanziano AL, Imbeth MAA, Sa AM, Cuzzi T, et al. Exfoliative erythroderma and palmoplantar hyperkeratosis associated with Majocchi's granuloma by *Trichophyton tonsurans* in a patient with AIDS. Rev Iberoam Micol. 2017;34(3):185–8.
92. Inaoki M, Nishijima C, Miyake M, Asaka T, Hasegawa YH, Anzawa K, et al. Case of dermatophyte abscess caused by *Trichophyton rubrum*: a case report and review of the literature. Mycoses. 2015;58(5):318–23.
93. Rouzaud C, Hay R, Chosidow O, Dupin N, Puel A, Lortholary O, et al. Severe dermatophytosis and acquired or innate immunodeficiency: a review. J Fungi (Basel). 2015;2(1):4.
94. Nazarian RM, Lilly E, Gavino C, Hamilos DL, Felsenstein D, Vinh DC, et al. Novel CARD9 mutation in a patient with chronic invasive dermatophyte infection (tinea profunda). J Cutan Pathol. 2020;47(2):166–700.
95. Cheikhrouhou F, Makni F, Masmoudi A, Sellami A, Turki H, Ayadi A. A fatal case of dermatomycoses with retropharyngeal abscess. Ann Dermatol Venereol. 2010;137(3):208–11.
96. Rouzaud C, Chosidow O, Brocard A, Fraitag S, Scemla A, Anglicheau D, et al. Severe dermatophytosis in solid organ transplant recipients: a French retrospective series and literature review. Transpl Infect Dis. 2018;20(1).
97. King D, Cheever LW, Hood A, Horn TD, Rinaldi MG, Merz WG. Primary invasive cutaneous *Microsporum canis* infections in immunocompromised patients. J Clin Microbiol. 1996;34(2):460–2.
98. Del Palacio A, Pereiro-Miguens M, Gimeno C, Cuétara MS, Rubio R, Costa R, et al. Widespread dermatophytosis due to *Microsporum* (*Trichophyton*) *gallinae* in a patient with AIDS—a case report from Spain. Clin Exp Dermatol. 1992;17(6):449–53.
99. Hadacek BM, Nassif A, Roux A, Desplaces N, Huerre M, De Bièvre C, et al. *Trichophyton tonsurans* dermatophyte granuloma in an HIV-1 infected patient. Br J Dermatol. 1999;140(4):762–3.
100. Kwon KS, Jang HS, Son HS, Oh CK, Kwon YW, Kim KH, et al. Widespread and invasive *Trichophyton rubrum* infection mimicking Kaposi's sarcoma in a patient with AIDS. J Dermatol. 2004;31(10):839–43.
101. Lowinger-Seoane M, Torres-Rodriguez JM, Madrenys-Brunet N, Aregall-Fusté S, Saballs P. Extensive dermatophytoses caused by *Trichophyton mentagrophytes* and *Microsporum canis* in a patient with AIDS. Mycopathologia. 1992;120(3):143–6.
102. Muñoz-Pèrez MA, Rodriguez-Pichardo A, Camacho F, Rios JJ. Extensive and deep dermatophytosis caused by *Trichophyton mentagrophytes* var. *interdigitalis* in an HIV-1 positive patient. J Eur Acad Dermatol Venereol. 2000;14(1):61–3.

103. Porro AM, Yoshioka MC, Kaminski SK, Palmeira Mdo C, Fisschman O, Alchorne MM. Disseminated dermatophytosis caused by *Microsporum gypseum* in two patients with the acquired immunodeficiency syndrome. Mycopathologia. 1997;137(1):9–12.
104. Tsang P, Hopkins T, Jimenez-Lucho V. Deep dermatophytosis caused by *Trichophyton rubrum* in a patient with AIDS. J Am Acad Dermatol. 1996;34(6):1090–1.
105. Grumach AS, de Querioz-Telles F, Migaud M, Lanternier F, Filho NR, Palma SM, et al. A homozygous CARD9 mutation in a Brazilian patient with deep dermatophytosis. J Clin Immunol. 2015;35(5):486–90.
106. Lanternier F, Cypowyj S, Picard C, Bustamante J, Lortholary O, Casanova JL, et al. Primary immunodeficiencies underlying fungal infections. Curr Opin Pediatr. 2013;25(6):736–47.
107. Zhang Y, Mijiti J, Huang C, Song Y, Wan Z, Li R, et al. Deep dermatophytosis caused by *Microsporum ferrugineum* in a patient with CARD9 mutations. Br J Dermatol. 2019;181 (5):1093–5.
108. Boudghene Stambouli O, Amrani N, Boudghene Stambouli K, Bouali F. Dermatophytic disease with deficit in CARD9: a new case with a brain impairment. J Mycol Méd. 2017;27 (2):250–3.
109. Nielsen J, Kofod-Olsen E, Spaun E, Larsen CS, Christiansen M, Mogensen TH. A STAT1-gain-of-function mutation causing Th17 deficiency with chronic mucocutaneous candidiasis, psoriasiform hyperkeratosis and dermatophytosis. BMJ Case Rep. 2015;2015: bcr2015211372.
110. Simpson JK, Fröbel P, Seneviratne SL, Brown M, Lowe DM, Grimbacher B, et al. Invasive dermatophyte infection with *Trichophyton interdigitale* is associated with prurigo-induced pseudoperforation and a signal transducer and activator of transcription 3 mutation. Br J Dermatol. 2018;179(3):750–4.
111. Nazzaro V, Blanchet-Bardon C, Badillet G, Puissant A. Hyperkeratosis, deafness and keratitis (KID syndrome). Study of 3 cases with dermatophytosis and review of the literature. G Ital Dermatol Venereol. 1988;123(1–2):5–14.
112. Machan M, Kestenbaum T, Fraga GR. Diffuse hyperkeratosis in a deaf and blind 48-year-old woman—quiz case. Diagnosis: keratitis-ichthyosis-deafness (KID) syndrome with secondary dermatophytosis. Arch Dermatol. 2012;148(10):1199–2000.
113. Babić-Erceg A, Barisić Z, Erceg M, Babić A, Borzić E, Zoranić V, et al. Dermatophytoses in Split and Dalmatia, Croatia, 1996–2002. Mycoses. 2004;47:297–9.
114. Saunte DM, Svejgaard EL, Haedersdal M, Frimodt-Møller N, Jensen AM, Arendrup MC. Laboratory-based survey of dermatophyte infections in Denmark over a 10-year period. Acta Derm Venereol. 2008;88:614–6.
115. Faure-Cognet O, Fricker-Hidalgo H, Pelloux H, Leccia MT. Superficial fungal infections in a French teaching hospital in Grenoble area: retrospective study on 5470 samples from 2001 to 2011. Mycopathologia. 2016;181:59–66.
116. Maraki S, Mavromanolaki VE. Epidemiology of dermatophytoses in Crete, Greece. Med Mycol J. 2016;57(4):e69–75.
117. Tsoumani M, Jelastopulu E, Bartzavali C, Vamvakopoulou S, Dimitracopoulos G, Anastassiou ED, et al. Changes of dermatophytoses in southwestern Greece: an 18-year survey. Mycopathologia. 2011;172:63–7.
118. Panasiti V, Devirgiliis V, Borroni RG, Mancini M, Curzio M, Rossi M, et al. Epidemiology of dermatophytic infections in Rome, Italy: a retrospective study from 2002 to 2004. Med Mycol. 2007;45(1):57–60.
119. Budak A, Bogusz B, Tokarczyk M, Trojanowska D. Dermatophytes isolated from superficial fungal infections in Krakow, Poland, between 1995 and 2010. Mycoses. 2013;56:422–8.
120. Drakensjo IT, Chryssanthou E. Epidemiology of dermatophyte infections in Stockholm, Sweden: a retrospective study from 2005–2009. Med Mycol. 2011;49(5):484–8.
121. Monod M, Jaccoud S, Zaugg C, Léchenne B, Baudraz F, Panizzon R. Survey of dermatophyte infections in the Lausanne area Switzerland. Dermatology. 2002;205:201–3.

122. Borman AM, Campbell CK, Fraser M, Johnson EM. Analysis of the dermatophyte species isolated in the British Isles between 1980 and 2005 and review of worldwide dermatophyte trends over the last three decades. Med Mycol. 2007;45:131–41.
123. Shimoyama H, Sei Y. 2016 epidemiological survey of dermatomycoses in Japan. Med Mycol J. 2019;60(3):75–82.
124. Ungpakorn R. Mycoses in Thailand: current concerns. Nippon Ishinkin Gakkai Zasshi. 2005;46(2):81–6.
125. Arenas R, Torres E, Amaya M, Rivera ER, Espinal A, Polanco M, et al. Emergence of *Microsporum audouinii* and *Trichophyton tonsurans* as causative organisms of tinea capitis in the Dominican Republic. Actas Dermosifiliogr. 2010;101:330–5.
126. López-Martínez R, Manzano-Gayosso P, Hernández-Hernández F, Bazán-Mora E, Méndez-Tovar LJ. Dynamics of dermatophytosis frequency in Mexico: an analysis of 2084 cases. Med Mycol. 2010;48(3):476–9.
127. Oke OO, Onayemi O, Olasode OA, Omisore AG, Oninla OA. The prevalence and pattern of superficial fungal infections among school children in Ile-Ife, South-Western Nigeria. Dermatol Res Pract. 2014;2014:842917.
128. Ayanbimpe GM, Taghir H, Diya A, Wapwera S. Tinea capitis among primary school children in some parts of Central Nigeria. Mycoses. 2008;51:336–40.
129. Dhib I, Fathallah A, Yaacoub A, Zemni R, Gaha R, Said MB. Clinical and mycological features of onychomycosis in Central Tunisia: a 22 years retrospective study (1986–2007). Mycoses. 2013;56(3):273–80.

Onychomycosis in the Twenty-First Century: An Update on Epidemiology and Diagnosis

Aditya K. Gupta, Maanasa Venkataraman, and Emma M. Quinlan

Abstract

Onychomycosis, a fungal infection of the nail, is the most common nail disease with a global prevalence of 5.5%. Caused by dermatophytes in the majority of countries in which it is studied, new evidence suggests that mixed infections and non-dermatophyte mold infections occur more often than previously thought, especially in warmer climates. Onychomycosis presents as a variety of symptoms depending on the pathogen and mode of invasion into the nail unit, both factors which also affect disease severity and optimal treatment. The risk of disease increases with age, as well as due to poor hygiene, genetic predisposition, history of *Tinea pedis*, diabetes, psoriasis, peripheral vascular disease, obesity, and immunosuppression. There are numerous techniques to diagnose onychomycosis, with direct microscopy and fungal culture as the gold standard. Other techniques such as histology and PCR are gaining favor due to faster and more accurate results. Mixed infections (dermatophytes with non-dermatophytes) require multiple confirmatory sampling and techniques to diagnose, as these fungi are commensals and contaminants. Proper and timely diagnosis of onychomycosis is required, as reinfection is common and more severe cases are less likely to respond to treatment.

A. K. Gupta (✉)
Division of Dermatology, Department of Medicine, University of Toronto, Toronto, ON, Canada

Mediprobe Research Inc., London, ON, Canada
e-mail: agupta@execulink.com

M. Venkataraman · E. M. Quinlan
Mediprobe Research Inc., London, ON, Canada

J.-P. Bouchara et al. (eds.), *Dermatophytes and Dermatophytoses*,
https://doi.org/10.1007/978-3-030-67421-2_3

Keywords

Onychomycosis · *Tinea unguium* · DLSO · Onychomycosis diagnosis · Onychomycosis epidemiology

1 Introduction

Onychomycosis is a fungal infection of the nail caused predominantly by dermatophytes, and less frequently by yeasts and non-dermatophyte molds (NDMs). The disease causes nail discoloration and thickening, with the most severe cases resulting in permanent disfiguration of the nail. Onychomycosis sufferers often feel shame and embarrassment over their condition, leading them to hide their feet and hands from both family and physicians, and shy away from certain activities in public. The disease is extremely difficult to treat, and often spreads between members of the same household and in public places such as pools and showers. Recognition of the clinical signs of the various types of onychomycosis and proper diagnosis of the infecting organism lead to earlier detection, appropriate treatment regimen choice, and better treatment outcomes.

2 Nail Anatomy

The anatomy and function of the nail directly influence its susceptibility to fungal infection. As useful tools that increase touch sensitivity and protect the fingertips, fingernails are frequently exposed to our external environment, increasing the chance of coming into contact with pathogenic fungi. Fingernails, along with fingers and hands, are often exposed to cleaners and soaps that kill competing bacteria, allowing fungal pathogens to colonize. Toenails are usually covered by socks and shoes in temperate climates, but in moist environments like sweaty socks and public showers, the anatomical barriers of the nail can be compromised, allowing both infectious and commensal fungi to invade the nail. In warmer climates, walking barefoot increases susceptibility to fungal infection.

To combat frequent exposure to physical damage and pathogens, the structure of the nail unit serves to protect the delicate tissues at the distal end of fingers and toes, while providing a barrier to environmental pathogens. The nail unit consists of 4 main components, namely: the nail plate, which is the hard, protective surface of the nail; the nail matrix, from which the nail plate grows; and the hyponychium and surrounding nail folds, which serve as barriers to intrusion from external organisms. The anatomy of the nail is depicted in Fig. 1 [1–4].

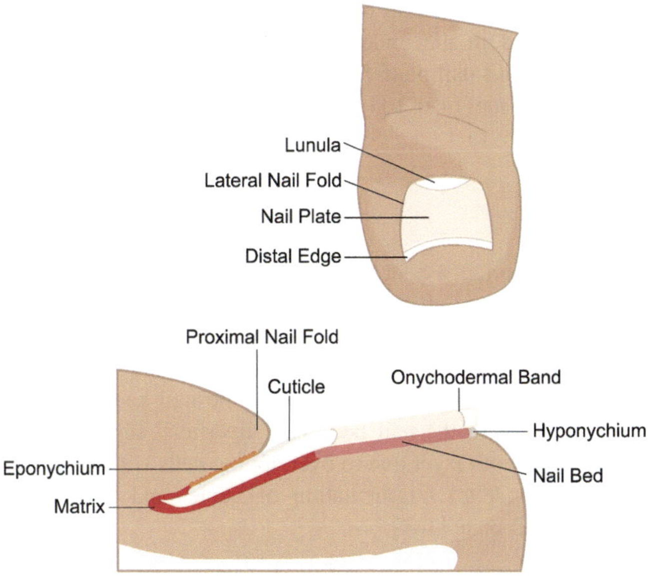

Fig. 1 Dorsal and lateral view of human nail anatomy

2.1 Nail Matrix

The nail matrix is the root of the nail unit and foundation of the nail plate. The distal end of the nail matrix opens into a crescent-shaped structure known as the lunula. The proximal nail folds cover the underlying nail matrix. The nail matrix is acanthotic; i.e., it has a thick epithelium due to thick stratum basal and stratum spinosum layers. However, the epithelium lacks a stratum granulosum layer. The nail matrix epithelium transitions into the thin nail bed at the distal edge of the lunula.

The nail matrix consists of melanocytes, keratinocytes, Langerhans cells, and Merkel cells. The keratinocytes undergo differentiation to become hard onychocytes and comprise the cells of the nail plate. As the basal cells differentiate, they flatten, lose their organelles, and become hard and keratinous in structure. The lunula is known as the keratogenous zone where the basal cells begin their differentiation process. However, unlike in the rest of the nail plate, at the lunula they retain their nuclei and form eosinophilic cytoplasm [1–4].

2.2 Nail Bed

The nail bed is the underlying component of the nail plate. It provides structural support to the nail unit. It forms the portion from the distal end of lunula to the distal end of the attached nail plate, i.e., the hyponychium. The nail bed epithelium is thin compared to that of the nail matrix and similar in structure, with an absence of the stratum granulosum layer. The nail bed is attached to the nail unit by longitudinal epidermal ridges found in the underside of the nail plate. The contribution of cornified cells to the hard nail plate from the basal cells of the nail bed epithelium is low compared to the nail matrix [1–4].

2.3 Nail Plate

The nail plate is an important aspect of the nail unit; it is the nail itself. The nail plate consists of about 196 rows of hard, cornified onychocytes which are devoid of nuclei, organelles, and keratohyalin granules [3]. It extends from the proximal nail fold to the hyponychium covering the nail matrix and the nail bed. It has longitudinal epidermal ridges that interdigitate with those of the nail bed, firmly attaching the ventral side of the plate to the nail bed. The superficial surface of the nail plate presents as a shiny, smooth layer due to overlapping cells without nuclei. The basal cells of the matrix undergo fragmentation of nuclei and lysis to form mature onychocytes of the nail plate [1–4].

2.4 Hyponychium

The hyponychium consists of the tissue found between the distal end of the nail bed and the distal nail groove beneath the free distal end of the nail plate. The epithelium of the hyponychium is different from that of the matrix and nail bed, with an additional stratum granulosum layer of which forms a thick cornified layer of cells that act as a physical barrier. Dermatophytes that cause distal and lateral subungual onychomycosis invade the hyponychium primarily before progressing into the nail bed and the nail plate [1–4].

2.5 Onychodermal Band

The onychodermal band is the small band between the nail bed epithelium and the hyponychium. The thickness of the band is between 0.5 and 1.5 mm. It has a different vascular supply than the rest of the nail unit, and works with the hyponychium to act as a barrier at the distal end of the nail [1–4].

2.6 Nail Folds

The nail unit is surrounded by the two lateral and single proximal nail folds or grooves of the skin. Similar to the hyponychium, the epithelium of the proximal and the lateral nail folds has an additional granulosum layer, and protects the nail bed and matrix [1–4].

2.7 Cuticle

The stratum corneum layer of the proximal nail fold epithelium forms the cuticle, a thin, colorless stretch of tissue attached to the nail plate, which acts as a physical barrier in the proximal end of the nail plate similar to the onychodermal band in the distal end. The dermatophytes causing proximal subungual onychomycosis primarily invade the cuticle [1–4].

2.8 Eponychium

The eponychium is the space between the cuticle and the proximal end of the nail plate [1–3].

3 Clinical Presentation of Onychomycosis

The general clinical features of onychomycosis include nail plate discoloration (yellow, brown, white, or black), subungual hyperkeratosis, onycholysis (lifting of the nail plate off the nail bed), and, rarely, paronychia. Zaias [5] proposed a classification of onychomycosis in 1972 based on the site of invasion in the nail and etiological agents. According to this classification, there are four types of onychomycosis: distal subungual onychomycosis (infection occurs primarily at the hyponychium and distal nail end), white superficial onychomycosis (superficial surface of the nail plate), proximal subungual onychomycosis (occurs primarily at proximal nail fold and eponychium), and *Candida* onychomycosis (entire nail plate and seen predominantly in fingernails) [5]. Baran [6, 7] modified this classification by accounting for the site and pattern of nail plate involvement and histopathology. The new, modern classification of onychomycosis has five types: Distal and lateral subungual onychomycosis (DLSO), superficial onychomycosis (SWO/SBO), proximal subungual onychomycosis (PSO), endonyx onychomycosis (EO), and total dystrophic onychomycosis (TDO) [6, 7].

3.1 Distal and Lateral Subungual Onychomycosis (DLSO)

Distal and lateral subungual onychomycosis (DLSO) is the most common clinical presentation found in patients with onychomycosis and 75% of DLSO cases are toenail infections [8]. DLSO could be a primary or secondary infection affecting onycholytic nails. The fungal pathogen disrupts the space between the hyponychium and the nail bed, invading through the distal end of the nail plate (see Fig. 2a). The infection spreads proximally along the nail bed causing subungual hyperkeratosis, eventually leading to onycholysis. DLSO is the most common presentation of a dermatophyte infection. Advanced infection involves the entire nail plate and nail

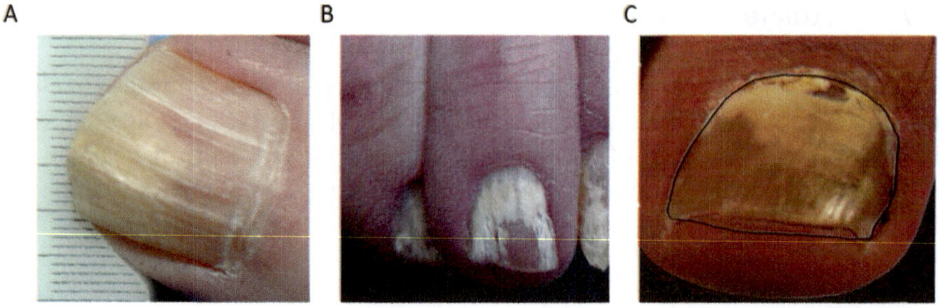

Fig. 2 Clinical types of onychomycosis. (**a**) Distal and lateral subungual onychomycosis (DLSO); (**b**) Superficial white onychomycosis (SWO); (**c**) Proximal subungual onychomycosis (PSO)

bed along with the surrounding skin, leading to paronychia (infection of the skin around the nails) [6, 9].

Trichophyton rubrum is the primary dermatophyte that causes typical DLSO. An infection caused by the non-dermatophyte mold *Neoscytalidium dimidiatum* (formerly *Scytalidium dimidiatum*) involves the surrounding nail folds along with the nail unit. Secondary DLSO following primary onycholysis or nail trauma in the fingers is usually caused by *Candida* or *Pseudomonas*, as these organisms do not have keratinolytic properties and invade nails that are already traumatized. Co-occurrence of *Tinea pedis* infection is often observed with DLSO [6, 9, 10].

3.2 Superficial Onychomycosis

Superficial White Onychomycosis (SWO). Superficial white onychomycosis (SWO) is a common superficial infection that affects either the dorsal surface or the undersurface of the nail plate and does not involve the nail bed. It is mostly caused by dermatophytes such as *Trichophyton mentagrophytes* or *Trichophyton interdigitale*. Infection by non-dermatophyte molds such as *Aspergillus* or *Fusarium* spp. invades deeply into the nail plate. The infection manifests as white flaky colonies on the nail plate (see Fig. 2b). It is not as prevalent as DLSO and onycholysis is not as common with SWO as with other clinical types. Although common overall, the co-occurrence of *Tinea pedis* and SWO is low compared to other clinical types of onychomycosis. SWO is commonly observed in toenails and rarely seen in fingernails. It may affect individuals who are immunosuppressed, such as patients with AIDS [6, 7, 9, 11]. SWO may involve the dorsal aspect of the fourth toenail especially when the third toe overlies the fourth toe. In this setting the organism is generally *T. mentagrophytes*.

Superficial Black Onychomycosis (SBO). Some non-dermatophyte molds such as *N. dimidiatum* or dermatophytes such as *T. rubrum* produce black colonies when they invade the nail plate. However, *T. rubrum* is also known to cause SWO in children [6, 7, 9, 11].

3.3 Proximal Subungual Onychomycosis (PSO)

Proximal subungual onychomycosis (PSO) is a rare manifestation and caused mostly by dermatophytes such as *T. rubrum*, followed by *Candida albicans* and non-dermatophyte molds. It is often associated with immunosuppressed individuals. The primary site of infection is the proximal nail fold, where it affects the nail matrix (see Fig. 2c). The infection spreads distally and may eventually corrode the entire nail plate. Infection by NDMs such as *Scopulariopsis brevicaulis* and *Fusarium* spp. manifests as white discoloration of the nail plate starting from the underside of the proximal nail plate along with periungual inflammation and purulent discharge. PSO can also be caused by a primary dermatophyte infection or by a secondary infection by *C. albicans* accompanying paronychia [6, 7, 11, 12].

3.4 Endonyx Onychomycosis (EO)

Endonyx onychomycosis is caused by the same dermatophytes that give rise to endothrix scalp infection such as *Trichophyton soudanense* and *Trichophyton violaceum*. The nail plate is infected with fungal hyphae. Pitts and splits are observed in the distal end of the nail plate and a milky white discoloration of the nail plate is observed, which are distinct characteristic features of this clinical type. EO is limited to the superficial surface and sometimes deeper layers of the nail plate without involvement of the nail bed or nail matrix. Hyperkeratosis of the nail bed and onycholysis are not generally observed in patients with EO [6, 7].

3.5 Total Dystrophic Onychomycosis (TDO)

Total dystrophic onychomycosis (TDO) is an advanced clinical representation of onychomycosis. Chronic disease cases may involve all 20 toenails. The entire nail plate crumbles and the nail bed hardens due to extreme dermatophyte infection. Secondary infections by *C. albicans* such as mucocutaneous candidiasis and by other NDMs often occur in nails with TDO [6, 10]. Recently, a 32-year-old woman was diagnosed with TDO who also had nail psoriasis. It was the first onychomycosis case reported to be caused by antifungal-resistant *Aspergillus clavatus* [13].

4 Epidemiology and Risk Factors

Onychomycosis is the most common nail disease in the world (50%), with a global prevalence of ~5.5% [14–20]. It is typically more common in the elderly population and prevalence increases with age: about 20.7%–40% of adults over the age of 60 are infected [20–23]. In contrast, prevalence is much lower in younger patients at 1.1%, as demonstrated by a retrospective study of children aged 12–18 years [19]. Various factors can increase susceptibility to onychomycosis infections such as poor hygiene, socioeconomic status, age, and numerous health conditions like diabetes, peripheral vascular disease, obesity, and immunosuppression [20, 24–26].

4.1 Epidemiology

Global occurrence. There is a geographical variation in the prevalence of onychomycosis based on several factors such as climate and rate of migration. For example, in North America, onychomycosis rates range from 8.7% to 13.8% [21, 27]. In these countries, onychomycosis is predominantly caused by dermatophytes, which cause 60%–90% of infections [15, 28, 29]; yeasts and NDMs cause 19% and 20%, respectively [27, 30]. The range of infection rates in Europe is broader, reported between 0.5% and 24% [15, 31–34]. In southern Europe, 18%–52% of infections are caused by dermatophytes and 24%–64% by yeasts [31, 34]. The wide ranges in the epidemiology data are due to sampling bias in population- and hospital-based studies. The mean prevalence of onychomycosis in North America and Europe is 4.3% based on population studies and 8.9% based on hospital/clinical studies [35]. In India, the prevalence ranges from 0.5% to 5% [36]. Studies from other parts of Asia, Latin, and South America report higher onychomycosis rates, particularly for those caused by NDMs and yeasts [15, 35]. The prevalence of onychomycosis in different geographical regions is depicted in Fig. 3 [20, 36, 37].

Rise of NDMs in different parts of the globe. Dermatophytes and yeasts account for similar rates of infection in the Middle Eastern and Asian countries (40%–48% and 43%–46% respectively), whereas NDMs account for 8%–11% of infection. Infections in Africa are mostly caused by yeasts [28, 38–40]. The high rates of onychomycosis infection in North America may be due to the movement of dermatophytes from other parts of the world caused by immigration and travel from other locations [28]. Dermatophyte infections are the most common, but

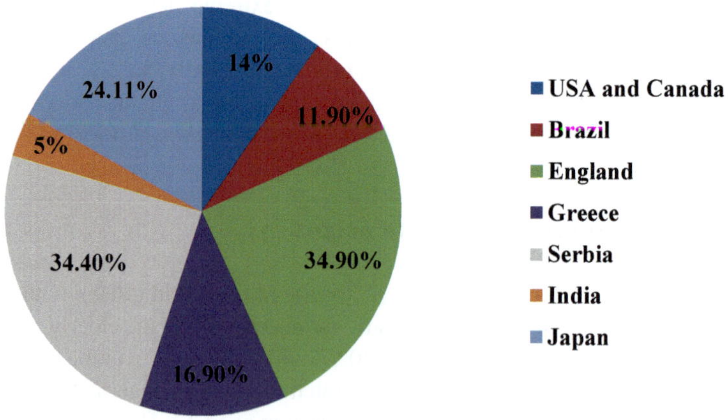

Fig. 3 Onychomycosis prevalence in different geographical regions

NDM infections are rising in frequency. Mixed infections are also becoming more common, representing about 3%–11% of infections [41, 42]. Earlier, NDMs had been ignored as nail contaminants if a dermatophyte was present, especially because their role in the presence of a dermatophyte was, and still remains, poorly understood [20, 43, 44]. NDM infections are also more difficult to diagnose because they are common nail and laboratory contaminants. As such, their treatment has not been studied sufficiently to garner effective strategies for disease management, especially in recurring cases [12].

4.1.1 Special Populations

Elderly. The elderly population is at risk of developing onychomycosis and the prevalence increases with age (~40% in patients >60 years of age) due to poor peripheral circulation, slower-growing nails, longer exposure to nail pathogens, several co-morbidities, poor compliance to medicines and treatments, etc. [20]. The global prevalence of onychomycosis in the elderly ranges from 20.7% to 28.1% based on several clinical and population studies [45–49]. According to a systematic review, the prevalence of dermatophyte onychomycosis in the elderly population is 10.28%. They also have a higher risk of contracting yeast onychomycosis nail infections (6.07%). The incidence of nail trauma or onycholysis is higher in people above 60 years of age, which in turn provides an ideal invasion portal for yeasts, especially *C. albicans* [46–49].

Diabetic patients. Patients with diabetes mellitus are prone to develop onychomycosis due to concomitant medications, compromised immune system, impaired wound healing, and foot ulcers [20]. About 15%–25% patients with diabetes develop a foot ulcer in their lifetime [50]. Sultana et al. [51] recently showed that the prevalence of dermatophyte onychomycosis in the diabetic population is high (35.18%–44.44%) compared to non-diabetic population (3%–6.06%) in Bangladesh. Similarly, Akkus et al. [52] showed that 34.9% of patients with diabetes mellitus have onychomycosis in Turkey. A two-center study by Gupta et al. [53] showed the prevalence of onychomycosis in type I diabetes patients visiting clinics in London, Ontario, Canada, and Boston, MA, USA, to be 13% compared to 9.10% in normal individuals without diabetes. The most common causative organisms of onychomycosis in diabetic patients are dermatophytes (8.75%), especially *T. rubrum*, followed by yeasts (3.97%), specifically *C. albicans*, and NDMs (1.68%) [46]. There is evidence that diabetes may increase treatment duration for onychomycosis [54].

Psoriatic patients. Common nail changes associated with psoriasis include onycholysis, paronychia, and subungual hyperkeratosis and are observed in 20%–78% of psoriatic patients [55]. Up to 47% of psoriatic patients also suffer from onychomycosis [56]. A 15-year retrospective study by Gallos et al. [57] on the prevalence of onychomycosis in psoriatic and non-psoriatic populations demonstrated that the occurrence of onychomycosis in the psoriatic population is 49.08% compared to 51.30% in the non-psoriatic population; i.e., there is no significant difference in the occurrence of onychomycosis in both the groups. A study by Romaszkiewicz et al. [58] showed similar results (23.53% vs. 22.37%, psoriatic and non-psoriatic populations, respectively).

In a hospital-based study in Israel, the prevalence of toenail onychomycosis in the psoriatic group was higher compared to the non-psoriatic control group (47.6% vs. 28.4%) in a span of 2 years [59]. There is higher prevalence of onychomycosis in psoriatic patients in Israel (47.6%) compared to North America, where it is only 13% [59, 60]. There is uncertainty on the relationship between psoriasis and onychomycosis and the nature of the causative organism(s). Some studies show high prevalence of onychomycosis in psoriatic patients while others report that there is no significant difference in the occurrence of onychomycosis between psoriatic and non-psoriatic populations. There could be a sampling bias in these studies or under reporting of incidences in psoriatic or non-psoriatic populations.

Immunosuppressed individuals. Patients with suppressed or compromised immunity are at greater risk of developing cutaneous fungal infections, both superficial and systemic. The factors that contribute to their vulnerability include low $CD4^+$ T cell count, poor cellular immune response, nail trauma, and history of *Tinea pedis* or onychomycosis. The global prevalence of onychomycosis in HIV-positive or autoimmune disease patients is 10%–44% [61, 62]. Onychomycosis in these populations is observed in tropical countries such as Thailand (10.2%) and more frequently in temperate regions such as Canada and southern Brazil (24% in Canada and 20% in Brazil) [62, 63].

Children. Tinea capitis (fungal infection of scalp) is the most commonly observed superficial dermatomycosis in children, with 90% of the cases involving *Trichophyton tonsurans*, superseding *Microsporum* spp., in the United States [64, 65]. Onychomycosis accounts for 20% of all superficial fungal infections of the skin, hair, and nail found in children; the prevalence of pediatric onychomycosis ranges from 0.35% to 5.5% [20, 66]. The worldwide prevalence of onychomycosis in children below 16 years of age is low compared to adults (0.5%–3%) [11, 67]. The prevalence of onychomycosis in children between 12 and 18 years of age was 1.1% in a retrospective study of 36,634 children in San Diego [68]. It ranges from 0% in the United States, Wales, and Finland to 2.6% in Guatemala [29, 69, 70]. The most common pathogens are dermatophytes, especially *T. rubrum* [46, 70]. There are several factors which contribute to the low occurrence in pediatric population compared to adults, such as faster nail plate growth, faster response to treatment, faster elimination of the fungal pathogen, low exposure to fungi and nail trauma, and good compliance (due to help from parents or caregivers) [70].

Romero et al. [71] studied the prevalence of onychomycosis in immunosuppressed children receiving chemotherapy and found it to be 1.3%. It is similar to other studies in pediatric populations despite the fact that immunosuppression is a risk factor for the development of onychomycosis [72–74]. The epidemiological data vary across different studies, but an overall increase in prevalence of onychomycosis in children is observed globally. Recently, children less than 2 years of age and infants are being diagnosed with onychomycosis. There was a case reported in 2007 by Bonifaz et al. [75] of an infant less than 11 weeks old being diagnosed with dermatophyte onychomycosis. A recent case of onychomycosis in the fingernails of a 28-day-old infant was reported by Subramanya et al. [76] in Nepal. The causative organism was found to be *C. albicans* and the infant was treated with oral fluconazole and topical amorolfine 5% nail lacquer once weekly for 3 months and was completely cured with normal growth of the nail plate [76].

Renal conditions. Patients with kidney transplants and those undergoing hemodialysis are at risk of developing superficial fungal infections. The duration of dialysis and co-morbidity of diabetes mellitus are the two important risk factors associated with renal transplant patients in developing onychomycosis [77]. Absence of a lunula is a characteristic commonly observed in patients with chronic renal failure. The worldwide prevalence of onychomycosis in hemodialysis, renal transplant, and chronic renal failure patients ranges from 5% to 39% and the majority of these infections are caused by dermatophytes, especially *T. rubrum*, and less commonly by yeasts and NDMs [46, 77–82].

4.2 Risk Factors

Risk factors that contribute to onychomycosis include age, genetic predisposition, employment, sports, recreational activities, immune-suppressed state and other medical conditions, nail trauma, and a history of *Tinea pedis* or athlete's foot [83, 84]. The presence and transmission of the disease is common in related family members and those sharing a household, especially dermatophyte toenail and foot infections [87, 88]. The risk of contracting onychomycosis may be autosomal-dominantly inherited with incomplete penetrance, as the human leukocyte antigen DR8 is more common in people with onychomycosis than unaffected populations [85, 86].

Age. The most common risk factor that contributes to onychomycosis is aging and is seen frequently in the elderly population. Onychomycosis occurs at 20.7%–28.1% and up to 40% in people aged 60 years and above [20, 26, 68]. As previously mentioned, there are several reasons for the increased occurrence in the elderly, such as low immune response, slow nail growth rate, co-morbidities such as diabetes or vascular diseases, nail trauma, and longer duration of infection, etc. [47].

Genetic predisposition. Genetic polymorphisms observed in human leukocyte antigen, i.e. HLA–DR8, predispose individuals to onychomycosis. It is seen in higher frequency in onychomycosis patients compared to normal individuals. Regulatory T cells (Tregs) are also found to be upregulated in patients with onychomycosis. This results in poor cellular immune response and decreases the capability to eliminate an infection [20, 90, 91]. Mutations in adaptor protein CARD9 and Dectin-1 receptor can potentially lead to mucocutaneous infections by *C. albicans* by inhibiting the production of IL-6 and -17, which trigger the adaptive immune response [91].

Tinea pedis. Tinea pedis is a fungal infection of the skin between the toes (*Tinea pedis interdigitalis*) and the bottom of the foot (*Tinea pedis plantaris*). More than half (~59%) of onychomycosis patients have co-existing *Tinea pedis* infection

[20, 92]. The factors that contribute to combined manifestation include infected footwear, diseases such as diabetes, poor vascular supply to the extremity, repeated trauma to the nails, and occupational lifestyle such as swimming [93]. The co-occurrence of onychomycosis and *Tinea pedis* is clinically significant and might lead to secondary bacterial infections, such as cellulitis [92]. Managing onychomycosis by simultaneously treating *Tinea pedis* infection might reduce relapse and re-infection rates [54].

5 Diagnosing Onychomycosis

5.1 Clinical Observation

The common clinical features typically include discoloration, thickening of the nail, and onycholysis [94]. It can also affect the adjacent skin, causing tenderness, redness, and swelling around the site of infection. Leaving it untreated causes the infection to progress and become more difficult to treat. This can cause discomfort and negatively impact the quality of life of patients, as they may be embarrassed about the appearance of their nails in social settings. It is important to identify the extent of nail unit involvement and the identity and viability of the fungal pathogen.

Clinical presentation can also determine site of sample collection for diagnosis—for example, with DLSO, samples would be collected from the most proximal area of involvement between the nail bed and overlying ventral aspect of the nail plate. PSO requires nail plate debridement to retrieve underlying nail debris at the most distal area of involvement, and SWO/SBO requires scrapings from the dorsal nail plate [29].

Common diagnostic techniques include potassium hydroxide (KOH) microscopy, fungal culture, histopathology, and PCR. Although a combination of techniques can be used, KOH-microscopy with culture is considered the gold standard [95, 96].

5.2 KOH Microscopy

Microscopy is largely used to confirm the presence and identification of fungal pathogens. 5%–40% KOH is used to dissolve large keratinocytes and flatten the nail segment, reducing reflection from cell borders. Then, light microscopy can be used to determine the presence of fungal elements such as hyphae. It is a rapid procedure performed in-office, but lacks sensitivity and will not determine fungal viability [6, 97]. Additionally, KOH can result in trapped air bubbles or fat droplets that lead to false-positive results as they resemble fungal structures [98]. The sensitivity of KOH is lower than histology and culture, but can be increased if combined with one or the other, from 48% to 89% or 74% if combined with PAS staining or culture, respectively [99]. It is recommended that microscopy be used with repeated isolations and identifications along with several subsequent cultures when culture and microscopy yield contradictory results (i.e., microscopy is positive and the culture is negative) [100]. Sodium sulfide (Na_2S) and sodium hydroxide (NaOH) are other commonly used clearing agents. Chlorazol Black is another stain used directly on debris collected from infected nails [6]. Parker Quink, also known as Parker's blue-black ink, is used in equal volume with KOH to identify NDM infections by highlighting NDM spores and mycelia [6].

5.3 Fluorescent Staining

Alternatively, Calcofluor White is a fluorescent stain used in equal volume with KOH for fungal identification [6]. It is a sensitive stain that can be used for detecting dermatophytes, yeasts, and NDMs [101]. During fluorescent microscope examinations, Calcofluor White is absorbed by fungal cell wall polysaccharides and enhances the visibility of fungal elements [102]. A negative stain does not necessarily mean the fungus is not present—it may have not absorbed enough dye to show fluorescence or a limited amount of hyphae was present in the sample [102]. This can also depend on the location of the nail sample, for example, in DLSO infections there can be fewer viable hyphae in distal areas of the nail, while higher concentrations are found in proximal portions of the nail bed [102, 103]. In studies, fluorescent staining produces more false-negative results when compared to other diagnostic techniques like KOH or culture [102, 104]. Regardless, microscopy and culture are helpful diagnostic tools—if direct microscopy is not used in diagnosis, approximately 53% of fungal infections would be missed, and if cultures were not used, 15% of infections would be missed [105]. In the United States, KOH is the most frequently used primary test because third-party payers require it, or medical professions were taught to use it and believe it is a good option for diagnosing onychomycosis [106].

5.4 Mycological Culture

Fungal culture is a gold standard diagnostic technique for laboratory testing because it can identify both the causative organism and its viability. It augments the diagnosis of onychomycosis following a visual confirmation of presence of fungi through direct microscopy. To prepare the nail sample for culture, the infected nail is cleaned with soap and water, then wiped with 70% alcohol, and clipped. The subungual debris is scraped off into a receptacle and cultured on Sabouraud dextrose agar (SDA) [6, 97]. SDA is a non-specific growth medium for dermatophytes, yeasts, and NDMs. Some of the dermatophytes and yeasts are slow-growing and could be superseded by the growth of NDMs. Therefore, the nail scrapings and debris are cultured in two separate SDA mediums, with and without cycloheximide. Cycloheximide encourages dermatophyte growth and kills NDMs, while agar without cycloheximide encourages NDM growth [6]. Both media contain chloramphenicol which inhibits concomitant bacterial growth. The cultures are maintained at 25–37 °C. Fungal colonies are identified based on their morphology, color, and other biochemical characteristics [6, 97]. The growth process may take up to 3 weeks or longer and tends to produce up to 40% false-negative rates [97]. Additionally, the fungal culture technique is not equipped to discern the fungal pathogen from a common laboratory or nail contaminant. However, its low cost and familiarity make it a favorable and gold standard diagnostic method.

5.5 Histopathology

Hematoxylin-eosin (H&E), Periodic acid-Schiff (PAS), and Grocott's methenamine stain (GMS) are used to visualize fungal elements and enhance hyphae by staining the fungal cell wall [6]. The nail scrapings or subungual debris are placed directly on paraffin and these scrapings are amplified in quality by chitin softening agents [6]. Immunofluorescence and Mayer's Mucicarmine staining could also be used. Histopathology alone cannot identify the causative organism or its viability, so it should be used along with fungal culture to improve the accuracy of clinical findings. PAS produces results within 24 h and can be more sensitive than KOH and fungal culture [98]. GMS may be superior to PAS, as one study demonstrated the former was able to visualize infective organisms in onychomycotic nails that were otherwise found negative by PAS [107].

5.6 Molecular Biology

Molecular biology is becoming more frequently used as a tool for the diagnosis of onychomycosis. PCR identifies fungal pathogens using specific dermatophyte primers. The target is a gene fragment of the fungal small ribosomal subunit 18s rRNA, an internal transcribed spacer region of ribosomal DNA, or the chitin synthase 1 (CHS1) gene [12, 20, 108, 109]. PCR can provide results within 48 h, but can have high false-positive rates due to the potential identification of a contaminant [97].

Real-time or quantitative PCR (qPCR) is a quantitative, sensitive technique which provides information on the fungal viability along with species identification. It amplifies the fungal DNA fragment and is detected by a fluorescent signal [97, 109]. qPCR takes longer than standard PCR, requires expensive laboratory equipment and trained personnel, and is not often used in a clinical setting. However, it is more sensitive than KOH microscopy and fungal culture, and can detect *T. rubrum* in samples that were initially reported negative by KOH and culture [110]. Commercial PCR kits for detecting fungi are being developed and show promising results [20].

5.7 Non-routine Techniques

Flow cytometry. Flow cytometry is a molecular biology technique which relies on granulosity, cell size, proteins, and DNA markers, and generates distinct profiles of fungi to identify the pathogen. The fungal mass is collected from the infected nail unit after treating with Tween 40. It is then sorted by a dual flow cytometer according to its size based on staining fungal DNA with propidium iodide (PI) and fungal protein with fluorescein isothiocyanate (FITC), which aids in distinguishing fungi. However, this technique requires large sample sizes to identify the causative organism and is expensive due to its specialized equipment and requirement of trained personnel [97, 111, 112].

Immunohistochemistry. Immunohistochemistry is a technique similar to histology since it uses cross sections of the nail sample. It is an upgraded technique where antibodies specific to the fungal pathogen are used to detect the causative organism. The labelled antibodies bind to the specific fungus and are detected by immunofluorescence, horse-radish peroxidase enzyme, or avidin-biotin complex. This diagnostic method is highly specific and eliminates false-positive contaminants. Accurate diagnosis could be achieved by combining immunohistochemistry along with other techniques such as imaging to estimate the viability of the pathogen [112, 113].

Confocal microscopy. Confocal microscopy is a non-invasive technique used in the real-time diagnosis of onychomycosis. The light source is an 830 nm laser which is operated in reflectance mode. It produces horizontal cross sections of 500 × 500μm in different depths from the nail plate to the nail bed. There are 2 commercial reflectance confocal microscopy (RCM) devices which are used to clinically diagnose patients with onychomycosis. It is an uncommon diagnostic tool and is limited to small cohort studies. The presence or absence of hyphae or other fungal elements is detected using the images produced from the nail sectioning. It is diagnosed positive if fungal hyphae are seen in three consecutive images. The sensitivity of this technique is 52%–60% [114]. It is an accurate method of detection, but cannot be used in routine clinical diagnosis because of several limitations, such as being unable to identify the causative organism, requiring highly-trained personnel, having a limited depth of view (200–300μm), and expense [68, 114, 115].

Optical coherence tomography. Optical coherence tomography (OCT) is another non-invasive, real-time imaging technique used for medical diagnosis. OCT images 1–2 mm beneath the tissue in vivo and is able to effectively image the nail plate, bed, and matrix. The OCT scans are analyzed to evaluate the morphological features of the nail and to detect fungal elements under the nail plate. Disturbed architecture is observed in patients with onychomycosis and fungal hyphae are visible. The surrounding, normal skin will produce homogenous bands [68, 116]. Abuzahra et al. [116] showed that OCT could be a novel diagnostic tool in detecting fungal pathogens in nail and the results were comparable to histopathology staining in 10 patients with onychomycosis. It provides reliable results, even in the case of previously tested nails with negative KOH microscopy or fungal culture. The sensitivity of this technique is up to 94%. Major limitations include poor resolution, high false-negative results, and the highly expensive device [68, 117, 118].

Dermoscopy. Dermoscopy is a non-invasive imaging technique used in the diagnosis of onychomycosis. It has an advantage over fungal culture due to its rapid, practical assessment of the nail plate, nail bed, and surrounding skin folds. It can distinguish between onycholytic nails with and without onychomycosis: a jagged appearance of the proximal edge with longitudinal spikes in the onycholytic area suggests fungal infection, compared to the linear appearance without longitudinal spikes in fungus-free onycholytic nails. Dermoscopy or videodermoscopic images aid in distinguishing onychomycosis from nail melanoma by visualizing nail pigmentation caused by the invading fungal pathogen. Absence of melanin and pigmentation are distinguishable characteristics of onychomycosis. This simple, rapid, practical method is proposed as the first line of diagnosis in onychomycosis replacing mycological examination [119–121].

5.8 Diagnosis of NDM Onychomycosis

As there is a rise in global NDM onychomycosis and mixed infections, accurate diagnosis is required. The diagnosis of NDMs is difficult and more challenging than dermatophyte onychomycosis. Originally, the Walshe and English (1976) criteria for diagnosing NDM onychomycosis were predominantly followed [122]; NDM is a true pathogen only when it is grown in actidione and actidione-free medium without the presence of a dermatophyte and visually confirmed by the presence of fungal elements in direct microscopy. Specifically, 5 out of 20 nail inocula should consistently grow an NDM to eliminate the uncertainty of the NDM as a common contaminant [122].

A recent study by Gupta et al. [123] empirically demonstrated that 5/20 positive inocula are not enough to confidently diagnose an NDM pathogen. The study used successive isolation procedures to verify NDM infections in untreated patients who were repeatedly followed up. They found that 4 colonies of 15 seeded nail fragments were predictive of NDM onychomycosis 23.2% of the time (concurrent to Walshe and English 5/20 inocula criteria), and 5 out of 15 inoculum predicted NDM onychomycosis 29.1% of the time. However, the diagnosis rate drastically increased to 89.7% when 15 of 15 colonies were counted [123]. If an NDM species is consistently isolated from a series of samples, the likelihood that the organism is a pathogen rather than a contaminant increases [12, 123].

Gupta et al. [12] established the next set of standards in NDM diagnosis. Accordingly, there are 6 main criteria in NDM diagnosis: direct microscopy confirmation of fungal elements using KOH, mycological culture, repeated isolation in culture, absence of dermatophyte growth in culture, inoculum counting, and histopathology. Three of 6 criteria are recommended as a standard for diagnosing NDM infection. The most frequently used criteria include mycological culture, repeated isolation, inoculum counting, and direct microscopy [12].

Shemer et al. [124] defined new criteria for diagnosing NDM onychomycosis. According to the authors, if NDM is observed in the first culture of a nail sample taken from a suspected patient, three nail samples should be retrieved from the patient in the subsequent visit and grown in culture. If all the 3 samples produced the same NDM, it is confirmed as the causative organism [124].

5.9 The "Gold Standard"—Not Anymore?

Direct microscopy confirmed with fungal culture is the gold standard diagnostic technique for onychomycosis. However, due to the high false-negative rates, this standard is currently being questioned. The false-negative rate of KOH microscopy is 5%–15% and that of fungal culture is ~40% or higher [29, 97, 125, 126]. Other techniques such as PAS staining and PCR are being used more frequently and are more sensitive than KOH microscopy and culture [127–129]. Both histology and PCR are rapid techniques compared to fungal culture which takes 7 days to 3 weeks to produce results. PCR and qPCR are effective tools in diagnosis as they provide information on fungal viability and the causative organism. It is suggested to use a combination of techniques to accurately and rapidly determine the pathogen. For example, if the KOH microscopy is positive, then one or more of the following procedures should be considered: culture, PCR, or histology (PAS staining). Gupta et al. [130] showed that it is cost-effective to obtain an accurate diagnosis before commencing therapy.

6 Measure of Disease Severity

There are three main definitions of cure for onychomycosis, based on the goals of stakeholders involved. Clinical cure is often sought out by patients because it is defined as 100% clear nail, and most patients are concerned with the cosmetic appearance of the nail. Mycological cure is defined by negative KOH microscopy and negative fungal culture. Complete cure is defined as clinical cure plus mycological cure [6]. Some clinical studies vary in their definitions of these terms; for example, mycological cure could be defined as only negative KOH microscopy, or clinical cure as $< 5\%$ nail involvement, although the use of these alternative definitions are discouraged. Complete cure (100% clear nail plus negative fungal culture and KOH microscopy) is the goal of physicians and regulatory bodies since they require both the elimination of the fungal pathogen and normal growth of healthy nail. However, in some cases of severe nail dystrophy such as TDO, the nails might never look normal even after achieving mycological cure [131].

Due to its effect on treatment outcome, disease severity should be evaluated in order to choose an appropriate treatment regimen. This is a subjective evaluation that may differ between disease studies, even among individual physicians. For example, the distinction between mild and moderate infection is not clear. Generally, mild onychomycosis is less than or equal to 20% nail plate involvement, moderate disease is 20–50% or 60% nail plate involvement, and severe disease is greater than 50% or 60% of nail plate involvement [20, 132, 133]. However, the infected area might not accurately predict disease severity, as very thick nails with minimal infected area could still have a poor prognosis [134]. Several researchers have embarked on the development of a grading system for evaluating nail involvement. Initially, the Scoring Clinical Index for Onychomycosis (SCIO) was developed to present the disease severity as a score based on clinical form, depth of nail unit involvement, and thickness of subungual hyperkeratosis [135, 136].

More recently, the Onychomycosis Severity Index (OSI) was created as a reproducible and objective numeric grading system based on involvement of the nail unit and the matrix proximity to infection. The scale separates the nail involvement into mild, moderate, and severe categories. This is useful for clinical trials, choosing a treatment regimen, and predicting treatment response [97, 134]. In addition, the OSI has demonstrated high statistical reliability in both nail experts and non-experts. When dermatologists observed live and photographed nails, results were consistent, proving it to be a simple scale to learn. In terms of scoring, nails with OSI scores of ≤ 5 are considered mild infections that are more likely to respond well to conventional therapies. Severe scores such as 16 through 35 may not respond well to treatment [134].

7 Conclusion

Onychomycosis is difficult to treat because its common risk factors (increased age, compromised immune system, frequent use of harsh cleansers, etc.) also negatively impact cure and reinfection rates. Wearing protective clothing such as sandals around pools and gloves around chemicals can help prevent onychomycosis. Understanding the epidemiology of the disease and its various causative organisms, along with using more efficient diagnostic tools, promotes early and accurate diagnosis, which can lead to more effective treatment, the specifics of which can be found in a later chapter.

References

1. de Berker D. Nail anatomy. Clin Dermatol. 2013;31(5):509–15.
2. Hordinsky MK, Sawaya ME, Scher RK. Atlas of hair and nails. London: Churchill Livingstone; 2000. 278 p.
3. Brahs AB, Bolla SR. Histology, Nail. In: StatPearls [Internet]. Treasure Island, FL: StatPearls; 2020. http://www.ncbi.nlm.nih.gov/books/NBK539733/. Accessed 6 Feb 2020.
4. Fernandez-Flores A, Saeb-Lima M, Martínez-Nova A. Histopathology of the nail unit. Rom J Morphol Embryol. 2014;55:235–56.
5. Zaias N. Onychomycosis. Arch Dermatol. 1972;105(2):263–74.
6. Baran R, editor. Baran & Dawber's diseases of the nails and their management. 5th ed. Hoboken, NJ: Wiley-Blackwell; 2019.
7. Baran R, Hay RJ, Tosti A, Haneke E. A new classification of onychomycosis. Br J Dermatol. 1998;139(4):567–71.
8. Vender RB, Lynde CW, Poulin Y. Prevalence and epidemiology of onychomycosis. J Cutan Med Surg. 2006;10(Suppl 2):S28–33.

9. Shirwaikar AA, Thomas T, Shirwaikar A, Lobo R, Prabhu KS. Treatment of onychomycosis: an update. Indian J Pharm Sci. 2008;70(6):710–4.
10. Gupta AK. Types of onychomycosis. Cutis. 2001;68(2 Suppl):4–7.
11. Piraccini BM, Alessandrini A. Onychomycosis: a review. J Fungi (Basel). 2015;1(1):30–43.
12. Gupta AK, Drummond-Main C, Cooper EA, Brintnell W, Piraccini BM, Tosti A. Systematic review of nondermatophyte mold onychomycosis: diagnosis, clinical types, epidemiology, and treatment. J Am Acad Dermatol. 2012;66(3):494–502.
13. Falahati M, Ghojoghi A, Abastabar M, Ghasemi Z, Farahyar S, Roudbary M, et al. The first case of total dystrophic onychomycosis caused by *Aspergillus clavatus* resistant to antifungal drugs. Mycopathologia. 2016;181(3–4):273–7.
14. Dubljanin E, Džamić A, Vujčić I, Grujičić SŠ, Arsenijević VA, Mitrović S, et al. Epidemiology of onychomycosis in Serbia: a laboratory-based survey and risk factor identification. Mycoses. 2017;60(1):25–32.
15. Gupta AK, Gupta G, Jain HC, Lynde CW, Foley KA, Daigle D, et al. The prevalence of unsuspected onychomycosis and its causative organisms in a multicentre Canadian sample of 30 000 patients visiting physicians' offices. J Eur Acad Dermatol Venereol. 2016;30 (9):1567–72.
16. Maraki S, Mavromanolaki VE. Epidemiology of onychomycosis in Crete, Greece: a 12-year study. Mycoses. 2016;59(12):798–802.
17. Gupta C, Jongman M, Das S, Snehaa K, Bhattacharya SN, Seyedmousavi S, et al. Genotyping and *in vitro* antifungal susceptibility testing of *Fusarium* isolates from onychomycosis in India. Mycopathologia. 2016;181(7–8):497–504.
18. Otašević S, Barac A, Pekmezovic M, Tasic S, Ignjatović A, Momčilović S, et al. The prevalence of *Candida* onychomycosis in Southeastern Serbia from 2011 to 2015. Mycoses. 2016;59(3):167–72.
19. Totri CR, Feldstein S, Admani S, Friedlander SF, Eichenfield LF. Epidemiologic analysis of onychomycosis in the San Diego pediatric population. Pediatr Dermatol. 2017;34(1):46–9.
20. Gupta AK, Versteeg SG, Shear NH. Onychomycosis in the 21st century: an update on diagnosis, epidemiology, and treatment. J Cutan Med Surg. 2017;21(6):525–39.
21. Elewski BE, Charif MA. Prevalence of onychomycosis in patients attending a dermatology clinic in northeastern Ohio for other conditions. Arch Dermatol. 1997;133(9):1172–3.
22. Vasconcellos C, Pereira CQM, Souza MC, Pelegrini A, Freitas RS, Takahashi JP. Identification of fungi species in the onychomycosis of institutionalized elderly. An Bras Dermatol. 2013;88(3):377–80.
23. Papini M, Piraccini BM, Difonzo E, Brunoro A. Epidemiology of onychomycosis in Italy: prevalence data and risk factor identification. Mycoses. 2015;58(11):659–64.
24. Haneke E, Roseeuw D. The scope of onychomycosis: epidemiology and clinical features. Int J Dermatol. 1999;38(Suppl 2):7–12.
25. Elewski BE, Tosti A. Risk factors and comorbidities for onychomycosis. J Clin Aesthet Dermatol. 2015;8(11):38–42.
26. Tosti A, Hay R, Arenas-Guzmán R. Patients at risk of onychomycosis—risk factor identification and active prevention. J Eur Acad Dermatol Venereol. 2005;19(Suppl 1):13–6.
27. Ghannoum MA, Hajjeh RA, Scher R, Konnikov N, Gupta AK, Summerbell R, et al. A large-scale North American study of fungal isolates from nails: the frequency of onychomycosis, fungal distribution, and antifungal susceptibility patterns. J Am Acad Dermatol. 2000;43 (4):641–8.
28. Gupta AK, Mays RR, Versteeg SG, Piraccini BM, Takwale A, Shemer A, et al. Global perspectives for the management of onychomycosis. Int J Dermatol. 2019;58(10):1128–9.
29. Elewski BE. Onychomycosis: pathogenesis, diagnosis, and management. Clin Microbiol Rev. 1998;11(3):415–29.
30. Welsh O, Vera-Cabrera L, Welsh E. Onychomycosis. Clin Dermatol. 2010;28(2):151–9.

31. Ioannidou DJ, Maraki S, Krasagakis SK, Tosca A, Tselentis Y. The epidemiology of onychomycoses in Crete, Greece, between 1992 and 2001. J Eur Acad Dermatol Venereol. 2006;20(2):170–4.
32. Sergeev AY, Ivanov OL, Sergeev YV, Vahlakov N, Sedova TN, Dudnik VS, et al. Epidemiology of onychomycosis in modern Russia: incidence is growing. Mycoses. 2002;45(S2):56.
33. Burzykowski T, Molenberghs G, Abeck D, Haneke E, Hay R, Katsambas A, et al. High prevalence of foot diseases in Europe: results of the Achilles Project. Mycoses. 2003;46 (11–12):496–505.
34. Vélez A, Linares MJ, Fenández-Roldán JC, Casal M. Study of onychomycosis in Córdoba, Spain: prevailing fungi and pattern of infection. Mycopathologia. 1997;137(1):1–8.
35. Sigurgeirsson B, Baran R. The prevalence of onychomycosis in the global population: a literature study. J Eur Acad Dermatol Venereol. 2014;28(11):1480–91.
36. Kayarkatte MN, Singal A, Pandhi D, Das S. Clinico-mycological study of onychomycosis in a tertiary care hospital—a cross-sectional study. Mycoses. 2020;63(1):113–8.
37. Shimoyama H, Sei Y. 2016 Epidemiological survey of dermatomycoses in Japan. Med Mycol J. 2019;60(3):75–82.
38. Scher RK, Rich P, Pariser D, Elewski B. The epidemiology, etiology, and pathophysiology of onychomycosis. Semin Cutan Med Surg. 2013;32(2 Suppl 1):S2–4.
39. Sehgal VN, Srivastava G, Dogra S, Chaudhary A, Adhikari T. Onychomycosis: an Asian perspective. Skinmed. 2010;8(1):37–45.
40. Thomas J, Jacobson GA, Narkowicz CK, Peterson GM, Burnet H, Sharpe C. Toenail onychomycosis: an important global disease burden. J Clin Pharm Ther. 2010;35(5):497–519.
41. Straten MRV, Balkis MM, Ghannoum MA. The role of nondermatophyte molds in onychomycosis: diagnosis and treatment. Dermatol Ther. 2002;15(2):89–98.
42. Moreno G, Arenas R. Other fungi causing onychomycosis. Clin Dermatol. 2010;28(2):160–3.
43. Ebihara M, Makimura K, Sato K, Abe S, Tsuboi R. Molecular detection of dermatophytes and nondermatophytes in onychomycosis by nested polymerase chain reaction based on 28S ribosomal RNA gene sequences. Br J Dermatol. 2009;161(5):1038–44.
44. Gupta A, Nakrieko K-A. Molecular determination of mixed infections of dermatophytes and nondermatophyte moulds in individuals with onychomycosis. J Am Podiatr Med Assoc. 2014;104(4):330–6.
45. Gupta AK. Onychomycosis in the elderly. Drugs Aging. 2000;16(6):397–407.
46. Gupta AK, Daigle D, Foley KA. The prevalence of culture-confirmed toenail onychomycosis in at-risk patient populations. J Eur Acad Dermatol Venereol. 2015;29(6):1039–44.
47. Loo DS. Onychomycosis in the elderly: drug treatment options. Drugs Aging. 2007;24 (4):293–302.
48. Gunduz T, Gündüz K, Degerli K, Limoncu ME. Epidemiological profile of onychomycosis in the elderly living in the nursing homes. Eur Geriatr Med. 2014;5(3):172–4.
49. Yoo JY, Suh MK, Ha GY, Kim H. A clinical and mycological study of onychomycosis in the elderly over 10 years (2001–2010). Korean J Med Mycol. 2014;19:83–92.
50. Gemechu FW, Seemant F, Curley CA. Diabetic foot infections. Am Fam Physician. 2013;88 (3):177–84.
51. Sultana S, Jaigirdar QH, Islam MA, Azad AK. Frequency of fungal species of onychomycosis between diabetic and non-diabetic patients. Mymensingh Med J. 2018;27(4):752–6.
52. Akkus G, Evran M, Gungor D, Karakas M, Sert M, Tetiker T. Tinea pedis and onychomycosis frequency in diabetes mellitus patients and diabetic foot ulcers. A cross sectional—observational study. Pak J Med Sci. 2016;32(4):891–5.
53. Gupta AK, Konnikov N, MacDonald P, Rich P, Rodger NW, Edmonds MW, et al. Prevalence and epidemiology of toenail onychomycosis in diabetic subjects: a multicentre survey. Br J Dermatol. 1998;139(4):665–71.
54. Gupta AK, Versteeg SG, Shear NH, Piguet V, Tosti A, Piraccini BM. A practical guide to curing onychomycosis: How to maximize cure at the patient, organism, treatment, and environmental level. Am J Clin Dermatol. 2019;20(1):123–33.

55. Chaowattanapanit S, Pattanaprichakul P, Leeyaphan C, Chaiwanon O, Sitthinamsuwan P, Kobwanthanakun W, et al. Coexistence of fungal infections in psoriatic nails and their correlation with severity of nail psoriasis. Indian Dermatol Online J. 2018;9(5):314–7.
56. Shemer A, Trau H, Davidovici B, Grunwald MH, Amichai B. Onychomycosis in psoriatic patients—rationalization of systemic treatment. Mycoses. 2010;53(4):340–3.
57. Gallo L, Cinelli E, Fabbrocini G, Vastarella M. A 15-year retrospective study on the prevalence of onychomycosis in psoriatic vs non-psoriatic patients: a new European shift from dermatophytes towards yeast. Mycoses. 2019;62(8):659–64.
58. Romaszkiewicz A, Bykowska B, Zabłotna M, Sobjanek M, Sławińska M, Nowicki RJ. The prevalence and etiological factors of onychomycosis in psoriatic patients. Postepy Dermatol Alergol. 2018;35(3):309–13.
59. Leibovici V, Hershko K, Ingber A, Westerman M, Leviatan-Strauss N, Hochberg M. Increased prevalence of onychomycosis among psoriatic patients in Israel. Acta Derm Venereol. 2008;88 (1):31–3.
60. Gupta AK, Lynde CW, Jain HC, Sibbald RG, Elewski BE, Daniel CR, et al. A higher prevalence of onychomycosis in psoriatics compared with non-psoriatics: a multicentre study. Br J Dermatol. 1997;136(5):786–9.
61. Ruíz-López P, Moreno-Coutiño G, Fernández-Martínez R, Espinoza-Hernández J, Rodríguez-Zulueta P, Reyes-Terán G. Evaluation of improvement of onychomycosis in HIV-infected patients after initiation of combined antiretroviral therapy without antifungal treatment. Mycoses. 2015;58(9):516–21.
62. Tuchinda P, Boonchai W, Prukpaisarn P, Maungprasat C, Suthipinittharm P. Prevalence of onychomycosis in patients with autoimmune diseases. J Med Assoc Thai. 2006;89 (8):1249–52.
63. Gupta AK, Taborda P, Taborda V, Gilmour J, Rachlis A, Salit I, et al. Epidemiology and prevalence of onychomycosis in HIV-positive individuals. Int J Dermatol. 2000;39 (10):746–53.
64. Andrews MD, Burns M. Common tinea infections in children. Am Fam Physician. 2008;77 (10):1415–20.
65. Ameen M. Epidemiology of superficial fungal infections. Clin Dermatol. 2010;28 (2):197–201.
66. Friedlander SF, Chan YC, Chan YH, Eichenfield LF. Onychomycosis does not always require systemic treatment for cure: a trial using topical therapy. Pediatr Dermatol. 2013;30 (3):316–22.
67. Rosen T, Friedlander SF, Kircik L, Zirwas MJ, Stein Gold L, Bhatia N, et al. Onychomycosis: epidemiology, diagnosis, and treatment in a changing landscape. J Drugs Dermatol. 2015;14 (3):223–33.
68. Lipner SR, Scher RK. Onychomycosis: clinical overview and diagnosis. J Am Acad Dermatol. 2019;80(4):835–51.
69. Kaur R, Kashyap B, Bhalla P. Onychomycosis—epidemiology, diagnosis and management. Indian J Med Microbiol. 2008;26(2):108–16.
70. Solís-Arias MP, García-Romero MT. Onychomycosis in children. A review. Int J Dermatol. 2017;56(2):123–30.
71. Garcia-Romero MT, Lopez-Aguilar E, Arenas R. Onychomycosis in immunosuppressed children receiving chemotherapy. Pediatr Dermatol. 2014;31(5):618–20.
72. Lange M, Roszkiewicz J, Szczerkowska-Dobosz A, Jasiel-Walikowska E, Bykowska B. Onychomycosis is no longer a rare finding in children. Mycoses. 2006;49(1):55–9.
73. Leibovici V, Evron R, Dunchin M, Westerman M, Ingber A. A population-based study of toenail onychomycosis in Israeli children. Pediatr Dermatol. 2009;26(1):95–7.
74. Gulgun M, Balci E, Karaoglu A, Kesik V, Babacan O, Fidanci MK, et al. Prevalence and risk factors of onychomycosis in primary school children living in rural and urban areas in Central Anatolia of Turkey. Indian J Dermatol Venereol Leprol. 2013;79(6):777–82.

75. Bonifaz A, Saúl A, Mena C, Valencia A, Paredes V, Fierro L, et al. Dermatophyte onychomycosis in children under 2 years of age: experience of 16 cases. J Eur Acad Dermatol Venereol. 2007;21(1):115–7.
76. Subramanya SH, Subedi S, Metok Y, Kumar A, Prakash PY, Nayak N. Distal and lateral subungual onychomycosis of the finger nail in a neonate: a rare case. BMC Pediatr. 2019;19 (1):168.
77. Kuvandik G, Çetin M, Genctoy G, Horoz M, Duru M, Akcali C, et al. The prevalence, epidemiology and risk factors for onychomycosis in hemodialysis patients. BMC Infect Dis. 2007;7:102.
78. Abdelaziz AM, Mahmoud KM, Elsawy EM, Bakr MA. Nail changes in kidney transplant recipients. Nephrol Dial Transplant. 2009;25(1):274–7.
79. Güleç AT, Demirbilek M, Seçkin D, Can F, Saray Y, Sarifakioglu E, et al. Superficial fungal infections in 102 renal transplant recipients: a case-control study. J Am Acad Dermatol. 2003;49(2):187–92.
80. Lamb FM, Ottonelli Stopiglia CD, Vetoratto G, Goldani JC, Scroferneker ML. Frequency of onychomycoses in chronic renal failure patients undergoing hemodialysis in Porto Alegre, Brazil. Acta Dermatovenerol Croat. 2013;21(1):19–23.
81. Anees M, Butt G, Gull S, Nazeer A, Hussain I, Ibrahim M. Factors affecting dermatological manifestations in patients with end stage renal disease. J Coll Physicians Surg Pak. 2018;28 (2):98–102.
82. Filho AMS, Ventura CG, Criado PR, Del Negro GB, Freitas RS, Luiz OC, et al. Hemodialysis and kidney transplantation as predisposing conditions to onychomycosis. Nephron. 2017;137 (1):38–46.
83. Avner S, Nir N, Henri T. Fifth toenail clinical response to systemic antifungal therapy is not a marker of successful therapy for other toenails with onychomycosis. J Eur Acad Dermatol Venereol. 2006;20(10):1194–6.
84. Walling HW, Sniezek PJ. Distribution of toenail dystrophy predicts histologic diagnosis of onychomycosis. J Am Acad Dermatol. 2007;56(6):945–8.
85. García-Romero MT, Arenas R. New insights into genes, immunity, and the occurrence of dermatophytosis. J Invest Dermatol. 2015;135(3):655–7.
86. Zaias N, Tosti A, Rebell G, Morelli R, Bardazzi F, Bieley H, et al. Autosomal dominant pattern of distal subungual onychomycosis caused by *Trichophyton rubrum*. J Am Acad Dermatol. 1996;34(2 Pt 1):302–4.
87. Ghannoum MA, Mukherjee PK, Warshaw EM, Evans S, Korman NJ, Tavakkol A. Molecular analysis of dermatophytes suggests spread of infection among household members. Cutis. 2013;91(5):237–45.
88. English MP. *Trichophyton rubrum* infection in families. Br Med J. 1957;1(5021):744–6.
89. Klaassen KMG, Dulak MG, van de Kerkhof PCM, Pasch MC. The prevalence of onychomycosis in psoriatic patients: a systematic review. J Eur Acad Dermatol Venereol. 2014;28(5):533–41.
90. Adams C, Athanasoula E, Lee W, Mahmudova N, Vlahovic TC. Environmental and genetic factors on the development of onychomycosis. J Fungi (Basel). 2015;1(2):211–6.
91. Gupta AK, Simpson FC, Brintnell WC. Do genetic mutations and genotypes contribute to onychomycosis? Dermatology (Basel). 2014;228(3):207–10.
92. Lipner SR, Scher RK. Management of onychomycosis and co-existing tinea pedis. J Drugs Dermatol. 2015;14(5):492–4.
93. de Berker D. Tinea pedis and onychomycosis. In: Onychomycosis [Internet]. Hoboken, NJ: Wiley; 2018. p. 21–30 . Accessed 6 Feb 2020. https://doi.org/10.1002/9781119226512.ch3.
94. Vlahovic TC. Onychomycosis: evaluation, treatment options, managing recurrence, and patient outcomes. Clin Pediatr Med Surg. 2016;33(3):305–18.
95. Elewski BE, Rich P, Pollak R, Pariser DM, Watanabe S, Senda H, et al. Efinaconazole 10% solution in the treatment of toenail onychomycosis: two phase III multicenter, randomized, double-blind studies. J Am Acad Dermatol. 2013;68(4):600–8.

96. Elewski BE, Aly R, Baldwin SL, González Soto RF, Rich P, Weisfeld M, et al. Efficacy and safety of tavaborole topical solution, 5%, a novel boron-based antifungal agent, for the treatment of toenail onychomycosis: results from 2 randomized phase-III studies. J Am Acad Dermatol. 2015;73(1):62–9.
97. Gupta AK, Mays RR, Versteeg SG, Shear NH, Piguet V. Update on current approaches to diagnosis and treatment of onychomycosis. Expert Rev Anti Infect Ther. 2018;16(12):929–38.
98. Reisberger E-M, Abels C, Landthaler M, Szeimies R-M. Histopathological diagnosis of onychomycosis by periodic acid–Schiff-stained nail clippings. Br J Dermatol. 2003;148 (4):749–54.
99. Wilsmann-Theis D, Sareika F, Bieber T, Schmid-Wendtner M-H, Wenzel J. New reasons for histopathological nail-clipping examination in the diagnosis of onychomycosis. J Eur Acad Dermatol Venereol. 2011;25(2):235–7.
100. US Food and Drug Administration. Medical devices and clinical trial design for the treatment or improvement in the appearance of fungally-infected nails—Draft Guidance for Industry and FDA Staff [Internet]. 2015 . http://www.fda.gov/downloads/MedicalDevices/DeviceRegulationandGuidance/GuidanceDocuments/UCM431312.pdf. Accessed 22 May 2015.
101. Weinberg JM, Koestenblatt EK, Tutrone WD, Tishler HR, Najarian L. Comparison of diagnostic methods in the evaluation of onychomycosis. J Am Acad Dermatol. 2003;49 (2):193–7.
102. Scherer WP, Kinmon K. Dermatophyte test medium culture versus mycology laboratory analysis for suspected onychomycosis. A study of 100 cases in a geriatric population. J Am Podiatr Med Assoc. 2000;90(9):450–9.
103. Queller JN, Bhatia N. The dermatologist's approach to onychomycosis. J Fungi (Basel). 2015;1(2):173–84.
104. Scherer WP, Scherer MD. A comparison of results from two mycology laboratories for the diagnosis of onychomycosis. J Am Podiatr Med Assoc. 2004;94(6):528–34.
105. Davies RR. Mycological tests and onychomycosis. J Clin Pathol. 1968;21(6):729–30.
106. Koshnick R, Lilly K, St Clair K, Finnegan M, Warshaw E. Use of diagnostic tests by dermatologists, podiatrists and family practitioners in the United States: pilot data from a cross-sectional survey. Mycoses. 2007;50(6):463–9.
107. D'Hue Z, Perkins SM, Billings SD. GMS is superior to PAS for diagnosis of onychomycosis. J Cutan Pathol. 2008;35(8):745–7.
108. Gupta AK, Ryder JE, Summerbell RC. Onychomycosis: classification and diagnosis. J Drugs Dermatol. 2004;3(1):51–6.
109. Emam SM, Abd El-salam OH. Real-time PCR: a rapid and sensitive method for diagnosis of dermatophyte induced onychomycosis, a comparative study. Alexandria J Med. 2016;52 (1):83–90.
110. Gupta AK, Zaman M, Singh J. Fast and sensitive detection of *Trichophyton rubrum* DNA from the nail samples of patients with onychomycosis by a double-round polymerase chain reaction-based assay. Br J Dermatol. 2007;157(4):698–703.
111. Gupta AK, Simpson FC. Diagnosing onychomycosis. Clin Dermatol. 2013;31(5):540–3.
112. Arrese JE, Piérard-Franchimont C, Greimers R, Piérard GE. Fungi in onychomycosis. A study by immunohistochemistry and dual flow cytometry. J Eur Acad Dermatol Venereol. 1995;4 (2):123–30.
113. Piérard GE, Arrese JE, De Doncker P, Piérard-Franchimont C. Present and potential diagnostic techniques in onychomycosis. J Am Acad Dermatol. 1996;34(2, Part 1):273–7.
114. Pharaon M, Gari-Toussaint M, Khemis A, Zorzi K, Petit L, Martel P, et al. Diagnosis and treatment monitoring of toenail onychomycosis by reflectance confocal microscopy: prospective cohort study in 58 patients. J Am Acad Dermatol. 2014;71(1):56–61.
115. Hongcharu W, Dwyer P, Gonzalez S, Anderson RR. Confirmation of onychomycosis by *in vivo* confocal microscopy. J Am Acad Dermatol. 2000;42(2, Part 1):214–6.

116. Abuzahra F, Spöler F, Först M, Brans R, Erdmann S, Merk HF, et al. Pilot study: optical coherence tomography as a non-invasive diagnostic perspective for real time visualisation of onychomycosis. Mycoses. 2010;53(4):334–9.
117. Olsen J, Lindsø Andersen P, Themstrup L, Jemec GBE, Saunte DML. Optical coherence tomography of onychomycosis: proposed terminology and a suggestion of practical usage. Arch Dermatol Res. 2020;312(1):51–8.
118. Optical Coherence Tomography Imaging of Onychomycosis [Internet]. Vivosight. 2012. https://vivosight.com/2012/06/18/optical-coherence-tomography-imaging-of-onychomycosis/. Accessed 7 Feb 2020.
119. Yorulmaz A, Yalcin B. Dermoscopy as a first step in the diagnosis of onychomycosis. Postepy Dermatol Alergol. 2018;35(3):251–8.
120. Piraccini BM, Balestri R, Starace M, Rech G. Nail digital dermoscopy (Onychoscopy) in the diagnosis of onychomycosis. J Eur Acad Dermatol Venereol. 2013;27(4):509–13.
121. Bet DL, dos Reis AL, Chiacchio ND, Belda Junior W. Dermoscopy and onychomycosis: guided nail abrasion for mycological samples. An Bras Dermatol. 2015;90(6):904–6.
122. English MP. Nails and fungi. Br J Dermatol. 1976;94(6):697–701.
123. Gupta AK, Cooper EA, MacDonald P, Summerbell RC. Utility of inoculum counting (Walshe and English criteria) in clinical diagnosis of onychomycosis caused by nondermatophytic filamentous fungi. J Clin Microbiol. 2001;39(6):2115–21.
124. Shemer A, Davidovici B, Grunwald MH, Trau H, Amichai B. New criteria for the laboratory diagnosis of nondermatophyte moulds in onychomycosis. Br J Dermatol. 2009;160(1):37–9.
125. Shemer A, Daniel R, Rigopoulos D, Farhi R, Babaev M. Variability in systemic treatment efficacy for onychomycosis: information that clinical studies do not impart to the office dermatologist. Skin Appendage Disord. 2018;4(3):141–4.
126. Gustafson E, Bakotic W, Bennett L, Page L, McCarthy L. DNA-based detection for onychomycosis correlates better to histopathology than does fungal culture. Dermatol Online J. 2019;25(7):13030/qt5bc2z46g.
127. Blake N, Zhu J, Hernandez G, Juliano PJ. A retrospective review of diagnostic testing for onychomycosis of the foot. J Am Podiatr Med Assoc. 2015;105(6):503–8.
128. Velasquez-Agudelo V, Cardona-Arias JA. Meta-analysis of the utility of culture, biopsy, and direct KOH examination for the diagnosis of onychomycosis. BMC Infect Dis. 2017;17(1):166.
129. Karaman BFO, Açıkalın A, Ünal İ, Aksungur VL. Diagnostic values of KOH examination, histological examination, and culture for onychomycosis: a latent class analysis. Int J Dermatol. 2019;58(3):319–24.
130. Gupta AK, Versteeg SG, Shear NH. Confirmatory testing prior to initiating onychomycosis therapy is cost-effective. J Cutan Med Surg. 2018;22(2):129–41.
131. Lipner SR, Scher RK. Long-standing onychodystrophy in a young woman. JAMA. 2016;316(18):1915–6.
132. Gupta AK, Sibbald RG, Andriessen A, Belley R, Boroditsky A, Botros M, et al. Toenail onychomycosis—a Canadian approach with a new transungual treatment: development of a clinical pathway. J Cutan Med Surg. 2015;19(5):440–9.
133. Gupta AK, Daigle D, Foley KA. Topical therapy for toenail onychomycosis: an evidence-based review. Am J Clin Dermatol. 2014;15(6):489–502.
134. Carney C, Tosti A, Daniel R, Scher R, Rich P, DeCoster J, et al. A new classification system for grading the severity of onychomycosis: onychomycosis severity index. Arch Dermatol. 2011;147(11):1277–82.
135. Gupta AK, Versteeg SG, Shear NH. A practical application of onychomycosis cure—combining patient, physician and regulatory body perspectives. J Eur Acad Dermatol Venereol. 2019;33(2):281–7.
136. Sergeev AY, Gupta AK, Sergeev YV. The Scoring Clinical Index for Onychomycosis (SCIO index). Skin Therapy Lett. 2002;7(Suppl 1):6–7.

Invasive Dermatophytoses: Clinical Presentations, Diagnosis, and Treatment

Murat Durdu and Macit Ilkit

Abstract

If not adequately treated, superficial dermatophytic infections may invade the dermis and subcutaneous tissue, resulting in invasive dermatophytoses. Invasive dermatophytic infections are broadly classified into Majocchi's granuloma, deep dermatophytosis, disseminated dermatophytosis, and mycetoma and pseudomycetoma. Most pseudomycetomas and mycetomas develop in immunocompetent individuals, while deep dermatophytosis and disseminated dermatophytosis usually develop in immunosuppressed individuals. Genetic tests for *CARD9* and *STAT3* mutations should be performed in patients with recurrent deep dermatophytosis. Majocchi's granuloma can occasionally be clinically indistinguishable from deep and disseminated dermatophytosis. Mycetoma should be considered if tumefactive tumoral lesions and sinuses are observed. In addition to histopathology, laboratory and imaging methods are used to differentiate different types of invasive dermatophytoses. Systemic antifungals constitute the first-line treatment of Majocchi's granuloma, deep dermatophytosis, and disseminated dermatophytosis. However, surgery is the preferred treatment approach for pseudomycetoma and mycetoma.

Keywords

Deep dermatophytosis · Disseminated dermatophytosis · Invasive dermatophytosis · Majocchi's granuloma · Mycetoma · Pseudomycetoma

M. Durdu
Department of Dermatology, Faculty of Medicine, Adana Hospital, Başkent University, Adana, Turkey

M. Ilkit (✉)
Division of Mycology, Department of Microbiology, Faculty of Medicine, University of Çukurova, Adana, Turkey

J.-P. Bouchara et al. (eds.), *Dermatophytes and Dermatophytoses*,
https://doi.org/10.1007/978-3-030-67421-2_4

1 Introduction

Superficial dermatophytic infections are the most common cutaneous fungal infections of humans worldwide [1]. It is estimated that over one billion people are acutely or chronically infected by ubiquitous dermatophytes [2]. Dermatophytes infect humans worldwide regardless of factors such as age, sex, ethnicity, and immune conditions [3]. The Leading International Fungal Education (LIFE) portal is an ongoing effort to characterize the epidemiological and ecological characteristics and the estimated fungal burden in each country, even for dermatophytoses [2]. In addition, advances in molecular diagnostic tools and dermatophyte virulence models have improved the current understanding of dermatophyte pathogenesis [4, 5]. However, more research is still required to characterize the mechanisms of infection in the human host and how dermatophytes cause an invasive disease [6].

Recent taxonomy of dermatophytes revealed the existence of at least seven different genera (*Arthroderma*, *Epidermophyton*, *Lophophyton*, *Microsporum*, *Nannizzia*, *Paraphyton*, and *Trichophyton*). These fungi have keratinophilic and keratinolytic properties and usually dwell and proliferate in the scalp, glabrous skin (the stratum corneum), and nails, which are keratin-rich structures [7]. If superficial dermatophytic infections are not adequately treated, they may invade the dermis and subcutaneous tissue and even spread to other tissues, resulting in invasive dermatophytoses [1]. However, the epidemiological characteristics of invasive dermatophytic infections are poorly understood.

Different types of invasive dermatophytic infections are recognized. Perifollicular dermatophyte invasion and perifollicular granulomatous inflammation were first described by Domenico Majocchi (1849–1929), who reported a case of dermal granuloma caused by the anthropophilic species *Trichophyton tonsurans*. Histopathological analysis indicated perifollicular granulomatous inflammation. This locally invasive dermatophytic infection was first described by him as "*granuloma tricofitico*" in 1883 [8]. Thereafter, perifollicular granulomatous infections caused by dermatophytes have been referred to as "Majocchi's granuloma" (MG) in the honor of Dr. Majocchi [9]. Later, based on dermatopathological examinations, it was discovered that dermatophytic infections are not limited to the perifollicular region but may also cause mycetoma and pseudomycetoma. In addition to these specific invasive dermatophytic infections, dermatophytes may also cause nonspecific dermal invasion, referred to as deep dermatophytosis. When a dermatophyte infection is not limited to the skin and spreads to other organs, it is called disseminated dermatophytosis. Therefore, it is important to distinguish these invasive dermatophytic infections in terms of disease prognosis and treatment approaches [1].

In this chapter, we explore and present the current state of knowledge on the rare and lesser-known clinical presentations of invasive dermatophytic infections, including etiological agents, pathogenesis, clinical findings, diagnostic methods, and treatment, other than the most common cutaneous dermatophyte infections.

2 Etiological Agents

Numerous dermatophytes cause MG. *Trichophyton rubrum* is the most frequently isolated fungal agent of MG in both immunocompetent and immunosuppressed individuals worldwide. *Trichophyton interdigitale*, *T. tonsurans*, *Trichophyton violaceum*, *Microsporum canis*, *Microsporum ferrugineum*, *Nannizzia gypsea*, and *Epidermophyton floccosum* are also reported (Table 1) [10]. Non-dermatophytic fungi, such as those belonging to the genera *Phoma*, *Aspergillus*, and *Malbranchea*, have been erroneously reported as other etiological agents of MG [11–13]. More importantly, such non-dermatophytic cases should be referred to as non-dermatophytic MG-like infections [1].

The most common agent of deep dermatophytosis (73.7%) is *T. rubrum*. In some cases, two agents are detected together [14]. The most common cause of disseminated dermatophytosis is *T. violaceum*. Furthermore, the most common site of involvement of mycetoma and pseudomycetoma is the scalp, and hence, the most common etiological agents are *M. ferrugineum* and *M. canis* [15, 16].

3 Pathogenesis

The keratin-rich stratum corneum, scalp, and nails provide the most suitable environment for dermatophytes [3]. Consequently, dermatophytes usually cause only superficial skin infections. When several predisposing conditions such as trauma and congenital or acquired immunodeficiencies occur, dermatophytes invade the dermis [9]. In this section, we provide information on the mechanisms of dermis invasion by dermatophytes.

Table 1 Dermatophytes detected in different invasive dermatophytic infections

Majocchi's granuloma	Deep dermatophytosis	Disseminated dermatophytosis	Mycetoma—pseudomycetoma
E. floccosum	*M. audouinii*	*T. rubrum*	*M. audouinii*
M. canis	*M. ferrugineum*	*T. verrucosum*	*M. canis*[a]
M. ferrugineum	*T. mentagrophytes*	*T. violaceum*[a]	*M. ferrugineum*[a]
N. gypsea	*T. rubrum*[a]		*N. gypsea*
T. interdigitale	*T. schoenleinii*		*T. mentagrophytes*
T. rubrum[a]	*T. violaceum*		*T. rubrum*
T. tonsurans			*T. schoenleinii*
T. violaceum			*T. soudanense*
			T. tonsurans
			T. verrucosum

[a]The most common etiological agent(s)

3.1 Trauma

Trauma is especially important for MG, pseudomycetoma, and mycetoma but not for deep and disseminated dermatophytoses. Although immunological factors play a role in each type of invasive dermatophytosis, they are quite important in the development of deep dermatophytosis and disseminated dermatophytosis [1].

The skin acts as a barrier to prevent microbial invasion. The effectiveness of the physical barrier varies depending on keratin production, the epidermal turnover rate, hydration level, amount of unsaturated transferrin in the sweat and serum, lipid composition of the stratum corneum, CO_2 levels, and the presence or absence of hair [17–20]. Skin scratching, depilation, and penetrating trauma disrupt the physical skin barrier, allowing the dermatophytes to easily enter the dermis [17, 21, 22]. Furthermore, dermatophyte cell morphology can change from a hypha to a spherical shape; the viability of dermal hyphae is higher than that of epidermal hyphae [23]. However, the dermal environment is more alkaline than the epidermis, and the dermatophyte-preferred substrate keratin is not present in the dermis [22]. Nevertheless, in addition to allowing dermatophyte entry into the dermis, trauma also allows the entry of keratin and necrotic material entry into the dermis, providing substrates for fungal growth. In addition, inflammatory response to dermatophyte invasion of the dermis increases the amount of dermal acidic mucopolysaccharides, thus reducing the pH of the dermis. All these changes provide a suitable dwelling environment for dermatophytes in the dermis [21, 22].

3.2 Congenital Immunodeficiencies

Dermatophytes stimulate both innate and adaptive immunity, leading to the release of certain antimicrobial peptides (AMPs) and cytokines, as well as dermal infiltration by neutrophils, macrophages, and lymphocytes. Numerous genes that predispose a host to fungal infections have been identified. By contrast, there are few known genetic defects that allow deep dermatophytosis [24].

The most important receptors for dermatophyte recognition are the C-type lectin receptors dectin-1 and dectin-2. They are pattern-recognition receptors expressed by neutrophils, dendritic cells, and monocytes. Dectin-1 recognizes β-glucan, and dectin-2 shows affinity for α-mannans (Fig. 1). Both receptors activate spleen tyrosine kinase (the Sky pathway). While the phosphorylation of Sky by dectin-1 leads to the activation of Ras protein-specific guanine nucleotide-releasing factor (Ras-GRF1) by phosphorylation, the phosphorylation of Sky by dectin-2 activates protein kinase C-delta (PKC-δ) [25]. The activation of both Ras-GRF1 and PKCγ leads to the stimulation of the transcription factor recruitment domain-containing protein 9 (*CARD9*) and increases the activity of NF-κB and AP-1, resulting in the release of AMPs, including human β-defensins, cathelicidin LL-37, ribonuclease-7, psoriasin, and dermcidin, and of various proinflammatory cytokines (TNF-α, IL-1β, IL-8, and IL-16). Keratinocytes also release some of these AMPs to prevent fungal invasion [18]. *CARD9* mutation facilitates fungal invasion by reducing Th17 cell

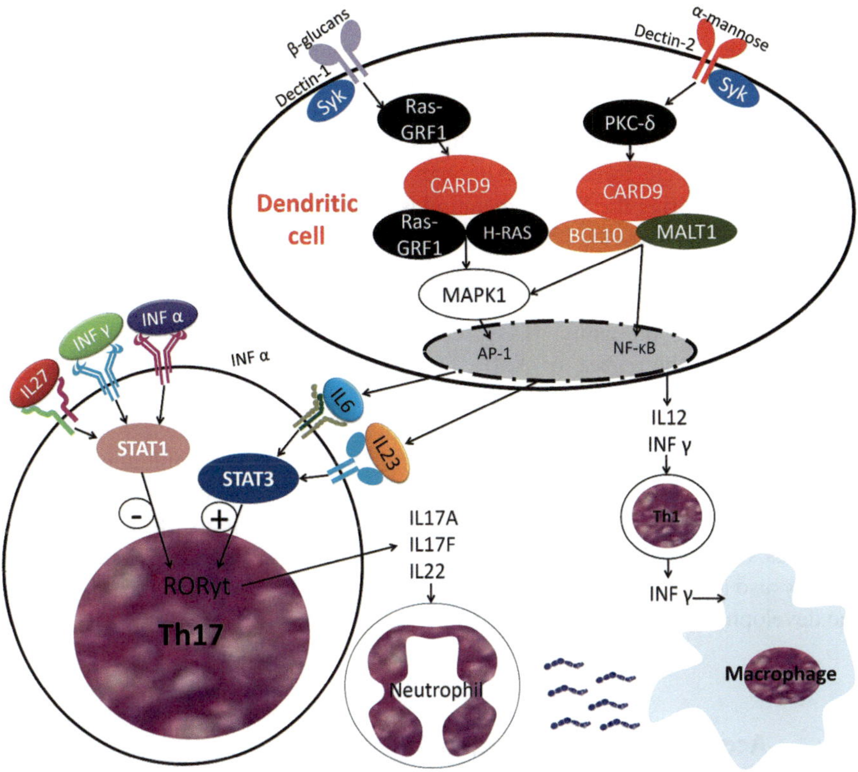

Fig. 1 Pathogenesis of invasive dermatophytosis. Dermatophytes are recognized by the mammalian cell dectin-1 and dectin-2 receptors. When these receptors are stimulated, transcription factor recruitment domain-containing protein 9 (CARD9) is activated via the spleen tyrosine kinase (Sky) pathway. Activation of NF-κB and AP-1 results in the release of some cytokines and antimicrobial peptides (AMPs). *CARD9* mutation predisposes to deep fungal infections. IL-6 and IL-23 activate Th17 lymphocytes by acting on another transcription factor, signal transducer, and activator of transcription 3 (STAT3), to induce transcription of retinoic-acid-receptor-related orphan nuclear receptor gamma (RORγt). Mutations causing loss of STAT3 function result in reduced Th17 differentiation and predispose to bacterial and fungal infections. Cytokines secreted by Th17 cells cause neutrophil chemotaxis and AMP release by keratinocytes

activity [26]. The mutation causes hyperimmunoglobulin E syndrome, hypereosinophilia, pneumonia, skeletal anomalies, and eczema and predisposes the host to fungal infections [25]. The most common fungal infection associated with *CARD9* mutation is candidiasis (40.3%). Other fungal infections in humans harboring this mutation are deep dermatophytosis (37.3%), phaeohyphomycosis (16.4%), invasive aspergillosis (3.0%), mucormycosis (1.5%), and protothecosis (1.5%). Treatment-resistant fungal infections develop in individuals with this mutation. Prophylactic antifungal treatment should be administered for months or a year after the eradication of fungi. Individuals harboring this mutation may die if appropriate treatment is not administered in a timely manner [27].

One important transcription factor in fungal immunology is the signal transducer and activator of transcription 3 (*STAT3*). This transcription factor is activated by IL-6, IL-21, and IL-23 receptors, and it induces the transcription of the retinoic acid receptor-related orphan nuclear receptor gamma t (RORγt). Mutations that cause the loss of *STAT3* function reduce Th17 differentiation, promoting the development of bacterial and fungal infections. Although the most common fungal infection in individuals harboring the *STAT3* mutation is candidiasis, deep dermatophytic infection has also been reported [26].

Although the Th1 response was considered the only response relevant to dermatophyte infection in the past, Th17 cells are currently recognized as important for anti-dermatophyte immunity. IL-17 secreted by Th17 cells stimulates neutrophil chemotaxis and the release of AMPs by keratinocytes. On the other hand, INF-γ produced by Th1 cells activates macrophages [28].

The immune response, in addition to trauma, plays a role in the development of mycetoma and pseudomycetoma. Neutrophil and macrophage infiltration initially develops as a response against dermatophytes, and it is followed by a Th2 response. Consequently, the local concentration of cytokines, such as IL-4 and IL-10, increases. Chitotriosidase gene polymorphisms that lead to decreased chitotriosidase activity also lead to reduced phagocytic activity of macrophages and neutrophils and the development of susceptibility to eumycetoma [29].

3.3 Acquired Immunodeficiencies

According to the most recent literature (published in the years 2011–2017), most MG patients (62%) are immunocompetent. Meanwhile, immunosuppressive treatments lead to MG susceptibility, especially in organ transplant patients [10]. Moreover, chemotherapy results in MG lesions that may recur after chemotherapy [30]. Another potential risk factor is steroid use. The use of systemic steroids has been reported in 27% of cases with MG [10]. MG may present as an exfoliative erythroderma in individuals with AIDS [31], and *Tinea incognito* and MG develop if a topical steroid treatment is mistakenly prescribed. In fact, topical steroid use is documented in 25% of reported cases of MG [32].

The majority (93%) of deep and disseminated dermatophytoses develop in immunosuppressed individuals. The most common reason is drug use for solid organ transplantation [33]. The use of systemic steroids to treat autoimmune disease or asthma also leads to deep dermatophytosis [34, 35]. Another cause of iatrogenic immunosuppression is the use of chemotherapy for various malignancies [36]. In addition, extensive deep dermatophytosis lesions have been reported in HIV-positive patients with cirrhosis associated with HBV and HCV [37]. Immunosuppression may develop not only because of iatrogenic factors but also as a result of genetic predisposition. Only a few cases of deep dermatophytosis without immunosuppression have been reported. However, genetic testing was not performed in these cases [38–40].

By contrast with deep dermatophytosis, most dermatophytic pseudomycetomas and mycetomas (85%) develop in immunocompetent individuals. Generally, localized lesions are observed in immunocompetent individuals, whereas widespread lesions may develop in immunosuppressed individuals. Tirado-González et al. [41] reported two cases of pseudomycetoma in immunosuppressed patients, one of whom showed diffuse hyperkeratotic plaques on the face, scalp, and neck. One of the two patients had IgM deficiency and lymphopenia (low serum CD3, CD4, and CD8 levels), while the other patient had diabetes mellitus, asthma, and systemic corticosteroid use. This indicates that immunosuppressive susceptibility should be investigated if widespread pseudomycetoma lesions are detected.

4 Clinical Findings

The clinical presentations of invasive dermatophytosis vary according to clinical type (Table 2). The most common type of invasive dermatophytosis is MG. Other forms include deep dermatophytosis, disseminated dermatophytosis, and mycetoma and pseudomycetoma. The clinical findings of mycetoma differ from those of other invasive dermatophytoses in that tumefactive tumoral lesions and sinuses are clinical indicators of mycetoma. However, nodular lesions are observed not only in MG but also in deep dermatophytosis and pseudomycetoma. Localizations of lesions also vary between the clinical types. While MG most frequently affects the lower extremities, pseudomycetomas and dermatophytic mycetomas are most commonly located on the scalp. Most people with deep dermatophytosis and disseminated dermatophytosis are immunosuppressed. We present the clinical findings in detail below.

4.1 Majocchi's Granuloma

Although MG can occur at any age (range: 3–87 years), the median age is 42 years. Men are affected more often than women (3:2). Unlike deep dermatophytosis, MG

Table 2 Type of lesion observed in invasive dermatophytic infections

	Majocchi's granuloma	Deep dermatophytosis	Disseminated dermatophytosis	Pseudomycetoma and mycetoma
Papule	+	+	+	–
Nodule	+	+	+	+
Plaque	+	+	+	+
Ulcer	+	+	+	+
Pustule	+	+	–	–
Abscess	+	+	–	+
Tumoral	–	+	–	+

+, positive; –, negative

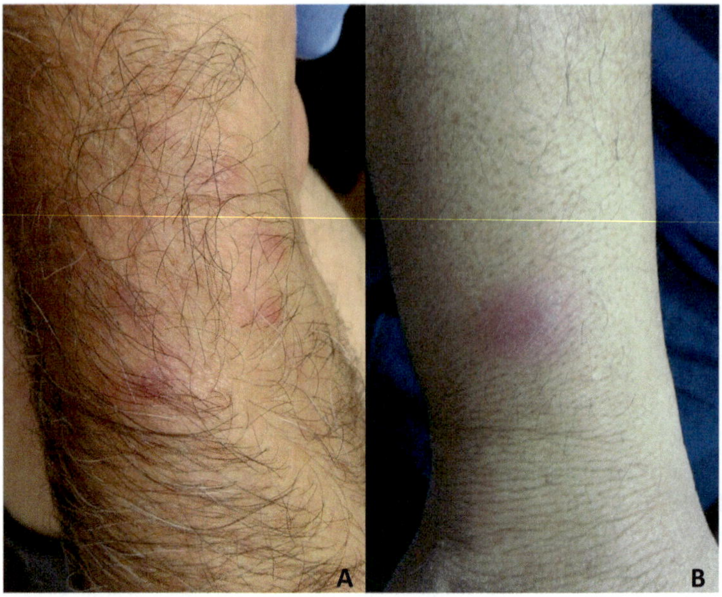

Fig. 2 Erythematous papulonodular lesions on the wrist (**a**) and erythematous plaque on the forearm (**b**) in two different Majocchi's granuloma patients receiving immunosuppressive therapy after renal transplantation

occurs in both immunosuppressed patients and healthy individuals [9, 10]. MG usually occurs in the presence of chronic dermatophytoses, such as *Tinea unguium* and *Tinea corporis* [42]. Other sources of dermatophytes are sexual contact or contact with animals [10]. The use of immunosuppressive treatment results in MG susceptibility, especially in organ transplant patients. The most common localizations of the lesions in adults are the lower extremities (37.5%). Other localizations are the upper extremities (27.7%), face and neck (15.2%), trunk (8%), genital region (7.1%), and the scalp (4.5%). The lower extremities are affected more frequently in the elderly (40.1%) than in children (22.2%), while head and neck involvement is more common in childhood (86%) than at other lifestages [9, 10].

Unlike superficial dermatophytoses, the clinical presentation of MG varies from papular lesion to abscess. Discrete or grouped papules and nodules occur either on the active border of erythematous plaques or by themselves (Fig. 2). Plaque, pustular, or papular lesions are observed in 15% of patients, while cellulitis-like erythematous patches or ulcerative lesions rarely develop [9]. Since the lesions can be asymptomatic, they can be pruritic or perceived as slight tenderness following the application of pressure. Painful lesions usually mimic bacterial infections, leading to the prescription of unnecessary antibiotic treatments. Painless chronic nodular lesions can also mimic other chronic infections, such as atypical mycobacterial infections, cutaneous leishmaniasis, and other deep fungal infections. If MG lesions occur on the face, they can be confused with granulomatous rosacea and granuloma faciale. Painful nodules on the leg imitate erythema nodosum, thrombophlebitis, and erythema induratum of Bazin [43]. Nodular lesions of MG share similarities with some tumoral diseases, such as lymphoma and Kaposi's sarcoma, especially in immunosuppressed patients [44, 45]. Keloid-like nodular lesions have also been reported [46]. If secondary bacterial infection does not develop, MG lesions do not suppurate. Lesions can be in the form of annular plaques and may mimic granuloma annulare, sarcoidosis, and annular elastolytic giant cell granuloma [21]. The lesions may also present as painful swelling of the vulva [47].

4.2 Deep Dermatophytosis

Deep dermatophytosis usually develop in adulthood (range: 18–91 years). They are twice as common in men as in women. Most cases (92.7%) of deep dermatophytosis occur in immunosuppressed patients. The most common localization of the lesions in adult patients is the lower extremities (36.6%). Other localizations are the upper extremities (4.9%), face and neck (9.8%), trunk (17.1%), genital region (14.6%), and scalp (17.1%). The most common findings are erythematous papules or nodules. Other findings include ulcers, erythematous plaques, and necrotic lesions. Although most lesions are asymptomatic, they may be pruritic or tender. The size of the ulcerated lesions varies. Although ulcerative lesions covered with a thick crust are usually 1–2 cm in size, giant ulcers with a diameter of 15 cm have also been reported [42, 48]. Because most patients are immunosuppressed, red or violaceous papulonodular lesions may be confused with Kaposi's sarcoma, pyogenic granuloma, bacillary angiomatosis, lymphoma, vasculitis, and other infections [49, 50]. Some papulonodular lesions show umbilical features and mimic cryptococcosis and molluscum contagiosum [51, 52]. Papules and nodules may mimic acute perforating collagenosis, both clinically and histopathologically [53]. Similar to MG, deep dermatophytosis lesions clinically cause cellulite-like plaques. Therefore, it is important to consider deep dermatophytosis in the differential diagnosis of cellulitis-like lesions, especially in immunosuppressed patients. Otherwise, unnecessary antibiotic use may cause a progression of the lesion [35]. Deep dermatophytosis lesions may mimic both MG and superficial dermatophytosis. Deep dermatophytosis has also been reported in *Tinea corporis*-like circinate and pustular plaques in an HIV-positive patient [37]. In immunosuppressed patients, lesions usually progress rapidly, and hundreds of nodular lesions may develop within a few days [54].

Deep dermatophytosis can cause abscesses and cystic lesions [55]. Forty-six cases of dermatophytic abscesses have been reported in the literature until 2016 [56]. Most abscess lesions develop in immunosuppressed patients (83%). Single or multiple (83%) abscess lesions have been reported on the lower extremities (54%), trunk (33%), upper extremities (33%), and head and face (22%) [57–59]. The size of abscesses also varies; while cystic lesions are usually a few centimeters in size, cysts up to 15 cm in size may develop [60].

4.3 Disseminated Dermatophytosis

In immunosuppressed patients, dermatophytes spread not only on the skin but also to other organs. Dermatophytes can spread via the lymphatic and hematogenous routes. Lymph node, liver, spleen, and bone involvement may be observed, together with plaques, infiltrative or ulcerative nodules, or papules. While more than half of the patients with *CARD9* mutation develop lymph node involvement, in 15%, the infection spreads to other organs (the bones, central nervous system, and digestive tract) [61, 62]. The involvement of other organs shows sepsis-like findings. Clinical findings vary according to the organ involved [63]. The mortality rate is high (23.5%) in patients with *CARD9* mutation [61].

4.4 Mycetoma and Pseudomycetoma

Mycetomas are a distinct form of invasive dermatophytoses, clinically characterized by tumefactive tumoral lesions and sinuses. Dermatophytes enter the dermis after trauma and form dermatophytic clusters referred to as grains. Africa is the most commonly reported location for eumycetoma, the mycetoma caused by fungal pathogens [64]. The median age of patients with mycetoma caused by dermatophytes is 20 years (range: 8–36 years). Importantly, the most common localization of eumycetoma is the foot, while eumycetoma caused by dermatophytes is most commonly localized to the scalp [65]. Patients with scalp involvement have a previous history of *Tinea capitis* [66]. Mycetoma begins as subcutaneous nodules, which are deeply located and drain pus to the dermis and epidermis, leading to sinus formation. The clusters (grains) formed by the fungus are located in draining sinuses. Grain color varies depending on the causative agent. *Microsporum audouinii* usually causes white or yellow grain structures, although black grain structures have been reported in few cases [67]. *Microsporum ferrugineum*, *Trichophyton mentagrophytes*, *T. rubrum*, *Trichophyton soudanense*, *Trichophyton schoenleinii*, and *T. tonsurans* lead to black grain formation [68–70]. In the chronic period, wood-like hard tumoral lesions develop. Secondary bacterial infection develops in approximately half of all cases [71].

Pseudomycetoma and mycetoma lesions differ in the following ways: (1) pseudomycetoma lesions are smaller than those of mycetoma, and they do not contain sinus structures; and (2) a history of trauma usually precedes pseudomycetoma [16]. Single or multiple nodular or tumoral lesions may develop in pseudomycetoma. Hyperkeratotic or verrucous lesions caused by pseudomycetoma have been reported [15, 41]. Pseudomycetoma lesions are most commonly located on the scalp, as in mycetomas caused by dermatophytes [72]. Involvement of the lower extremities may also be observed in pseudomycetoma [15].

5 Diagnosis

In addition to histopathology, some laboratory and imaging methods can be used to differentiate the different types of invasive dermatophytosis (Table 3).

5.1 KOH Examination

Diagnosis of invasive dermatophytoses requires confirmation that the causative agent is a dermatophyte. Direct microscopic examination after clearing with KOH is the most commonly used method for the diagnosis of dermatophytic infection in the dermatology clinic. It allows the rapid and inexpensive detection of fungal elements such as hyphae and arthrospores. However, the positivity of KOH examination in invasive dermatophytic infections is not as high as that in superficial dermatophytic infections because of the deep localization of dermatophytes. KOH testing yields negative results in approximately one quarter of MG cases [9]. Therefore, negative KOH testing does not necessarily exclude the possibility of a dermatophyte infection. For the detection of dermatophytes, fungal culture and other molecular methods should also be used. Furthermore, KOH examination cannot be used to distinguish between the types of invasive dermatophytic infections [1].

5.2 Cytology

Cytological samples can be obtained by an imprint method or by fine-needle aspiration. The specimens are stained with hematoxylin–eosin (HE), May–Grünwald–Giemsa, Papanicolaou, or periodic acid–Schiff (PAS) stains. This inexpensive and rapid diagnostic method can reveal the presence of dermatophytes as well as dermal invasion. Dermal scraping or fine-needle aspiration cytology can reveal the Splendore–Hoeppli phenomenon, characterized by radially distributed eosinophilic material around the microorganism, and grain formation, indicative of mycetoma. Although cytological examination may reveal granuloma formation, it cannot be used to determine whether granulomatous infiltration is perifollicular or not. Furthermore, cytology can be used not only for skin samples but also for the diagnosis of extracutaneous organ involvement. Thus, cytology helps in differentiating disseminated dermatophytosis from other invasive dermatophytoses [73].

5.3 Fungal Culture

Fungal culture helps to detect dermatophytes and to determine their antifungal susceptibility. Sabouraud glucose agar containing chloramphenicol and cycloheximide is the most convenient and inexpensive medium for culturing dermatophytes.

Table 3 Laboratory and imaging techniques for differentiating clinical types of invasive dermatophytic infections

Diagnostic method	Visualization of fungal elements	Dermal invasion	Perifollicular granuloma	Grain	Splendore–Hoeppli phenomenon	Extracutaneous involvement
KOH examination	+	−	−	−	−	−
Cytology	+	+	−		+	+
Fungal culture	+	−	−	−	−	+
Histopathology	+	+	+	+	+	+
Radiological methods						
Ultrasonography	−	+	−	−	−	−
Magnetic resonance imaging	+	+	−	+	−	−
Computed tomography	−	−	−	−	−	−
Reflectance confocal microscopy	+	+	+	+	−	−

+, positive; −, negative

However, fungal culture cannot be used to distinguish superficial dermatophytic infection from invasive dermatophytic infection [1].

5.4 Histopathology

The gold standard method for the differentiation of invasive dermatophytosis is histopathological examination. Although HE staining performs well for determining the type of inflammation, it may be inadequate for the detection of hyphae and spores. To improve the detection rate of fungal elements, PAS and Grocott-Gomori's methenamine silver (GMS) staining should also be performed. In addition, HE-stained preparations can be examined under an immunofluorescence microscope, as fungi are autofluorescent [74]. In the case of a doubtful diagnosis, especially in immunosuppressed patients, it is important to identify the fungal species using conventional or molecular-based techniques, such as polymerase chain reaction or matrix-assisted laser desorption ionization time-of-flight mass spectrometry [65]. These molecular diagnostic methods provide a more rapid and reliable identification compared to conventional methods [7, 39, 75, 76].

5.4.1 Majocchi's Granuloma

Unlike deep dermatophytosis, acanthosis occurs in most individuals with MG, to varying degrees. Histopathological examination of MG reveals perifollicular granulomatous inflammation [21]. Perifollicular inflammation is also observed in kerion celsi. In kerion lesions, however, the cellular infiltrate spreads to the interfollicular areas and is rich in neutrophils. Unlike kerion, MG lesions show chronic inflammation with lymphocytes, macrophages, epithelioid cells, and scattered multinucleated giant cells [10]. Capillary proliferation, vascular dilation, and erythrocyte extravasation are detected in most patients. Although intradermal edema and fibrosis are usually observed, their intensity varies considerably. Mucicarmine stain can be used to detect dermal mucin deposition. Fibrinoid accumulation within the vessel is rarely detected [21]. If there is a history of trauma, disruption of the follicle and intradermal keratin structures can also be detected [77].

Depending on the immunity of the individual, some differences might be observed in MG histopathology. Epidermal acanthosis and granulomatous reactions are less prominent, while tissue necrosis is more prominent, in immunosuppressed individuals than in immunocompetent individuals. The amount of dermal hyphae and arthrospores increases with increasing immunosuppression [21].

Unlike superficial dermatophytoses, the appearance of hyphae and arthrospores in these individuals differs. Arthrospores within the multinuclear giant cells are thicker (up to 6μm in diameter) than intrafollicular arthrospores (up to 2μm in diameter). Single or multiple budding may be observed. Both the thickness and shape of the hyphae vary. In some cases, hyphae and spores are seen clustered as in mycetoma [77].

5.4.2 Deep Dermatophytosis

Histopathological examination of deep dermatophytosis reveals less extensive epidermal changes than those observed during superficial dermatophytosis. Acanthosis and pseudoepitheliomatous hyperplasia are observed in one third of patients with deep dermatophytosis [42]. Other epidermal changes are epidermal necrosis, thickening and elongation of the rete ridges, parakeratosis, and spongiosis [36]. The main histopathological findings of deep dermatophytosis involve the dermal region [10]. The type and severity of dermal infiltration vary, and acute or chronic dermal infiltrates are not limited to the perifollicular area. In such patients, inflammatory cell infiltration is apparent in the dermal region, as well as the subcutaneous tissue [42]. Granuloma formation and multinuclear giant cells are observed in more than half of the patients with deep dermatophytosis [62]. Dermal infiltration sometimes contains numerous eosinophils and plasma cells [42]. Vascular thrombosis and extensive hemorrhage throughout the reticular dermis are also reported [36]. Furthermore, the dermal infiltrate can be surrounded by fibrotic collagen [39]. Acquired reactive perforating collagenosis-like degenerated collagen structures excreted from the epidermis have been reported for a diabetic patient with *STAT3* mutation [53]. Hyphae and arthrospores should be detected for histopathological diagnosis of deep dermatophytosis. If fungal structures are not detected by HE staining, other confirmatory tests, such as PAS and GMS staining, can be performed [34].

5.4.3 Disseminated Dermatophytosis

Histopathological findings of disseminated dermatophytosis are similar to those of deep dermatophytosis. Acute or chronic inflammatory infiltrate and dermal hyphae and arthrospores are observed. Unlike deep dermatophytosis, this inflammatory reaction is observed in the lymph nodes and other organs, besides the skin. Tuberculosis-like caseating granuloma has also been reported [62].

5.4.4 Dermatophytic Mycetoma and Pseudomycetoma

In addition to dermatophytes, more than 30 fungal species cause eumycetoma. Colored granules are large enough to be detected macroscopically (0.5–2.0 mm in diameter). These granules contain septate hyphae (4–5μm thick) and spores. Eumycetomas can be distinguished from actinomycetomas by Gram staining. While filamentous bacteria that cause actinomycosis are stained positive, eumycetomas appear gram negative. Granulomatous inflammation other than grains is nonspecific. While lymphoid cells, plasma cells, histiocytes, and fibroblasts are observed early in the infection, fibroblast density increases late in the infection. Positively stained pigment structures are detected by Fontana–Masson staining [78].

The Splendore–Hoeppli phenomenon, i.e., localization of eosinophilic material radially around the hyphae and spores, is a characteristic histopathological finding of pseudomycetoma. The eosinophilic material is thought to consist of an antigen–antibody complex, major basic protein, and the remainder of inflammatory cells [79]. The inflammatory reaction involves eosinophils, histiocytes, and numerous foreign body-type and Langhans-type giant cells. The Splendore–Hoeppli phenomenon is observed not only in dermatophyte infections but also in other fungal infections (sporotrichosis, zygomycosis, *Malassezia* folliculitis, candidiasis, aspergillosis, and blastomycosis), certain bacterial infections (botryomycosis, nocardiosis, and actinomycosis), and some parasitic diseases (strongyloidiasis, schistosomiasis, and cutaneous larva migrans). To distinguish between these diseases, the causative agent should be identified [80]. Furthermore, if the causative agent cannot be isolated, this finding can be confused with other types of eosinophilic material, such as flame figures, actinomycotic granules, asteroid bodies, tophaceous lesions, and keratin granuloma [79].

5.5 Radiological Methods

Radiological methods are mainly used for the diagnosis of eumycetoma. Eumycetomas cause a few large ($\geq$ 10 mm in diameter) cavities in the bone, while actinomycoses cause a number of small cavities. These cavities are seen as a central hyperechoic area (dot) surrounded by hypoechoic tissue (circle) in ultrasonographic examination. The sensitivity and specificity of magnetic resonance imaging for the early diagnosis of mycetoma are higher than those of computed tomography. The diagnostic radiological finding for mycetoma is a "dot-in-circle" sign. Magnetic resonance imaging is important for early diagnosis and for the evaluation of the response to treatment [29].

5.6 Reflectance Confocal Microscopy

Reflectance confocal microscopy is a noninvasive, painless diagnostic method, in which thin-section images of the skin are acquired without biopsy. Although the diagnostic value of reflectance confocal microscopy for superficial dermatophyte infections has been established, no data on the diagnostic value in invasive dermatophyte infections are available [81]. However, reflectance confocal microscopy cannot be used to distinguish dermatophytes from other fungi causing infections. For this distinction, mycological cultures should be performed when fungal structures are detected. Furthermore, the high cost of this noninvasive diagnostic method limits its routine practice [82].

6 Treatment

Before determining the most appropriate treatment for invasive dermatophytosis, it is necessary to determine the disease type. This is particularly important in immunosuppressed patients. While the main treatment for MG, deep dermatophytosis, and disseminated dermatophytosis consists of systemic antifungals, surgical treatment is the first choice for mycetoma and pseudomycetoma.

6.1 Majocchi's Granuloma

Since MG was described, different therapeutic approaches have been used for treatment. The original treatments were oral potassium iodide, local X-radiation, and topical 2-dimethylamino-6-(β-diethylaminoethoxy)-benzothiazole (Asterol®). The approach has changed with the discovery of antifungal drugs, which can be used for topical and systemic applications. Topical treatments suffice for the treatment of superficial dermatophytosis. Systemic therapy has also been often used to treat MG, and it initially involved griseofulvin (250–500 mg/day) and ketoconazole (200 mg/day) [83–86]. However, new antifungals have been developed after certain toxic effects of these antifungal drugs were observed. Terbinafine (250 mg/day) has been used for treating approximately half of the MG cases reported in the last 30 years. Oral itraconazole (100–200 mg/day) has been the therapy of choice in approximately a quarter of reported cases [9]. Patients who failed to improve upon systemic itraconazole or griseofulvin administration and whose status improved with terbinafine treatment have been reported [12, 87, 88]. On the other hand, a patient with MG resistant to systemic terbinafine and itraconazole treatment was reported; this patient was treated with voriconazole (200 mg, twice daily) [89]. Antifungal therapy should be continued until the lesions are completely resolved. Depending on the severity of the disease, the duration of MG treatment varies from 1 to 6 months [39, 90].

Another factor to consider when choosing antifungals is drug interactions. Drug interactions are particularly important for patients receiving immunosuppressive therapy. Itraconazole interacts with cyclosporine, dexamethasone, fluticasone, methylprednisolone, rapamycin (sirolimus), and tacrolimus. Especially in elderly patients, azole antifungals also interact with a number of drugs used for treating heart disease and hypertension [91]. Terbinafine is more advantageous in terms of drug interactions than azole antifungals [92]. Nevertheless, postinflammatory pigmentation, atrophic scarring, and alopecia may develop despite appropriate antifungal treatments [45, 87, 93].

Surgical treatment is suggested for solitary or discrete lesions [94]. Moreover, cryotherapy can be used as an adjuvant therapeutic method if MG does not respond to antifungal treatments [95].

6.2 Deep and Disseminated Dermatophytoses

Systemic antifungal treatment should be used for deep dermatophytosis, e.g., triazoles (fluconazole, itraconazole, voriconazole, and posaconazole) and terbinafine [33, 42]. The response to antifungal treatment depends on the immunity of the patient and the immunosuppressive drug used by the patient. Rouzaud et al. [33] reported eight cases of deep dermatophytosis in solid organ transplant patients and reviewed the appropriate literature. Approximately half of the patients who receive solid organ transplants and develop deep dermatophytosis respond completely to the treatment. When the dose of the immunosuppressive drug is reduced to control fungal infection, graft rejection develops in a quarter of patients. In addition, concomitant complications accompanying immunosuppression may result in death (17%). A patient who did not improve with amphotericin B but responded well to fluconazole was also reported [36]. Generally, treatment should continue until complete recovery. In patients with *CARD9* deficiency, recurrence may be observed when the treatment is discontinued. Therefore, maintenance therapy is recommended to prevent these recurrences (oral fluconazole, 100–200 mg/day) [96]. Complete recovery can only be achieved with allogeneic bone marrow transplantation in patients with *CARD9* deficiency [97].

6.3 Mycetoma and Pseudomycetoma

Surgical excision is the best treatment option for mycetoma and pseudomycetoma with solitary lesions. If multiple lesions are present, long-term (6 months to 1 year) antifungal therapy is required [41]. Itraconazole (200–400 mg/day), terbinafine (250–500 mg/day), or griseofulvin is generally preferred for long-term antifungal therapy. Antifungal treatment alone results in a reduction in lesion size, but a complete response is usually not observed [71]. Recovery with liposomal amphotericin B was also reported, but recurrence occurred within 6 months [67]. Antifungal therapy should be continued until 3 months after clinical and radiological improvement is achieved. For clinical improvement, swelling, pain, and movement limitation should be reduced; purulent discharge should be stopped; and sinus tracts should be closed. For radiological improvement, osteomyelitis findings should regress. This period usually lasts at least 1 year, but some patients may require 2 years of treatment. After the discontinuation of the therapy, patients should be followed for at least 2 years [98].

If no response to antifungal therapy is apparent, surgical treatment is recommended. Recurrence may occur (20–90%) after amputation. Therefore, itraconazole treatment should be continued after surgery. Both clinical and radiological findings should be evaluated to assess treatment response in patients with mycetoma [71].

7 Conclusions

Currently, four different clinical types of invasive dermatophytoses are recognized. It is very important to distinguish these different types in terms of response to treatment and prognosis. Histopathological examination is the gold standard method for differentiating invasive dermatophytoses, and fungal culture or molecular methods should be used to identify the causative species. Notably, MG, pseudomycetoma, and mycetoma may also develop in immunocompetent individuals. Furthermore, deep dermatophytosis and disseminated dermatophytosis often accompany immunosuppression scenarios. Genetic susceptibility should be investigated if a recurrent deep dermatophytosis infection is apparent. Systemic antifungal drugs are the best treatment choice for deep dermatophytosis and MG, while surgical treatment is the best option for mycetoma and pseudomycetoma. The risk of death in patients with disseminated dermatophytosis despite treatment should also be considered.

Conflict of Interest The authors report no conflicts of interest. The authors alone are responsible for the content and the writing of this paper.

Ethical Approval This article does not describe any studies with human participants or animals performed by the authors.

References

1. Durdu M, Kandemir H, Ilkit M, de Hoog GS. Changing concepts and current definition of Majocchi's granuloma. Mycopathologia. 2020;185(1):187–92.
2. Bongomin F, Gago S, Oladele RO, Denning DW. Global and multi-national prevalence of fungal diseases-estimate precision. J Fungi (Basel). 2017;3(4):E57.
3. Ilkit M, Durdu M. Tinea pedis: the etiology and global epidemiology of a common fungal infection. Crit Rev Microbiol. 2015;41(3):374–88.
4. Persinoti GF, Martinez DA, Li W, Döğen A, Billmyre RB, Averette A, et al. Whole-genome analysis illustrates global clonal population structure of the ubiquitous dermatophyte pathogen *Trichophyton rubrum*. Genetics. 2018;208(4):1657–69.
5. Kaplan E, Gonca S, Kandemir H, Döğen A, Hilmioğlu-Polat S, Ilkit M, et al. Genes encoding proteolytic enzymes fungalysin and subtilisin in dermatophytes of human and animal origin: a comparative study. Mycopathologia. 2020;185(1):137–44.
6. Achterman RR, White TC. A foot in the door for dermatophyte research. PLoS Pathog. 2012;8(3):e1002564.
7. de Hoog GS, Dukik K, Monod M, Packeu A, Stubbe D, Hendrickx M, et al. Toward a novel multilocus phylogenetic taxonomy for the dermatophytes. Mycopathologia. 2017;182(1–2):5–31.
8. Majocchi D. Sopra una nuova tricofizia (Granuloma tricofitico), Studi clinici e micologici. Bull R Acad Med Roma. 1883;9:220–3.
9. Ilkit M, Durdu M, Karakaš M. Majocchi's granuloma: a symptom complex caused by fungal pathogens. Med Mycol. 2012;50(5):449–57.
10. Boral H, Durdu M, Ilkit M. Majocchi's granuloma: current perspectives. Infect Drug Resist. 2018;11:751–60.
11. Saadat P, Kappel S, Young S, Abrishami M, Vadmal MS. *Aspergillus fumigatus* Majocchi's granuloma in a patient with acquired immunodeficiency syndrome. Clin Exp Dermatol. 2008;33(4):450–3.
12. Sequeira M, Burdick AE, Elgart GW, Berman B. New-onset Majocchi's granuloma in two kidney transplant recipients under tacrolimus treatment. J Am Acad Dermatol. 1998;38(3):486–8.
13. Govind A, L'Etoile N, Vasquez G. The first reported case of Majocchi's granuloma with *Malbranchea* sp. in an immunocompetent patient. Case Rep Infect Dis. 2017;2017:9196528.
14. Chai ZT, Oh CC, Chan MM, Foo M, Pang SM. Deep dermal fungal infection in an Asian renal transplant recipient. Int J Dermatol. 2017;56(3):269–71.
15. Berg JC, Hamacher KL, Roberts GD. Pseudomycetoma caused by *Microsporum canis* in an immunosuppressed patient: a case report and review of the literature. J Cutan Pathol. 2007;34(5):431–4.

16. Botterel F, Romand S, Cornet M, Recanati G, Dupont B, Bourée P. Dermatophyte pseudomycetoma of the scalp: case report and review. Br J Dermatol. 2001;145(1):151–3.
17. Dahl MV. Dermatophytosis and the immune response. J Am Acad Dermatol. 1994;31(3 Pt 2): s34–41.
18. Brasch J. Pathogenesis of tinea. J Dtsch Dermatol Ges. 2010;8(10):780–6.
19. Svejgaard EL. Immunologic investigations of dermatophytoses and dermatophytosis. Semin Dermatol. 1985;4:201–21.
20. Marsh TW, Artis WM. Host defence mechanisms and the superficial fungal infections. Dermatol Clin. 1984;2:67–79.
21. Smith KJ, Neafie RC, Skelton HG 3rd, Barrett TL, Graham JH, Lupton GP. Majocchi's granuloma. J Cutan Pathol. 1991;18(1):28–35.
22. Elgart ML. Tinea incognito: an update on Majocchi's granuloma. Dermatol Clin. 1996;14 (1):51–5.
23. Nishiyama C, Miyaji M, Saheki M, Morioka S. Studies on the parasitic forms of *Trichophyton rubrum* isolated from patients with granuloma trichophyticum using the "agar-implantation method". J Dermatol. 1985;12:325–8.
24. Abdel-Rahman SM. Genetic predictors of susceptibility to dermatophytoses. Mycopathologia. 2017;182(1–2):67–76.
25. Yoshikawa FS, Yabe R, Iwakura Y, de Almeida SR, Saijo S. Dectin-1 and dectin-2 promote control of the fungal pathogen *Trichophyton rubrum* independently of IL-17 and adaptive immunity in experimental deep dermatophytosis. Innate Immun. 2016;22(5):316–24.
26. Wang X, Li R. *STAT3* and inherited susceptibility to invasive dermatophytosis. Br J Dermatol. 2018;179(3):567–8.
27. Vaezi A, Fakhim H, Abtahian Z, Khodavaisy S, Geramishoar M, Alizadeh A, et al. Frequency and geographic distribution of *CARD9* mutations in patients with severe fungal infections. Front Microbiol. 2018;9:2434.
28. Heinen MP, Cambier L, Fievez L, Mignon B. Are Th17 cells playing a role in immunity to dermatophytosis? Mycopathologia. 2017;182(1–2):251–61.
29. Verma P, Jha A. Mycetoma: reviewing a neglected disease. Clin Exp Dermatol. 2019;44 (2):123–9.
30. Lourdes LS, Mitchell CL, Glavin FL, Schain DC, Kaye FJ. Recurrent dermatophytosis (Majocchi's granuloma) associated with chemotherapy-induced neutropenia. J Clin Oncol. 2014;32(27):e92–4.
31. Lyra MR, Muniz Álvarez B, Lanziano AL, Imbeth MAA, Sá AM, Cuzzi T, et al. Exfoliative erythroderma and palmoplantar hyperkeratosis associated with Majocchi's granuloma by *Trichophyton tonsurans* in a patient with AIDS. Rev Iberoam Micol. 2017;34(3):185–8.
32. Li FQ, Lv S, Xia JX. Majocchi's granuloma after topical corticosteroids therapy. Case Rep Dermatol Med. 2014;2014:507176.
33. Rouzaud C, Chosidow O, Brocard A, Fraitag S, Scemla A, Anglicheau D, et al. French mycoses study group. Severe dermatophytosis in solid organ transplant recipients: a French retrospective series and literature review. Transpl Infect Dis. 2018;20(1). https://doi.org/10.1111/tid.12799.
34. Nir-Paz R, Elinav H, Pierard GE, Walker D, Maly A, Shapiro M, et al. Deep infection by *Trichophyton rubrum* in an immunocompromised patient. J Clin Microbiol. 2003;41 (11):5298–301.
35. Lowther AL, Somani AK, Camouse M, Florentino FT, Somach SC. Invasive *Trichophyton rubrum* infection occurring with infliximab and long-term prednisone treatment. J Cutan Med Surg. 2007;11(2):84–8.
36. Smith KJ, Welsh M, Skelton H. *Trichophyton rubrum* showing deep dermal invasion directly from the epidermis in immunosuppressed patients. Br J Dermatol. 2001;145(2):344–8.
37. Muñoz-Pèrez MA, Rodriguez-Pichardo A, Camacho F, Rios JJ. Extensive and deep dermatophytosis caused by *Trichophyton mentagrophytes* var. *interdigitalis* in an HIV-1 positive patient. J Eur Acad Dermatol Venereol. 2000;14(1):61–3.

38. Gong JQ, Liu XQ, Xu HB, Zeng XS, Chen W, Li XF. Deep dermatophytosis caused by *Trichophyton rubrum*: report of two cases. Mycoses. 2007;50(2):102–8.
39. Su H, Li L, Cheng B, Zhu J, Zhang Q, Xu J, Zhu M. *Trichophyton rubrum* infection characterized by Majocchi's granuloma and deeper dermatophytosis: case report and review of published literature. Mycopathologia. 2017;182(5–6):549–54.
40. Sommer S, Barton RC, Wilkinson SM, Merchant WJ, Evans EG, Moore MK. Microbiological and molecular diagnosis of deep localized cutaneous infection with *Trichophyton mentagrophytes*. Br J Dermatol. 1999;141(2):323–5.
41. Tirado-González M, Ball E, Ruiz A, Rodriguez Y, Goudet CE, Finkel O, et al. Disseminated dermatophytic pseudomycetoma caused by *Microsporum* species. Int J Dermatol. 2012;51 (12):1478–82.
42. Kershenovich R, Sherman S, Reiter O, Huss SR, Didkovsky E, Mimouni D, et al. A unique clinicopathological manifestation of fungal infection: a case series of deep dermatophytosis in immunosuppressed patients. Am J Clin Dermatol. 2017;18(5):697–704.
43. Wilson JW, Plunkett OA, Gregersen A. Nodular granulomatous perifolliculitis of the legs caused by *Trichophyton rubrum*. AMA Arch Dermatol Syphilol. 1954;69(3):258–77.
44. Brod C, Benedix F, Röcken M, Schaller M. Trichophytic Majocchi granuloma mimicking Kaposi sarcoma. J Dtsch Dermatol Ges. 2007;5(7):591–3.
45. Kim JE, Won CH, Chang S, Lee MW, Choi JH, Moon KC. Majocchi's granuloma mimicking Kaposi sarcoma in a heart transplant patient. J Dermatol. 2011;38(9):927–9.
46. Molina-Leyva A, Perez-Parra S, Garcia-Garcia F. Case for diagnosis. An Bras Dermatol. 2014;89(5):839–40.
47. Chang SE, Lee DK, Choi JH, Moon KC, Koh JC. Majocchi's granuloma of the vulva caused by *Trichophyton mentagrophytes*. Mycoses. 2005;48(6):382–4.
48. Grumach AS, de Queiroz-Telles F, Migaud M, Lanternier F, Filho NR, Palma SM, et al. A homozygous *CARD9* mutation in a Brazilian patient with deep dermatophytosis. J Clin Immunol. 2015;35(5):486–90.
49. Seçkin D, Arikan S, Haberal M. Deep dermatophytosis caused by *Trichophyton rubrum* with concomitant disseminated nocardiosis in a renal transplant recipient. J Am Acad Dermatol. 2004;51(Suppl. 5):S173–6.
50. Dan P, Rawi R, Hanna S, Reuven B. Invasive cutaneous *Trichophyton shoenleinii* infection in an immunosuppressed patient. Int J Dermatol. 2011;50(10):1266–9.
51. Gönül M, Saraçlı MA, Demiriz M, Gül U. Deep *Trichophyton rubrum* infection presenting with umbilicated papulonodules in a cardiac transplant recipient. Mycoses. 2013;56(3):361–4.
52. Ergen EN, Donaldson SL, Stratton MS, Pavlidakey PG, Elewski BE. Image gallery: Molluscum contagiosum-like facial lesions in a patient with a renal transplant: deep dermatophytosis due to *Trichophyton rubrum*. Br J Dermatol. 2018;178(5):e328.
53. Simpson JK, Fröbel P, Seneviratne SL, Brown M, Lowe DM, Grimbacher B, et al. Invasive dermatophyte infection with *Trichophyton interdigitale* is associated with prurigo-induced pseudoperforation and a signal transducer and activator of transcription 3 mutation. Br J Dermatol. 2018;179(3):750–4.
54. Lestringant GG, Lindley SK, Hillsdon-Smith J, Bouix G. Deep dermatophytosis to *Trichophyton rubrum* and *T. verrucosum* in an immunosuppressed patient. Int J Dermatol. 1988;27(10):707–9.
55. Kobayashi M, Ishida E, Yasuda H, Yamamoto O, Tokura Y. Tinea profunda cysticum caused by *Trichophyton rubrum*. J Am Acad Dermatol. 2006;54(Suppl 2):S11–3.
56. Inaoki M, Nishijima C, Miyake M, Asaka T, Hasegawa Y, Anzawa K, et al. Case of dermatophyte abscess caused by *Trichophyton rubrum*: a case report and review of the literature. Mycoses. 2015;58(5):318–23.
57. Patel G, Mills C. Tinea faciei due to *Microsporum canis* abscess formation. Clin Exp Dermatol. 2000;25(8):608–10.

58. Okata-Karigane U, Hata Y, Watanabe-Okada E, Miyakawa S, Ota M, Uzawa Y, et al. Subcutaneous abscesses caused by *Trichophyton rubrum* in the unilateral groin of an immunocompromised patient: a case report. Med Mycol Case Rep. 2018;21:16–9.
59. Kim SH, Jo IH, Kang J, Joo SY, Choi JH. Dermatophyte abscesses caused by *Trichophyton rubrum* in a patient without pre-existing superficial dermatophytosis: a case report. BMC Infect Dis. 2016;16:298.
60. Toyoki M, Hase M, Hirasawa Y, Umeyama T, Miyazaki Y, Kano R, et al. A giant dermatophyte abscess caused by *Trichophyton rubrum* in an immunocompromised patient. Med Mycol J. 2017;58(2):E63–6.
61. Pilmis B, Puel A, Lortholary O, Lanternier F. New clinical phenotypes of fungal infections in special hosts. Clin Microbiol Infect. 2016;22(8):681–7.
62. Tejasvi T, Sharma VK, Sethuraman G, Singh MK, Xess I. Invasive dermatophytosis with lymph node involvement in an immunocompetent patient. Clin Exp Dermatol. 2005;30(5):506–8.
63. Wu LC, Sun PL, Chang YT. Extensive deep dermatophytosis cause by *Trichophyton rubrum* in a patient with liver cirrhosis and chronic renal failure. Mycopathologia. 2013;176(5–6):457–62.
64. van de Sande WW. Global burden of human mycetoma: a systematic review and meta-analysis. PLoS Negl Trop Dis. 2013;7(11):e2550.
65. Diongue K, Boye A, Bréchard L, Diallo MA, Dione H, Ndoye NW, et al. Dermatophytic mycetoma of the scalp due to an atypical strain of *Microsporum audouinii* identified by MALDI-TOF MS and ITS sequencing. J Mycol Med. 2019;29(2):185–8.
66. Chiapello LS, Dib MD, Nuncira CT, Nardelli L, Vullo C, Collino C, et al. Mycetoma of the scalp due to *Microsporum canis*: hystopathologic, mycologic, and immunogenetic features in a 6-year-old girl. Diagn Microbiol Infect Dis. 2011;70(1):145–9.
67. West BC, Kwon-Chung KJ. Mycetoma caused by *Microsporum audouinii.* First reported case. Am J Clin Pathol. 1980;73(39):447–54.
68. Baylet R, Camain R, Juminer B, Faye I. *Microsporum ferrugineum* Ota, 1921, an agent of scalp mycetomas in dark Africa. Pathol Biol. 1973;21(1):5–12.
69. Develoux M, Ndiaye B, Kane A, Diousse P, Huerre M. Mycetoma of the scalp and the neck in Dakar. J Mycol Med. 1999;9(1):65–7.
70. Strobel M, Ndiaye B, Marchand JP, Ravisse P, Basset M. Dermatophyte mycetoma of the scalp. Ann Dermatol Venereol. 1980;107(12):1181–4.
71. Zijlstra EE, van de Sande WW, Welsh O, Mahgoub el S, Goodfellow M, Fahal AH. Mycetoma: a unique neglected tropical disease. Lancet Infect Dis. 2016;16(1):100–12.
72. Colwell AS, Kwaan MR, Orgill DP. Dermatophytic pseudomycetoma of the scalp. Plast Reconstr Surg. 2004;113(3):1072–3.
73. Durdu M. Diagnostic approach to granulomatous diseases. In: Durdu M, editor. Cutaneous cytology and Tzanck smear test. Cham: Springer; 2018. p. 125–64.
74. Elston DM. Fluorescence of fungi in superficial and deep fungal infections. BMC Microbiol. 2001;1:21.
75. Mirhendi H, Makimura K, de Hoog GS, Rezaei-Matehkolaei A, Najafzadeh MJ, Umeda Y, et al. Translation elongation factor 1-α gene as a potential taxonomic and identification marker in dermatophytes. Med Mycol. 2015;53(3):215–24.
76. Rezaei-Matehkolaei A, Mirhendi H, Makimura K, de Hoog GS, Satoh K, Najafzadeh MJ, et al. Nucleotide sequence analysis of beta tubulin gene in a wide range of dermatophytes. Med Mycol. 2014;52(7):674–88.
77. Mikhail GR. *Trichophyton rubrum* granuloma. Int J Dermatol. 1970;9(1):41.
78. Kramer SC, Ryan M, Bourbeau P, Tyler WB, Elston DM. Fontana-positive grains in mycetoma caused by *Microsporum canis*. Pediatr Dermatol. 2006;23(59):473–5.
79. Hussein MR. Mucocutaneous Splendore-Hoeppli phenomenon. J Cutan Pathol. 2008;35(11):979–88.
80. Gopinath D. Splendore-Hoeppli phenomenon. J Oral Maxillofac Pathol. 2018;22(29):161–2.

81. Hui D, Xue-cheng S, Ai-e X. Evaluation of reflectance confocal microscopy in dermatophytosis. Mycoses. 2013;56(2):130–3.
82. Navarrete-Dechent C, Bajaj S, Marghoob AA, Marchetti MA. Rapid diagnosis of tinea incognito using handheld reflectance confocal microscopy: a paradigm shift in dermatology? Mycoses. 2015;58(6):383–6.
83. Kanaan IC, Santos TB, Kac BK, Souza AM, Cerqueira AM. Majocchi's granuloma—case report. An Bras Dermatol. 2015;90(2):251–3.
84. Kranseler J, Gupta D. Erythematous annular plaques on a 9-year-old's arms and chest. Clin Pediatr. 2017;57(4):488–90.
85. Lewkowicz C, Lakhiani C. Girl's papular rash will not respond to treatment. Int J Contemp Pediatr. 2017;34:42–4.
86. Seddon ME, Thomas MG. Invasive disease due to *Epidermophyton floccosum* in an immunocompromised patient with a Behçet's syndrome. Clin Infect Dis. 1997;25(1):153–4.
87. Rallis E, Katoulis A, Rigopoulos D. Pubic Majocchi's granuloma unresponsive to itraconazole successfully treated with oral terbinafine. Skin Appendage Disord. 2015;1(3):111–3.
88. Gupta AK, Prussick R, Sibbald RG, Knowles SR. Terbinafine in the treatment of Majocchi's granuloma. Int J Dermatol. 1995;34(7):489.
89. Liu HB, Liu F, Kong QT, Shen YN, Lv GX, Liu WD, Sang H. Successful treatment of refractory Majocchi's granuloma with voriconazole and review of published literature. Mycopathologia. 2015;180(3–4):237–43.
90. Bressan AL, Silva RS, Fonseca JC, Alves Mde F. Majocchi's granuloma. An Bras Dermatol. 2011;86(4):797–8.
91. Gupta AK, Versteeg SG, Shear NH. Common drug-drug interactions in antifungal treatments for superficial fungal infections. Expert Opin Drug Metab Toxicol. 2018;14(49):387–98.
92. Bonifaz A, Tirado-Sánchez A, Mercadillo P, Fierro-Arias L. Treatment of Majocchi granuloma. A retrospective study of 36 cases. J Dermatol Treat. 2021;32(2):264–65.
93. Drakensjö IT, Vassilaki I, Bradley M. Majocchi's granuloma caused by *Trichophyton mentagrophytes* in 2 immunocompetent patients. Actas Dermosifiliogr. 2017;108(1):e6–8.
94. Ma L, Li R, Yu J, Wang C, Wan Z, Wang X. Majocchi's granuloma in a liver transplant recipient caused by *Trichophyton* spp., phenotypically consistent with *Trichophyton rubrum* var. *raubitschekii*. Med Mycol. 2009;47(3):312–6.
95. Pietrzak A, Tomasiewicz K, Kanitakis J, Paszkowski T, Dybiec E, Donica H, et al. *Trichophyton mentagrophytes*-associated Majocchi's granuloma treated with cryotherapy. Folia Histochem Cytobiol. 2012;50(3):486–9.
96. Corvilain E, Casanova JL, Puel A. Inherited *CARD9* deficiency: invasive disease caused by ascomycete fungi in previously healthy children and adults. J Clin Immunol. 2018;38 (6):656–93.
97. Queiroz-Telles F, Mercier T, Maertens J, Sola CBS, Bonfim C, Lortholary O, et al. Successful allogenic stem cell transplantation in patients with inherited *CARD9* deficiency. J Clin Immunol. 2019;39(5):462–9.
98. Bendl BJ, Mackey D, Al-Saati F, Sheth KV, Ofole SN, Bailey TM. Mycetoma in Saudi Arabia. J Trop Med Hyg. 1987;90(2):51–9.

Unusual Dermatophytosis Presentations and New Emerging Dermatophytes Species

Marie-Pierre Hayette and Rosalie Sacheli

Abstract

Dermatophytes are causing superficial mycosis in animals and humans. Depending on the geophilic, zoophilic or anthropophilic origin of the fungus but also on the immunological status of the patient, symptomatology can widely differ. Nevertheless, each species is currently associated with typical clinical manifestations, even if atypical localizations and/or clinical pictures are sometimes also reported. Diagnostic tools applied to species identification have been changing since the last two decades with the more frequent use of molecular methods currently considered nowadays as reference methods for species identification. It becomes obvious that the algorithm used for the distinction of closely related species needs to combine phenotypic and genomic methods. All these different points are discussed and the most recent novel species causing or involved in human dermatophytosis are reported.

Keywords

Dermatophytes · Unusual dermatophytosis · New species

1 Introduction

Dermatophytes are keratinophilic fungi, which cause superficial infections of the nails, skin, and hair. They belong to the Arthrodermataceae family that is divided into nine genera that are *Trichophyton*, *Epidermophyton*, *Nannizzia*, *Lophophyton*,

---

M.-P. Hayette (✉) · R. Sacheli
Department of Clinical Microbiology, National Reference Centre for Mycosis, University Hospital of Liège, Liège, Belgium
e-mail: mphayette@chuliege.be; R.Sacheli@chuliege.be

J.-P. Bouchara et al. (eds.), *Dermatophytes and Dermatophytoses*,
https://doi.org/10.1007/978-3-030-67421-2_5

Paraphyton, *Microsporum*, *Arthroderma*, *Guarromyces*, and *Ctenomyces* following the recent nomenclature revision by De Hoog et al. [1].

Dermatophytosis varies with several factors such as the climate, the geographic area, the contact with animals, and the age of the patient. The dermatophyte species encountered can also change over the time, with some species becoming less or more common like *Trichophyton rubrum*, which emerged after the second world war, or the zoophilic *Trichophyton benhamiae*, which is now frequently isolated in Western countries from children infected by Guinea pigs [2–5]. In Europe, the most frequent etiological agent for onychomycosis is *T. rubrum*, followed by *Trichophyton interdigitale*. For *Tinea capitis*, anthropophilic species are predominating in many countries, mostly due to populations of immigrant origin. Indeed, anthropophilic species are mostly isolated in prepubescent children from African immigrant families living in big European or American cities [6]. *Trichophyton tonsurans* is currently the most important etiological agent of *Tinea capitis* in the USA and UK. Its frequency increases in France, while in other European countries or some cities *Microsporum audouinii* (Belgium) and *Trichophyton soudanense* (Paris, France) are still predominant [7, 8]. Meanwhile, *Tinea capitis* due to zoophilic species result from close contact with infected pets among which cats are mostly involved causing *Microsporum canis* infections that are sometimes difficult to cure [9]. For skin infections, anthropophilic and zoophilic strains may be in cause and particularly *T. rubrum*, the *Trichophyton mentagrophytes* series, *Epidermophyton floccosum*, or *M. canis* [2, 3, 10]. In some rare cases, geophilic species may also be the etiologic agents [11]. Most dermatophyte species have become adapted to human or animals. Although they can infect other hosts, each dermatophyte tends to be associated with a particular host or group of hosts and is not maintained at long term in other ones. Regarding the identification, dermatophytes are mainly identified by microscopy performed on Sabouraud agar medium, which can be combined with molecular biology methods such as pan-dermatophytes PCR or sequencing of the internal transcribed spacer (ITS) 1 and 2 regions of ribosomal DNA [12–14]. However, if more and more laboratories include PCR assays for dermatophyte identification to decrease the dependence to skilled technicians and the turnaround time to results, it is critical to refer to reference laboratories because in some cases, new or atypical species are involved. The most frequent dermatophytes isolated in dermatomycosis still belong to the three genera *Epidermophyton, Trichophyton*, and *Microsporum*. However, some geophilic and some rare zoophilic species causing human disease are distributed into the genera *Arthroderma, Paraphyton, Lophophyton*, and *Nannizzia*.

In this review, we focused our attention on new dermatophyte species, which have been recently described as a cause of dermatomycosis in humans, on dermatophyte species infrequently implicated in dermatomycosis, and on unusual clinical manifestations caused by known dermatophytes.

2 New Species of Dermatophytes

2.1 Geophilic Dermatophytes Isolated from Human Infection

Only a few new dermatophyte species causing human infections or isolated from human body parts have been described in the last 20 years thanks to molecular biology techniques and mating experiments. All these new species are related or belong to geophilic dermatophytes, which are more susceptible to genetic variation than zoophilic or anthropophilic ones. Six new species recently described and related to clinical cases are presented under the names used in the last taxonomy revision and listed in Table 1 as: *Arthroderma chiloniense* [15], *Nannizzia perplicata* [16], *Nannizzia aenigmaticum* [17], *Arthroderma onychocola* [18], *Paraphyton mirabile* [19], and *Arthroderma eboreum* [20, 21].

2.1.1 *Arthroderma chiloniense*

Arthroderma chiloniense was isolated in Germany from a 68-year-old female presenting with scaly skin lesions [15]. Skin samples were collected; however, no fungal element was ever detected by microscopic examination. Fungal culture led to the isolation of a presumptive dermatophyte. Subsequent detailed investigations with conventional morphological and physiological methods and a phylogenetic analysis of combined sequences of LSU (D1/D2 domains) and ITS regions of rDNA led to the description of a new geophilic species named *Arthroderma chiloniense* sp. nov., EMBL accession no. LT992885. This species lacks specific microscopic features and can be confused with *T. rubrum* or *T. interdigitale* except the disposition of microconidia on hyphae, which were more similar to *Chrysosporium* conidia. Because no fungal element could be seen and no further isolation of the fungus was achieved, the strain was, however, supposed to be a transient colonizer and the lesions due to autoimmune disorder. This highlights the importance of the direct microscopy in the definition of infection.

2.1.2 *Nannizzia perplicata*

A novel dermatophyte species was recently isolated from skin scales of a female patient in UK presenting with itching rash of the wrist and arm [16]. Her principal risk factor was long-term corticosteroid use for underlying Lupus autoimmune syndrome explaining probably the inflammatory lesion. Skin scales revealed positive hyphae and grew a dermatophyte that could be reisolated 5 months later despite oral and systemic 3 months terbinafine treatment. In vitro resistance to terbinafine was demonstrated by a high minimum inhibitory concentration (MIC) of 16μg/mL explaining the treatment failure. The treatment was shifted to itraconazole 400 mg per day plus clotrimazole cream for 2 weeks. The isolate was referred to the UK National Reference Mycology Laboratory due to atypical morphology and for MIC determination. Culture on Sabouraud agar revealed white to pale buff colonies with a radially folded inner zone harboring a cerebriform center and nonpigmented reverse. Microscopy was consistent with species of the geophilic *Trichophyton terrestre* complex. The use of multilocus phylogenetic analysis using 6 loci allowed the

Table 1 New species isolated from human infection/colonization in the past 15 years

New species	Year of first publication	Country of isolation	Site of isolation	Sex / age	Infection confirmed	Dermatophyte with similar features	Reference (s)
Arthroderma chiloniense	2019	Germany	Trunk	F/68	No	*T. rubrum/T. interdigitale*	[15]
Nannizzia perplicata	2018	United Kingdom	Wrist/arm	F/44	Yes	*T. terrestre*	[16]
Nannizzia aenigmaticum	2014	Czech Republic	Wrist	F/46	Yes	*T. tonsurans/T. equinum*	[17]
Arthroderma onychocola	2014	Czech Republic	Toenail	F/46	Yes	*T. rubrum/T. interdigitale*	[18, 22]
Paraphyton mirabile	2012	The Netherlands	Toenail	ND	No	*P. cookei*	[19, 24]
Arthroderma eboreum	2005	Germany	Foot	F/14	Yes	*T. terrestre*	[20, 21, 25]

F female, *M* male, *Nd* no data, *T Trichophyton*, *P Paraphyton*

identification of a novel species within the *Nannizzia gypsea* (formerly *Microsporum gypseum*) complex and close to *Nannizzia nana* (formerly *Microsporum nanum*). This novel species surprisingly matched at 99% with a previously called "*Microsporum* sp." isolated from a leg lesion in Czech Republic, attesting the presence of the dermatophyte outside UK and confirming the pathogenic or opportunistic character of this geophilic species. This patient worked regularly with compost and potted plants, which can explain the source of this soil-related geophilic species.

2.1.3 *Nannizzia aenigmaticum*

This species (formerly *Microsporum aenigmaticum)* was recently isolated from a scaling lesion of the wrist of a 46-year-old Czech woman. The lesion was large (40 mm diameter), erythematous, well delineated with scaling at the periphery but was nonitchy. Direct microscopy showed hyphae compatible with a dermatophyte, and a culture on Sabouraud Dextrose agar (SDA) with and without cycloheximide yielded a colony, which was first identified as *T. tonsurans/Trichophyton equinum* [17]. The patient presented several risk factors for a dermatophyte infection: she was hunting, took care from horses, owned a dog breeding station, and worked frequently in her garden. She often visited swimming pools and saunas. Topical treatment based on clotrimazole 1% was successful. The isolate was sent to the reference laboratory in Prague for further characterization where molecular studies based on ITS rDNA and β-tubulin encoding gene analysis were used for species identification. Phylogenetic analysis led to the positioning of the isolate among the *N. gypsea* clade [17]. The morphological characteristics are as follows: (1) a slower development than other members of *N. gypsea* clade, (2) growth at 37 °C, and (3) the reverse of the colonies was yellow to orange with the time. The species produced numerous chlamydospores and aggregates of hyphae and racket filaments. Molecular data clearly distinguished this species from known *Microsporum* species. Again, sending the strain to a Reference laboratory allowed the identification of a novel dermatophyte species, which was further renamed as *N. aenigmaticum*. The most frequently isolated geophilic species causing human infection are members of the *N. gypsea* clade, with *N.* gypsea and *Nannizzia fulva* (formerly *Microsporum fulvum*) being the most frequent. However, these species are sometimes difficult to distinguish by simple microscopy and physiological characteristics. Therefore, the use of DNA sequencing techniques, and perhaps also in the next future matrix-assisted laser desorption ionization–time of flight (MALDI-TOF) mass spectrometry (MS), is recommended for a reliable identification to the species level [17–19].

2.1.4 *Arthroderma onychocola* (Formerly *Trichophyton onychocola*)

Arthroderma onychocola is a presumed geophilic dermatophyte first isolated in 2011 from a toenail in a Czech patient with a history of *T. rubrum* infection [18]. Another similar case was observed in Denmark (strain isolated in 2013) in a Kurdish descent presenting onychomycosis of the great toenail and having a past history (2004) of *T. rubrum* infection at the same site. Both strains could be successfully mated, and the sexual stage showed all the characteristics of *Arthroderma* teleomorph

[18]. *Arthroderma onychocola* is classified into the basal clades of the phylogenic tree of dermatophytes, which includes almost only geophilic species. Colonies growing on SGA are white yellowish to white greenish yellow, plane, floccose, and raised at the center, crateriform, with a light orange to red-brown color at the reverse. On SGA, this species shows numerous microconidia, which are clavate to piriform, smooth-walled, and born singly on short lateral branches or at the end of hyphae or sessile on the hyphae. No macroconidia are observed. Numerous spiral hyphae are often present. Thick-walled chlamydospores (mostly in chain) are observed only on Malt Agar Extract (MAE) medium. Optimal temperature is between 18 °C and 25 °C, and no growth is observed above 34 °C [22]. However, because geophilic species are rarely pathogens, one should be cautious when reporting a geophilic dermatophyte as causal agent of dermatomycosis. In the first case involving the Czech patient, it was not possible to confirm the causal role of the fungus. Indeed, due to the noncompliance of the patient who did not come back for control, the fungus was isolated only once and direct microscopic examination of the unique sample obtained from this patient was negative. In the second case, however, the isolate was probably the causal agent of relapsing onychomycosis because the culture yielded only this species and direct examination revealed hyphae and spores. It would have been, however, interesting to use a pan-dermatophyte PCR assay to exclude a potential coinfection with the previous *T. rubrum* isolated in the past [3, 23]. A misidentification of the first so-called *T. rubrum* cannot be excluded. Macroscopical and microscopical features of this species are described in Fig. 1a, b.

2.1.5 *Paraphyton mirabile* (Formerly *Microsporum mirabile*)

This species was first isolated from a toenail in The Netherlands but the clinical case was not published. It has also been isolated from asymptomatic animals like a domestic dog in Germany and from an alpine chamois in Italy [24]. The phenotypic characteristics include production of large, thick-walled, multicelled conidia, resembling those of *Paraphyton cookei* except the strains are highly pleomorphic. Because the isolates had been deposited in reference centers, comparison of morphology, physiology, mating tests, and analysis of ITS and partial β-tubulin (BT2) gene sequences led to the recognition of a novel species that was classified among the geophilic *P. cookei* clade [19]. So far, no other clinical case has been reported. The pathogenicity of this novel species is therefore not confirmed and reports of other well-documented clinical cases are needed before conclusion. Macroscopical and microscopical features of this species are described in Fig. 1c, d.

2.1.6 *Arthroderma eboreum*

Arthroderma eboreum was firstly isolated in an African patient living in Germany for the last 14 years and going back occasionally to Ivory coast, his homeland. The patient was HIV positive and was suffering from scaly and itchy moccasin-type *Tinea pedis* that lasted since about 10 years. Scales taken from his foot grew rapidly on Sabouraud-gentamicin agar (SGA). The morphological features of the colony were similar on different culture media, and these are described as follows: an off-white spreading colony with a granular to powdery surface and a radiating

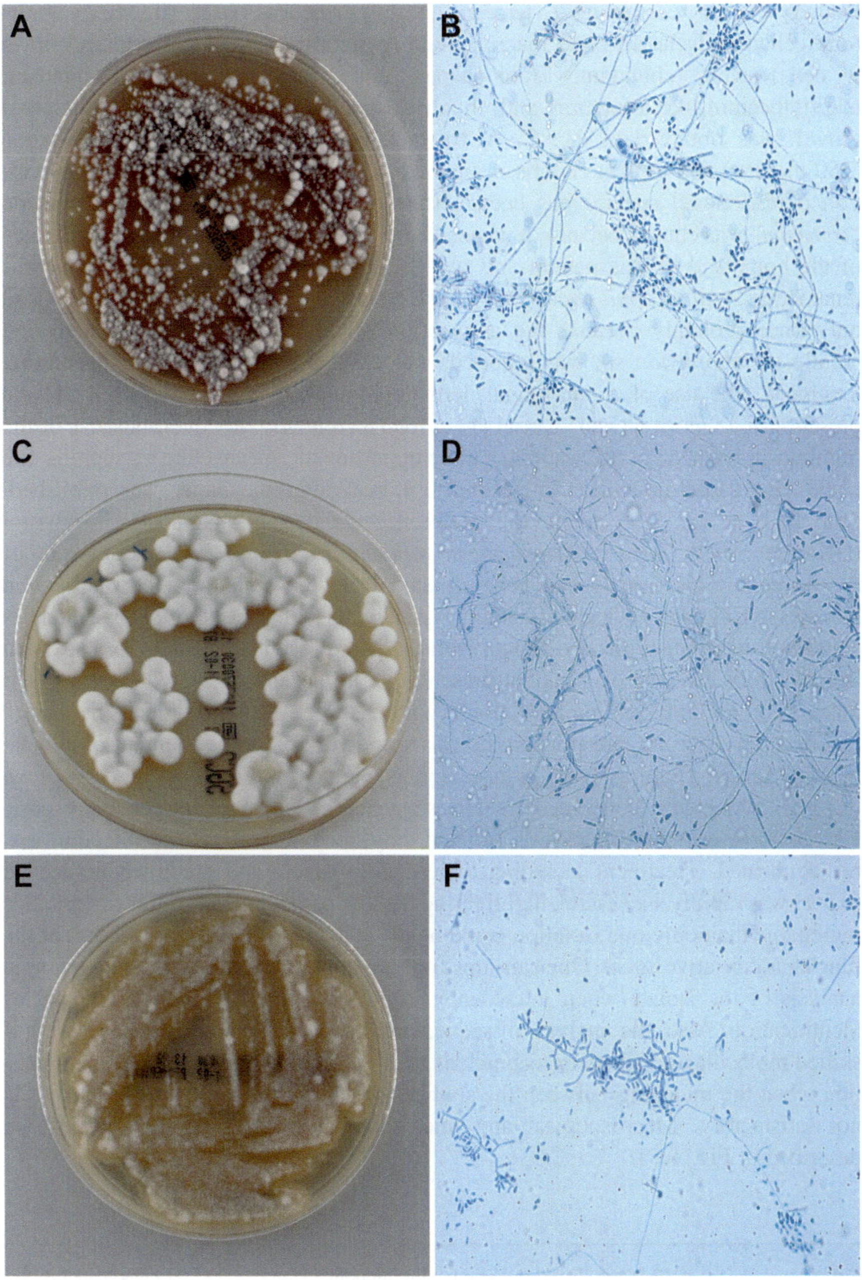

Fig. 1 Macroscopic (**a**, **c**, and **e**) and microscopic (**b**, **d**, and **f**) features of some new species of dermatophytes. *Arthroderma onychocola* IHEM 26867 on Sabouraud chloramphenicol agar, 28 °C, 21 days (**a** and **b**). (**a**) Surface of the colonies is white and slightly powdery; red-brown pigment is visible. (**b**) Numerous clavate to tear-shaped microconidia are observed; rare macroconidia are also visible such as in the middle of the photograph (magnification 20 × 40). *Paraphyton mirabile*

feathery margin resembling *T. terrestre*. No pigment was observed. Microconidia (one celled and clavate) and macroconidia (cylindrical or club-shaped and up to 9 cells) were abundant. Hook-shaped and spirals were also abundant. Cleistothecium-like structures and chlamydospores appeared in old cultures. No growth was obtained at 37 °C. A 10-week course associating oral terbinafine (250 mg/day) and topical ciclopirox olamine was successful. The detailed description of this novel species has been performed by Brash et al. [20]. The name "*eboreum*" was chosen because it means "ivory" in Latin, which was a reminiscence of the homeland of this patient and of the color of the colony. The species was renamed as *Arthroderma eboreum* by De Hoog et al. [1]. The species was isolated only once probably because the association of positive direct examination and culture (*Trichophyton* sp. was reported) was enough for the clinician for starting treatment. The use of a pan-fungal dermatophyte PCR could have excluded the presence of a "more typical" causal agent for *Tinea pedis*. However, because of the immunosuppression of this patient, a more opportunistic species like a geophilic one could also cause infection. Interestingly, 1 year after the paper was published, Campbell et al. [21] from the Reference Center for Mycosis in Bristol (UK) described former *Arthroderma olidum* as being the teleomorph of *A. eboreum*. This fungus had been isolated from the soil of badger and rabbit burrows in England. Molecular methods based on the comparison of ITS1 sequences could not distinguish the two fungal isolates. More recently, *A. eboreum* has been isolated from the skin of a patient presenting with eczema-like erythematous skin macules on the arm and inguinal region with a 2-week evolution [25]. The patient reported traveling in Greece where he was woodworking at the moment of clinical manifestations. Cultures from skin scrapings yielded a fungus with *A. eboreum* features, which was confirmed by ITS sequencing. However, direct microscopy was negative and the fungus was isolated only once. Therefore, the infection was not confirmed. Treatment including topical clotrimazole and prednisolone acetate (0.5%) was rapidly successful, highlighting the low pathogenicity of this fungus. It is regrettable than only one isolation could be made hampering the involvement of the fungus as causative agent. One can stress out the importance of performing genomic analysis of the isolate when microscopy is not characteristic enough for species identification. Also, the analysis of sequencing results must be performed by high-skilled molecular scientists or technicians to avoid approximated species identification when the matching between the studied isolate and the reference sequences is not satisfactory. Macroscopical and microscopical features of this species are described in Fig. 1e, f.

Fig. 1 (continued) IHEM 24407 on Sabouraud chloramphenicol agar, 28 °C, 21 days (**c** and **d**). (**c**) Surface of the colonies is white and curved; reverse is slightly pigmented in red. (**d**) Lactophenol blue staining shows abundant pear shaped and clavate microconidia. *Arthroderma eboreum* IHEM 21959 on Sabouraud chloramphenicol agar, 28 °C, 21 days (**e** and **f**). (**e**) Surface of the colonies is white and slightly powdery; reverse is nonpigmented. (**f**) Lactophenol blue staining shows numerous clavate microconidia (magnification 20 × 40)

2.2 Novel Geophilic Dermatophyte Species Isolated in the Human Environment

New species have been identified through screening studies of the soil in the vicinity of human activities but not as human pathogens. Three new *Ctenomyces* species have been isolated from soil in China in places with high human activity: *C. albus* sp. nov., *C. obovatus* sp. nov., and *C. peltricolor* sp. nov. [26]. To remind, the main diagnostic criteria of *Ctenomyces* are verrucose conidia, which are thick-walled, lightly pigmented, commonly with ampulliform swellings, and mostly longer than 8 μm. Similarly, one new species of the genus *Nannizzia*, *N. graeserae*, has been isolated in India from a barbershop in Buldhana district after systematic culture screening of the soil [27]. This species yields a unique morphology of its macroconidia, which are rough-walled and clavate to cylindrical. Identification was performed by sequence analysis of ITS and 28S rDNA regions. Systematic screening of soil is not a current use, but it can lead to the discovery of new species potentially pathogenic for humans of animals.

3 Rarely Identified Dermatophytes Causing Infection in Humans

Some publications attest of the recent involvement of geophilic or zoophilic dermatophytes never seen previously in human infections. It generally concerns species, which are rarely isolated from soil or animals. However, it is important to recognize them in order to screen properly the environment of the patient and prevent reinfection.

3.1 Rare Cases of Dermatophytosis Caused by Zoophilic Species

In 2012, Sitterle et al. [28] described the first case of human infection due to the zoophilic species, *Trichophyton bullosum*. It concerned an infection in a 21-year-old male who had a skin lesion on his forearm. Clinical examination revealed the presence of one erythemato-squamous lesion with undefined border. The lesion was pruritic but noninflammatory. The transmission was probably due to contact with infected donkeys in a rural area in France. The patient was treated by topical (ketoconazole 2% cream) and systemic therapy (griseofulvin 1 g/day) during 2 months, and complete remission was obtained. Direct examination of skin scrapings showed septate hyphae. A dermatophyte was isolated in culture and thereafter characterized by ITS (ITS1 and ITS2) sequencing. It was closely related to an African strain of *T. benhamiae* (formerly *Arthroderma benhamiae*) and grouped in a zoophilic cluster with *Trichophyton verrucosum*, *Trichophyton erinacei*, and *T. benhamiae*. Further studies are needed to know if this species can be another anamorph of *T. benhamiae*. The colonies were slow growing, glabrous, and cream-yellow, and reverse was nonpigmented. Microscopy revealed the

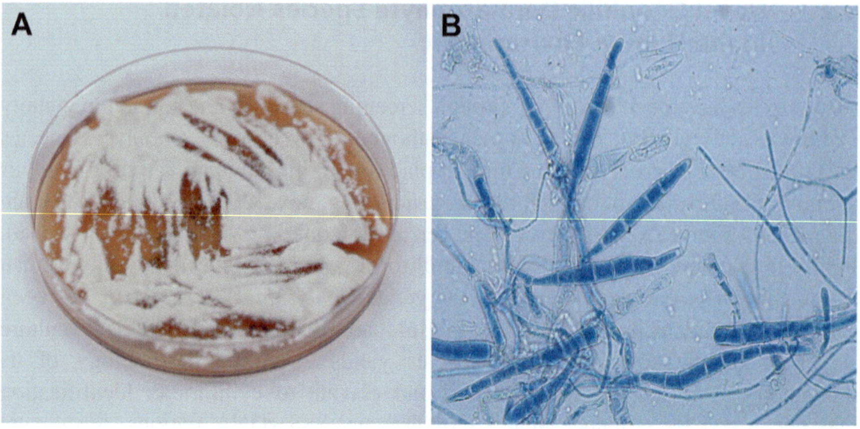

Fig. 2 Macroscopic (**a**) and microscopic (**b**) features of *Lophophyton gallinae*. *Lophophyton gallinae* IHEM 13565 cultivated on Sabouraud chloramphenicol agar, 28 °C, 21 days. (**a**) Surface of the colonies is white and fluffy, and reverse is pigmented in red. (**b**) Lactophenol blue staining shows elongated septate macroconidia (2–8 cells) (magnification 20 × 40)

presence of thick hyphae irregularly fragmented with chlamydospores in chains resembling to those of *T. verrucosum* strains. This species was described for the first time by Lebasque in 1933 and reported as a zoophilic dermatophyte isolated from cutaneous lesions in horses in Tunisia, Syria, and Sudan. More recently, it has been isolated from a horse infection in Czech Republic, and this case represents the first report in Europe [29]. However, the origin of the infection was unknown. *Trichophyton bullosum* is very rarely isolated in horses and is probably underreported because microscopy looks like *T. verrucosum.* However, it can be easily confirmed by molecular sequencing, and contrary to *T. verrucosum,* it does not cause inflammatory lesions. To note, the name "*bullosum*" comes from the morphological aspect of the culture that appears like an air bubble in a viscous liquid [28]. Since it can cause infections in humans, it would be of interest to follow the incidence of infections due to *T. bullosum* in domestic horses.

Lophophyton gallinae (formerly *Microsporum gallinae*) is another rarely described zoophilic species causing infections in humans. In 2011, Miyasato et al. [30] described the first case of *Tinea corporis* in Japan. A 96-year-old healthy man was infected. The patient had a long working history as a breeder of fighting cocks. Lesions consisted of two erythematous macules, which appeared after the man was being bitten by a cock. *Lophophyton gallinae* was identified as the infectious agent based on the morphology and ITS sequencing. Macroscopical and microscopical features of this species are described in Fig. 2a, b.

Trichophyton erinacei (formerly *T. mentagrophytes* var. *erinacei*) is frequently found in hedgehogs, and some cases of transmission to humans have been described [31]. Indeed, in some countries, hedgehogs became popular as household pets. This is the case in Korea where a case of *Tinea manuum* caused by this species has been described. This concerned a 15-year-old girl with a pruritic scaly erythematous patch and pustules on her left hand. She kept a hedgehog at home. A fungus was recovered and morphologic aspects suggested *T. erinacei*. ITS sequencing confirmed the similarity with *T. erinacei* DNA [32]. Another similar case of *Tinea manuum* caused by this dermatophyte was described in Japan [33]. In 2011, a case of unusual *Tinea faciei* was described again in Korea and 1 year later in Chile [34, 35].

Another closely related species of *T. erinacei* has been published in 2010 by Contet-Audonneau [31, 36, 37] who proposed the name of *T. mentagrophytes* var. *porcellae* because it is mostly transmitted by Guinea pigs and present in breeding farms [36]. It causes erythematous and inflammatory lesions of the exposed part of the body in young adults and children. This species has been renamed as *T. benhamiae*. It harbors a yellow and a white phenotype and is increasingly isolated in Europe since 2008 mainly in children. However, the first human case report has been described in Japan by Nakamura et al. [38]. The yellow phenotype, which can be confused with *M. canis*, is downy with a pleated mycelium, a lemon-orange color and poor sporulation. The white phenotype is powdery to floccose with a yellow orange to brown reverse and rapidly sporulates, harboring numerous spherical to clavate microconidia, sparse macroconidia, and rare spirals. It can be confused with *T. mentagrophytes* and is more rarely isolated than the yellow phenotype. A 9 years French retrospective study reports this species as the third most frequent zoophilic species isolated from 2008 to 2013, [39] while it is the second most frequent zoophilic dermatophyte after *M. canis* isolated in Japan and the first one in Germany in a study performed from 2010 to 2013 [4, 40]. Because this species is now present all over the world, easy and cheap diagnostic methods such as the short subculture on CandiSelect™ 4 medium (Bio-Rad, France) allows the differentiation with *M. canis* [39]. However, more sophisticated methods such as MALDI-TOF/MS or dermatophyte PCR can also be successfully applied [12, 41].

In 1997, a case of *Tinea corporis* due to a zoophilic strain of *N. nana* has been described in Madras, India [42]. A 16-year-old boy working in a poultry farm presented pustular scaly circinate lesions on the back, leg, and chest. Histopathological section of a biopsy reported the presence of arthroconidia in the stratum corneum. In culture, a white granular colony with a yellow pigment grew. Even rare, this kind of infection in human has been reported in several countries such as Poland [43], the USA [44], the Netherlands [45], China [46], and Africa [47]. It concerned cases of *Tinea barbae, Tinea corporis*, and *Tinea capitis*. More recently, one case of onychomycosis due to *N. nana* has been described in a retrospective study conducted in Guatemala [48]. This concerned a 35-year-old woman with no predisposing risk factor. She presented a total dystrophic onychomycosis. In 2019, a Mexican team described two cases of siblings infected by this zoophilic species. The authors described the clinical presentation in a previously healthy 8 year-old boy, living in a pig-farm area. He presented a 1-month history of a tumoral lesion localized on the scalp with hair loss, scales, multiple pustules, and abscesses that drained seropurulent material and pseudoalopecic areas with few short hairs. All these symptoms were accompanied by intense pain. KOH direct examination showed ectothrix type infection. In culture, beige-colored colonies with a brown pigment on the reverse side were observed. By microscopy, multiple small egg-shaped macroconidia of 12–15 μm with 1–3 septa and few sessile microconidia, corresponding to *N. nana,* were observed. The 6-year-old sister of the boy developed an extended *Tinea corporis* due to the same fungus. She presented disseminated dermatophytosis on the left cheek, trunk, and upper limbs, with low to moderate pruritus; several 2–3 cm erythematous, scaly, and round plaques with raised borders and scarce crusts were observed. Both cases received treatment with griseofulvin for 40 days for the boy associated with oral prednisolone and 20 days for the girl associated with 2% cream. There was clinical and mycological cure in both cases [49].

3.2 Rare Cases of Dermatophytosis Due to Geophilic Dermatophytes

Nannizzia gypsea, *Nannizzia incurvata*, and *N. fulva* represent the most frequent geophilic dermatophytes causing human dermatophytosis. However, *N. incurvata* and *N. fulva* are rarely reported, which is probably due to misidentification with *N. gypsea sensu stricto* because of very similar morphology. A recent publication reports not more than 30 clinical cases due to *N. incurvata* published until now [50]. Generalization of the use of MALDI-TOF/MS or ITS sequencing (particularly ITS1 in this case) will probably reassess the role of this group of geophilic dermatophytes in human dermatophytosis as it was the case for an isolate causing *Tinea corporis* isolated from a young man in Iran. Culture of skin scrapings grew a dermatophyte, which was identified as *N. gypsea* according to its morphological aspects. PCR amplification of ITS1–5.8S rDNA-ITS2 region and subsequent ITS-RFLP and sequencing identified *N. fulva* as the true causative agent [51].

Microsporum racemosum is a rare geophilic dermatophyte, which was isolated the first time by Borelli in 1965 from the hair of a wild rat in South America [52]. Rare cases of human infection due to this species have been described. The first case of dermatophytosis concerned a 14-year-old girl in Venezuela who developed one typical lesion of *Tinea corporis* on her right forearm. She had been probably contaminated by rats which were inhabitants of the closet where her clothes were stored [53]. In 1976, a doubtful case was reported in the USA. It concerned a 79-year-old man who had numerous excoriations of the forehead, scalp, and nape of the neck [54]. However, the patient did not respond to antifungal treatment and the fungus grew only once. It was considered to be transient carriage probably due to large gardening activities. Another case was reported from the USA in 1978 in a 51-year-old patient. He had a circinate maculopapular lesion on the dorsum of the right hand, which was vesicular with an erythematous border, pruritic and developed rapidly to 5 cm in diameter. The patient denied any trauma and contact with soil, nursery material, or potted plants [55]. In 1999, the first case of human infection due to *M. racemosum* was reported from Europe. It concerned a 60-year-old Spanish female who developed onychomycosis. Identification of the isolate was confirmed by ITS1/ITS2 and 5.8S rDNA sequencing [56].

4 Unusual Clinical Presentations of Dermatophytosis

Each species of dermatophyte has its specific clinical manifestations, which are characteristic. However, sometimes the picture is different, making the recognition of the species more difficult and doubtful. That is why in these cases, microscopy should be combined with reference methods before to conclude. Examples of unusual manifestations are described and summarized in Table 2.

4.1 Unusual Presentations Due to Geophilic Dermatophytes

Nannizzia gypsea currently causes *Tinea corporis* on exposed parts of the body. Lesions are generally pruritic and inflammatory. However, this species can lead to unusual clinical manifestations in humans. In 2001, Luque et al. [57] described an atypical *Tinea corporis* caused by *N. gypsea* in a 39-year-old male in Argentina. The man developed a seborrheic dermatitis with erythematous vesicular and granulomatous skin lesions with clear lightly infiltrated edges on his lower legs. The skin surrounding the lesions was scaly and pruritic. In 2006, an 18-year-old healthy female presented with skin lesion on her abdomen. Lesion showed concentric annular erythematous rings with a fine scaly border. Scales were cultured, and cinnamon brown colonies identified as *N. gypsea* grew after 5 days [58]. In 2017, a rare case of dermatophyte infection of the glabrous skin caused by *N. gypsea* has been described in an immunocompetent woman. A 22-year-old Malagasy female who reported close contact with cats, presented a single round lesion with a peripheral, active, squamous, and pruriginous inflammatory bead. Morphological species identification was confirmed by sequencing the ITS region [59]. Many other unusual clinical presentations caused by *N. gypsea* have been described such as circumscribed scleroderma, crusty circular plaques, white paint-like dots, psoriasis-like lesions, or dystrophic onychomycoses [60–64]. Also, presentations can be different if the patient has an immunodeficiency like in the case presented by Polili et al. [65]. This case concerned a 37-year-old male with HIV and HCV infections. He presented annular erythematous scaly plaques with inflammatory advancing borders located on the upper lip, nose, and side of the neck. A few months after a successful treatment with itraconazole, a relapse of skin lesions occurred concomitantly with a central nervous system lesion causing seizures. Microscopy from new skin scrapings was positive, and culture grew again *N. gypsea*. However, the brain lesion was not biopsied and the fungal etiology could not be confirmed. Intravenous cotrimoxazole, 70 mg/kg daily, associated with liposomal amphotericin B, 3 mg/kg daily, was started; at the ninth week of therapy, when 3 mg/kg liposomal amphotericin B was still administered 3 times weekly for maintenance, lower limb edemas ensued, the cutaneous lesions had cleared, and his cerebral lesion was still detectable on a control CT scan, although much reduced in size [65].

Table 2 Atypical clinical manifestations caused by dermatophytes

Dermatophyte species	Type of infection	Clinical manifestations	Age	Sex	Country of isolation	Reference
Geophilic						
Nannizzia nana	Onychomycosis	Total dystrophic onychomycosis	35	F	Guatemala	Martinez et al. (2014) [48]
Microsporum racemosum	Onychomycosis	Yellow, purulent exudate, and a slight perilesional erythema when the thumb became painful	60	M	Spain	Garcia-Martos et al. (1999) [56]
Nannizzia gypsea	Onychomycosis	Dystrophic onychomycosis and progressive alteration of the nail of the right big toe over a 2-year period, after sustaining trauma in a fall from a horse	35	F	Italy	Romano et al. (2006) [62]
Nannizzia gypsea	*Tinea corporis*	Seborrheic dermatitis with erythematous vesicular and granulomatous skin lesions with clear lightly infiltrated edges on the lower legs	39	M	Argentina	Luque et al. (2001) [57]
Nannizzia gypsea	*Tinea corporis*	Round lesion with peripheral active squamous and pruriginous inflammatory beads	22	F	Madagascar	Soankasina et al. (2018) [59]
Zoophilic						
Trichophyton equinum	Onychomycosis	Progressive scaling of the skin of the left thumb and forefinger, the distal part of the thumbnail had turned yellow	70	M	Finland	Huovinen et al. (1998) [98]
Microsporum canis	*Tinea favosa*	Itchy scaling of the scalp and hair loss, 6 cm wide plaque covered with yellowish scale, favic invasion of the hair follicle	8	F	Chicago, USA	Krunic et al. (2007) [66]
Microsporum canis	Blepharitis and eyelid *Tinea*	Erythematous edematous plaque associated with blepharitis and chalazion in the right upper eyelid	3	F	Spain	Calles-Monar et al. (2018) [74]
Trichophyton mentagrophytes	*Tinea blepharo-ciliaris* and *Tinea barbae*	Edema extended up to the right eye with purulent hemorrhagic secretions and hair loss on the eyelid, associated with *Tinea barbae*	24	M	Romania	Burulana et al. (2015) [73]

Trichophyton mentagrophytes	*Tinea faciei*	Perinasal erythematous papules progressively extending with numerous papules and nodules forming erythematous and scaly plaques localized on the face	47	F	Italy	Calcaterra et al. (2013) [71]
Trichophyton mentagrophytes type VII	*Tinea genitalis*	Erythematous scaly plaques on pubis, labia majora, and proximal penile shaft (+ one case of *Tinea barbae profunda* with abscesses)	Various	M/F	South East Asia	Luchsinger et al. (2015) [77]
Trichophyton mentagrophytes type VIII	*Tinea corporis*	Extensive dermatophytosis of the back, buttocks, chest, and groins	6 months	F	Bahrain	Nenoff et al. (2019) [80]
Trichophyton simii	*Tinea corporis*	Extensive erythematous lesions on the trunk	9 months	F	Iran	Ansari et al. (2016) [76]
Anthropophilic						
Trichophyton tonsurans	*Tinea pedis*	Hyperkeratosis, erythema, and scaling on the feet	47	M	Denmark	Carlsen et al. (2013) [84]
Trichophyton tonsurans	Onychomycosis	Recurrent onychomadesis, repeated episodes of shedding and regrowth of the affected nail, and skin lesions on the right hand and forearm, no immunosuppression	9 months	M	India	Khanna et al. (2013) [100]
Trichophyton violaceum	Onychomycosis	Disto-lateral subungual onychomycosis and hyperkeratosis of the first toenail	41	M	Egypt	Mapelli et al. (2012) [99]
Trichophyton schoenleinii	Onychomycosis	Dystrophic lesions in toenails	49	M	Poland	Macura et al. (2012) [101]
Microsporum audouinii	Kerion celsi	Scaly pruriginous plaques on the scalp	3	F	Portugal	Fernandes et al. (2013) [85]
Trichophyton rubrum	*Tinea incognito*	*Tinea incognito* mimicking syphilid, erythema on the palm, soles, glans penis, and foreskin	56	M	China	Tan et al. (2014) [88]

(continued)

Table 2 (continued)

Dermatophyte species	Type of infection	Clinical manifestations	Age	Sex	Country of isolation	Reference
Deep infections						
Trichophyton verrucosum	Face and neck	Severe sycosis barbae vegetating purulent secreting, painful infiltrations with nodules, and papules on the lower third of his face and on the neck.	80	M	Germany	Wollina et al. (2018) [90]
Trichophyton rubrum	Mandible	Majocchi's granuloma in a 45-year-old immunocompetent man with erythema, papules, and nodules on the submandibular area	45	M	China	Su et al. (2017) [91]
Trichophyton rubrum	Deep infection	Multiple fungal abscesses in the lower extremities of an immunocompromised patient	66	M	China	Dai et al. (2019) [92]
Microsporum ferrugineum	Head skin lesions	Boil-like lesion on the head, large elevated mass with purulent exudation	27	F	China	Zhang et al. (2019) [93]
Trichophyton interdigitale	Groin	Invasive dermatophytic infection, enlarging right-sided groin mass and widespread itchy eruption in a patient with a mutation in Stat3	45	M	United Kingdom	Simpson et al. (2018) [97]

F female, *M* male

4.2 Unusual Presentations Due to Zoophilic Dermatophytes

In some cases, zoophilic strains may also cause atypical infections, as it has been described in 2006 for *M. canis* causing *Tinea favosa* [66]. This case described in the USA concerned an 8-year-old Caucasian female with a history of itchy scaling of the scalp and hair loss. The girl had a dog sleeping with her. No history of immunosuppression was known. The fungal isolate was identified only by microscopy. In general, the anthropophilic *Trichophyton schoenleinii* is responsible for this kind of clinical manifestation with permanent alopecia. Some cases have also been reported with *Trichophyton quinckeanum* (formerly *T. mentagrophytes* var. *quinckeanum*) and *Trichophyton violaceum* [67, 68]. Because of the absence of molecular assay, a misidentification, however, is not excluded.

Nannizzia praecox (formerly *Microsporum praecox*) is another zoophilic dermatophyte mostly transmitted by horses to human. Infections due to *N. praecox* are particularly rare in human as only 29 cases have been reported around the world until 1944. In 2010, Alanio et al. [50] described a rare case of dermatophytosis in a French 28-year-old horse rider. The observed lesion was located on the right external malleolus. Microscopic examination of skin scrapings revealed the presence of a dermatophyte species and molecular sequencing confirmed the identification as *N. praecox* [69].

In 2007, Sanchez-Castellanos et al. [51] described a case of *Tinea incognito* due to the zoophilic strain *T. mentagrophytes.* A 2-year-old girl presented a 3 months history of pruritic facial eruption that had begun as perinasal erythematous papules and then extended with numerous papules and nodules forming erythematous and scaly plaques localized on the face. Fungal culture from scales yielded granular and powdery colonies with a yellow brown pigment. Microscopic examination identified the fungus as *T. mentagrophytes* [70].

In Italy, another case of rosacea-like *Tinea incognito* due to the zoophilic *T. mentagrophytes* was described by Calcaterra et al. [52] in 2013. This concerned a 47-year-old woman with a 3 months history of erythematous papules with rare pustules on the face. There was a progressive exacerbation of the eruption by the application of topical corticosteroids. The absence of classical features of ringworm made the initial diagnostic difficult. The patient had no remarkable history before this episode. Culture on Mycosel agar let grow powdery and granular colonies with a yellowish color at the reverse [71].

Tinea imbricata is an uncommon described dermatophytosis caused by the anthropophilic dermatophyte *Trichophyton concentricum*. It is characterized by widespread, annular, concentric, squamous lesions. *Tinea imbricata* is endemic in three geographical areas: Southwest Pacific, Southeast Asia, and Central and South America. However, this clinical picture is exceptionally rare in travelers returning from endemic areas. In 2015, a case of *Tinea pseudo-imbricata* caused by *T. mentagrophytes* has been described in India. *Tinea pseudo-imbricata* is characterized by concentric scaly rings mimicking *Tinea imbricata* but is caused by other species than *T. concentricum*. The case reported concerned an 18-month-old baby. The diagnosis of *Tinea pseudo-imbricata* was made on the basis of the morphology of the lesions and confirmed by the isolation of *T. mentagrophytes* in culture [72].

In 2015, a (zoophilic) strain of *T. mentagrophytes* has been responsible for *Tinea blepharo-ciliaris* associated with *Tinea barbae* in Romania. This clinical manifestation in eyes is rare by dermatophytes. The infection was present in an adult male of 24 years old living in a rural environment, with no history of trauma, immunosuppression, or corticosteroid treatment. First diagnosis was conjunctivitis, the edema extended up to the right eye with purulent hemorrhagic secretions and hair loss on the eyelid. The man presented also a lesion on the cheek with a decreased hair presence. Ecto-endothrix hair invasion was seen after KOH clearing on hairs of the cheek and of the eyelid. *Trichophyton mentagrophytes* was identified as the etiological agent by culture and ITS1 sequencing. Sequence analysis showed, however, 100% similarity with *T. interdigitale*, *Arthroderma vanbreuseghemii*, and zoophilic strain of *T. mentagrophytes,* which is not surprising as the ITS sequencing cannot differentiate between these closely related species [73].

In 2018, an eyelid *Tinea* with blepharitis due to *M. canis* has been described by Calles Monar and Juarez Martin [74]. The case is presented in a 3-year-old girl with an erythematous edematous plaque associated with blepharitis and chalazion in the right upper eyelid. The culture of the eyelid scrape showed *M. canis* as the etiological agent.

In 2016, a rare case of *Tinea corporis* in a 9-month-old female with extensive erythematous lesions on the trunk has been described in Iran. The girl was living in a farm. Morphological features of the isolate resulted in the identification of *T. interdigitale*. For further identification, the ITS1 and 2 regions of rDNA were sequenced and the presumed *T. interdigitale* isolate was finally identified as *Trichophyton simii*. It is an exotic zoophilic dermatophyte that is supposed to be restricted to the Indian subcontinent, where it causes dermatophytosis lesions on animals such as monkeys, cattle, poultry, Guinea baboons, dogs, and cats [75, 76].

A new genotype of the zoophilic species *T. mentagrophytes* was recently described by Luchsinger et al. [77] in Switzerland. This strain was responsible for several cases of sexually transmitted infections. It was isolated from seven patients who had visited Southeast Asia. These patients reported on having had sexual intercourse 1–2 weeks earlier during their trip in Asia. Clinical presentation was quite similar in all patients: sharply demarcated erythematous scaly plaques on pubis, labia majora, or proximal penile shaft, and in one patient, additional involvement of the scrotum. Two female patients showed additional plaques on the gluteal region and one of them on the neck. Kupsch et al. [78] in 2019 described numerous cases caused by this new genotype developing as an epidemic outbreak. A total of 43 patients, mostly suffering from highly inflammatory, painful, and persistent infection of the pubogenital region, were observed between January 2016 and July 2017 in Berlin. Fungal culture and sequencing of the Ef1-alpha locus were applied; a phylogenetic analysis was then done based on these sequences. In 37 cases, a new genotype of *T. mentagrophytes* (referred as *T. mentagrophytes* VII or "Thailand genotype") was isolated as the etiological agent and sequencing revealed identical sequences for all tested isolates. Of these patients, only two reported animal contacts with rabbits and cats, and four had a history of travel in Thailand or Southeast Asia shortly before the symptoms appeared; none of the others reported a recent stay abroad. Regular intimate shaving was documented in 16 of the patients. Eight patients reported visiting a fitness club regularly. Even if no contact with sex workers has been reported in this study, the hypothesis of a direct human-to-human infection via sexual contact is supported by the fact that the sexual partners of some patients were also infected in five cases. This attests of the interhuman transmission (possibly sexually transmitted by contact) of this new strain circulating known in Germany and probably also in other European countries. Mycologists should be aware of the existence of this genotype.

Wendrock Shiga et al. [79] reported in 2017 a case of *Tinea barbae profunda* caused by the new variant of *T. mentagrophytes* described above (type VII). The pathology followed a journey to Thailand. A businessman was affected by a fulminant abscessing infection of the upper lip and beard area. Fungal mycelium was observed in the tissue biopsy. *Trichophyton mentagrophytes* was grown in culture, and ITS sequencing was done. The patient had contact with Thaï female sex workers who must be considered as the source of infection taking into account previously similar cases.

In 2019, Nenoff et al. [80] described a human infection due to the recently characterized zoophilic *T. mentagrophytes* genotype VIII in a young baby. This genotype is responsible for the current Indian epidemic caused by a terbinafine-resistant strain. The described infection concerned a 6-month-old female infant from Bahrain visiting Germany with her family who consulted for extended dermatophytosis of the back, buttocks, chest, and groins. Mycological examination performed in Germany identified the "Indian type" zoophilic dermatophyte *T. mentagrophytes*. The newly described genotype VIII within this species was identified by sequencing of the ITS region of rDNA. This strain was also characterized to be resistant to terbinafine. A significant percentage of these Indian *T. mentagrophytes* strains are resistant to terbinafine in vitro because of point mutations in the squalene epoxidase (SQLE) gene. Authors discuss the risk of dissemination to other countries due to globalization [81]. As it is the case with the example above with the young baby, the strain initially from India could have circulate in Arabia and then in Europe with travels. Nenoff and his team [82] comment in a recent paper that the predominant organism responsible for the "epidemic" in India is the zoophilic *T. mentagrophytes* genotype VIII and not the anthropophilic *T. interdigitale* as described by Chowdhary et al. [83] in recent studies about the epidemic Indian strain. Chowdhary et al. [83] answered to that affirmation suggesting perspectives on misidentification of *T. interdigitale/T. mentagrophytes* using ITS sequencing as these two species are molecularly much closer to each other than was anticipated on the basis of clinical and phenotypic data. The author suggests that description of the distinct Indian genotype VIII of *T. mentagrophytes,* recognized by ITS sequencing, needs to be evaluated using neotype *T. mentagrophytes* (IHEM 4268^{NT}), which was omitted by Nenoff et al. [82] in their phylogenetic analysis. Additionally, they suggest the use a multilocus phylogeny approach to describe a new genotype. Finally, polyphasic studies including clinical, molecular, ecological, and life cycle data are needed to establish the validity of *T. mentagrophytes* genotype VIII with certitude [83].

4.3 Unusual Clinical Presentation Due to Anthropophilic Dermatophytes

In 2013, Carlsen and Menné [84] described a surprising case of *Tinea pedis* due to *T. tonsurans*. This anthropophilic species is generally associated with *Tinea capitis* or *Tinea corporis* in judokas. Typically, dermatophytes causing *Tinea pedis* in Europe are *T. rubrum*, *T. interdigitale*, and *E. floccosum*. The case of *Tinea pedis* due to *T. tonsurans* was described in a Danish Caucasian man of 46 years old. Hyperkeratosis, erythema, and scaling were observed bilaterally on the feet. No sports activity or travel history was noted for this patient. Mycological examination of feet scrapings was performed, and arthroconidia were visualized. *Trichophyton tonsurans* was identified based on morphological features.

In 2013, a case of kerion celsi caused by *M. audouinii* has been described in Portugal. Kerion celsi occurs generally in individuals infected by zoophilic dermatophytes. The case concerned a 3-year-old Caucasian girl with unremarkable past medical history. Direct microscopic examination revealed *Microsporum* endo-ectothrix parasitism of some hairs. Positive culture was identified as *M. audouinii* by conventional and molecular methods [85]. In 2015, an unusual strain of *M. audouinii* causing *Tinea corporis* in Europe has been described. This concerned a boy who lived in Germany and most likely acquired his infection during a trip on a farm in Poland. The strain showed features of *M. canis* (abundance of rough-walled macroconidia, growth on rice, and positive hair perforation), as well as of *M. audouinii* (white thallus and long macroconidia with central constriction). ITS sequencing showed 99.9% similarity with a *M. audouinii* reference strain. This case shows that some strains of *M. audouinii* can present atypical morphological and physiological features [86]. A similar case of *Tinea corporis* was previously described by the same team in a young German woman who stayed in Ghana. In that case, the strain showed morphological aspects concordant with *M. audouinii* but a positive hair perforation test made its identification with conventional methods equivocal. A genetic analysis finally defined the strain as *M. audouinii* [87].

A case of *Tinea incognito* caused by *T. rubrum* mimicking syphilid was described in 2014 by Tan et al. [88]. A 56-year-old man under chemotherapy presented with a 2-week history of erythema on the left palm, soles, glans penis, and the foreskin with no itching and pain. Initially, syphilid was suspected. However, both toluidine red unheated serum test and *Treponema pallidum* particle agglutination assay were negative. Microscopy showed hyphae in all sites, and skin culture revealed *T. rubrum* infection.

4.4 Deep Infections Caused by Dermatophytes

Deep dermatophytosis is very uncommon and is mostly encountered in immunosuppressed patients [89]. In 2018, Wollina et al. [90] described a deep facial mycosis due to *T. verrucosum.* This concerned an 80-year-old male patient with severe sycosis barbae. He was treated with systemic antibiotics without any improvement. His medical history was positive for diabetes mellitus treated by oral antidiabetics and insulin, chronic obstructive pulmonary disease, arterial hypertension, and absolute arrhythmia. Surprisingly, no animal contact or activity as a farmer was reported for this man. Sequencing of the ITS2 region of rDNA revealed a 100% accordance with *T. verrucosum.*

Deep infections caused by *T. rubrum* are rarely reported especially in immunocompetent people. In 2017, Su et al. [91] described a case of deep infection caused by *T. rubrum* in a 45-year-old man with no significant immunodeficiency. This patient had a history of onychomycosis on the toenails without any regular treatment for 6 years. He had erythema, papule, and nodules on the submandibular area, neck, and chest for almost 1 year. Direct microscopic examination of pyogenic fluid was positive, and the fungal cultures showed the macroscopic and microscopic features of *T. rubrum*. After DNA sequencing, the strain was confirmed as *T. rubrum.*

A case of an unusual presentation of *T. rubrum* infection causing multiple fungal abscesses in the lower extremities of an immunocompromised patient has been described by Dai et al. [92] in 2019. A 66-year-old male who received immunosuppressive drugs for 7 years developed numerous subcutaneous nodules in the lower extremities. The purulent fluid obtained from the cyst was positive for *T. rubrum.*

A deep dermatophytosis caused by *Microsporum ferrugineum* has been described recently in a girl presenting mutations in *CARD9*. A 27-year-old female developed a boil-like lesion on her head more than 10 years ago, without any other trauma. Later, the lesion became a large mass. Physical examination revealed an elevated mass (10×15 cm in diameter) with a clear border on her head with purulent exudation. Skin biopsy revealed intense inflammatory cell infiltration with irregular septate hyphae in multinuclear giant cells. From tissue culture, *M. ferrugineum* was isolated, with a white and downy colony, and identification was further confirmed by ITS sequencing. *CARD9* mutations in this patient have been evaluated. Sanger sequencing of the CARD9 coding exons indicated two compound heterozygous mutations in exons 6 and 8 [93]. Autosomal recessive CARD9 deficiency has been already shown to be associated with deep dermatophytosis. Indeed, patients with autosomal recessive *CARD9* mutations are predisposed to recurrent mucocutaneous and invasive fungal infections with *Candida* spp., dermatophytes and phaeohyphomycetes [94, 95].

In 2018, Simpson et al. [96] reported the first case of an invasive dermatophyte infection associated with abrupt onset of a prurigo-induced pseudoperforation of the skin and a loss-of-function mutation in signal transducer and activator of transcription 3 (STAT3). This mutation is able to induce inherited susceptibility to fungal infections in seemingly 'immunocompetent' individuals. STAT3 mutation has been shown to be associated with oral fungal infections [97]. A 45-year-old man of Indian origin presented a 10-week history of an enlarging right-sided groin mass and a widespread itchy eruption. Fungal hyphae were observed on direct examination. Fungal cultures did not permit to identify any organism, and panfungal PCR was performed permitting the identification of *T. interdigitale* [96].

4.5 Unusual Dermatophyte Species Causing Onychomycosis

In 2014, a retrospective study in Guatemala showed that *N. gypsea* could be responsible for onychomycoses. *Nannizzia gypsea* caused in 6 cases a disto-lateral subungual onychomycosis and in one case a total dystrophic onychomycosis. None of the patients had *Tinea* affecting other parts of their body such as *Tinea corporis* or *Tinea capitis* [48].

In 1998, a fingernail onychomycosis caused by *T. equinum* has been described in Finland. This concerned a 70-years-old farmer breeding horses. Lesions described were a progressive scaling of the skin of his left thumb and forefinger. The distal part of the thumbnail had simultaneously turned yellow. Direct examination and fungal culture showed hyphae and oval microconidia, suggesting *T. equinum* infection. Later samples were taken from the horse, and culture gave a similar isolate of *T. equinum* [98].

In 2011, a case of toenail onychomycosis due to *T. violaceum* was described in Italy. This dermatophyte species is typically responsible for *Tinea capitis* with endothrix parasitism and sometimes *Tinea corporis*. Onychomycoses due to this species are rarely reported. This concerned a 41-year-old man coming from Egypt who presented a disto-lateral subungual onychomycosis and hyperkeratosis of the first toenail. No remarkable clinical history was recorded for the patient. Mycological examination of the nail showed mycelial elements. Colonies with a wrinkled surface, red-brown color, and diffusible brown pigment grew. Microscopic examination showed distorted hyphae with reflexive branchings, characteristics of *T. violaceum* [99].

In 2013, a case of recurrent onychomadesis due to *T. tonsurans* in a 9-month-old baby was described in India. The boy had redness, discoloration, and thickening of the fingernails on the right hand since 2 weeks of age. He had a history of repeated episodes of shedding and regrowth of the affected nail. Skin lesions appeared on the right hand and forearm. There was no evidence of sepsis or immunosuppression of the infant. Observation of skin scrapings and nails revealed multiple hyphae suggestive of a dermatophytosis. Fungal cultures revealed *T. tonsurans* as the etiological agent. This was the first report of onychomycosis due to *T. tonsurans* in a young baby [100].

Trichophyton schoenleinii is an anthropophilic dermatophyte species that may be responsible for *Tinea capitis* (favus) but is less implicated in onychomycosis. In 2012, a case of onychomycosis due to *T. schoenleinii* was described in Poland. The patient who was a 49-year-old carpenter and brick-layer, presented dystrophic lesions in toenails, lasting for many years, and less severe lesions in fingernails. The mycological examination revealed characteristic brain-shaped colonies penetrating the medium and an image typical for *T. schoenleinii* in microscopy [101].

5 Conclusion

This review of the literature shows that dermatophytes can cause a lot of clinical manifestations that can highly differ from those usually described for specific species. Dermatophyte transmission from animal or soil to human is frequently described. Therefore, dermatologists should make a complete anamnesis of the patient and information about contacts with pets/soil at home or at work in order to contribute to the identification of the right dermatophyte species. It is also crucial to promote a correct identification of the etiological agent by using molecular biology methods as reference methods based on ITS sequencing and in some doubtful cases, on multilocus phylogenic analysis. The latter techniques are usually performed by high-skilled reference laboratories, which should be regularly solicited. Indeed, because of the societal changes, mycologists should play a role of sentinels for detecting new trends in the epidemiology of dermatophytosis. To conclude, despite the preponderant role of sequencing methods, the awareness of a skilled mycologist plays a major role and can make the difference in alerting the scientific community about new circulating variants or atypical species.

References

1. de Hoog GS, Dukik K, Monod M, Packeu A, Stubbe D, Hendrickx M, et al. Toward a novel multilocus phylogenetic taxonomy for the dermatophytes. Mycopathologia. 2017;182 (1–2):5–31.
2. Seebacher C, Bouchara JP, Mignon B. Updates on the epidemiology of dermatophyte infections. Mycopathologia. 2008;166(5–6):335–52.
3. Hayette M-P, Sacheli R. Dermatophytosis, trends in epidemiology and diagnostic approach. Curr Fungal Infect Rep. 2015;9(3):164–79.
4. Nenoff P, Uhrlass S, Kruger C, Erhard M, Hipler UC, Seyfarth F, et al. *Trichophyton* species of *Arthroderma benhamiae*—a new infectious agent in dermatology. J Dtsch Dermatol Ges. 2014;12(7):571–81.
5. Rippon JW. The changing epidemiology and emerging patterns of dermatophyte species. Curr Top Med Mycol. 1985;1:208–34.
6. Hay RJ. Tinea capitis: current status. Mycopathologia. 2017;182(1–2):87–93.
7. Swinne D, Watelle M, Suetens C, Mertens K, Fonteyne PA, Nolard N. A one-year survey of candidemia in Belgium in 2002. Epidemiol Infect. 2004;132(6):1175–80.
8. Gits-Muselli M, Benderdouche M, Hamane S, Mingui A, Feuilhade de Chauvin M, Guigue N, et al. Continuous increase of *Trichophyton tonsurans* as a cause of tinea capitis in the urban area of Paris, France: a 5-year-long study. Med Mycol. 2017;55(5):476–84.
9. Aneke CI, Otranto D, Cafarchia C. Therapy and antifungal susceptibility profile of *Microsporum canis*. J Fungi (Basel). 2018;4(3):107.
10. Nenoff P, Kruger C, Ginter-Hanselmayer G, Tietz HJ. Mycology—an update. Part 1: dermatomycoses: causative agents, epidemiology and pathogenesis. J Dtsch Dermatol Ges. 2014;12(3):188–209; quiz 10, 188–211; quiz 2.

11. García-Martos P, Ruiz-Aragón J, García-Agudo L, Linares M. Dermatophytoses due to *Microsporum gypseum*: report of eight cases and literature review. Rev Iberoam Micol. 2004;21(3):147–9.
12. Hayette MP, Seidel L, Adjetey C, Darfouf R, Wery M, Boreux R, et al. Clinical evaluation of the DermaGenius® Nail real-time PCR assay for the detection of dermatophytes and *Candida albicans* in nails. Med Mycol. 2019;57(3):277–83.
13. Kupsch C, Ohst T, Pankewitz F, Nenoff P, Uhrlass S, Winter I, et al. The agony of choice in dermatophyte diagnostics-performance of different molecular tests and culture in the detection of *Trichophyton rubrum* and *Trichophyton interdigitale*. Clin Microbiol Infect. 2016;22 (8):735.e11–7.
14. Verrier J, Monod M. Diagnosis of dermatophytosis using molecular biology. Mycopathologia. 2017;182(1–2):193–202.
15. Brasch J, Beck-Jendroschek V, Voss K, Yurkov A, Graser Y. *Arthroderma chiloniense* sp. nov. isolated from human stratum corneum: description of a new *Arthroderma* species. Mycoses. 2019;62(1):73–80.
16. Borman AM, Szekely A, Fraser M, Lovegrove S, Johnson EM. A novel dermatophyte relative, *Nannizzia perplicata* sp. nov., isolated from a case of tinea corporis in the United Kingdom. Med Mycol. 2018. https://doi.org/10.1093/mmy/myy099.
17. Hubka V, Dobiasova S, Dobias R, Kolarik M. *Microsporum aenigmaticum* sp. nov. from *M. gypseum* complex, isolated as a cause of tinea corporis. Med Mycol. 2014;52(4):387–96.
18. Hubka V, Nissen CV, Jensen RH, Arendrup MC, Cmokova A, Kubatova A, et al. Discovery of a sexual stage in *Trichophyton onychocola*, a presumed geophilic dermatophyte isolated from toenails of patients with a history of *T. rubrum* onychomycosis. Med Mycol. 2015;53 (8):798–809.
19. Choi JS, Graser Y, Walther G, Peano A, Symoens F, de Hoog S. *Microsporum mirabile* and its teleomorph *Arthroderma mirabile*, a new dermatophyte species in the *M. cookei* clade. Med Mycol. 2012;50(2):161–9.
20. Brasch J, Gräser Y. *Trichophyton eboreum* sp. nov. isolated from human skin. J Clin Microbiol. 2005;43(10):5230–7.
21. Campbell CK, Borman AM, Linton CJ, Bridge PD, Johnson EM. *Arthroderma olidum*, sp. nov. a new addition to the *Trichophyton terrestre* complex. Med Mycol. 2006;44(5):451–9.
22. Hubka V, Cmokova A, Skorepova M, Mikula P, Kolarik M. *Trichophyton onychocola* sp. nov. isolated from human nail. Med Mycol. 2014;52(3):285–92.
23. Elavarashi E, Kindo AJ, Kalyani J. Optimization of PCR-RFLP directly from the skin and nails in cases of dermatophytosis, targeting the ITS and the 18S ribosomal DNA regions. J Clin Diagn Res. 2013;7(4):646–51.
24. Peano A, Tizzani P, Gallo M, Molinar Min A, Rambozzi L, Meneguz P. Dermatophytosis due to *Trichophyton verrucosum* in a chamois (*Rupicapra rupicapra*). Eur J Wildl Res. 2008;54 (1):153–6.
25. Keller MC, French LE, Hofbauer GFL, Bosshard PP. Erythematous skin macules with isolation of *Trichophyton eboreum*—infection or colonisation? Mycoses. 2013;56(3):373–5.
26. Zhang ZY, Han YF, Chen WH, Liang ZQ. Phylogeny and taxonomy of three new *Ctenomyces* (Arthrodermataceae, Onygenales) species from China. MycoKeys. 2019;47:1–16.
27. Rahul Sharma YS. *Nannizzia graeserae* sp nov, a new dermatophyte of geophilic clade isolated from vicinity of a barbershop in India. Kavaka. 2018;50:14–20.
28. Sitterle E, Frealle E, Foulet F, Cabaret O, Cremer G, Guillot J, et al. *Trichophyton bullosum*: a new zoonotic dermatophyte species. Med Mycol. 2012;50(3):305–9.
29. Lyskova P, Hubka V, Cmokova A, Kolarik M, Petricakova A, Dobias R. Equine dermatophytosis due to *Trichophyton bullosum*, a poorly known zoophilic dermatophyte masquerading as *T. verrucosum*. Mycopathologia. 2015;180(5–6):407–19.
30. Miyasato H, Yamaguchi S, Taira K, Hosokawa A, Kayo S, Sano A, et al. Tinea corporis caused by *Microsporum gallinae*: first clinical case in Japan. J Dermatol. 2011;38(5):473–8.

31. Sabou M, Boulanger N, Ganel S, Filali F, Lipsker D, Glatz I, et al. Émergence d'un dermatophyte proche de *Trichophyton mentagrophytes* var. *erinacei* et inféodé au cochon d'inde: description de sept cas diagnostiqués au CHU de Strasbourg entre novembre 2010 et octobre 2011. J Mycol Med. 2012;22(1):114.
32. Rhee DY, Kim MS, Chang SE, Lee MW, Choi JH, Moon KC, et al. A case of tinea manuum caused by *Trichophyton mentagrophytes* var. *erinacei*: the first isolation in Korea. Mycoses. 2009;52(3):287–90.
33. Mochizuki T, Takeda K, Nakagawa M, Kawasaki M, Tanabe H, Ishizaki H. The first isolation in Japan of *Trichophyton mentagrophytes* var. *erinacei* causing tinea manuum. Int J Dermatol. 2005;44(9):765–8.
34. Lee DW, Yang JH, Choi SJ, Won CH, Chang SE, Lee MW, et al. An unusual clinical presentation of tinea faciei caused by *Trichophyton mentagrophytes* var. *erinacei*. Pediatr Dermatol. 2011;28(2):210–2.
35. Concha M, Nicklas C, Balcells E, Guzman AM, Poggi H, Leon E, et al. The first case of tinea faciei caused by *Trichophyton mentagrophytes* var. *erinacei* isolated in Chile. Int J Dermatol. 2012;51(3):283–5.
36. Charlent AL, Bloch M, Nowak N, Machouart M, Debourgogne A, Cavignaux R, et al. *Trichophyton mentagrophytes* variété *porcellae* nouvel agent d'épidémie de teigne chez le cochon d'inde: enquête dans des animaleries. J Mycol Med. 2014;24(2):e76.
37. Contet-Audonneau N, Leyer C. Émergence d'un dermatophyte transmis par le cochon d'Inde et proche de *Trichophyton mentagrophytes* var. *erinacei*: *T. mentagrophytes* var. *porcellae*. J Mycol Med. 2010;20(4):321–5.
38. Nakamura Y, Kano R, Nakamura E, Saito K, Watanabe S, Hasegawa A. Case report. First report on human ringworm caused by *Arthroderma benhamiae* in Japan transmitted from a rabbit. Mycoses. 2002;45(3–4):129–31.
39. Sabou M, Denis J, Boulanger N, Forouzanfar F, Glatz I, Lipsker D, et al. Molecular identification of *Trichophyton benhamiae* in Strasbourg, France: a 9-year retrospective study. Med Mycol. 2018;56(6):723–34.
40. Brasch J, Wodarg S. Morphological and physiological features of *Arthroderma benhamiae* anamorphs isolated in northern Germany. Mycoses. 2015;58(2):93–8.
41. Uhrlass S, Wittig F, Koch D, Kruger C, Harder M, Gaajetaan G, et al. Do the new molecular assays-microarray and realtime polymerase chain reaction-for dermatophyte detection keep what they promise? Hautarzt. 2019;70(8):618–26. Article in German.
42. Ranganathan S, Menon T, Balajee SA. Isolation of *Microsporum nanum* from a patient with tinea corporis in Madras, India. Mycoses. 1997;40(5–6):229–30.
43. Ratka P, Slusarczyk E, Sloboda T, Kusmierski W. Case of tinea barbae profunda caused by *Microsporum nanum*. Przegl Dermatol. 1983;70(5–6):549–52. Article in Polish.
44. Sinski JT, Flouras K. A survey of dermatophytes isolated from human patients in the United States from 1979 to 1981 with chronological listings of worldwide incidence of five dermatophytes often isolated in the United States. Mycopathologia. 1984;85(1–2):97–120.
45. Ratka P. Microsporid gonitis caused by *Microsporum nanum*. Mycopathologia. 1985;92(1):45–7.
46. Wu SX, Zhu WY, Xia MY, Huang L. Tinea capitis caused by *Microsporum nanum*. Chin Med J (Engl). 1987;100(9):709–11.
47. Ponnighaus JM, Warndorff D, Port G. *Microsporum nanum*—a report from Malawi (Africa). Mycoses. 1995;38(3–4):149–50.
48. Martinez E, Ameen M, Tejada D, Arenas R. *Microsporum* spp. onychomycosis: disease presentation, risk factors and treatment responses in an urban population. Braz J Infect Dis. 2014;18(2):181–6.
49. Bonifaz A, Cordoba-Garcia B, Simancas-Llanos T, Hernandez MA, Martinez-Herrera E, Tirado-Sanchez A. Dermatophytosis caused by *Nannizzia nana* in two siblings. Rev Iberoam Micol. 2019;36(1):30–3.

50. Rezaei-Matehkolaei A, Makimura K, Graser Y, Seyedmousavi S, Abastabar M, Rafiei A, et al. Dermatophytosis due to *Microsporum incurvatum*: notification and identification of a neglected pathogenic species. Mycopathologia. 2016;181(1):107–13.
51. Nouripour-Sisakht S, Rezaei-Matehkolaei A, Abastabar M, Najafzadeh MJ, Satoh K, Ahmadi B, et al. *Microsporum fulvum*, an ignored pathogenic dermatophyte: a new clinical isolation from Iran. Mycopathologia. 2013;176(1–2):157–60.
52. Borelli D. *Microsporum racemosum* nova species. Acta Med Venez. 1965:148–51.
53. Albornoz MB. Primer caso de tinea corporis por el *Microsporum racemosum* (Dante Borelli, 1965). Dermatol Venez. 1972;11(2):310–8.
54. Daum V, McCloud D. *Microsporum racemosum*: first isolation in the United States. Mycopathologia. 1976;59(3):183–5.
55. Rippon JW, Andrews TW. *Microsporum racemosum*. Second clinical isolation from the United States and the Chicago area. Mycopathologia. 1978;64(3):187–90.
56. Garcia-Martos P, Gene J, Sole M, Mira J, Ruiz-Henestrosa R, Guarro J. Case of onychomycosis caused by *Microsporum racemosum*. J Clin Microbiol. 1999;37(1):258–60.
57. Luque AG, Biasoli MS, Sortino MA, Lupo SH, Bussy RF. Atypical tinea corporis caused by *Microsporum gypseum* in a subject with acquired immune deficiency syndrome. J Eur Acad Dermatol Venereol. 2001;15(4):374–5.
58. Sun PL, Ho HT. Concentric rings: an unusual presentation of tinea corporis caused by *Microsporum gypseum*. Mycoses. 2006;49(2):150–1.
59. Soankasina AH, Rakotozandrindrainy N, Andrianteloasy S, Zafindraibe NJ, Rasamoelina T, Rafalimanana C, et al. Dermatophyte infection caused by *Nannizzia gypsea*: a rare case report from Madagascar. Med Mycol Case Rep. 2018;20:7–9.
60. Miranda MF, de Brito AC, Zaitz C, de Carvalho TN, Carneiro FR. *Microsporum gypseum* infection showing a white-paint-dot appearance. Int J Dermatol. 1998;37(12):956–7.
61. Vollekova A, Tothova I, Kolibasova K, Sevcekova A, Baumgartner J. Atypical infection caused by *Microsporum gypseum* mimicking circumscribed scleroderma. Bratisl Lek Listy. 1996;97(3):162–4. Article in Slovak.
62. Romano C, Ghilardi A, Fimiani M. Dystrophic onychomycosis due to *Microsporum gypseum*. Mycoses. 2006;49(4):335–7.
63. Romano C, Asta F, Massai L. Tinea incognito due to *Microsporum gypseum* in three children. Pediatr Dermatol. 2000;17(1):41–4.
64. Qianggiang Z, Limo Q, Jiajun W, Li L. Report of two cases of tinea infection with scutula-like lesions caused by *Microsporum gypseum*. Int J Dermatol. 2002;41(6):372–3.
65. Polilli E, Fazii P, Ursini T, Fantini F, Di Masi F, Tontodonati M, et al. Tinea incognito caused by *Microsporum gypseum* in a patient with advanced HIV infection: a case report. Case Rep Dermatol. 2011;3(1):55–9.
66. Krunic AL, Cetner A, Tesic V, Janda WM, Worobec S. Atypical favic invasion of the scalp by *Microsporum canis*: report of a case and review of reported cases caused by *Microsporum* species. Mycoses. 2007;50(2):156–9.
67. Joly J, Delage G, Auger P, Ricard P. Favus: twenty indigenous cases in the province of Quebec. Arch Dermatol. 1978;114(11):1647–8.
68. Niczyporuk W, Krajewska-Kulak E, Lukaszuk C. Tinea capitis favosa in Poland. Mycoses. 2004;47(5–6):257–60.
69. Alanio A, Romand S, Penso-Assathiany D, Foulet F, Botterel F. *Microsporum praecox*: molecular identification of a new case and review of the literature. Mycopathologia. 2011;171(1):61–5.
70. Sanchez-Castellanos ME, Mayorga-Rodriguez JA, Sandoval-Tress C, Hernandez-Torres M. Tinea incognito due to *Trichophyton mentagrophytes*. Mycoses. 2007;50(1):85–7.
71. Calcaterra R, Fazio R, Mirisola C, Baggi L. Rosacea-like tinea incognito due to *Trichophyton mentagrophytes* vr. *mentagrophytes*. Acta Dermatovenerol Croat. 2013;21(4):263–4.
72. Kumar S, Kamra N, Mahajan BB. Tinea pseudomibricata due to *Trichophyton mentagrophytes*: a rare case report. J Pak Assoc Dermatol. 2015;25(2):151–3.

73. Buruiana AM, Mihali CV, Popescu C. Sequence-based identification of a zoophilic strain of *Trichophyton interdigitale* in a rare case of Tinea blepharo-ciliaris associated with Tinea barbae. Mycopathologia. 2015;180(5–6):403–6.
74. Calles Monar PS, Juarez MA. Eyelid tinea with blepharitis due to *Microsporum canis*. Arch Soc Esp Oftalmol. 2018;93(10):491–3.
75. Gugnani HC, Wattal BL, Sandhu RS. Dermatophytes and other keratinophilic fungi recovered from small mammals in India. Mykosen. 1975;18(12):529–38.
76. Ansari S, Hedayati MT, Nouripour-Sisakht S, Rezaei-Matehkolaei A, Jannesar R, Mohammadi H, et al. A 9-month-old girl from Iran with extensive erythematous plaques due to *Trichophyton simii*, a zoophilic dermatophyte. Mycopathologia. 2016;181(5–6):451–5.
77. Luchsinger I, Bosshard PP, Kasper RS, Reinhardt D, Lautenschlager S. Tinea genitalis: a new entity of sexually transmitted infection? Case series and review of the literature. Sex Transm Infect. 2015;91(7):493–6.
78. Kupsch C, Czaika VA, Deutsch C, Gräser Y. *Trichophyton mentagrophytes*—a new genotype of zoophilic dermatophyte causes sexually transmitted infections. J Dtsch Dermatol Ges. 2019;17(5):493–501.
79. Wendrock-Shiga G, Mechtel D, Uhrlass S, Koch D, Kruger C, Nenoff P. Tinea barbae profunda due to *Trichophyton mentagrophytes* after journey to Thailand: case report and review. Hautarzt. 2017;68(8):639–48. Article in German.
80. Nenoff P, Verma SB, Vasani R, Burmester A, Hipler UC, Wittig F, et al. The current Indian epidemic of superficial dermatophytosis due to *Trichophyton mentagrophytes*—a molecular study. Mycoses. 2019;62(4):336–56.
81. Suss A, Uhrlass S, Ludes A, Verma SB, Monod M, Kruger C, et al. Extensive tinea corporis due to a terbinafine-resistant *Trichophyton mentagrophytes* isolate of the Indian genotype in a young infant from Bahrain in Germany. Hautarzt. 2019;70(11):888–96. Article in German.
82. Nenoff P, Verma SB, Uhrlass S, Burmester A, Graser Y. A clarion call for preventing taxonomical errors of dermatophytes using the example of the novel *Trichophyton mentagrophytes* genotype VIII uniformly isolated in the Indian epidemic of superficial dermatophytosis. Mycoses. 2019;62(1):6–10.
83. Chowdhary A, Singh A, Singh PK, Khurana A, Meis JF. Perspectives on misidentification of *Trichophyton interdigitale/Trichophyton mentagrophytes* using internal transcribed spacer region sequencing: urgent need to update the sequence database. Mycoses. 2019;62(1):11–5.
84. Carlsen BC. A case of tinea pedis due to *Trichophyton tonsurans*. Intern J Vet Med Res Rep. 2013;2013:204515. https://doi.org/10.5171/2013.204515.
85. Fernandes S, Amaro C, da Luz MM, Inacio J, Araujo T, Vieira R, et al. Kerion caused by *Microsporum audouinii* in a child. Med Mycol Case Rep. 2013;2:52–4.
86. Brasch J, Muller S, Graser Y. Unusual strains of *Microsporum audouinii* causing tinea in Europe. Mycoses. 2015;58(10):573–7.
87. Brasch J, Hugel R, Lipowsky F, Graser Y. Tinea corporis caused by an unusual strain of *Microsporum audouinii* that perforates hair *in vitro*. Mycoses. 2010;53(4):360–2.
88. Tan Y, Lin L, Feng P, Lai W. Dermatophytosis caused by *Trichophyton rubrum* mimicking syphilid: a case report and review of literature. Mycoses. 2014;57(5):312–5.
89. Rouzaud C, Hay R, Chosidow O, Dupin N, Puel A, Lortholary O, et al. Severe dermatophytosis and acquired or innate immunodeficiency: a review. J Fungi (Basel). 2015;2(1):4.
90. Wollina U, Hansel G, Uhrlass S, Kruger C, Schonlebe J, Hipler UC, et al. Deep facial mycosis due to *Trichophyton verrucosum*-molecular genetic identification of the dermatophyte in paraffin-embedded tissue-case report and review of the literature. Mycoses. 2018;61(3):152–8.
91. Su H, Li L, Cheng B, Zhu J, Zhang Q, Xu J, et al. *Trichophyton rubrum* infection characterized by Majocchi's granuloma and deeper dermatophytosis: case report and review of published literature. Mycopathologia. 2017;182(5–6):549–54.
92. Dai Y, Xia X, Shen H. Multiple abscesses in the lower extremities caused by *Trichophyton rubrum*. BMC Infect Dis. 2019;19(1):271.

93. Zhang Y, Mijiti J, Huang C, Song Y, Wan Z, Li R, et al. Deep dermatophytosis caused by *Microsporum ferrugineum* in a patient with CARD9 mutations. Br J Dermatol. 2019;181(5):1093–5.
94. Lanternier F, Pathan S, Vincent QB, Liu L, Cypowyj S, Prando C, et al. Deep dermatophytosis and inherited CARD9 deficiency. N Engl J Med. 2013;369(18):1704–14.
95. Alves de Medeiros AK, Lodewick E, Bogaert DJ, Haerynck F, Van Daele S, Lambrecht B, et al. Chronic and invasive fungal infections in a family with CARD9 deficiency. J Clin Immunol. 2016;36(3):204–9.
96. Simpson JK, Frobel P, Seneviratne SL, Brown M, Lowe DM, Grimbacher B, et al. Invasive dermatophyte infection with *Trichophyton interdigitale* is associated with prurigo-induced pseudoperforation and a signal transducer and activator of transcription 3 mutation. Br J Dermatol. 2018;179(3):750–4.
97. Abusleme L, Diaz PI, Freeman AF, Greenwell-Wild T, Brenchley L, Desai JV, et al. Human defects in STAT3 promote oral mucosal fungal and bacterial dysbiosis. JCI Insight. 2018;3(17):e122061.
98. Huovinen S, Tunnela E, Huovinen P, Kuijpers AF, Suhonen R. Human onychomycosis caused by *Trichophyton equinum* transmitted from a racehorse. Br J Dermatol. 1998;138(6):1082–4.
99. Mapelli ET, Colombo L, Crespi E, Menni S. Toenail onychomycosis due to *Trichophyton violaceum* complex. (An unusual, emerging localization of this anthropophilic dermatophyle). Mycoses. 2012;55(2):193–4.
100. Khanna D, Goel A, Kedar A, Manchanda V, Agarwal S. *Trichophyton tonsurans* induced recurrent onychomadesis in a very young infant. Pediatr Dermatol. 2013;30(3):390–1.
101. Macura AB, Krzysciak P, Skora M, Gniadek A. Case report: onychomycosis due to *Trichophyton schoenleinii*. Mycoses. 2012;55(2):e18–9.

Non-dermatophyte Dermatoses Mimicking Dermatophytoses in Animals

Didier Pin

Abstract

Dermatophytoses in animals are fungal diseases of the skin caused by dermatophyte fungi of the genus *Arthroderma*, *Lophophyton*, *Microsporum*, *Nannizzia*, and *Paraphyton* or *Trichophyton*. The infection is generally follicular causing one or many circular areas of alopecia with variable erythema, scaling and crusting, the so-called ringworm. Thus, the primary differential diagnoses are follicular infections, such as bacterial folliculitis and demodicosis. Although dermatophyte folliculitis is the most common lesion of dermatophytoses in animals, other presentations may be observed according to the host species and the dermatophyte involved: dermatophyte folliculitis or ringworm, scaling and crusting in dermatophytosis due to *Nannizzia persicolor* (formerly *Microsporum persicolor*), nodule in the case of kerion or mycetoma, matted hairs, seborrheic dermatosis, or miliary dermatitis in cats, generalized exfoliative dermatoses in dogs, cats, and horses, superficial nonfollicular pustules, papules and macules in the Devon Rex cat, pruritic dermatophytoses in dogs, cats, and horses, and onychomycosis in dogs, cats, and horses. Since manifestations of dermatophytosis are highly variable, particularly in the cat, dermatophytosis should be considered in the case of any annular, papular, nodular, or pustular dermatoses, alopecic or not, sometimes pruritic, and nodular dermatoses as well. Dermatophytoses affect other species such as cattle, sheep, goats, pigs, rodents, rabbits, hedgehog, nonhuman primates, bats, camelids, marine mammals, and birds and must also be differentiated from non-dermatophyte dermatoses mimicking dermatophytoses in these animals.

Keywords

Animal · Dermatophytosis · Dermatology · Differential diagnosis

D. Pin (✉)
Univ Lyon, VetAgro Sup, UPSP ICE, Marcy-l'Etoile, France
e-mail: didier.pin@vetagro-sup.fr

J.-P. Bouchara et al. (eds.), *Dermatophytes and Dermatophytoses*,
https://doi.org/10.1007/978-3-030-67421-2_6

1 Introduction

Dermatophytoses in animals are fungal diseases of the skin caused by dermatophyte fungi of the genus *Arthroderma*, *Lophophyton*, *Microsporum*, *Nannizzia*, and *Paraphyton* or *Trichophyton*, which use keratin as a source of nutrients. Arthrospores cannot penetrate intact skin, but minor skin trauma or damage from moisture can be adequate to allow infection to develop. Once infection is established, dermatophytes proliferate in keratinous tissues and invade dead parts of hair, skin, and nails. The incubation period is approximately 1–3 weeks.

2 Typical and Rare Presentations of Dermatophytoses

Although they are highly variable, clinical signs are explained by the keratinous tropism of dermatophytes. Dermatophytes invade and proliferate into keratinous tissues, e.g., the horny layer or *stratum corneum* of the epidermis, hair shafts, or claws, and do not generally penetrate the living tissues. Although dermatophyte folliculitis or ringworm is the most common lesion of dermatophytoses in animals, other presentations may be observed according to the host species and the dermatophyte involved [1].

2.1 Dermatophyte Folliculitis or Ringworm

In a large number of cases, dermatophytes invade hair shafts and the *stratum corneum* of the epidermis and produce the classical lesion called ringworm or dermatophyte folliculitis. Ringworm or dermatophyte folliculitis is characterized by a regular and circular, well-delimited area of peripherally expanding alopecia. Discrete erythema and scales are often associated as well as some crusts due to a moderate exudation. Less commonly, follicular papules and pustules may be present. Sometimes, hair regrowth is observed in the center of the lesion. Ringworm may be single or multifocal.

2.2 Scaling and Crusting in Dermatophytosis Due to *Nannizzia persicolor* (Formerly *Microsporum persicolor*)

As *N. persicolor* invades only the *stratum corneum*, it produces scaling and crusting but no alopecia, at least at the beginning of the infection. Hunting dogs and dogs known to frequently encounter small rodents are predisposed. In half of the cases, pruritus is present [2]. Lesions classically consist of erythema and papules, with a mild to severe associated scaling, and alopecia after a certain time of evolution. They may be very localized or generalized, and in the latter case, they are initially located on the face (on the bridge of the nose, chin, pinnae, and periocular region) before generalization.

2.3 Nodule in the Case of Kerion or Mycetoma

Generally, the inflammation is moderate, nonexudative, without pus. However, in some cases of infection caused by *Nannizzia gypsea* (formerly *Microsporum gypseum*) or *Trichophyton mentagrophytes*, but also, more rarely, *Microsporum canis*, the inflammation is severe and exudative, with pus, and induces a nodular lesion called kerion.

Exceptionally, some dermatophytes, particularly *M. canis*, have been associated with a nodular dermatosis, called mycetoma, particularly in Persian cats. Mycetoma is characterized by one or more cutaneous or subcutaneous nodules that are often ulcerated and discharging and caused by the development of a granulomatous dermatitis or panniculitis containing granules, which are formed of branching hyphae or pseudohyphae of the fungus [3, 4].

2.4 Matted Hairs or Seborrheic Dermatosis in Cats

In some cases of dermatophytosis in the cat, alopecia is absent and matted hairs or seborrheic dermatosis with mild hypotrichosis is the main clinical feature [5].

2.5 Miliary Dermatitis in Cats

Miliary dermatitis, which is a common and often pruritic, papulopustular, and papulocrustous skin syndrome in cats due to a variety of causes, may be due to *M. canis* infection in some cases [6].

2.6 Generalized Exfoliative Dermatoses in Dogs, Cats, and Horses

After a chronic evolution or if a predisposing factor exists, such as young or old age, breed predisposition in the Persian cat or Yorkshire terrier dog or concomitant glucocorticoid treatment in the cat, dog, or horse, the alopecia could generalize in some individuals. The complete absence of hairs is then associated with an erythematous, lichenified, hyperpigmented, and scaly skin. In such cases, a marked lymph node involvement may occur [7, 8].

2.7 Superficial Nonfollicular Pustules in Dogs and Horses

An acantholytic dermatophytosis, associated with *T. mentagrophytes*, *N. persicolor*, or *Trichophyton equinum* infection, characterized by nonfollicular pustules, epidermal collarettes, scales, and crusts, has been described in the dog and horse [9, 10].

2.8 Papules and Macules in the Devon Rex Cat

In Devon Rex cats, dermatophytosis can be characterized by multiple hyperpigmented, sometimes crusted, papules, and macules [11].

2.9 Pruritic Dermatophytoses in Dogs, Cats, and Horses

Dermatophytoses classically are not pruritic, except in the cat in which dermatophytosis is frequently associated with sometimes severe pruritus [12], as are some cases in the dog or in the horse [7, 13].

As mentioned above, dermatophytosis due to *N. persicolor* may be pruritic.

2.10 Onychomycosis Due to Dermatophytes in Dogs, Cats, and Horses

Conversely to humans, dermatophyte onychomycosis is rare in animals and it has been described rarely in the dog [14] and exceptionally in the cat [15–18] and the horse [19, 20]. In dogs and cats, one or a few digits are usually affected. The claw is somewhat opaque and mottled irregularly with whitish macules [15, 16]. In horses suffering from onychomycosis, the horn defects range from brittle hooves with sand cracks to horn fissures and large crumbling areas around the nail holes [20].

Because the infection is very often follicular in animals, the most common clinical signs are hair loss, erythema in white skin areas, scaling, and crusting. Rare lesions are macules, crusted papules, pustules, epidermal collarettes, scales or nodule, and onychodystrophy as well, sometimes associated with pruritus. Thus, dermatophytoses are very pleomorphic diseases and it is true to say "if it looks like ringworm, it is probably not ringworm" (Danny Scott). One should add, "if it does not look like ringworm, it could be" (Didier-Noël Carlotti).

In addition to these well-known presentations of dermatophytoses in dogs, cats, and horses, dermatophytoses are observed in numerous other species, such as the cattle, sheep, goats, the pig, the rodents, rabbit, and hedgehog, the nonhuman primates, the Chiroptera or bats, the camelids, the marine mammals, and the birds with varied presentations that must be differentiated from no less varied non-dermatophyte dermatoses mimicking dermatophytoses more or less specific to these species).

3 Non-dermatophyte Dermatoses Mimicking Dermatophytoses in Animals

Although numerous other dermatoses mimic the classical ringworm lesion, dermatophytosis is often missed because of the protean nature of the clinical signs. In most studies, the incidence of dermatophytosis is low, e.g., accounting for only

0.26–5.6% of all cases examined of canine or feline skin diseases [6]. Nevertheless, due to its highly contagious tendency, dermatophytosis should be considered in the differential diagnosis of any annular, papular, or pustular eruption.

3.1 In the Dog

The prevalence of dermatophytoses is lower in the dog than in the cat. The four species the most frequently encountered in dogs are *M. canis* (40–80% of cases), *N. gypsea* (up to 25% of cases), rarely *N. persicolor*, and *T. mentagrophytes* (also called formerly *Arthroderma vanbreuseghemii*) (5–35% of cases). More rarely, other dermatophytes can be identified in dogs such as *Trichophyton erinacei* and *Trichophyton verrucosum*.

Since manifestations of dermatophytosis are highly variable in the dog, e.g., dermatophyte folliculitis or ringworm, scaling and crusting of the face, kerion or mycetoma, generalized exfoliative dermatosis, superficial nonfollicular pustules, pruritic dermatosis, and onychomycosis, dermatophytosis should be considered in the case of any annular, papular, nodular, or pustular dermatoses, alopecic or not, and of nodular dermatoses as well. The main differential diagnoses, bacterial and demodectic folliculitis, and the other non-dermatophyte dermatoses mimicking dermatophytoses are listed in Table 1.

3.2 In the Cat

Microsporum canis is responsible for more than 95% of the cases of dermatophytosis in the cat. Other agents, much rarer, are *N. gypsea*, *N. persicolor*, and *T. mentagrophytes* (*A. vanbreuseghemii*). Long-haired cats, particularly Persian cats, are strongly predisposed to develop *M. canis* infection, especially in catteries (probably due to a genetic predisposition). On the contrary, short hair European cats more commonly develop *T. mentagrophytes* infection [21].

Feline dermatophytosis has a plethora of clinical presentations ranging from classic focal or multifocal alopecia with broken hairs, scaling and minimal inflammation to more inflammatory lesions as miliary dermatitis, matted hairs, or seborrheic dermatitis, exfoliative generalized dermatitis, tail seborrhea, mycetoma, pruritic dermatitis, onychomycosis, and a recently described presentation in the Devon Rex cat, characterized by erythematous crusted papules and macules. So, the differential diagnosis is wide and includes all skin diseases listed in Table 2.

These presentations may be associated with pruritus or not. Table 3 lists the non-dermatophyte dermatoses mimicking dermatophytoses in the cat and indicates their relative frequency and if pruritus is associated.

Table 1 Non-dermatophyte dermatoses mimicking dermatophytoses in the dog and their clinical presentations

Group of dermatoses	Dermatoses and their clinical presentations
Folliculitis	Bacterial and demodectic folliculitis are the main differentials. In the dog, bacterial folliculitis is much more common than demodicosis and dermatophytosis. Furthermore, follicular papules or pustules, epidermal collarettes, and a moth-eaten appearance of the hair coat are often associated with bacterial folliculitis, while comedones and a sharp delimitation of edges of alopecic areas are associated with demodicosis
Nodular skin diseases	Bacterial, mycobacterial, parasitic, and fungal nodules, foreign-body granuloma, sterile granulomas or pyogranulomas, and neoplasms such as histiocytoma, mastocytoma, fibroma, or plasmacytoma
Autoimmune dermatoses	Affecting the face as pemphigus foliaceus of mucocutaneous lupus erythematosus. Pemphigus foliaceus is characterized by nonfollicular pustules, crusts, and alopecia. Pemphigus foliaceus and discoid lupus erythematosus frequently produce facial scaly, papulopustular, or crusty lesions often affecting the planum nasale unlike dermatophytosis Characterized by alopecic circular areas of alopecia as *alopecia areata* and pseudopelade. In *alopecia areata* and pseudopelade, the alopecic skin appears otherwise normal
Generalized exfoliative dermatoses	Endocrinopathies, leishmaniasis, exfoliative lupus erythematosus, granulomatous sebaceous adenitis, and skin lymphoma
Pruritic dermatoses	Ectoparasitoses, pyodermas, *Malassezia* dermatitis, and allergies
Claw disorders	Bacterial infections, neoplasms, candidiasis, blastomycosis, geotrichosis, cryptococcosis, symmetric lupoid onychodystrophy, pemphigus foliaceus, pemphigus vulgaris, bullous pemphigoid, and epidermolysis bullosa In general, onychomycosis is characterized by an asymmetric paronychia or onychodystrophy of one digit or multiple digits of one paw [6]

3.3 In the Horse

Since *Trichophyton equinum* is the most common agent of dermatophytosis in horses throughout the world, other less frequently encountered dermatophytes are *T. mentagrophytes*, *T. verrucosum*, *Microsporum equinum*, and *N. gypsea* [8]. Dermatophytosis is common in horses, but, as in other host species, it is overdiagnosed when diagnosis relies exclusively on clinical signs.

Because the infection is generally follicular in horses, the most common clinical sign is one or many circular areas of alopecia with variable erythema, scaling, and crusting and the primary differential diagnoses are follicular infections, such as bacterial folliculitis, demodicosis, and eosinophilic folliculitis (Table 4). In horses, demodicosis is much rarer than dermatophytosis and bacterial folliculitis. The differential diagnosis also includes other skin diseases listed in Table 4.

Table 2 Non-dermatophyte dermatoses mimicking dermatophytoses in the cat and their clinical presentations

Clinical presentation	Diseases
Focal or multifocal alopecia	Demodicosis, bacterial folliculitis, *alopecia areata*, or pseudopelade [62]. Inversely to the dog, dermatophytosis in cats is more common than demodicosis and bacterial folliculitis
Miliary dermatitis	Ectoparasitoses (cheyletiellosis, trombiculiasis, pediculosis, and flea infestation), bacterial folliculitis, allergic dermatoses (flea allergy dermatitis and nonflea-induced hypersensitivity dermatitis) [63], and pemphigus foliaceus
Matted hairs or seborrheic dermatitis	*Malassezia* spp. overgrowth or dermatitis [64, 65], demodicosis, *Lynxacarus radovskyi* infestation [66], hyperthyroidism and other endocrinopathies, defect of grooming habits due to obesity or psychogenic trouble, dietary deficiencies, *Proisotoma* spp. springtails infestation [67], and sebaceous gland dysplasia [68]
Exfoliative generalized dermatitis	Thymoma-associated exfoliative dermatitis, nonthymoma-associated exfoliative dermatitis [69], epitheliotropic T-cell lymphoma, systemic lupus erythematosus or chronic cutaneous lupus erythematosus [70], pseudopelade, sebaceous adenitis, and drug reaction
Tail seborrhea	Tail gland hyperplasia [71], flea allergy dermatitis, *Malassezia* spp. dermatitis or overgrowth, and demodicosis
Nodules	Non-dermatophytic fungal mycetomas, other infectious or foreign-body granulomas, sterile panniculitis, and various neoplasms
Onychitis or paronychia	Bacterial paronychia, pemphigus foliaceus, squamous cell carcinoma, and lupus erythematosus [6, 18]
Erythematous crusted papules and macules in Devon Rex and Sphynx cats	Feline urticaria pigmentosa [11]

As initially, papules with tufted erected hairs in annular areas or more rarely, an urticarial-like eruption may be observed in dermatophytosis, other causes of urticaria or urticarial-like eruption must be ruled out (Table 4).

In some horses infected with *T. equinum*, the development of the dermatophyte is accompanied by an acantholysis of the epidermis and follicular wall [10]. Fungal filaments are found on the surface of the skin and in the wall of the follicles but generally not in hairs. This presentation must be differentiated from pemphigus foliaceus.

In horses, dermatophytoses can be pruritic, particularly in the case of *T. mentagrophytes* infection, in which multiple small bumps in "grain of millet" appear with tufted hairs, which rapidly turn into scabs and multiple hair loss areas of

Table 3 Pruritic and nonpruritic non-dermatophyte dermatoses mimicking dermatophytoses in the cat and their estimated frequencies

Pruritus	Diseases	Frequency
Presence	Hypersensitivity disorders	Very frequent
	Parasitic skin diseases	Frequent
	Malassezia dermatitis	Frequent
	Bacterial skin diseases	Frequent
	Generalized form of pemphigus foliaceus	Rare
	Mucinotic mural folliculitis	Very rare
	FeLV-associated dermatosis	Very rare
Absence	Acne	Very frequent
	Physiological preauricular alopecia	Frequent
	Bacterial folliculitis	Frequent
	Localized form of pemphigus foliaceus	Frequent
	Seborrhea due to obesity or old age	Frequent
	Tail seborrhea	Rare
	Demodicosis	Rare
	Systemic lupus erythematosus	Rare
	Feline urticaria pigmentosa	Rare
	Cutaneous lupus erythematosus	Very rare
	Alopecia areata and pseudopelade	Very rare
	Nonthymoma-associated exfoliative dermatosis	Very rare
	Thymoma-associated exfoliative dermatosis	Very rare
	Epitheliotropic cutaneous lymphoma	Very rare
	Sebaceous gland dysplasia	Very rare
	Paraneoplastic alopecia	Very rare
	Endocrinopathies (hyperthyroidism, hypothyroidism, naturally occurring and iatrogenic hypercortisolism, and diabetes mellitus)	Rare to very rare

a few millimeters in diameter with pityriasiform scales [13]. Thus, other causes of pruritus must be ruled out in horses.

Rarely, dermatophytoses in horses could generalize and therefore must be differentiated from other causes of generalized, more or less scaly, dermatoses (Table 4).

3.4 In Cattle, Sheep, and Goats

In cattle, as well as in goats and sheep, the most frequently encountered dermatophyte is *T. verrucosum*, with *T. mentagrophytes*, *T. equinum*, *N. gypsea*, *Nannizzia nana* (formerly *Microsporum nanum*), and *M. canis* less commonly isolated. Dermatophytosis is common in cattle, but infrequent in goats and sheep.

In cattle, lesions are typically circular areas of alopecia of 10–50 mm in diameter, either covered with thin farinaceous desquamations or with thick crusty lamellar scales difficult to pull out of the skin. Rarely, they are characterized by severe

Table 4 Non-dermatophyte dermatoses mimicking dermatophytoses in the horse and their clinical presentations

Clinical presentation	Differential diagnosis
One or many circular areas of alopecia with variable erythema, scaling, and crusting	Bacterial folliculitis, demodicosis, and eosinophilic folliculitis as well as dermatophilosis (Fig. 1), pemphigus foliaceus, and *alopecia areata*. In horses, demodicosis is much rarer than dermatophytosis and bacterial folliculitis
Urticaria or urticarial-like eruption	Insect allergies, vasculitis, infectious and sterile folliculitis, pemphigus foliaceus, lymphoma, and amyloidosis
Pustules due to acantholysis	Pemphigus foliaceus
Pruritic dermatoses	*Culicoides* hypersensitivity [72], various fly bites-induced dermatoses, sarcoptic mange, psoroptic mange, trombiculiasis, *Dermanyssus gallinae* infestation, pediculosis [73], bacterial pyoderma [74], and pemphigus foliaceus [75]
Generalized alopecia	Pemphigus foliaceus, systemic or chronic cutaneous lupus erythematosus, *alopecia areata*, effluvium, sarcoidosis, cutaneous lymphoma, and multisystemic eosinophilic epitheliotropic disease
Onychomycosis	Cases due to *Scopulariopsis brevicaulis* [20] and from other causes of coronary band and hoof disorders [8]
Nodules: kerion and dermatophytic mycetoma	Other infectious or sterile granulomas, panniculitis, and various neoplasms

suppuration and ulceration. They particularly occur on the head, neck, and pelvis but can spread all over the body. Pruritus is variable [22].

In goats, lesions vary from circular to diffuse areas of alopecia, scaling, erythema, and yellowish crusts and are most commonly seen on the face, pinnae, neck, and limbs. Pruritus is rare [22].

Fig. 1 Numerous circular crusted lesions of the back in the case of dermatophilosis in a horse (Dermatologie, VetAgro Sup, Marcy l'Etoile, France)

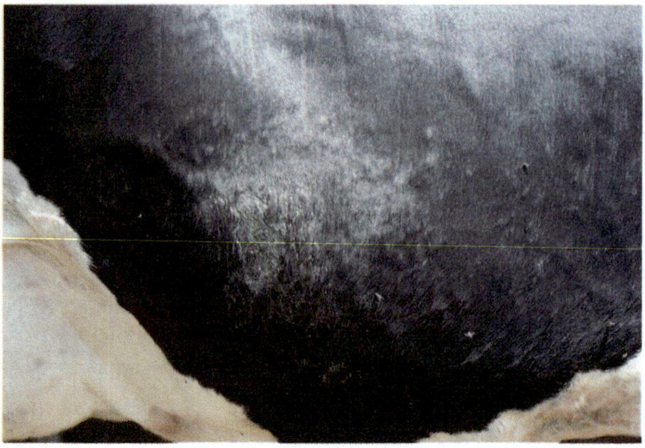

Fig. 2 Round area of hair loss and tufted erected hairs in the case of staphylococcal folliculitis in a cow (Dermatologie, VetAgro Sup, Marcy l'Etoile, France)

In sheep, lesions are characterized by circular areas of alopecia and thick grayish crusts and are most commonly seen on the face, neck, thorax, and back [22].

The main differential diagnosis includes staphylococcal folliculitis (Fig. 2), dermatophilosis, zinc-responsive dermatosis, and ectoparasitoses such as chorioptic mange, psoroptic mange, sarcoptic mange, and demodicosis. *Alopecia areata* must be ruled out in cattle, as well as pemphigus foliaceus in cattle [23] and goats [24, 25].

3.5 In the Pig

In swine, *N. nana* is the most common cause of dermatophytosis, with *T. mentagrophytes*, *T. verrucosum*, and *M. canis* being less often encountered.

Lesions start by small well-circumscribed macules, papules, and annular areas of red to brown discoloration of the skin and superficial dry and brown crusts. They spread out and sometimes converge forming large flat plaques with irregular borders [22, 26]. They could be located on any part of the animal's body but are especially common on the jowls, behind the ears, and on the trunk. Alopecia and pruritus are rare.

The main differential diagnosis includes staphylococcal skin disease also known as greasy pig disease (Fig. 3), staphylococcal folliculitis, and dermatophilosis, as well as pityriasis rosea or porcine juvenile pustular psoriasiform dermatitis, psoriasiform parakeratosis, zinc deficiency, early-stage swinepox, *alopecia areata*, and, if pruritus is present, sarcoptic mange and fly bites [26].

3.6 In Rodents, Rabbit, and Hedgehog

The most encountered dermatophyte in domestic and wild rodents and in rabbits is *T. mentagrophytes*. *Microsporum canis* is also frequent in rabbit pets and farms. Other encountered dermatophytes are *N. gypsea* and *Paraphyton cookei* (formerly *Microsporum cookei*) [27].

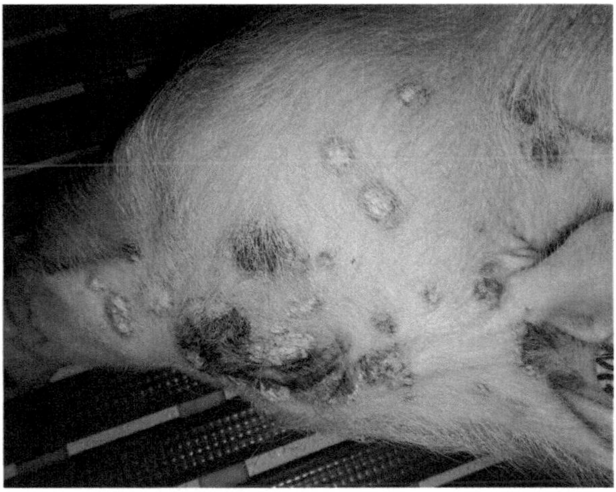

Fig. 3 Numerous circular alopecic crusted lesions of the face in the case of superficial extensive pyoderma in a pig (Dermatologie, VetAgro Sup, Marcy l'Etoile, France)

In guinea pigs, typical lesions are circumscribed areas of alopecia with erythema, a marked scaling and crusting. They are commonly located on the muzzle, around the eyes, and on the forehead or ears and could affect, in severe cases, the back and limbs. Dermatophytosis may be pruritic or not and may affect the nails with severe onychogryphosis and onychodystrophy [21, 27]. In this species, the differential diagnosis mainly includes demodicosis and, in the case of associated pruritus, *Trixacarus caviae*, *Chirodiscoides caviae*, and lice infestations.

Dermatophytosis is rare in rats and mice, and the condition is usually not pruritic limiting the differential diagnosis to poorly pruritic ectoparasite infestations and seborrheic dermatoses observed in older rats [6].

In rabbits, dermatophytosis is common. Lesions are characterized by patchy alopecia, broken hairs, erythema, and yellow or silvery crusts, initially located on the bridge of the nose, eyelids, and pinnae. Owing to pruritus, lesions can be spread to the feet and other areas of the body. The differential diagnosis includes sarcoptic mange, psoroptic mange, cheyletiellosis (Fig. 4), flea infestation, other causes of pododermatitis or moist dermatitis, syphilis, and sebaceous adenitis.

3.7 In Nonhuman Primates

Dermatophytosis is rarely reported in nonhuman primates [28]. *Microsporum canis* and *Trichophyton rubrum* have been isolated from lesions on great apes characterized by well-circumscribed alopecic circular or ring-shaped areas, generalized scaliness, and patchy hair loss to generalized alopecia [29]. Dermatophytosis due to *T. mentagrophytes* has been diagnosed in a Coquerel's sifaka (*Propithecus coquereli*) and in Capuchin monkeys (*Cebus nigrivitatus*) characterized by crusting and alopecia of the skin of the face and neck or generalized, associated with mild pruritus [30, 31]. A case of dermatophytosis due to *M. canis* in a rhesus monkey (*Macaca mulatta*) has been reported [32], and a case

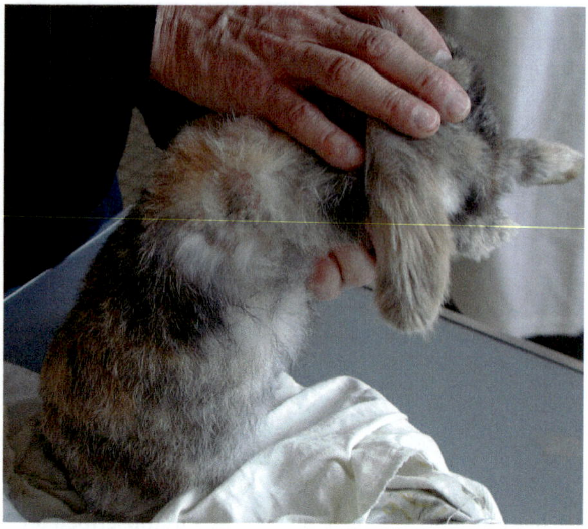

Fig. 4 Areas of scaling, crusting, and alopecia of the back in the case of cheyletiellosis in a rabbit (Dermatologie, VetAgro Sup, Marcy l'Etoile, France)

due to *Trichophyton* spp. has been identified in a group of L'Hoest's monkeys (*Cercopithecus lhoesti*) [33]. Pruritus is not usually a clinical feature of dermatophyte infection in primates.

The differential diagnosis includes all causes of acquired alopecia, scaling, and crusting associated or not with pruritus such as ectoparasitoses (particularly scabies and demodicosis), *Strongyloides stercoralis* or *Pelodera strongyloides* dermatitis in Old World monkeys and great apes (gibbons, chimpanzees, gorillas, and orangutans) [34], *Psorergates pitheci* dermatitis [35], *Dunnalges lambrechti*, or *Rosalialges cruciformis* infections [36], bacterial infections, nutritional deficiencies, or hormonal imbalances (hypothyroidism and hyperadrenocorticism), color-linked follicular dysplasias, autoimmune diseases (*alopecia areata*, pseudopelade, and sebaceous adenitis), and hypersensitivities [33, 37].

3.8 In Chiroptera (Bats)

Dermatophytes have been observed in hibernating bats [38, 39], but have not been linked to disease probably because they cause nonlethal infections and have gone undetected. Recently, *Arthroderma redellii* (formerly *Trichophyton redellii*) infection has been diagnosed in hibernating bats [40]. This dermatophytic dermatosis is characterized by white fungal growth on the ears, legs, wings, tail, or uropatagium. The muzzle often lacks clinical signs of infection. Ten years ago, the white-nose syndrome, a deadly disease caused by the fungus *Pseudogymnoascus destructans*, has been described, [41, 42] which is characterized by very similar lesions as those of *A. redellii* infection, but associates lesions on the muzzle in addition to other areas of unfurred skin.

Other non-dermatophyte dermatoses mimicking dermatophytoses in bats are idiopathic alopecia syndrome [43], parasitic diseases caused by mites belonging to the families *Macronyssidae*, *Myobiidae*, and *Spinturnicidae*, found on the skin, and other less common ectoparasites, bat flies, fleas, and bat bugs [29]. Many ectoparasitic infections in bats are self-limiting.

3.9 In Camelids

The camelids include the South American Camelids, with two genuses, *Llama* and *Vicugna*, and the Old World Camels, with one genus *Camelus*. Dermatophytosis is rare and tends to be self-limiting in South American Camelids but common in Old World Camels and caused by *T. mentagrophytes*, *T. verrucosum*, *Trichophyton schoenleinii*, *Trichophyton sarkisovii*, *Trichophyton dankaliense*, *N. gypsea*, and *M. canis*. Lesions most commonly occur on short-haired areas (head, face, distal legs, and perineum) and vary from annular areas of alopecia and scaling to thick grayish crusts. In general, dermatophytosis is not pruritic in camelids. The differential diagnosis includes cutaneous bacterial infections such as bacterial folliculitis or dermatophilosis, demodicosis, sterile eosinophilic folliculitis, and zinc-responsive dermatitis [44, 45].

3.10 In Marine Mammals

Marine mammals include Cetacea (Order *Cetacea*), with whales (Suborder *Mysticeti*), and porpoises and dolphins (Suborder *Odontoceti*), Pinnipeds (Order *Carnivora*, Suborder *Caniformia*), with *Otariidae* (eared seals), *Phocidae* (earless seals), *Odobenidae* (walruses), and sirenians (Order *Sirenia*), with manatees (*Trichechidae*), and dugongs (Dugongidae).

Dermatophytosis in Cetacea is rare and due to some *Microsporum* or *Trichophyton* species. The most frequent lesions are nodules as in a captive Atlantic bottlenose dolphin with nodules on the trunk [46]. Differential diagnosis includes dermatitis due to *Fusarium* infection, which is characterized by raised, firm, erythematous, cutaneous nodules, most prominent on the head, trunk, and caudal portions of the body [47], and lacaziosis (formerly lobomycosis), a chronic granulomatous skin disease affecting bottlenose dolphins, caused by the yeast-like organism *Lacazia loboi* (formerly *Loboa loboi*). The lesions are generally limited to the skin and present as a chronic proliferative and ulcerated dermatitis and cellulitis [48]. Amazingly, as humans are the only other species known to be infected by this organism, this disease is potentially zoonotic [49]. Differential diagnosis also includes viral skin diseases such as mucosal and cutaneous papillomas or fibropapillomas due to papillomavirus infection, or ring or pinhole to black, and punctiform or stippled lesions on any part of the body due to poxvirus, herpesvirus, or calicivirus infections. Localized or diffuse bacterial granulomatous dermatitis or panniculitis due to *Erysipelothrix rhusopathiae*, *Nocardia asteroides*, *Nocardia brasiliensis*, or *Mycobacterium marinum* must also be excluded.

Dermatophytosis in pinnipeds is due to some *Trichophyton* species, notably *T. mentagrophytes*, *Epidermophyton floccosum*, *M. canis*, or *N. gypsea*, and is characterized by oval areas of inflammation of varying sizes, depilation or pustular dermatitis, alopecia, depigmentation, hyperemia, and hyperkeratosis, or nodules on the face and nose, the flippers, the tail, and the axillary region. Non-dermatophyte dermatomycoses mimicking dermatophytoses in pinnipeds are: (1) fusariosis in California sea lions, gray seals, and harbor seals characterized by firm, raised papules found mainly on the face, trunk, and flippers; (2) dermatitis caused by *Candida albicans* characterized by alopecia and scaling, less often ulcers, at mucocutaneous junctions, around nail beds, and in the axillae or the interdigital folds; and (3) dermatitis caused by *Malassezia pachydermatis* characterized by multiple wheals (2–3 mm in diameter) on both flanks and the chest without any pruritus. Nonmycotic dermatoses mimicking dermatophytoses must be included in the differential diagnosis such as vesicular stomatitis and dermatitis of the glabrous skin of the flippers or dermatitis caused by calicivirus, poxvirus, and herpesvirus infections, demodicosis characterized by follicular pustules around the genitalia and flippers, pustular folliculitis caused by *Staphylococcus* sp. or *Corynebacterium* sp. infection mainly developed on skin after trauma or abrasions and alopecia secondary to pruritus caused by lice (*Anoplura* sp.) [29, 50–59].

Only two cases of skin disease have been reported in Manatees (order *Sirenia*). One was a captive Manatee, which developed erosions on the skin of the nose, face, flippers, and tail and skin scrapings of the lesions revealed *E. floccosum* [60]. The second was papillomatosis affecting a group of captive Manatees with lesions typical of cutaneous papillomatosis due to a specific papillomavirus [61].

3.11 In Birds

Dermatophytosis is rare in birds and caused by *Lophophyton gallinae* (formerly *Trichophyton gallinae*). The infection, which is observed in chicken, turkey, duck, quail, and canary, is also called favus. It is characterized by white scaly or crusty lesions on the comb and on the skin of the head and neck and loss of feathers. Feather loss is due to the colonization of the keratinized shaft of the feathers [27].

The differential diagnosis includes poxvirus infection for the lesions on the comb, lice (Fig. 5), *Dermanyssus gallinae*, or *Neocnemidocoptes laevis* var. *gallinae* infestations, and choline deficiency for the feather loss.

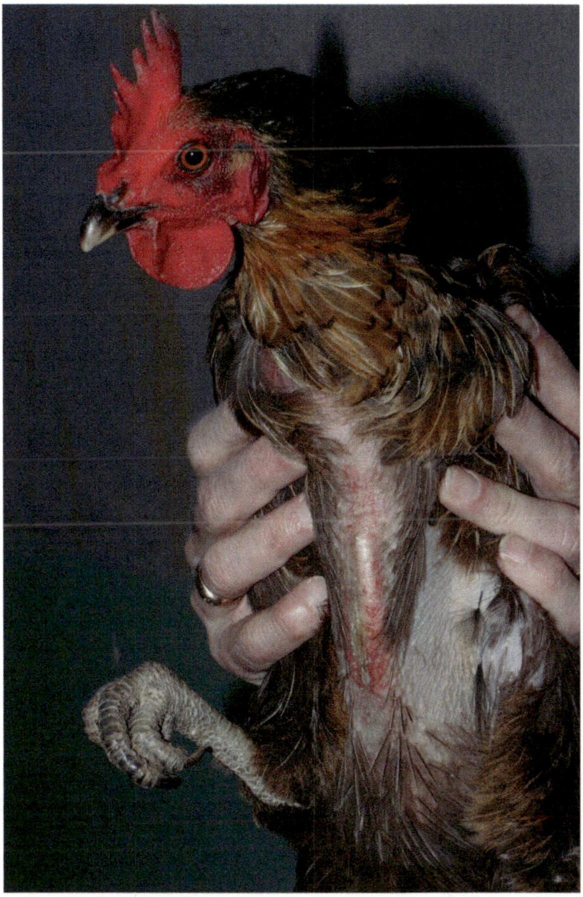

Fig. 5 Feather loss in the case of lice infestation in a hen (Dermatologie, VetAgro Sup, Marcy l'Etoile, France)

4 Conclusion

In all species, non-dermatophyte dermatoses mimicking dermatophytoses are numerous. It is then necessary that the clinician realizes a thorough and step-by-step diagnostic approach including history, clinical signs, and complementary examinations such as direct examination of plucked hairs, Wood's lamp examination of lesions, fungal culture or PCR, and histopathological examination of skin biopsies.

References

1. Pin D. Non-dermatophyte dermatoses mimicking dermatophytoses in animals. Mycopathologia. 2017;182(1,2):113–26.
2. Carlotti D-N, Bensignor E. Dermatophytosis due to *Microsporum persicolor* (13 cases) or *Microsporum gypseum* (20 cases) in dogs. Vet Dermatol. 1999;10:17–27.

3. Bourdin M, Destombes P, Parodi AL, Drouhet E, Secretain G. Première observation d'un mycétome à *Microsporum canis* chez un chat. Rec Méd Vét. 1975;151:475–80.
4. Pin D, Carlotti D-N. Un cas de mycétome dermatophytique à *Microsporum canis* chez un chat Persan. Prat Méd Chir Anim Comp. 2000;35:105–12.
5. Carlotti D-N, Pin D. Aspects cliniques et histopathologiques, diagnostic différentiel et traitements des dermatophytoses chez les carnivores domestiques. Ann Méd Vét. 2002;147:85–96.
6. Miller WH Jr, Griffin CE, Campbell KL. Muller and Kirk's small animal dermatology. 7th ed. Elsevier Mosby: St. Louis; 2013.
7. Mignon B. Dermatophytosis. In: Guaguère E, Prélaud P, editors. A practical guide to canine dermatology: Merial: Italy; 2008. p. 151–64.
8. Scott DW, Miller WH Jr. Equine dermatology. St. Louis: Elsevier Saunders; 2011. p. 172–83.
9. Poisson L, Mueller RS, Olivry T. Dermatophytose pustuleuse cornéophilique canine évoquant un pemphigus foliacé. Prat Méd Chir Anim Comp. 1998;33:229–34.
10. Scott DW. Marked acantholysis associated with dermatophytosis due to *Trichophyton equinum* in two horses. Vet Dermatol. 1994;5:105–10.
11. Colombo S, Scarampella F, Ordeix L, Roccabianca P. Dermatophytosis and papular eosinophilic/mastocytic dermatitis (urticarial pigmentosa-like dermatitis) in three Devon Rex cats. J Fel Med Surg. 2012;14:498–502.
12. Moriello KA, DeBoer DJ. Dermatophytosis. In: Guaguère E, Prélaud P, editors. A practical guide to feline dermatology: Merial: Italy; 1999. p. 4.1–4.11.
13. Guillot J, Ribot X, Chermette R. Connaître et savoir traiter les formes atypiques de dermatophytose équine. Nouv Prat Vét Eq. 2008;16:33–5.
14. Scott DW, Miller WH Jr. Disorders of the claw and clawbed in dogs. Comp Cont Educ. 1992;14:1448–58.
15. LaTouche CJ. Some clinical and microscopic features of *Microsporum canis* Bodin infection of the skin and its appendages as it occurs in the cat. Vet Rec. 1953;65:680–1.
16. LaTouche CJ. Onychomycosis in cats infected by *Microsporum canis* Bodin. Vet Rec. 1955;67:578–9.
17. Carlotti D-N. Claw diseases in dogs and cats. Eur J Comp Anim Pract. 1999;9:21–33.
18. Scott DW, Miller WH Jr. Disorders of the claw and clawbed in cats. Comp Cont Educ. 1992;14:449–57.
19. Keller M, Krehon S, Stanek C, Rosengarten R. Keratinopathogenic mould fungi and dermatophytes in healthy and diseased hooves of horses. Vet Rec. 2000;147:619–22.
20. Apprich V, Spergser J, Rosengarten R, Hinterhofer C, Stanek C. Scanning electron microscopy and fungal culture of hoof horn from horses suffering from onychomycosis. Vet Dermatol. 2010;21:335–40.
21. Drouot S, Mignon B, Fratti M, Roosje P, Monod M. Pets as the main source of two zoonotic species of the *Trichophyton mentagrophytes* complex in Switzerland, *Arthroderma vanbreuseghemii* and *Arthroderma benhamiae*. Vet Dermatol. 2009;20:13–8.
22. Scott DW. Large animal dermatology. Philadelphia: WB Saunders Comp; 1988. p. 487.
23. Mosca M, Fantini O, Druart S, Arnoult A, Ledoux D, Becker C, et al. Pemphigus foliaceus in a cow. Vet Dermatol. 2015;26:310.
24. Jackson PGG, Lloyd S, Jeffries AR. Pemphigus foliaceus in a goat. Vet Rec. 1984;114:479.
25. Scott DW, Smith MC, Smith CA. Pemphigus foliaceus in a goat. Agri-Practise. 1984;5:38–45.
26. Gourreau J-M, Drolet R, Martineau G-P, Morvan H, Pastoret P-P, Pin D, et al. Atlas of porcine dermatology. Paris: OIE, World Organisation for Animal Health; 2015. p. 456.
27. Chermette R, Ferreiro L, Guillot J. Dermatophytoses in animals. Mycopathologia. 2008;166:385–405.
28. Bernstein JA, Didier PJ. Nonhuman primate dermatology: a literature review. Vet Dermatol. 2009;20:145–56.
29. Fowler ME, Miller RE. Zoo and wild animal medicine. 5th ed. St. Louis: Saunders; 2003.

30. Phair K, Larsen RS, Wack R. Dermatophytosis (*Trichophyton mentagrophytes*) in a Coquerel's sifaka (*Propithecus coquereli*). J Zoo Wildl Med. 2011;42:759–62.
31. Bagnall BG, Grünberg W. Generalized *Trichophyton mentagrophytes* ringworm in Capuchin monkeys (*Cebus nigrivitatus*). Br J Dermatol. 1972;87:565–70.
32. Baker HJ, Bradford LG, Montes LE. Dermatophytosis due to *Microsporum canis* in a rhesus monkey. J Am Vet Med Assoc. 1971;159:1607–11.
33. Keeble EJ, Neuber A, Hume L, Goodman G. Medical management of *Trichophyton* dermatophytosis using a novel treatment regimen in L'Hoest's monkeys (*Cercopithecus lhoesti*). Vet Rec. 2010;167:862–4.
34. Toft JD. The pathoparasitology of nonhuman primates: a review. In: Benirschle, editor. Primates. New York: Springer-Verlag; 1986.
35. Zumpt F, Till W. The mange-causing mites of the genus *Psorergates* (Acarina: *Myobiidae*) with description of a new species from a South African monkey. Parasitology. 1955;45:269–74.
36. Lavoipierre MMJ. A note on the family Psoralgidae (Acari: Sarcoptiformes) together with a description of two new genera and two new species parasitic on primates. Acarologia. 1964;6:342–52.
37. Novak MA, Meyer JS. Alopecia: possible causes and treatments in captive nonhuman primates. Comp Med. 2009;59:18–26.
38. Courtin F, Stone WB, Risatti G, Gilbert K, van Kruiningen JV. Pathologic findings and liver elements in hibernating bats with white-nose syndrome. Vet Pathol. 2010;47:214–9.
39. Vanderwolf KJ, McAlpine DF, Malloch D, Forbes GJ. Ectomycota associated with hibernating bats in eastern Canadian caves prior to the emergence of white-nose syndrome. Northeast Nat. 2013;20:115–30.
40. Lorch JM, Minnis AM, Meteyer CU, Redell JA, White JP, Kaarakka HM, et al. The fungus *Trichophyton redellii* sp. nov. causes skin infections that resemble white-nose syndrome of hibernating bats. J Wildl Dis. 2015;51(1):36–47.
41. Blehert DS, Hicks AC, Behr M, Meteyer CU, Berlowski-Zier BM, Buckles EL, et al. Bat white-nose syndrome: an emerging fungal pathogen? Science. 2009;323:227.
42. Lorch JM, Meteyer CU, Behr MJ, Boyles JG, Cryan PM, Hicks AC, et al. Experimental infection of bats with *Geomyces destructans* causes white-nose syndrome. Nature. 2011;480:376–9.
43. Bello-Gutierrez J, Suzan G, Hidalgo-Mihart MG, Salas G. Alopecia in bats from Tabasco, Mexico. J Wildl Dis. 2010;46(3):1000–4.
44. Scott DW. Color atlas of farm animal dermatology. 2nd ed. Hoboken: Wiley-Blackwell; 2018. p. 295–329.
45. Scott DW, Vogel JW, Fleis RI, Miller WH Jr, Smith MC. Skin diseases in the alpaca (*Vicugna pacos*): a literature review and retrospective analysis of 68 cases (Cornell University 1997–2006). Vet Dermatol. 2011;22:2–16.
46. Hoshina T, Sigiura Y. On skin disease and nematode parasite of a dolphin *Tursiops truncatus* (Montagu, 1821). Scient Whale Res Inst (Tokyo). 1956;1:133–7.
47. Frasca S, Dunn JL, Cookev JC, Buck JD. Mycotic dermatitis in an Atlantic white-sided dolphin, a pygmy sperm whale, and two harbor seals. J Am Vet Med Assoc. 1996;208:727–9.
48. Durden WN, St Leger J, Stolen M, Mazza T, Londono C. Lacaziosis in bottlenose dolphins (*Tursiops truncatus*) in the Indian River Lagoon, Florida, USA. J Wildl Dis. 2009;45 (3):849–56.
49. Symmers W. A possible case of Lobo's disease acquired in Europe from a bottle-nosed dolphin (*Tursiops truncatus*). Bull Soc Pathol Exot. 1983;76:777–84.
50. Vedros NA, Quinlivan J, Cranford R. Bacterial and fungal flora of wild northern fur seals (*Callorhinus ursinus*). J Wildl Dis. 1982;18:447–56.
51. Fansworth RJ, McKeever PJ, Fletcher JA. Dermatomycosis in a harbor seal caused by *Microsporum canis*. J Zoo Anim Med. 1975;6:26–8.
52. Nakagaki K, Hata K, Iwata E, Takeo K. *Malassezia pachydermatis* isolated from a South American sea lion (*Otaria byronia*) with dermatitis. J Vet Med Sci. 2000;62:901–3.

53. Guillot J, Petit T, Degorce-Rubiales F, Guého E, Chermette R. Dermatitis caused by *Malassezia pachydermatis* in a California sea lion (*Zalophus californianus*). Vet Rec. 1998;142:311–2.
54. Phillips PH, Davenport PK, Schultz DJ. *Microsporum gypseum* dermatophytosis in captive Australian sea lions (*Neophoca cinerea*). J Zoo Anim Med. 1986;17:136–8.
55. Pollock CG, Rohrbach B, Ramsay EC. Fungal dermatitis in captive pinnipeds. J Zoo Wildl Med. 2000;31:374–8.
56. Tanaka E, Kimura T, Wada S, Hatai K, Sonoda S. Dermatophytosis in a Steller sea lion (*Eumetopias jubatus*). J Vet Med Sci. 1994;56:551–3.
57. Sós E, Molnár V, Lajos Z, Koroknai V, Gál J. Successfully treated dermatomycosis in California sea lions (*Zalophus californianus*). J Zoo Wildl Med. 2013;44:462–5.
58. Dailey DM, Nutting WB. *Demodex zalophi* sp. nov. (Acari: *Demodicidae*) from *Zalophus californianus*, the California sea lion. Acarologia. 1980;21:423–8.
59. Fowler ME, Miller RE. Zoo and wild animal medicine. 6th ed. St. Louis: Saunders; 2010.
60. Dilbone RP. Mycosis in a manatee. J Am Vet Med Assoc. 1965;147:1095.
61. Woodruff RA, Bonde RK, Bonilla JA, Romero CH. Molecular identification of a Papilloma virus from cutaneous lesions of captive and free-ranging Florida Manatees. J Wildl Dis. 2005;41:437–41.
62. Olivry T, Power H, Woo JC, Moore PF, Tobin DJ. Anti-isthmus autoimmunity in a novel feline acquired alopecia resembling pseudopelade of humans. Vet Dermatol. 2000;11:261–70.
63. Favrot C, Steffan J, Seewald W, Hobi S, Linek M, Marignac G, et al. Establishment of diagnostic criteria for feline nonflea-induced hypersensitivity dermatitis. Vet Dermatol. 2012;23:45–50.e11.
64. Mauldin EA, Morris DO, Goldschmidt MH. Retrospective study: the presence of *Malassezia* in feline skin biopsies. A clinicopathological study. Vet Dermatol. 2002;13:7–14.
65. Ordeix L, Galeotti F, Scarampella F, Dedola C, Bardagi M, Romano E, et al. *Malassezia* spp. overgrowth in allergic cats. Vet Dermatol. 2007;18:316–23.
66. Munro R, Munro HMC. *Lynxacarus* on cats in Fiji. Aust Vet J. 1979;55:90.
67. Beccati M, Gallo MG, Chiavassa E, Peano A. A case of apparent infestation by *Proisotoma* spp. springtails (Collembola: *Isotomidae*) in a cat. Vet Dermatol. 2012;23:157–61.
68. Yager JA, Gross TL, Shearer D, Rothstein E, Power H, Sinke JD, et al. Abnormal sebaceous gland differentiation in 10 kittens ('sebaceous gland dysplasia') associated with generalized hypotrichosis and scaling. Vet Dermatol. 2012;23:136–44.
69. Linek M, Rüfenacht S, Brachelente C, von Tscharner C, Favrot C, Wilhelm S, et al. Nonthymoma-associated exfoliative dermatitis in 18 cats. Vet Dermatol. 2015;26:40–5, e12–13.
70. Wilhelm S, Grest P, Favrot C. Two cases of feline exfoliative dermatitis and folliculitis with histological features of cutaneous lupus erythematosus. Tierarztliche Praxis Ausgabe. 2005;33:364–9.
71. Ruth J. Case presentation #1. Comp Cont Educ Pract Vet. 2009:208–10.
72. Pin D. Connaître et traiter les affections cutanées allergiques chez le cheval. Nouv Prat Vét Eq. 2008;16:7–14.
73. Franc M. Connaître et traiter les principales dermatoses parasitaires chez les équidés. Nouv Prat Vét Eq. 2008;16:15–20.
74. Depecker M, Pin D. Connaître et traiter les pyodermites chez les équidés. Nouv Prat Vét Eq. 2008;16:28–32.
75. Besson B, Cadoré J-L, Pin D. Démarche diagnostique et traitement du pemphigus foliacé chez les équidés. Nouv Prat Vét Eq. 2008;16:37–40.

Part III

Pathogenesis of Dermatophytosis

Experimental Models of Dermatophytosis

Emilie Faway, Marc Thiry, Bernard Mignon, and Yves Poumay

Abstract

Dermatophytosis is a superficial fungal infection of the keratinized structures of the host. Since the last decade, this mycosis became an important health concern due to an increasing prevalence and to the limited number and efficacy of available treatments. Several experimental models have then been developed in order to improve knowledge about this infection and to design new therapeutic strategies. This chapter presents the variety of dermatophytosis experimental models and their contribution in the understanding of mechanisms used by dermatophytes to adhere and to invade the host tissue. Their support to study the establishment of effective antifungal defenses by the host is also summarized. The usefulness of these models for testing the efficacy of antifungal compounds is finally discussed.

Keywords

Dermatophytosis · *Trichophyton rubrum* · *Microsporum canis* · Experimental models · Skin equivalents · Reconstructed human epidermis · Adhesion · Invasion · Host responses · Antifungal efficacy

E. Faway · Y. Poumay (✉)
URPHYM-NARILIS, Faculty of Medicine, University of Namur, Namur, Belgium
e-mail: yves.poumay@unamur.be

M. Thiry
Cell and Tissue Biology Unit, GIGA-Neurosciences, University of Liège, Liège, Belgium

B. Mignon
FARAH, Faculty of Veterinary Medicine, University of Liège, Liège, Belgium

J.-P. Bouchara et al. (eds.), *Dermatophytes and Dermatophytoses*,
https://doi.org/10.1007/978-3-030-67421-2_7

Abbreviations

AMP	Antimicrobial peptides
CFU	Colony-forming unit
GM-CSF	Granulocyte-macrophage colony-stimulating factor
IFNγ	Interferon-gamma
IL	Interleukin
Lap	Leucine aminopeptidase
Mep	Metalloprotease
NET	Neutrophil extracellular trap
PAS	Periodic acid-Schiff
PMN	Polymorphonuclear neutrophils
RFE	Reconstructed feline epidermis
RHE	Reconstructed human epidermis
Sub	Subtilisin protease
TGF	Transforming growth factor
TLR	Toll-like receptors
TNFα	Tumor necrosis factor alpha

1 Introduction

Dermatophytosis is the most common fungal disease in the world with a current prevalence estimated between 20% and 25% of the general population [1–4], but reaching more than 50% among some populations at risk such as diabetic patients or sport practitioners [5, 6]. This infection is caused by specific filamentous and keratinolytic fungi named dermatophytes. Several species are identified and classified into three groups according to their natural host or environment [7]:

- Geophilic species, for example *Nannizzia gypsea* and *Nannizzia fulva* (previously known as *Microsporum gypseum* and *Microsporum fulvum*, respectively), generally feed on keratinized wastes found in the environment and frequently remain nonpathogenic.
- Zoophilic species preferentially infect specific animal hosts but can also infect other animal species including humans. For example, the preferred hosts for *Microsporum canis* and *Trichophyton benhamiae* (previously known as *Arthroderma benhamiae*) are, respectively, cats and rodents.
- Anthropophilic species, as for example *Trichophyton rubrum* and *Trichophyton interdigitale*, exclusively infect humans.

Among all dermatophyte species able to cause infections in humans, *T. rubrum* is responsible for a vast majority of dermatophytosis in humans: indeed, depending on the geographical areas, between 50% and 90% of cases are due to *T. rubrum* [2, 8].

In immunocompetent patients, dermatophytosis affects the keratinized structures of the host, namely hairs, nails, and epidermis, resulting in superficial lesions with erythema, dryness, and desquamation as clinical symptoms [9, 10]. The severity of lesions partly depends on the species involved, zoophilic and geophilic dermatophytes usually inducing more severe inflammatory responses in human hosts in comparison with anthropophilic species that are well adapted to humans as a consequence of more efficient mechanisms for immunoevasion [11–13]. Anyway, lesions resulting from dermatophytosis cause pain and discomfort and their anesthetic appearance is clearly responsible for individual shame and low self-esteem in concerned patients, along with reduced social interactions and quality of life [14–16].

Despite their overwhelming prevalence, dermatophytosis has been yet poorly studied. Thus, many questions regarding the mechanisms deployed by dermatophytes to invade host tissues remain to be addressed. Similarly, other questions concerning the establishment of an adequate immune response in infected host require further investigation. Moreover, while antifungal agents currently available are effective against dermatophytes, these drugs have several limitations such as extended duration of the treatment, associated toxicity, especially encountered when oral administration is needed, and finally emergence of resistant strains [17].

During the last decades, several in vivo, in vitro, and ex vivo experimental models of dermatophytosis have been developed in order to answer outstanding questions. This chapter aims to describe various models useful for studying dermatophytosis, with a special focus on skin equivalent models, to depict their current contribution in the understanding of the multiple steps involved in host infection by dermatophytes, as well as to present their potential use in performing efficacy testing of antifungal compounds.

2 Experimental Models of Dermatophytosis

The increase in dermatophytosis prevalence observed since the last decade has gradually raised the interest of the scientific community for the study of this infection. This has led to the development and diversification of experimental models of dermatophytosis over the last few years.

2.1 Aleurioconidia or Arthroconidia as Infective Elements Initiating Dermatophytosis

Spores are quiescent fungal unicellular particles characterized by high mechanical resistance and low metabolic activity. They represent the initial stage of fungal development, able to reactivate, germinate, and produce new mycelium when environmental conditions are favorable [18]. Spores are physiologically produced by fungi to ensure survival in adverse conditions, as well as dispersion in the environment of individuals which can adhere to host tissue before starting a new infectious process. Dermatophytes are able to produce two kinds of spores: aleurioconidia arising terminally or laterally from the hyphae (microconidia corresponding to one-celled aleurioconidia and macroconidia to several successive cells detaching together at the end of hyphae), and arthroconidia resulting from fragmentation of hyphae. Transmission and scanning electron microscopy analysis performed on aleurioconidia and arthroconidia adhering over corneocytes in suspension revealed that the cell wall of arthroconidia is thicker than that of aleurioconidia, and that germination of arthroconidia occurs faster after adhesion [19]. Moreover, arthroconidia appear more resistant to certain antifungal drugs (e.g., fluconazole, griseofulvin, itraconazole) than microconidia [20]. In addition to these intrinsic differences, aleurioconidia, although they are extensively produced by dermatophytes cultured in vitro, have never been observed in vivo on lesions [21]. Conversely, arthroconidia are efficiently produced in vivo by dermatophytes.

In order to create adequate models for dermatophytosis, the use of arthroconidia as infective elements seems therefore more appropriate. In practice, arthroconidia can be produced in vitro following the procedure described by Tabart et al. [22]. Practically, after approximately 2-week growth on nutrient-rich agar, the fungal mass is recovered by scrapping, and cultured for additional 2–3 weeks in conducive conditions that combine a nutrient-poor culture medium and a 12% CO_2 atmosphere. Unicellular elements, corresponding to arthroconidia, are finally isolated after agitation and filtration of the fungal material. However, aleurioconidia are used in most studies (e.g., [12, 23–26]), because of the ease and speed of their production in vitro: only 2-week culture on nutrient-rich agar before agitation and filtration steps. To date, only a few studies have been performed using arthroconidia (e.g., [19, 27–31]).

2.2 In Vitro, Ex Vivo, and In Vivo Models of Dermatophytosis

In the past, simple models of dermatophytosis were developed to describe the early steps of the infection, for instance, dermatophytes adhering to human corneocytes, either isolated after skin scrapping or in suspension, or invading sheets of cornified layer collected by the tape-stripping method [19, 30, 31]. More recently, models using nail or hair fragments infected by dermatophytes allowed the investigation of the mechanisms used by these fungi to degrade keratin [23] or permitted to characterize the expression of virulence factors by dermatophytes [24]. Although such models are suitable to monitor dermatophyte adhesion, growth, and, to a lesser extent, invasion, the absence of living keratinocytes impedes any study about the host tissue responses. Complementarily, in vitro models of keratinocytes [12, 26, 32, 33], polymorphonuclear neutrophils (PMN) [28], or macrophages [34] cultured in the presence of dermatophytes were designed to overcome those limitations. In vitro models allow the evaluation of expression and release of cytokines and antimicrobial peptides (AMP), together with the expression of toll-like receptors (TLR) or of co-stimulatory molecules by those cell types. However, even keratinocytes cultured as monolayers cannot be used to accurately model the epidermal adhesion and invasion processes followed by dermatophytes since they lack keratinized material.

Indeed, an ideal model to study dermatophytosis should allow simultaneous analysis of the infection steps (i.e., adhesion, germination, and tissue invasion) used by dermatophytes on one hand, and analysis of the host responses that trigger the recruitment and activation of the antifungal immune defenses on the other. In addition, such a relevant model might provide an appropriate tool to perform efficacy and toxicity assays of antifungal compounds. Ex vivo models of skin explants and in vivo animal models easily fulfill these requirements and appear therefore as promising solutions. To date, skin explants from several animal species (e.g., cats, Guinea pigs, mice) as well as from human beings have already been used to study adhesion and invasion by dermatophytes [35–39] or to characterize their expression of potential virulence factors [24]. Nonetheless, the use of skin explants, especially from humans, is quite restricted by limited availability, poor standardization of samples (for instance, regarding hairiness or body area), and large variability between the donors.

Animal models, mainly based on Guinea pigs or mice experimentally infected by dermatophytes, have been used to study steps of infection (e.g., [35, 40, 41]) or to analyze the establishment of host immune responses (e.g., [25, 29, 42]). Several animal models used to study dermatophytosis have been recently reviewed in details by Cambier et al. [43]. Nevertheless, an accurate model of dermatophytosis should mimic infection by a specific dermatophyte on tissue from an adapted host. Thereby, studying the infection of animal tissue by typical anthropophilic dermatophytes is particularly irrelevant. Indeed, there is no report of natural infection of nonhuman epidermal tissue by anthropophilic species and experimental trials devoted to infect animal tissue by the same species remain complicated because anthropophilic dermatophytes are poorly adapted to adhere and invade nonhuman tissues. Since the anthropophilic *T. rubrum* species is the most common dermatophyte involved in human infections [2, 8], the availability of an accurate model of infection for this dermatophyte species is required. Despite the technical limitations mentioned, models of *T. rubrum* dermatophytosis using mouse [44, 45], Guinea pig [46–48], or rat [49, 50] have nonetheless been designed. Interestingly, repeated applications of spores, previous abrasive treatments, or injection of corticosteroids prior to and after the infection are clearly required in order to obtain significant lesions. Anyway, even if they are of a considerable interest to gain knowledge about activation of the immune system in the host organism in response to dermatophyte infection, one must keep in mind that critical differences between nonhuman and human epidermis might influence data interpretation. Furthermore, the animal models have obvious ethical concerns. Therefore, there is a strong case for the development of alternative models.

2.3 Skin Equivalents to Create Dermatophytosis Models

Human skin equivalents obtained in culture, including reconstructed epidermis alone or epidermis reconstructed on a dermal equivalent generally made of collagen lattice with fibroblasts, are currently the closest in vitro models that mimic cutaneous human tissues. In such models, human keratinocytes cultured at the air–liquid interface, in an environment containing appropriate growth factors and elevated Ca^{++} concentration, undergo a complete program of differentiation that creates a stratified reconstructed tissue covered by a keratinized layer, quite similar to the human epidermis [51]. Infection models developed on skin equivalents allow studies of interactions between host tissue and pathogens, i.e., analysis of the infection process, identification of responses triggered in keratinocytes, and characterization of potential alterations induced in the tissue function. Notably, studies about cutaneous infections by the yeast *Candida albicans* [52], *Staphylococcus aureus* bacteria [53], or helminths [54] have already proven the usefulness of skin equivalents to investigate the mechanisms underlying infection.

Several reports already described infection of skin equivalents by dermatophytes. In 1995, a model involving infection by *Trichophyton mentagrophytes* arthroconidia of reconstructed human epidermis (RHE) has demonstrated for the first time the ability of dermatophytes to adhere onto reconstructed tissues and to invade their cell layers [55]. Later on, a model of infection by *M. canis* arthroconidia on reconstructed feline epidermis (RFE) was created to allow investigation of the adhesion mechanisms used by this dermatophyte on its natural preferred host [22]. More recently, two models of dermatophytosis based on commercially reconstructed skin tissue EpiDerm (MatTek) [26] or EpiSkin® [56] were reported. The first model was explored to evaluate the release of cytokines by keratinocytes during the infection process by several dermatophyte species, including the anthropophilic *T. rubrum*. Characterization of the signaling pathways simultaneously involved was also undertaken [26]. By mean of morphological analysis, the second model of dermatophytosis on EpiSkin® illustrated the different steps of the infection process by *T. rubrum* [56]. Although both models brought interesting insight and information about dermatophytosis and its pathogenesis, they used aleurioconidia as infective elements. Since aleurioconidia are only produced by dermatophytes cultured in vitro and have never been observed in vivo [21], those models cannot be considered fully representative of the in vivo conditions of infection.

Our team designed a model of dermatophytosis using RHE grown on polycarbonate filter and then infected by *T. rubrum* arthroconidia [27]. Practically, infection of RHE is initiated by topical application of *T. rubrum* arthroconidia suspended in phosphate-buffered saline to reach a final density of 1700 arthroconidia per cm^2. After 4 h of exposure, washes are performed in order to eliminate non-adherent arthroconidia from the apical surface of RHE and to expose them again to the air–liquid interface. Infected RHE are then cultured during four additional days. Morphological analysis of infected RHE revealed that arthroconidia were able to rapidly adhere to the surface of corneocytes and to produce hyphae that progressively invaded the cornified layer (Fig. 1a). When infected RHE were maintained in culture

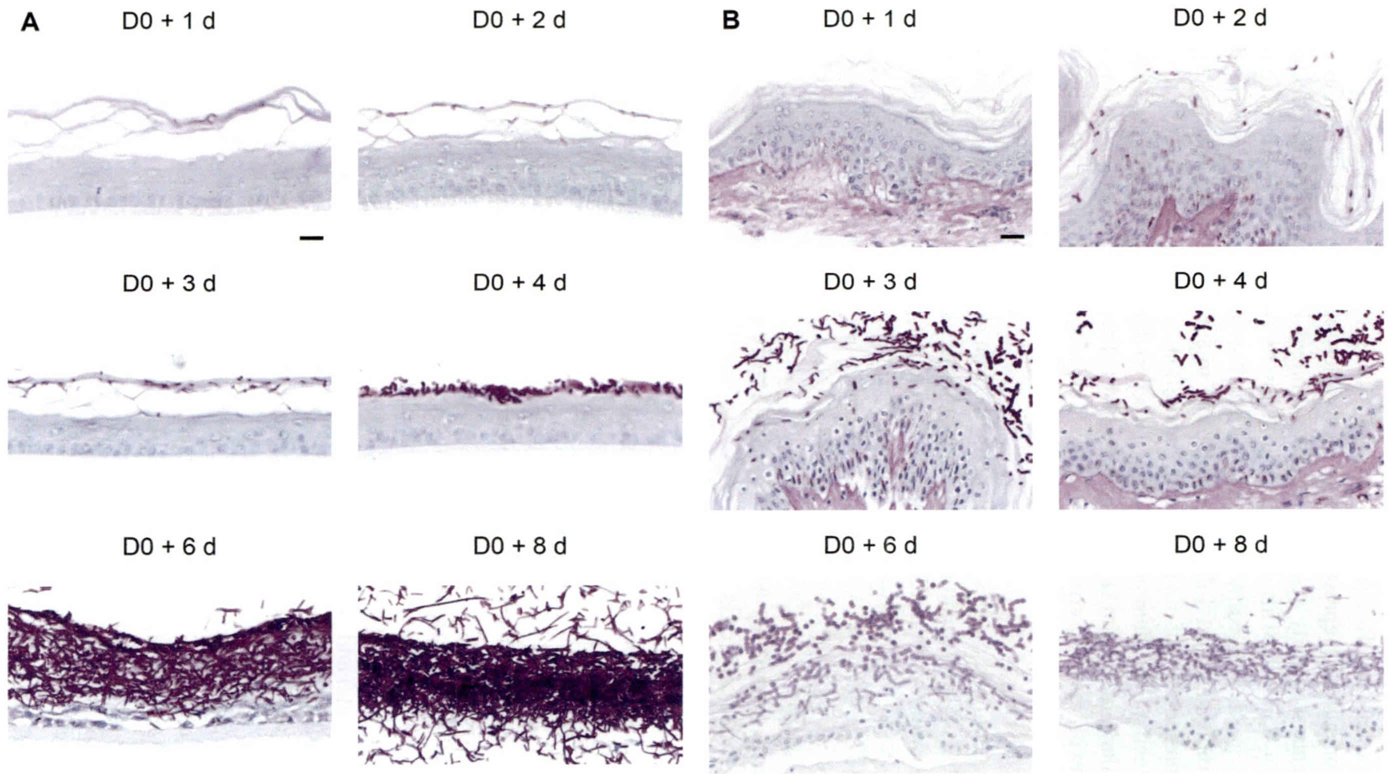

Fig. 1 In vitro infection of reconstructed human epidermis or of human skin explants by *Trichophyton rubrum* dermatophytes. Periodic acid-Schiff (PAS) staining with hemalun counterstaining of histological sections performed on infected (**a**) reconstructed human epidermis (RHE) or on (**b**) human skin explants, chemically fixed and processed for histology, one (D0 + 1 day), two (D0 + 2 days), three (D0 + 3 days), four (D0 + 4 days), six (D0 + 6 days), or eight (D0 + 8 days) days after exposure to *T. rubrum* arthroconidia. Histological sections of RHE were pretreated with α-amylase before the PAS staining. Normal human skin explants were obtained from abdominal plastic surgery after informed consent of patient (Drs. B. Bienfait and J.S. Blairvacq, Clinique St. Luc, Namur-Bouge, Belgium). Scale bars = 20 μm

for longer periods, hyphae progressed deeper between layers of the epidermis until entire invasion of the RHE, resulting in its total disorganization. This excessive invasion is for sure not representative of in vivo lesions since the progression of fungal elements is usually restricted to the cornified layer in naturally infected normal human skin [10]. This difference can likely be explained by the absence of immune cells in RHE-based models. Therefore, culture of infected RHE is systematically interrupted at latest 4 days after exposure to arthroconidia in order to keep the model representative of in vivo lesions. In addition, using the protocol developed for RHE infection, we demonstrated that infection by *T. rubrum* arthroconidia happens similarly on human skin explant (Fig. 1b). This model based on infection of RHE was explored to describe the progressive steps of infection [27], as well as to identify the cellular responses of host keratinocytes and the alterations of the barrier in the infected RHE [57, 58].

The aforementioned models of dermatophytosis on skin equivalent have proven their usefulness in the evaluation of infectious processes and of human keratinocytes responses, as well as in the identification of signaling pathways triggered during these responses. In addition, such models are valuable to test efficacy of antifungal compounds [27, 55, 59].

2.4 Complementary Models for the Study of Experimental Dermatophytosis

Every model described earlier allows advances in the study of dermatophytosis, but each of them is also more appropriate for the evaluation of certain aspects of the problem (Table 1). Thus, choosing a model depends essentially on the question addressed.

Adhesion is the initial step of infection during which fungal cells adhere to host substrate (i.e., keratinized structures). Ex vivo models of corneocytes in suspension and cornified layer sheets, or nail and hair fragments remain easy and adequate for the study of adhesion processes without any particular interference. Adhesion can also be studied using skin explants, skin equivalents, or animal models, but activation of host living cells (e.g., through production of AMP or recruitment of other immune components) and desquamation may influence the adhesion process. Invasion of the host tissue and consequences onto the epidermal barrier function can be evaluated in a similar manner using skin explants, skin equivalents, or animal models.

About the understanding of host tissue responses, through investigation of expression and release of cytokines, AMP, or other factors, in vitro, ex vivo, and in vivo models provide different and complementary information. In vitro models of cells cultured in suspension, as monolayers or inside RHE, allow monitoring the activation of specific cell types (e.g., keratinocytes, PMN, macrophages) independently of the others. Conversely, evaluation of the tissue reaction as a whole, including the activation of keratinocytes and of immune cells, either innate or adaptive, requires the use of in vivo animal models. For their part, ex vivo models

Table 1 Use of different experimental models to study specific aspects of the pathogenesis of dermatophytosis

	Adhesion	Invasion and effect on the epidermal barrier	Activation of the immune system	Activation of individual cell type	Antifungal efficacy
Ex vivo models					
Corneocytes, cornified layer sheets, nail and hair fragments	+				±
Skin explant	+	+	±	±	+
In vitro models					
Culture of keratinocytes, PMN, or macrophages				+	
Skin equivalents reconstructed in culture	+	+		+	+
In vivo models					
Animal models	+	+	+		+

using skin explants can be used to study the local responses given by either keratinocytes or cells of the innate immunity located in the epidermis and dermis of the explant.

Finally, testing efficacy of antifungal compounds is performed using ex vivo model of skin explant, in vitro model of skin equivalent, or in vivo animal model. Indeed, these three models cover the analysis of complete infection processes by dermatophytes and allow to evaluate the efficacy of antifungal compounds, whatever their mode of action. To a lesser extent, ex vivo models using corneocytes in suspension, cornified layer sheets, or nail and hair fragments, can be used to monitor the efficacy of compounds targeting the adhesion processes deployed by dermatophytes.

3 Experimental Models to Study Infectious Processes During Dermatophytosis

Dermatophytes infect host tissues through three successive steps: adhesion, germination, and invasion (Fig. 2a). In reaction to tissue invasion by dermatophytes, host cells become activated, most likely in order to initiate antifungal defenses and counteract the infection. Mechanisms involved during the infectious process as well as cellular responses triggered in host tissue can be both investigated by mean of experimental models of dermatophytosis, as mentioned above and depicted in details hereunder.

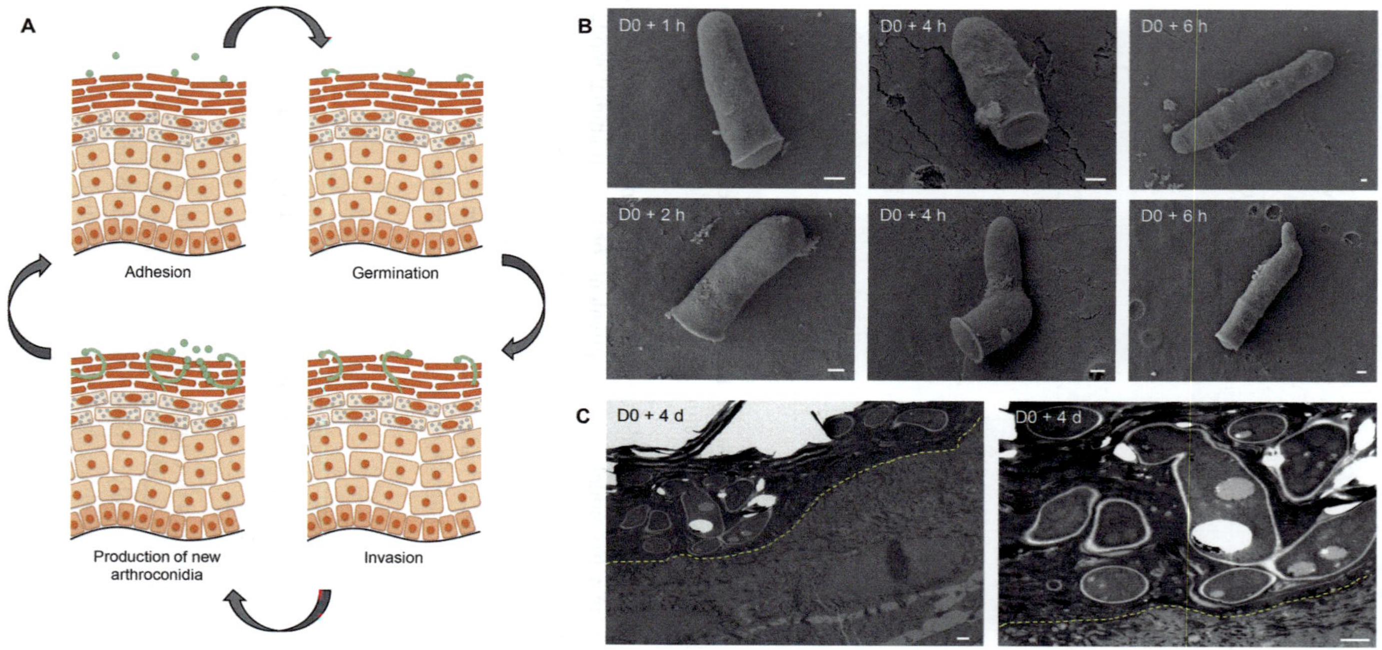

Fig. 2 Dermatophytes infect epidermis through four steps. (**a**) Schematic representation of the different steps observed during infection of epidermis by dermatophytes: adhesion of arthroconidia to tissue surface, germination of arthroconidia to become elongating hyphae, invasion of the cornified layer by hyphae, and finally production of new arthroconidia by hyphal fragmentation. (**b**) Adhesion and germination of *Trichophyton rubrum* arthroconidia over the surface of reconstructed human epidermis (RHE) observed by scanning electron microscopy one (D0 + 1 h), two (D0 + 2 h), four (D0 + 4 h), or six (D0 + 6 h) hours after exposure of the tissue to arthroconidia. Scale bars = 1 μm. (**c**) Invasion through the cornified layer of RHE by hyphae observed by transmission electron microscopy 4 days after infection (D0 + 4 days) by *T. rubrum* arthroconidia. Yellow dotted lines indicate the limits between cornified and granular layers. Scale bars = 1 μm

3.1 Adhesion to the Host Tissue

Adhesion is the first step of tissue infection by dermatophytes and involves close contact between arthroconidia and the surface of the host epidermis. Notably, analysis by electron microscopy have revealed that arthroconidia or microconidia from *T. mentagrophytes* and *T. interdigitale*, inoculated on human skin explants, produce fibrils that connect them to the epidermal surface [30, 38, 39].

3.1.1 Adhesion Is an Early Process in the Infection Development

Several studies performed on different experimental models of dermatophytosis have characterized the adhesion kinetics of dermatophytes to the host tissue. By light microscopy, Zurita and Hay [19] observed arthroconidia from three different *Trichophyton* species (i.e., *T. rubrum*, *T. interdigitale*, and *T. quinckeanum*) adhering to corneocytes in suspension. Their observations revealed that adhesion occurs as soon as after 2 h of contact, reaching a maximum after 4 h. Moreover, those authors report increased adhesion of *T. interdigitale* and *T. rubrum* to corneocytes from plantar skin, whereas the adhesion to corneocytes isolated from forearm skin is weaker, suggesting that the strength of dermatophyte adhesion varies upon body location. Accordingly, arthroconidia from *T. interdigitale* or *T. mentagrophytes* adhere to cornified layer sheets as soon as 1 h after tissue exposure and their adhesion increasingly occurred during the first 6 h of contact [30]. Other studies using microscopy further showed that adhesion of *T. mentagrophytes* on ex vivo models based on human nail fragments [60] or on human skin explants [38] started, respectively, 6 or 12 h after infection. Finally, adhesion of *M. canis* arthroconidia to the surface of RFE can be observed under a fluorescent light microscope after labeling arthroconidia with Calcofluor White. In this case, adhesion was shown starting within 2 h and increased up to 6 h after infection [61].

More recently, adhesion assays based on colony-forming units (CFU) counting have been developed. In one method, arthroconidia adhering to skin explants from human or other animal species were recovered by scrapping and were seeded over Sabouraud agar for a few days at 27 °C in order to assess the number of CFU which were adherent [35, 36]. This kind of assay demonstrated strong adhesion of *M. canis* arthroconidia over skin explants after 4 h of exposure. On the other hand, a second method developed using an in vitro model of dermatophytosis on RHE, rather assessed the number of non-adherent arthroconidia in a defined number laid over the epidermal tissue [27]. Practically, non-adherent arthroconidia were recovered by several washes of the tissue surface before being seeded over Sabouraud agar and counting of CFU as above. Knowing the initial number of arthroconidia entering in contact with the RHE, the number of adherent arthroconidia was determined by simple calculation. By using this method, it was shown that arthroconidia of *T. rubrum* start to adhere to RHE as soon as 1 h after infection and that *T. rubrum* adhesion increases for the first 24 h of exposure.

Altogether, these data suggest that adhesion is an early process in infection development and that it increases over time.

3.1.2 Surface Molecules and Secreted Proteases Are Required for Adhesion

Currently, the precise mechanisms used by dermatophytes to adhere to the host tissues are still incompletely understood. However, pieces of information have been obtained by the mean of experimental models of dermatophytosis. Interestingly, adhesion seems to rely on complex processes that simultaneously involve molecules expressed at the surface of arthroconidia and secreted proteases (reviewed by Baldo et al. [62]).

In such context, the study of interactions between *T. mentagrophytes* or *T. rubrum* with mutant Chinese ovary epithelial cells expressing various terminal carbohydrates on their cell surface revealed that mannose and galactose-binding proteins are present on the surface of microconidia [63, 64]. Accordingly, Bitencourt et al. [65] observed that *T. rubrum* aleurioconidia express a gene encoding an adhesin-like protein when they are cocultured with human keratinocytes. As other fungal adhesins [66–68], this adhesin-like protein contains a central domain with a tandem repeat sequence whose length and number of repeats likely influence adhesion ability of the different dermatophyte species or strains.

In addition, dermatophytes express a huge number of proteases, including subtilisin proteases (Sub), metalloproteases (Mep), and leucine aminopeptidases (Lap), which seem all implied in adhesion and invasion processes [62, 69]. Since the proteases Sub1, Sub3, Sub4, Lap1, Lap2, and Mep4 are highly expressed by dermatophytes when they are cultured in a protein-rich medium [40, 41, 65, 70, 71], they were long ago considered as virulence factors. Recent studies though have shown that proteases expressed by dermatophytes differ during in vivo infection: for instance, Sub6 is the main protease expressed by *T. benhamiae* during experimental infection on Guinea pig [40, 41], and by *T. rubrum* inside onychomycosis lesions naturally induced in humans [72]. Besides, even if Sub3 is not the main protease expressed in vivo, Sub3 was detected by immunohistochemistry in the skin of Guinea pig experimentally infected with *M. canis* [73]. Such a difference between in vitro and in vivo studies underlines the importance of suitable experimental models.

Proteases of the subtilisin family, especially Sub3, have been shown to have a major role in dermatophyte adhesion. Indeed, adhesion of *M. canis* arthroconidia to RFE surface is reduced by the chymostatin inhibitor which targets serine-proteases including subtilisins, and by specific antibody against Sub3, although in a lesser extent [61]. In addition, the adhesion of a *M. canis* strain invalidated for Sub3 by RNA silencing [74] to cutaneous explants from humans or other animals is decreased by comparison with a control strain of *M. canis* expressing Sub3 [35, 36]. Using an experimental model of dermatophytosis on RHE [27], we confirmed the involvement of subtilisin proteases in the adhesion of *T. rubrum* to the epidermal surface (Fig. 3). First, the inhibiting activity of chymostatin on that of Sub3 was verified, either in the absence and in the presence of the RHE, thanks to an assay based on an artificial substrate for this protease (*N*-succinyl-Ala-Ala-Pro-Phe-P-nitroanilide Sigma cat. no. S7388) (Fig. 3a). In addition, the viability of RHE was controlled when exposed to chymostatin (Fig. 3b). Infection of RHE was then

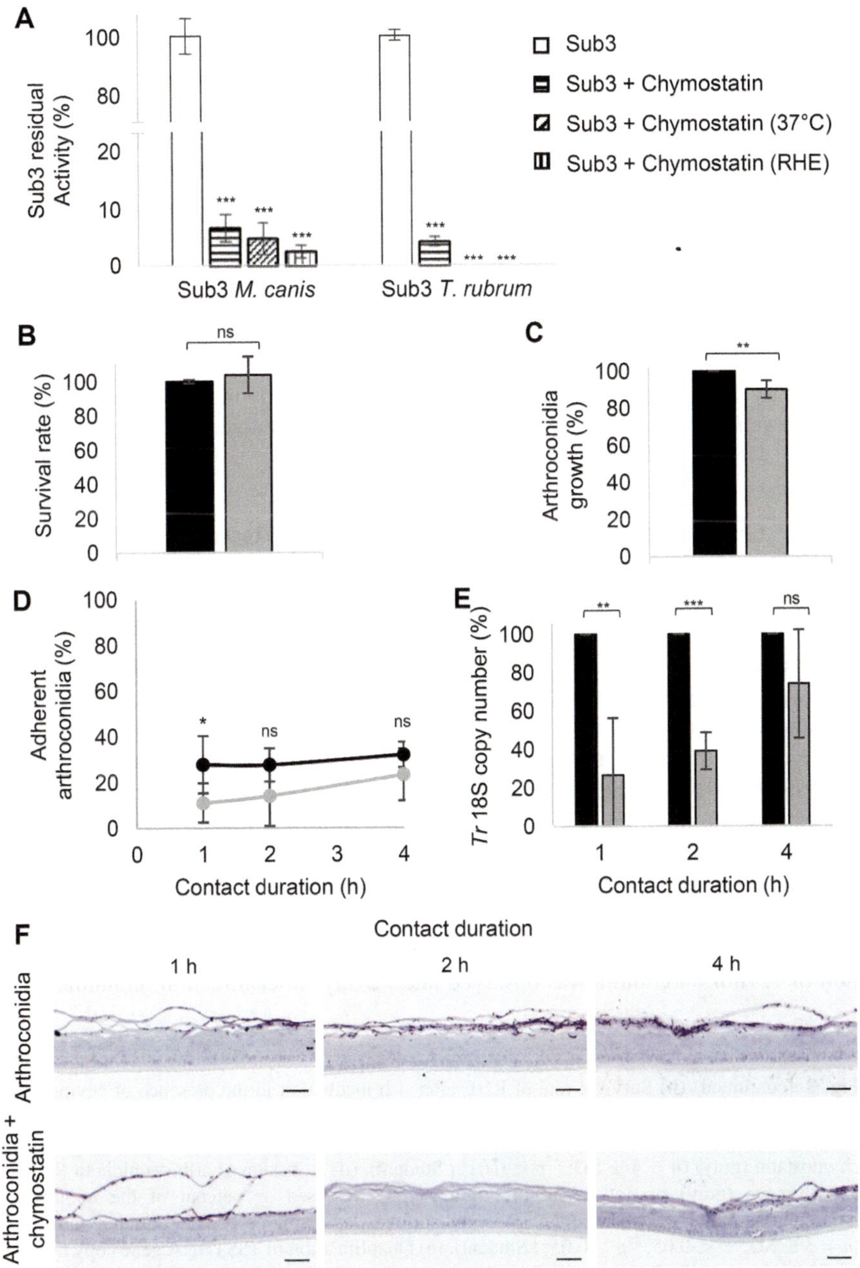

Fig. 3 Serine proteases are involved in the adhesion process of arthroconidia to host epidermis. (**a**) Residual activity of recombinant subtilisin-3 (Sub3) serine proteases from *Trichophyton rubrum* or *Microsporum canis* alone (white), in the presence of 100 μM chymostatin either under optimal conditions (horizontal hatchings), or after 4 h incubation at 37 °C (diagonal hatchings) or after 4 h incubation in topical application on reconstructed human epidermis (RHE) at 37 °C (vertical hatchings) ($n = 3 \pm$ SD; $^{***}p < 0.001$ in comparison to the activity of Sub3 alone; ANOVA1).

performed by topical exposure to *T. rubrum* arthroconidia, in the presence or not of chymostatin (100 μM), followed by washes of their surface in order to eliminate non-adherent arthroconidia after 1, 2, or 4 h of infection. Non-adherent arthroconidia recovered by washing were numbered as described above after seeding on Sabouraud agar, taking into account the effect of chymostatin on the growth of arthroconidia (Fig. 3c), in order to calculate the percentage of adherent arthroconidia (Fig. 3d). Adhesion is slightly reduced by the presence of chymostatin, especially when contact duration was as short as 1 h. This decrease in adhesion resulted, 4 days after infection, in a decrease in the number of fungi detected on the epidermal surface, as assessed by PCR quantification of the copy number of *T. rubrum* ribosomal DNA 18S (Fig. 3e) and by histological analysis after periodic acid-Schiff (PAS)-staining of infected RHE (Fig. 3f). Therefore, proteases of the subtilisin family appear clearly involved in the adhesion of dermatophytes to the host tissue, even if mechanisms promoting adhesion are still unclear.

3.2 Germination and Invasion of the Host Tissue

After adherence to host tissue, arthroconidia initiate germination. Once arthroconidia adhere to the host tissue, they are able to perceive that this environment is favorable for their growth and they initiate the germination step during which they may swell before producing germ tubes that will elongate to become hyphae [75]. Germination is initiated when arthroconidia produce a germ tube with a length at least equivalent to the one of the initial arthroconidia. In ex vivo models based on corneocytes in suspension [19] or on sheets of cornified layer [31], the germination of arthroconidia, respectively, from *T. interdigitale* and *T. mentagrophytes*, was observed after 4 h of incubation. Simultaneously though, aleurioconidia of *T. interdigitale* adherent to suspended corneocytes do not exhibit germination yet [19]. In accordance with a delayed germination of aleurioconidia, *T. mentagrophytes* microconidia inoculated on human skin explant seemed to start germination after 24 h [38], while germination of *T. rubrum* conidia was observed later, 2 days after infection on human skin

---

Fig. 3 (continued) (**b**) Survival rate of RHE after 4 h incubation in the presence of chymostatin (gray) compared with control untreated RHE (black) ($n = 3 \pm$ SD; $^{ns}p \geq 0.05$; *t* Student). (**c**) Growth of arthroconidia on Sabouraud agar in untreated conditions (black) or in the presence of chymostatin (gray) ($n = 4 \pm$ SD; $^{**}p < 0.01$; *t* Student). (**d**) Adhesion of arthroconidia to RHE in the presence (gray) or not (black) of chymostatin expressed as percent of the number of arthroconidia initially applied on the epidermal surface, depending on contact duration with RHE ($n = 3 \pm$ SD; $^{*}p < 0.05$, $^{ns}p \geq 0.05$; *t* Student). (**e**) Quantification of 18S rDNA gene copy number of *T. rubrum* by qPCR after total DNA extraction from infected RHE performed 4 days after infection by arthroconidia in the presence (gray) or not (black) of chymostatin, depending on the duration of contact with RHE ($n = 3 \pm$ SD; $^{**}p < 0.01$, $^{***}p < 0.001$, $^{ns}p \geq 0.05$; *t* Student). (**f**) Periodic acid-Schiff (PAS) staining with α-amylase pretreatment and hemalun counterstaining of histological sections prepared from RHE 4 days after their infection by *T. rubrum* arthroconidia, after 1, 2, or 4 h of exposure in the presence or not of chymostatin. Scale bars = 50 μm

equivalent [56]. By scanning electron microscopy performed on infected RHE (Fig. 2b), one can observe *T. rubrum* arthroconidia adhering to the tissue surface as early as after 1 h of contact. Later on, arthroconidia that become slightly swollen and germinate can be seen after 4 h. Elongating septate hyphae formed from germinated arthroconidia are observed after 6 h and become able to penetrate the host tissue during the invasion step.

During invasion, hyphae progress through the host tissue while continuing to elongate. Initial microscopic observations suggested that dermatophytes invade the host tissue by progressing through intercellular spaces without causing extensive damages to the cells. Indeed, 72 h after infection, *T. mentagrophytes* arthroconidia were observed invading sheets of cornified layer [31] or nail fragments [60] by progressing into intercellular spaces. In accordance, 3 days after infection, hyphae from *T. mentagrophytes* were found penetrating human skin explants between cells of the cornified layer and separating them [38]. In the dermatophytosis model on RHE developed by our team [27], we also observed using transmission electron microscopy analysis of infected tissues that hyphae are present in intercellular spaces between corneocytes 4 days after exposure to *T. rubrum* arthroconidia (Fig. 2c). These results suggest that dermatophytes are able to degrade intercellular junctions (e.g., corneodesmosomes or tight junctions) and/or the extracellular lipid matrix. In addition, these observations also suggest that hyphal elongation is orientated according to the physical and topographical features of the substrate in order to facilitate its invasion, a phenomenon known as "thigmotropism" [76]. This ability of dermatophytes to orientate their growth was also described by Perera et al. [77] who observed the growth of *Epidermophyton floccosum*, *M. canis*, and *T. mentagrophytes* on artificial membranes.

In addition to the above-described properties, dermatophytes secrete several proteases that exhibit specificity toward hard keratin substrates [69], rendering these fungi able to invade host tissue directly through corneocytes. Accordingly, scanning electron microscopy revealed penetration of corneocytes by *T. mentagrophytes* hyphae, resulting in damage to the corneocyte surface 21 h after the infection of cornified layer sheets [31]. Moreover, Jensen et al. [78] observed hyphae inside the corneocytes on human skin biopsy from *Tinea corporis* lesion. By infecting nail or hair fragments with various mutants of *T. benhamiae*, the critical role of the enzyme cysteine dioxygenase Cdo1 and of the sulfite efflux pump SSu1 in the degradation of keratinized structures by dermatophytes could be demonstrated [23]. Actually, thanks to Cdo1 and Ssu1, sulfite are produced by dermatophytes from environmental cysteine and secreted. As a reducing agent, sulfite can then cleave keratin-stabilizing cysteine bonds, rendering keratin more accessible for its degradation by proteases. Growth of strains deficient for one of the two (Cdo1 or Ssu1) proteins was indeed impaired on hair or nail fragments. Besides, Sub3, which seems involved in adhesion processes, is not required for invasion of the host tissue by dermatophytes. Indeed, arthroconidia from a *M. canis* strain invalidated for Sub3 produced hyphae able to invade the epidermis of experimentally infected Guinea pigs when the adhesion deficiency is artificially compensated by poloxamer 407, increasing the remanence of arthroconidia at the infection site [35].

It therefore appears that dermatophytes invade the host tissue by progressing simultaneously between and through corneocytes. Anyway, invasion of the cornified layer leads to alterations of the epidermal barrier, as observed on skin biopsy collected from *Tinea corporis* lesions [78]. Alterations of the epidermal barrier in response to dermatophyte infection of the in vitro RHE model were characterized by measurement of transepithelial electrical resistance, by assessing the outside-in permeability, together with the inside-out permeability [58]. While the two former assays confirmed loss of integrity in the epidermal barrier 4 days after the infection of RHE, the latter assay indicated that this loss can be partly due to the disorganization of tight junctions.

Finally, it is important to note that the infection by dermatophytes in vivo remains superficially localized in the keratinized structures (i.e., cornified layer of the epidermis, nails, and hairs), their progression into living tissues being probably blocked by the host immune system [7, 10]. However, in dermatophytosis models based on skin equivalent [27, 56] or on skin explant [38], the invasion reaches deeper epidermal layers in a few days, leading to the disorganization of the host tissue. This can be explained by the total absence of immune cells in skin equivalents, whereas cells from the innate immunity are sometimes present in skin explants, even if adaptive immunity is nonetheless missing. Those observations highlight the importance to monitor the development of infection in experimental models in order to perform assays and measurements when invasion is similar to that observed in vivo during natural infection.

3.3 Host Responses Against Dermatophytosis

Host responses against infecting dermatophytes include both activation of local cells and recruitment and activation of immune cells. Additionally, some dermatophytes also deploy several mechanisms to evade or silence host immune response.

3.3.1 Recruitment of the Immune System Through Activation of Local Cells

Experimental animal models are necessary to study the entire recruitment and activation of the host immune system in response to dermatophytosis. Firstly, Green et al. [79] demonstrated the requirement of cell-mediated immunity to fight against dermatophytosis by observing that nude mice experimentally infected by *T. mentagrophytes* were unable to heal. In accordance, Calderon and Hay [80] showed that transfer of T cells from mice acutely infected by *T. quinckeanum* into naive mice resulted in their protection against infection by this fungus. Conversely, serum transfer did not confer any protection. Later, secretion of interferon-gamma (IFNγ) and recruitment of macrophages were detected in the skin of wild-type mice experimentally infected by *T. rubrum*, whereas IFNγ or interleukin (IL)-12 knockout mice presented higher fungal burdens and lower macrophage recruitment than wild-type animals [45]. Those observations suggested that the Th1 immune response, implying IFN-γ production and macrophages activation, was the effective response allowing to control and resolve infection by dermatophytes [81, 82].

However, emerging data tend to show that the Th17 immune response could also be required for the control of dermatophyte infection [25, 29]. Indeed, using mouse models based on *T. quinckeanum* [83], *T. mentagrophytes* [84], or *T. benhamiae* [42], it was shown that PMN were recruited in large numbers to the infection site. Secretion of cytokines involved in the Th17 response (i.e., IL-6, IL-17A, IL-23, and transforming growth factor (TGF)-β) was also reported during experimental infection by *T. mentagrophytes* and *T. benhamiae* on mouse models [42, 84]. The involvement of the Th17 immune response in dermatophytosis clearance was finally clearly demonstrated in a recent study comparing *M. canis* infection on wild-type or IL-17RA or IL-17A/F-deficient mice: while it remains superficially localized in the cornified layer of wild-type mice, *M. canis* extensively colonizes the epidermis when the Th17 pathway is dysfunctional [25].

Interestingly, the recent report by Heinen et al. [29] reconciled all data since they observed that T cells isolated from the skin-draining lymph nodes of mice experimentally infected by *T. benhamiae* exhibit both Th1 and Th17 differentiation as assessed through cytokines production (IL-17A, IL-22, IFNγ) and transcription factors mRNA expression (retinoic acid receptor-related orphan receptor γt and T-box transcription factor). In addition, they showed that fungal clearance and clinical recovery are lower in IFNγ and IL-17A double-deficient mice than in IFNγ or IL-17A single-deficient mice, suggesting complementary roles of Th1 and Th17 immune responses.

On another hand, some studies performed on simpler models, such as cells cultured either in suspension or in monolayers, as well as skin equivalents reconstructed in culture, allowed to monitor the specific activation of various cell types in response to dermatophytosis. Keratinocytes are the first cells to encounter dermatophytes and to react to their presence by production of cytokines and AMP. Indeed, human keratinocytes cultured as monolayers, in the presence of different dermatophyte species, exhibit release of several pro-inflammatory cytokines such as IL-1α, IL-1β, IL-8, and tumor necrosis factor alpha (TNFα), growth factors like granulocyte-macrophage colony-stimulating factor (GM-CSF), and AMP including β-defensins-2 and -3 and protein S100A7 [12, 13, 26, 32, 33]. Similarly, mRNA expression and release of pro-inflammatory cytokines and AMP by keratinocytes embedded in skin equivalents are observed in such models exposed to infection by dermatophytes [26, 58]. Responses triggered in keratinocytes appear dependent on the dermatophyte species selected for epidermal infection, the release of cytokines being generally higher in the presence of zoophilic species, such as *T. benhamiae* or *T. mentagrophytes*, than in the presence of anthropophilic dermatophytes, i.e., *T. rubrum* or *Trichophyton tonsurans* [12, 13]. This reflects the high or low inflammatory levels observed in cutaneous lesions induced in vivo by the different species of dermatophytes and pledges again for the right model selection when studying specific dermatophytosis. Besides, the mRNA expression of TLR-2, -4, and -6 by keratinocytes cultured as monolayers was modulated during stimulation by *T. rubrum* [85].

By analysis of PMN maintained in culture, the release of IL-1β, IL-8, and TNFα and the production of neutrophil extracellular traps (NET) by this cell type were observed when PMN are incubated in the presence of *T. benhamiae* or *M. canis* [26, 86]. Overexpression of TLR-2 and -4 mRNA was further characterized in PMN exposed to *M. canis* [87]. In addition, release of TNFα and IL-1β by macrophages challenged by *T. rubrum* was measured on a simple culture model [34, 88]. Finally, IL-6, IL-8, IL-12, and TNFα were produced by dendritic cells upon contact with *T. benhamiae* or *M. canis* ([86]; Tabart, unpublished data).

Considering all these data, the antifungal response by the host tissue during epidermal dermatophytosis has been summarized in Fig. 4. Initially, keratinocytes detect the presence of dermatophytes through the recognition of specific fungal motifs by TLR, mainly TLR-2, -4, and -6. Activated keratinocytes start producing various pro-inflammatory cytokines, including IL-1α, IL-1β, IL-8, and TNFα, which become upon release responsible for the recruitment of PMN and macrophages. Simultaneously, dendritic cells also detect dermatophytes via TLR or dectin receptors and subsequently participate to inflammation by their own secretion of pro-inflammatory cytokines. Thereafter, dendritic cells mature to become antigen-presenting cells and migrate into the lymph nodes in order to induce the differentiation of Th1 and Th17 cells, respectively, through the production of IL-12 or IL-6. Activated T cells then migrate to the infection site and secrete specialized cytokines: IFNγ by Th1 cells, IL-17 and IL-22 by Th17 cells. IFNγ stimulates macrophages which then become able to produce pro-inflammatory cytokines and to phagocyte small fungal elements. In the same time, the production by keratinocytes of PAM, responsible for direct antimicrobial effects, and GM-CSF is induced by IL-17 and IL-22. Finally, PMN are activated by GM-CSF and participate to the clearance of dermatophytes by phagocytosis, production of NET, and secretion of pro-inflammatory cytokines.

3.3.2 Modulation of Host Immune Response by Dermatophytes

Certain dermatophytes might dispose of mechanisms to evade or silence the immune response, causing chronic and low inflammatory lesions. Particularly, the anthropophilic *T. rubrum* species exhibits inhibition of macrophage functions and an induced secretion of anti-inflammatory proteins by macrophages and keratinocytes. Indeed, exposure of cultured macrophages to *T. rubrum* resulted in down-modulation of the major histocompatibility complex class II, in reduction of expression of co-stimulatory molecules, and in a upregulated release of IL-10 [34]. Moreover, *T. rubrum* further induced expression and release by keratinocytes of the anti-inflammatory protein encoded by TNFα-stimulated gene 6 [58]. Altogether, properties of *T. rubrum* which favor its adaptation to human host may explain why this species induces lower inflammatory lesions than zoophilic or geophilic dermatophytes.

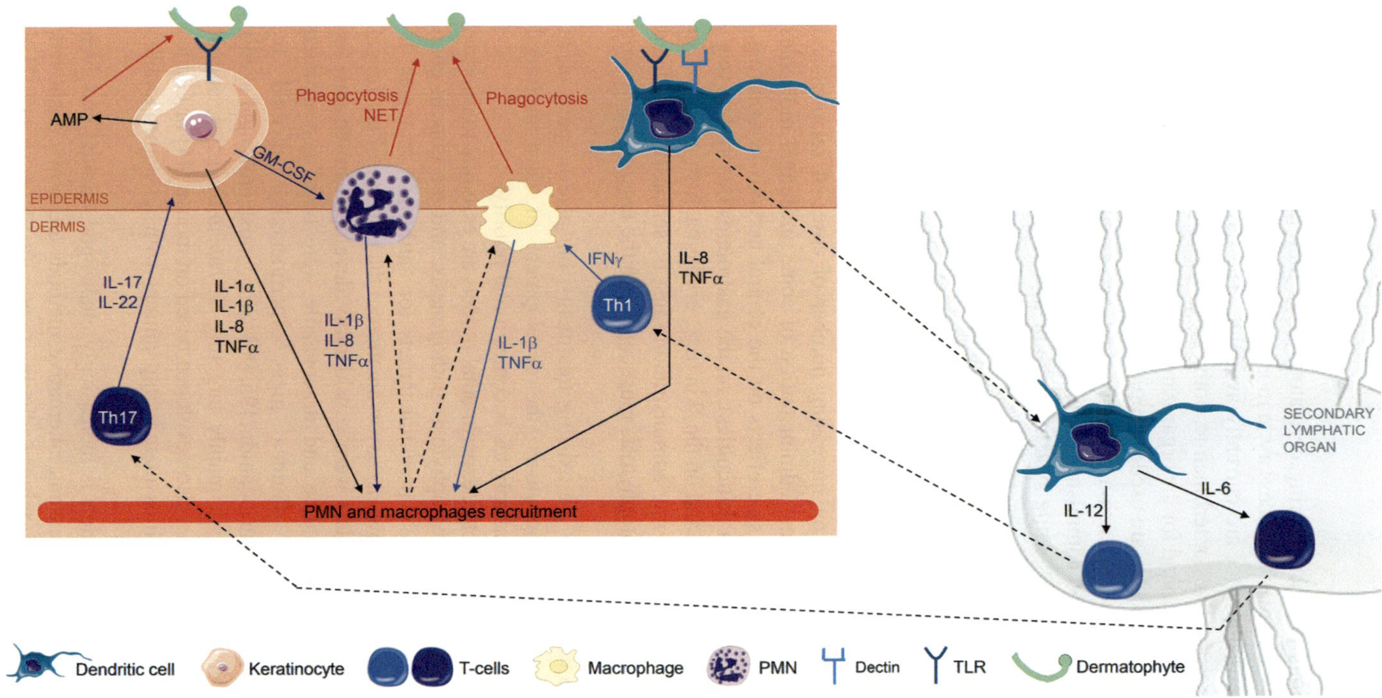

Fig. 4 Innate and adaptive immune responses are induced during dermatophytosis. Host keratinocytes and dendritic cells detect the presence of dermatophytes invading the epidermis through specific receptors, i.e., toll-like receptors (TLR) and dectin. They then produce and release antimicrobial peptides (AMP) and pro-inflammatory cytokines which initiate the recruitment of macrophages and polymorphonuclear neutrophils (PMN). Dendritic cells migrate to secondary lymphatic organs where they induce the differentiation, activation, and recruitment of Th1 and Th17 cells. Th1 cells, through the release of IFNγ, improve the recruitment and activation of macrophages. In the same time, IL-17 and IL-22, produced by Th17 cells, enhance the release of AMP and cytokines by keratinocytes and induce the release of GM-CSF activating PMN. Finally, macrophages and PMN, through phagocytosis and neutrophil extracellular traps (NET) formation, as well as AMP by their direct antimicrobial activities, are together responsible for the control and the eventual resolution of infection

4 Experimental Models to Evaluate Efficacy of Antifungal Compounds

Although skin lesions produced during dermatophytosis may heal spontaneously within a few months, treatment is necessary in most cases. Several antifungal agents effective against dermatophytes are currently available. The most frequently used molecules are azole derivatives or terbinafine, both acting by inhibition of ergosterol synthesis, an essential compound for the organization of normal membrane in fungal cells [89]. Antifungal molecules are used for topical application or systemic administration, depending on the extension, severity, and accessibility of the lesion [17, 90]. For instance, local administration is chosen to treat simple lesions of *Tinea corporis* or *Tinea pedis*, while systemic administration is often required for nail or hair lesions, as well as for severe or recalcitrant cutaneous lesions.

Despite the current availability of effective antifungal agents, the management of dermatophytosis must face several problems [17]. Firstly, the treatment of dermatophytosis may become long-lasting and thus expensive, leading too often to reduced compliance in patients who generally stop taking medication as soon as the clinical symptoms improve. Premature interruption of antifungal treatment contributes to the occurrence of an increasing number of cases of relapses and reinfections. Secondly, the systemic administration of some antifungal molecules is unfortunately accompanied by side effects on the central nervous system (e.g., headache, dizziness, concentration difficulties), on the gastrointestinal system (e.g., nausea, diarrhea, abdominal pain), and on the skin (e.g., erythema, rash, pruritus). In addition, toxicity for liver of systemic administration must be seriously considered [91]. Finally, treatments for dermatophytosis must cope with the emergence of drug resistance [92–94].

In view of problems linked to the existing treatments, but also because of an increasing incidence of dermatophytosis, the development of new antifungal compounds effective against dermatophytes becomes a real priority.

As explained above in this chapter, experimental models are essential to improve the current understanding of dermatophytosis pathogenesis, and are thus in good position to help in identifying new potential targets for antifungal strategies. In addition, experimental models allow an easy evaluation of the efficacy of antifungal compounds. Notably, animal models were considerably developed in the past for this purpose; for instance, terbinafine and itraconazole were both demonstrated efficient by the analysis of Guinea pigs experimentally infected by *T. mentagrophytes* or *M. canis* aleurioconidia [95, 96]. Since most antifungal agents are developed for human treatment, mainly targeting *T. rubrum* species, animal models of *T. rubrum* infection were also developed based on mouse [44] or Guinea pigs [46, 48], despite the difficulties in infecting animal with an anthropophilic dermatophyte (see Sect. 2.2).

Because the use of animal models is increasingly challenged, limited and even banned for ethical and political reasons, alternative methods are now considered. Thereby, ex vivo models of dermatophytes adhering to cornified layer sheet [97] or to corneocytes in suspension [19] were used to monitor the impact of photodynamic therapy or of antifungal molecules (i.e., ketoconazole, itraconazole, and griseofulvin) on adherence between dermatophytes and host cells. Furthermore, skin equivalent models were proved to provide valid methods to evaluate antifungal efficacy by using reference antifungal compounds such as terbinafine [55] or miconazole [27, 59]. Using a model of dermatophytosis on RHE, our group lately identified PD169316, a well-known specific inhibitor of the human p38 mitogen-activated protein kinase, as a potential antifungal agent effective against dermatophytes [58]. Indeed, PD169316 exhibits some direct effect on growth of fungi and thereby interferes with the infection of RHE by *T. rubrum* arthroconidia. Finally, a model of infection based on silkworm was recently developed by injection of *T. mentagrophytes* aleurioconidia in this invertebrate, in order to evaluate antifungal agents against dermatophytosis [98].

Therefore, each model appears as useful experimental device to identify and evaluate the efficacy of new antifungal molecules while, except for sheets of cornified layer, assessing potential toxicity for the treated tissue. In addition, one must keep in mind that arthroconidia are the infective elements produced in vivo by dermatophytes [21] and are more resistant than microconidia to some antifungal compounds [20]. In consequence, experimental models using arthroconidia as infective elements (e.g., [27, 55, 59]) must be considered as more appropriate to perform efficacy assay of antifungal molecules.

5 Conclusion

This chapter summarized experimental models for the investigations of both the infection process deployed by dermatophytes and the development of the host immune response. A better understanding of the pathogenesis of dermatophytosis will allow the identification of potential therapeutic targets and the subsequent development of new antifungal compounds. Since each type of model, whether ex vivo, in vitro, or animal model, holds characteristics bringing intrinsic advantages and limitations that make each of them more or less appropriate for precise investigations, the choice of one or another model must always consider the purpose of the study. For example, while simple in vitro models of keratinocytes cultured as monolayers remain useful to study particular responses of this host cell type exposed to dermatophytes, they are certainly less appropriate to investigate the recruitment of host immune cells and the activation of immunity. Conversely, animal models are suitable to evaluate the entire host immune responses, but do not allow in-depth focus on the activation of individual cells. In conclusion, a complete understanding of dermatophytosis will still depend upon the deployment of complementary experimental models.

References

1. Seebacher C, Bouchara J, Mignon B. Updates on the epidemiology of dermatophyte infections. Mycopathologia. 2008;166:335–52.
2. Zhan P, Liu W. The changing face of dermatophytic infections worldwide. Mycopathologia. 2017;182:77–86.
3. Havlickova B, Czaika V, Friedrich M. Epidemiological trends in skin mycoses worldwide. Mycoses. 2008;51:2–15.
4. Hayette M, Sacheli R. Dermatophytosis, trends in epidemiology and diagnostic approach. Curr Fungal Infect Rep. 2015;9:164–79.
5. Saunte D, Holgersen J, Haedersdal M, Strauss G, Bitsch M, Svendsen O, et al. Prevalence of toe nail onychomycosis in diabetic patients. Acta Derm Venereol. 2006;86:425–8.
6. Eckhard M, Lengler A, Liersch J, Bretzel R, Mayser P. Fungal foot infections in patients with diabetes mellitus – results of two independent investigations. Mycoses. 2007;50:14–9.
7. Weitzman I, Summerbell R. The dermatophytes. Clin Microbiol Rev. 1995;8:240–59.
8. Lee W, Kim S, Jang Y, Lee S, Kim D, Bang Y, et al. Increasing prevalence of *Trichophyton rubrum* identified through an analysis of 115,846 cases over the last 37 years. J Korean Med Sci. 2015;30:639–43.
9. Degreef H. Clinical forms of dermatophytosis (ringworm infection). Mycopathologia. 2008;166:257–65.
10. Nenoff P, Krüger C, Ginter-Hanselmayer G, Schulte-Beerbühl R, Tietz H. Mycology – an update. Part 2: Dermatomycoses: clinical picture and diagnosis. J Dtsch Dermatol Ges. 2014;12:749–77.
11. Vermout S, Tabart J, Baldo A, Mathy A, Losson B, Mignon B. Pathogenesis of dermatophytosis. Mycopathologia. 2008;166:267–75.
12. Shiraki Y, Ishibashi Y, Hiruma M, Nishikawa A, Ikeda S. Cytokine secretion profiles of human keratinocytes during *Trichophyton tonsurans* and *Arthroderma benhamiae* infections. J Med Microbiol. 2006;55:1175–85.
13. Tani K, Adachi M, Nakamura Y, Kano R, Makimura K, Hasegawa A, et al. The effect of dermatophytes on cytokine production by human keratinocytes. Arch Dermatol Res. 2007;299:381–7.
14. Drake L, Scher R, Smith E, Faich G, Smith S, Hong J, et al. Effect of onychomycosis on quality of life. J Am Acad Dermatol. 1998;38:702–4.
15. Szepietowski J, Reich A. Stigmatisation in onychomycosis patients: a population-based study. Mycoses. 2009;52:343–9.
16. Nenoff P, Krüger C, Ginter-Hanselmayer G, Tietz H. Mycology – an update. Part 1: Dermatomycoses: causative agents, epidemiology and pathogenesis. J Dtsch Dermatol Ges. 2014;12:188–209.
17. Gupta A, Foley K, Versteeg S. New antifungal agents and new formulations against dermatophytes. Mycopathologia. 2017;182:127–41.
18. Deacon J. Fungal biology. 4th ed. London: Blackwell; 2006.
19. Zurita J, Hay R. Adherence of dermatophyte microconidia and arthroconidia to human keratinocytes *in vitro*. J Invest Dermatol. 1987;89:529–34.
20. Coelho L, Aquino-Ferreira R, Leite Maffei C, Martinez-Rossi N. *In vitro* antifungal drug susceptibilities of dermatophytes microconidia and arthroconidia. J Antimicrob Chemother. 2008;62:758–61.
21. Rashid A. Arthroconidia as vectors of dermatophytosis. Cutis. 2001;67:23.
22. Tabart J, Baldo A, Vermout S, Nusgens B, Lapiere C, Losson B, et al. Reconstructed interfollicular feline epidermis as a model for *Microsporum canis* dermatophytosis. J Med Microbiol. 2007;56:971–5.
23. Grumbt M, Monod M, Yamada T, Hertweck C, Kunert J, Staib P. Keratin degradation by dermatophytes relies on dioxygenase and a sulfite efflux pump. J Invest Dermatol. 2013;133:1550–5.

24. Teixeira de Aguiar Peres N, Gomes da Silva L, da Silva Santos R, Jacob T, Persinoti G, Rocha L, et al. *In vitro* and *ex vivo* infection models help assess the molecular aspects of the interaction of *Trichophyton rubrum* with the host milieu. Med Mycol. 2016;54:420–7.
25. Burstein V, Guasconi L, Beccacece I, Theumer M, Mena C, Prinz I, et al. IL-17-mediated immunity controls skin infection and T Helper 1 response during experimental *Microsporum canis* dermatophytosis. J Invest Dermatol. 2018;138:1744–53.
26. Achterman R, Moyes D, Thavaraj S, Smith A, Blair K, Withe T, et al. Dermatophytes activate skin keratinocytes via mitogen-activated protein kinase signaling and induce immune responses. Infect Immun. 2015;83:1705–14.
27. Faway E, Cambier L, Mignon B, Poumay Y, Lambert de Rouvroit C. Modeling dermatophytosis in reconstructed human epidermis: a new tool to study infection mechanisms and to test antifungal agents. Med Mycol. 2017;55:485–94.
28. Cambier L, Mathy A, Baldo A, Bagut E, Tabart J, Antoine N, et al. Feline polymorphonuclear neutrophils produce pro-inflammatory cytokines following exposure to *Microsporum canis*. Vet Microbiol. 2013;162:800–5.
29. Heinen M, Cambier L, Antoine N, Gabriel A, Gillet L, Bureau F, et al. Th1 and Th17 immune responses act complementarily to optimally control superficial dermatophytosis. J Invest Dermatol. 2019;139:626–37.
30. Aljabre S, Richardson M, Scott E, Rashid A, Shankland G. Adherence of arthroconidia and germlings of anthropophilic and zoophilic varieties of *Trichophyton mentagrophytes* to human corneocytes as an early event in the pathogenesis of dermatophytosis. Clin Exp Dermatol. 1993;18:231–5.
31. Aljabre S, Richardson M, Scott E, Shankland G. Germination of *Trichophyton mentagrophytes* on human stratum corneum *in vitro*. J Med Vet Mycol. 1992;30:145–52.
32. Nakamura Y, Kano R, Hasegawa A, Watanabe S. Interleukin-8 and tumor necrosis factor alpha production in human epidermal keratinocytes induced by *Trichophyton mentagrophytes*. Clin Diag Lab Immunol. 2002;9:935–7.
33. Firat Y, Simanski M, Rademacher F, Schröder L, Brasch J, Harder J. Infection of keratinocytes with *Trichophytum rubrum* induces epidermal growth factor-dependent RNase 7 and human beta-defensin-3 expression. PLoS Onc. 2014;9:e93941.
34. Campos M, Russo M, Gomes E, Almeida S. Stimulation, inhibition and death of macrophages infected with *Trichophyton rubrum*. Microbes Infect. 2006;8:372–9.
35. Baldo A, Mathy A, Tabart J, Camponova P, Vermout S, Massart L, et al. Secreted subtilisin Sub3 from *Microsporum canis* is required for adherence to but not for invasion of the epidermis. Br J Dermatol. 2010;162:990–7.
36. Bagut E, Baldo A, Mathy A, Cambier L, Antoire N, Cozma V, et al. Subtilisin Sub3 is involved in adherence of *Microsporum canis* to human an animal epidermis. Vet Microbiol. 2012;160:413–9.
37. Baldo A, Chevigné A, Dumez M, Mathy A, Power P, Tabart J, et al. Inhibition of the keratinolytic subtilisin protease Sub3 from *Microsporum canis* by its propeptide (proSub3) and evaluation of the capacity of proSub3 to inhibit fungal adherence to feline epidermis. Vet Microbiol. 2012;159:479–84.
38. Duek L, Kaufman G, Ulman Y, Berdicevsky I. The pathogenesis of dermatophyte infections in human skin sections. J Infect. 2004;48:175–80.
39. Kaufman G, Horwitz B, Duek L, Ullman Y, Berdicevsky I. Infection stages of the dermatophyte pathogen *Trichophyton*: microscopic characterization and proteolytic enzymes. Med Mycol. 2007;45:149–55.
40. Staib P, Zaugg C, Mignon B, Weber J, Grumbt M, Pradervand S, et al. Differential gene expression in the pathogenic dermatophyte *Arthroderma benhamiae in vitro* versus during infection. Microbiology. 2010;156:884–95.
41. Tran T, De Col N, Feuermann M, Schmid-Slegert E, Bagut E, Mignon B, et al. RNA sequencing-based genome reannotation of the dermatophyte *Arthroderma benhamiae* and

characterization of its secretome and whole gene expression profile during infection. mSystems. 2016;1:e00036–16.
42. Cambier L, Weatherspoon A, Defaweux V, Bagut E, Heinen M, Antoine N, et al. Assessment of the cutaneous response during *Arthroderma benhamiae* and *A. vanbreuseghemii* infection using an experimental mouse model. Br J Dermatol. 2014;170:625–33.
43. Cambier L, Heinen M, Mignon B. Relevant animal models in dermatophyte research. Mycopathologia. 2017;182:229–40.
44. Baltazar L, Werneck S, Carneiro H, Gouveia L, de Paula T, Byrro R, et al. Photodynamic therapy efficiently controls dermatophytosis caused by *Trichophyton rubrum* in a murine model. Br J Dermatol. 2015;172:801–4.
45. Baltazar L, Santos P, de Paula T, Rashid M, Cisalpino P, Souza D, et al. IFN-γ impairs *Trichophyton rubrum* proliferation in a murine model of dermatophytosis through the production of IL-1β and reactive oxygen species. Med Mycol. 2014;52:293–302.
46. Li Z, Guo X, Dawuti G, Aibai S. Antifungal activity of ellagic acid *in vitro* and *in vivo*. Phytother Res. 2015;29:1019.
47. Chen X, Shen Y, Lü G, Liu W. Establishing an experimental Guinea pig model of dermatophytosis using *Trichophyton rubrum*. Acta Acad Med Sin. 2008;30:599–602.
48. Mei Y, Dai X, Yang W, Xu X, Liang Y. Antifungal activity of chitooligosaccharides against the dermatophyte *Trichophyton rubrum*. Int J Biol Macromol. 2015;77:330–5.
49. Kumar N, Goindi S. Statistically designed nonionic surfactant vesicles for dermal delivery of itraconazole: characterization and *in vivo* evaluation using a standardized Tinea pedis infection model. Int J Pharm. 2014;472:224–40.
50. Kumar N, Goindi S. D-optimal experimental approach for designing topical microemulsion of itraconazole: characterization and evaluation of antifungal efficacy against a standardized Tinea pedis infection model in Wistar rats. Eur J Pharm Sci. 2015;67:97–112.
51. Poumay Y, Dupont F, Marcoux S, Leclercq-Smekens M, Hérin M, Coquette A. A simple reconstructed human epidermis: preparation of the culture model and utilization in *in vitro* studies. Arch Dermatol Res. 2004;296:203–11.
52. Green C, Cheng G, Chandra J, Mukherjee P, Ghannoum M, Hoyer L. RT-PCR detection of *Candida albicans ALS* gene expression in the reconstituted human epithelium (RHE) model of oral candidiasis and in model biofilms. Microbiology. 2004;150:267–75.
53. Bäsler K, Galliano M, Bergmann S, Rohde H, Wladykowski E, Vidal-y-Sy S, et al. Biphasic influence of *Staphylococcus aureus* on human epidermal tight junctions. Ann N Y Acad Sci. 2017;1405:53–70.
54. Jannasch M, Groeber F, Brattig N, Unger C, Walles H, Hansmann J. Development and application of three-dimensional skin equivalents for the investigation of percutaneous worm invasion. Exp Parasitol. 2015;150:22–30.
55. Rashid A, Edward M, Richardson M. Activity of terbinafine on *Trichophyton mentagrophytes* in human living skin equivalent model. J Med Vet Mycol. 1995;33:229–33.
56. Liang P, Huang X, Yi J, Chen Z, Ma H, Ye C, et al. A *Trichophyton rubrum* infection model based on the reconstructed human epidermis – Episkin®. Chin Med J (Engl). 2016;129:54–8.
57. Faway E, Lambert de Rouvroit C, Poumay Y. *In vitro* models of dermatophyte infection to investigate epidermal barrier alterations. Exp J Dermatol. 2018;27:915–22.
58. Faway E, Cambier L, De Vuyst E, Evrard C, Thiry M, Lambert de Rouvroit C, et al. Responses of reconstructed human epidermis to *Trichophyton rubrum* infection and impairment of infection by the inhibitor PD169316. J Invest Dermatol. 2019;139(10):2080–2089.e6.
59. Tabart J, Baldo A, Vermout S, Losson B, Mignon B. Reconstructed interfollicular feline epidermis as a model for the screening of antifungal drugs against M*icrosporum canis*. Vet Dermatol. 2008;19:130–3.
60. Rashid A, Scott E, Richardson M. Early events in the invasion of the human nail plate by *Trichophyton mentagrophytes*. Br J Dermatol. 1995;133:932–40.

61. Baldo A, Tabart J, Vermout S, Mathy A, Collard A, Losson B, et al. Secreted subtilisins of *Microsporum canis* are involved in adherence of arthroconidia to feline corneocytes. J Med Microbiol. 2008;57:1152–6.
62. Baldo A, Monod M, Mathy A, Cambier L, Bagut E, Defaweux V, et al. Mechanisms of skin adherence and invasion by dermatophytes. Mycoses. 2011;55:218–23.
63. Esquenazi D, de Souza W, Sales Alviano C, Rozental S. The role of surface carbohydrates on the interaction of microconidia of *Trichophyton mentagrophytes* with epithelial cells. Immunol Med Microbiol. 2003;35:113–23.
64. Esquenazi D, Sales Alviano C, de Souza W, Rozental S. The influence of surface carbohydrates during *in vitro* infection of mammalian cells by the dermatophyte *Trichophyton rubrum*. Res Microbiol. 2004;155:144–53.
65. Bitencourt T, Macedo C, Franco M, Assis A, Komoto T, Stehling E, et al. Transcription profile of *Trichophyton rubrum* conidia grown on keratin reveals the induction of an adhesin-like protein gene with a tandem repeat pattern. BMC Genomics. 2016;17:249.
66. Verstrepen K, Klis F. Flocculation, adhesion and biofilm formation in yeasts. Mol Microbiol. 2006;60:5–15.
67. Brandhorst TT, Roy R, Wüthrich M, Nanjappa S, Filutowicz H, Galles K, et al. Structure and function of a fungal adhesin that binds heparin and mimics thrombospondin-1 by blocking T cell activation and effector function. PLoS Pathog. 2013;9:e1003464.
68. Hung CY, Yu JJ, Seshan KR, Reichard U, Cole GT. A parasitic phase-specific adhesin of *Coccidioides immitis* contributes to the virulence of this respiratory fungal pathogen. Infect Immun. 2002;70:3443–56.
69. Monod M. Secreted proteases from dermatophytes. Mycopathologia. 2008;166:285–94.
70. Mignon B, Swinnen M, Bouchara J, Hofinger M, Nikkels A, Pierard G, et al. Purification and characterization of a 31.5 kDa keratinolytic subtilisin-like serine protease from *Microsporum canis* and evidence of its secretion in naturally infected cats. Med Mycol. 1998;36:395–404.
71. Leng W, Liu T, Wang J, Li R, Jin Q. Expression dynamics of secreted protease genes in *Trichophyton rubrum* induced by key host's proteinaceous components. Med Mycol. 2009;47:759–65.
72. Méhul B, Gu Z, Jomard A, Laffet G, Feuilhade M, Monod M. Sub6 (Tri r 2), an onychomycosis marker revealed by proteomics analysis of *Trichophyton rubrum* secreted proteins in patient nail samples. J Invest Dermatol. 2016;136:331–3.
73. Mignon B, Leclipteux T, Focant C, Nikkels A, Piérard G, Losson B. Humoral and cellular immune response to a crude exo-antigen and purified keratinase of *Microsporum canis* in experimentally infected Guinea pigs. Med Mycol. 1999;37:123–9.
74. Vermout S, Tabart J, Baldo A, Monod M, Losson B, Mignon B. RNA silencing in the dermatophyte *Microsporum canis*. FEMS Microbiol Lett. 2007;275:38–45.
75. Liu T, Zhang Q, Wang L, Yu L, Leng W, Yang J, et al. The use of global transcriptional analysis to reveal the biological and cellular events involved in distinct development phases of *Trichophyton rubrum* conidial germination. BMC Genomics. 2007;8:100.
76. Almeida M, Brand A. Thigmo responses: the fungal sense of touch. Microbiol Spectr. 2017;5 https://doi.org/10.1128/microbiolspec.FUNK-0040-2016.
77. Perera T, Gregory D, Marshall D, Gow N. Contact-sensing by hyphae of dermatophytic and saprophytic fungi. J Med Vet Mycol. 1997;35:289–93.
78. Jensen J, Pfeiffer S, Akaki T, Schröder J, Kleine M, Neumann C, et al. Barrier function, epidermal differentiation, and human β-defensin 2 expression in Tinea corporis. J Invest Dermatol. 2007;127:1720–7.
79. Green F, Lee KW, Balish E. Chronic *T. mentagrophytes* dermatophytosis of Guinea pig skin grafts on nude mice. J Invest Dermatol. 1982;79:125–9.
80. Calderon R, Hay R. Cell-mediated immunity in experimental murine dermatophytosis. II. Adoptive transfer of immunity to dermatophyte infection by lymphoid cells from donors with acute or chronic infections. Immunology. 1984;53:465–72.
81. Almeida S. Immunology of dermatophytosis. Mycopathologia. 2008;166:277–83.

82. Mignon B, Tabart J, Baldo A, Mathy A, Losson B, Vermout S. Immunization and dermatophytes. Curr Opin Infect Dis. 2008;21:134–40.
83. Hay R, Calderon R, Mackenzie C. Experimental dermatophytosis in mice: correlation between light and electron microscopic changes in primary, secondary and chronic infections. Br J Exp Pathol. 1988;69:703–16.
84. Nakamura Y, Nishibu A, Yasoshima M, Tanoue C, Yoshida N, Hatta J, et al. Analysis of *Trichophyton* antigen-induced contact hypersensitivity in mouse. J Dermatol Sci. 2012;66:144–53.
85. Garcia-Madrid L, Huizar-Lopez M, Flores-Romo L, Islas-Rodriguez A. *Trichophyton rubrum* manipulates the innate immune functions of human keratinocytes. Cent Eur J Biol. 2011;6:902–10.
86. Heddergott C, Bruns S, Nietzsche S, Leonhardt I, Kurzai O, Kniemeyer O, et al. The *Arthroderma benhamiae* hydrophobin HypA mediates hydrophobicity and influences recognition by human immune effector cells. Eukaryot Cell. 2012;11:673–82.
87. Cambier L, Heinen M, Bagut E, Antoine N, Mignon B. Overexpression of TLR-2 and TLR-4 mRNA in feline polymorphonuclear neutrophils exposed to *Microsporum canis*. Vet Dermatol. 2016;27:78–e22.
88. Yoshikawa F, Ferreira L, de Almeida S. IL-1 signaling inhibits *Trichophyton rubrum* conidia development and modulates the IL-17 response *in vivo*. Virulence. 2015;6:449–57.
89. Ghannoum M, Rice L. Antifungal agents: mode of action, mechanisms of resistance, and correlation of these mechanisms with bacterial resistance. Clin Microbiol Rev. 1999;12:501–17.
90. Gupta A, Cooper E. Update in antifungal therapy of dermatophytosis. Mycopathologia. 2008;166:353–67.
91. Raschi E, Poluzzi E, Koci A, Caraceni P, De Ponti F. Assessing liver injury associated with antimycotics: concise literature review and clues from data mining of the FAERS database. World J Hepatol. 2014;6:601–12.
92. Osborne C, Leitner I, Favre B, Ryder N. Amino acid substitution in *Trichophyton rubrum* squalene epoxidase associated with resistance to terbinafine. Antimicrob Agents Chemother. 2005;49:2840–4.
93. Mukherjee P, Leidich S, Isham N, Leitner I, Ryder N, Ghannoum M. Clinical *Trichophyton rubrum* strain exhibiting primary resistance to terbinafine. Antimicrob Agents Chemother. 2003;47:82–6.
94. Salehi Z, Shams-Ghahfarokhi M, Razzaghi-Abyaneh M. Antifungal drug susceptibility profile of clinically important dermatophytes and determination of point mutations in terbinafine-resistant isolates. Eur J Clin Microbiol Infect Dis. 2018;37:1841–6.
95. Ghannoum M, Hossain M, Long L, Mohamed S, Reyes G, Mukherjee P. Evaluation of antifungal efficacy in an optimized animal model of *Trichophyton mentagrophytes*-dermatophytosis. J Chemother. 2004;16:139–44.
96. Saunte D, Hasselby J, Brillowska-Dabrowska A, Frimodt-Moller N, Svejgaard E, Linnemann D, et al. Experimental Guinea pig model of dermatophytosis: a simple and useful tool for the evaluation of new diagnostics and antifungals. Med Mycol. 2008;46:303–13.
97. Smijs T, Bouwstra J, Schuitmaker H, Talebi M, Pavel S. A novel *ex vivo* skin model to study the susceptibility of the dermatophyte *Trichophyton rubrum* to photodynamic treatment in different growth phases. J Antimicrob Chemother. 2007;59:433–40.
98. Ishii M, Matsumoto Y, Yamada T, Abe S, Sekimizu K. An invertebrate infection model for evaluating anti-fungal agents against dermatophytosis. Sci Rep. 2017;7:12289.

Transcriptome of Host–Dermatophyte Interactions Using Infection Models

Ana Lúcia Fachin, Monise Fazolin Petrucelli, Tamires Aparecida Bitencourt, Felipe Garcia Nishimura, Gabriela Gonzalez Segura, Bruna Aline Michelotto Cantelli, Mariana Heinzen de Abreu, and Mozart Marins

Abstract

Dermatophytes are able to invade keratinized structures such as skin, nails, and hair, causing superficial infections. However, the number of deep dermatophytoses has been increasing among immunodepressed and diabetic patients, as well as among individuals without immunodeficiency. Models for the study of superficial infection use culture medium containing protein substrates and fragments of human nails or keratinocyte/skin fragment cultures. On the other hand, new models should be explored for the study of deep infections, including monocyte/macrophage cultures. Large-scale gene expression techniques (transcriptome analysis) are a molecular approach to investigate the mechanisms involved in the host–fungus interactions and to highlight new antifungal targets. Furthermore, studies of the metabolome, proteome, and fungal and human microRNAs should be performed to gain a better understanding of the mechanisms underlying pathogenesis and the host or fungal response.

A. L. Fachin (✉) · M. F. Petrucelli · F. G. Nishimura · G. G. Segura · B. A. M. Cantelli · M. H. de Abreu · M. Marins
Biotechnology Unit, University of Ribeirão Preto, UNAERP, Ribeirão Preto, Brazil
e-mail: afachin@unaerp.br; mmarins@gmb.bio.br

T. A. Bitencourt
Biotechnology Unit, University of Ribeirão Preto, UNAERP, Ribeirão Preto, Brazil

Department of Genetics, Ribeirão Preto Medical School, University of São Paulo, Ribeirão Preto, Brazil

Department of Biochemistry and Immunology, Ribeirão Preto Medical School, University of São Paulo, Ribeirão Preto, Brazil

J.-P. Bouchara et al. (eds.), *Dermatophytes and Dermatophytoses*,
https://doi.org/10.1007/978-3-030-67421-2_8

Keywords

Antifungals · Superficial infection · Deep infection · Infection model · Transcriptome · *Trichophyton rubrum*

1 Introduction

Dermatophytes are pathogenic fungi that cause infections in animals and particularly in humans. They comprise a group of filamentous fungi that are able to invade keratinized tissues such as skin, hair, and nails [1], causing superficial infections called dermatophytoses [2].

Dermatophyte fungi comprise 52 species that degrade keratin and are divided into nine genera: *Trichophyton*, *Microsporum*, *Epidermophyton*, *Arthroderma*, *Lophophyton*, *Nannizzia*, *Ctenomyces*, *Guarromyces*, and *Paraphyton* [3]. Among the fungi that cause dermatophytoses in humans, the filamentous fungus *Trichophyton rubrum* is the main causative agent of infections of the skin, nails, and body [4]. Dermatophytoses are transmitted through direct contact with an infected human or animal, or indirectly by fomites. Clinical characterization of the disease depends on the fungal species and anatomical site involved. These fungal infections receive the name *Tinea* followed by the site involved, for example, *Tinea corporis* for body infections and *Tinea unguium* for nail infections [5, 6].

According to the World Health Organization, dermatophytes affect about 25% of the world population and 30–70% of adults are asymptomatic carriers of these diseases [7]. Dermatophytes generally cause superficial infections. However, there have been an increasing number of reports of dermatophytes that invade deep layers such as the dermis and hypodermis and that can spread to the lymph nodes, brain, and bloodstream, causing deep infections [8]. Some conditions predispose to the establishment of this infection in the patient, including an immunosuppressive state, obesity, diabetes, and advanced age. Other factors associated with this disease are environmental factors and genetic predisposition. Environmental factors often influence the susceptibility to infection; for example, humidity and temperature favor fungal growth. Regarding genetic factors, mutations in specific genes can modify gene expression or alter the conformational structure of proteins responsible for the host's innate and adaptive immunity [9].

Invasive fungal infections are characterized by high rates of morbidity and mortality in patients with a weakened immune system. In this respect, cases of deep infections caused by dermatophytes have been reported in the literature, for example, in a patient with hepatitis C, liver cirrhosis, and subcutaneous nodules [10]. A patient with deep dermatophytosis caused by *T. rubrum* developed complications caused by pre-existing pathologies and died [11]. Other examples include patients infected with human immunodeficiency virus (HIV) [12], transplant recipients and patients undergoing chemotherapy [13], and patients with a history of diabetes mellitus [14]. A patient without significant immunodeficiency with a history of onychomycosis of the toe nails exhibited erythema, papules, and nodules in the submandibular area, neck, and chest [15]. Fungal abscesses were reported in a patient using immunosuppressive drugs [16]. A patient without immunodeficiency but with a history of hypertension and angina showed red patches with secretion [17]. All of these cases were diagnosed as deep infection caused by *T. rubrum*. A case of dermatophytosis caused by *Trichophyton interdigitale* was reported in a man with a 2-year history of fungal infection who lost his vision and had onychomycosis in the fingernails of the left hand [18].

In an attempt to improve diagnostic techniques and the treatments for deep infections, sequencing of the patient genome revealed a correlation between the presence of severe dermatophytoses and deficiency of caspase recruitment domain-containing signaling protein (CARD9). This protein is responsible for activating and recruiting T cells of the immune system [19, 20]. In a case study, a healthy individual was admitted to the hospital with skin lesions and localized dermatitis on the neck and face. A skin biopsy showed septate and pigmented hyphae but the fungal species was not characterized. In view of the absence of a history of immunosuppression, DNA was extracted from the patient and CARD9 deficiency was detected. Despite antifungal therapy and surgical procedures, the patient died after 38 days of hospitalization [21]. In another case of deep infection, the patient had an erythematous plaque in the left groin and a 6-year history of superficial ulcers. A biopsy showed the presence of septate hyphae and the lesions were found to be caused by *T. rubrum*. Improvement of the lesions was observed after 4 months of antifungal treatment. CARD9 deficiency was also detected in this patient [22].

2 Molecular Approaches to Dermatophytes: Advances and Limitations

In the 1990s, the genome sequencing of several species was started, including the human genome, which enabled great and rapid progress in science, and was widely used [23]. Indeed, molecular studies of dermatophytes were facilitated by the sequencing and publication of the genome of 24 species of this fungal class (http://www.broadinstitute.org/annotation/genome/dermatophyte_comparative/MultiHome.html), permitting the discovery of new therapeutic targets for the possible development of more effective antifungal agents [24].

Several techniques are used for the study of gene and protein expression, such as northern blotting, reverse transcription-quantitative PCR (RT-qPCR), and western blotting [25]. However, most of these techniques permit to analyze only a small number of previously selected genes and proteins. Thus, large-scale expression techniques are gaining space in scientific research as they provide global expression profiles of genes in a given sample. These techniques started with the development of the SAGE (serial analysis of gene expression) method and array techniques [26]. cDNA microarrays led to the first large-scale transcriptome studies. This technique consists of immobilizing specific DNA probes of genes of interest on a matrix and hybridizing them to their corresponding fluorescence-labeled cDNA. The fluorescence intensity of each labeled cDNA reveals the abundance of the transcript in the genome. Despite its efficacy, the microarray technique has some limitations, including cross-reactions of the probes between the pathogen and host and the need to separately analyze the RNA of either organism. In addition, the knowledge of transcriptomes is restricted to the identification of mRNAs, tRNAs, and rRNAs [27].

The RNA-seq technique is nowadays used for the analysis of differential gene expression [28]. The use of transcriptome technologies based on large-scale sequencing has led to great advances in the understanding of biological processes in different organisms [29]. Differential gene expression is important in different areas, such as the study of novel antifungal drugs and antibiotics that can modulate the expression of genes important for the infection process of a fungus or bacterium [30], or genes involved in multidrug resistance pathways [30–32].

Considering that host–fungi interaction does not happen in an isolated manner, another interesting technique is dual gene expression assays (dual RNA-seq), in which both the pathogen and the infected host are analyzed to obtain a better understanding of the infection process, observing pathways and genes that are modulated during this process [33]. With the emergence of this new sequencing technology, dual RNA-seq finds wide application in the study of interactions of parasitism, commensalism, and mutualism between various species [33] and has been used to better understand the interaction that occurs between dermatophytes and the host [34, 35].

Furthermore, the new molecular approaches revealed novel target genes using genetic analysis methods. Some genes potentially involved in the virulence and response to drugs of dermatophytes were studied in mutants and/or strains with silenced genes. These genes include the gene encoding malate synthase *ACUE* and *KU70* in *Arthroderma benhamiae*, now called *Trichophyton benhamiae* [36], the gene encoding the ABC transporter *Tru*MDR2 in *T. rubrum* [37], the gene for the transcription factor pacC in *T. rubrum* [38], the *KU80*, *areA*, and *Trim4* genes in *T. mentagrophytes* [39, 40], the *areA* and *sub3* genes in *Microsporum canis* [41, 42] and, more recently, the genes encoding the StuA/SteA, and HacA transcription factors in *T. benhamiae* and *T. rubrum*, respectively [43, 44].

Another important aspect in the study of dermatophytes is to clarify the functions of microRNAs in cellular processes. MicroRNAs are small non-coding RNAs (containing 20–30 nucleotides) that control the expression of genes, normally inhibiting protein translation. This mechanism of action has been associated with the immune response to a variety of pathogens, especially modulating the intensity of the inflammatory response [45].

The microRNA database (mirBASE) contains more than 2.8 million sequences of different species and this number has been increasing exponentially. Studies have demonstrated that more than 60% of genes contain at least one microRNA-binding site and that these small molecules are part of the organism in animals, plants, fungi, and viruses [46].

The existence of microRNA-like RNAs (milRNAs) in fungi was reported in *Neurospora crassa*, *Sclerotinia sclerotiorum*, *Cryptococcus neoformans*, *Fusarium oxysporum*, *Metarhizium anisopliae*, *Trichoderma reesei*, *Aspergillus fumigatus*,

Aspergillus flavus, and *Penicillium marneffei* [47–54]. Knowledge of these small regulatory RNAs and how they participate in the regulation of gene expression in fungi is still lacking. Wang et al. [55] performed a first integrated analysis of milRNAs and mRNAs between conidial and mycelial stages to investigate the roles of microRNAs in regulating the expression of target genes in *T. rubrum*. A total of 158 conserved milRNAs and 12 novel milRNAs were identified in that study, corresponding to 5470 target genes that are involved in various essential biological pathways. MicroRNA studies may inspire further research on the regulatory mechanism of milRNAs in *T. rubrum* as well as other dermatophytes [55].

3 Molecular Studies and Development of Models That Mimic Superficial Infection

The new molecular approaches have resulted in several discoveries regarding microbial pathogenesis in different research areas. However, in contrast to other fungi of clinical interest such as *Candida albicans*, *A. fumigatus*, or *C. neoformans*, these methods have been poorly exploited for genetic analyses in dermatophytes. Consequently, little is known about the pathogenicity of dermatophytes at the molecular level [56].

The study of the pathogenicity of dermatophytes is usually complex, involving several still unknown fungal mechanisms. It is important to note that the adaptation and response of dermatophytes are associated with the type of host of the infecting species, producing variable clinical signs ranging from a chronic to a more inflammatory dermatophytosis. Thus, the study of the pathogenicity of dermatophytes does not only involve pathogenic attributes of the fungus but also the host's immune response mechanisms [56].

Despite the high global prevalence of dermatophytoses, knowledge of the molecular mechanisms underlying the host–pathogen interaction is still scarce, possibly because of technical difficulties of the models that mimic this interaction and the lack of genetic tools for the study of these organisms [57]. The animal model most used for the study of dermatophytoses is the Guinea pig; however, this model is more suitable for studying zoophilic dermatophytes and is not effective for investigation of the host–pathogen relationship of anthropophilic dermatophyte species [57].

Different models have been developed to overcome this problem and provide a better understanding of dermatophytoses. Commercially available human epithelial tissues may represent an alternative for studying the early stages of infection such as adherence and invasion [57]. Other methods such as fungal growth in culture media containing protein substrates that mimic infection and coculture of the pathogen with human keratinocyte lines are excellent tools for the analysis of gene expression, providing a better understanding of how this interaction occurs at the molecular level [34, 58–60].

Our research group has associated large-scale gene expression techniques with models that mimic superficial infection. We compared the protease expression profile using the *T. rubrum* microarray grown in a culture medium containing keratin

Table 1 Protease expression profile using the *T. rubrum* microarray grown in a culture medium containing keratin with the Dual RNAseq of *T. rubrum* cocultured in HaCat keratinocytes

Gene product name	Log_2 fold change
Culture medium containing keratin	
Subtilisin-like protein (Sub1)	13.73
Subtilisin-like protein (Sub6)	12.45
Leucine aminopeptidase 2	10.74
Subtilisin-like protein (Sub3)	9.66
Leucine aminopeptidase 2	6.62
Metalloproteinase (Mep3)	5.81
Keratinocytes	
Dipeptidyl peptidase V	2.16
Leucine aminopeptidase 2	1.29

[61] with the Dual RNAseq of *T. rubrum* cocultured in HaCat keratinocytes [34]. The microarray transcripts showed the modulation of 142 transcripts in the presence of keratin. Regarding protease activity, the growth of *T. rubrum* conidia on keratin induced six protease genes. The results of dual RNA-seq showed the modulation of 70 genes and induction of two protease genes (Table 1).

4 Transcriptome Analysis for the Study of the Host–Fungus Relationship: Use of Culture Medium Containing Protein Substrates

In view of the need for in-depth assessment of the pathophysiological features of infections caused by dermatophytes and for expanding our knowledge of these microorganisms at the molecular level, techniques have been applied to trace their transcriptional profile under conditions related to pathogenicity. Based on these data, potential gene candidates for virulence factors were identified and new insights into the mechanisms underlying cellular adaptation and the success of fungal infection were obtained [24, 56].

Given the importance of keratin for the parasitic life of dermatophytes, most studies investigating virulence factors analyzed the response of dermatophytes during their growth on keratin as the main nutrient source [61–63]. In parallel, other studies used different molecules found in the host such as elastin, lipids [64], and collagen [61, 65, 66]. The in vitro growth of dermatophytes on keratin sources is associated with the secretion of different keratinases, many of them characterized in the last decades [67]. Different from other fungal species, dermatophytes secrete multiple endoprotease members of the S8 (subtilisin, SUB) and M36 (fungalysins, MEP) families [68–70]. In fact, proteases are key factors for the degradation of keratin and pathogenicity of these microorganisms. Comparative analysis of the genome of dermatophytes and other fungi identified proteases as one of the four gene classes most enriched in the genome of this group of fungi, highlighting their role in the success of infection process and niche preference of these microorganisms [71].

The first studies assessing the response of dermatophytes during the interaction with host molecules investigated the individual role of secreted proteases using different infection models. Analyses conducted during the growth of dermatophytes such as *T. rubrum* and *T. benhamiae* on keratin and soy medium confirmed the induction of the SUB and MEP genes, as well as genes encoding exoproteases such as leucine aminopeptidases (LAP) and dipeptidyl peptidases (DPP) [62, 72]. In addition to proteases, these studies also showed the induction of genes encoding heat shock proteins, transporters, metabolic enzymes, transcription factors, and hypothetical proteins with unknown functions, as well as of a gene encoding an enzyme of the glyoxylate cycle. In a recent study from our group that assessed the transcription profile of *T. rubrum* grown on keratin or elastin using an oligonucleotide microarray approach covering about 70% of the genome of this fungus, we demonstrated the modulation of 15 protease-coding genes, with expressive induction of the *mep4* and *lap1* genes. In addition, genes encoding heat shock proteins such as Hsp 70 like-protein, Hsp 88-like protein, and Hsp 90 like-protein were modulated [58]. The function of heat shock proteins in the establishment of dermatophyte infections and in the regulation of protease secretion has been documented [73, 74].

The metabolic profile of *T. rubrum* was also evaluated by RNA-seq analysis during the growth of the fungus on keratin compared to glucose. That study showed important repression of essential genes involved in nitrogen metabolism accompanied by the induction of a urea transporter. In addition, the activation of alternative pathways for energy production, such as the glyoxylate cycle and of lipid metabolism, was demonstrated [75]. Using the same approach, another study compared the modulation of genes involved in cell wall synthesis during the growth of *T. rubrum* on keratin *versus* glucose. The data obtained showed that the presence of keratin promotes the consistent induction of genes encoding cell wall-associated proteins such as glycosyl hydrolases, glucanases and an alpha-1,2-mannosyltransferase, as well as a gene encoding a hydrophobin [76].

Assays assessing the growth of *T. rubrum* on keratin or elastin/collagen sources that simulate superficial and deep infection, respectively, demonstrated similar induction of gene products on the two protein sources [65, 66]. In addition, transcriptome data of *T. rubrum* grown on elastin showed that proteases and lipases are equally important during the growth of the dermatophyte [58]. Within this context, it is worth mentioning that skin is rich in lipids and that the *T. benhamiae* and *T. verrucosum* genomes contain 16 lipase genes [58, 77].

Analysis of the transcriptional profile of *T. rubrum* conidia grown on keratin or elastin substrates also demonstrated a quiescent status associated with the maintenance of conidial dormancy, which is an important adaptive mechanism in the early stages of infection. That study revealed the modulation of genes related to the survival and germination of conidia, including genes involved in the cellular respiration chain, signal transduction, and lipid metabolism. In addition, one protein was characterized as a putative adhesin of dermatophytes, which is potentially involved in the host–fungus interaction [61]. The gene coding for this adhesin was induced in culture medium containing keratin and in the presence of antifungal agents, especially fluconazole, indicating an adaptive response to the stress caused by this drug

[78]. Previous proteome data of *T. rubrum* conidia revealed that the maintenance of dormancy is a complex and elaborate process that involves mechanisms associated with reorganization of the cell wall, maintenance of energy metabolism, and the response to different environmental stresses. Taken together, these data contribute to the understanding of the cellular and molecular features of these conidia and indicate new paths to the development of therapeutic strategies [79].

Besides the addition of host molecules in liquid medium, ex vivo infection models using human nail fragments have been proposed to evaluate the expression of genes that are important for the establishment and maintenance of dermatophyte infections and to compare profiles of gene expression obtained during the growth on different keratin substrates [63, 80]. These studies have shown that the modulation of genes is substrate- and time-dependent and that differences in the physiochemical characteristics of the substrates as well as nutrient availability in each protein source influence metabolic responses and gene activation [63]. Finally, infection models provide indications of potential candidates for virulence factors and drug targets [81].

5 Transcriptome Analysis of Dermatophytes in Superficial Infection Models: Importance of Keratinocytes for the Immune Response to Dermatophytes

The human skin is an important organ for the immune system, especially because of the production of soluble factors such as cytokines, chemokines, and antimicrobial peptides by infiltrated leukocytes, keratinocytes, and fibroblasts [82], once again highlighting the importance of these cells as key players in the defense reactions against dermatophytes.

During the infection process in dermatophytoses, keratinocytes are the first cells to come in contact with the fungus. Thus, in addition to their function as structural cells of the epidermis, keratinocytes also participate directly in the defense against these fungal infections and activate other cells of the immune system [83]. Despite these defense mechanisms, dermatophytes selectively reach the epithelial layer using keratin as a nutrient source. Thus, the infection is generally restricted to the stratum corneum [82]. Considering the important role of keratinocytes in the immune defense, molecular studies identifying the genes and pathways that are modulated when dermatophytes interact with human keratinocytes during the infection process are interesting to gain a better understanding of how this host–fungus interaction occurs. In addition, such studies will allow the identification of new target genes for the development of more selective antifungal agents.

Petrucelli et al. [34] were the first to evaluate the transcriptional profile in cocultures of the dermatophyte *T. rubrum* and a human keratinocyte cell line (HaCat) using the dual RNAseq technique. The authors highlighted the importance of the modulation of genes involved in metabolic flexibility and nutrient assimilation by the fungus during the infection process, including genes involved in the glyoxylate cycle, a carboxylic acid transporter gene, and the *ERG6* gene. In addition, the authors demonstrated the induction of genes encoding keratinolytic proteases (LAP2 and DPP V), which are important virulence factors for this dermatophyte. With respect to keratinocytes, genes involved in the repair of the epithelial barrier, genes increasing cell migration, and genes encoding compounds with antimicrobial activity were found to be modulated [34].

The importance of keratinolytic proteases during dermatophyte–keratinocyte interactions has also been explored through comparative and functional genomic analysis of the dermatophytes *T. benhamiae* and *Trichophyton verrucosum*, which cause highly inflammatory infections in humans. These authors cocultured *T. benhamiae* conidia with human keratinocytes and demonstrated the modulation of the genes that encode carboxypeptidase S1 and DPP V to be important for infection [77].

The identification of these proteases is fundamental for the study of the infection process of dermatophytes. After adherence to the host tissue, dermatophytes need to sense and adapt to the pH of the host and obtain nutrients for their survival. The latter are obtained through the secretion of proteolytic enzymes that cleave macromolecules, which are assimilated by the fungus as an energy source. The secretion of these enzymes, particularly keratinolytic enzymes, is considered an important virulence factor of these fungi [24].

The large-scale transcriptional sequencing also was used to evaluate alterations in the gene modulation of HaCat keratinocytes when infected with *T. rubrum* (anthropophilic species) and *Nannizzia gypsea* (geophilic species, formerly called *Microsporum gypseum*). A total of 686 differentially expressed human genes were identified when HaCat cells were infected with *T. rubrum* and *N. gypsea*. Of these, only 51 genes were common to the two species, 568 were specifically expressed in cells infected with the geophilic dermatophyte, and 67 were specifically expressed in cells infected with the anthropophilic dermatophyte. The data reported in that study indicate that different dermatophyte species elicit a different cell response in keratinocytes, a fact that would explain the differences in the clinical manifestations of infections caused by anthropophilic and geophilic dermatophytes [35].

The different responses of keratinocytes to the type of infecting dermatophyte species are important factors during the host–dermatophyte interaction since anthropophilic species usually cause chronic infections accompanied by low-grade inflammation, while infections caused by zoophilic and geophilic species are acute and highly inflammatory [84]. Within this context, the expression of genes encoding several proinflammatory cytokines and chemokines (IL-1β, IL-1A, TNF, IL-6, IL-8, IL-23A, and CXCL1), peptides with antimicrobial activity (HBD2, HBD3, S100A7, and RNASE7), and the pattern recognition receptor TLR2 was observed when keratinocytes were incubated with the zoophilic dermatophyte *T. benhamiae* [82], again demonstrating the importance of these cells as key players in the defense reactions against dermatophytes.

Table 2 shows an overview of the main interleukins released during the infection process caused by *T. rubrum* and other dermatophytes, using different models that mimic the infection.

Table 2 Overview of the main anti-inflammatory and pro-inflammatory interleukins released during infections caused by dermatophytes

IL	Dermatophyte	Infection model	Function	Reference
Anti-inflammatory activity				
TGF-β	*T. benhamiae*	Animal infection model (mice)	It promotes the development of Th17 cells and activates the transcription factors RORγt and STAT3	[100]
IL-4	*T. benhamiae*	Superficial infection in human keratinocyte cell line	It acts in the differentiation of *naive* Th cells into Th2 cells	[103]
IL-10	*T. rubrum*	Resident and peritoneal macrophage cell lines	It acts in the inhibition of macrophages and dendritic cells and in the control of cell-mediated immunity	[92]
IL-13	*T. benhamiae*	Superficial infection in human keratinocyte cell line	Central mediator of physiological changes induced by allergic inflammation in tissues	[103]
Pro-inflammatory activity				
INF-γ	*T. rubrum*	Animal infection model (mice)	It assists in the recruitment of macrophages and neutrophils, and is associated with decreased fungal load	[93]
TNF-α	*T. rubrum*	Resident and peritoneal macrophage cells line	It acts in the activation and differentiation of macrophages	[92]
IL-1β	*T. rubrum*	Murine bone marrow macrophages	It acts in the control of fungal proliferation. When associated with ROS, it acts directly on the inflammatory process	[104]
IL6	*T. rubrum*	Neutrophils and macrophages collected from patients diagnosed with generalized chronic dermatophytosis	It acts to protect against infection by activating the immune system	[105]
IL-8	*T. tonsurans*	Superficial infection in human keratinocyte cell line	It acts as a potent neutrophil chemotactic	[103]
IL-12	*T. mentagrophytes*	Animal infection model (Guinea pig)	T cells: Th1 differentiation. NK cells and T cells: IFN-γ synthesis,	[106]

(continued)

Table 2 (continued)

IL	Dermatophyte	Infection model	Function	Reference
			increased cytotoxic activity	
IL-16	*T. tonsurans*	Superficial infection in human keratinocyte cells line	It acts as a chemoattractant for CD4 + T cells	[103]
IL-17	*T. rubrum*	Superficial infection in human keratinocyte cell line	It stimulates the differentiation of Th17 cells	[107]
IL-18	*T. tonsurans* and *T. benhamiae*	Superficial infection in human keratinocyte cell line	It acts mainly on NK cells and T cells, causing the synthesis of INF-γ	[103]
IL-23	*T. mentagrophytes*	Mouse model of inflammation induced by trichophytin	It acts mainly on NK cells maintaining IL-17 producing T cells. It is produced by macrophages and dendritic cells	[108]

Considering that each dermatophyte species elicits a different cellular response and since models mimicking the interaction between these dermatophyte species and their respective hosts are still scarce, keratinocyte models are promising by increasing the molecular understanding of the infection process. The conclusions drawn from these cell models of host–fungus interaction are still limited since they do not consider the interference of the cells studied with professional immune cells, only providing results regarding the reactions of the cell types evaluated. Nevertheless, we believe that these results contribute to elucidating the immune response mechanisms of keratinocytes to dermatophytes [82].

In addition to infection models using monolayer keratinocyte cultures, another strategy has emerged for the study of dermatophyte adherence and invasion: the in vitro use of human epidermis reconstructed from normal cultured keratinocytes for basic and preclinical studies. However, like monolayer cells, this model also has limitation such as the absence of immune system cells, impairing the investigation of the immune response in dermatophytosis [85].

The development of sequencing associated to superficial infection model has resulted in a better understanding of the host pathways involved in the combat of the pathogen, besides evidence of new targets using antifungal [86]. One example is the study conducted by Petrucelli et al. [87], which identified 277 differentially expressed genes in cocultures of the dermatophyte *T. rubrum* with the HaCat keratinocyte cell line in the presence of terbinafine. Approximately 28% of these genes have not yet been studied in *T. rubrum* but were investigated in other dermatophytes. That study may therefore contribute to elucidating the functions of these genes in this species and the mechanisms underlying the responses to antifungal agents. In addition, the modulation of different genes that are important for the biosynthesis and transport of ergosterol, such as *ERG1*, *ERG5*, *ERG11*, *CYP51*, and *CYP61*, was observed [87].

6 New Possibilities for Studying Deep Infections

Models of immortalized cells, generally human (THP1) [88] or murine macrophages (J774) [89, 90], can be used to simulate deep infections caused by dermatophytes. Another possibility are models using primary cell lines (extracted from the patient's blood) [91] and some studies have used mice as an in vivo infection model [92, 93]. These models can be used to carry out assays that investigate the interactions between the pathogen and its host in order to facilitate the understanding of the infection process in deep dermatophytoses. However, such studies are scarce in the literature and are slow to emerge as demonstrated by the increasing number of cases of deep infection and lack of adequate treatment.

Upon entering the human organism, pathogens through their receptors activate specific pathways in response to this situation, generating an infectious process. Generally the interleukins studied, which are responsible for the immune response, are pro-inflammatory interleukins, such as IL-6 [94], IL-17 [95, 96], TNF-α [97, 98], and IL-1β [99]. High levels of TGF-β and IL-22 were also found in the immune response against dermatophytes [100].

In an attempt to clarify how dermatophytes circumvent the immune system, survive inside phagocytic cells and, consequently, cause deep infections, studies in the field of fungal immunology increased with aiming at the identification of the immunological mechanism behind the fungal infection process. In this context, the glyoxylate cycle is a determining factor for the survival of microorganisms in niches with limited nutrient availability. Based on data mainly obtained for *Candida albicans*, the activation of this cycle permits the synthesis of glucose from lipids and other alternative carbon sources [101, 102]. These pathways can also be active in infections caused by *T. rubrum*, since Cantelli et al. [59] and Komoto et al. [60] verified the induction of genes encoding enzymes involved in the glyoxylate cycle after the coculture of the dermatophyte *T. rubrum* with HaCat keratinocytes.

7 Future Perspectives

Large-scale approaches have contributed to clarifying the host–fungus relationship and response to antifungal agents using microarrays, RNAseq, and dualRNAseq. Future perspectives include improving superficial infection models and focusing on the development of more appropriate models to study deep infections because of the increasing number of fungal infections in immunodepressed patients. In this sense, a schematic representation that illustrates the main achievements about the dermatophyte infection process and new roads for investigation is shown in Fig. 1. Additionally, metabolomic and proteomic studies should be performed. Research on

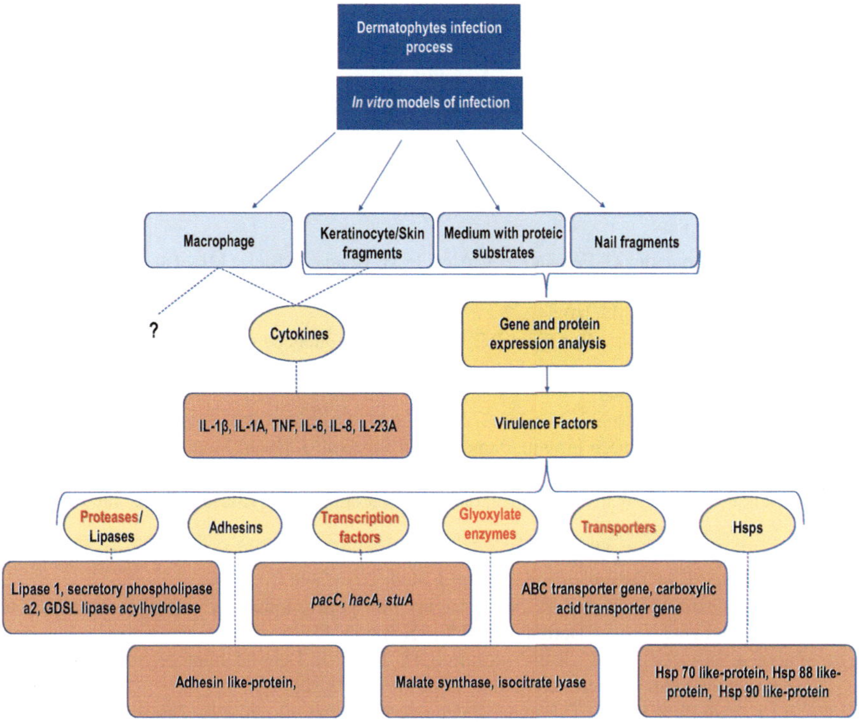

Fig. 1 A summary of the main achievements about the dermatophyte infection process using different models of infection (keratinocytes strain line, human nail fragments, medium supplemented with proteic sources). The concomitant use of models of infection and approaches to assess gene modulation led to identification of crucial dermatophyte virulence factors such as the genes encoding some heat shock proteins (Hsps), proteases, lipases, enzymes from alternative metabolism, adhesins, and transcription factors. Herein, virulence factors highlighted in red color are related to virulence factors validated by functional genomics. Also, the interleukins profiles were mainly obtained from models of infection that used cell lines such as keratinocytes and macrophages. New roads for investigation are represented by the question mark related to use of macrophages to assess more aspects for deep infections caused by dermatophytes

fungal and human microRNAs related to the response to infection and antifungal therapy also has a great potential to be explored.

Acknowledgments Research in the laboratories of authors has been supported by grants from Fundação de Amparo à Pesquisa do Estado de São Paulo (FAPESP) Process: 2016/22701-9 and 2019/10514-8 and Postdoctoral fellowship 2015/23435-8 granted to TAB, CAPES (PhD fellowships granted to G.G.S., M.H.A., M.F.P., B.A.M.C. and Master degree fellowship granted to F.G.N.) and Postdoctoral fellowship (T.A.B.).

References

1. Khurana A, Sardana K, Chowdhary A. Antifungal resistance in dermatophytes: recent trends and therapeutic implications. Fungal Genet Biol. 2019;132:103255.
2. Gnat S, Łagowski D, Nowakiewicz A, Zięba P. The host range of dermatophytes, it is at all possible? Phenotypic evaluation of the keratinolytic activity of *Trichophyton verrucosum* clinical isolates. Mycoses. 2019;62(3):274–83.
3. Mercer DK, Stewart CS. Keratin hydrolysis by dermatophytes. Med Mycol. 2019;57 (1):13–22.
4. Antuori A, Fernández G, Fernández A, Alcaide M, Boada A, Bielsa MI, et al. Epidemiology of dermatophytic infections between 2008 and 2017 in Barcelona, Spain. Enferm Infecc Microbiol Clin. 2019;37(10):642–7.
5. Achterman RR, Smith AR, Oliver BG, White TC. Sequenced dermatophyte strains: growth rate, conidiation, drug susceptibilities, and virulence in an invertebrate model. Fungal Genet Biol. 2011;48(3):335–41.
6. Hube B, Hay R, Brasch J, Veraldi S, Schaller M. Dermatomycoses and inflammation: the adaptive balance between growth, damage, and survival. J Mycol Med. 2015;25(1):e44–58.
7. Sai BS, Tejashree A, Veeranna S, Karthik KM. Speciation and *in vitro* activity of four antifungal drugs against clinical isolates of dermatophytes by E-test method. Int J Sci Res. 2019;8(6):16–8.
8. Lanternier F, Pathan S, Vincent QB, Liu L, Cypowyj S, Prando C, et al. Deep dermatophytosis and inherited CARD9 deficiency. N Engl J Med. 2013;369(18):1704–14.
9. García-Romero MT, Arenas R. New insights into genes, immunity, and the occurrence of dermatophytosis. J Invest Dermatol. 2015;135(3):655–7.
10. Warycha MA, Leger M, Tzu J, Kamino H, Stein J. Deep dermatophytosis caused by *Trichophyton rubrum*. Dermatol Online J. 2011;17(10):21.
11. Wu LC, Sun PL, Chang YT. Extensive deep dermatophytosis cause by *Trichophyton rubrum* in a patient with liver cirrhosis and chronic renal failure. Mycopathologia. 2013;176 (5–6):457–62.
12. Costa JEF, Neves RP, Delgado MM, Lima-Neto RG, Morais VMS, Coêlho MRCD. Dermatophytosis in patients with human immunodeficiency virus infection: clinical aspects and etiologic agents. Acta Trop. 2015;150:111–5.
13. Kershenovich R, Sherman S, Reiter O, Huss SR, Didkovsky E, Mimouni D, et al. A unique clinicopathological manifestation of fungal infection: a case series of deep dermatophytosis in immunosuppressed patients. Am J Clin Dermatol. 2017;18(5):697–704.
14. Kim SH, Jo IH, Kang J, Joo SY, Choi JH. Dermatophyte abscesses caused by *Trichophyton rubrum* in a patient without pre-existing superficial dermatophytosis: a case report. BMC Infect Dis. 2016;16:298.
15. Su H, Li L, Cheng B, Zhu J, Zhang Q, Xu J, et al. *Trichophyton rubrum* infection characterized by Majocchi's granuloma and deeper dermatophytosis: case report and review of published literature. Mycopathologia. 2017;182(5–6):549–54.
16. Dai Y, Xia X, Shen H. Multiple abscesses in the lower extremities caused by *Trichophyton rubrum*. BMC Infect Dis. 2019;19:271.
17. Seo J-K, Jeong K-H, Shin MK, Choi JS, Lee M-H. Dermal infection with *Trichophyton rubrum* in an immunocompetent patient. Ann Dermatol. 2019;31(Suppl):S32–3.
18. Zhang M, Jiang L, Li F, Xu Y, Lv S, Wang B. Simultaneous dermatophytosis and keratomycosis caused by *Trichophyton interdigitale* infection: a case report and literature review. BMC Infect Dis. 2019;19:983.
19. Rouzaud C, Hay R, Chosidow O, Dupin N, Puel A, Lortholary O, et al. Severe dermatophytosis and acquired or innate immunodeficiency: a review. J Fungi (Basel). 2016;2(1):4.
20. Zhong X, Chen B, Yang L, Yang Z. Molecular and physiological roles of the adaptor protein CARD9 in immunity review-article. Cell Death Dis. 2018;9:52.

21. Vaezi A, Mardani M, Fakhim H, Yaghoobi MH, Abtahian Z, Nasri E, et al. Severe disseminated phaeohyphomycosis in a patient with inherited CARD9 deficiency. Arch Clin Infect Dis. 2018;13(6):e84006.
22. Huang C, Peng Y, Zhang Y, Li R, Wan Z, Wang X. Deep dermatophytosis caused by *Trichophyton rubrum*. Lancet Infect Dis. 2019;19(12):1380.
23. Schena M, Shalon D, Davis RW, Brown PO. Quantitative monitoring of gene expression patterns with a complementary DNA microarray. Science. 1995;270(5235):467–70.
24. Martinez-Rossi NM, Peres NTA, Rossi A. Pathogenesis of dermatophytosis: sensing the host tissue. Mycopathologia. 2017;182(1–2):215–27.
25. Hovhannisyan H, Gabaldón T. Transcriptome sequencing approaches to elucidate host–microbe interactions in opportunistic human fungal pathogens. Curr Top Microbiol Immunol. 2019;422:193–235.
26. Velculescu VE, Zhang L, Vogelstein B, Kinzler KW. Serial analysis of gene expression. Science. 1995;270(5235):484–7.
27. Westermann AJ, Gorski SA, Vogel J. Dual RNA-seq of pathogen and host. Nat Rev Microbiol. 2012;10(9):618–30.
28. Wang Z, Gerstein M, Snyder M. RNA-Seq: a revolutionary tool for transcriptomics. Nat Rev Genet. 2009;10(1):57–63.
29. Westermann AJ, Barquist L, Vogel J. Resolving host–pathogen interactions by dual RNA-seq. PLoS Pathog. 2017;13(2):e1006033.
30. Martinez-Rossi NM, Peres NTA, Rossi A. Antifungal resistance mechanisms in dermatophytes. Mycopathologia. 2008;166(5–6):369–83.
31. Kathiravan MK, Salake AB, Chothe AS, Dudhe PB, Watode RP, Mukta MS, et al. The biology and chemistry of antifungal agents: a review. Bioorg Med Chem. 2012;20:5678–98.
32. Martinez-Rossi NM, Bitencourt TA, Peres NTA, Lang EAS, Gomes EV, Quaresemin NR, et al. Dermatophyte resistance to antifungal drugs: mechanisms and prospectus. Front Microbiol. 2018;9:1108.
33. Wolf T, Kämmer P, Brunke S, Linde J. Two's company: studying interspecies relationships with dual RNA-seq. Curr Opin Microbiol. 2018;42:7–12.
34. Petrucelli MF, Peronni K, Sanches PR, Komoto TT, Matsuda JB, da Silva Junior WA, et al. Dual RNA-seq analysis of *Trichophyton rubrum* and HaCat keratinocyte co-culture highlights important genes for fungal-host interaction. Genes (Basel). 2018;9(7):362.
35. Deng W, Liang P, Zheng Y, Su Z, Gong Z, Chen J, et al. Differential gene expression in HaCaT cells may account for the various clinical presentation caused by anthropophilic and geophilic dermatophytes infections. Mycoses. 2020;63(1):21–9.
36. Grumbt M, Defaweux V, Mignon B, Monod M, Burmester A, Wöstemeyer J, et al. Targeted gene deletion and *in vivo* analysis of putative virulence gene function in the pathogenic dermatophyte *Arthroderma benhamiae*. Eukaryot Cell. 2011;10(6):842–53.
37. Fachin AL, Ferreira-Nozawa MS, Maccheroni W, Martinez-Rossi NM. Role of the ABC transporter *Tru*MDR2 in terbinafine, 4-nitroquinoline N-oxide and ethidium bromide susceptibility in *Trichophyton rubrum*. J Med Microbiol. 2006;55(8):1093–9.
38. Ferreira-Nozawa MS, Silveira HCS, Ono CJ, Fachin AL, Rossi A, Martinez-Rossi NM. The pH signaling transcription factor PacC mediates the growth of *Trichophyton rubrum* on human nail *in vitro*. Med Mycol. 2006;44(7):641–5.
39. Yamada T, Makimura K, Hisajima T, Ishihara Y, Umeda Y, Abe S. Enhanced gene replacements in *Ku80* disruption mutants of the dermatophyte, *Trichophyton mentagrophytes*. FEMS Microbiol Lett. 2009;298(2):208–17.
40. Yamada M, Yawata K, Orino Y, Ueda S, Isogai Y, Taguchi G, et al. *Agrobacterium tumefaciens*-mediated transformation of antifungal lipopeptide producing fungus *Coleophoma empetri* F-11899. Curr Genet. 2009;55(6):623–30.
41. Baldo A, Mathy A, Tabart J, Camponova P, Vermout S, Massart L, et al. Secreted subtilisin Sub3 from *Microsporum canis* is required for adherence to but not for invasion of the epidermis. Br J Dermatol. 2010;162(5):990–7.

42. Yamada T, Makimura K, Abe S. Isolation, characterization, and disruption of *dnr1*, the *areA/nit-2*-like nitrogen regulatory gene of the zoophilic dermatophyte, *Microsporum canis*. Med Mycol. 2006;44(3):243–52.
43. Kröber A, Etzrodt S, Bach M, Monod M, Kniemeyer O, Staib P, et al. The transcriptional regulators SteA and StuA contribute to keratin degradation and sexual reproduction of the dermatophyte *Arthroderma benhamiae*. Curr Genet. 2017;63(1):103–16.
44. Bitencourt TA, Lang EAS, Sanches PR, Peres NTA, Oliveira VM, Fachin AL, et al. HacA governs virulence traits and adaptive stress responses in *Trichophyton rubrum*. Front Microbiol. 2020;11:193.
45. Ma C, Li Y, Li M, Deng G, Wu X, Zeng J, et al. MicroRNA-124 negatively regulates TLR signaling in alveolar macrophages in response to mycobacterial infection. Mol Immunol. 2014;62(1):150–8.
46. Axtell MJ, Westholm JO, Lai EC. Vive la différence: biogenesis and evolution of microRNAs in plants and animals. Genome Biol. 2011;12(4):221.
47. Chen R, Jiang N, Jiang Q, Sun X, Wang Y, Zhang H, et al. Exploring microRNA-like small RNAs in the filamentous fungus *Fusarium oxysporum*. PLoS One. 2014;9(8):e104956.
48. Jiang N, Yang Y, Janbon G, Pan J, Zhu X. Identification and functional demonstration of miRNAs in the fungus *Cryptococcus neoformans*. PLoS One. 2012;7(12):e52734.
49. Kang K, Zhong J, Jiang L, Liu G, Gou CY, Wu Q, et al. Identification of microRNA-like RNAs in the filamentous fungus *Trichoderma reesei* by Solexa sequencing. PLoS One. 2013;8(10):e76288.
50. Lau SKP, Chow WN, Wong AYP, Yeung JMY, Bao J, Zhang N, et al. Identification of microRNA-like RNAs in mycelial and yeast phases of the thermal dimorphic fungus *Penicillium marneffei*. PLoS Negl Trop Dis. 2013;7(8):e2398.
51. Lee HC, Li L, Gu W, Xue Z, Crosthwaite SK, Pertsemlidis A, et al. Diverse pathways generate microRNA-like RNAs and dicer-independent small interfering RNAs in fungi. Mol Cell. 2010;38(6):803–14.
52. Özkan S, Mohorianu I, Xu P, Dalmay T, Coutts RHA. Profile and functional analysis of small RNAs derived from *Aspergillus fumigatus* infected with double-stranded RNA mycoviruses. BMC Genomics. 2017;18(1):416.
53. Zhou J, Fu Y, Xie J, Li B, Jiang D, Li G, et al. Identification of microRNA-like RNAs in a plant pathogenic fungus *Sclerotinia sclerotiorum* by high-throughput sequencing. Mol Gen Genomics. 2012;287(4):275–82.
54. Zhou Q, Wang Z, Zhang J, Meng H, Huang B. Genome-wide identification and profiling of microRNA-like RNAs from *Metarhizium anisopliae* during development. Fungal Biol. 2012;116(11):1156–62.
55. Wang L, Xu X, Yang J, Chen L, Liu B, Liu T, et al. Integrated microRNA and mRNA analysis in the pathogenic filamentous fungus *Trichophyton rubrum*. BMC Genomics. 2018;19(1):933.
56. Grumbt M, Monod M, Staib P. Genetic advances in dermatophytes. FEMS Microbiol Lett. 2011;320:79–86.
57. Achterman RR, White TC. A foot in the door for dermatophyte research. PLoS Pathog. 2012;8(3):e1002564.
58. Bitencourt TA, Macedo C, Franco ME, Rocha MC, Moreli IS, Cantelli BAM, et al. Trans-chalcone activity against *Trichophyton rubrum* relies on an interplay between signaling pathways related to cell wall integrity and fatty acid metabolism. BMC Genomics. 2019;20(1):411.
59. Cantelli BAM, Bitencourt TA, Komoto TT, Beleboni RO, Marins M, Fachin AL. Caffeic acid and licochalcone A interfere with the glyoxylate cycle of *Trichophyton rubrum*. Biomed Pharmacother. 2017;96:1389–94.
60. Komoto TT, Bitencourt TA, Silva G, Beleboni RO, Marins M, Fachin AL. Gene expression response of *Trichophyton rubrum* during coculture on keratinocytes exposed to antifungal agents. Evid Based Complement Alternat Med. 2015;2015:180535.

61. Bitencourt TA, Macedo C, Franco ME, Assis AF, Komoto TT, Stehling EG, et al. Transcription profile of *Trichophyton rubrum* conidia grown on keratin reveals the induction of an adhesin-like protein gene with a tandem repeat pattern. BMC Genomics. 2016;17:249.
62. Zaugg C, Monod M, Weber J, Harshman K, Pradervand S, Thomas J, et al. Gene expression profiling in the human pathogenic dermatophyte *Trichophyton rubrum* during growth on proteins. Eukaryot Cell. 2009;8(2):241–50.
63. Peres NT, Silva LG, Santos Rda S, Jacob TR, Persinoti GF, Rocha LB, et al. *In vitro* and *ex vivo* infection models help assess the molecular aspects of the interaction of *Trichophyton rubrum* with the host milieu. Med Mycol. 2016;54(4):420–7.
64. Maranhão FCA, Silveira HCS, Rossi A, Martinez-Rossi NM. Isolation of transcripts overexpressed in the human pathogen *Trichophyton rubrum* grown in lipid as carbon source. Can J Microbiol. 2011;57(4):333–8.
65. Leng W, Liu T, Wang J, Li R, Jin Q. Expression dynamics of secreted protease genes in *Trichophyton rubrum* induced by key host's proteinaceous components. Med Mycol. 2009;47 (7):759–65.
66. Kaufman G, Berdicevsky I, Woodfolk JA, Horwitz BA. Markers for host-induced gene expression in *Trichophyton* dermatophytosis. Infect Immun. 2005;73(10):6584–90.
67. Monod M. Secreted proteases from dermatophytes. Mycopathologia. 2008;166(5–6):285–94.
68. Jousson O, Léchenne B, Bontems O, Mignon B, Reichard U, Barblan J, et al. Secreted subtilisin gene family in *Trichophyton rubrum*. Gene. 2004;339:79–88.
69. Kaplan E, Gonca S, Kandemir H, Döğen A, Hilmioğlu-Polat S, Ilkit M, et al. Genes encoding proteolytic enzymes fungalysin and subtilisin in dermatophytes of human and animal origin: a comparative study. Mycopathologia. 2020;185(1):137–44.
70. Jousson O, Léchenne B, Bontems O, Capoccia S, Mignon B, Barblan J, et al. Multiplication of an ancestral gene encoding secreted fungalysin preceded species differentiation in the dermatophytes *Trichophyton* and *Microsporum*. Microbiology. 2004;150(2):301–10.
71. Martinez DA, Oliver BG, Gräser Y, Goldberg JM, Li W, Martinez-Rossi NM, et al. Comparative genome analysis of *Trichophyton rubrum* and related dermatophytes reveals candidate genes involved in infection. MBio. 2012;3(5):e00259–12.
72. Staib P, Zaugg C, Mignon B, Weber J, Grumbt M, Pradervand S, et al. Differential gene expression in the pathogenic dermatophyte *Arthroderma benhamiae in vitro versus* during infection. Microbiology. 2010;156(3):884–95.
73. Neves-da-Rocha J, Bitencourt TA, Oliveira VM, Sanches PR, Rossi A, Martinez-Rossi NM. Alternative splicing in heat shock protein transcripts as a mechanism of cell adaptation in *Trichophyton rubrum*. Cell. 2019;8(10):1206.
74. Martinez-Rossi NM, Jacob TR, Sanches PR, Peres NTA, Lang EAS, Martins MP, et al. Heat shock proteins in dermatophytes: current advances and perspectives. Curr Genomics. 2016;17 (2):99–111.
75. Martins MP, Rossi A, Sanches PR, Bortolossi JC, Martinez-Rossi NM. Comprehensive analysis of the dermatophyte *Trichophyton rubrum* transcriptional profile reveals dynamic metabolic modulation. Biochem J. 2020;477(5):873–85.
76. Martins MP, Silva LG, Rossi A, Sanches PR, Souza LDR, Martinez-Rossi NM. Global analysis of cell wall genes revealed putative virulence factors in the dermatophyte *Trichophyton rubrum*. Front Microbiol. 2019;10:2168.
77. Burmester A, Shelest E, Glöckner G, Heddergott C, Schindler S, Staib P, et al. Comparative and functional genomics provide insights into the pathogenicity of dermatophytic fungi. Genome Biol. 2011;12(1):R7.
78. de Abreu MH, Bitencourt TA, Franco ME, Moreli IS, Cantelli BAM, Komoto TT, et al. Expression of genes containing tandem repeat patterns involved in the fungal-host interaction and in the response to antifungals in *Trichophyton rubrum*. Mycoses. 2020;63:610–6.
79. Leng W, Liu T, Li R, Yang J, Wei C, Zhang W, et al. Proteomic profile of dormant *Trichophyton rubrum* conidia. BMC Genomics. 2008;9:303.

80. Bitencourt TA, Oliveira FB, Sanches PR, Rossi A, Martinez-Rossi NM. The prp4 kinase gene and related spliceosome factor genes in *Trichophyton rubrum* respond to nutrients and antifungals. J Med Microbiol. 2019;68(4):591–9.
81. Mercer DK, Stewart CS, Miller L, Robertson J, Duncan VMS, O'Neil DA. Improved methods for assessing therapeutic potential of antifungal agents against dermatophytes and their application in the development of NP213, a novel onychomycosis therapy candidate. Antimicrob Agents Chemother. 2019;63(5):e02117–8.
82. Hesse-Macabata J, Morgner B, Morgenstern S, Grimm MO, Elsner P, Hipler UC, et al. Innate immune response of human epidermal keratinocytes and dermal fibroblast to *in vitro* incubation of *Trichophyton benhamiae* DSM 6916. J Eur Acad Dermatol Venereol. 2019;33 (6):1177–88.
83. Brasch J. Pathogenesis of tinea. J Dtsch Dermatol Ges. 2010;8(10):780–6.
84. Heinen M-P, Cambier L, Fievez L, Mignon B. Are Th17 cells playing a role in immunity to dermatophytosis? Mycopathologia. 2017;182(1–2):251–61.
85. Faway É, Lambert de Rouvroit C, Poumay Y. *In vitro* models of dermatophyte infection to investigate epidermal barrier alterations. Exp Dermatol. 2018;27(8):915–22.
86. Ryder NS. Terbinafine: mode of action and properties of the squalene epoxidase inhibition. Br J Dermatol. 1992;126:2–7.
87. Petrucelli MF, Matsuda JB, Peroni K, Sanches PR, Silva WA, Beleboni RO, et al. The transcriptional profile of *Trichophyton rubrum* co-cultured with human keratinocytes shows new insights about gene modulation by terbinafine. Pathogens. 2019;8(4):274.
88. Hennen J, Blömeke B. Keratinocytes improve prediction of sensitization potential and potency of chemicals with THP-1 cells. ALTEX. 2017;34(2):279–88.
89. Kitahara N, Morisaka H, Aoki W, Takeda Y, Shibasaki S, Kuroda K, et al. Description of the interaction between *Candida albicans* and macrophages by mixed and quantitative proteome analysis without isolation. AMB Express. 2015;5(1):127.
90. Stavrum R. Modulation of transcriptional and inflammatory responses in murine macrophages by the *Mycobacterium tuberculosis* mammalian cell entry (Mce)1 complex. PLoS One. 2011;6 (10):e26295.
91. Santiago K, Bomfim GF, Criado PR, Almeida SR. Monocyte-derived dendritic cells from patients with dermatophytosis restrict the growth of *Trichophyton rubrum* and induce CD4-T cell activation. PLoS One. 2014;9(11):e110879.
92. Campos MRM, Russo M, Gomes E, Almeida SR. Stimulation, inhibition and death of macrophages infected with *Trichophyton rubrum*. Microbes Infect. 2006;8(2):372–9.
93. Baltazar L d M, Santos PC, de Paula TP, Rachid MA, Cisalpino PS, Souza DG, et al. IFN-γ impairs *Trichophyton rubrum* proliferation in a murine model of dermatophytosis through the production of IL-1β and reactive oxygen species. Med Mycol. 2014;52(3):293–302.
94. Tanaka T, Narazaki M, Kishimoto T. IL-6 in inflammation, immunity, and disease. Cold Spring Harb Perspect Biol. 2014;6(10):a016295.
95. Sparber F, Dolowschiak T, Mertens S, Lauener L, Clausen BE, Joller N, et al. Langerin+ DCs regulate innate IL-17 production in the oral mucosa during *Candida albicans*-mediated infection. PLoS Pathog. 2018;14(5):e1007069.
96. Conti HR, Gaffen SL. IL-17-mediated immunity to the opportunistic fungal pathogen *Candida albicans*. J Immunol. 2015;195(3):780–8.
97. Bradley J. TNF-mediated inflammatory disease. J Pathol. 2008;214(2):149–60.
98. Yoshikawa FS, De Almeida SR. The role of phagocytes and NETs in dermatophytosis. Mycopathologia. 2017;182(1–2):263–72.
99. Chu J-Q, Shi G, Fan Y-M, Choi I-W, Cha G-H, Zhou Y, et al. Production of IL-1β and inflammasome with up-regulated expressions of NOD-like receptor related genes in *Toxoplasma gondii*-infected THP-1 macrophages. Korean J Parasitol. 2016;54(6):711–7.
100. Cambier L, Weatherspoon A, Defaweux V, Bagut ET, Heinen MP, Antoine N, et al. Assessment of the cutaneous immune response during *Arthroderma benhamiae* and

A. vanbreuseghemii infection using an experimental mouse model. Br J Dermatol. 2014;170 (3):625–33.
101. Lorenz MC, Fink GR. The glyoxylate cycle is required for fungal virulence. Nature. 2001;412 (6842):83–6.
102. Cheah HL, Lim V, Sandai D. Inhibitors of the glyoxylate cycle enzyme ICL1 in *Candida albicans* for potential use as antifungal agents. PLoS One. 2014;9(4):e95951.
103. Shiraki Y, Ishibashi Y, Hiruma M, Nishikawa A, Ikeda S. Cytokine secretion profiles of human keratinocytes during *Trichophyton tonsurans* and *Arthroderma benhamiae* infections. J Med Microbiol. 2006;55(9):1175–85.
104. Tóth EJ, Boros É, Hoffmann A, Szebenyi C, Homa M, Nagy G, et al. Interaction of THP-1 monocytes with conidia and hyphae of different *Curvularia* strains. Front Immunol. 2017;8:1369.
105. de Sousa MGT, Santana GB, Criado PR, Benard G. Chronic widespread dermatophytosis due to *Trichophyton rubrum*: a syndrome associated with a *Trichophyton*-specific functional defect of phagocytes. Front Microbiol. 2015;6:801.
106. Shi Y, Niu Q, Yu X, Jia X, Wang J, Lin D, et al. Assessment of the function of *SUB6* in the pathogenic dermatophyte *Trichophyton mentagrophytes*. Med Mycol. 2016;54(1):59–71.
107. Rincon M. Interleukin-6: from an inflammatory marker to a target for inflammatory diseases. Trends Immunol. 2012;33(11):571–7.
108. Nakamura T, Yoshida N, Nishibu A, Mochizuki T. A mouse model of *Trichophyton* inflammation based on trichophytin-induced contact hypersensitivity. Med Mycol J. 2019;60 (3):65–70.

Genetic Predictors of Susceptibility to Dermatophytosis

Susan M. Abdel-Rahman

1 Introduction

Dermatophyte species infect humans of every age, race, gender, and socioeconomic status on every continent that man inhabits. Yet, despite their widespread nature, not everyone exposed to these pathogens will suffer infection (Fig. 1). For those that do, infections can present totally devoid of clinical manifestations, with symptoms that are mild and indolent, or with acutely inflammatory lesions. When treated, infections may completely resolve, partially respond, or persist recalcitrantly for years on the infected host. An intuitive rationale for these disparities is that something innate to the host determines the nature of their relationship with these ubiquitous organisms. Consequently, investigators have spent over seven decades exploring pedigrees and polymorphisms to discover whether genetics influence the receptivity of the host to the pathogen, the nature and severity of infection, or the likelihood of response to treatment. In this chapter, we review the existing data that examine a link between genetics and infection with consideration for the putative underlying mechanisms that support these findings.

2 Observations in Support of an Association

Dermatophytoses are as commonplace as the family of pathogens that cause them [1–7]. However, active infections are not guaranteed to result following exposure to infectious fungal elements. The binding of fungal elements to keratinized host

S. M. Abdel-Rahman (✉)
Division of Pediatric Clinical Pharmacology, Toxicology, and Therapeutic Innovation, Children's Mercy Hospital, Kansas City, MO, USA

Department of Pediatrics, University of Missouri-Kansas City, School of Medicine, Kansas City, MO, USA
e-mail: srahman@cmh.edu

J.-P. Bouchara et al. (eds.), *Dermatophytes and Dermatophytoses*,
https://doi.org/10.1007/978-3-030-67421-2_9

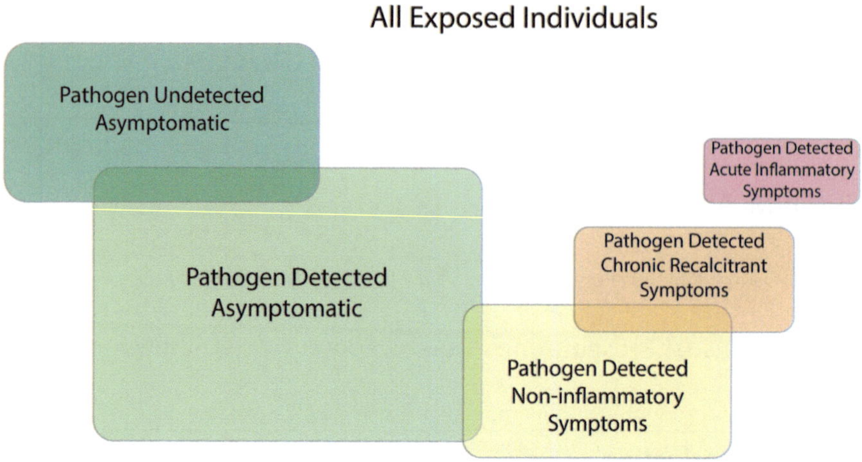

Fig. 1 Varying clinical presentations seen with dermatophyte infections, in relative proportion, as observed in the general population

tissues is not a passive process [8–14], and the adhesion of spores does not, in and of itself, prompt germination and the subsequent cascade of events that lead to active infection [10, 11, 15]. In fact, dermatophytes can be recovered from human skin structures in the absence of clinical disease indicators and can persist on their hosts for months to years without causing symptomatic infection [16–19]. Moreover, the prevalence of asymptomatic infection can exceed that of symptomatic diseases by an order of magnitude [20, 21]. Though chronic carriers appear to be at higher risk for infection, binding of the organism, while necessary, is not sufficient for infection [22].

These findings prompt one to consider why it is that some individuals fail to acquire the pathogen, even in high-risk settings, some acquire the pathogen but remain free of symptoms, and others develop active infections. One explanation could be that the natural course of dermatophyte infections does not lend itself to the cross-sectional sampling strategies favored by most epidemiologic studies. Consequently, such studies, by the nature of their design, may simply fail to identify individuals who will eventually progress to develop active disease. However, there exists compelling evidence that supports the findings above. Using molecular fingerprinting strategies, a longitudinal study following preschool-aged children every month for 2 years was clearly able to discriminate children who never acquired the dermatophyte, from those that intermittently acquired and lost different genetic strains of the pathogen, and those that acquired and sustained infection with the same genetic strain over the entire observation period [21].

Another explanation for the findings may be that the symptoms of infection are mistakenly attributed to other conditions such as eczema, atopic dermatitis, or seborrheic dermatitis [22, 23]. Finally, it may be that different genetic strains with different capacities for infection are resident within the population being evaluated.

Within singular host communities, studies have observed marked genotypic and phenotypic diversity among dermatophyte strains where the secreted proteins responsible for disrupting surface structures, liberating nutrients, and modulating host response span orders of magnitude [15, 24–32]. However, these studies also report identical genetic strains causing different infection patterns in related sibling pairs who share not only the same home but the same school environment [21]. Collectively, these studies lend support to the assertion that, in some fashion, the constitution of the host influences the nature and extent of the relationship that is established with the dermatophytes they encounter.

3 In Search of Evidence

Some of the earliest investigations in which a genetic basis for susceptibility to dermatophyte infections was inferred arise from studies examining infection rates between family members who were genetically related and those having married into the family [33–36]. The problem with these associative studies is that they are confounded by the nature of contact that occurs in a shared environment which has been shown to influence infection rates in populations with no familial relationship [37–43]. As such, these early studies were largely dismissed by investigators arguing that their observations are not practically relevant [34, 44].

The introduction of pedigree analyses into these studies did little to advance evidence of a genetic basis for susceptibility to dermatophyte infections. Autosomal dominant, autosomal recessive, and no discernible pattern of inheritance for susceptibility to dermatophyte infections have all found their way into the medical literature curbing the value of these studies [45–49]. A notable limitation of these pedigree studies is the assumption that susceptibility is a monogenic trait independent of non-inherited factors.

More recently, the search for a link between genetics and infection was broadened by the way of a genome-wide association study in a cohort of children for whom the frequency of dermatophyte infection was characterized with repeated mycologic sampling over the course of several years [50]. This study identified that over 65% of the variability observed in infection rates could be accounted for by sequence variations in 21 genes (Table 1). These included genes involved in leukocyte recruitment, activation, and migration (*SEMA6A*, *ROBO1*, *SLIT3*, *cd99L2*, *CSMD1*, *GAB2*), extracellular matrix formation, integrity and remodeling (*FBLN5*, *FBN2*, *MFAP4*, *SMOC2*, *PCDH7*, *MMP3*, *ADAM 12*), epidermal development, maintenance, and wound repair (*FGF1*, *MAPK8*, *IGF1R*), and skin homeostasis, and host–pathogen interaction (*LASS4*, *GALP*, *KAL1*, *FibCD1*) (Fig. 2) [51–94]. While these genes appear to discriminate the likelihood of chronic dermatophyte infections in the specific pediatric subgroup studied [50], confirming a causal relationship will require replication in an independent cohort of children.

Table 1 Genes identified for their association with cutaneous fungal infections

Gene locus investigated		Putative mechanistic role	Ref.[a]
ADAM12	A disintegrin and metalloprotease 12	• Involved with remodeling of the ECM under normal and pathological conditions	[51, 52]
AIRE	Autoimmune regulator	• Regulates expression of autoantigens and negative selection of autoreactive T-cells	
apoE	Apolipoprotein E	• Proposed to modulate microbial growth by interfering with pathogen-mediated cytokine signaling	
CARD9	Caspase recruitment domain family, member 9	• Mediates cross talk between pattern recognition cells and downstream signaling; selected alleles associated with hyperkeratotic skin conditions	[53]
cd99L2	CD99-like 2	• Extravasation of neutrophils and monocytes into inflamed tissue including skin	[54, 55]
CLEC7A	C-type lectin domain family 7 member A	• Recognition of fungal cell wall β-glucans	
CsmDI	Cub and sushi multiple domains-1	• Complement activation; recently identified as candidate gene associated with psoriasis	[56, 57]
DOCK8	Dedicator of cytokinesis 8	• Involved with intracellular signaling; associated with immunodeficiency and Hyper-IgE recurrent infection syndrome	
FBLN5	Fibulin5	• Essential component of the ECM and elastin fiber network under normal homeostatic conditions and during periods of remodeling and wound healing; mutations in FBLN5 disrupt the elastic properties of the skin	[58–60]
FBN2	Fibrillin-2	• Essential components of the ECM and elastin fiber network; expression markedly upregulated in sclerotic skin diseases	[61–63]
FGF1	Fibroblast growth factor 1	• Epidermal development, maintenance, and wound repair	
FibCD1	Fibrinogen C domain containing 1	• Binds chitin from invading fungi and parasites; highly expressed in the gastrointestinal tract	[64, 65]
GAB2	Grb2-associated binder 2	• Participates in the activation of T- and B-cells via cytokine- and growth factor-mediated signaling; involved in mast cell signaling and function in response to pathogenic stimuli	[66, 67]
GALP	Galanin-like peptide	• A neuropeptide expressed in dermal microvasculature; members of this gene family regulate cutaneous blood flow and demonstrate the ability to inhibit the budding and growth of *Candida albicans*	[68–70]
HLA	Human leukocyte antigens	• Involved in immune regulation; various alleles identified protecting against or predisposing to dermatophytosis	
ICAM-1	Intercellular adhesion molecule 1	• Plays a role in the trans-endothelial migration of leukocytes	

(continued)

Table 1 (continued)

Gene locus investigated		Putative mechanistic role	Ref.[a]
IGF1R	Insulin-like growth factor 1 receptor	• Responds to IGF1 secreted from dermal fibroblasts; dermatologic changes stimulated by *Propionibacterium acnes* appear to be mediated through the IGF/IGF1R system	[71–73]
IL-17F	Interleukin 17 F	• Stimulates the production of several cytokines including IL-6, IL-8, CSF2/GM-CSF	
IL-17RA	Interleukin 17 receptor A	• Binds the pro-inflammatory IL-17A inducing the maturation of neutrophils	
KAL1	Kallmann syndrome 1	• Innervates itch perception; differentially regulated in atopic dermatitis; ortholog of a canine antimicrobial peptide that is active on skin	[74, 75]
Lass4	Longevity assurance gene 4	• Involved in the formation of n-acetylsphinganine from sphinganine; maintains the water permeability and barrier function of the skin; the substrate on which LASS4 acts (sphinganine) is capable of inhibiting mycelial development and growth for a number of cutaneous pathogens including *T. tonsurans*	[76–79]
MAPK8	Mitogen-activated protein kinase	• Involved in TLR signaling in the skin; implicated in the pathogenesis of psoriasis	[80]
MATN2	Matrilin-2	• Thought to be involved in the formation of filamentous networks in the ECM	
MFAP4	Microfibril-associated protein 4	• Essential components of the ECM and elastin fiber network under normal homeostatic conditions and during periods of remodeling and wound healing	[81]
MMP3	Matrix metalloproteinase 3	• Involved with remodeling of the ECM	[82–84]
MPN2	Marapsin-2	• Serine endopeptidase; possible role in epidermal development, maintenance, and wound repair	
PCDH7	BH-protocadherin	• Involved with cell–cell recognition; upregulated during keratinocyte differentiation	[85, 86]
PTPN22	Protein tyrosine phosphatase nonreceptor 22	• Regulation of T-cell and B-cell antigen receptor signaling	
ROBO1	Roundabout receptor 1	• Functions cooperatively with SLIT to influence the directional migration of monocytes and lymphocytes	[87–89]
SEMA6A	Semaphorin 6A	• Guides the migration of antigen-presenting Langerhans cells out of the skin	[90, 91]
SMOC2	Secreted modular calcium-binding protein-2	• Involved with melanocyte function; GWAS studies have identified an association between SMOC2 and vitiligo which is associated with	[92–94]

(continued)

Table 1 (continued)

Gene locus investigated		Putative mechanistic role	Ref.[a]
		lower hydration and altered permeability in the stratum corneum	
SLIT3	Slit homolog 3	• Functions cooperatively with ROBO to influence the directional migration of monocytes and lymphocytes	[87–89]
STAT1/3	Signal transducer and activator of transcription	• Nuclear transcription factors that are activated in response to interferons, cytokines, and other growth factors	
TLR3	Toll-like receptor 3	• Microbial recognition and activation of innate immunity	
Tribl	Tribbles homolog 1	• Binding to signaling proteins of the MAPK pathway; mediates proliferation, apoptosis, and differentiation in cells	

ECM extracellular matrix, *TLR* Toll-like receptor, *GWAS* genome-wide association studies, *MAPK* mitogen-activated protein kinase

[a]Additional references are provided for some genes to supplement the text

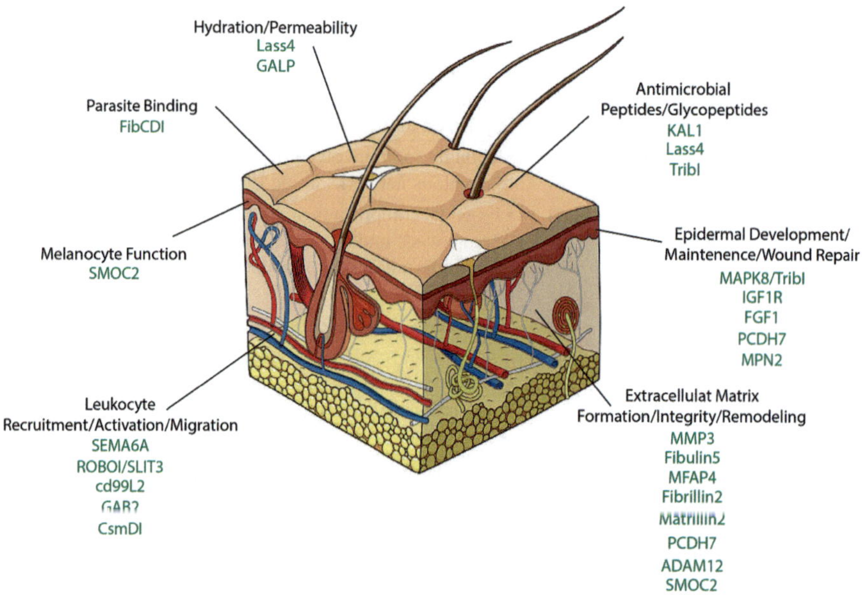

Fig. 2 Putative role of various genes in the skins' defense against dermatophytes (image of skin architecture reproduced with permission © Alexander Pokusay/stock.adobe.com)

4 Observations in Support of Causation

Among the first molecular determinants explored for their relationship with susceptibility to dermatophyte infections were the human leukocyte antigens (HLA) which play a key role in immune regulation. Though early serological testing revealed no association between class I or class II HLA antigens and the presence of infection [95], a more densely populated, controlled investigation suggested that HLA A26 and A33 were observed at higher frequencies in patients with chronic infection [96]. Using polymerase chain reaction, subsequent investigations were able to offer evidence that specific HLA haplotypes or alleles might confer susceptibility to (e.g., DQB1*06, DR8), or protection against (e.g., DR53, DR4, B14, DR6), dermatophyte infections [97–100]. Unfortunately, most of these studies were largely underpowered to detect differences between cases and controls (after controlling for multiple comparisons) even when focused on genetically homogenous populations (e.g., Ashkenazi Jews). Nevertheless, the assertion of a link between inherited immune dysregulation and dermatophyte infections persisted with empiric observations that patients with atopic conditions (e.g., atopic dermatitis, asthma, allergy) experienced higher rates of infection [101–104].

It is only in recent decades that specific gene defects, many of which interfere with innate or adaptive immune responses, have been linked with susceptibility to superficial fungal infections (Table 1). However, the largest fund of knowledge does not come from studies of dermatophytoses but of chronic mucocutaneous candidiasis (CMC). Since there are so few studies centered solely on the dermatophytes, the remainder of this chapter references findings from CMC as they may offer insight into host defense pathways responsible for controlling cutaneous fungal disease.

Several inherited immunodeficiencies feature cutaneous fungal infections, though dermatophytes are infrequently represented among these. However, widespread dermatophyte infection has been observed in the presence of severe combined immunodeficiency (SCID) accompanied by lymphopenia, hypogammaglobulinemia, and reduced complement C3 and C4 [105]. CMC is a hallmark feature of polyglandular autoimmune syndrome type 1 otherwise known as autoimmune polyendocrinopathy with candidiasis and ectodermal dystrophy (APECED). Mutations in the autoimmune regulator gene (*AIRE*) cause this rare autosomal-recessive disease where autoantibodies contribute to markedly reduced levels of IL-17 and IL-22 [106–108]. CMC has also been associated with a mutation in protein tyrosine phosphatase nonreceptor 22 (*PTPN22*) which shares many features with APECED [109].

Superficial fungal infections accompany sequence variations that produce global impairments in T cell function. Mutations in a guanine nucleotide exchange factor involved with intracellular signaling (Dedicator Of Cytokinesis 8, *DOCK8*) have been associated with susceptibility to mucocutaneous fungal infections stemming from defects in CD4+ T helper 17 (T_H17) cell differentiation [110–112]. CMC has also been linked to mutations that disrupt cytokine production and response. These include an autosomal dominant inherited missense mutation in the IL-17F which impairs production of this inflammatory cytokine, and an autosomal recessive nonsense mutation in the interleukin 17 receptor A (*IL-17RA*) gene which results in a failure to express the cytokine receptor and a complete lack of response to IL-17 [113]. In this report, the IL-17F mutation was identified in two healthy familial relations suggesting incomplete clinical penetrance [113]. Evaluations in double-knockout mice confirm the importance of IL-17 and other T_H17 derived cytokines in conferring protection against dermatophytes [114].

Mutations in the Signal Transducer and Activator of Transcription (*STAT*) genes which encode nuclear transcription factors activated in response to interferons, cytokines, and growth factors also predispose cutaneous fungal infections. Both CMC and concurrent, severe dermatophytoses have been reported in patients with a *STAT1* missense mutation that diminishes T_H cell-mediated immunity via inhibition of IL-12 and IL-23 signaling [115, 116]. A separate *STAT1* gain-of-function mutation, which impairs T_H17 response, has also been associated with CMC and extensive *Tinea pedis* [117, 118]. Additional *STAT3* mutations, including *STAT3* mosaicism, have been linked with CMC [119, 120]. Notably, changes in the cutaneous microbiome accompany *STAT1* and *STAT3* mutations which offers a secondary mechanism by which the immune response to cutaneous fungal pathogens may be perturbed [121].

When considering the innate immune system, genes involved with pathogen recognition (e.g., toll-like receptors and c-type lectin receptors) have been implicated in the pathogenesis of cutaneous fungal infections. Mutations in toll-like receptors (TLRs), arguably the most recognizable pathogen recognition receptors, are widely linked with the risk of disseminated infections. However, only TLR3, more precisely the sequence variation L412F which confers blunted cytokine release in response to stimulation with TLR ligands, has been associated with CMC [122, 123]. In contrast, a premature stop codon in C-Type Lectin Domain Family 7 Member A (CLEC7A, Dectin-1) has been linked with both recurrent candidal and dermatophyte infections [124]. The corresponding failure to recognize fungal β-glucans appears to impair the cytokine-mediated response of mononuclear cells [125–127]. The related CLEC6A (Dectin-2) appears to function in a similar role against *Candida* species; however, a clinical link between sequence variations in CLEC6A and susceptibility to dermatophyte infections has not been established.

Also playing a role in cutaneous fungal infections are genes involved in the cross talk between pathogen recognition receptors and downstream transcription factors which drive the expression of cytokines and co-stimulatory molecules that promote proliferation and differentiation of T_H cells, specifically T_H1 and T_H17 cells. For example, reports of missense and nonsense mutations in Caspase Recruitment Domain Family Member 9 (*CARD9*) have been associated with dermatophyte infections which are chronic and spread beyond the superficial keratinized tissues [128–130].

There are also genetic associations where the mechanism is less well understood. In patients with dermatophytosis, 2 of 6 apolipoprotein E (apoE) genotypes appear to be present at frequencies higher (ε23) or lower (ε33) than observed in uninfected controls [131]. Finally, there are associations with cutaneous fungal infections that, while likely genetic in nature, have only been described at the cellular or protein level including lower circulating Intercellular Adhesion Molecule 1 (ICAM-1) levels [132]; increased CD4+CD25+ T regulatory cell numbers [133]; higher numbers of activated T lymphocytes (CD3+HLA-DR+) [134]; and higher total IgE and dermatophyte specific IgG4 [135–137].

We would be remiss not to highlight that far more immunomodulatory proteins than mentioned in this chapter have been implicated in the pathogenesis of dermatophyte infections [138, 139]. However, the naturally occurring sequence variations identified to date which both alter their expression/activity and influence susceptibility to infection are largely restricted to those referenced above.

5 Conclusions

Most of the causally associated sequence variations associated with the risk of cutaneous fungal infections are sufficiently rare that they are of limited relevance in explaining the overwhelmingly large fraction of the population who acquire infection with no clinical indication of immune dysregulation. The heterogeneous nature of our human population with their comparably diverse integumentary barriers [140–146] argue that predisposition to, or protection from, infection is likely the cumulative result of minor changes in multiple genes or gene pathways working in concert with environmental risk factors. Additional genome-wide studies conducted in large, well-phenotyped populations may identify genes that mediate infection risk and, by extension, the gene targets against which therapies can be directed to disrupt host–pathogen interactions. Until such time as these studies are published, a definitive understanding of the genes involved remains elusive.

References

1. Gould WL. Ringworm of the feet. JAMA. 1931;96:1300.
2. Barlow JE, Chaattaway FW. The parasitism of the ringworm group of fungi. AMA Arch Derm. 1958;77:399–405.
3. Kamalam A, Thambiah AS. *Trichophyton violaceum* infection in an Indian school. Int J Dermatol. 1976;15:136–9.
4. Chandrasekaran N, Kamalam A, Thambiah AS. Tinea capitis in an Anglo-Indian school. Trop Geogr Med. 1981;33:253–5.
5. Omar AA. Ringworm of the scalp in primary-school children in Alexandria: infection and carriage. East Mediterr Health J. 2000;6:961–7.
6. Woldeamanuel Y, Mengistu Y, Chryssanthou E, Petrini B. Dermatophytosis in Tulugudu Island, Ethiopia. Med Mycol. 2005;43:79–82.
7. Woldeamanuel Y, Leekassa R, Chryssanthou E, Menghistu Y, Petrini B. Prevalence of tinea capitis in Ethiopian schoolchildren. Mycoses. 2005;48:137–41.
8. Zurita J, Hay RJ. Adherence of dermatophyte microconidia and arthroconidia to human keratinocytes *in vitro*. J Invest Dermatol. 1987;89:529–34.
9. Aljabre SH, Richardson MD, Scott EM, Shankland GS. Germination of *Trichophyton mentagrophytes* on human stratum corneum *in vitro*. J Med Vet Mycol. 1992;30:145–52.

10. Aljabre SH, Richardson MD, Scott EM, Rashid A, Shankland GS. Adherence of arthroconidia and germlings of anthropophilic and zoophilic varieties of *Trichophyton mentagrophytes* to human corneocytes as an early event in the pathogenesis of dermatophytosis. Clin Exp Dermatol. 1993;18:231–5.
11. Rashid A, Scott E, Richardson MD. Early events in the invasion of the human nail plate by *Trichophyton mentagrophytes*. Br J Dermatol. 1995;133:932–40.
12. Esquenazi D, Alviano CS, de Souza W. RozentalS. The influence of surface carbohydrates during *in vitro* infection of mammalian cells by the dermatophyte *Trichophyton rubrum*. Res Microbiol. 2004;155:144–53.
13. Duek L, Kaufman G, Ulman Y, Berdicevsky I. The pathogenesis of dermatophyte infections in human skin sections. J Infect. 2004;48:175–80.
14. Vermout S, Tabart J, Baldo A, Mathy A, Losson B, Mignon B. Pathogenesis of dermatophytosis. Mycopathologia. 2008;166:267–75.
15. Tsuboi R, Ogawa H, Bramono K, Richardson MD, Shankland GS, Crozier WJ, et al. Pathogenesis of superficial mycoses. J Med Vet Mycol. 1994;32(Suppl 1):91–104.
16. Cornbleet T. Cultures from the skin of apparently normal feet. Arch Derm Syphilol. 1926;13:670–1.
17. Ive FA. The carrier stage of tinea capitis in Nigeria. Br J Dermatol. 1966;78:219–21.
18. Neil G, Hanslo D, Buccimazza S, Kibel M. Control of the carrier state of scalp dermatophytoses. Pediatr Infect Dis J. 1990;9:57–8.
19. Williams JV, Honig PJ, McGinley KJ, Leyden JJ. Semiquantitative study of tinea capitis and the asymptomatic carrier state in inner city school children. Pediatrics. 1995;96:265–7.
20. Strickler A, Friedman R. Symptomatic and asymptomatic ringworm of the feet. Arch Derm Syphilol. 1931;24:430.
21. Abdel-Rahman SM, Simon S, Wright KJ, Ndjountche L, Gaedigk A. Tracking *Trichophyton tonsurans* through a large urban childcare center: defining infection prevalence and transmission patterns by molecular stain typing. Pediatrics. 2006;118:2365–73.
22. Mitchell JH. Further studies on ringworm of the hands and feet. Arch Derm Syphilol. 1933;5:174–97.
23. Honig PJ, Smith LR. Tinea capitis masquerading as atopic or seborrheic dermatitis. J Pediatr. 1979;94:604–5.
24. Rippon JW, Varadi DP. The elastase of pathogenic fungi and actinomycetes. J Invest Dermatol. 1968;50:54–8.
25. Page WJ, Stock JJ. Phosphate-mediated alteration of the *Microsporum gypseum* germination protease specificity for substrate: enhanced keratinase activity. J Bacteriol. 1974;117:422–31.
26. Meevootisom V, Niederpruem DJ. Control of exocellular proteases in dermatophytes and especially *Trichophyton rubrum*. Sabouraudia. 1979;17:91–106.
27. Siesenop U, Bohm KH. Comparative studies on keratinase production of *Trichophyton mentagrophytes* strains of animal origin. Mycoses. 1995;38:205–9.
28. Muhsin TM, Aubaid AH, Al-Duboon AH. Extracellular enzyme activities of dermatophytes and yeast on solid media. Mycoses. 1997;40:465–9.
29. Abdel-Rahman SM. Polymorphic exocellular protease expression in clinical isolates of *Trichophyton tonsurans*. Mycopathologia. 2001;150:117–20.
30. Abdel-Rahman SM, Sugita T, Gonzalez-Gonzalez G, Ellis D, Arabatzis M, Vella-Zahra L, et al. Divergence among an international population of *Trichophyton tonsurans* isolates. Mycopathologia. 2010;169:1–13.
31. Abdel-Rahman SM, Farrand N, Schuenemann E, Stering TK, Preuett B, Magie R, et al. Prevalence of infections with *Trichophyton tonsurans* in school children (the CAPITIS study). Pediatrics. 2010;125:966–73.
32. Gnat S, Łagowski D, Nowakiewicz A, Zięba P. Phenotypic characterization of enzymatic activity of clinical dermatophyte isolates from animals with and without skin lesions and humans. J Appl Microbiol. 2018;125(3):700–9.

33. Hodges RS. Ringworm of the nails: a preliminary report of sixteen cases of onychomycosis with a cultural study of twelve of these cases due to Trichophytons. Arch Derm Syphilol. 1921;4:1–26.
34. Lewis GM, Spoor H. Trichophyton purpureum infection (generalized). Arch Derm Syphilol. 1953;68:354–5.
35. English MD. *Trichophyton rubrum* infection in families. Br Med J. 1957;1:744–6.
36. Many H, Derbes VJ, Friedman L. *Trichophyton rubrum*: exposure and infection within household groups. Arch Dermatol. 1960;82:6–9.
37. Lewis SM, Lewis BG. Nosocomial transmission of *Trichophyton tonsurans* tinea corporis in a rehabilitation hospital. Infect Control Hosp Epidemiol. 1997;18:322–5.
38. Poisson DM, Rousseau D, Defo D, Estève E. Outbreak of tinea corporis gladiatorum, a fungal skin infection due to *Trichophyton tonsurans*, in a French high level judo team. Euro Surveill. 2005;10:187–90.
39. Viguié-Vallanet C, Serre M, Masliah L, Tourte-Schaefer C. Épidémie de teignes à *Trichophyton tonsurans* dans une école maternelle de la régionparisienne. Ann Dermatol Venereol. 2005;132:432–8.
40. Ergin S, Ergin C, Erdogan BS, Kaleli I, Evliyaoglu D. An experience from an outbreak of tinea capitis gladiatorum due to *Trichophyton tonsurans*. Clin Exp Dermatol. 2006;31:212–4.
41. Shiraki Y, Hiruma M, Hirose N, Sugita T, Ikeda S. A nationwide survey of *Trichophyton tonsurans* infection among combat sport club members in Japan using a questionnaire form and the hairbrush method. J Am Acad Dermatol. 2006;54:622–6.
42. Shroba J, Olson-Burgess C, Preuett B, Abdel-Rahman S. Molecular strain typing identifies a large outbreak of *Trichophyton tonsurans* among healthcare workers in a pediatric hospital. Am J Infect Control. 2009;37:43–8.
43. Ilkit M, Saracli MA, Kurdak H, Turac-Bicer A, Yuksel T, Karakas M, et al. Clonal outbreak of *Trichophyton tonsurans* tinea capitis gladiatorum among wrestlers in Adana, Turkey. Med Mycol. 2010;48:480–5.
44. Hecht R, Sulzberger MB, Baer RL, Hecht R. Common fungous infections of the feet and groins: negligible role of exposure causing attacks. Arch Derm Syphilol. 1942;45:670–5.
45. Zaias N, Tosti A, Rebell G, Morelli R, Bardazzi F, Bieley H, et al. Autosomal dominant pattern of distal subungual onychomycosis caused by *Trichophyton rubrum*. J Am Acad Dermatol. 1996;34:302–4.
46. Bonifaz A, Araiza J, Koffman-Alfaro S, Paredes-Solis V, Cuevas-Covarrubias S, Rivera MR. Tinea imbricata: autosomal dominant pattern of susceptibility in a polygamous indigenous family of the Nahuatl zone in Mexico. Mycoses. 2004;47:288–91.
47. Serjeantson S, Lawrence G. Autosomal recessive inheritance of susceptibility to tinea imbricata. Lancet. 1977;1:13–5.
48. Ravine D, Turner KJ, Alpers MP. Genetic inheritance of susceptibility to tinea imbricata. J Med Genet. 1980;17:342–8.
49. Hay RJ, Reid S, Talwat E, Macnamara K. Endemic tinea imbricata—a study on Goodenough Island, Papua New Guinea. Trans R Soc Trop Med Hyg. 1984;78:246–51.
50. Abdel-Rahman SM, Preuett BL. Genetic predictors of susceptibility to cutaneous fungal infections: a pilot genome wide association study to refine the candidate gene search. J Dermatol Sci. 2012;67:147–52.
51. Harsha A, Stojadinovic O, Brem H, Sehara-Fujisawa A, Wewer U, Loomis CA, et al. ADAM12: a potential target for the treatment of chronic wounds. J Mol Med. 2008;86:961–9.
52. Seifert O, Bayat A, Geffers R, Dienus K, Buer J, Löfgren S, et al. Identification of unique gene expression patterns within different lesional sites of keloids. Wound Repair Regen. 2008;16:254–65.
53. Furumoto H, Nakamura K, Imamura T, Hamamoto Y, Shimizu T, Muto M, et al. Association of apolipoprotein allele epsilon 2 with psoriasis vulgaris in Japanese population. Arch Dermatol Res. 1997;289:497–500.

54. Schenkel AR, Dufour EM, Chew TW, Sorg E, Muller WA. The murine CD99-related molecule CD99-like 2 (CD99L2) is an adhesion molecule involved in the inflammatory response. Cell Commun Adhes. 2007;14:227–37.
55. Bixel MG, Li H, Petri B, Khandoga AG, Khandoga A, Zarbock A, et al. CD99 and CD99L2 act at the same site as, but independently of, PECAM-1 during leukocyte diapedesis. Blood. 2010;116:1172–84.
56. Kraus DM, Elliott GS, Chute H, Horan T, Pfenninger KH, Sanford SD, et al. CSMD1 is a novel multiple domain complement-regulatory protein highly expressed in the central nervous system and epithelial tissues. J Immunol. 2006;176:4419–30.
57. Sun LD, Cheng H, Wang ZX, Zhang AP, Wang PG, Xu JH, et al. Association analyses identify six new psoriasis susceptibility loci in the Chinese population. Nat Genet. 2010;42:1005–9.
58. Yanagisawa H, Schluterman MK, Brekken RA. Fibulin-5, an integrin-binding matricellular protein: its function in development and disease. J Cell Commun Signal. 2009;3:337–47.
59. Auer-Grumbach M, Weger M, Fink-Puches R, Papic L, Fröhlich E, Auer-Grumbach P, et al. Fibulin-5 mutations link inherited neuropathies, age-related macular degeneration and hyperelastic skin. Brain. 2011;134:1839–52.
60. Langton AK, Sherratt MJ, Griffiths CE, Watson RE. Differential expression of elastic fibre components in intrinsically aged skin. Biogerontology. 2012;13:37–48.
61. Olivieri J, Smaldone S, Ramirez F. Fibrillin assemblies: extracellular determinants of tissue formation and fibrosis. Fibrogenesis Tissue Repair. 2010;3:24.
62. Brinckmann J, Hunzelmann N, Kahle B, Rohwedel J, Kramer J, Gibson MA, et al. Enhanced fibrillin-2 expression is a general feature of wound healing and sclerosis: potential alteration of cell attachment and storage of TGF-beta. Lab Investig. 2010;90:739–52.
63. Samuel CS, Sakai LY, Amento EP. Relaxin regulates fibrillin 2, but not fibrillin 1, mRNA and protein expression by human dermal fibroblasts and murine fetal skin. Arch Biochem Biophys. 2003;411:47–55.
64. Schlosser A, Thomsen T, Moeller JB, Nielsen O, Tornøe I, Mollenhauer J, et al. Characterization of FIBCD1 as an acetyl group-binding receptor that binds chitin. J Immunol. 2009;183:3800–9.
65. Thomsen T, Moeller JB, Schlosser A, Sorensen GL, Moestrup SK, Palaniyar N, et al. The recognition unit of FIBCD1 organizes into a noncovalently linked tetrameric structure and uses a hydrophobic funnel (S1) for acetyl group recognition. J Biol Chem. 2010;285:1229–38.
66. Xie ZH, Ambudkar I, Siraganian RP. The adapter molecule Gab2 regulates Fc epsilon RI-mediated signal transduction in mast cells. J Immunol. 2002;168:4682–91.
67. Nishida K, Yamasaki S, Hasegawa A, Iwamatsu A, Koseki H, Hirano T. Gab2, via PI-3K, regulates ARF1 in FcεRI-mediated granule translocation and mast cell degranulation. J Immunol. 2011;187:932–41.
68. Schmidhuber SM, Santic R, Tam CW, Bauer JW, Kofler B, Brain SD. Galanin-like peptides exert potent vasoactive functions *in vivo*. J Invest Dermatol. 2007;127:716–21.
69. Kofler B, Berger A, Santic R, Moritz K, Almer D, Tuechler C, et al. Expression of neuropeptide galanin and galanin receptors in human skin. J Invest Dermatol. 2004;122:1050–3.
70. Lang R, Gundlach AL, Kofler B. The galanin peptide family: receptor pharmacology, pleiotropic biological actions, and implications in health and disease. Pharmacol Ther. 2007;115:177–207.
71. Lewis DA, Travers JB, Somani AK, Spandau DF. The IGF-1/IGF-1R signaling axis in the skin: a new role for the dermis in aging-associated skin cancer. Oncogene. 2010;29:1475–85.
72. Isard O, Knol AC, Ariès MF, Nguyen JM, Khammari A, Castex-Rizzi N, et al. *Propionibacterium acnes* activates the IGF-1/IGF-1R system in the epidermis and induces keratinocyte proliferation. J Invest Dermatol. 2011;131:59–66.
73. Miura M, Sasaki M, Mizukoshi K, Shibasaki M, Izumi Y, Shimosato G, et al. Peripheral sensitization caused by insulin-like growth factor 1 contributes to pain hypersensitivity after tissue injury. Pain. 2011;152:888–95.

74. Tengara S, Tominaga M, Kamo A, Taneda K, Negi O, Ogawa H, et al. Keratinocyte-derived anosmin-1, an extracellular glycoprotein encoded by the X-linked Kallmann syndrome gene, is involved in modulation of epidermal nerve density in atopic dermatitis. J Dermatol Sci. 2010;58:64–71.
75. Wingate KV, Torres SM, Silverstein KAT, Hendrickson JA, Rutherford MS. Expression of endogenous antimicrobial peptides in normal canine skin. Vet Dermatol. 2009;20:19–26.
76. Hicks AA, Pramstaller PP, Johansson A, Vitart V, Rudan I, Ugocsai P, et al. Genetic determinants of circulating sphingolipid concentrations in European populations. PLoS Genet. 2009;5(10):e1000672.
77. Jarrold B, Mullins L, Bimder R, Osborne R. Expression profiles of stratum corneum lipid metabolism pathways associated with intrinsic and extrinsic aging (Abstract P824). J Am Acad Dermatol. 2009;60:AB28.
78. Jarrold B, Tiesman J, Robinson M, Binder R, Osborne R. Hexamidine, a protease inhibitor, promotes stratum corneum lipid biomarkers *in vitro* (Abstract). J Am Acad Dermatol. 2010;62:AB1.
79. Bibel DJ, Aly R, Shah S, Shinefield HR. Sphingosines: antimicrobial barriers of the skin. Acta DermVenereol. 1993;73:407–11.
80. Ogawa E, Okuyama R, Egawa T, Nagoshi H, Obinata M, Tagami H, et al. p63/p51-induced onset of keratinocyte differentiation via the c-Jun N-terminal kinase pathway is counteracted by keratinocyte growth factor. J Biol Chem. 2008;283:34241–9.
81. Plasari G, Calabrese A, Dusserre Y, Gronostajski RM, McNair A, Michalik L, et al. Nuclear factor I-C links platelet-derived growth factor and transforming growth factor β1 signaling to skin wound healing progression. Mol Cell Biol. 2009;29:6006–17.
82. Clemmensen A, Andersen KE, Clemmensen O, Tan Q, Petersen TK, Kruse TA, et al. Genome-wide expression analysis of human *in vivo* irritated epidermis: differential profiles induced by sodium lauryl sulfate and nonanoic acid. J Invest Dermatol. 2010;130:2201–10.
83. McFarland KL, Glaser K, Hahn JM, Boyce ST, Supp DM. Culture medium and cell density impact gene expression in normal skin and abnormal scar-derived fibroblasts. J Burn Care Res. 2011;32:498–508.
84. Russell SB, Russell JD, Trupin KM, Gayden AE, Opalenik SR, Nanney LB, et al. Epigenetically altered wound healing in keloid fibroblasts. J Invest Dermatol. 2010;130:2489–96.
85. Banno T, Adachi M, Mukkamala L, Blumenberg M. Unique keratinocyte-specific effects of interferon-γ that protect skin from viruses, identified using transcriptional profiling. Antivir Ther. 2003;8:541–54.
86. Taylor JM, Street TL, Hao L, Copley R, Taylor MS, Hayden PJ, et al. Dynamic and physical clustering of gene expression during epidermal barrier formation in differentiating keratinocytes. PLoS One. 2009;4(10):e7651.
87. Geutskens SB, Hordijk PL, van Hennik PB. The chemorepellent Slit3 promotes monocyte migration. J Immunol. 2010;185:7691–8.
88. Wu JY, Feng L, Park HT, Havlioglu N, Wen L, Tang H, et al. The neuronal repellent Slit inhibits leukocyte chemotaxis induced by chemotactic factors. Nature. 2001;410:948–52.
89. Denk AE, Braig S, Schubert T, Bosserhoff AK. Slit3 inhibits activator protein 1-mediated migration of malignant melanoma cells. Int J Mol Med. 2011;28:721–6.
90. Gautier G, de Saint-Vis B, Sénéchal B, Pin JJ, Bates EE, Caux C, et al. The class 6 semaphorin SEMA6A is induced by interferon-gamma and defines an activation status of Langerhans cells observed in pathological situations. Am J Pathol. 2006;168:453–65.
91. Klostermann A, Lutz B, Gertler F, Behl C. The orthologous human and murine semaphorin 6A-1 proteins (SEMA6A-1/Sema6A-1) bind to the enabled/vasodilator-stimulated phosphoprotein-like protein (EVL) via a novel carboxyl-terminal zyxin-like domain. J Biol Chem. 2000;275:39647–53.
92. Birlea SA, Gowan K, Fain PR, Spritz RA. Genome-wide association study of generalized vitiligo in an isolated European founder population identifies SMOC2, in close proximity to IDDM8. J Invest Dermatol. 2010;130:798–803.

93. Alkhateeb A, Al-Dain Marzouka N, Qarqaz F. *SMOC2* gene variant and the risk of vitiligo in Jordanian Arabs. Eur J Dermatol. 2010;20:701–4.
94. Liu J, Man WY, Lv CZ, Song SP, Shy YJ, Elias PM, et al. Epidermal permeability barrier recovery is delayed in vitiligo-involved sites. Skin Pharmacol Physiol. 2010;23:193–200.
95. Svejgaard E, Jakobsen B, Svejgaard A. HLA studies in chronic dermatophytosis caused by *Trichophyton rubrum*. Acta Derm Venereol. 1983;63:254–5.
96. Ahmed AR, Schreiber P, Aiello J, Tiwari JL, Terasaki PI. A preliminary report on the role of some immunologic factors in persistence of chronic tinea pedis. Clin Exp Dermatol. 1985;10:45–50.
97. Zaitz C, Campbell I, Moraes JR, Moraes ME, Gouvea C, Romero M, et al. HLA-associated susceptibility to chronic onychomycosis in Brazilian Ashkenazic Jews. Int J Dermatol. 1996;35:681–2.
98. Sadahiro A, Moraes JRF, Moraes MEH, Romero M, de Lima Gouvea NA, Gouvea CS, et al. HLA in Brazilian Ashkenazic Jews with chronic dermatophytosis caused by *Trichophyton rubrum*. Braz J Microbiol. 2004;35:69–73.
99. Asz-Sigall D, López-García L, Vega-Memije ME, Lacy-Niebla RM, García-Corona C, Ramírez-Rentería C, et al. HLA-DR6 association confers increased resistance to *T. rubrum* onychomycosis in Mexican mestizos. Int J Dermatol. 2010;49:1406–9.
100. García-Romero MT, Granados J, Vega-Memije ME, Arenas R. Analysis of genetic polymorphism of the HLA-B and HLA-DR loci in patients with dermatophytic onychomycosis and in their first-degree relatives. Acta Dermosifiliogr. 2012;103:59–62.
101. Jones HE, Reinhardt JH, Rinaldi MG. Immunologic susceptibility to chronic dermatophytosis. Arch Dermatol. 1974;110:213–20.
102. Rajka G, Barlinn C. On the significance of the trichophytin reactivity in atopic dermatitis. Acta Derm Venereol. 1979;59:45–7.
103. Hay RJ, Shennan G. Chronic dermatophyte infections. II. Antibody and cell-mediated immune responses. Br J Dermatol. 1982;106:191–8.
104. Woodfolk JA. Allergy and dermatophytes. Clin Microbiol Rev. 2005;18:30–43.
105. Jiménez-Puya R, Vázquez-Bayo C, Rodriguez-Bujaldón A, Gómez-García F, Moreno-Giménez JC. Extensive tinea in a patient with severe combined immunodeficiency. Pediatr Dermatol. 2009;26:213–4.
106. Kisand K, Bøe Wolff AS, Podkrajsek KT, Tserel L, Link M, Kisand KV, et al. Chronic mucocutaneous candidiasis in APECED or thymoma patients correlates with autoimmunity to Th17-associated cytokines. J Exp Med. 2010;207:299–308.
107. Kisand K, Lilic D, Casanova JL, Peterson P, Meager A, Willcox N. Mucocutaneous candidiasis and autoimmunity against cytokines in APECED and thymoma patients: clinical and pathogenetic implications. Eur J Immunol. 2011;41:1517–27.
108. Pedroza LA, Kumar V, Sanborn KB, Mace ME, Niinikoski H, Nadeau K, et al. Autoimmune regulator (AIRE) contributes to Dectin-1-induced TNF-α production and complexes with caspase recruitment domain-containing protein 9 (CARD9), spleen tyrosine kinase (Syk), and Dectin-1. J Allergy Clin Immunol. 2012;129:464–72.
109. Nahum A, Bates A, Sharfe N, Roifman CM. Association of the lymphoid protein tyrosine phosphatase, R620W variant, with chronic mucocutaneous candidiasis. J Allergy Clin Immunol. 2008;122:1220–2.
110. Korn T, Bettelli E, Oukka M, Kuchroo VK. IL-17 and Th17 cells. Annu Rev Immunol. 2009;27:485–517.
111. Chu EY, Freeman AF, Jing H, Cowen EW, Davies J, Su HC, et al. Cutaneous manifestations of DOCK8 deficiency syndrome. Arch Dermatol. 2012;148:79–84.
112. Al Khatib S, Keles S, Garcia-Lloret M, Karakoc-Aydiner E, Reisli I, Artac H, et al. Defects along the T(H)17 differentiation pathway underlie genetically distinct forms of the hyper IgE syndrome. J Allergy Clin Immunol. 2009;124:342–8.
113. Puel A, Cypowyj S, Bustamante J, Wright JF, Liu L, Lim HK, et al. Chronic mucocutaneous candidiasis in humans with inborn errors of interleukin-17 immunity. Science. 2011;332:65–8.

114. Verma AH, Gaffen SL. Dermatophyte immune memory is only skin-deep. J Invest Dermatol. 2019;139(3):517–9.
115. van de Veerdonk FL, Plantinga TS, Hoischen A, Smeekens SP, Joosten LA, Gilissen C, et al. STAT1 mutations in autosomal dominant chronic mucocutaneous candidiasis. N Engl J Med. 2011;365:54–61.
116. Smeekens SP, Plantinga TS, van de Veerdonk FL, Heinhuis B, Hoischen A, Joosten LA, et al. STAT1 hyperphosphorylation and defective IL12R/IL23R signaling underlie defective immunity in autosomal dominant chronic mucocutaneous candidiasis. PLoS One. 2011;6(12):e29248.
117. Nielsen J, Kofod-Olsen E, Spaun E, Larsen CS, Christiansen M, Mogensen TH. A STAT1-gain-of-function mutation causing Th17 deficiency with chronic mucocutaneous candidiasis, psoriasiform hyperkeratosis and dermatophytosis. BMJ Case Rep. 2015;2015:bcr2015211372. https://doi.org/10.1136/bcr-2015-211372.
118. Mekki N, Ben-Mustapha I, Liu L, Boussofara L, Okada S, Cypowyj S, et al. IL-17 T cells' defective differentiation *in vitro* despite normal range *ex vivo* in chronic mucocutaneous candidiasis due to STAT1 mutation. J Invest Dermatol. 2014;134:1155–7.
119. Milner JD, Brenchley JM, Laurence A, Freeman AF, Hill BJ, Elias KM, et al. Impaired T(H)17 cell differentiation in subjects with autosomal dominant hyper-IgE syndrome. Nature. 2008;452:773–6.
120. Hsu AP, Sowerwine KJ, Lawrence MG, Davis J, Henderson CJ, Zarember KA, et al. Intermediate phenotypes in patients with autosomal dominant hyper-IgE syndrome caused by somatic mosaicism. J Allergy Clin Immunol. 2013;131:1586–93.
121. Smeekens SP, Huttenhower C, Riza A, van de Veerdonk FL, Zeeuwen PL, Schalkwijk J, et al. Skin microbiome imbalance in patients with STAT1/STAT3 defects impairs innate host defense responses. J Innate Immun. 2014;6:253–62.
122. Nahum A, Dadi H, Bates A, Roifman CM. The L412F variant of Toll-like receptor 3 (TLR3) is associated with cutaneous candidiasis, increased susceptibility to cytomegalovirus, and autoimmunity. J Allergy Clin Immunol. 2011;127:528–31.
123. Nahum A, Dadi H, Bates A, Roifman CM. The biological significance of TLR3 variant, L412F, in conferring susceptibility to cutaneous candidiasis, CMV and autoimmunity. Autoimmun Rev. 2012;11:341–7.
124. Ferwerda B, Ferwerda G, Plantinga TS, Willment JA, van Spriel AB, Venselaar H, et al. Human dectin-1 deficiency and mucocutaneous fungal infections. N Engl J Med. 2009;361:1760–7.
125. Saijo S, Fujikado N, Furuta T, Chung SH, Kotaki H, Seki K, et al. Dectin-1 is required for host defense against *Pneumocystis carinii* but not against *Candida albicans*. Nat Immunol. 2007;8:39–46.
126. Taylor PR, Tsoni SV, Willment JA, Dennehy KM, Rosas M, Findon H, et al. Dectin-1 is required for beta-glucan recognition and control of fungal infection. Nat Immunol. 2007;8:31–8.
127. Hise AG, Tomalka J, Ganesan S, Patel K, Hall BA, Brown GD, et al. An essential role for the NLRP3 inflammasome in host defense against the human fungal pathogen *Candida albicans*. Cell Host Microbe. 2009;5:487–97.
128. Glocker EO, Hennigs A, Nabavi M, Schäffer AA, Woellner C, Salzer U, et al. A homozygous CARD9 mutation in a family with susceptibility to fungal infections. N Engl J Med. 2009;361:1727–35.
129. Engelhardt KR, Grimbacher B. Mendelian traits causing susceptibility to mucocutaneous fungal infections in human subjects. J Allergy Clin Immunol. 2012;129:294–305.
130. Lanternier F, Pathan S, Vincent QB, Liu L, Cypowyj S, Prando C, et al. Deep dermatophytosis and inherited CARD9 deficiency. N Engl J Med. 2013;369:1704–14.
131. Tursen U, Kaya TI, Eskandari G, Bokecli E, Muslu N, Camdeviren H, et al. Apolipoprotein E gene polymorphism and serum lipids in patients with superficial fungal disease. Yonsei Med J. 2004;45:375–9.

132. Zuccarello D, Salpietro DC, Gangemi S, Tosacano V, Merlino MV, Briuglia S, et al. Familial chronic nail candidiasis with ICAM-1 deficiency: a new form of chronic mucocutaneous candidiasis. J Med Genet. 2002;39:671–5.
133. Kaya TI, Eskandari G, Guvenc U, Gunes G, Tursen U, BurakCimen MY, et al. CD4+CD25+ Treg cells in patients with toenail onychomycosis. Arch Dermatol Res. 2009;301:725–9.
134. Maleszka R, Adamski Z, Dworacki G. Evaluation of lymphocytes subpopulations and natural killer cells in peripheral blood of patients treated for dermatophyte onychomycosis. Mycoses. 2001;44:487–92.
135. Jones HE. The atopic-dermatophytosis syndrome. Acta Derm Venereol Suppl. 1980;92:81–5.
136. Leibovici V, Evron R, Axelrod O, Westerman M, Shalit M, Barak V, et al. Imbalance of immune responses in patients with chronic and widespread fungal skin infection. Clin Exp Dermatol. 1995;20:390–4.
137. Tigga RA, Das S, Bhattacharya SN, Saha R, Pandhi D, Datt S, Rai G. Burden of chronic dermatophytosis in a tertiary care hospital: interaction of fungal virulence and host immunity. Mycopathologia. 2018;183(6):951–9.
138. de Hoog S, Monod M, Dawson T, Boekhout T, Mayser P, Gräser Y. Skin fungi from colonization to infection. Microbiol Spectr. 2017;5(4) https://doi.org/10.1128/microbiolspec.FUNK-0049-2016.
139. Hube B, Hay R, Brasch J, Veraldi S, Schaller M. Dermatomycoses and inflammation: the adaptive balance between growth, damage, and survival. J Mycol Med. 2015;25(1):e44–58.
140. Fuchs E. Keratins and the skin. Annu Rev Cell Dev Biol. 1995;11:123–53.
141. Deedrick DW. Hairs, fibers, crime, and evidence. Forensic Sci Commun. 2000;2(3)
142. Franbourg A, Hallegot P, Baltenneck F, Toutain C, Leroy F. Current research on ethnic hair. J Am Acad Dermatol. 2003;48:S115–9.
143. Saferstein R. Criminalistics: an introduction to forensic science. 8th ed. Upper Saddle River, NJ: Pearson Education Inc.; 2004.
144. Rogers MA, Winter H, Langbein L, Bleiler R, Schweizer J. The human type I keratin gene family: characterization of new hair follicle specific members and evaluation of the chromosome 17q21.2 gene domain. Differentiation. 2004;72:527–40.
145. Rogers MA, Edler L, Winter H, Langbein L, Beckmann I, Schweizer J. Characterization of new members of the human type II keratin gene family and a general evaluation of the keratin gene domain on chromosome 12q13.13. J Invest Dermatol. 2005;124:536–44.
146. Jablonski NG. The evolution of human skin and skin color. Annu Rev Anthropol. 2004;33:585–623.

Part IV

Epidemiology of Dermatophytes and Dermatophytosis

The *Trichophyton rubrum* Complex

Ann Packeu, Dirk Stubbe, and Marijke Hendrickx

Abstract

The most important species of the *Trichophyton rubrum* complex are *T. rubrum*, causing mainly skin and nail infections, and *Trichophyton violaceum*, and *Trichophyton soudanense*, which are mostly associated with *Tinea capitis*. Despite their close similarity, recent polyphasic studies confirm their position as valid and independent species with distinct clinical outcomes and geographical distributions. Combining clinical manifestations with results obtained from physiological and morphological tests can be useful for mycologists in the identification of these closely related species. However, sequencing of the internal transcribed spacer (ITS) regions of ribosomal DNA remains the gold standard for correct identification of isolates belonging to this species complex.

Keywords

Dermatophytes · Dermatophytosis · *Tinea* · Taxonomy · Epidemiology

A. Packeu (✉)
Sciensano, Service of Mycology and Aerobiology, Brussels, Belgium
e-mail: ann.packeu@sciensano.be

D. Stubbe · M. Hendrickx
Sciensano, Service of Mycology and Aerobiology, Brussels, Belgium

BCCM/IHEM Fungi Collection, Sciensano, Service of Mycology and Aerobiology, Brussels, Belgium
e-mail: dirk.stubbe@sciensano.be; marijke.hendrickx@sciensano.be

J.-P. Bouchara et al. (eds.), *Dermatophytes and Dermatophytoses*,
https://doi.org/10.1007/978-3-030-67421-2_10

1 Taxonomical Developments in the *Trichophyton rubrum* Complex

Trichophyton rubrum (Castell.) Sabour 1911 and *Trichophyton violaceum* Sabour. ex E. Bodin 1902 are the two core species of what is now considered the *T. rubrum* complex. *Trichophyton violaceum* was described in 1902 by Raymond Sabouraud who isolated this species from a patient returning from Sudan [1]. Unlike other *Trichophyton* species known at the time, he noticed that the cultures were slow growing and developing a violet to aubergine color, hence the species epithet *violaceum*. It was the Italian doctor Aldo Castellani, while working in Ceylon (currently Sri Lanka), who was the first to describe *Epidermophyton rubrum* in 1910 which was isolated from two cases of *Tinea cruris* [2]. The main distinguishing character for this new dermatophyte species was the development of red pigmentation on Sabouraud culture medium. The following year, in 1911, Raymond Sabouraud transferred the species to the genus *Trichophyton*. Between 1910 and 1950, numerous new species names of dermatophytes were introduced, many of which were later considered superfluous. In 1958, the Botanical Code for Nomenclature demanded the designation of a type specimen when describing a new species. This most likely had a stabilizing effect on the dermatophyte taxonomy, which nonetheless, remained complicated. In a 1995 review of dermatophytes, Weitzman and Summerbell [3] gave a synopsis of the accepted dermatophyte species which included *T. rubrum*, *T. violaceum*, *Trichophyton soudanense* Joyeux 1912, *Trichophyton gourvilii* Catanei 1933, *Trichophyton yaoundei* G. Cochet and Doby-Dub 1957 (nom. inv.), *Trichophyton megninii* R. Blanch 1895, *Trichophyton raubitschekii* J. Kane, Salkin, Weitzman and Smitka 1981, and *Trichophyton kanei* Summerb 1987, although they mentioned that some mycologists considered the latter two species as synonyms of *T. rubrum*. In 2000, Gräser et al. [4] published the first molecular phylogenetic analysis of the *T. rubrum* complex. They incorporated the type specimens of *Trichophyton fischeri* J. Kane 1977, *Trichophyton fluviomuniense* Pereiro 1968, *T. kanei*, *Trichophyton kuryangei* Vanbreus. and S.A. Rosenthal 1961, *T. raubitschekii*, *Trichophyton rodhainii* Vanbreus 1949, *Trichophyton cerebriforme* Sabour 1910, and *T. violaceum* var. *indicum* Acton and McGuire 1929. A neotype was assigned to *T. rubrum* (CBS 392.58) and *T. violaceum* (CBS 374.92). They also included strains considered as authentic strains of the species *T. yaoundei* and *Trichophyton pervesii* Catanei 1937, and strains identified as *Trichophyton circonvolutum* Sabour 1910, *Trichophyton glabrum* Sabour 1910, *T. gourvilii*, *T. megninii*, *Trichophyton pedis* M. Ota 1922, *T. rubrum* var. *nigricans* Frágner 1966, and *T. soudanense*. In their conclusion, they proposed a drastic reduction in names, keeping only *T. rubrum* and *T. violaceum*. This sudden simplification did not appear to be completely accepted by the medical community. Especially *T. soudanense* continued to be considered as a relevant species in clinical surveys and case studies [5–8]. In 2008, Gräser et al. [9] proposed an updated taxonomy for the dermatophytes, in which they listed—next to *T. violaceum* and *T. rubrum*—also a *T. rubrum* "African population" (instead of *T. soudanense*). They also mentioned the distinct characteristics of strains identified

Table 1 Historical overview of accepted species related to the *T. rubrum* complex

Sabouraud [13]	Ota and Langeron [14]	Emmons [15]	Ajello [16]	Matsumoto and Ajello [17]
T. circonvolutum	*Sabouraudites ruber*	*T. megninii*	*T. gourvilii*	*T. gourvilii*
T. glabrum	*S. violaceus*	*T. rubrum*	*T. megninii*	*T. megninii*
T. rosaceum	*T. circonvolutum*	*T. violaceum*	*T. rubrum*	*T. rubrum*
T. vinosum	*T. megninii*		*T. soudanense*	*T. soudanense*
T. violaceum	*T. pedis*		*T. violaceum*	*T. violaceum*
			T. yaoundei (nom. inv.)	*T. yaoundei* (nom. inv.)
Weitzman and Summerbell [3]	**Gräser et al.** [4]	**Gräser et al.** [9]	**de Hoog et al.** [10]	**Su et al.** [11] **Packeu et al.** [12]
T. gourvilii	*T. rubrum*	*T. rubrum*	*T. rubrum*	*T. rubrum*
T. kanei	*T. violaceum*	*T. rubrum* "Afr. pop."	*T. soudanense*	*T. soudanense*
T. megninii		*T. violaceum*	*T. violaceum*	*T. violaceum*
T. raubitschekii				(*T. kuryangei*)
T. rubrum				(*T. megninii*)
T. soudanense				
T. violaceum				
T. yaoundei (nom. inv.)				

Nom. inv. invalid name, *Afr. pop.* African population

as *T. megninii*, but still regarded this species name as a synonym of *T. rubrum* based on their phylogenetic results. The latest phylogenetic revision of the *Arthrodermataceae*, in 2017, by de Hoog et al. [10], took into account the "one fungus one name" principle, and is currently the reference for the taxonomy of dermatophytes. Although they accepted *T. soudanense* as a species (assigning a neotype IHEM 19751), they remark that it might be indistinguishable from *T. violaceum* and that species borderlines should further be investigated. In two recent publications, Su et al. [11] and Packeu et al. [12], both in 2019, zoomed in on the *T. rubrum* complex and performed extensive polyphasic analyses, incorporating many strains of African origin. Their results corroborated the status of *T. soudanense* as a species in its own right. A historical overview of accepted taxa connected to the *T. rubrum* complex is given in Table 1 [3, 4, 9–17]. It shows that also *T. megninii*, *T. gourvilii*, and *T. yaoundei* (nom. inv.) have been accepted as species for a long time. Following Packeu et al. [12], *T. gourvilii* should be considered a synonym of *T. soudanense*, and strains identified as *T. yaoundei* are in fact white variants of *T. violaceum.* On the other hand, *T. megninii* and *T. kuryangei* appear to be distinct entities, though this conclusion was based only on a few strains and merits additional research. As these two species are rather rare and geographically restricted (*T. megninii*: Portugal and Sardinia; *T. kuryangei*: central Africa), we will discuss in this chapter only the three main species of the *T. rubrum* complex, being *T. rubrum*, *T. violaceum*, and *T. soudanense*.

2 The Big Three of the Complex

Since the 1950s, *T. rubrum* is widely reported and the most prevalent species involved in chronic skin and nail infections. Important risk factors are a family history of *Tinea pedis*, onychomycosis and advanced age [18]. Its universal predominance can be explained by a shift in lifestyle of the population, such as the use of modern occlusive footwear, causing humidity, and leading to maceration, thus promoting the emergence of *Tinea pedis* and toenail onychomycosis [19, 20], and the common use of sport facilities, pools, or showers [21]. Moreover, the development of new antifungal drugs such as griseofulvin in 1958, leading to the disappearance of both *Microsporum audouinii* and *Trichophyton schoenleinii* from central Europe, can also be linked to the rise of *T. rubrum* infections [22]. On the contrary, *T. violaceum* and *T. soudanense* which are predominant in arid climate zones of western China, Africa, and the Middle East, mostly cause *Tinea capitis* with clinical appearance varying from asymptomatic carriage to kerion, favus, scalp penetration, and black dot infections [23–26]. Both species share the same prevalent sites of isolation, i.e., scalp and hair (penetrating the hair shaft—endothrix) (Fig. 1), but show different geographical distributions. While *T. violaceum* is reported as a major causal agent of *Tinea capitis* in the Middle East and in East African countries, *T. soudanense* is predominantly observed in West African countries such as Mali

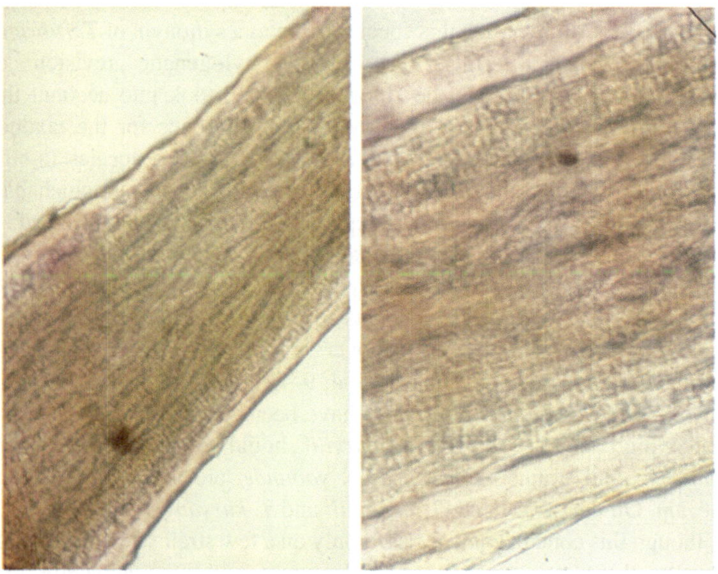

Fig. 1 Endothrix colonization of the hair

and Senegal. The appearance of these two species in outbreaks of *Tinea capitis* in several European countries is linked to human migration [23]. Clarification of the taxonomy of this group is crucial for correct identifications in routine analyses and to allow detailed epidemiological studies.

3 How to Recognize the Three Main Taxa?

3.1 Molecular Characterization

Nomenclature of the members of the *T. rubrum* complex has a long history. Synonymy of different species, based on a molecular phylogeny, was first proposed in 2000 by Gräser et al. [4] who reclassified all members as *T. rubrum* and *T. violaceum. Trichophyton soudanense* was considered a synonym of *T. violaceum*. However, results from Ohst et al. [27] in 2004 obtained via microsatellite-typing grouped isolates of *T. soudanense* with *T. rubrum* rather than with *T. violaceum*. A 2008 study of Gräser et al. [28] listed isolates of *T. soudanense* this time as part of an "African population" of *T. rubrum*. Meanwhile, the name *T. soudanense* remained to be used all over the world in routine clinical analysis. *Tinea capitis* reports from New Zealand [29], USA [7], Mali [30], and Italy [31], all (morphologically) identified and reported both *T. soudanense* and *T. violaceum* as the causal agents. This illustrates their distinctive phenotypes and the wide acceptance of both species concepts in the medical community.

Despite their close genetic similarity, two recent polyphasic studies confirm the position of *T. rubrum*, *T. soudanense*, and *T. violaceum* as valid and independent species [11, 12] as proposed in 2017 by de Hoog et al. [10]. In order to delineate these taxonomic entities, the combination of molecular, phenotypic, and physiological characterization was needed.

Mainly sequencing of the internal transcribed spacer (ITS) regions of ribosomal DNA (rDNA) revealed the taxonomy of these three closely related species. Several studies already confirmed that the ITS regions provide sufficient phylogenetic resolution to support the different lineages within the genus *Trichophyton*. Adding the β-tubulin sequence data does not provide any additional support [10, 12, 32–34]. Indeed, studies performed by Su et al. [11] and Packeu et al. [12] using ITS alignments showed the existence of three species in the *T. rubrum* complex which are interpreted as *T. rubrum*, *T. soudanense*, and *T. violaceum*. In certain strains identified as *T. soudanense*, Packeu et al. [12] have discovered a deletion of 37 base pairs in the otherwise conserved region of ITS1 without resulting in a separate position in the multilocus sequence analysis. Currently other molecular methods, such as amplified fragment length polymorphism (AFLP), and Matrix-Assisted Laser Desorption/Ionization Time-Of-Flight Mass Spectrometry (MALDI-TOF MS) do not have enough discriminatory power to accurately distinguish isolates of these three different species [35].

An alternative and interesting hypothesis was suggested in 2018 by Zhan et al. [32]. In this study, it was proposed that *T. violaceum* could be a phenotypically different strain of *T. rubrum*, showing epigenetic changes when growing on either the scalp or naked skin. Whole-genome sequencing was applied on a clinical strain of both species and draft sequences were compared regarding proteases and adhesins, which are key virulence factors for dermatophytes. The genome size of *T. violaceum* (23.4 Mb) appeared to be larger than that of *T. rubrum* (22.3 Mb). Gene families of secreted proteases appeared to be conserved throughout the members of the *Arthrodermataceae* family. However, the clinical difference of both species might partially be explained by the differences in the production of adhesins which determine host tissue adherence. While both species have 17 putative adhesins in common, four are specific for *T. rubrum* and eight for *T. violaceum*. Performing multilocus phylogeny and genome comparison underlined once again the close affinity between both dermatophyte species. Nevertheless, further exploration of the genomes, has proven useful to uncover and understand the exact biological relationships between the species of the *Trichophyton rubrum* species complex.

3.2 Phenotypical Characterization

Over a large period of time, diverse morphological studies on the members of the complex have been performed. Striking is the slower growth of the colonies of *T. soudanense* and *T. violaceum* in comparison with isolates of *T. rubrum*. This latter species displays colonies which are fluffy to cottony, white, sometimes becoming pink when aging and with a wine-red to olive reverse (Fig. 2a). In contrast, strains identified as *T. soudanense* are usually slightly wrinkled, and predominantly yellow to orange with yellow-orange to brownish reverse producing star-like colonies with irregular margins, while others can be more cottony with more regular, smooth margins (Fig. 2b). Strains of the *T. violaceum* group are white/purple, white/beige, or pink, with a purple to brown reverse (Fig. 2c). Microscopically, no microconidia are observed in isolates of *T. violaceum* (Fig. 3c), while these are generally present in *T. rubrum* and *T. soudanense* (Fig. 3a, b). The microconidia from the latter species are clavate to piriform and reflexive branching is mostly observed (Fig. 3b).

Physiological tests revealed that all members of the complex were positive for hydrolysis of urea at 24 °C [11]. Only isolates belonging to *T. soudanense* and *T. violaceum* were able to hydrolyze Bromocresol Purple-Milk solids-Glucose Agar (BCP) when incubated at 37 °C, and this property was usually associated with the presence of asymmetrical chlamydospores [9, 12]. Based on the results obtained via the lipolysis of Tween 80, *T. violaceum* has a higher lipolytic ability than the other two species [11]. Since Tween 80 is derived from polyethoxylated sorbitan and oleic acid, it can be used as an indicator of the production of lipolytic enzymes [36]. The reduced lipolytic ability of *T. rubrum* can be linked to the fact that it infects glabrous skin resulting in the direct degradation of the epidermis of the skin in contrast to *T. violaceum* which enters the more lipid-rich central hair shaft [37]. As proposed by Su et al. [11], the limited hydrolytic capacity of *T. soudanense* may be associated with a dryer hair-type prevalent in Africa, explaining the different patterns between

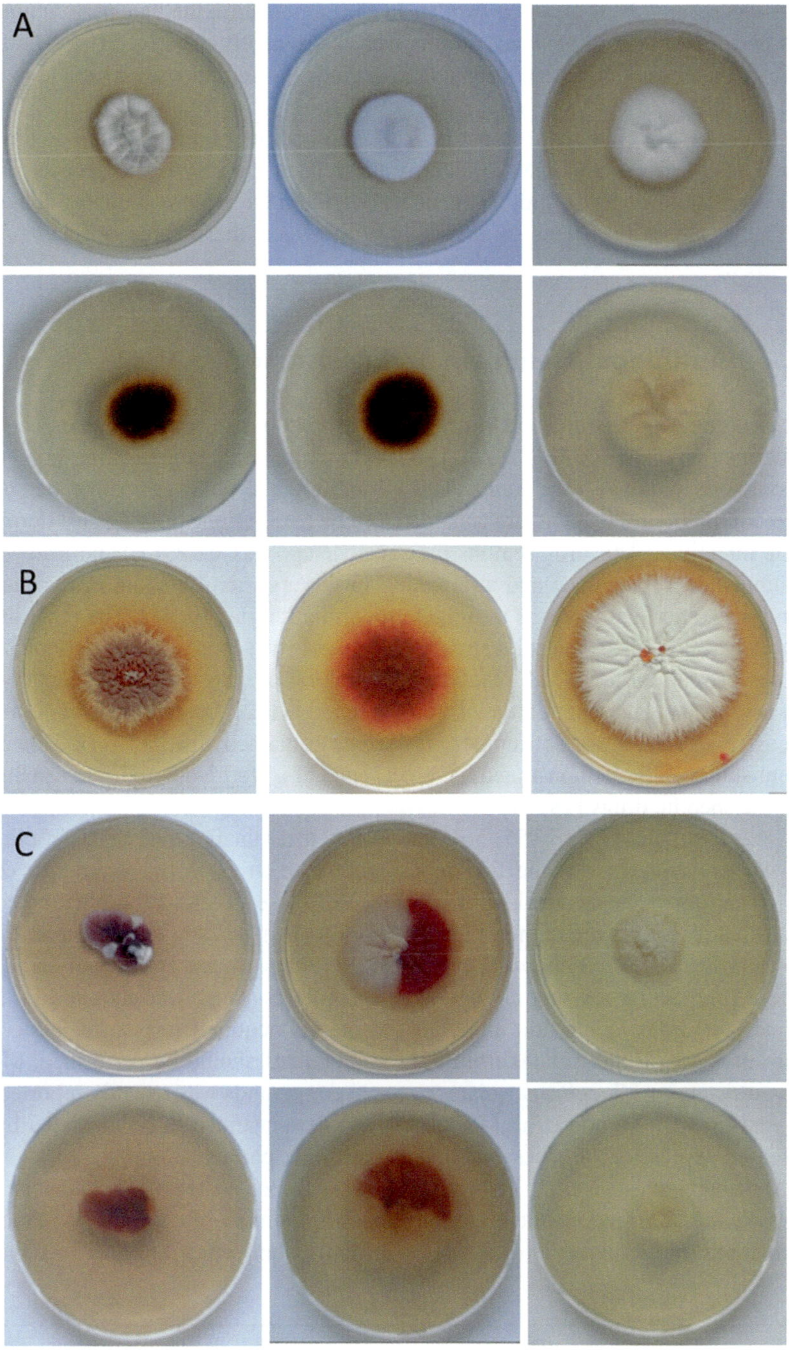

Fig. 2 Colony characteristics of *Trichophyton rubrum* (**a**), *Trichophyton soudanense* (**b**), and *Trichophyton violaceum* (**c**). *Trichophyton rubrum* (**a**) displays colonies which are fluffy to cottony, white, sometimes becoming pink when aging and with a wine-red to olive reverse. Colonies of *Trichophyton soudanense* (**b**) are slightly wrinkled and predominantly yellow to orange with

both species involved in *Tinea capitis*. In the same study, the hair perforation test was performed and all analyzed strains were consistently negative. These findings indicate that these fungi are not capable to degrade the keratinous hair cuticle in vitro [11, 38, 39].

Clinical manifestations, morphological examination, and the results of physiological tests can be combined to identify these three closely related species. However, for routine diagnostics sequencing of the ITS regions remains the gold standard, and is sufficient to discriminate between these three species of the complex.

4 Antifungal Treatment and Resistance

The antifungal agents (oral and topical) that are predominantly used for the treatment of dermatophytosis are the azoles and allylamines, both targeting the ergosterol biosynthesis pathway. The latter class of antifungals, with terbinafine being the most commonly used agent, are considered to be the first-line drug against dermatophyte infections. This drug inhibits the squalene epoxidase (SQLE) and leads to an accumulation of squalene and a depletion in the ergosterol content of the plasma membrane, and finally results in fungal growth inhibition [40, 41]. The other group of widely used drugs, the azoles such as fluconazole, itraconazole, and voriconazole, act downstream of the SQLE reaction by inhibition of lanosterol 14-α-demethylase thus resulting in an accumulation of sterol precursors [42]. Despite the long-term and extensive use of these antifungals, resistance remained low [43–47]. However, recent studies have demonstrated the emergence of a worldwide reduced sensitivity to these specific drugs [48–54].

Several studies [50, 51, 55, 56] demonstrated terbinafine resistance in clinical *T. rubrum* isolates, with the first report published in 2003 by Mukherjee et al. [54]. The molecular mechanism of resistance of these isolates to terbinafine could be attributed to point mutations in the SQLE target gene leading to single amino acid substitutions at 1 of 4 positions (Leu393, Phe397, Phe415, and His440) of the SQLE protein [51]. So far it is not clear how terbinafine resistance was acquired in these isolates. In order to provide more insight in this acquired resistance, molecular analysis of sequential isolates from the same patients should be performed. Moreover, resistance of *T. rubrum* isolates to azole compounds was recently reported for the first time by Monod et al. [48] and it was demonstrated to be mediated by an increased expression of genes encoding some efflux pumps.

Antifungal resistance has not yet been reported for isolates of *T. violaceum* and *T. soudanense*. However, there are an increasing number of cases of resistance in dermatophytes reported worldwide. Currently, an epidemic of recalcitrant

Fig. 2 (continued) yellow-orange to brownish reverse producing star-like colonies with irregular margins. Other strains can be more cottony with more regular, smooth margins. Isolates of *Trichophyton violaceum* are white/purple, white/beige, or pink, with a purple to brown reverse

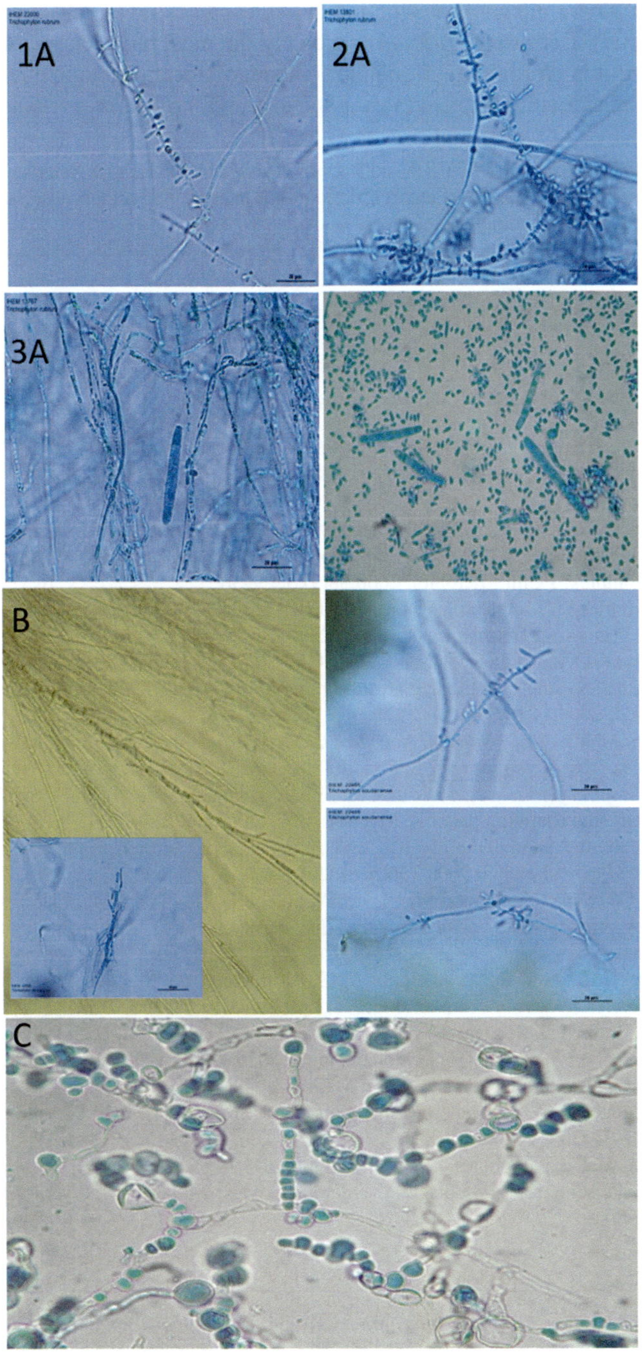

Fig. 3 Microscopic morphological characteristics of *Trichophyton rubrum* (**1a**—IHEM 2200, **2a**—IHEM 13801, **3a**—IHEM 13767), *Trichophyton soudanense* IHEM 22465 (**b**), and *Trichophyton violaceum* (**c**). Microscopically, clavate to piriform microconidia are observed

T. mentagrophytes infections is ongoing in India and caused by one specific "Indian genotype" [57]. Because *T. rubrum* is globally the most prevalent dermatophyte at the moment, it is worthwhile to monitor epidemiological developments also in the *T. rubrum* species complex, and to continue to scan for putative-resistant genotypes.

Acknowledgments The BCCM/IHEM collection is funded by the Belgian Science Policy Office.
Conflict of Interest: The authors report no conflicts of interest. The authors alone are responsible for the content and the writing of this paper.
Ethical Approval: This article does not describe any studies with human participants or animals performed by the authors.

References

1. Bodin E. Les champignons parasites de l'homme. Paris, France: Masson et Cie; 1902.
2. Castellani A. Observation on a new species found in tinea cruris. Br J Dermatol. 1910;5:148–50.
3. Weitzman I, Summerbell R. The dermatophytes. Clin Microbiol Rev. 1995;8(2):240–59.
4. Gräser Y, Kuijpers AF, Presber W, de Hoog GS. Molecular taxonomy of the *Trichophyton rubrum* complex. J Clin Microbiol. 2000;38(9):3329–36.
5. Hay RJ, Robles W, Midgley G, Moore MK. Tinea capitis in Europe: new perspective on an old problem. J Eur Acad Dermatol. 2001;15(3):229–33.
6. Adou-Bryn K-D, Kouame K, Ouhon J, Yapo C-G, Aka B-R, Silue S-K, et al. Fungi of skin, hair and nails in 219 samples at the free health centre of Pont in Treichville (Abidjan, Ivory Cost). J Mycol Med. 2004;14(2):89–92.
7. Magill SS, Manfredi L, Swiderski A, Cohen B, Merz WG. Isolation of *Trichophyton violaceum* and *Trichophyton soudanense* in Baltimore, Maryland. J Clin Microbiol. 2007;45(2):461–5.
8. Ginter-Hanselmayer G, Weger W, Ilkit M, Smolle J. Epidemiology of tinea capitis in Europe: current state and changing patterns. Mycoses. 2007;50(Suppl 2):6–13.
9. Gräser Y, Scott J, Summerbell R. The new species concept in dermatophytes – a polyphasic approach. Mycopathologia. 2008;166(5–6):239–56.
10. de Hoog GS, Dukik K, Monod M, Packeu A, Stubbe D, Hendrickx M, et al. Toward a novel multilocus phylogenetic taxonomy for the dermatophytes. Mycopathologia. 2017;182 (1–2):5–31.
11. Su H, Packeu A, Ahmed SA, Al-Hatmi AMS, Blechert O, Ilkit M, et al. Species distinction in the *Trichophyton rubrum* complex. J Clin Microbiol. 2019;57(9):e00352–19.
12. Packeu A, Stubbe D, Roesems S, Goens K, Van Rooij P, de Hoog S, et al. Lineages within the *Trichophyton rubrum* complex. Mycopathologia. 2020;185(1):123–36.
13. Sabouraud R. Les teignes. Paris, France: Masson et Cie; 1910.
14. Ota M, Langeron M. Nouvelle classification des dermatophytes. Ann Parasitol Hum Comp. 1923;1(4):305–36.
15. Emmons CW. Dermatophytes. Arch Derm Syphilol. 1934;30(3):337–62.
16. Ajello L. Present day concepts of the dermatophytes. Mycopathol Mycol Appl. 1962;17 (4):315–24.

Fig. 3 (continued) for *Trichophyton rubrum* (**a**) and *Trichophyton soudanense* (**b**). A typical feature for this latter species is the presence of reflexive branching (**b**). No microconidia are observed for isolates of *Trichophyton violaceum*, but the species is able to hydrolyze BCP when incubated at 37 °C, with the typical production of asymmetrical chlamydospores (**c**)

17. Matsumoto T, Ajello L. Current taxonomic concepts pertaining to the dermatophytes and related fungi. Int J Dermatol. 1987;26(8):491–9.
18. Ilkit M, Durdu M. Tinea pedis: the etiology and global epidemiology of a common fungal infection. Crit Rev Microbiol. 2015;41:374–88.
19. Lee WJ, Kim SL, Jang YH, Lee SJ, Kim DW, Bang YJ, et al. Increasing prevalence of *Trichophyton rubrum* identified through an analysis of 115.846 cases over the last 37 years. J Korean Med Sci. 2015;30(5):639–43.
20. Daniel CR, Jellinek NJ. The pedal fungus reservoir. Arch Dermatol. 2006;142:1344–6.
21. Nenoff P, Krüger C, Ginter-Hanselmayer G, Tietz HJ. Mycology – an update. Part 1: Dermatomycoses: causative agents, epidemiology and pathogenesis. J Dtsch Dermatol Ges. 2014;12:188–210.
22. Ginter-Hanselmayer G, Seebacher C. Treatment of tinea capitis – a critical appraisal. J Dtsch Dermatol Ges. 2011;9:109–14.
23. Hayette MP, Sacheli R. Dermatophytosis, trends in epidemiology and diagnostic approach. Curr Fungal Infect Rep. 2015;9:164–79.
24. Ilkit M, Demirhindi H. Asymptomatic dermatophyte scalp carriage: laboratory diagnosis, epidemiology and management. Mycopathologia. 2008;165:61–71.
25. Ilkit M. Favus of the scalp: an overview and update. Mycopathologia. 2010;170:143–54.
26. John AM, Schwartz RA, Janniger CK. The kerion: an angry tinea capitis. Int J Dermatol. 2018;57:3–9.
27. Ohst T, de Hoog S, Presber W, Stavrakieva V, Gräser Y. Origins of microsatellite diversity in the *Trichophyton rubrum-T. violaceum* clade (Dermatophytes). J Clin Microbiol. 2004;42 (10):4444–8.
28. Gräser Y, Fröhlich J, Presber W, de Hoog S. Microsatellite markers reveal geographic population differentiation in *Trichophyton rubrum*. J Med Microbiol. 2007;56(Pt 8):1058–65.
29. Lamb SR, Rademaker M. Tinea due to *Trichophyton violaceum* and *Trichophyton soudanense* in Hamilton, New Zealand. Australas J Dermatol. 2001;42(4):260–3.
30. Coulibaly O, Kone AK, Niaré-Doumbo S, Goïta S, Gaudart J, Djimdé AA, et al. Dermatophytosis among schoolchildren in three eco-climatic zones of Mali. PLoS Negl Trop Dis. 2016;10(4):e0004675.
31. Farina C, Fazii P, Imberti G, Lombardi G, Passera M, Andreoni S. *Trichophyton violaceum* and *T. soudanense*: re-emerging pathogens in Italy, 2005–2013. Italian Association of Clinical Microbiology (AMCLI) Dermatophytes' Study Group; AMCLI Dermatophytes' Study Group. New Microbiol. 2015;38(3):409–15.
32. Zhan P, Dukik K, Li D, Sun J, Stielow JB, Gerrits van den Ende B, et al. Phylogeny of dermatophytes with genomic character evaluation of clinically distinct *Trichophyton rubrum* and *T. violaceum*. Stud Mycol. 2018;89:153–75.
33. Kong F, Tong Z, Chen X, Sorrell T, Wang B, Wu Q, et al. Rapid identification and differentiation of *Trichophyton* species, based on sequence polymorphisms of the ribosomal internal transcribed spacer regions, by rolling-circle amplification. J Clin Microbiol. 2008;46 (4):1192–9.
34. Irinyi L, Lackner M, de Hoog GS, Meyer W. DNA barcoding of fungi causing infections in humans and animals. Fungal Biol. 2016;120(2):125–36.
35. Persinoti G, Martinez DA, Li W, Dögen A, Billmyre RB, Averette A, et al. Whole-genome analysis illustrates global clonal population structure of the ubiquitous dermatophyte pathogen *Trichophyton rubrum*. Genetics. 2018;208:1657–69.
36. Slifkin M. Tween 80 opacity test responses of various *Candida* species. J Clin Microbiol. 2000;38:4626–8.
37. Ro BI, Dawson TL. The role of sebaceous gland activity and scalp microfloral metabolism in the etiology of seborrheic dermatitis and dandruff. J Investig Dermatol Symp Proc. 2005;10:194–7.
38. Ates A, Ozcan K, Ilkit M. Diagnostic value of morphological, physiological and biochemical tests in distinguishing *Trichophyton rubrum* from *Trichophyton mentagrophytes* complex. Med Mycol. 2008;46:811–22.

39. Salkin IF, Hollick GE, Hurd NJ, Kemna ME. Evaluation of human hair sources for the *in vitro* hair perforation test. J Clin Microbiol. 1985;22:1048–9.
40. Ryder NS. Terbinafine: mode of action and properties of the squalene epoxidase inhibition. Br J Dermatol. 1992;126(39):2–7.
41. Valachovic M, Garaiova M, Holic R, Hapala I. Squalene is lipotoxic to yeast cells defective in lipid droplet biogenesis. Biochem Biophys Res Commun. 2016;469(4):1123–8.
42. Van den Bossche H, Marichal P, Gorrens J, Coene MC, Willemsens G, Bellens D, et al. Biochemical approaches to selective antifungal activity. Focus on azole antifungals. Mycoses. 1989;32(1):35–52.
43. Ghannoum MA, Chaturvedi V, Espinel-Ingroff A, Pfaller MA, Rinaldi MG, Lee-Yang W, et al. Intra and interlaboratory study of a method for testing the antifungal susceptibilities of dermatophytes. J Clin Microbiol. 2004;42:2977–9.
44. Jessup CJ, Warner J, Isham N, Hasan I, Ghannoum MA. Antifungal susceptibility testing of dermatophytes: establishing a medium for inducing conidial growth and evaluation of susceptibility of clinical isolates. J Clin Microbiol. 2000;38:341–4.
45. Badali H, Mohammadi R, Mashedi O, de Hoog GS, Meis JF. *In vitro* susceptibility patterns of clinically important *Trichophyton* and *Epidermophyton* species against nine antifungal drugs. Mycoses. 2015;58:303–7.
46. Afshari MA, Shams-Ghahfarokhi M, Razzaghi-Abyaneh M. Antifungal susceptibility and virulence factors of clinically isolated dermatophytes in Tehran, Iran. Iran J Microbiol. 2016;8:36–46.
47. Indira G. *In vitro* antifungal susceptibility testing of 5 antifungal agents against dermatophytic species by CLSI (M38-A) microdilution method. Clin Microbial. 2014;3:145.
48. Monod M, Feuermann M, Salamin K, Fratti M, Makino M, Alshahni MM, et al. *Trichophyton rubrum* azole resistance mediated by a new ABC transporter, TruMDR3. Antimicrob Agents Chemother. 2019;63(11):e00863–19.
49. Singh A, Masih A, Khurana A, Singh PK, Gupta M, Hagen F, et al. High-terbinafine resistance in *Trichophyton interdigitale* isolates in Delhi, India harbouring mutations in the squalene epoxidase gene. Mycoses. 2018;61:477–84.
50. Rudramurthy SM, Shankarnarayan SA, Dogra S, Shaw D, Mushtaq K, Paul RA, et al. Mutation in the squalene epoxidase gene of *Trichophyton interdigitale* and *Trichophyton rubrum* associated with allylamine resistance. Antimicrob Agents Chemother. 2018;62(5):e02522–17.
51. Yamada T, Maeda M, Alshahni MM, Tanaka R, Yaguchi T, Bontems O, et al. Terbinafine resistance of *Trichophyton* clinical isolates caused by specific point mutations in the squalene epoxidase gene. Antimicrob Agents Chemother. 2017;61:e00115–7.
52. Digby SS, Hald M, Arendrup MC, Hjort SV, Kofoed K. Darier disease complicated by terbinafine-resistant *Trichophyton rubrum*: a case report. Acta Derm Venereol. 2017;97 (1):139–40.
53. Schosler L, Andersen LK, Arendrup MC, Sommerlund M. Recurrent terbinafine-resistant *Trichophyton rubrum* infection in a child with congenital ichthyosis. Pediatr Dermatol. 2018;35:259–60.
54. Mukherjee PK, Leidich SD, Isham N, Leitner I, Ryder NS, Ghannoum MA. Clinical *Trichophyton rubrum* strain exhibiting primary resistance to terbinafine. Antimicrob Agents Chemother. 2003;47:82–6.
55. Osborne CS, Leitner I, Hofbauer B, Fielding CA, Favre B, Ryder NS. Biological, biochemical, and molecular characterization of a new clinical *Trichophyton rubrum* isolate resistant to terbinafine. Antimicrob Agents Chemother. 2006;50:2234–6.
56. Osborne CS, Leitner I, Favre B, Ryder NS. Amino acid substitution in *Trichophyton rubrum* squalene epoxidase associated with resistance to terbinafine. Antimicrob Agents Chemother. 2005;49:2840–4.
57. Nenoff P, Verma SB, Vasani R, Burmester A, Hipler UC, Wittig F, et al. The current Indian epidemic of superficial dermatophytosis due to *Trichophyton mentagrophytes* – a molecular study. Mycoses. 2018;62:336–56.

Microsporum ferrugineum: The Renaissance of a Forgotten Dermatophyte

Silke Uhrlaß, Cassian Sitaru, Christine Scholz, Matthias Gebhardt, Anja Baunacke, Claudia Friedrichs, Iris Ranke-Greve, Ulrike Cleffmann, Martin Schaller, Sybille Schmidt, Daniela Koch, Hanna Mütze, Franziska Wittig, Constanze Krüger, Nadine Rahmig, Uta-Christina Hipler, and Pietro Nenoff

Abstract

In Germany, the anthropophilic dermatophyte *Microsporum ferrugineum* has been rarely isolated for the last 50 years. Currently, started in 2016, *M. ferrugineum* strains were occasionally identified although this dermatophyte is difficult to differentiate from *Microsporum audouinii* or *Microsporum canis*.

S. Uhrlaß · D. Koch · H. Mütze · F. Wittig · C. Krüger · P. Nenoff (✉)
Labor für medizinische Mikrobiologie, Rötha, Germany
e-mail: S.Uhrlass@Mykologie-Experten.de; koch@mykologie-experten.de; franziska-wittig92@web.de; krueger@mykologie-experten.de; nenoff@mykologie-experten.de

C. Sitaru · C. Scholz
Klinik für Dermatologie und Venerologie, Universitätsklinikum Freiburg, Freiburg, Germany
e-mail: cassian.sitaru@mail.medizin.uni-freiburg.de; christine.scholz@uniklinik-freiburg.de

M. Gebhardt
Hautarztpraxis Dr. Matthias Gebhardt, Zwickau, Germany
e-mail: gebderm@t-online.de

A. Baunacke
Hautarztpraxis Baunacke, Radeberg, Germany

C. Friedrichs
Medizinisches Labor Ostsachsen, Mikrobiologie Görlitz, Görlitz, Germany
e-mail: c.friedrichs@labor-ostsachsen.de

I. Ranke-Greve · U. Cleffmann
Hautarztpraxis Dr. Stephan Bartels, Iris Ranke-Greve und Ulrike Cleffmann, Göttingen, Germany
e-mail: ranke-greve@hautarzt-goettingen.de; praxis@hautarzt-goettingen.de

M. Schaller · S. Schmidt
Universitätshautklinik Tübingen, Tübingen, Germany
e-mail: martin.schaller@med.uni-tuebingen.de; sybille@schsy.de

N. Rahmig · U.-C. Hipler
Klinik für Hautkrankheiten, Universitätsklinikum Jena, Jena, Germany
e-mail: Nadine.Rahmig@med.uni-jena.de; Christina.Hipler@med.uni-jena.de

J.-P. Bouchara et al. (eds.), *Dermatophytes and Dermatophytoses*,
https://doi.org/10.1007/978-3-030-67421-2_11

Furthermore, a small outbreak of infections was observed in Germany. From July 2016 until April 2019, 19 patients with *M. ferrugineum* infections were diagnosed all over Germany. This included 15 both culture and molecular confirmed cases, and four cases exclusively diagnosed based on sequencing of fungal DNA. Cultural analysis revealed slowly growing colonies with white thallus and peripheral yellow-brownish submerged hyphae bundles. The reverse side of the furrowed colonies showed cream color to yellow staining. Microscopically, big spherical and oval double-walled intercalary-localized chlamydospores, typical "bamboo" hyphae, and acute-angled branched hyphae were observed. Fungal culture material from all isolates was identified by polymerase chain reaction (PCR), Sanger sequencing of the internal transcribed spacer (ITS) region, and/or the translation elongation factor (*TEF*)-*1α* gene. Results were referred to the *M. ferrugineum* CBS 497.48 strain (Centraalbureau voor Schimmelcultures CBS, Utrecht, The Netherlands, www.westerdijkinstitute.nl). Patients were children and adolescents under 18 years, mainly males. Suggested source of infection was martial sports, e.g., wrestling, judo, and boxing. Surprisingly, a significant part of affected patients were Germans of Russian descent. A migrant 3-year-old boy from Afghanistan suffering from *Tinea capitis* was also among the patients. Another strain was isolated from a 10-year-old wrestler with suspected *Tinea corporis*. There was no migration background or contact to foreigners, the boy did not stay abroad. The mycological challenge is the cultural identification of *M. ferrugineum* due to the morphological similarity not only to *M. canis*, but also to *M. audouinii*, and *Trichophyton verrucosum*. Phylogenetic analysis of ITS region of ribosomal DNA and the *TEF-1α* gene was performed using MEGAX, the statistical maximum likelihood method, and the Tamura-Nei substitution model. Bootstrapping was performed with 1000 replicates. The alignment of the ITS sequences (partial 18.S, ITS1, 5.8S, ITS2, partial 28.S) with known sequences deposited at the Genbank database available at the National Center for Biotechnology Information (NCBI), Bethesda, Maryland, U.S., clearly identified the *M. ferrugineum* strains after cultural analysis. The phylogenetic analysis of the dermatophytes—the dendrogram of fungal strains—demonstrated the genetic differences between *M. ferrugineum* strains and *M. audouinii* or *M. canis*. The three species could be clearly distinguished from each other. In particular, sequencing of the *TEF-1α* gene allowed a better differentiation between *M. ferrugineum* and *M. audouinii* or *M. canis* than sequencing of the ITS 2 region.

Meanwhile, the *M. ferrugineum* strain 208361/2016 is deposited at the Deutsche Sammlung von Mikroorganismen und Zellkulturen (DSMZ, German Collection of Microorganisms and Cell cultures) in Braunschweig, Germany (DSM no. 103785). The DNA sequences of the *M. ferrugineum* strains were deposited as MF173061—ITS—and as MF173060—*TEF-1α* gene sequence—at the Genbank database. In Germany, the isolation of *M. ferrugineum* was not described during the last decades. Nowadays, an infection with this forgotten dermatophyte needs to be considered, in particular for migrants and asylum seekers. The species identification of *M. ferrugineum* represents a challenge for the dermatologist. If there are morphological and microscopic features suspicious for this rare *Microsporum* species, molecular identification of the causative pathogen is essential. For oral therapy of *Tinea capitis* due to *M. ferrugineum*, griseofulvin should be applied, alternatively itraconazole or fluconazole.

1 Introduction

Microsporum ferrugineum was initially described by Ota in 1921 as a dermatophyte causing *Tinea capitis* in Japan and surrounding areas [1]. Once classified, it belonged to the genus *Trichophyton*; however, after the observation of typical rough-walled macroconidia, the fungus was finally classified in the genus *Microsporum* [2]. In Germany and Europe, the anthropophilic dermatophyte *M. ferrugineum* has been rarely isolated for the last 50 years. However, it was first isolated in Germany in 1970 by Garske and Nuber [3]. The famous German mycologist Luise Krempl-Lamprecht from Munich published the next case in 1982 [4]. Currently, started in 2016, *M. ferrugineum* strains were occasionally identified although this dermatophyte is difficult to differentiate from *Microsporum audouinii* and *Microsporum canis*. Furthermore, a small outbreak of infections was observed in Germany.

2 Methods

2.1 Patients

Based both from routine diagnostics of the laboratory Mölbis, Germany, and from cultures sent for fungal species identification, 19 patients (all out-patients) with dermatomycoses and *Tinea capitis* due to *M. ferrugineum* have been diagnosed all over Germany in the last 3 years (Table 1). From these 19 patients, we were able to detect altogether 25 isolates (mainly cultures, but also DNA samples) of *M. ferrugineum*.

The majority of the patients was of male ($n = 16$) gender, there was one girl, only, among the patients. For the remaining patients, the gender was not told. The age range was from three to 13 years. The kids suffered predominantly from *Tinea capitis* ($n = 13$). Three children had *Tinea corporis*, two *Tinea faciei*. At least, altogether six children were of Russian German origin, but living for years in Germany. Two kids were migrants, one of them originating from Afghanistan.

2.2 Conventional Cultural Diagnostics

In mycological routine diagnostics, scrapings from centrifugal lesions of the free skin and hair roots from lesions of the capillitium were cultured on Sabouraud 4% dextrose agar (Sifin, Berlin, Germany), and, additionally, on cycloheximide (Actidione®)-containing Sabouraud dextrose agar (Becton Dickinson, Heidelberg,

Table 1 *Microsporum ferrugineum* strains identified or detected from skin swabs/skin scrapings in the lab Mölbis, Germany, from July 2016 until April 2019

No.	Lab no. Mölbis	Gender	Age in years	Geographic region in Germany	Material and diagnostic investigation	Kind of *Tinea*	Patient's history	Date	Identification by sequencing
1	208361/16	Male	3	Göttingen, Lower Saxony	Skin scrapings with fungal culture growth and PCR	*Tinea capitis*	Migrant (Refugee)	June 2016	ITS, *TEF-1α*
	209665/16	Male	3	Göttingen, Lower Saxony	Skin scrapings and hair with fungal culture growth and PCR	*Tinea capitis*		June 2016	
2	210168/16	Male	10	Saxony	Skin scrapings with fungal culture growth and PCR	*Tinea corporis*	Wrestler	July 2016	ITS, *TEF-1α*
3	205548/17	Male	8	Zwickau, Saxony	Fungal culture and PCR	*Tinea corporis*		April 2017	ITS, *TEF-1α*
4	215735/17	Male	4	Zwickau, Saxony	Fungal culture and PCR		Martial arts Brother of patient 3	October 2017	ITS, *TEF-1α*
	216800/17	Male	4	Zwickau, Saxony	Skin scrapings and PCR	*Tinea capitis*		October 2017	
5	216660/17	Male	13	Pulsnitz, Saxony	Fungal culture and PCR	*Tinea corporis* (side of trunk)	Football with Arabic children	October 2017	ITS, *TEF-1α*
6	219889/17	Male	4	Schneckengrün, Saxony	Skin scrapings with fungal culture growth and PCR	*Tinea corporis* (trunk)		December 2017	ITS, *TEF-1α*
7	201048/18	Male	10	Schorndorf, Baden-Wurttemberg	Skin swab and PCR		Martial arts, box fight From Ukraine	January 2018	
	202557/18	Male	10	Schorndorf, Baden-Wurttemberg	Skin scrapings with fungal culture growth and PCR	*Tinea faciei* (forehead)		February 2018	ITS, *TEF-1α*

8	207888/18	Male	9	Schorndorf, Baden-Wurttemberg	Skin swab and PCR		Brother of patient 7 (202557/2018)	May 2018	
9	205957/18	Male		Tübingen, Baden-Wurttemberg Prof. Martin Schaller	Fungal culture and PCR	*Tinea capitis*	Judoka having developed infection after visiting his homeland in Africa	March/ April 2018	ITS, *TEF-1α*
10	205958/18	Male		Tübingen, Baden-Wurttemberg Prof. Martin Schaller	Fungal culture and PCR	*Tinea capitis*	Judoka?	March/ April 2018	ITS, *TEF-1α*
11	207736/18	Male	10	Greifswald, Mecklenburg-Western Pomerania Laboratory Rostock	Fungal culture and PCR			May 2018	ITS, *TEF-1α*
12	208287/18	Male	10	Bayreuth, Bavaria Lab Dr. Dostmann Leipzig	Skin scrapings with fungal culture growth and PCR	*Tinea capitis*	Migrant	May 2018	*TEF-1α*
13	210807/18	Female	6	Zella-Mehlis, Thuringia	Skin scrapings and PCR	*Tinea faciei* (forehead)	Sister of patient 14 Russian German, martial arts	June 2018	*TEF-1α*
	213906/18	Female	6	Zella-Mehlis Thuringia	Dandruff and PCR	*Tinea capitis*		August 2018	*TEF-1α*

(continued)

Table 1 (continued)

No.	Lab no. Mölbis	Gender	Age in years	Geographic region in Germany	Material and diagnostic investigation	Kind of *Tinea*	Patient's history	Date	Identification by sequencing
14	210808/18	Male	10	Zella-Mehlis, Thuringia	Dandruff and PCR	*Tinea capitis*	Brother of patient 13 Russian German, martial arts	June 2018	*TEF-1α*
	216487/18	Male	10	Zella-Mehlis, Thuringia	Skin scrapings from dandruff, control investigation, and PCR	*Tinea capitis*		September 2018	
15	216124/18	Male	11	Oelsnitz, Saxony	Skin scrapings and PCR	*Tinea capitis*	Wrestler	September 2018	*TEF-1α*
16	202108/19	Male		Freiburg, Baden-Wurttemberg	Fungal culture and PCR	*Tinea capitis*	Wrestler	February 2019	ITS, *TEF-1α*
17	600152/19			Freiburg, Baden-Wurttemberg	Fungal culture and PCR	*Tinea capitis*	Capillitium/ hair roots Wrestler Russian German	March 2019	ITS, *TEF-1α*
18	600178/19	Male		Freiburg, Baden-Wurttemberg	Fungal culture and PCR	*Tinea capitis*	Wrestler, brother of patient 17 (600152/19) Russian German	April 2019	ITS, *TEF-1α*
19	600179/19			Freiburg, Baden-Wurttemberg	Fungal culture and PCR	*Tinea capitis*	Wrestler Child, Russian German	2018 April 2019	ITS, *TEF-1α*
	600180/19			Freiburg, Baden-Wurttemberg	Fungal culture and PCR	*Tinea capitis*	Wrestler Child, Russian German Under oral terbinafine treatment	April 2019	ITS, *TEF-1α*

All strains were identified by sequencing the ITS regions and/or *TEF-1α* gene and comparison with sequences deposited at the NCBI database
ITS internal transcribed spacer, *TEF-1α gene* translation elongation factor-1α gene

Germany). In case of fast-growing flat fungal colonies with a white or beige and partly yellowish color and a submerse periphery in the agar that are typical for *M. ferrugineum*, the fungi were further analyzed. Microscopic Lactophenol Cotton Blue preparations were performed from these colonies. Identification was confirmed by DNA sequencing.

2.3 Molecular Biological Diagnostics

For confirmation of the suspected dermatophyte species, Sanger sequencing of the ribosomal DNA (rDNA), mainly the regions ITS 1, 5.8S rRNA, ITS 2, and the translation elongation factor (*TEF*)-*1α* gene were performed for all strains to identify dermatophytes at a species level after extracting DNA from fungal cultures [5–7]. This required PCR amplification of a ~900 bp DNA fragment using universal primers that bind to flanking pan-fungal sequence regions. The following primers were used for sequencing of the ITS region of the rDNA: V9G 5′-TTACGTCCCTGCCCTTTGTA-3′ and LSU266 5′-GCATTCCCAAACAACTCGACTC-3′.

The length of the analyzed region in the *TEF-1α* gene varied from 709 to 769 nucleotides among the various dermatophyte species. Primers EF-DermF EF1a-F 5′CACATTAACTTGGTCGTTATCG 3′ and EF-DermR 5′CATCCTTGGAGATACCAGC3′ were used for sequencing [5].

3 Results

3.1 Clinical Characterization of Patients Affected by Dermatophytosis

From July 2016 until April 2019, in total 19 patients with *M. ferrugineum* infections were diagnosed all over Germany. This included 15 both culture and molecular confirmed cases, and four cases exclusively diagnosed based on sequencing of fungal DNA.

Patients were children and adolescents under 18 years, mainly males (Table 1). They were found in the following geographical regions of the country: Lower Saxony (1 patient), Saxony (6), Baden-Wurttemberg (8), Thuringia (2), Mecklenburg-Western Pomerania (1), and Bavaria (1).

3.2 Clinical Summary—Selected Case Reports

3.2.1 Patient 1

A 3-year-old Afghan boy (Patient no. 1, see Table 1) together with his family—refugees and asylum seekers—immigrated to Germany. The first 2 months, the family lived in a refugee's camp. The boy developed gray-patch *Tinea capitis*. From skin scrapings from the scalp grew fast, white to yellow in subculture, flat,

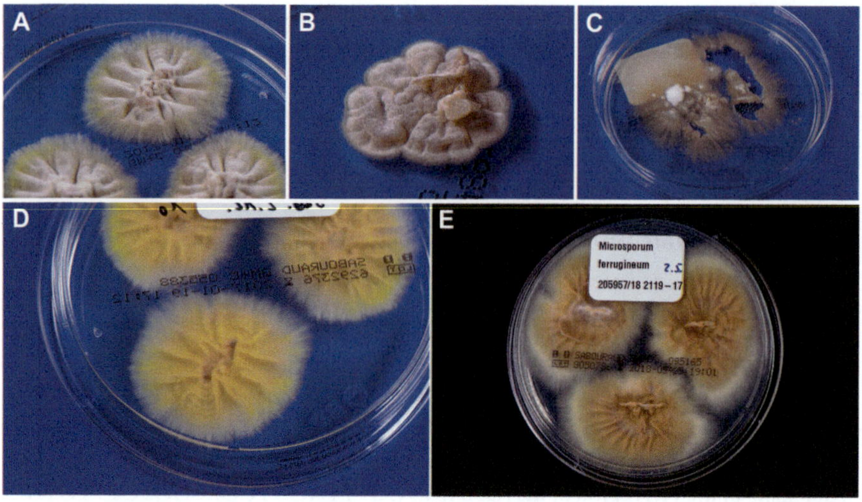

Fig. 1 Colony morphology of *Microsporum ferrugineum* on Sabouraud dextrose agar. (**a**) Subculture on agar plate with yellow, flat, central-folded colonies with submerged hyphae bundles at the periphery. Isolate from a 3-year-old boy from Afghanistan (Patient no. 1, Table 1). (**b**) Slow growing cerebral and verrucous colonies (subculture) with beige, peripheral submerged hyphae bundles. Isolate from a male patient, practicing judo (Patient no. 10, Table 1). (**c**) Slowly, but flat and radiating growing dermatophyte with partly white thallus and peripheral beige brownish submerged hyphae bundles. Primary culture of an isolate of an 8-year-old boy (Patient no. 3, Table 1). (**d**) The reverse side of the furrowed colonies showed bright yellow staining. Isolate from a 3-year-old boy from Afghanistan (Patient no. 1, Table 1). (**e**) The reverse side of the furrowed colonies showed dark yellow to brown staining. Isolate from a male patient with *Tinea capitis* developed after visiting his homeland in Africa (Patient no. 9, Table 1)

central-folded colonies with submerged hyphae bundles at the periphery (isolate lab no. 208361/2016 and 209665/16, Fig. 1a). The reverse side of the furrowed colonies showed yellow staining (Fig. 1d). Microscopically, bamboo and racquet hyphae were detected, from which acute-angled branched hyphae developed (Fig. 2a). Big spherical and oval double-walled intercalary-localized chlamydospores were observed as typical feature of both *M. ferrugineum* and *M. audouinii.*

3.2.2 Patient 2

Another strain was isolated from a 10-year-old wrestler (Patient no. 2, see Table 1) with suspected *Tinea corporis*. There was no migration background or contact to foreigners, the boy did not stay abroad.

3.2.3 Patient 3

An 8-year-old boy (Patient no. 3, see Table 1) from Saxony suffered from *Tinea corporis* due to *M. ferrugineum*. Clinical information and patient's history were not available for this child. However, other boys from the same town in Saxony showed dermatophytoses due to *M. ferrugineum* as well. At least, one of them, the brother of the here described boy, was active in martial arts.

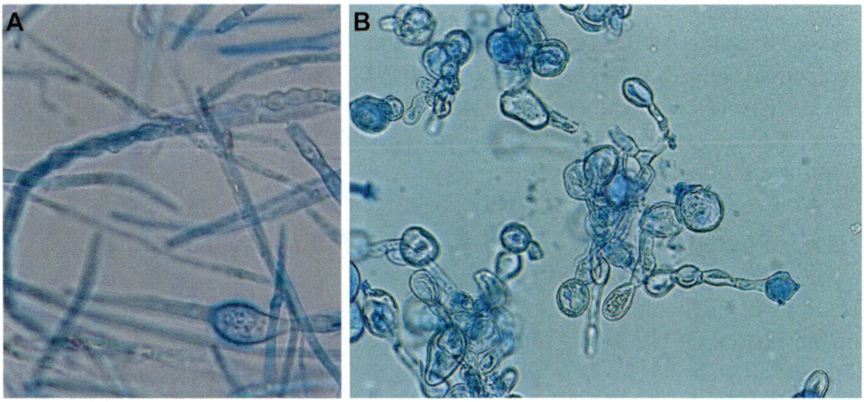

Fig. 2 Microscopic features of *Microsporum ferrugineum* on Sabouraud dextrose agar. (**a**) Bamboo, but also racquet hyphae appeared together with big spherical oval double-walled intercalary-localized chlamydospores (Lactophenol Cotton Blue staining). (**b**) Microscopic picture of big spherical oval double-walled chlamydospores (Lactophenol Cotton Blue staining)

3.3 Mycological Characterization

3.3.1 Cultural Features of *Microsporum ferrugineum*

The mycological challenge is the cultural identification of *M. ferrugineum* due to the morphological similarity not only to *M. canis*, but also to *M. audouinii*, and *Trichophyton verrucosum*. Cultural analysis of the here presented isolates revealed both fast (Fig. 1a), but also slow (Fig. 1b) growing colonies with a white to grayish and beige thallus and peripheral yellow-brownish submerged hyphae bundles. The reverse side of the furrowed colonies showed cream to yellow staining (Fig. 1e). The isolates showed different growth on various dermatophyte agar, i.e., several strains grew on agar no. 1–7, one strain on all agars and one isolate on all except agar no. 6. Morphological-based differentiation of all isolates revealed *Microsporum* species, e.g., *M. audouinii* or *M. ferrugineum*.

3.3.2 Microarchitecture of *Microsporum ferrugineum*

Microscopically, big spherical and oval double-walled intercalary-localized round and oval chlamydospores (Fig. 2b), together with typical "bamboo" hyphae were observed. The strains showed also racquet hyphae, which are typically intrinsic for *M. audouinii*. While fungi developed acute-angled branched hyphae, neither microconidia, nor macroconidia were found in all analyzed strains.

3.3.3 Molecular Identification of *Microsporum ferrugineum*

The sequence of each strain was aligned to sequences of reference strains from the databases. Based on the principle of similarity search (BLASTn search), individual strains were identified down to the species level by using the validated Online Dermatophyte Database of the Westerdijk Fungal Biodiversity Institute (formerly

Centraalbureau voor Schimmelcultures CBS, Utrecht, The Netherlands, "www.westerdijkinstitute.nl"). The reference strain used for identification of the dermatophytes was *M. ferrugineum* CBS 497.48. The comparison of the obtained ITS 2 sequences with sequences deposited at the Genbank database available at the National Center for Biotechnology Information (NCBI), Bethesda, Maryland, U.S., confirmed the species identification.

3.3.4 Phylogenetic Tree of *Microsporum ferrugineum* and Other *Microsporum* Species

The phylogenetic analysis of the dermatophytes—the dendrogram of fungal strains—based on ITS region of rDNA and *TEF-1α* gene sequences allows genetic differentiation between *M. ferrugineum* and the closely related *M. canis* and *M. audouinii* (Fig. 3a, b and Table 2). The three species could be clearly distinguished from each other. In particular, sequencing of the *TEF-1α* gene allowed a better differentiation of *M. ferrugineum* and *M. audouinii* or *M. canis* than sequencing of the ITS 2 region.

3.4 Deposition of the Isolates in Gene Databases and Strain Collections

Both ITS region and *TEF-1α* gene sequences of some strains were deposited at the Genbank database, the ISHAM-ITS-Database and at the "Fungal MLST Database". The DNA sequences of strain 208361/2016 (Patient no. 1, see Table 1) were deposited as MF173061—ITS region—and as MF173060—TEF-1α gene—at the Genbank database. In addition, this strain was deposited at the "Deutsche Sammlung von Mikroorganismen und Zellkulturen" (DSMZ, German Collection of Microorganisms and Cell cultures) in Braunschweig, Germany, with the DSM no. 103785.

4 Discussion

4.1 Clinical Symptoms and Predisposing Factors

The anthropophilic dermatophyte *M. ferrugineum* causes an ectothrix infection of hair roots at the scalp and an inflammatory erythematosquamous *Tinea corporis*. Recently, a deep dermatophytosis caused by *M. ferrugineum* was reported in a patient from China [8]. As a mutation of the Caspase recruitment domain protein 9 (CARD9) which is responsible for immunodeficiency predisposes individuals to suffer from severe dermatophytosis caused by fungi of the genus *Trichophyton*, this mutation might also serve as a predisposing factor for this infection [8]. Kosanke et al. [9] identified the Mating Type (MAT) locus idiomorphs of *M. ferrugineum* as

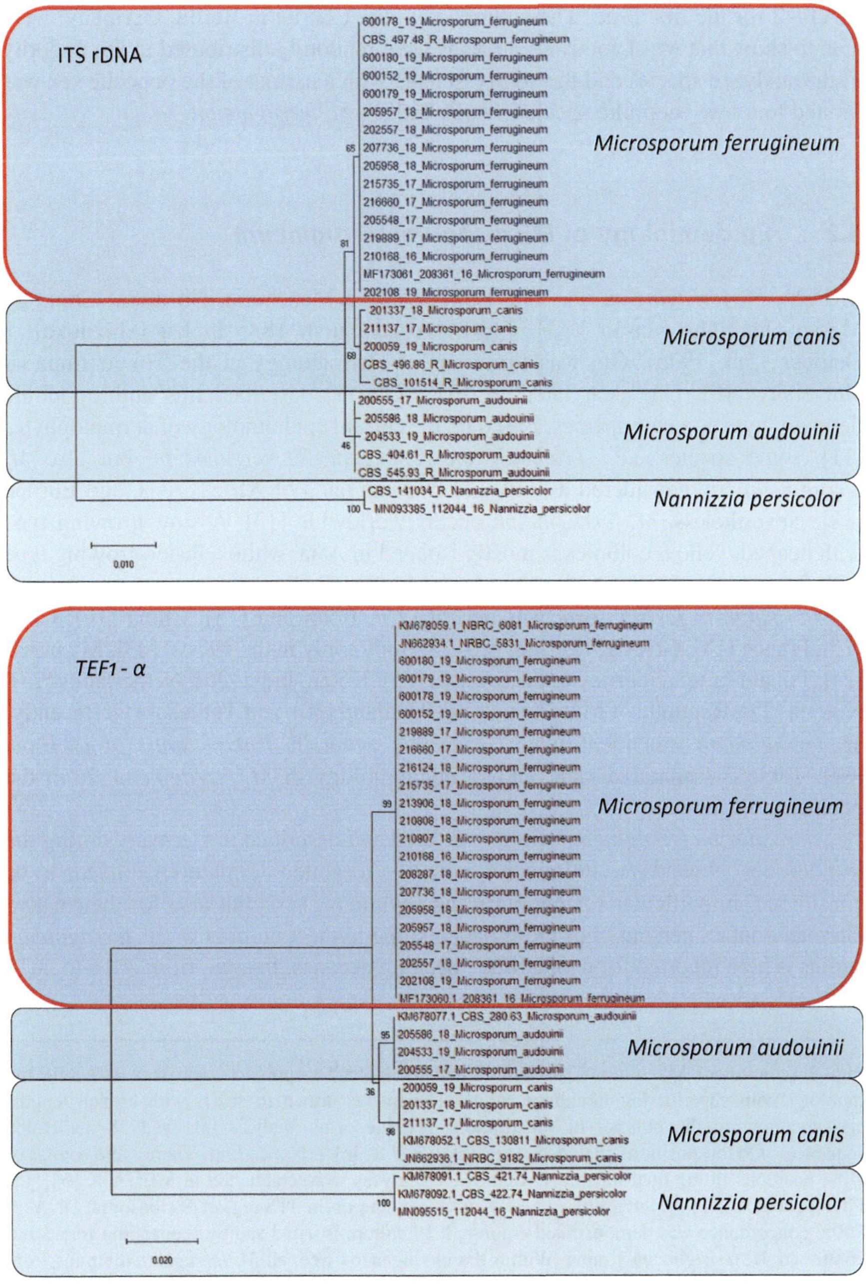

Fig. 3 The phylogenetic analysis of the dermatophytes—the dendrogram of fungal strains—based on ITS region of rDNA and of the *TEF-1α* gene. Statistical Method: Evolutionary analysis by Maximum Likelihood method. The evolutionary history was inferred by using the Maximum Likelihood method and Tamura-Nei model [63]. The tree with the highest log likelihood (−907.70) is shown. The percentage of trees in which the associated taxa clustered together are shown next to the branches. Initial tree(s) for the heuristic search were obtained automatically by applying Neighbor-Join and BioNJ algorithms to a matrix of pairwise distances estimated using the

MAT1-2 for the first time. The authors from the Charité in Berlin, Germany, were able to show that MAT locus idiomorphs were randomly distributed in the majority of the analyzed species and the ability to mate with a partner of the opposite sex was limited to a few zoophilic species, hence barring *M. ferrugineum*.

4.2 Epidemiology of *Microsporum ferrugineum*

In 1921, *M. ferrugineum* was first isolated in the Manchuria, Northeast China by Masao Ota ([Ōta Masao 太田 正雄], *August first, 1885 in Izu [Shizuoka]; † October 15th, 1945). Ota was Professor of Dermatology at the Tokyo Imperial University [10]. One year later, in 1922, Ota [1] described this anthropophilic dermatophyte as a new species. Based on the current epidemiology of dermatophytes [11], some species, i.e., *Trichophyton violaceum*, *T. verrucosum*, but also *M. ferrugineum* are considered as endemic in some parts of Africa, Asia, and Europe [12]. Nevertheless, *M. ferrugineum* occurs worldwide [13]. A slow-growing type with heaped, yellow colonies is mostly isolated in Asia, while a faster-growing type with flat, white colonies is prevalent in the Balkan [14].

Microsporum ferrugineum was described in Botswana [15], China [16], Cuba [17], France [18], Greece [19], Iran [20–22], Italy (only in the 1970s) [23], Myanmar [24], Poland (after a journey to the tropics) [25], Korea, India [26], Netherlands [27], Nigeria [28], Republic of Macedonia [29], Thailand [30], and Tunisia [31]. Recently, *M. ferrugineum* was identified as causative agent in *Tinea capitis* in Ethiopia [32]. Table 3 summarizes the known epidemiology of *M. ferrugineum* from the beginning till now.

Microsporum ferrugineum was not isolated and described in Germany during the last decades. Nowadays, an infection with this forgotten dermatophyte needs to be considered, in particular for migrants and asylum seekers, but also for their native German contact persons. For these patients, the highest number of *M. ferrugineum* strains is here reported for a long time, meaning decades. In total, from 2016 to 2019

Fig. 3 (continued) Maximum Composite Likelihood (MCL) approach, and then selecting the topology with superior log likelihood value. The tree is drawn to scale, with branch lengths corresponding to the number of substitutions per site. This analysis involved 28 nucleotide sequences. Codon positions included were 1st + 2nd + 3rd + Noncoding. There were a total of 1086 positions in the final dataset. Evolutionary analyses were conducted in MEGA X [64]. (**a**) Phylogenetic tree of *M. ferrugineum* based on sequencing of the ITS region of ribosomal DNA. A 100% concordance was demonstrated among all 19 culture-isolated and by sequencing found and confirmed *M. ferrugineum* strains. Within the phylogenetic tree, all *M. ferrugineum* strains from Germany are forming their own clade which is clearly discriminated from *M. canis* and *M. audouinii*. Rooted with *Nannizzia persicolor*. (**b**) Phylogenetic tree of *M. ferrugineum* based on sequencing of the *TEF-1α* gene. A 100% concordance was demonstrated among all 19 culture-isolated and by sequencing found and confirmed *M. ferrugineum* strains. Within the phylogenetic tree, all *M. ferrugineum* strains from Germany are forming their own clade which is clearly discriminated from *M. canis* and *M. audouinii*. Rooted with *Nannizzia persicolor*

Table 2 Strains used to generate the phylogenetic tree

Species	Strain	Sequencing region/gene	Genbank accession number
M. audouinii	Lab Mölbis 200555/17	ITS, *TEF-1α*	
M. audouinii	Lab Mölbis 205586/18	ITS, *TEF-1α*	
M. audouinii	Lab Mölbis 204533/19	ITS, *TEF-1α*	
M. audouinii	CBS 545.93	ITS	NR_144883
M. audouinii	CBS 404.61	ITS	MF926387
M. audouinii	CBS 280.63	*TEF-1α*	KM678077
M. canis	Lab Mölbis 211137/17	ITS, *TEF-1α*	
M. canis	Lab Mölbis 201337/18	ITS, *TEF-1α*	
M. canis	Lab Mölbis 200059/19	ITS, *TEF-1α*	
M. canis	CBS 496.86	ITS	MH861991
M. canis	CBS 101514	ITS	KT155672
M. canis	CBS 130811	*TEF-1α*	KM678052
M. canis	NRBC 9182	*TEF-1α*	JN662936
M. ferrugineum	Lab Mölbis Table 1	ITS, *TEF-1α*	
M. ferrugineum	Lab Mölbis 208361/16	ITS	MF173061
M. ferrugineum	Lab Mölbis 208361/16	*TEF-1α*	MF173060
M. ferrugineum	CBS 497.48	ITS	NR_155417
M. ferrugineum	NRBC 5831	*TEF-1α*	JN662934
M. ferrugineum	NBRC 6081	*TEF-1α*	KM678059
Nannizzia persicolor	Lab Mölbis 112044/16	ITS	MN093385
Nannizzia persicolor	Lab Mölbis 112044/16	*TEF-1α*	MN095515
Nannizzia persicolor	CBS 141034	ITS	MH378249
Nannizzia persicolor	CBS 871.70	ITS	MH859989
Nannizzia persicolor	CBS 421.74	*TEF-1α*	KM678091
Nannizzia persicolor	CBS 422.74	*TEF-1α*	KM678092

CBS Centraalbureau voor Schimmelcultures (now Westerdijk Institute), Utrecht, The Netherlands, *ITS* internal transcribed spacer, *TEF-1α gene* translation elongation factor-1α gene

Table 3 Summary of interesting published reports on *Microsporum ferrugineum* cases

No.	Year published	Country	Patients and dermatophytosis due to *Microsporum ferrugineum*	Reference
1	1922	Manchuria, Northeast China	*Tinea pedis*	[1]
2	1950	Congo	136 *M. ferrugineum* strains	[35]
3	1953	Former Czechoslovakia	One strain	[42]
4	1959	France	Outbreak of *M. ferrugineum*	[43]
5	1966	France	Four cases of *M. ferrugineum* infections	[44]
6	1970	Germany	First patient in Germany	[3]
7	1970	Germany	*Tinea capitis*	[45]
8	1971	Former Czechoslovakia	Combined nail infection due to *M. ferrugineum* and *Trichophyton rubrum*	[46]
9	1972	Serbia (formerly Yugoslavia)	*Tinea*	[47]
10	1973	West Africa	*Tinea capitis profunda* (Mycetoma of the scalp)	[48]
11	1975	Former Czechoslovakia	*Tinea capitis*	[49]
12	1979	France	*Tinea capitis* in a single immigrant from West Africa	[18]
13	1981	India	*Tinea capitis*	[26]
14	1981	India (Jamnagar, Gujarat)	*Tinea cruris/corporis*	[50]
15	1982	Germany	9-year-old German boy, and a 7-year-old Korean boy, adopted by this family, and the mother were infected by the same fungus *M. ferrugineum*	[4]
16	1984	USA	Single case, only	[51]
17	1984	Puerto Rico	Single case, only	[52]
18	1984	USA	Single cases, only	[53]
19	1987	Japan	Unusual skin manifestations of a dermatophyte infection with peripheral hyperkeratosis with central yellowish pus in a 30-year-old female	[54]
20	1989	Cuba	*Tinea corporis*	[17]
21	1991	Qatar	*Tinea capitis*	[55]
22	1993	Italy	Sporadically isolated in the 1970–1979 period, but disappeared in the years 1980–1989	[23]
23	1993	Taiwan	25-year-old man with several prominent subcutaneous masses in the occipital region of the scalp diagnosed as *Tinea capitis* and *Tinea corporis* due to *M. ferrugineum*	[56]
24	1996	Thailand	*Microsporum ferrugineum* was the major pathogen of an outbreak of *Tinea capitis*	[30]

25	1999	Poland	Fungal skin infection due to *M. ferrugineum* in persons returning from the tropics. *Microsporum* genus was isolated from persons returning from East, West, and Central Africa	[25]
26	2000	Austria	Veterinary medicine *Microsporum ferrugineum* was isolated as keratinopathogenic fungus from hooves of horses	[57]
27	2003	Myanmar	Two cases of *Tinea capitis* in a boy and his mother due to *M. ferrugineum*, successfully treated by itraconazole	[24]
28	2005	Thailand	Altogether 81 patients with *Tinea capitis* (49 boys and 32 girls) were treated by an average dose of itraconazole of 4.5 mg/kg/day. Continuous and longer treatment was significantly better than pulse therapy	[58]
29	2006	Iran	A 42-year-old man was examined for *Tinea faciei* due to *M. ferrugineum*	[20]
30	2007	Nigeria	*Microsporum ferrugineum* was here isolated for the first time as rare cause of dermatophytoses in Nigeria	[28]
31	2008	Nigeria	*Microsporum ferrugineum* was the second prevalent (17%) dermatophyte in *Tinea capitis* among primary school children in Anambra state of Nigeria	[59]
32	2008	Nigeria	The etiological agent of *Tinea capitis* in the study population was *Trichophyton soudanense* (30.6%), followed by *M. ferrugineum* (7.7%), and *Microsporum audouinii* (7.7%)	[60]
33	2008	China	*Microsporum ferrugineum* was among agents of *Tinea capitis* in school children from Western China	[16]
34	2009	Tunisia	*Microsporum ferrugineum* as cause of *Tinea* sycosis associated with circinate herpes of the hand in a 60-year-old man	[31, 61]
35	2010	Republic of Macedonia	Rare isolation of *M. ferrugineum* in dermatophytosis in Macedonia	[29]
36	2012	Iran	Two patients with *Tinea* due to *M. ferrugineum*	[21]
37	2012	Turkey	Identification of rare macroconidia-producing dermatophytic fungi, like *M. ferrugineum*, by real-time PCR	[62]
38	2013	Iran	Rare infection due to *M. ferrugineum* in Tehran, Iran	[22]
39	2017	Ethiopia	In rural southern Ethiopia, two out of 88 culture proven *Tinea capitis* patients were caused by *M. ferrugineum*	[32]
40	2019	Germany	In Germany, between 2016 and 2019, altogether 25 strains or DNA isolates were collected from 19 pediatric patients	The here presented own investigation

in Germany, 25 *M. ferrugineum* strains and DNA extracts from 19 different patients were diagnosed at the species level by DNA sequencing. More than half of the identified *M. ferrugineum* strains were sent as fungal cultures to our lab from other microbiological labs, dermatological offices, and clinics in order to confirm the suspected rare dermatophyte, which was finally done by DNA sequencing.

4.3 Source of Infections Caused by *Microsporum ferrugineum*

The suggested source of infection for some patients discussed in this study is martial sports, e.g., wrestling, judo, and boxing. Mainly children and adolescents were affected. For example, one strain was isolated from a 10-year-old wrestler showing symptoms of *Tinea corporis*. Only this patient had no migration background or contact to foreigners, the boy did not stay abroad. Since further children did wrestling sports, *M. ferrugineum* might also cause *Tinea gladiatorum capitis* and *corporis* in addition to *Trichophyton tonsurans* [33]. Surprisingly, a significant part of affected patients were Germans of Russian descent living already for several years in Germany. As a 3-year-old boy who migrated from Afghanistan suffered from *Tinea capitis*, migration movements might further introduce *M. ferrugineum* to Germany. Hence, a similar epidemiology could be expected in other European countries.

4.4 Species Identification of *Microsporum ferrugineum*

In young patients with migration background, fungal isolates with morphological features including a *Microsporum*-like thallus in combination with microscopically observed chlamydospores indicate rare *Microsporum* species like *M. audouinii* and *M. ferrugineum*. In the Atlas of Clinical Fungi [34], two colony forms are distinguished: A slow-growing type with heaped, yellow colonies, which is mostly isolated in Asia, and a faster-growing type with flat, white colonies that is prevalent in the Balkan. Already in 1930, Vanbreuseghem [35] reported on 136 strains of *Trichophyton* (today *Microsporum*) *ferrugineum* (Ota 1921), Langeron and Milochevitch, in the formerly Belgian Congo. He described a white variant of *M. ferrugineum*. The variety of colony morphology complicates a correct species differentiation.

Weitzman and Rosenthal described *M. ferrugineum* as follows: "*M. ferrugineum* was found to be urease positive in urea broth (most isolates were negative on urea agar); all produced light-colored colonies on Lowenstein-Jensen medium; spreading colonies on lactose agar and failed to perforate hair *in vitro* or to produce reflexive branching. Most isolates manifested poorer to no growth at 37 degrees C compared to room temperature and all but one failed to decompose casein and tyrosine" [36]. *Microsporum ferrugineum* is rarely forming macroconidia and, therefore, cannot be easily identified. In a study from Turkey, the production of macroconidia on nine common laboratory media was investigated [37]. In particular, Borelli's lactrimel agar (BLA) apparently improved macroconidia production after 15 days and was the most useful nutrient agar medium to induce these phenotypic characters in daily practice.

Thus, the species identification needs to include molecular methods. PCR diagnostics based on the topoisomerase II gene [38] are not able to distinguish between *M. canis*, *M. audouinii*, and *M. ferrugineum*. Moreover, the differentiation between *M. canis* and *M. ferrugineum* is not possible using real-time PCR systems based on the ITS region of rDNA, although the two species can be defined from *M. audouinii* [39].

Kobylak et al. [40] from Poland developed a diagnostic assay—both traditional and real-time PCR—using specific primers based on differences in the β-tubulin gene enabling detection of *M. canis*- and *M. audouinii*/*M. ferrugineum*-DNA from a pure culture. However, a recently developed microarray test (EUROArray Dermatomycosis, Euroimmun, Lübeck, Germany), is able to correctly identify *M. ferrugineum* both, directly from clinical material (skin scrapings) and from fungal culture material [41]. In order to ensure a clear identification of *M. ferrugineum*, DNA sequencing is required, as it was used for this German epidemiological study.

4.5 Therapy

For oral therapy of *Tinea capitis* due to *M. ferrugineum*, griseofulvin should be given, alternatively itraconazole or fluconazole. Additional topical therapy is mandatory, antidermatophyte effective antifungals, like amorolfine, azoles (clotrimazole, bifonazole, miconazole, sertaconazole), ciclopirox olamine, or terbinafine, can be used.

Acknowledgments We thank Esther Klonowski, biologist from Leipzig, for excellent support in preparing and formatting the manuscript. The photographer Uwe Schossig, Leipzig, has provided impressive photographs of the fungal cultures.

References

1. Ota M. Sur deux espèces nouvelles de dermatophytes en Mandchourie—*Microsporum ferrugineum* et T. pedis n. sp. Bull Soc Pathol Exot. 1922:588–96.
2. Nishimoto K. Remarkable works and cases in the history of medical mycology in Japan. Med Mycol J. 2017;58:J29–33.
3. Garske I, Nuber D. *Trichophyton ferrugineum* infection simulating microsporosis. Z Haut Geschlechtskr. 1970;45:543–6.
4. Krempl-Lamprecht L, Wagner N, Seitz R. First isolation of *Microsporum ferrugineum* (Ota 1921) in Munich. Mykosen. 1982;25:469–79.
5. Mirhendi H, Makimura K, De Hoog GS, Rezaei-Matehkolaei A, Najafzadeh MJ, Umeda Y, et al. Translation elongation factor 1-α gene as a potential taxonomic and identification marker in dermatophytes. Med Mycol. 2015;53:215–24.
6. Kargl A, Kosse B, Uhrlaß S, Koch D, Krüger C, Eckert K, et al. Hedgehog fungi in a dermatological office in Munich: case reports and review. Hautarzt. 2018;69:576–85.

7. Uhrlaß S, Schroedl W, Mehlhorn C, Krüger C, Hubka V, Maier T, et al. Molekulare Epidemiologie von *Trichophyton quinckeanum*—ein zoophiler Dermatophyt im Aufwind. J Dtsch Dermatol Ges. 2018;16:21–33.
8. Zhang Y, Mijiti J, Huang C, Song Y, Wan Z, Li R, et al. Deep dermatophytosis caused by *Microsporum ferrugineum* in a patient with CARD9 mutations. Br J Dermatol. 2019;181:1093–5.
9. Kosanke S, Hamann L, Kupsch C, Moreno Garcia S, Chopra A, Gräser Y. Unequal distribution of the mating type (MAT) locus idiomorphs in dermatophyte species. Fungal Genet Biol. 2018;118:45–53.
10. Ono T. Prof. Masao Ota. Med Mycol J. 2017;58:J71–5.
11. De Hoog GS, Dukik K, Monod M, Packeu A, Stubbe D, Hendrickx M, et al. Toward a novel multilocus phylogenetic taxonomy for the dermatophytes. Mycopathologia. 2017;182:5–31.
12. Zhan P, Liu W. The changing face of dermatophytic infections worldwide. Mycopathologia. 2017;182:77–86.
13. Uhrlaß S, Ranke-Greve I, Cleffmann U, Koch D, Krüger C, Rahmig N, et al. Tinea capitis and tinea corporis due to *Microsporum ferrugineum*—a forgotten dermatophyte in Europe. Mycoses. 2017;60(Suppl S2):234. (Abstract).
14. Kwong-Chung KJ, Benett JW. Medical mycology. Philadelphia: Lea & Febiger; 1992.
15. Thakur R. Spectrum of dermatophyte infections in Botswana. Clin Cos Inves Dermatol. 2015;8:127–33.
16. Deng S, Bulmer GS, Summerbell RC, de Hoog GS, Hui Y, Gräser Y. Changes in frequency of agents of tinea capitis in school children from Western China suggest slow migration rates in dermatophytes. Med Mycol. 2008;46:421–7.
17. Valencia León G, Tió Polledo L. Isolation of *Microsporum ferrugineum* in Cuba. Presentation of a case. Rev Cubana Med Trop. 1989;41:290–8.
18. Maleville J, Moulinier C, Boineau D, Marc-Antoine H, Mollard S, Guillet G, et al. Tinea capitis: 73 cases observed in Bordeaux (author's transl). Ann Dermatol Venereol. 1979;106:869–72.
19. Ieronymaki A, Papadogiannakis E, Kargadouri A, Poutouri E, Papadogeorgaki E, Chryssou SE. Tinea capitis in the area of Athens, Greece: a five-year retrospective analysis. Poster P2394, 28th ECCMID.2018, Madrid, Spain (Abstract).
20. Mahmoudabadi AZ. First case of *Microsporum ferrugineum* from Iran. Mycopathologia. 2006;161:337–9.
21. Rezaei-Matehkolaei A, Makimura K, de Hoog GS, Shidfar MR, Satoh K, Najafzadeh MJ, et al. Multilocus differentiation of the related dermatophytes *Microsporum canis*, *Microsporum ferrugineum* and *Microsporum audouinii*. J Med Microbiol. 2012;61:57–63.
22. Rezaei-Matehkolaei A, Makimura K, De Hoog GS, Shidfar MR, Zaini F, Eshraghian M, et al. Molecular epidemiology of dermatophytosis in Tehran, Iran, a clinical and microbial survey. Med Mycol. 2013;51:203–7.
23. Terragni L, Lasagni A, Oriani A. Dermatophytes and dermatophytoses in the Milan area between 1970 and 1989. Mycoses. 1993;36:313–7.
24. Ishizaki S, Ito H, Hanyaku H, Harada T. Two cases of tinea capitis by *Microsporum ferrugineum* believed infected in Myanmar. Nippon Ishinkin Gakkai Zasshi. 2003;44:203–7.
25. Jeske J, Lupa S, Seneczko F, Kamerys J, Malinowska B. Dermatophyte infections in persons returning from the tropics. Wiad Parazytol. 1999;45:395–400.
26. Khosa RK, Girgla HS, Hajini GH, Sharma BM, Singh GM. Study of dermatomycoses. Int J Dermatol. 1981;20:130–2.
27. Korstanje MJ, Staats CC. Fungal infections in the Netherlands. Prevailing fungi and pattern of infection. Dermatology (Basel). 1995;190:39–42.
28. Ngwogu AC, Otokunefor TV. Epidemiology of dermatophytoses in a rural community in Eastern Nigeria and review of literature from Africa. Mycopathologia. 2007;164:149–58.
29. Starova A, Balabanova-Stefanova M, V'lckova-Laskoska M. Dermatophytes in republic of Macedonia. Prilozi. 2010;31:317–26.

30. Wisuthsarewong W, Chaiprasert A, Viravan S. Outbreak of tinea capitis caused by *Microsporum ferrugineum* in Thailand. Mycopathologia. 1996;135:157–61.
31. Neji S, Makni F, Sellami H, Cheikhrouhou F, Sellami A, Ayadi A. First case of *Microsporum ferrugineum* from Tunisia. Mycopathologia. 2009;167:351–3.
32. Pérez-Tanoira R, Marín I, Berbegal L, Prieto-Pérez L, Tisiano G, Cuadros J, et al. Mycological profile of tinea capitis in schoolchildren in rural southern Ethiopia. Med Mycol. 2017;55:262–8.
33. Mayser P, Handrick W, Nenoff P. [Sports-associated dermatophytoses: an overview]. Hautarzt. 2016;67:680–8.
34. De Hoog GS, Guarro J, Gené J, Ahmed S, Al Hatmi AMS, Figueras MJ, et al. Atlas of clinical fungi: online version 2019, Utrecht, The Netherlands. http://www.clinicalfungi.org/.2019. Accessed 18 Jan 2020.
35. Vanbreuseghem R. Study of 136 strains of *Trichophyton ferrugineum* (Ota 1921), Langeron and Milochevitch 1930, and of its white variety isolated in Belgian Congo. Ann Parasitol Hum Comp. 1950;25:485–92.
36. Weitzman I, Rosenthal S. Studies in the differentiation between *Microsporum ferrugineum* Ota and *Trichophyton soudanense* Joyeux. Mycopathologia. 1984;84:95–101.
37. Ilkit M, Gümral R, Döğen A. Borelli's lactritmel agar induces conidiation in rare-macroconidia producing dermatophytic fungi. Med Mycol. 2012;50:735–9.
38. Winter I, Uhrlaß S, Krüger C, Herrmann J, Bezold G, Winter A, et al. Molecular biological detection of dermatophytes in clinical samples when onychomycosis or tinea pedis is suspected. A prospective study comparing conventional dermatomycological diagnostics and polymerase chain reaction. Hautarzt. 2013;64:283–9.
39. Wittig F, Uhrlaß S, Krüger C, Nenoff P. Nachweis von pathogenen Dermatophyten mittels multiplex real-time PCR. Derm Prakt Dermatol. 2019;25:158–69.
40. Kobylak N, Bykowska B, Kurzyk E, Nowicki R, Brillowska-Dąbrowska A. PCR and real-time PCR approaches to the identification of *Arthroderma otae* species *Microsporum canis* and *Microsporum audouinii/Microsporum ferrugineum.* J Eur Acad Dermatol Venereol. 2016;30:1819–22.
41. Uhrlaß S, Wittig F, Koch D, Krüger C, Harder M, Gaajetaan G, et al. Do the new molecular assays-microarray and real-time polymerase chain reaction-for dermatophyte detection keep what they promise? Hautarzt. 2019;70:618–26.
42. Langer J. [First isolation of the *Microsporon ferrugineum* in Czechoslovakia]. Cesk Dermatol 1953;28:446–52.
43. Bureau Y, Jarry, Barriere, Charpentier. Results obtained by griseofulvin in an epidemic of tinea caused by "*Microsporum ferrugineum*". Bull Soc Fr Dermatol Syphiligr. 1959;5:665–8.
44. Coudert J, Fleurette J, Battesti MR, Cadi-Soussi M. 4 Autochtonal cases of microspory due to *Microsporum ferrugineum.* Extension of the endemic to the south-east region. Bull Soc Fr Dermatol Syphiligr. 1966;73:154–6.
45. Gromzig H, Rieth H. Should fungus diseases of the scalp caused by *Microsporum ferrugineum* be called microsporosis or by a different name? Mykosen. 1970;13:563.
46. Otcenásek M, Sich J, Dvorák J. [Combined dermatophytic nail infection due to *Microsporum ferrugineum* and *Trichophyton rubrum*]. Mykosen. 1971;14:111–3.
47. Pap J. [1st cases of isolation of *Microsporum* (Tr.) *ferrugineum* in Vojvodina]. Med Pregl. 1972;25:5–9.
48. Baylet R, Camain R, Juminer B, Faye I. *Microsporum ferrugineum* OTA, 1921, an agent of scalp mycetomas in Dark Africa. Pathol Biol. 1973;21:5–12.
49. Kubec K, Stáva Z. [Tinea capitis caused by *Microsporum ferrugineum.* Case report]. Cesk Dermatol. 1975;50:327–30.
50. Poria VC, Samuel A, Acharya MK, Tilak SS. Dermatomycoses in and around Jamnagar. Indian J Dermatol Venereol Leprol. 1981;47:84–7.
51. Sinski JT, Flouras K. A survey of dermatophytes isolated from human patients in the United States from 1979 to 1981 with chronological listings of worldwide incidence of five dermatophytes often isolated in the United States. Mycopathologia. 1984;85:97–120.

52. Vázquez M, Sánchez JL. A clinical and mycologic study of tinea corporis and pedis in Puerto Rico. Int J Dermatol. 1984;23:550–1.
53. Sinski JT, Kelley LM. A survey of dermatophytes from human patients in the United States from 1985 to 1987. Mycopathologia. 1991;114:117–26.
54. Shoji A, Kitajima J, Hamada T. *Microsporum ferrugineum* infection with unusual clinical features. J Dermatol. 1987;14:476–82.
55. el-Benhawi MO, Fathy S, Moubasher AH, Alem NS. Mycologic study of tinea capitis in Qatar. Int J Dermatol. 1991;30:204–5.
56. Chen AW, Kuo JW, Chen JS, Sun CC, Huang SF. Dermatophyte pseudomycetoma: a case report. Br J Dermatol. 1993;129:729–32.
57. Keller M, Krehon S, Stanek C, Rosengarten R. Keratinopathogenic mould fungi and dermatophytes in healthy and diseased hooves of horses. Vet Rec. 2000;147:619–22.
58. Wisuthsarewong W, Chaiprasert A. Treatment of tinea capitis caused by *Microsporum ferrugineum* with itraconazole. J Med Assoc Thail. 2005;88(Suppl 8):S72–9.
59. Emele FE, Oyeka CA. Tinea capitis among primary school children in Anambra state of Nigeria. Mycoses. 2008;51:536–41.
60. Ayanbimpe GM, Taghir H, Diya A, Wapwera S. Tinea capitis among primary school children in some parts of central Nigeria. Mycoses. 2008;51:336–40.
61. Neji S, Makni F, Cheikhrouhou F, Sellami A, Sellami H, Marreckchi S, et al. Epidemiology of dermatophytoses in Sfax, Tunisia. Mycoses. 2009;52:534–8.
62. Yüksel T, Ilkit M. Identification of rare macroconidia-producing dermatophytic fungi by real-time PCR. Med Mycol. 2012;50:346–52.
63. Tamura K, Nei M. Estimation of the number of nucleotide substitutions in the control region of mitochondrial DNA in humans and chimpanzees. Mol Biol Evol. 1993;10:512–26.
64. Kumar S, Stecher G, Li M, Knyaz C, Tamura K. MEGA X: molecular evolutionary genetics analysis across computing platforms. Mol Biol Evol. 2018;35:1547–9.

Trichophyton mentagrophytes ITS Genotype VII from Thailand

Pietro Nenoff, Gudrun Wendrock-Shiga, Dirk Mechtel, Katja Schubert, Regina Jarsumbeck, Elke Lusmöller, Rudolf Stadler, Gabriele Ginter-Hanselmayer, Hans-Jürgen Tietz, Roland Pfüller, Helena Dröge, Daniela Koch, Franziska Wittig, Constanze Krüger, Bartosz Malisiewicz, Gökhan Mutluer, and Silke Uhrlaß

Abstract

Currently, in Germany, Switzerland, and Austria, a new and until now not known entity of inflammatory and abscessing dermatophyte infection is observed. In

P. Nenoff (✉) · D. Koch · F. Wittig · C. Krüger · S. Uhrlaß
Labor für Medizinische Mikrobiologie, Partnerschaft Dr. Constanze Krüger & Prof. Pietro Nenoff, Rötha, Germany
e-mail: nenoff@mykologie-experten.de; koch@mykologie-experten.de; franziska-wittig92@web.de; krueger@mykologie-experten.de; S.Uhrlass@Mykologie-Experten.de

G. Wendrock-Shiga
Hautarztpraxis, Oelsnitz, Germany

D. Mechtel
Klinik für Hautkrankheiten und Allergologie, Heinrich-Braun-Kliniken gGmbH, Zwickau, Germany
e-mail: dirk.mechtel@hbk-zwickau.de

K. Schubert
Hautarztpraxis Dresden, Dresden, Germany
e-mail: kontakt@meine-hautaerztin.de

R. Jarsumbeck
Medizinisches Labor Ostsachsen, Niederlassung Dresden, Dresden, Germany
e-mail: r.jarsumbeck@labor-ostsachsen.de

E. Lusmöller · R. Stadler
Universitätsklinik für Dermatologie, Venerologie, Allergologie und Phlebologie, Johannes Wesling Klinikum Minden, Minden, Germany
e-mail: Elke.LusmoellerPortodaRocha@muehlenkreiskliniken.de; rudolf.stadler@ruhr-uni-bochum.de

G. Ginter-Hanselmayer
Department of Dermatology and Venerology, Medical University of Graz, Graz, Austria
e-mail: gabriele.ginter@klinikum-graz.at; gabriele.ginter@medunigraz.at; h.hanselmayer@aon.at

J.-P. Bouchara et al. (eds.), *Dermatophytes and Dermatophytoses*,
https://doi.org/10.1007/978-3-030-67421-2_12

particular, these deep infections of the skin and their appendages include *Tinea pubogenitalis*, *Tinea cruris*, *Tinea barbae*, *Tinea corporis*, and *Tinea capitis*. Some of the patients came back from a journey to Thailand or South East Asia where transmission occurred. But, independently from a journey to abroad, the dermatophyte might be transferred in Germany, too. In every case, this deep infection is caused by the zoophilic dermatophyte *Trichophyton mentagrophytes* of the newly described Internal Transcribed Spacer (ITS) genotype VII. From 24 patients with dermatomycoses, altogether 25 isolates of *T. mentagrophytes* were isolated in the last 6 years. Among them were both male and female patients, the age range was from 1 to 50 years. The infection was affecting the pubogenital area, the capillitium, the face, or the trunk. Most of these infections were acquired in Germany without a stay abroad. A total of 25 dermatophyte isolates—all belonging to *T. mentagrophytes*—could be characterised. Species identification was confirmed for all isolates by sequencing of the ITS region of rDNA genes. Molecular relationship of the 24 isolates in comparison to already known genotypes within this species was depicted in a phylogenetic tree. The sequences obtained for these isolates are forming their own phylogenetic cluster, corresponding to a new genotype, the ITS genotype VII ("Thai variant"). This genotype is clearly set up from other clusters of *T. mentagrophytes*, e. g. zoophilic strains isolated from animals and human dermatophytoses, and from *T. mentagrophytes* genotype VIII "India". This genotype, as the other zoophilic *T. mentagrophytes* genotypes, can also be distinguished from the anthropophilic *Trichophyton interdigitale*.

Besides a direct sexual transmission and non-sexual human-to-human contagion, an indirect transmission due to sports or fitness training is suspected. As major predisposing factor for *Tinea* pubogenital, intimate shaving has to be considered. For treatment, oral antifungal agents should be used, first of all terbinafine, alternatively itraconazole or fluconazole. In particular, *Tinea genitalis profunda* may require oral terbinafine treatment for several months.

H.-J. Tietz
Mycoclinic Berlin, Berlin, Germany
e-mail: tietz@mycoclinic.de

R. Pfüller
Labor Dr. Eicke und Kollegen, EicLab GmbH, Abteilung Mikrobiologie, Berlin, Germany
e-mail: r.pfueller@mvz-weissensee.de

H. Dröge
Hautarztpraxis Helena und Henning Dröge, Berlin, Germany
e-mail: info@hautarzt-moabit.de

B. Malisiewicz
Klinik für Dermatologie, Venerologie und Allergologie, Universitätsklinikum, Goethe-Universität, Frankfurt am Main, Germany
e-mail: Bartosz.Malisiewicz@kgu.de

G. Mutluer
Klinik für Dermatologie und Phlebologie, Katharinen-Hospital, Unna, Germany

The question arises if the isolates of *T. mentagrophytes* genotype VII recently detected in Europe, especially in Germany, do possess higher virulence when compared to other genotypes of *T. mentagrophytes*. Laboratory investigations on pathogenicity and virulence factors of this dermatophyte should be initiated to answer to this speculation.

1 Introduction

Currently, in Germany, Switzerland, and Austria, a new and until now not known entity of inflammatory and abscessing dermatophyte infection is observed [1, 2]. In particular, these deep infections of the skin and their appendages include *Tinea genitalis*, *Tinea cruris*, *Tinea barbae*, *Tinea corporis*, and *Tinea capitis* [3]. Some of the patients came back from a journey to Thailand or South East Asia where transmission occurred. But, independently from a journey to abroad, the dermatophyte might be transferred in Germany, too [4]. In every case, this deep infection is caused by *Trichophyton mentagrophytes* of the newly described Internal Transcribed Spacer (ITS) genotype VII [5]. Here, we report on 24 patients with cutaneous infections due to this fungus throughout Germany and Austria during the last 6 years.

2 Patients

Based both from routine diagnostics of the laboratory Mölbis, Germany, and from cultures sent for fungal species identification, 24 patients with dermatomycoses due to *T. mentagrophytes* of ITS genotype VII ("Thailand") have been diagnosed all over Germany ($n = 23$) and Austria ($n = 1$) in the last 6 years (Table 1). From these patients, we were able to detect altogether 25 isolates of the *T. mentagrophytes* ITS genotype VII.

About half of the patients were of male ($n = 9$) and female ($n = 9$) gender. For the remaining patients, the gender was not told. The age range was from one (two siblings were diagnosed with dermatophytoses) to 50 years. Frequently, patients suffered from *Tinea pubogenitalis* ($n = 6$). Three children of 1 ($n = 2$) and 4 years suffered from *Tinea capitis* or both *Tinea capitis* and *Tinea corporis*. Four foreigners living in Germany, mostly from Syria, were among the patients. One German patient was of Turkish origin, he was fallen ill with a *Tinea pubogenitalis* after a journey to Turkey.

The mostly deep fungal skin infection affected the pubogenital area, the capillitium, the face, or the trunk. The dermatophyte was predominantly acquired in Germany, without a stay abroad.

Table 1 Overview of the 24 patients with dermatomycoses due to *T. mentagrophytes* of ITS genotype VII ("Thailand") diagnosed all over Germany ($n = 23$) and Austria ($n = 1$) in the last 6 years

Patient No.	Laboratory No.	Sending institution	Geographic site/ town/country	Gender	Age (years)	Clinical picture	Material	Transmission, treatment	Comment	Identification
1	205520/14	Department of Dermatology, University Graz, Austria	Oberwart, Austria	Male	40	*Tinea partim profunda mons pubis*	Fungal culture for confirmation	Intercourse with unknown partner in Thailand, condom tored	Pre-treatment: amoxicillin/ clavulanic acid and fluconazole. No animal contact, strong pain	ITS
2	216715/15	Diagnosticum Dr. Scholz and Partner, Neukirchen, Chemnitz, Germany	Ehrenfriedersdorf, Germany	Male	33	/	Skin scrapings			ITS
3	206565/16	Dr. Gudrun Wendrock-Shiga, Dermatological Office	Oelsnitz, Saxon Vogtland, Germany	Female	/	*Tinea faciei*	Fungal culture for confirmation	Infection transmitted from her husband who suffered from *Tinea barbae* after a stay in Thailand and contact with prostitutes [3]	Indirect transmission from Thailand via husband's *Tinea barbae* [3]	ITS, *TEF1-α*
4	206727/16	Dr. Kastl	Ammersbek, Germany	Female	50	/	Skin scrapings			ITS, *TEF1-α*
5	210363/16	Dr. Regina Jarsumbeck, Medical Laboratory East Saxoniae/Dr. Katja Schubert, Dermatological Office Dresden	Dresden, Germany	Female	27	*Tinea pubogenitalis*	Fungal culture for confirmation	*Tinea corporis* and *Tinea pubogenitalis* after a trip to Egypt, journey with her partner [2]		ITS, *TEF1-α*

6	211577/16	Dr. Meixner, Dept. of Dermatology, Ludwig Maximilians University Munich	München, Germany	/	/	/	Fungal culture for confirmation			ITS, *TEF1-α*
7	212014/16	Dr. Lindner and PD Dr. Ziemer, Dermatological Office Erfurt	Erfurt, Germany	Female	30	Suspected eczema, *Tinea corporis*	Skin scrapings			ITS, *TEF1-α*
8	213177/16	Prof. H.-J. Tietz, Mycoclinic Berlin	Berlin, Germany	/	/	/	Fungal culture for confirmation			ITS, *TEF1-α*
9	213178/16	Prof. H.-J. Tietz, Mycoclinic Berlin	Berlin, Germany	/	/	/	Fungal culture for confirmation			ITS, TEF1-α
10	215003/16	Charité Berlin	Berlin, Germany	Female	/	Genital mycosis	Fungal culture for confirmation			ITS, *TEF1-α*
11	218904/16	Dr. Maier, Dermatological Office Staßfurt	Egeln, Germany	Male	30	*Tinea barbae*	Skin scrapings			ITS, *TEF1-α*
12	200128/17	Clinic for Skin Diseases, Johannes Wesling Clinics Minden	Hille, Germany	Male	29	*Tinea profunda pubogenitalis*, *Tinea cruris*	Fungal culture for confirmation	No stay abroad, intercourse with his wife, only. Painful deep infection in the pubogenital area. No fever, no systemic signs of infection. Treatment at the dermatological ward in the Skin Clinic as in-patient. Cefuroxime intravenously, later on supplemented with oral clindamycin. Topical antifungals (clotrimazole), prednicarbate + clotrimazole,	Foreigner, in Germany living patient originating from Syria	ITS, *TEF1-α*

(continued)

Table 1 (continued)

Patient No.	Laboratory No.	Sending institution	Geographic site/ town/country	Gender	Age (years)	Clinical picture	Material	Transmission, treatment	Comment	Identification
								ciclopirox olamine, start with terbinafine 250 mg orally		
13	204544/17	Prof. Stadler, Clinic for Skin Diseases, Johannes Wesling Clinics Minden	Minden, Germany	Female	31	*Tinea corporis* and *Tinea capitis*	Fungal culture for confirmation	Mother of patients 14 and 16. Pregnant (36 week of pregnancy). Centrifugal, erosive lesions abdominal. Suspicion of scabies, permethrin and methylprednisolone aceponate, topically. Topical antifungal therapy by ciclopirox olamine. After childbirth, papular, pustular, nodular lesions developed pubogenital. After stopping breastfeeding, treatment with oral terbinafine 250 mg, and ciclopirox olamine. Healing		ITS, *TEF1-α*
14	204545/17	Prof. Stadler, Clinic for Skin Diseases, Johannes Wesling Clinics Minden	Minden, Germany	Female	4	*Tinea corporis* and *Tinea capitis*	Fungal culture for confirmation	Daughter of patient 13 and sister of patient 16. Erythematosquamous lesions at the trunk, hair loss areas and dandruff at the capillitium. Topical ciclopirox olamine cream, oral 125 mg terbinafine daily for 3 months, healing	Ciclopirox olamine treatment, but further positive fungal cultures	ITS, *TEF1-α*
15	204546/17	Prof. Stadler, Clinic for Skin Diseases, Johannes Wesling Clinics Minden	Hille, Germany	Female	21	*Tinea corporis*	Fungal culture for confirmation		Foreigner	ITS, *TEF1-α*

16	204547/17	Prof. Stadler, Clinic for Skin Diseases, Johannes Wesling Clinics Minden	Minden, Germany	Male	3 months old	*Tinea capitis*	Fungal culture for confirmation	Son of patient 13 and brother of patient 14. Newborn with pustules and erythematosquamous lesions left occipital. Topical application of ciclopirox olamine. Oral treatment with terbinafine, 3 mg/kg body weight (in gelatine), daily for 2 weeks, then once weekly until healing (for 2 months)	Suspected diagnosis was *Candida* infection of the skin	ITS, *TEF1-α*
17	216656/17	Tropical Medicine, Out-Patients Department, University Düsseldorf, Germany	Düsseldorf, Germany	Female	22	/	Skin scrapings, no cultural fungal detection, sequencing from extracted DNA from skin scrapings	After stay in the tropics	Foreigner	ITS
18	219080/17	Dermatological Office Dr. med. Götz Itschert	Pinneberg, Germany	Male	49	*Tinea genitalis* and *Tinea cruris profunda*	Fungal culture for confirmation	No contact with Thailand women		ITS, *TEF1-α*
19	205874/18	Surrounding of Essen, Ruhr District, Germany	Essen, Germany	/	/	*Tinea genitalis*	Fungal culture for confirmation			ITS, *TEF1-α*
20	205953/18	Prof. M. Schaller, Department of Dermatology, University Tübingen	Tübingen, Germany	Male	/	*Tinea corporis* (breast)	Fungal culture for confirmation	Patient originated from Republic of Kosovo		ITS, *TEF1-α*
21	250002/18	Dr. Mutluer Gökhan, Katharinen	Unna, Germany	Male	/	*Tinea genitalis*	Fungal culture for confirmation		German patient of Turkish	ITS, *TEF1-α*

(continued)

Table 1 (continued)

Patient No.	Laboratory No.	Sending institution	Geographic site/ town/country	Gender	Age (years)	Clinical picture	Material	Transmission, treatment	Comment	Identification
		Hospital, Skin Clinic, Unna							origin, after journey to Turkey	
22	901140/18	Dr. med. Bartosz Malisiewicz, Department of Dermatology, University Frankfurt/Main	Frankfurt/Main, Germany	Male	24	*Tinea*	Skin scrapings		Foreigner	ITS, *TEF1-α*
23	204511/19 204512/19	Dr. Roland Pfüller, EicLab GmbH, Mikrobiologie, Jacobsohnstr. 20, 13086 Berlin, Germany	Berlin, Germany	Male	41	*Tinea corporis* (stomach) and *Tinea manuum*	Fungal culture for confirmation			ITS, *TEF1-α*
24	600012/19	Dermatological Office Dr. Dröge, Berlin, and Dr. Roland Pfüller, EicLab GmbH, Mikrobiologie, Jacobsohnstr. 20, 13086 Berlin, Germany	Berlin, Germany	Male	39	*Tinea corporis* and *Tinea faciei*	Fungal culture for confirmation	Foreigner. Dermatophytosis of the groins (initially *Tinea cruris*), trunk, and spreading to the face after visiting a friend 10 days ago. Now stay abroad, never in Thailand, no animal contact	No intercourse?	ITS, *TEF1-α*

3 Description of Selected Cases of Dermatophytoses Due to *Trichophyton mentagrophytes* Genotype VII

3.1 Case Report: Patient 6 with *Tinea pubogenitalis profunda* After a Trip to Egypt

After a holiday trip to Egypt, a 28-year-old woman developed erythematosquamous, pustular, and purulent abscessing lesions at the mons pubis, in the groins, and the underbelly [2]. Because of the suspicion of a dermatomycosis, topical antifungal treatment was started. Microbiological diagnostics revealed, in addition to *Staphylococcus aureus* and group B-streptococci, a fast-growing fungus producing white-yellowish pigmented, granular colonies. Morphological identification resulted in *Trichophyton mentagrophytes* var. *asteroides*, a zoophilic dermatophyte. Treatment of the kerion-like dermatomycosis of the pubogenital area with oral terbinafine was started for 6 weeks with hesitant recovery. A short-time change to fluconazole, and antibiotic treatment with sultamicillin did not bring any progress. Finally, full recovery was achieved after a 4.5 months treatment with terbinafine 250 mg per day orally. Brocq's pseudopélade resulted as final condition. Precise species identification using sequencing of the ITS regions of ribosomal DNA (rDNA) genes yielded for the dermatophyte isolate, a 100% agreement with sequences of zoophilic *T. mentagrophytes* isolates deposited at the National Center for Biotechnology Information (NCBI), Bethesda, MD. The DNA sequence of this isolate matches, indeed, with *T. mentagrophytes* isolates currently found in Germany from patients returning from Thailand and suffering from abscessing *Tinea genitalis*. An animal source of infection was not identified. An indirect transmission was suggested, probably in the wellness and fitness sector of the hotel in Egypt. As predisposing factor shaving of the intimate area, which was maintained even during the entire duration of treatment has to be considered.

3.2 Case Report: Patient 12 with *Tinea pubogenitalis profunda* in a Syrian Refugee

A 29-year-old Syrian refugee, living for already 2 years in Germany, suffered from severe and extensive pustular and abscessing pubogenital infection, which not improved after treatment by oral cefuroxime. No fever or other systemic infection signs occurred. He never travelled to Thailand or South East Asia. The patient developed highly inflammatory deep red partly erosive papules, pustules, nodules, and plaques on erythematous ground, both in the mons pubis area, and also in the groins, scrotal, and at the underbelly (Fig. 1). Intercourse with his wife, only, was declared. No unprotected sexual contacts were reported. Due to elevated inflammatory signs in the blood serum (C-reactive protein), and to suspected bacterial skin infection (erysipelas and intertrigo), antibiotics (cefuroxime 1.5 g, three times a day) were applied intravenously. Paracetamol and ibuprofen were given due to pain and strong inflammation of the infected area. Histologically, the haematoxylin-eosin

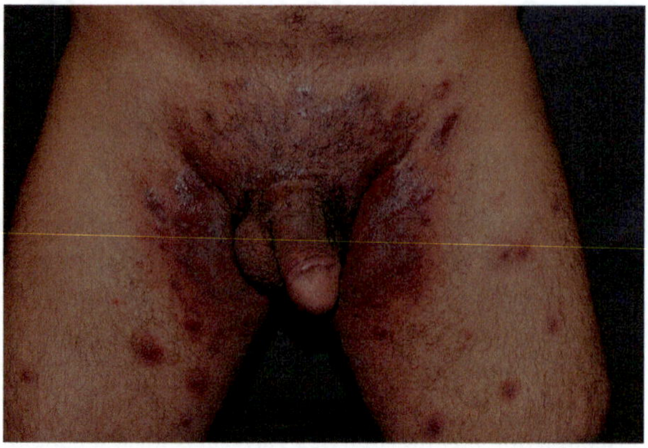

Fig. 1 Abscessing *Tinea cruris* and *Tinea genitalis profunda* in a 29-year-old Syrian migrant in Germany (Patient 12, see Table 1). Causative agent was *Trichophyton mentagrophytes*. Sequencing of the ITS revealed genotype VII "Thai variant". The patient, however, never visited Thailand. Way of transmission was a sexually one

staining revealed an inflammatory infiltrate of preferably neutrophil granulocytes, lymphocytes, and histiocytes. The Periodic acid-Schiff (PAS) reaction revealed fungal hyphae preferably in the deepest parts of the follicles, in particular in the ruptured hair follicles. Oral terbinafine 250 mg daily was started. Topical antifungals were added, first clotrimazole 2% + prednicarbate 0.25% cream, later treatment was switched to ciclopirox olamine twice daily. Due to newly increased inflammatory laboratory signs, intravenous cefuroxime (1.5 g, three times per day) was given again for 10 days. For better skin tissue penetration, oral clindamycin (600 mg, three times per day) was added. Terbinafine 250 mg orally was further given, both under inpatient conditions, but also after the patient has leaved the hospital. A slow improvement of the *Tinea pubogenitalis profunda* and the suspected bacterial cutaneous infection was achieved.

3.3 Case Report: Patients 13, 14, and 16 with *Tinea genitalis, Tinea capitis,* and *Tinea corporis* in a Family

A pregnant woman (36 weeks of pregnancy, mother of patients 14 and 16) developed centrifugal, erosive lesions of the abdominal skin. As scabies was suspected, topical permethrin and methylprednisolone aceponate cream were administered. Due to the microbiological detection of *T. mentagrophytes*, the topical therapy of the *Tinea corporis* was switched to ciclopirox olamine cream. After childbirth, papular, pustular, and nodular lesions have spread over the pubogenital region. The woman stopped breastfeeding, and treatment with oral terbinafine 250 mg was started, together with ciclopirox olamine cream. Complete healing of the *Tinea pubogenitalis* and *Tinea cruris* was achieved. The *T. mentagrophytes* isolate was later on identified as belonging to the cluster ITS genotype VII.

A 3-month-old male newborn (daughter of the mentioned woman) developed white pustules on erythematous base, together with left occipital erythematosquamous lesions. Topical application of ciclopirox olamine was started. After cultural detection of *T. mentagrophytes*, oral terbinafine was given (as Off-

Label-use for treatment of *Tinea capitis* in Germany), 3 mg/kg body weight (in gelatine capsules), once daily for 2 weeks, then once per week until healing of the *Tinea capitis* (altogether for 2 months).

The 4-year-old girl, child of patient 13, showed erythematosquamous lesions at the trunk, and hair loss areas, and dandruff at the capillitium. From skin scrapings, *T. mentagrophytes* ITS genotype VII was detected. Topical ciclopirox olamine cream and oral 125 mg terbinafine (Off-Label-use in Germany) daily for 3 months lead to complete healing.

Later on, both *T. mentagrophytes* isolates from the kids were identified as belonging to the cluster of ITS genotype VII.

The sexual partner of their mother (i.e. the biological father of both children) had a history of skin lesions at the mons pubis and in the genital area, too. He was the suspected source of infection of the whole family.

4 Conventional Mycological Diagnostics and Features of *Trichophyton mentagrophytes* Genotype VII

4.1 Mycological Diagnostics

Trichophyton mentagrophytes isolates from patients with highly inflammatory deep infections of the pubogenital area and *Tinea cruris* were investigated based on their cultural features, but also by molecular methods. Besides conventional PCR-ELISA [6], sequencing of the ITS region of the fungal rDNA for species and ITS genotype identification was performed for all dermatophyte isolates [7, 8]. Mostly, fungal cultures sent from all over the country were included in this survey. The patients suffered mainly from inflammatory dermatophytoses of the genital area, sometimes from other kinds of dermatophytoses, e.g. *Tinea cruris*, *Tinea capitis*, and *Tinea corporis*.

4.2 Cultural Features of *Trichophyton mentagrophytes* ITS Genotype VII

Cultural analysis of the here presented isolates revealed fast-growing fungi with white colonies, typical for *T. mentagrophytes* (Fig. 2a). In 4–5 days grew flat, radiating colonies with white to beige-stained granular, partly floccose surface (Fig. 3a, c). The centre of the thallus shows sometimes slight yellow pigmentation.

On cycloheximide (Actidione®)-containing Sabouraud dextrose agar (Becton Dickinson, Heidelberg, Germany), the reverse of the colonies was yellow to brown coloured. On common Sabouraud 4% dextrose agar (Sifin, Berlin, Germany), the reverse colony side often showed a strong brown-red to mahogany-coloured pigmentation, characteristic for this species (Figs. 2b and 3b, d).

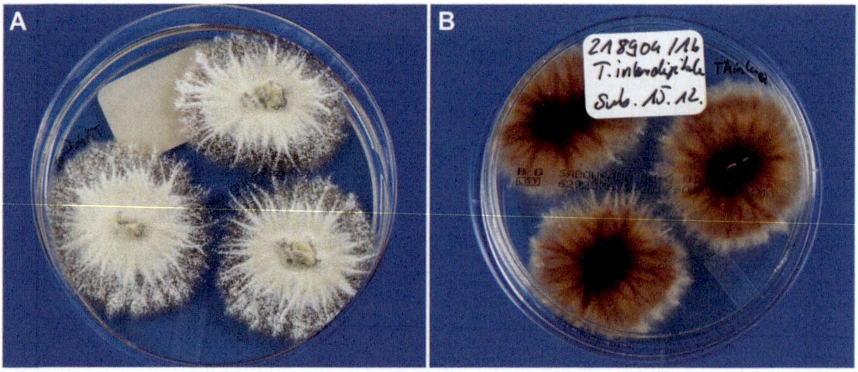

Fig. 2 *Trichophyton mentagrophytes* ITS genotype VII "Thai variant", isolated from a 30-year-old male patient with *Tinea barbae* (Patient 11, see Table 1). (**a**) Subculture with white to yellowish flat, granular, fast-growing colonies on Sabouraud 4% dextrose agar. (**b**) Reverse side of the colonies showing with strong brownish red pigment

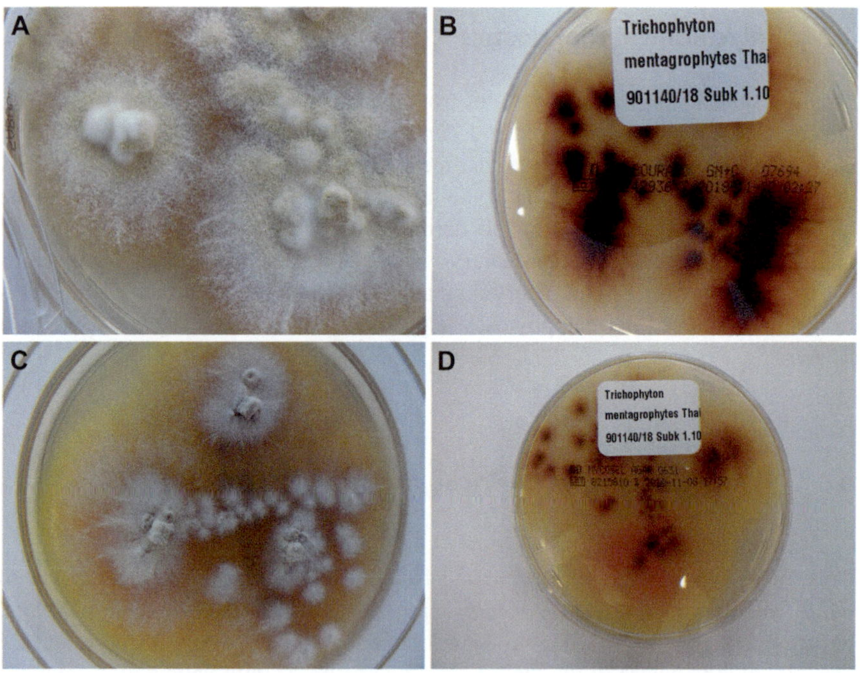

Fig. 3 *Trichophyton mentagrophytes* ITS genotype VII "Thai variant". Isolate from a 24-year-old male patient with *Tinea genitalis profunda* (Patient 22, see Table 1). (**a**) Typical flat white granular colonies with a yellow centre on Sabouraud 4% dextrose agar. (**b**) Reverse side with typical brownish red pigmentation. (**c**) Typical flat white granular colonies on cycloheximide-containing Sabouraud agar (Mycosel®). (**d**) Brown-stained reverse side

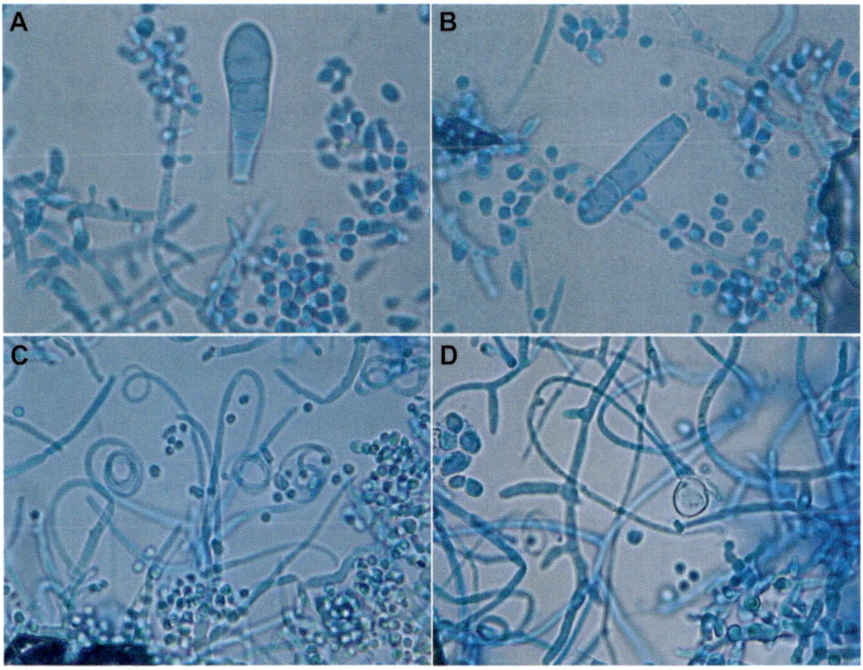

Fig. 4 Microscopic features of *Trichophyton mentagrophytes* ITS genotype VII "Thai variant". Lactophenol Cotton Blue staining. (**a**) Clavate and thick-walled macroconidia. (**b**) Cigar-shaped thick-walled macroconidia and round and sessile microconidia. (**c**) Spiral hyphae and microconidia. (**d**) Chlamydospores

4.3 Microarchitecture of *Trichophyton mentagrophytes* ITS Genotype VII

Microscopically, round and elongated microconidia, cigar-shaped macroconidia, and spiral hyphae were detected (Fig. 4a–d). Due to these morphological characteristics, the dermatophyte was identified as the zoophilic species *T. mentagrophytes*. One investigator diagnosed the former *asteroides* variety of *T. mentagrophytes* [9, 10]. Based on macroscopic morphology and microarchitecture, the secure identification of the distinct genotype within the species *T. mentagrophytes* was not possible [11].

5 Molecular Biological Characteristics of *Trichophyton mentagrophytes* ITS VII

5.1 PCR-ELISA for Molecular Identification of Dermatophytes

DNA from either the skin scrapings or fungal isolates (for identification of submitted fungal cultures) was extracted according to the manufacturer's protocol using the QIAamp® DNA Mini Kit (Qiagen, Hilden, Germany). Samples were analysed using a validated and standardised *in-house*-developed enzyme linked immunoassay (PCR-ELISA) to detect dermatophyte DNA [12]. Specific probes detecting the following relevant dermatophytes were used: *Trichophyton rubrum*, *T. interdigitale*/*T. mentagrophytes*, *Microsporum canis*, and *Trichophyton benhamiae* (formerly referred to as *Trichophyton* anamorph of *Arthroderma benhamiae*).

All samples—DNA extracted either from skin scrapings or from fungal cultures—were positive in the PCR-ELISA for *T. interdigitale*/*T. mentagrophytes*. As the differentiation between *T. interdigitale* and the *T. mentagrophytes* complex was not possible by PCR-ELISA, the ITS regions of rDNA genes and the translation elongation factor (TEF)1-α gene were sequenced.

5.2 Sequencing of the ITS Regions of rDNA Genes

For confirmation of the suspected dermatophyte species, Sanger sequencing of the ITS regions of rDNA genes (mainly the regions ITS 1, 5.8S rRNA, ITS 2) and *TEF1-α* gene was performed for all isolates [13–15]. This required PCR amplification of a ~900 bp DNA fragment using universal primers that bind to flanking pan-fungal sequence regions: V9G (5′-TTACGTCCCTGCCCTTTGTA-3′) and LS266 (5-′-GCATTCCCAAACAACTCGACTC-3′).

The length of the analysed region in the *TEF1-α* gene varied from 709 to 769 nucleotides among the various dermatophyte species. Primers EF1a-F (5-′-CACATTAACTTGGTCGTTATCG-3′) and EF1a-R (5-′-CATCCTTGGAGATACCAGC-3′) were used for sequencing [13].

5.3 Species Identification and Phylogenetic Analysis of *Trichophyton mentagrophytes* Genotypes

Species identification was confirmed for the 25 isolates by sequencing of the ITS regions of rDNA genes. Molecular relationship of these 25 isolates with other genotypes within the species *T. mentagrophytes*, and with closely related dermatophytes, was depicted in a phylogenetic tree/dendrogram of the sequences (Fig. 5a). The 25 isolates belonged to the same cluster, called ITS genotype VII and referred to as "Thailand variant" or "Thai variant". The isolates of *T. mentagrophytes* ITS genotype VII were forming their own phylogenetic cluster. This genotype was clearly set up from other already known genotypes of *T. mentagrophytes*,

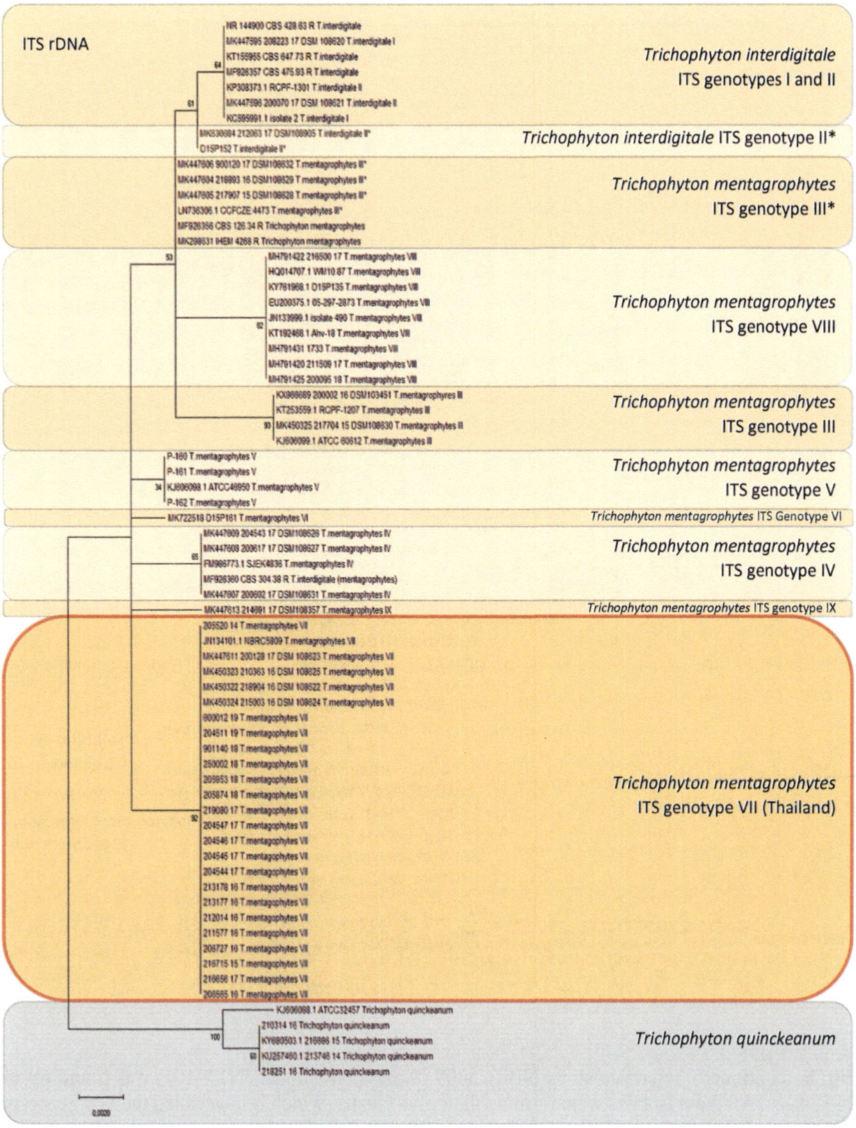

Fig. 5 Phylogenetic analysis of the dermatophyte isolates based on the ITS regions of rDNA and the *TEF1-α* gene (Table 2). The evolutionary history was inferred by using the Maximum Likelihood method and Tamura-Nei model [33]. The tree with the highest log likelihood (−907.70) is shown. The percentages of trees in which the associated taxa clustered together are shown next to the branches. Initial tree for the heuristic search was obtained automatically by applying Neighbor-Join and BioNJ algorithms to a matrix of pairwise distances estimated using the Maximum Composite Likelihood (MCL) approach, and then selecting the topology with superior log likelihood value. The tree is drawn to scale, with branch lengths measured in the number of substitutions per site. This analysis involved 28 nucleotide sequences. Codon positions included were first + second + third + Non-coding. There were a total of 1086 positions in the final dataset. Evolutionary analyses were conducted in MEGA X [34]. (**a**) Phylogenetic tree of *T. mentagrophytes* based on sequencing of the ITS regions of rDNA genes. By sequencing, a 100% concordance with the

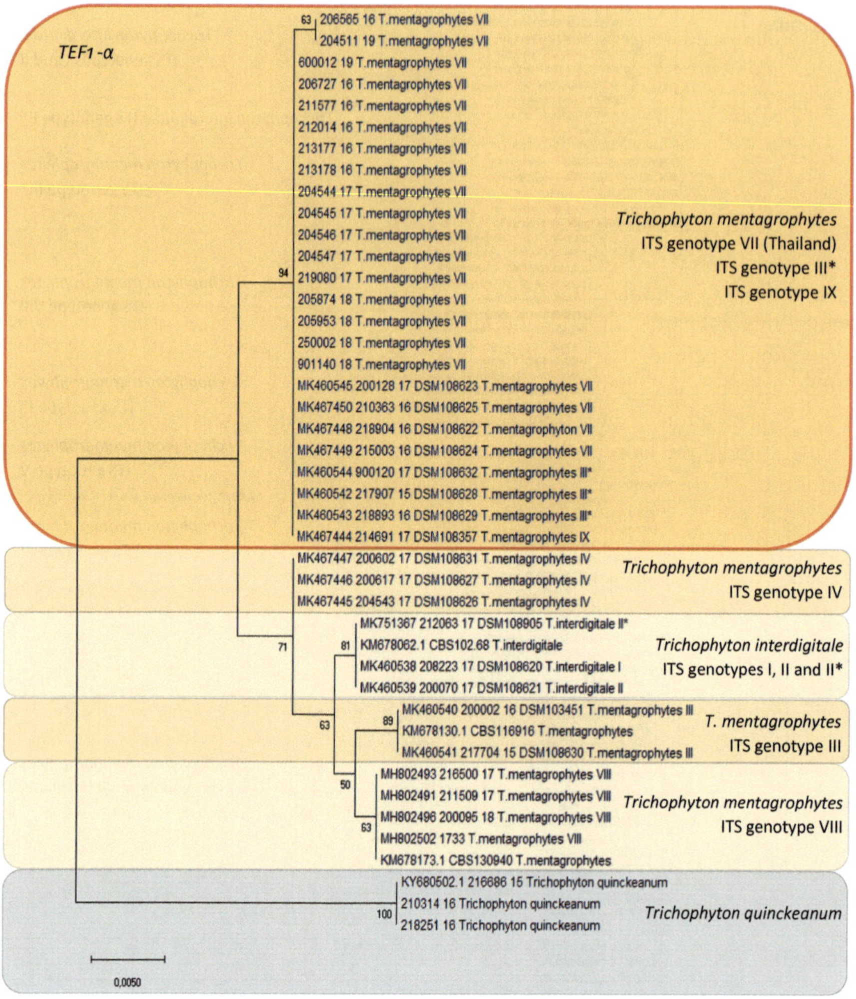

Fig. 5 (continued) reference strain NBRC5809 (Accession number JN134101) was found for all 24 isolates. All these isolates were forming their own cluster, which is now called the ITS genotype VII (Thai variant) of *T. mentagrophytes*. These isolates (*T. mentagrophytes* genotype VII or clade) were clearly discriminated from already known *T. mentagrophytes* genotypes, e. g. II, V, VIII. Rooted with *Trichophyton quinckeanum*. (**b**) Phylogenetic tree of *T. mentagrophytes* based on sequencing of the *TEF1-α* gene. Within the phylogenetic tree, all *T. mentagrophytes* ITS genotype VII strains from Germany were forming their own clade which is clearly discriminated from the other, above-mentioned *T. mentagrophytes* genotypes. Rooted with *T. quinckeanum*

e.g. zoophilic strains isolated from human dermatophytoses and from animals, including a snow leopard at a zoo garden, and from *T. mentagrophytes* ITS genotype

VIII from India. The anthropophilic *T. interdigitale* can be distinguished clearly from zoophilic *T. mentagrophytes* clusters.

Phylogenetic tree based on sequencing of the *TEF1-α* gene revealed a 100% concordance of all 25 isolates belonging to genotype VII of *T. mentagrophytes* (Fig. 5b). Within the phylogenetic tree, all *T. mentagrophytes* isolates from this study were forming their own clade which was clearly differentiated from other *T. mentagrophytes* genotypes, and from *T. interdigitale.*

The origin of all reference strains and clinical or animal isolates and their sequences used here for comparison are presented in Table 2.

6 Discussion

6.1 *Tinea genitalis*

Tinea pubogenitalis or *Tinea genitalis* represents a rare type of dermatophytosis which, however, is increasingly being diagnosed [16]. The mons pubis is affected, but also the outer regions to the penis shaft and the labia together with the groins and underbelly. *Tinea pubogenitalis* is a more superficial erythrosquamous scaling type. However, an emergence of strong inflammatory dermatomycoses of the genital area as *Tinea genitalis profunda* ranging to kerion Celsi is observed. Recently, 30 patients (14–63 years of age, 11 men and 19 women) with *Tinea pubogenitalis* were reported in an epidemiological survey from Austria and Germany [16]. Causative agents were mainly zoophilic dermatophytes: *M. canis*, *T. mentagrophytes*, *T. benhamiae*, and *Trichophyton verrucosum.* Rarely, anthropophilic fungi, as *T. rubrum*, and *Trichophyton tonsurans*, were isolated from *Tinea pubogenitalis.* The authors of this survey focussed on patient's history, which should include questions concerning contact to pets, physical activities (fitness), and travel. Genital shaving and concurrent *Tinea pedis* and onychomycosis were found as main predisposing factors.

6.2 Abscessing *Tinea genitalis* and *Tinea barbae* Due to *Trichophyton mentagrophytes* in Germany as New Entity of Mainly Sexually Transmitted Infection

In Germany, Austria, and Switzerland, an increasing number of male and female patients presenting with abscessing and purulent *Tinea pubogenitalis* and *Tinea barbae* due to zoophilic and anthropophilic dermatophytes can be observed [16, 17]. Transmission occurs, however, not always via a zoophilic way from animals, but rather by human-to-human transfer of the infectious agent. Some of the patients report on a previous journey to South-East-Asia, where they had sexual contacts.

First, in 2015, Luchsinger et al. [1] from Switzerland reported on *Tinea genitalis* as a probably new entity of sexually transmitted infection due to *T. interdigitale* (what is now *T. mentagrophytes*) after a journey to South-East-Asia. The authors

Table 2 Reference strains and clinical isolates used to generate the phylogenetic tree based on sequencing the ITS region of rDNA genes and the *TEF1-α* gene

Species	Strain	Locus or gene	GenBank accession number
T. interdigitale	CBS 647.73 R	ITS rDNA	KT155955
T. interdigitale	CBS 475.93 R	ITS rDNA	MF926357
T. interdigitale	CBS 428.63 R	ITS rDNA	NR_144900
T. interdigitale ITS genotype II	200070_17_DSM_108621	ITS rDNA	MK447596
		TEF1-α	MK460539
T. interdigitale ITS genotype II	RCPF-1301	ITS rDNA	KP308373
T. interdigitale ITS genotype I	Isolate 2	ITS rDNA	KC595991
T. interdigitale ITS genotype I	208223_17_DSM_108620	ITS rDNA	MK447595
		TEF1-α	MK460538
T. interdigitale	CBS102.68	*TEF1-α*	KM678062
T. interdigitale ITS genotype II*	212063_17_DSM_108905	ITS rDNA	MK630684
		TEF1-α	MK751367
T. interdigitale ITS genotype II*	D15P152	ITS rDNA	
T. mentagrophytes	CBS 126.34 R	ITS rDNA	MF926356
T. mentagrophytes	IHEM 4268 R	ITS rDNA	MK298631
T. mentagrophytes	CBS 116916	*TEF1-α*	KM678130
T. mentagrophytes ITS genotype III	200002_16_DSM_103451	ITS rDNA	KX866689
		TEF1-α	MK460540
T. mentagrophytes ITS genotype III	217704_15_DSM_108630	ITS rDNA	MK450325
		TEF1-α	MK460541
T. mentagrophytes ITS genotype III	ATCC 60612	ITS rDNA	KJ606099
T. mentagrophytes ITS genotype III	RCPF 1207	ITS rDNA	KT253559
T. mentagrophytes ITS genotype III*	900120_17_DSM_108632	ITS rDNA	MK447606
		TEF1-α	MK460544
T. mentagrophytes ITS genotype III*	218893_16_DSM_108629	ITS rDNA	MK447604
		TEF1-α	MK460543
T. mentagrophytes ITS genotype III*	217907_15_DSM_108628	ITS rDNA	MK447605
		TEF1-α	MK460542
T. mentagrophytes ITS genotype IV	200617_17_DSM_108627	ITS rDNA	MK447608
		TEF1-α	MK467446
T. mentagrophytes ITS genotype IV	SJEK4836	ITS rDNA	FM986773
T. mentagrophytes ITS genotype IV	204543_17_DSM_108626	ITS rDNA	MK447609
		TEF1-α	MK467445
T. mentagrophytes ITS genotype IV	200602_17_DSM_108631	ITS rDNA	MK447607
		TEF1-α	MK467447

(continued)

Table 2 (continued)

Species	Strain	Locus or gene	GenBank accession number
T. interdigitale (mentagrophytes)	CBS 304.38 R	ITS rDNA	MF926360
T. mentagrophytes ITS genotype V	ATCC 46950	ITS rDNA	KJ606098
T. mentagrophytes ITS genotype V	P-160	ITS rDNA	
T. mentagrophytes ITS genotype V	P-161	ITS rDNA	
T. mentagrophytes ITS genotype V	P-162	ITS rDNA	
T. mentagrophytes ITS genotype VI	D15P161	ITS rDNA	MK722518
T. mentagrophytes ITS genotype VII	NBRC5809	ITS rDNA	JN134101
T. mentagrophytes ITS genotype VII	200128_17_DSM_108623	ITS rDNA	MK447611
		TEF1-α	MK460545
T. mentagrophytes ITS genotype VII	210363_16_DSM_108625	ITS rDNA	MK450323
		TEF1-α	MK467450
T. mentagrophytes ITS genotype VII	218904_16_DSM_108622	ITS rDNA	MK450322
		TEF1-α	MK467448
T. mentagrophytes ITS genotype VII	215003_16_DSM_108624	ITS rDNA	MK450324
		TEF1-α	MK467449
T. mentagrophytes ITS genotype VII	600012 19	ITS rDNA, *TEF1-α*	
T. mentagrophytes ITS genotype VII	204511 19	ITS rDNA, *TEF1-α*	
T. mentagrophytes ITS genotype VII	901140 18	ITS rDNA, *TEF1-α*	
T. mentagrophytes ITS genotype VII	250002 18	ITS rDNA, *TEF1-α*	
T. mentagrophytes ITS genotype VII	205953 18	ITS rDNA, *TEF1-α*	
T. mentagrophytes ITS genotype VII	205874 18	ITS rDNA, *TEF1-α*	
T. mentagrophytes ITS genotype VII	219080 17	ITS rDNA, *TEF1-α*	
T. mentagrophytes ITS genotype VII	216656 17	*ITS rDNA*	
T. mentagrophytes ITS genotype VII	204547 17	ITS rDNA, *TEF1-α*	
T. mentagrophytes ITS genotype VII	204546 17	ITS rDNA, *TEF1-α*	
T. mentagrophytes ITS genotype VII	204545 17	ITS rDNA, *TEF1-α*	

(continued)

Table 2 (continued)

Species	Strain	Locus or gene	GenBank accession number
T. mentagrophytes ITS genotype VII	204544 17	ITS rDNA, *TEF1-α*	
T. mentagrophytes ITS genotype VII	213178 16	ITS rDNA, *TEF1-α*	
T. mentagrophytes ITS genotype VII	213177 16	ITS rDNA, *TEF1-α*	
T. mentagrophytes ITS genotype VII	212014 16	ITS rDNA, *TEF1-α*	
T. mentagrophytes ITS genotype VII	211577 16	ITS rDNA, *TEF1-α*	
T. mentagrophytes ITS genotype VII	206727 16	ITS rDNA, *TEF1-α*	
T. mentagrophytes ITS genotype VII	206565 16	ITS rDNA, *TEF1-α*	
T. mentagrophytes ITS genotype VII	216715 15	ITS rDNA	
T. mentagrophytes ITS genotype VII	205520 14	ITS rDNA	
T. mentagrophytes ITS genotype VIII	WM10.87	ITS rDNA	HQ014707
T. mentagrophytes ITS genotype VIII	Ahv-18	ITS rDNA	KT192468
T. mentagrophytes ITS genotype VIII	D15P135	ITS rDNA	KY761968
T. mentagrophytes ITS genotype VIII	05-297-2873	ITS rDNA	EU200375
T. mentagrophytes ITS genotype VIII	Isolate 490	ITS rDNA	JN133999
T. mentagrophytes ITS genotype VIII	216500 17	ITS rDNA	MH791422
		TEF1-α	MH802493
T. mentagrophytes ITS genotype VIII	211509 17	ITS rDNA	MH791420
		TEF1-α	MH802491
T. mentagrophytes ITS genotype VIII	200095 18	ITS rDNA	MH791425
		TEF1-α	MH802496
T. mentagrophytes ITS genotype VIII	1733	ITS rDNA	MH791431
		TEF1-α	MH802502
T. mentagrophytes	CBS 130940	*TEF1-α*	KM678173
T. mentagrophytes ITS genotype IX	214691_17_DSM_108357	ITS rDNA	MK447613
		TEF1-α	MK467444
Trichophyton quinckeanum	213746 14	ITS rDNA	KU257460
Trichophyton quinckeanum	210314 16	ITS rDNA, *TEF1-α*	

(continued)

Table 2 (continued)

Species	Strain	Locus or gene	GenBank accession number
Trichophyton quinckeanum	216686 15	ITS rDNA	KY680503
		TEF1-α	KY680502
Trichophyton quinckeanum	ATCC 32457	ITS rDNA	KJ606088
Trichophyton quinckeanum	218251 16	ITS rDNA, *TEF1-α*	

GenBank accession number of the nucleotide sequences used in this study is available at the NCBI (National Centre for Biotechnology Information, Bethesda, MD)
ITS internal transcribed spacer, *rDNA* ribosomal DNA, *TEF1-α* translation elongation factor 1-α, *CBS* CentraalBureau voor Schimmelcultures (Westerdijk Fungal Biodiversity Institute, Utrecht, The Netherlands), *DSM* Deutsche Sammlung von Mikroorganismen und Zellkulturen, Braunschweig, Germany, *NCPF* The National Collection of Pathogenic Fungi (Bristol, UK), *IHEM* Institute of Hygiene and Epidemiology-Mycology section (Sciensano, Brussels, Belgium), *ATCC* American Type Culture Collection (Manassas, VA)

presented a case series of seven patients (two female and five male patients, age range 18–55 years) with *Tinea* in the genital region emerging after sexual intercourse in South-East-Asia. In six patients, *T. interdigitale* (today classified as *T. mentagrophytes*) was detected. Three patients suffered from a severe inflammatory reaction of the cutaneous-subcutaneous tissue and two of them were hospitalised due to severe pain.

The first German report of an infection due to *T. mentagrophytes* genotype VII is on a patient who came back from Thailand with a foudroyant abscessing and painful *Tinea barbae* [3]. As a diagnosis of herpes and sycosis barbae was evoked, initially oral acyclovir, followed by ciprofloxacin and ampicillin plus sulbactam, was given, without any effect. Histologically, in Grocott–Gomori's methenamine silver stain, fungal mycelium was apparent. Terbinafine 250 mg was started together with topically 1% ciclopirox olamine containing cream. Later, fungal culture grew a dermatophyte which could be identified by sequencing of the ITS regions of rDNA as *T. mentagrophytes* genotype VII ("Thai variant"). The dermatophyte was transferred to the patient's wife causing *Tinea faciei.* The index patient had contact with Thai female sex workers who must be considered as a source of infection of the dermatophytosis. There was, like in nearly all other patients with this infection, no animal contact, neither in Thailand, nor in Germany.

The same genotype of *T. mentagrophytes* (ITS genotype VII) could be isolated in an Australian returning from Southeast Asia, and this infection was transmitted via sexual contact [18]. The infection was the first reported case of Majocchi's granuloma (MG) in an Australian returned traveller. By DNA sequencing, the former ITS type IV variant, a type reported as having a zoophilic origin, was identified. This corresponds to the current ITS genotype VII of *T. mentagrophytes* [19, 20]. The strain was subsequently transmitted via sexual contact to another intimate partner of the patient. Both patients were successfully treated with systemic antifungals. The first patient was treated with oral terbinafine 500 mg daily, from day +28 after diagnosis until day +140, and then with terbinafine 250 mg daily together with 50 mg itraconazole until day +196. Pseudopelade with scarring and alopecia areata resulted after therapy. The second patient got terbinafine 250 mg daily for 6 weeks.

7 Intra Species Classification of *Trichophyton mentagrophytes* and *Trichophyton interdigitale*

Due to the new taxonomy of dermatophytes, the former *T. mentagrophytes* complex is now separated in *T. mentagrophytes* (geophilic and zoophilic strains), *T. benhamiae* (zoophilic species) and *T. interdigitale* (anthropophilic species) [21]. Based on sequencing of the ITS regions of rDNA genes, currently at least ten different genotypes are distinguished within the species *T. mentagrophytes* (eight genotypes), and *T. interdigitale* (two genotypes) [19]. These genotypes correspond significantly with distinct geographic regions (e.g. genotype VIII and India), an animal source of infection, or a clinical picture (genotype VIII and *Tinea cruris*) [22, 23]. The here isolated *T. mentagrophytes* isolates are belonging to ITS genotype VII. Most of these isolates were recovered in Germany, in one case after travelling to Southeast Asia, but predominately after transmission in Germany. A sexual pathway of infection has to be considered in most patients with *Tinea cruris* and *Tinea genitalis* [24, 25].

7.1 Dermatophytosis Due to *Trichophyton mentagrophytes* Genotype VII in Germany

A first huge case series of abscessing dermatophytoses of the pubogenital area due to *T. mentagrophytes* ITS genotype VII in Germany was recently published [5]. Between January 2016 and July 2017, 43 patients attended the Skin Clinic of the Charité, Berlin, Germany; all of them were suffering from mostly highly inflammatory, painful, and persistent infections of the pubogenital region. Sequencing of fungal DNA revealed that 37 out of the 43 patients were suffering from an infection due to the new genotype VII of *T. mentagrophytes*. Interestingly, all patients were from Berlin or its region. Most of the patients had no history of travel. Only four out of 18 patients with available history have been on vacation in Thailand or in Southeast Asia a short time before the infection developed. Main predisposing factor in 16 out of 18 patients was regular intimate shaving.

Mating type analysis which was performed in Berlin revealed the Mat1-2 locus (HMG) in all strains, while the α-box was not present in any of the isolates. By sequencing of the squalene epoxidase (SQLE) gene of the Berlin *T. mentagrophytes* VII isolates, no one of the four described point mutations involved in treatment failure of terbinafine could be detected [26, 27].

Nearly all patients were treated with oral terbinafine (250 mg daily), and either ciclopirox olamine or miconazole-containing cream was used for topical treatment. In a single case, fluconazole was first used, but it was changed to terbinafine. In two patients, due to treatment failure, terbinafine was switched to itraconazole. Duration of oral antifungal treatment was 4–15 weeks, in average 7 weeks.

7.2 Dermatophytosis Due to *Trichophyton mentagrophytes* Genotype VII Are Present Throughout Germany

Finally, a total of 25 dermatophyte strains—all belonging to *T. mentagrophytes*, isolated from altogether 24 patients, could be characterised in the present survey. Species identification was confirmed for all strains by sequencing of the ITS regions of rDNA genes. Nucleotide sequences of this variant were deposited in the GenBank database. Although abscessing *Tinea genitalis* was the most frequent clinical manifestation, other kinds of *Tinea* were diagnosed. These were *Tinea cruris*, *Tinea corporis*, but also *Tinea capitis*. The latter mentioned dermatophytoses of the capillitium affected exclusively children—a newborn child of 3 months, and two 1- and 4-year-old toddlers. Transmission in these children took place from the likewise with the dermatophyte infected mother who was finally initially infected by her intimate partner.

7.3 Transmission and Predisposing Factors

The infection chain of the mostly purulent dermatophytosis due to *T. mentagrophytes* genotype VII from Southeast-Asia, e.g. Thailand, probably reaching from sex worker to the finally affected European (and Australian) patients, has to be considered. Besides a direct sexual transmission, an indirect infection due to sports or fitness training is suspected. Currently, most infections caused by the here presented *T. mentagrophytes* genotype that occurred in Germany, were probably transmitted in this country, without a history of travelling abroad, in particular Southeast Asia. A sexual contact pathway, however, has to be considered. Likewise, as major predisposing factor for *Tinea pubogenitalis*, intimate shaving has to be considered [28].

7.4 *Trichophyton mentagrophytes* in Thailand

In Southeast Asia, in particular in Thailand, no abscessing pubogenital infections or other kinds of *Tinea* due to *T. mentagrophytes* have been reported so far including in sex workers. There are very few reports on *T. mentagrophytes* dermatophytoses in Thailand, and most of them were published a long time ago [29, 30]. In a recent study from the Dermatological Clinic of the Hospital for Tropical Medicine at Mahidol University, Bangkok, only one case due to *T. mentagrophytes* was identified among 27 patients diagnosed with dermatophytoses during a 1-year period [31]. More frequently, *M. canis* was detected. In vitro susceptibility testing of the isolates revealed no resistance against terbinafine, itraconazole, fluconazole, griseofulvin, and the topical antifungal ciclopirox olamine.

More recently, in a survey on superficial fungal infection of the feet at the Thai naval rating cadets training in Naval rating school, Sattahip, Thailand, superficial fungal infections were detected using microscopical preparation and fungal culture in 57 out of 788 participants, giving a point prevalence of 7.2% [32]. *Tinea pedis* was diagnosed in 54 participants with the leading causative organism being *T. mentagrophytes* (52.8%). Species identification was done using conventional differentiation methods, not by molecular means. At least a significant part of these *T. mentagrophytes* strains might be considered as anthropophilic *T. interdigitale* according to the current dermatophyte taxonomy and classification.

7.5 Treatment

For treatment of *Tinea profunda* due to *T. mentagrophytes* ITS genotype VII, oral antifungal agents should be used, first of all terbinafine, alternatively itraconazole, or in single cases only fluconazole. In particular, *Tinea genitalis profunda* requires oral terbinafine treatment for at least 4 weeks, but sometimes a longer duration of treatment may be needed, up to 4.5 months. Topical antifungal agents and antiseptics should be used, too. In case of mixed fungal and bacterial infections, oral antibiotics are necessary. In one patient, oral antifungal therapy was changed from terbinafine to fluconazole due to side effects of terbinafine (hair loss).

7.6 Conclusion

Tinea pubogenitalis due to *T. mentagrophytes* genotype VII represents a new clinical entity. Infection is acquired not via animal contact, but in every case by human-to-human transmission, preferably following sexual contacts in Germany, rarely abroad, e.g. Southeast Asia, but probably also in other countries, e.g. Egypt and Turkey. The question arises if the zoophilic strains of *T. mentagrophytes* genotype VII "Thai variant" recently emerging in Germany do possess a higher virulence when compared to other genotypes of *T. mentagrophytes*. Laboratory investigations on pathogenicity and virulence factors of this dermatophyte should be initiated to answer this speculation.

References

1. Luchsinger I, Bosshard PP, Kasper RS, Reinhardt D, Lautenschlager S. Tinea genitalis: a new entity of sexually transmitted infection? Case series and review of the literature. Sex Transm Infect. 2015;91:493–6.
2. Nenoff P, Schubert K, Jarsumbeck R, Uhrlaß S, Krüger C. Tinea genitalis profunda due to *Trichophyton mentagrophytes* after a journey to Egypt. Akt Dermatol. 2017;43:146–53.

3. Wendrock-Shiga G, Mechtel D, Uhrlaß S, Koch D, Krüger C, Nenoff P. Tinea barbae profunda due to *Trichophyton mentagrophytes* after journey to Thailand: case report and review. Hautarzt. 2017;68:639–48.
4. Nenoff P, Uhrlaß S, Wendrock-Shiga G, Mechtel D, Schubert K, Jarsumbeck R, et al. *Trichophyton mentagrophytes* genotype Thailand Type 1 as causative pathogen of abscessing dermatophytoses – tinea genitalis, tinea capitis and tinea barbae – in Germany. Med Mycol. 2018;56:24. (Abstract).
5. Kupsch C, Czaika V-A, Deutsch C, Gräser Y. *Trichophyton mentagrophytes* – a new genotype of zoophilic dermatophyte causes sexually transmitted infections. J Dtsch Dermatol Ges. 2019;17:493–501.
6. Beifuss B, Bezold G, Gottlöber P, Borelli C, Wagener J, Schaller M, et al. Direct detection of five common dermatophyte species in clinical samples using a rapid and sensitive 24-h PCR-ELISA technique open to protocol transfer. Mycoses. 2011;54:137–45.
7. Wiegand C, Mugisha P, Mulyowa GK, Elsner P, Hipler UC, Gräser Y, et al. Identification of the causative dermatophyte of tinea capitis in children attending Mbarara Regional Referral Hospital in Uganda by PCR-ELISA and comparison with conventional mycological diagnostic methods. Med Mycol. 2017;55:660–8.
8. Kupsch C, Ohst T, Pankewitz F, Nenoff P, Uhrlaß S, Winter I, et al. The agony of choice in dermatophyte diagnostics-performance of different molecular tests and culture in the detection of *Trichophyton rubrum* and *Trichophyton interdigitale*. Clin Microbiol Infect. 2016;22:735.e11–7.
9. Nenoff P, Krüger C, Ginter-Hanselmayer G, Tietz HJ. Mycology – an update. Part 1: Dermatomycoses: causative agents, epidemiology and pathogenesis. J Dtsch Dermatol Ges. 2014;12:188–209.
10. Nenoff P, Herrmann J, Gräser Y. *Trichophyton mentagrophytes* sive *interdigitale*? A dermatophyte in the course of time. J Dtsch Dermatol Ges. 2007;5:198–202.
11. De Hoog GS, Guarro J, Gené J, Ahmed S, Al Hatmi AMS, Figueras MJ, et al. Atlas of clinical fungi: online version 2019, Utrecht, The Netherlands. http://www.clinicalfungi.org/.2019. Accessed 18 Jan 2020.
12. Winter I, Uhrlaß S, Krüger C, Herrmann J, Bezold G, Winter A, et al. Molecular biological detection of dermatophytes in clinical samples when onychomycosis or tinea pedis is suspected. A prospective study comparing conventional dermatomycological diagnostics and polymerase chain reaction. Hautarzt. 2013;64:283–9.
13. Mirhendi H, Makimura K, De Hoog GS, Rezaei-Matehkolaei A, Najafzadeh MJ, Umeda Y, et al. Translation elongation factor 1-α gene as a potential taxonomic and identification marker in dermatophytes. Med Mycol. 2015;53:215–24.
14. Kargl A, Kosse B, Uhrlaß S, Koch D, Krüger C, Eckert K, et al. Hedgehog fungi in a dermatological office in Munich: case reports and review. Hautarzt. 2018;69:576–85.
15. Uhrlaß S, Schroedl W, Mehlhorn C, Krüger C, Hubka V, Maier T, et al. Molecular epidemiology of *Trichophyton quinckeanum* – a zoophilic dermatophyte on the rise. J Dtsch Dermatol Ges. 2018;16:21–32.
16. Ginter-Hanselmayer G, Nenoff P, Kurrat W, Propst E, Durrant-Finn U, Uhrlaß S, et al. Tinea in the genital area: a diagnostic and therapeutic challenge. Hautarzt. 2016;67:689–99.
17. Wagenknecht D, Simon JC, Treudler R, Nenoff P. Pubogenitale abszedierende Tinea profunda nach Intimrasur mit Nachweis mit Nachweis eines seltenen Erregers. J Dtsch Dermatol Ges. 2018;16:597–9.
18. Gallo JG, Woods M, Graham RM, Jennison AV. A severe transmissible Majocchi's granuloma in an immunocompetent returned traveler. Med Mycol Case Rep. 2017;18:5–7.
19. Nenoff P, Verma SB, Uhrlaß S, Burmester A, Gräser Y. A clarion call for preventing taxonomical errors of dermatophytes using the example of the novel *Trichophyton mentagrophytes* genotype VIII uniformly isolated in the Indian epidemic of superficial dermatophytosis. Mycoses. 2019;62:6–10.

20. Taghipour S, Pchelin IM, Zarei-Mahmoudabadi A, Ansari S, Katiraee F, Rafiei A, et al. *Trichophyton mentagrophytes* and *T. interdigitale* genotypes are associated with particular geographic areas and clinical manifestations. Mycoses. 2019;62(11):1084–91.
21. De Hoog GS, Dukik K, Monod M, Packeu A, Stubbe D, Hendrickx M, et al. Toward a novel multilocus phylogenetic taxonomy for the dermatophytes. Mycopathologia. 2017;182:5–31.
22. Pchelin IM, Zlatogursky VV, Rudneva MV, Chilina GA, Rezaei-Matehkolaei A, Lavnikevich DM, et al. Reconstruction of phylogenetic relationships in dermatomycete genus *Trichophyton* Malmsten 1848 based on ribosomal internal transcribed spacer region, partial 28S rRNA and beta-tubulin genes sequences. Mycoses. 2016;59:566–75.
23. Uhrlaß S, Pchelin IM, Rezaei-Matehkolaei A, Verma SB, Vasani R, Koch D, et al. Molecular epidemiology of *Trichophyton mentagrophytes* – a new approach. J Dtsch Dermatol Ges. 2019;17:159. (Abstract).
24. Nenoff P, Shyam B, Verma SB, Vasani R, Uhrlaß S. Die alarmierende Epidemie exzessiver, Therapie-refraktärer, rezidivierender und Steroid-modifizierter Dermatophytosen durch *Trichophyton mentagrophytes* in Indien. J Dtsch Dermatol Ges. 2019;17:99–100. (Abstract).
25. Nenoff P, Verma SB, Vasani R, Burmester A, Hipler UC, Wittig F, et al. The current Indian epidemic of superficial dermatophytosis due to *Trichophyton mentagrophytes* – a molecular study. Mycoses. 2019;62:336–56.
26. Singh A, Masih A, Khurana A, Singh PK, Gupta M, Hagen F, et al. High terbinafine resistance in *Trichophyton interdigitale* isolates in Delhi, India harbouring mutations in the squalene epoxidase gene. Mycoses. 2018;61:477–84.
27. Süß A, Uhrlaß S, Verma SB, Ludes A, Monod M, Krüger C, et al. Extensive tinea corporis due to a terbinafine resistant *Trichophyton mentagrophytes* isolate of the 'Indian genotype' in a young infant from Bahrain in Germany. Hautarzt. 2019;70:888–96.
28. Ginter-Hanselmayer G, Nenoff P. Dermatomycoses. In: Clinically relevant mycoses. 1st ed. Cham: Springer; 2019. https://doi.org/10.1007/978-3-319-92300-0_10.
29. Ekavithayopas A, Phoovijitre B, Puapan Y, Jongsatien S. Outbreaks of *Trichophyton mentagrophytes* infection in a premature baby ward. J Med Assoc Thail. 1988;71(Suppl 3):38–40.
30. Imwidthaya S, Thianprasit M. A study of dermatophytoses in Bangkok (Thailand). Mycopathologia. 1988;102:13–6.
31. Muangkaew W, Wongsuk T, Luplertlop N. Common dermatophytes and *in vitro* anti-fungal susceptibility testing in patients attending the Dermatological Clinic at the Hospital for Tropical Medicine, Bangkok. New Microbiol. 2017;40:175–9.
32. Ongsri P, Bunyaratavej S, Leeyaphan C, Pattanaprichakul P, Ongmahutmongkol P, Komoltri C, et al. Prevalence and clinical correlation of superficial fungal foot infection in Thai naval rating cadets. Mil Med. 2018;183:e633–7.
33. Tamura K, Nei M. Estimation of the number of nucleotide substitutions in the control region of mitochondrial DNA in humans and chimpanzees. Mol Biol Evol. 1993;10:512–26.
34. Kumar S, Stecher G, Li M, Knyaz C, Tamura K. MEGA X: molecular evolutionary genetics analysis across computing platforms. Mol Biol Evol. 2018;35:1547–9.

Trends in Epidemiology of Dermatophytes in Iran

Maral Gharaghani, Simin Taghipour, Forough Shamsizadeh, Mahdi Abastabar, Saham Ansari, Ahmad Heidari, Ali Zarei-Mahmoudabadi, and Ali Rezaei-Matehkolaei

Abstract

The epidemiological trends of dermatophyte infections (dermatophytosis or *Tinea*) in Iran from 1960 to 2018 are characterized here. A thorough search in different literature databases was performed together with statistical analyses. Our findings stress that mycological and clinical aspects of dermatophytosis in Iran have notably been revolutionized over time. *Tinea capitis* has remained as the dominant *Tinea* infection in preschool and school-age children throughout the country though its occurrence has markedly subsided after 1980s. In adults, *Tinea pedis* (22%), *Tinea corporis* (19.6%), and *Tinea cruris* (18.4%) were the dominant forms of infection and currently a gradual increase in their incidence is of note. Some clinical types more frequently occurred in some age categories and were caused by some particular species. *Tinea capitis* more often affected patients under 20 years old, while other *Tinea* infections were more frequent in middle-aged and older adults. While *Trichophyton schoenleinii* and *Trichophyton violaceum* species accounted for a significant part of dermatophytic infections

M. Gharaghani · F. Shamsizadeh · A. Zarei-Mahmoudabadi · A. Rezaei-Matehkolaei (✉)
Infectious and Tropical Diseases Research Center, Health Research Institute, Ahvaz Jundishapur University of Medical Sciences, Ahvaz, Iran

S. Taghipour
Department of Medical Parasitology and Mycology, Faculty of Medicine, Shahrekord University of Medical Sciences, Shahrekord, Iran

M. Abastabar
Invasive Fungi Research Center, Mazandaran University of Medical Sciences, Sari, Iran

S. Ansari
Department of Medical Parasitology and Mycology, School of Medicine, Shahid Beheshti University of Medical Sciences, Tehran, Iran

A. Heidari
Congenital Malformations Research Center, Golestan University of Medical Sciences, Gorgan, Iran

J.-P. Bouchara et al. (eds.), *Dermatophytes and Dermatophytoses*,
https://doi.org/10.1007/978-3-030-67421-2_13

in the past, *Epidermophyton floccosum*, *Trichophyton rubrum*, *Trichophyton tonsurans*, and the *Trichophyton mentagrophytes* / *Trichophyton interdigitale* species group are currently predominating in Iran. A more striking mycological aspect of dermatophytosis, however, is the emergence of infection with less-known zoophilic species, e.g. *Trichophyton benhamiae, Trichophyton eriotrephon, Trichophyton simii*, and *Microsporum ferrugineum,* all of which having been identified by ITS-rDNA sequencing. Terbinafine, itraconazole, luliconazole, lanoconazole, and efinaconazole showed the best in vitro efficacy against dermatophyte isolates from Iran.

1 Introduction

Dermatophytosis (also known as *Tinea*), which is a public health issue throughout the world, is principally superficial mycosis caused by a group of filamentous fungi known as "dermatophytes" [1, 2]. These fungi grow on the keratinized tissues of skin (in various body sites), nail and hair of human and animals, and spread in an eccentric pattern, leading to small to extended red circular lesions as rings and therefore termed as "ringworm." Depending on the anatomical site of involved skin, the resulted infections are called *Tinea corporis*, *Tinea cruris*, *Tinea pedis*, and *Tinea manuum*. Infections of the skin and hair shafts of the beard and scalp areas are termed as *Tinea barbae* and *Tinea capitis*, respectively, while involvement of the nails is known as *Tinea unguium* [2, 3]. Based on the newest taxonomy of dermatophytes, this group contains more than 50 species distributed in seven genera: *Trichophyton*, *Microsporum*, *Epidermophyton*, *Lophophyton*, *Paraphyton, Nannizzia*, and *Arthroderma* [4]. Geographically, the distribution profiles of pathogenic dermatophytes and relevant infections are known to change over time due to immigration, tourism activities, socioeconomic and sanitary conditions [3, 5]. These infections are not life-threatening and, perhaps for this reason, they were more important in view of aesthetics in the past. But, during the last decade there have been reports highlighting the emergence and remarkable increase in the number of recalcitrant *Tinea* cases with therapeutic failure and repeated recurrences, hence the treatment of dermatophytoses and in vitro antifungal susceptibility testing on the causative fungi became of particular importance [6]. Such variation in the face of *Tinea* infections led to a real apprehension among dermatologists. Accordingly, regular and retrospective sequential surveys on dermatophytosis and relevant agents as well as in vitro testing of anti-dermatophyte drugs can provide keystones for the upcoming epidemiological investigations on these fungi, and appropriate decision for clinical management of relevant infections.

The first documented report on the prevalence and the etiologic agents of *Tinea capitis* in Iran returns to the work by Asgari et al. [7], during 1964–1966. But the first and the only critical review on the epidemiological aspects of dermatophytosis in Iran was conducted by Khosravi et al. [8] in 1994. In this study, a total of 12,150 cases of suspected dermatophytosis were investigated between 1986 and 1991 and the infection was confirmed by direct microscopy and culture in 9345 of the cases (76.9%). Since that time, multiple regional studies focused on the epidemiology and etiology of dermatophytosis in different parts of the country have been reported [8–113]. Herein, we aimed to restructure the epidemiological trends of dermatophytoses and antifungal susceptibility aspects of these infections by a comprehensive nation-wide study during the time period of 1960 to 2018. An analysis of the frequency of all *Tinea* infections in regards to the causative species, location, and time period is reported.

2 Material and Methods

2.1 Data Collection Methods

A literature search was performed in the ISC, Google Scholar, Scopus, PubMed, and Web of science (ISI) databases by using the following key search terms: "Dermatophytosis," "*Tinea*," "Dermatophytes," "Antifungal Susceptibility Testing" (AFST), and "Iran." The key words were corroborated by the Mesh dictionary, by using the Boolean operator "AND" to link them. To recover the reports lost in our quest, we also scrutinized the references listed in the selected publications. All studies were checked out for information, e.g. place (province) and time of study (as decade), age, and gender of patients if specified, type of dermatophyte infection and the causative species. The extracted data were transferred to the IBM SPSS Statistics program (version 22). The period of analysis was designated from 1960 to 2018.

2.2 Inclusion Criteria

The eligible criteria designated for inclusion of the studies in the review were as follows: (1) publications in English or Persian (Farsi); (2) publications with enough mycological evidences for diagnosis of dermatophytosis and identification of dermatophyte agents including direct microscopy, culture, and morphological identification or sequencing of the internal transcribed spacer (ITS) regions of ribosomal DNA (rDNA). By contrast, a number of studies were excluded from the analysis due to some reasons: (1) studies conducted exclusively on dermatomycoses other than dermatophytosis or relying solely on dermatological findings; (2) studies in which the AFST was not performed according to the Clinical Laboratory Standards Institute (CLSI) M38-A2 guidelines [114] or the European Committee on Antimicrobial Susceptibility Testing (EUCAST) EDef 9.1 guidelines [115]; and (3) studies with discrepant or inexplicable results. In 2017, a new taxonomy for dermatophytes was proposed by De Hoog et al. [4] and we followed this taxonomy for dermatophytes species in this review.

2.3 Statistical Analysis

The differences in the distribution of some species between various infections and time periods of occurrence were compared by using chi-square (χ^2) test. Also, the distribution of each *Tinea* infection between various age groups was calculated by chi-square (χ^2) test. Two sides p value was calculated and the level of statistical significance was set at $p < 0.05$.

3 Results

The circumstance of data collection and entrance in this review was illustrated in Fig. 1. Of the 135 publications identified in this search, 106 studies from 24 provinces met our inclusion criteria and entered into the analysis among which a large part was performed in Tehran, the capital of Iran (33 studies), followed by East Azerbaijan ($n = 11$), Khuzestan ($n = 9$), Mazandaran ($n = 9$), and Isfahan ($n = 8$). In Fig. 2 the location of different provinces (cities) in the country and the provinces of realization in the studies was illustrated. Thirty eight studies were performed from 1960 to 1999 and 68 studies within 2000–2018. Altogether, of

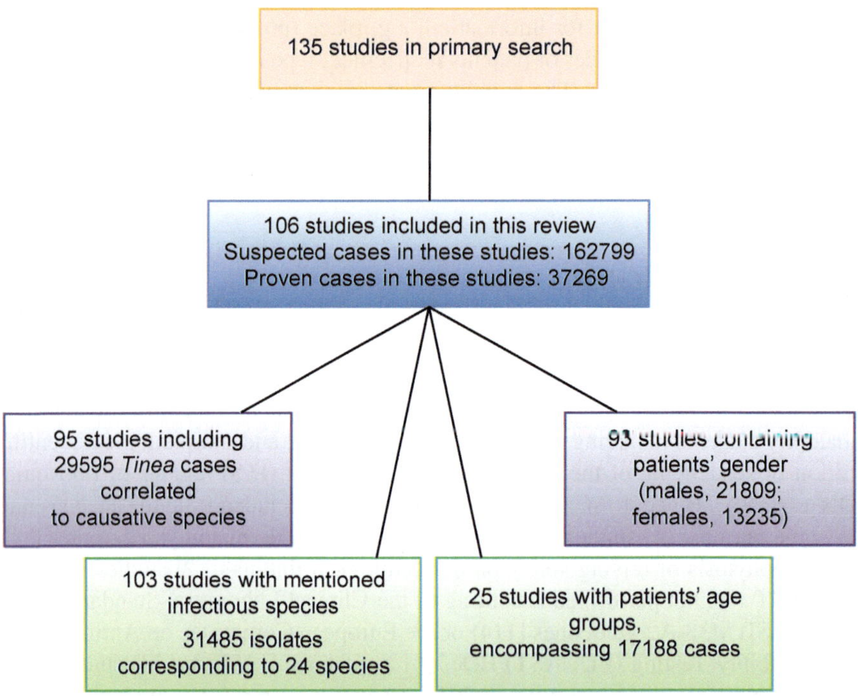

Fig. 1 The flow diagram explaining the steps involved in collection and processing of data in this review

Fig. 2 The location of different provinces in Iran. The yellow highlighted locations corresponding to provinces and cities with eligible studies included in this review

162,799 suspected cases, 37,269 cases (22.9%) were confirmed for dermatophytosis by direct microscopic examination with 10% KOH and/or culture. In Table 1 the distribution of all dermatophytoses in Iran in regards to the time of occurrence was illustrated. Most cases of dermatophytic infections (n = 18,950; 50.8%) were reported within 2000–2009. *Tinea pedis* (22%) was the dominant form of dermatophytosis followed by *Tinea capitis* (20.3%), *Tinea corporis* (19.6%), *Tinea cruris* (18.4%), *Tinea unguium* (12.3%), *Tinea manuum* (7%), and *Tinea barbae* (0.4%). In 93 studies, the distribution of *Tinea* infections was correlated to the gender: 21089 individuals were males and 13,234 were females (M/F ratio = 1.59).

The evolution and frequency of pathogenic dermatophytes species in Iran during the last six decades are depicted in Table 2. A total of 31,485 isolates affiliated to 24 species were detected as causative agents of dermatophytoses in 103 reviewed studies which were, in decreasing order, *Epidermophyton floccosum* (n = 6098; 19.4%), *Trichophyton rubrum* (n = 5272), *Trichophyton mentagrophytes* / *Trichophyton interdigitale* species group (*TMTI*SG) (n = 4896), *Trichophyton*

Table 1 Distribution of dermatophytic infections in Iran according to the time of occurrence

Tinea	1960–69	1970–79	1980–89	1990–99	2000–2009	2010–18	Total
Tinea pedis (N)[a]	None	116	1502	147	5322	1113	8200
% within *Tinea*[b]		1.4	18.3	1.8	64.9	13.6	100
% within period		70.3[c]	14.5[c]	5.4[c]	28.1[c]	22.3[c]	22.0[d]
Tinea capitis (N)[a]	67	1	3965	1170	1848	508	7559
% within *Tinea*[b]	0.9	0	52.5	15.5	24.4	6.7	100
% within period	100[c]	0.6[c]	38.2[c]	42.9[c]	9.8[c]	10.2[c]	20.3[d]
Tinea corporis (N)[a]	None	2	2080	435	3584	1214	7315
% within *Tinea*[b]		0	28.4	5.9	49	16.6	100
% within period		1.2[c]	20[c]	16[c]	18.9[c]	24.4[c]	19.6[d]
Tinea cruris (N)[a]	None	27	1397	325	3937	1167	6853
% within *Tinea*[b]		0.4	20.4	4.7	57.4	17	100
% within period		16.4[c]	13.5[c]	11.9[c]	20.8[c]	23.4[c]	18.4[d]
Tinea unguium (N)[a]	None	19	516	475	3131	436	4577
% within *Tinea*[b]		0.4	11.3	10.4	68.4	9.5	100
% within period		11.5[c]	5[c]	17.4[c]	16.5[c]	8.8[c]	12.3[d]
Tinea manuum (N)[a]	None	None	855	162	1062	536	2615
% within *Tinea*[b]			32.7	6.2	40.6	20.5	100
% within period			8.2[c]	5.9[c]	5.6[c]	10.8[c]	7.0[d]
Tinea barbae (N)[a]	None	None	66	12	66	6	150
% within *Tinea*[b]			44	8	44	4	100
% within period			0.6[c]	0.4[c]	0.3[c]	0.1[c]	0.4[d]
Total (N)[a]	67	165	10,381	2726	18,950	4980	37,269
% within period	0.2[c]	0.4[c]	27.9[c]	7.3[c]	50.8[c]	13.4[c]	100[d]

The different clinical forms of dermatophytic infections are presented in decreasing order of frequency on the whole study period (1960 to 2018)
[a]Number of cases during the decade
[b]Frequency within all cases of that *Tinea* among the whole study period
[c]Frequency among all *Tinea* infections during the decade
[d]Frequency within all cases of dermatophytic infections on the whole study period

verrucosum ($n = 3610$), *Trichophyton tonsurans* ($n = 2871$), *T. interdigitale* ($n = 2459$), *Microsporum canis* ($n = 2365$), *Trichophyton violaceum* ($n = 1877$), *Trichophyton schoenleinii* ($n = 1532$), *Nannizzia gypsea* (formerly *Microsporum gypseum*; $n = 235$), *T. mentagrophytes* ($n = 78$), *Trichophyton erinacei* ($n = 64$), *Trichophyton benhamiae* ($n = 36$), *Lophophyton gallinae* (formerly *Trichophyton gallinae*), *Trichophyton soudanense* and *Microsporum audouinii* (15 cases each one), *Microsporum ferrugineum* ($n = 12$), *Trichophyton concentricum* ($n = 4$), *Arthroderma uncinatum* (formerly *Trichophyton ajelloi*), *Trichophyton eriotrephon, Trichophyton simii*, *Nannizzia fulva* (formerly *Microsporum fulvum*), *Nannizzia nana* (formerly *Microsporum nanum*), and *Nannizzia persicolor* (formerly *Microsporum persicolor*), two cases each, and *Nannizzia incurvata* and *Paraphyton cookei* (formerly *Microsporum cookei*), one case each. A total of 22 isolates were reported as unknown species. Altogether, in 29,595 cases from 95 studies, the species causing dermatophytosis were correlated to the type of *Tinea* infection (Table 3). While *TMTI*SG ($n = 2117$), *T. rubrum* ($n = 1966$), *T. interdigitale* ($n = 876$), and *E. floccosum* ($n = 799$) were respectively determined as the commonest species causing *Tinea pedis, T. tonsurans* ($n = 1429$), *T. verrucosum* ($n = 1332$), *TMTI*SG ($n = 845$), *M. canis* ($n = 714$), *E. floccosum* ($n = 710$) and *T. rubrum* ($n = 622$) were the main species causing *Tinea corporis*. The species causing other *Tinea* infections and their frequencies are presented in Table 3. A statistically significant difference was found in distribution of some species including *T. rubrum, T. interdigitale, T. schoenleinii, T. violaceum, T. tonsurans, T. benhamiae, T. verrucosum, M. canis, E. floccosum*, and *N. gypsea* between different clinical conditions and time periods ($p < 0.05$). Concerning the correlation of each *Tinea* infection with the patients' age, only a total of 17,188 *Tinea* cases from 25 out of the 106 reviewed publications could be sorted into six age groups (Table 4). Regarding AFST, only 10 publications were eligible for analysis.

4 Discussion

In this review, the epidemiological and mycological features of dermatophytosis in Iran were investigated according to the reports from 1960s to 2018. All reviewed studies were found to be descriptive (but not population based) investigations focused on outpatients attending the mycological laboratories in university-hospital clinics, or on skin, hair, and nail samples referred for direct examination and fungal culture. Except *Tinea capitis* which more often arose in the 1980s, the detection rates of *Tinea* infections were the highest in 2000–2009, likely because the vast majority of reviewed studies ($n = 46$) and then dermatophytic cases ($n = 18{,}950$) was from this period. It is known that dermatophytosis more often occurs in men [8]. Consistent with this fact, in almost all reports from Iran, the rate of infection was greater in males except for three studies from Guilan [57, 100] and Tehran [105] that found dermatophytosis affecting predominantly females. With regard to the age of patients,

Table 2 Spectrum of dermatophytes species causing human dermatophytosis in Iran during the last six decades

Species	1960–69	1970–79	1980–89	1990–99	2000–09	2010–18	Total
E. floccosum (N)[a]	None	14	1443	195	3309	1137	6098
% within species[b]		0.2	23.7	3.2	54.3	18.6	100
% within period		13.5[c]	14.2[c]	8.1[c]	23.8[c]	23.6[c]	19.4[d]
T. rubrum (N)[a]	None	2	1340	211	2821	898	5272
% within species[b]		< 0.1	25.4	4	53.5	17	100
% within period		1.9[c]	13.2[c]	8.8[c]	20.3[c]	18.7[c]	16.7[d]
*TMTI*SG (N)[a]	None	71	776	396	2458	1195	4896
% within species[b]		1.5	15.8	8.1	50.2	24.4	100
% within period		68.3[c]	7.6[c]	16.5[c]	17.7[c]	24.9[c]	15.6[d]
T. verrucosum (N)[a]	None	1	1476	534	1361	238	3610
% within species[b]		< 0.1	40.9	14.8	37.7	6.6	100
% within period		1[c]	14.5[c]	22.2[c]	9.8[c]	5[c]	11.5[d]
T. tonsurans (N)[a]	1	1	204	106	2206	353	2871
% within species[b]	< 0.1	< 0.1	7.1	3.7	76.8	12.3	100
% within period	1.7[c]	1[c]	2[c]	4.4[c]	15.8[c]	7.3[c]	9.1[d]
T. interdigitale (N)[a]	None	None	1128	18	564	655	2365
% within species[b]			47.7	0.8	23.8	27.7	100
% within period			11.1[c]	0.7[c]	4.1[c]	13.6[c]	7.5[d]
M. canis (N)[a]	None*	None*	1702*	223*	348*	184*	2457
% within species[b]			69.3	9.1	14.2	7.5	100
% within period			16.7[c]	9.3[c]	2.5[c]	3.8[c]	7.8[d]
T. violaceum (N)[a]	48*	3*	943*	333*	496*	54*	1877
% within species[b]	2.6	0.2	50.2	17.7	26.4	2.9	100
% within period	81.4[c]	2.9[c]	9.3[c]	13.9[c]	3.6[c]	1.1[c]	6.0[d]
T. schoenleinii (N)[a]	1*	12*	997*	364*	130*	28*	1532
% within species[b]	0.1	0.8	65.1	23.8	8.5	1.8	100
% within period	1.7[c]	11.5[c]	9.8[c]	15.1[c]	0.9[c]	0.6[c]	4.9[d]
N. gypsea (N)[a]	None	None	91	19	102	23	235
% within species[b]			38.7	8.1	43.4	9.8	100
% within period			0.9[c]	0.8[c]	0.7[c]	0.5[c]	0.7[d]
T. mentagrophytes (N)[a]	None	None	None	None	78	None	78
% within species[b]					100		100
% within period					0.6[c]		0.2[d]
T. erinacei (N)[a]	None	None	60	None	None	4	64
% within species[b]			93.8			6.3	100
% within period			0.6[c]			0.1[c]	0.2[d]
T. benhamiae (N)[a]	None*	None*	None*	None*	None*	32*	36
% within species[b]						88.9	100
% within period						0.7[c]	0.1[d]
A. uncinatum (N)[a]	None	None	2	None	None	None	2
% within species[b]			100				100

(continued)

Table 2 (continued)

Species	1960–69	1970–79	1980–89	1990–99	2000–09	2010–18	Total
% within period			< 0.1[c]				< 0.1[d]
L. gallinae (N)[a]	None	None	13	None	None	2	15
% within species[b]			86.7			13.3	100
% within period			0.1[c]			< 0.1[c]	< 0.1[d]
M. audouinii (N)[a]	None	None	None	1	14	None	15
% within species[b]				6.7	93.3		100
% within period				< 0.1[c]	0.1[c]		< 0.1[d]
M. ferrugineum (N)[a]	None	None	9	None	3	None	12
% within species[b]			75		25		100
% within period			0.1[c]		< 0.1[c]		< 0.1[d]
N. fulva (N)[a]	None	None	None	None	None	2	2
% within species[b]						100	100
% within period						< 0.1[c]	< 0.1[d]
N. incurvata (N)[a]	None	None	None	None	None	1	1
% within species[b]						100	100
% within period						< 0.1[c]	< 0.1[d]
N. nana (N)[a]	None	None	2	None	None	None	2
% within species[b]			100				100
% within period			< 0.1[c]				< 0.1[d]
N. persicolor (N)[a]	None	None	None	None	1	1	2
% within species[b]					50	50	100
% within period					< 0.1[c]	< 0.1[c]	< 0.1[d]
P. cookei (N)[a]	None	None	None	None	1	None	1
% within species[b]					100		100
% within period					< 0.1[c]		< 0.1[d]
T. concentricum (N)[a]	None	None	None	None	4	None	4
% within species[b]					100		100
% within period					< 0.1[c]		< 0.1[d]
T. eriotrephon (N)[a]	None	None	None	None	None	2	2
% within species[b]						100	100
% within period						< 0.1[c]	< 0.1[d]
T. simii (N)[a]	None	None	None	None	1	1	2
% within species[b]					50	50	100
% within period					< 0.1[c]	< 0.1[c]	< 0.1[d]
T. soudanense (N)[a]	9	None	2	3	None	1	15
% within species[b]	60		13.3	20		6.7	100

(continued)

Table 2 (continued)

Species	1960–69	1970–79	1980–89	1990–99	2000–09	2010–18	Total
% within period	15.3[c]		< 0.1[c]	0.1[c]		< 0.1[c]	< 0.1[d]
Unknown species (N)[a]	None	None	None	4	18	None	22
% within species[b]				18.2	81.8		100
% within period				0.2[c]	0.1[c]		0.1[d]
Total (N)	59	104	10,188	2407	13,919	4808	31,485
% within the whole study period	0.2[c]	0.3[c]	32.4[c]	7.6[c]	44.2[c]	15.3[c]	100[d]

Results are related to the 103 studies which reported frequency of each isolated species. Species involved are presented in decreasing order of frequency on the whole study period (1960 to 2018). * *P* value < 0.0001

A. Arthroderma, E. Epidermophyton, L. Lophophyton, M. Microsporum, N. Nannizzia, P. Paraphyton, T. Trichophyton, TMTISG *Trichophyton mentagrophytes/Trichophyton interdigitale* species group

[a]Number of cases during the decade

[b]Frequency of this species among the whole study period

[c]Frequency among all dermatophyte isolates during the decade

[d]Frequency within all cases of dermatophytic infections on the whole study period

in 25 reviewed publications with the same age ranges, some clinical presentations were found to occur predominantly in some age categories and to be caused by some particular species. As shown in Table 4, *Tinea capitis* more often affected patients under 20 years old, whereas other infections affected more frequently middle-aged and older adults. The species causing infection in Iranian people during the last 58 years, e.g. *E. floccosum, T. rubrum,* TMTISG, *T. verrucosum, M. canis, T. tonsurans, T. interdigitale*, and *T. violaceum,* in fact are currently the main agents of dermatophytoses worldwide though their distribution pattern differs inside the country and also with other parts of the world. Herein, the evolution of human dermatophytoses and of dermatophytes species pertaining to the time of occurrence is discussed. To our knowledge, this review is the largest Iranian dataset on the epidemiological aspects of human dermatophytosis.

4.1 *Tinea pedis*

Tinea pedis (also known as athlete foot) is the fungal infection of the soles of the feet and interdigital spaces by a dermatophyte species that can lead to *Tinea unguium* in 50–60% of cases [116, 117]. The infection is also entitled with "pedal fungus reservoir" due to its extension to other body sites causing *Tinea manuum, Tinea unguium*, or even *Tinea cruris* [118]. Autoinoculation from *Tinea pedis* and onychomycosis is responsible for *Tinea corporis, Tinea faciei,* and rarely for

Table 3 Distribution of species causing infection in Iran according to the clinical manifestation

Species	*Tinea corporis*	*Tinea capitis*	*Tinea pedis*	*Tinea cruris*	*Tinea unguium*	*Tinea manuum*	*Tinea barbae*	Total
E. floccosum (N)[a]	710	4	799*	4121*	172	152	2	5960
% within species[b]	11.9	0.1	13.4	69.1	2.9	2.6	< 0.1	100
% within *Tinea*[c]	11	0.1	13.3	72.2	6.4	6.8	1.4	20.1
T. rubrum (N)[a]	622*	78*	1966*	864*	1250*	410*	1*	5191
% within species[b]	12	1.5	37.9	16.6	24.1	7.9	< 0.1	100
% within *Tinea*[c]	9.6	1.2	32.6	15.1	46.4	18.4	0.7	17.5
*TMTISG*** (N)[a]	845	365	2117	337	773	267	14	4718
% within species[b]	17.9	7.7	44.9	7.1	16.4	5.7	0.3	100
% within *Tinea*[c]	13.1	5.7	35.1	5.9	28.7	12	10.1	15.9
T. verrucosum (N)[a]	1332*	1262*	99*	27*	85*	545*	89*	3439
% within species[b]	38.7	36.7	2.9	0.8	2.5	15.8	2.6	100
% within *Tinea*[c]	20.7	19.8	1.6	0.5	3.2	24.5	64	11.6
M. canis (N)[a]	714*	1423*	45*	51*	20*	83*	25*	2361
% within species[b]	30.2	60.3	1.9	2.2	0.8	3.5	1.1	100
% within *Tinea*[c]	11.1	22.4	0.7	0.9	0.7	3.7	18	8
T. tonsurans (N)[a]	1429*	651*	50*	33*	26*	192*	None*	2381
% within species[b]	60	27.3	2.1	1.4	1.1	8.1		100
% within *Tinea*[c]	22.2	10.2	0.8	0.6	1	8.6		8
T. interdigitale (N)[a]	327*	112*	876*	245*	250*	518*	3*	2331
% within species[b]	14	4.8	37.6	10.5	10.7	22.2	0.1	100
% within *Tinea*[c]	5.1	1.8	14.5	4.3	9.3	23.3	2.2	7.9
T. violaceum (N)[a]	250*	1335*	49*	12*	63*	37*	3*	1749
% within species[b]	14.3	76.3	2.8	0.7	3.6	2.1	0.2	100
% within *Tinea*[c]	3.9	21	0.8	0.2	2.3	1.7	2.2	5.9
T. schoenleinii (N)[a]	59*	931*	5*	3*	25*	2*	1*	1026

(continued)

Table 3 (continued)

Species	*Tinea corporis*	*Tinea capitis*	*Tinea pedis*	*Tinea cruris*	*Tinea unguium*	*Tinea manuum*	*Tinea barbae*	Total
% within species[b]	5.8	90.7	0.5	0.3	2.4	0.2	0.1	100
% within *Tinea*[c]	0.9	14.6	0.1	0.1	0.9	0.1	0.7	3.5
N. gypsea (N)[a]	102*	78*	4*	4*	9*	8*	1*	206
% within species[b]	49.5	37.9	1.9	1.9	4.4	3.9	0.5	100
% within *Tinea*[c]	1.6	1.2	0.1	0.1	0.3	0.4	0.7	0.7
T. erinacei (N)[a]	None	60	1	1	None	1	None	63
% within species[b]		95.2	1.6	1.6		1.6		100
% within *Tinea*[c]		0.9	< 0.1	< 0.1		< 0.1		0.2
T. mentagrophytes (N)[a]	19	13	7	3	7	3	None	52
% within species[b]	36.5	25	13.5	5.8	13.5	5.8		100
% within *Tinea*[c]	0.3	0.2	0.1	0.1	0.3	0.1		0.2
L. gallinae (N)[a]	8	7	None	None	None	None	None	15
% within species[b]	53.3	46.7						100
% within *Tinea*[c]	0.1	0.1						0.1
M. audouinii (N)[a]	2	10	2	1	None	None	None	15
% within species[b]	13.3	66.7	13.3	6.7				100
% within *Tinea*[c]	< 0.1	0.2	< 0.1	< 0.1				0.1
T. benhamiae (N)[a]	14*	2*	2*	1*	2*	7*	None*	28
% within species[b]	50	7.1	7.1	3.6	7.1	25		100
% within *Tinea*[c]	0.2	< 0.1	< 0.1	< 0.1	0.1	0.3		0.1
A. uncinatum (N)[a]	2	None	None	None	None	None	None	2
% within species[b]	100							100
% within *Tinea*[c]	< 0.1							< 0.1
M. ferrugineum (N)[a]	1	10	1	None	None	None	None	12
% within species[b]	8.3	83.3	8.3					100

% within *Tinea*[c]	< 0.1	0.2	< 0.1					< 0.1
N. fulva (N)[a]	1	None	None	None	None	None	None	1
% within species[b]	100							100
% within *Tinea*[c]	< 0.1							< 0.1
N. incurvata (N)[a]	1	None	None	None	None	None	None	1
% within species[b]	100							100
% within *Tinea*[c]	< 0.1							< 0.1
N. nana (N)[a]	None	2	None	None	None	None	None	2
% within species[b]		100						100
% within *Tinea*[c]		< 0.1						< 0.1
N. persicolor (N)[a]	2	None	None	None	None	None	None	2
% within species[b]	100							100
% within *Tinea*[c]	< 0.1							< 0.1
P. cookei (N)[a]	None	1	None	None	None	None	None	1
% within species[b]		100						100
% within *Tinea*[c]		< 0.1						< 0.1
T. concentricum (N)[a]	2	None	None	None	None	2	None	4
% within species[b]	50					50		100
% within *Tinea*[c]	< 0.1					0.1		< 0.1
T. eriotrephon (N)[a]	1	1	None	None	None	None	None	2
% within species[b]	50	50						100
% within *Tinea*[c]	< 0.1	< 0.1						< 0.1
T. simii (N)[a]	2	None	None	None	None	None	None	2
% within species[b]	100							100
% within *Tinea*[c]	<0.1							< 0.1
T. soudanense (N)[a]	None	14	None	None	None	None	None	14
% within species[b]		100						100

(continued)

Table 3 (continued)

Species	*Tinea corporis*	*Tinea capitis*	*Tinea pedis*	*Tinea cruris*	*Tinea unguium*	*Tinea manuum*	*Tinea barbae*	Total
% within *Tinea*[c]		0.2						< 0.1
Unknown species (N)[a]	5	None	None	4	11	None	None	20
% within species[b]	25			20	55			100
% within *Tinea*[c]	0.1			0.1	0.4			0.1
Total	6448	6358	6023	5707	2693	2227	139	29,595
% within the whole study period	21.8	21.5	20.4	19.3	9.1	7.5	0.5	100

Results are related to the 95 studies which reported frequency of each species according to the type of *Tinea*. Species involved are presented in decreasing order of frequency on the whole study period (1960 to 2018). * P value < 0.0001

A. Arthroderma, *E. Epidermophyton*, *L. Lophophyton*, *M. Microsporum*, *N. Nannizzia*, *P. Paraphyton*, *T. Trichophyton*, *TMTI*SG *Trichophyton mentagrophytes*/*Trichophyton interdigitale* species group

[a]Number of cases during the whole study period

[b]Frequency within all cases of *Tinea* during the whole study period

[c]Frequency among all species causing that *Tinea* infection during the whole study period

Table 4 Distribution of each *Tinea* infection in patients according to the age categories

Age group (year-old)	*Tinea capitis*	*Tinea corporis*	*Tinea cruris*	*Tinea pedis*	*Tinea manuum*	*Tinea unguium*	*Tinea barbae*	Total
1–9	3344 (84.5%)	495 (12.5%)	43 (1.1%)	19 (0.5%)	30 (0.8%)	26 (0.7%)	0	3957 (100%)
10–19	1628 (61.1%)	575 (21.6%)	246 (9.2%)	49 (1.8%)	134 (5%)	33 (1.2%)	1 (0%)	2666 (100%)
20–29	93 (2.3%)	1790 (45.2%)	1257 (31.7%)	426 (10.8%)	255 (6.4%)	124 (3.1%)	15 (0.4%)	3960 (100%)
30–39	29 (1.2%)	516 (21.5%)	661 (27.6%)	586 (24.5%)	366 (15.3%)	202 (8.4%)	35 (1.5%)	2395 (100%)
40–49	18 (0.8%)	285 (13.1%)	396 (18.2%)	763 (35%)	442 (20.3%)	249 (11.4%)	27 (1.2%)	2180 (100%)
$\geq$ 50	20 (1%)	183 (9%)	257 (12.7%)	803 (39.6%)	304 (15%)	460 (22.7%)	3 (0.1%)	2030 (100%)
Total	5132	3844	2860	2646	1531	1094	81	17,188

Results are related to the 25 studies which reported frequency of each type of *Tinea* according to patients' age

Tinea capitis in the elderly. Besides, due to the long duration of treatment and the potential for recurrence of infection, *Tinea pedis* is yet postulated as a matter of concern for public health [119]. Because of health improvement in the mid-twentieth century, expansion of immigration and change in the human lifestyle, such as use of occlusive footwear, the incidence of *Tinea pedis* has increased. Unlike *Tinea capitis* or *Tinea unguium*, we found no worthwhile survey with regular intervals and specific focus on foot dermatophytosis as a distinct entity in Iran. However, integrated with other *Tinea* conditions, the epidemiological aspects of *Tinea pedis* were the matter of investigation in different geographical parts of Iran [27, 37]. In this review, *Tinea pedis* accounted for 22% of all dermatophytoses with an incidence peak in 2000–2009 (28% of all *Tinea* cases during this decade), and 64.9% of all cases of *Tinea pedis* also occurred in 2000s. Study of dermatophytosis among some military personnel in south of Iran was the first Iranian report addressing the mycological features of *Tinea pedis* [120]. In that study dermatophytosis was diagnosed in 145 out of 290 individuals who had lesions suggestive of dermatophytosis, and among them *Tinea pedis* was the major clinical form ($n = 116$). Though somewhat old, that study resonated that clinical manifestations of *Tinea* can vary depending on the occupation, and in certain occupational groups where the individuals have to stand or wear the boat (shoes) for a long time, dermatophytosis more often appears as *Tinea pedis* [120]. *Tinea pedis* was then found with a frequency of 11.9% in Isfahan from 1983 to 1984, with *T. mentagrophytes* as the main causative species [37]. In two later studies within 1983–1994, dermatophytosis of the feet accounted for 8.9% and 5% of all dermatophytoses in Isfahan and Hamadan, respectively [27, 72]. During 1986 to 1991, frequency of *Tinea pedis* in Iran increased to 15.2% in the study by Khosravi et al. [8], and it ranked the third among the different clinical presentations of dermatophytosis. Generally, before the new century *Tinea pedis* was a less frequent condition in Iran. At the beginning of the twenty-first century and parallel to infrequency of *Tinea capitis*, a meaningful upward shift in the prevalence of *Tinea pedis* occurred, especially in the north of the country. Similar to other *Tinea* infections, most *Tinea pedis* cases were recorded within 2000–2009 (Table 1). In Isfahan, the infection took a gradual rising trend after 2007 with an incidence peak in 2009 [31]. During 2000 to 2017, dermatophytic infection of the feet appeared as the dominant *Tinea* presentation in Tehran [19, 21, 46, 79, 121], Mazandaran [75] and Hamadan [55]. The frequency of infection increased from 20% [84] to 32.1% [75] in Mazandaran (North of Iran) from 2003 to 2014, and it reached 43.4% in Tehran during 2008–2010 [21] which is the highest incidence rate of *Tinea pedis* from Iran till now. Similarly, in the next investigations during 2010–2014, the infection remained as the most common presentation of dermatophytosis in this city [19, 79, 121]. In the only study performed in Qazvin, *Tinea pedis* was the third prevalent manifestation of dermatophytosis [30]. A review by Seebacher et al. [5] showed that *Tinea pedis* accompanied with toenail involvement accounts for more than 50% of *Tinea* infections in central Europe.

As for the pathogenic agents, Emami et al. [120] found *T. mentagrophytes* as the main species causing *Tinea pedis* in 1974. Likewise, *T. mentagrophytes* (51.7%) was the major causative species in Isfahan during 1983 to 1994 [37, 72], but at that time, *T. interdigitale* was not recognized as a distinct species. So it is very likely that majority of those *T. mentagrophytes* isolates actually were *T. interdigitale.* In the nationwide study by Khosravi et al. [8], *T. mentagrophytes* var. *interdigitale* (now called *T. interdigitale*) (40%), *T. rubrum* (32%) and *E. floccosum* (24%) were the leading causes of *Tinea pedis* in Iran between 1986 and 1991. In Qazvin, during 2004–2006, *Tinea pedis* was mainly caused by *E. floccosum*, *T. rubrum*, and *T. mentagrophytes* var. *interdigitale* [30]. In general, *T. rubrum*, *T. interdigitale*, and *E. floccosum* are the predominant species causing *Tinea pedis* worldwide [3, 5]. As concluded from Table 3, *TMT*ISG, *T. rubrum*, *T. interdigitale*, and *E. floccosum* were the dominant pathogens causing foot infection during the last 58 years in Iran, a statistically notable finding and congruent with the global pattern. Recently, a thorough sequence-based study on the human clinical isolates morphologically identified as *TMT*ISG indicated that *Tinea pedis* is mainly caused by *T. interdigitale* rather than *T. mentagrophytes*, and this finding was statistically significant [122]. Likewise, many isolates from foot clinical materials that were morphologically identified as *T. mentagrophytes* were, in fact, *T. interdigitale* in ITS sequencing. Members of the genera *Microsporum*, *Arthroderma*, *Nannizzia*, *Lophophyton*, and *Paraphyton* were less frequently reported as agents of *Tinea pedis* in Iran. Given the increase in the rate of host factors such as elderly population and predisposing diseases like diabetes, the growing incidence of foot infection is predicted.

4.2 *Tinea capitis*, an Everlasting Condition in Iran

Tinea capitis (scalp ringworm) is an infection of the hair, scalp, and annexes (eyebrows and eyelashes) caused by dermatophytes, mainly of the genera *Trichophyton*, *Microsporum*, and *Nannizzia* [123]. The most important concerns regarding dermatophytosis of the scalp are its high propensity for transmissibility and the long duration of the treatment even with new antifungals [124]. About five decades ago, *Tinea capitis* was known as "*Tinea* of elementary students" in Iran and the spectrum of dermatophyte species causing *Tinea capitis* encompassed all ecological categories of geophilic, zoophilic, and anthropophilic species [8, 37, 125]. In this review, *Tinea capitis* was found as the second most frequent form of dermatophytosis but nowadays scalp infection is clearly less common in Iran. As shown in Table 1, *Tinea capitis* was more prevalent in the twentieth century but, similar to the global trend, the incidence of scalp infection sharply decreased after 2010 through the country. This may be the result of improvement in child's standards and of greater availability of oral antifungals such as terbinafine and itraconazole.

The oldest study regarding dermatophytosis in Iran, conducted by Asgari et al. [113], was a preliminary screening on the prevalence and etiological agents of *Tinea capitis* in 14 different cities of Iran during 1964 to 1966. In this study, three distinct geographic foci were identified: in decreasing order of importance, the Caspian Sea coast with 25%, the central plateau of Iran with 7.9% and the Persian Gulf coast with 2.4%. In February to March 1970, 1891 children aged 6 to 24 years from Bandar Abbas and Minab cities, South of Iran, were screened by Asgari et al. [7] to define the prevalence of *Tinea capitis*. They found a prevalence of 3.3% with dominance of endothrix-type *Tinea capitis* (95%), and neither ectothrix-type infection nor female infection was detected. Failure to diagnose infection among females was attributed to the fact that women and girls refused to be investigated for scalp infection due to cultural issues. Besides, infection was more common in age group ranging from 5 to 14 years. As inferred from Table 1, ringworm of the scalp had an increasing trend in prevalence between 1960 and 2000 with 52.5% of all *Tinea capitis* cases detected during 1980s and the highest rate of scalp infection in the 1990s (42.9% of all *Tinea* infections in this decade). *Tinea capitis* was reported as the most common type of dermatophytosis in studies performed in Hamadan, Isfahan and East Azerbaijan during 1971 to 2014 [27, 31, 37, 47, 48, 50, 72, 125] but also, with lower frequency, from all parts of Iran. In the nationwide study conducted by Khosravi et al. [8] between 1986 and 1991, *Tinea capitis* was also noted as the predominant type of dermatophyte infection (35.6%).

The prevalence rate of scalp ringworm in this reviewed series changed from 72% in a study in 1983 [37] to 0.8% 30 years later [21]. However, in spite of this considerable decrease in incidence, *Tinea capitis* remains as a never-ending problem in most parts of the country. In a recent study conducted in Mashhad Northern Iran, *Tinea capitis* accounted for 32.5% of all dermatophytosis [65], and among them, 53% were endothrix infections caused by *T. violaceum* and 28% favus due to *T. schoenleinii*. Regarding the age distribution, *Tinea capitis* is known to primarily, but not exclusively, affect prepubertal children [3, 123]. Consistent with this fact, the patients with scalp infection in Iran ranged from under one year old to more than 50 years old with majority of them in the age groups of 1–9 and 10–19 years.

Regarding the causal agents of *Tinea capitis*, they were not static over the time. While *M. canis* and *M. audouinii* were the predominant species in Europe and Mediterranean countries, *T. tonsurans* appeared in the Americas and is currently the most frequent species causing scalp infection worldwide [124, 126]. As shown in Table 3, spectrum of species causing scalp infection in Iran encompassed all seven genera of dermatophytes. Infection mostly occurred by *M. canis*, followed by *T. violaceum*, *T. verrucosum*, *T. schoenleinii*, and *T. tonsurans*. The earliest survey on the causative agents of scalp infection in Iran was conducted by Asgari and Sateri [7] in 1970. They found *T. violaceum* (80%), *T. soudanense* (15%), and *T. tonsurans* and *T. schoenleinii* (1.6% each) as agents of *Tinea capitis* in Bandar Abbas. In 1980s, *T. verrucosum*, *T. schoenleinii*, and *T. violaceum* were the dominant agents of *Tinea capitis* in Isfahan [37]. Till 1990s, *T. verrucosum* remained as the dominant agent of infection in Isfahan, but *M. canis* appeared as the second common species during this period [72]. In Hamadan during 1991–1992 *T. schoenleinii* and *T. verrucosum* were equally the most common agents of *Tinea capitis* [27] while in the new century, *T. tonsurans* appeared as the dominant agent of infection in this city and in Tehran as well [46, 55]. In the national study conducted by Khosravi et al. [8], *M. canis* was found as the most prevalent agent of *Tinea capitis*, followed by

T. verrucosum and *T. violaceum.* Along with *T. verrucosum*, *T. schoenleinii* was endemic in the rural areas of Iran [8], but in the last decade, the rate of infection by this species decreased progressively and it is now eradicated from most parts of the country; the latest cases of *T. schoenleinii* infection were reported in 2017 by Salari et al. [34]. Generally, overview of the infection pattern shows that over the last decades, the dominant zoophilic species, e.g. *M. canis* and *T. verrucosum,* have been replaced by anthropophilic species, especially *T. tonsurans* [19]. The recent molecular-based studies from Northern and central cities, e.g. Sari, Rasht, and Arak, underlined this changing in the clinical-etiological profile of *Tinea capitis* with a gradual increase in the rate of endothrix-type *Tinea capitis* by *T. tonsurans* [57, 75, 78, 127]. On the other hand, in two recent similar studies from Southwestern cities like Ahvaz, Shiraz, and Yasuj, ectothrix-type scalp infections emerged with *T. interdigitale* (currently *T. mentagrophytes*) as the main causative agent [29, 109]. Until the last taxonomic changes from de Hoog et al. [4], all strains of the *TMTI*SG were gathered under the term *T. interdigitale* which did not discriminate the zoophilic and anthropophilic variants of the species. However, based on the current taxonomy of dermatophytes, it is known that almost all of those Iranian strains causing *Tinea capitis* were in fact zoophilic species of *T. mentagrophytes.* Similar pictures were described from Tehran. The predominant species such as *T. violaceum* and *M. canis* causing *Tinea capitis* in the late 1990s and early 2000s [42, 46, 60] were gradually replaced by *T. tonsurans* and *T. mentagrophytes* [19, 79, 121]. *Nannizzia gypsea* was the most common geophilic species causing *Tinea capitis,* so that 38% of all infections by this species were scalp infections (Table 3). Generally, *Tinea capitis* was the most common dermatophytic condition before twenty-first century in Iran, but its occurrence decreased considerably during the last two decades. Additionally, the pattern of the causative species has progressively changed with a substantial increase in the incidence of *T. tonsurans* and *T. mentagrophytes* and a remarkable decline in the rate of infection by *T. schoenleinii* after 2000s (Table 3).

4.3 *Tinea corporis*

Tinea corporis is a worldwide distributed infection that affects the skin of trunk, arms, legs, and neck. In a recent review on human dermatophytosis, this type of infection was found as the predominant *Tinea* in African adults [126]. It was stated that *Tinea corporis* is the main clinical form of dermatophytosis in the Middle East [3]. Consistent with this, 65.6% of all *Tinea corporis* cases were detected in the twenty-first century (during 2000–2018). In this review, *Tinea corporis* was the third most common presentation of dermatophytic infections with the highest rate of incidence in 2010s. As it can be inferred from Table 1, 24.4% of all dermatophytosis cases between 2010 and 2018 were *Tinea corporis.* To our knowledge, the oldest retrieved data regarding *Tinea corporis* also came from the work of Emami et al. [120] in 1974. They found this type of dermatophytosis as the lowest frequent type (1.4%) among affected military personnel. In five distinct surveys from center and west of the country between 1983 and 2018, cases of *Tinea corporis* constituted the second most frequent type of dermatophytosis after scalp infection [27, 31, 37, 55, 72]. Likewise, in East Azerbaijan till early 1990s *Tinea corporis* ranked second among all forms of dermatophytosis after *Tinea capitis* but during 1997–98, it appeared as the main manifestation of dermatophytosis [128]. Similarly, *Tinea corporis* ranked the first in frequency (31.4%) among all *Tinea* forms over the period of 1999–2001 in Tehran [60]. The infection also appeared as the main *Tinea* infection in Khorasan [65], Khuzestan [109], and Kermanshah [18] with prevalence rates of 33%, 37.2, and 38.5%, respectively.

All dermatophytes from all ecological categories cause *Tinea corporis*, but the main species varies geographically. In Africa, the most prevalent causative species were found to be *T. violaceum, T. rubrum*, and *M. canis* [126]. In this review, *T. verrucosum* and *TMTI*SG were respectively the dominant species involved in *Tinea corporis* in central provinces of Isfahan and Yazd during 1983–2015 [31, 37, 72, 104]. Likewise, in three different surveys from west of Iran between 1991 and 2015, *T. verrucosum* was reported as the species predominantly causing infection [18, 27, 55]. In different surveys from Tehran, *E. floccosum* [42, 60, 79], *T. tonsurans* [19, 46], *T. interdigitale* [21, 25], *M. canis* [129], and *T. rubrum* [121] were involved as the dominant species causing infection. In Northern provinces, the infection more often occurred by *T. verrucosum* [128], *T. mentagrophytes* [65, 84], and *T. rubrum* [57]. In South part of the country, *TMTI*SG was found as the leading species [29, 82, 109]. Overall, as pointed out in Table 3, *T. tonsurans* (22.1%), *T. verrucosum* (20.6%), *TMTI*SG (13.1%), *M. canis* (11.07%), and *E. floccosum* were the main species involved in *Tinea corporis* in Iran ($p < 0.05$).

4.4 *Tinea cruris*

The dermatophytic infection of the skin in genitals, inner tights, and buttocks is termed as *Tinea cruris* (also jock itch and *Tinea inguinalis*). The infection is globally distributed with different prevalence rates ranging from 2.5 to 52% [130]. It seems that the prevalence rate of *Tinea cruris* was low before the new century in most parts of Iran. To our knowledge, the first Iranian study addressing the mycological features of *Tinea cruris* was performed by Emami et al. [120]. They found *Tinea cruris* as the second type of *Tinea* among military personnel in some regions in the south of Iran. In Isfahan, during 1983 to 1984, a frequency of 7.3% was reported, with *E. floccosum* as the only causative agent [37]. Afterward, during 1983 to 1993 the infection ranked fourth among all clinical forms of dermatophyte infections (6.8%) [72]. In four investigations from North West of the country between 1985 and 1997, *Tinea cruris* rated 1.3–5.3% in term of frequency [48, 50, 110, 128]. In other studies over the period of 2001–2011, *Tinea cruris* was reported as the most prevalent *Tinea* with frequency up to 47% [30, 57, 58, 82, 84, 100]. As shown in Table 1, the highest rate of groin infection (57.4%) was from 2000–2009. But it should be stressed that such rate of infection was not solely the result of widespread occurrence of *Tinea cruris* during 2000s, but also due to a publication bias with multitude of studies focused on dermatophytosis in this decade (46 out of 106 reviewed publications). *Tinea cruris* is well known to be more common in adolescents and young adults [130]. Congruent with this, the highest rate of infection in Iran was described in 20–39 years old individuals (Table 4).

From a mycological perspective, the most common species causing *Tinea cruris* worldwide in 1970s were *T. mentagrophytes* and *E. floccosum*, but since the 1980s they have been replaced by *T. rubrum* and now this species is responsible for the majority of the cases [5, 130]. In Iran, however, *E. floccosum* has been settled as the main pathogen causing groin infection from the past till now [8, 37, 72, 109]. As shown in Table 3, 69.1% of all *E. floccosum* strains were isolated from groin infection and 72.2% of all *Tinea cruris* cases were caused by this species, and this finding was statistically significant ($p < 0.05$). Historically, in 1974 *T. mentagrophytes* and *E. floccosum* were reported as the most common cause of *Tinea corporis* among the personnel of military services in South of Iran [120]. Fourteen years later, Sadri et al. [59] found *E. floccosum* as the most frequent species causing *Tinea cruris* in patients referred to the Mycological Laboratory, Razi Hospital, in Tehran during a 3-month period. Between 1983 and 2014 *E. floccosum,* with notable difference, exceeded other species such as *TMTI*SG, *T. rubrum*, and *M. canis* as the agent of groin infection in center, North East, and South of the country [30, 31, 37, 58, 65, 72, 82, 109]. In two different surveys in Hamadan, west of the country, *T. verrucosum* and *E. floccosum* were the main agents of jock itch between 1991 to 2016 with almost the same ratio [27, 55], whereas the dominant species causing infection were *E. floccosum* and *T. rubrum* in the north of Iran [57, 75, 84], and *E. floccosum*, *T. rubrum* and *TMTI*SG in Tehran [19, 21, 25, 46, 60, 79]. Finally, when data regarding the age and gender of patients were available, the infection more often occurred in adults and males. Generally, *Tinea cruris* remains as a common *Tinea* infection in Iran with predominance of *E. floccosum* as the etiological agent.

4.5 *Tinea unguium*

The term *Tinea unguium* (or dermatophyte onychomycosis) encompasses any fungal infection of the finger and toe nails by dermatophytes [131, 132]. *Tinea unguium* together with nail infections by yeasts and non-dermatophyte molds is responsible for about half of all the nail abnormalities and it still remains as an unsolved and even rising problem in many parts of the world [117, 131, 133]. In Australia, the incidence of toenail dermatophytosis increased from 13.8% during 1961–64 to the striking rate of 54.8% in 1995–96 [134]. Though not life-threatening, onychomycosis still remains a serious global disease due to some negative social consequences [132]. Meanwhile, unlike to skin infections by dermatophytes and due to factors intrinsic to the nail, dermatophytosis of the nails is widely difficult to eradicate with antifungal therapy [117]. Today, seven decades after introduction of griseofulvin, *Tinea unguium* along with *Tinea pedis* and *Tinea capitis* remains a never-ending story in dermatology [5]. It is estimated that the global prevalence of *Tinea unguium* will increase as predisposing conditions such as diabetes, senility, and smoking are becoming more common [117]. Like in other parts of the world, *Tinea unguium* was found to be common in Iran, with a prevalence rate varying from 1.5% [27] to 21.3% [21]. The first report of *Tinea unguium* in Iran comes from the study of Ardehali [76] in 1973 with frequency rate of 38%. In the 1980s and early 1990s, *Tinea unguium* was the less common form of dermatophytosis in most parts of Iran. In Isfahan, Hamadan, and East Azerbaijan frequency of the infection during 1983 to 1998 varied from 1.3 to 2.6% [27, 37, 48, 72, 128]. In the nationwide study by Khosravi et al. [8] between 1986 and 1991, nail infection accounted for 2.6% of all dermatophytoses and it was the lowest common among all dermatophytic infections. Aghamirian and Ghiasian [28] investigated over a 4-year period (2004–2007) the prevalence of onychomycosis in Qazvin in 308 suspected patients and found a rate of 40.2%, with 50% of these infections caused by dermatophytes. In the unique investigation of dermatophytosis from Khorramabad, West of Iran (2007–2008), nail infection ranked second among all *Tinea* infections (25.5%) [35]. As concluded from Table 1, similar to *Tinea cruris*, *Tinea unguium* had a sinusoidal trend in prevalence during the last six decades. But except the report by Ardehali [76], review of the literature indicated that dermatophyte infection of the finger and toe nails was uncommon or less prevalent before 2000s and majority of *Tinea unguium* cases occurred during the period of 2000–2009 (Table 1). Compared to *Tinea capitis* which gradually diminished over the time, and similar to *Tinea pedis*, frequency of nails dermatophytosis has increased over the time, and with reference to the change in the lifestyle and to growth in the number of populations at greater risk for onychomycosis, such rising trend in the frequency was predictable. However, compared to the period of 2000 to 2009, frequency of the infection has decreased to 8.8% in 2009–2018 (Table 1). Although this may imply the improvement in social awareness of fungal nail infection, such frequency is yet remarkable due to its negative social outcomes.

The species most often isolated from *Tinea unguium* are *T. rubrum*, *T. interdigitale* and *E. floccosum,* worldwide [117]. Consistent with the global trend, *T. rubrum* and *TMTI*SG constituted the major causative agents of *Tinea unguium* in this review (Table 3) and this finding was statistically significant. Surprisingly, in the oldest available report from Iran by Ardehali [76], *T. schoenleinii* was incriminated as the leading agent (63.1%) of onychomycosis. The other species in decreasing order were *T. violaceum* (15.8%), *T. rubrum* (10.5%), *T. tonsurans*, and *T. verrucosum* (5.3% each). In 93.6% of the cases due to *T. schoenleinii* and *T. violaceum*, an accompanying scalp infection was detected and the patients were mostly the peasants who had nail infection in left hands. Khosravi et al. [8] during 1980s showed that *T. rubrum*, with notable difference, was the leading cause of nail infection compared to *T. interdigitale* and *E. floccosum*. In another study by Khosravi and Mansouri [73], *T. interdigitale*, *T. violaceum*, and *T. rubrum* were found to be the foremost cause of *Tinea unguium* in Tehran from 1996 to 1997. In Isfahan, from 1983 to 2014, the causative agents were found to be *TMTI*SG, *E. floccosum* and *T. rubrum* [31, 72, 107]. In Kerman, South of the country, Salari et al. [34] recently identified *T. interdigitale* as the main agent (64.7%) of nail infection. As illustrated in Table 3, nearly half of the cases of *Tinea unguium* in Iran were due to *T. rubrum*. Since 2000s, *T. rubrum* has remarkably overtaken the *TMTI*SG as the agent of nail and foot mycoses, particularly in Tehran and Northern half of the country, e.g. Khorasan and Mazandaran [21, 46, 65, 75]. In Tehran, during two independent studies from 2000 to 2011, *T. rubrum* was isolated from 64–74% of all *Tinea unguium* cases [21, 46]. Similarly, in Qazvin, two different investigations between 2004 and 2007 demonstrated the predominance of *T. rubrum* followed by *T. interdigitale* and *E. floccosum* [28, 30]. Similar to the scenario for *Tinea pedis*, the recent comprehensive sequence-based study by Taghipour et al. [122], on the human clinical isolates of *TMTI*SG revealed that *Tinea unguium* is also more commonly caused by *T. interdigitale* rather than *T. mentagrophytes*.

Recently, some mycological studies from Tehran, Isfahan, Kermanshah and Ahvaz demonstrated a shift in the etiology of onychomycosis from dermatophytes towards non-dermatophyte molds (NDMs) and yeasts [18, 80, 107, 135, 136]. The frequency of nail infections due to yeasts such as *Candida* species and NDMs in these investigations varied from 62 to 83.1%. It is known that *Tinea unguium* is a gender- and age-related infection and its prevalence peaked in males and elderly individuals. Furthermore, nail infection by dermatophytes is much more abundant in toe nails than in finger nails [117, 133]. In agreement with these facts, in 18 studies with retrievable data, we found that dermatophytic onychomycosis more often occurred in toe nails (54.7%) and male patients (M/F ratio: 1.1), aged of 50 years and above. Generally, the increase in life expectancy and in the number of diabetic individuals has resulted in an increased frequency of particular infections such as *Tinea unguium* and *Tinea pedis* [3].

4.6 *Tinea manuum*

Tinea manuum is the term applied for dermatophyte infection of the intertriginous web spaces, palmar and dorsum surfaces of the hands which typically occurs in adults. The infection may also be seen together with *Tinea pedis* and the so-called 'two-feet-one-hand syndrome implies to the unilateral infection coexistent with bilateral plantar *Tinea pedis* [137]. But, compared to *Tinea pedis*, only a few investigations have been published regarding the clinical and mycological perspective of *Tinea manuum*. According to our dataset, the first study in Iran with information on the clinico-mycological features of *Tinea manuum* was performed by Chadeganipour et al. [72] in Isfahan between 1983 and 1993. In their study *Tinea manuum* accounted for only 2.6% of dermatophytosis cases. Subsequently, Omidynia et al. [27] during their survey of dermatophytosis in Hamadan between 1991 and 1992 found *Tinea manuum* as the third form of infection (10.4%) in term of abundance. After that, Khosravi et al. [8], in their nationwide study during 1986 to 1994, reported the frequency of *Tinea manuum* as 9.7% and *T. interdigitale* as the main causative species. In another surveillance, during 1991 to 1993 hand dermatophytosis accounted for 12% of all dermatophytosis in Tabriz, North West of Iran [48]. In different reports, the frequency of *Tinea manuum* varied from 2.1% to 18.7% in Tehran [21, 25, 46, 79, 129, 138] or from 4% to 16.4% in Mazandaran [75, 84], whereas it was of 17.5% in Mashhad [65]. But, the highest frequency of *Tinea manuum* throughout the country (35.5%) was reported from Kerman by Mosavi et al. [16] during 2007–2011. While the frequency of *Tinea manuum* in Iran was found to be 7% in this review, in different reports from Australia, Poland, China, and Syria, it ranged from 3.4 to 16% [134, 139–141]. On the whole, our data support the conclusion that *Tinea manuum* occurs uncommonly worldwide. Regarding its etiologic agents, *Tinea manuum* is known to be caused mainly by *T. rubrum* throughout the world [134, 137, 142], but the causative agents may vary geographically. The survey of all eligible reports from Iran shows that especially before 2000, *T. verrucosum* was the dominant species causing hand infection but after that *T. interdigitale, T. rubrum, TMTI*SG and *T. tonsurans* emerged as the main involved species. But, given the marked reduction in the isolation rate of *T. verrucosum* after 2009 (Table 2), more occurrence of hand infection by human-adapted species like *T. rubrum* and *T. interdigitale* is expectable.

4.7 *Tinea barbae*

Tinea barbae, also called sycosis, is a dermatophytic infection of the skin and hairs of the beard and moustache in men [143–145]. Though rare, these disorders impose major complications and healthcare burdens. Besides, given that the clinical appearance of infection simulate to the folliculitis or pseudo-folliculitis barbae, misdiagnosis is highly usual [145]. Nevertheless, few investigations have rated the clinical and mycological features of this infection. Congruent with this view, our review also revealed that the infection has taken less attention in Iran, with only 12 studies reporting a total of 150 cases. This rarity among our dataset made it statistically difficult to take a comprehensive assessment on clinical and mycological aspects of the infection. In our dataset, the frequency of the infection regularly declined after its highest value (44%) in 1980–89, except the period of 2000–2009 where there was a publication bias as for other *Tinea* presentations. During 2010 to 2018 only six episodes of infection were detected. Notably, this finding was expected because such trend was observed in some recent reports from Iran [21, 25, 109]. Overall, dermatophytosis of the beard was formerly more common but after substantial progresses in lifestyle and sanitary habits of men, especially application of single-use razors, its occurrence took apparently a descending trend worldwide [109, 145]. Overview of the few available literature indicated that most cases of *Tinea barbae* cases were reported as small series and caused by zoophilic or anthropophilic species [143–146]. Bonifaz et al. [144] diagnosed 9 cases of *Tinea barbae* in Mexico over an 18-year retrospective study (1983–2001). Infection was more common in the 60s of life and mainly resulted from *T. rubrum* and *M. canis*. The most recent cross-sectional study on *Tinea barbae* was performed in a major Tertiary Dermatology Department in Lisbon, Portugal by Duarte et al. [145]. Overall, seven cases all of which caused by *T. rubrum* were detected over an 11-years period (2008 to 2018), with a 46% decrease in the frequency of *Tinea barbae* compared to similar survey performed in the past decade. A high rate of 43% misdiagnosis at the first clinical inspection by a dermatologist was reported which accentuates the role of mycological examinations for confirmation of the diagnosis. As to the etiology, *T. verrucosum, M. canis*, and *T. mentagrophytes* are the dominant zoophilic agents, while *T. rubrum* is the more common anthropophilic species [143–145]. The highest frequency of beard infection in Iran (6.4%) backs to a study in Lorestan, Southwest of Iran. Sepahvand et al. [35] in that scrutiny, found 11 cases of *Tinea barbae* among which a dermatophyte strain was isolated from culture in 5 cases, i.e. *T. violaceum* in 4 cases and *E. floccosum* in the last one. In Isfahan, during 1983–1993, 6 cases of beard infection by *T. violaceum, M. canis*, and *T. verrucosum* were detected [72]. Likewise, 6 cases of infection due to *T. verrucosum* and *TMTI*SG were found over the period of 2004 to 2014 in Kerman [34]. Generally, in the few reported cases among our series, the most common dermatophytes involved were the zoophilic species *T. verrucosum* (64%), *M. canis* (18%), *TMTI*SG (10.1%), followed by the anthropophilic species *T. violaceum* and *T. interdigitale* (one 2.2% each).

4.8 Trends of Change of Infectious Species

Analysis of the literature reveals that the spectrum of dermatophyte species can be altered throughout the time and may depend on the variation in clinical manifestation of dermatophytosis in the respective regions [5]. This was reflected in this review of dermatophytosis in Iran. As shown in Table 2, favus by *T. schoenleinii* was prevalent for 3–4 decades before 2000s, but afterward its occurrence strikingly decreased and it has almost disappeared today in Iran. This phenomenon may likely be due to the effectivity of griseofulvin in pediatric *Tinea capitis* and improvement in the sanitary condition of children during two recent decades. In some recent sequence-based investigations from cities like Tehran, Ahvaz, Shiraz and Yasuj, *T. interdigitale* was reported as the most common species causing *Tinea* infections [21, 25, 109, 147]. As indicated before, in those papers from 2013–2016, all anthropophilic and zoophilic strains of the current *T. mentagrophytes/T. interdigitale* complex were gathered under the term *T. interdigitale*. Given the recent taxonomy proposed by De Hoog et al. [4], differentiation of *T. mentagrophytes* and *T. interdigitale* can be accomplished by ITS sequencing. As stated by Taghipour et al. [122], due to the lack of sequence-based investigations in the past and to the taxonomic instability across dermatophytes species, the true estimation on the rate and spectrum of *Tinea* infections by either *T. mentagrophytes* or *T. interdigitale* in Iran is difficult. In limited number of studies with detailed morphological evaluation, *T. mentagrophytes* var. *interdigitale* (currently an invalid name) had predominantly been isolated from *Tinea pedis* and *Tinea unguium* while *T. mentagrophytes* var. *mentagrophytes* was mainly reported from *Tinea corporis* and *Tinea capitis* [8, 31]. The recent comprehensive survey on the GenBank entries accompanied with practical assessment by Taghipour et al. [122] indicated that *T. mentagrophytes* and *T. interdigitale* are associated with particular clinical manifestations and a great number of Iranian *T. interdigitale* strains, especially those isolated from infections other than foot and nail, should be reclassified as *T. mentagrophytes*. By contrast, majority of submitted *T. mentagrophytes* which had been isolated from *Tinea unguium* and *Tinea pedis* were, in fact, *T. interdigitale*. Then, estimation of the real incidence of foot and other infections by *T. interdigitale* and *T. mentagrophytes* in previous studies (whether from Iran or abroad) when there were no available ITS sequences connected to the isolates, is very difficult.

In terms of frequency of infectious species, *E. floccosum* was reported as the dominant agent of dermatophytosis in some parts of the country from the past till now [19, 30, 37, 57, 60, 79, 84] and now it is the most prevalent anthropophilic species in the South and West regions of Iran [18, 82, 109]. As a whole, it has also been found as the main species causing groin infection in Iran which was statistically significant [25, 30, 31, 37, 57, 60, 82, 109] and currently it is the third agent of *Tinea pedis* and *Tinea unguium* after *T. interdigitale* and *T. rubrum* [5]. Of note, the species very rarely invade the hair but four episodes of scalp infection by *E. floccosum* were found in our dataset. Surveillance of trend of infections by *E. floccosum* in this review supports the conclusion of Seebacher et al. [5] that contrary to some European countries like Germany, incidence of infection by this anthropophilic species in many Islamic countries is increasing.

The anthropophilic dermatophyte *T. rubrum* accounted as the second taxon causing dermatophytosis in Iran. *Trichophyton rubrum*, in general, is a complex comprising three taxa including *T. rubrum*, *T. violaceum*, and *T. soudanense* that despite their relative genetic similarity can be considered as separate species with distinct geographical distributions and clinical preferences [148]. *Trichophyton rubrum* has emerged since the twentieth century as a cause of chronic *Tinea corporis* from the endemic areas in South Asia. In the middle of the twentieth century, it has been the most incriminated species causing mainly *Tinea pedis* and *Tinea unguium* in Europe and currently it is the main entity causing dermatophytosis worldwide [149]. This species appeared in Iran in the late of the twentieth century, with a very low frequency [37, 76, 150]. For example, in 1973 it accounted for only 10.6% of *Tinea unguium* cases and its rarity was attributed to its non-indigenous nature [76]. In 1987 and 1996 the frequency of this species among clinical isolates of dermatophytosis in Hamadan and Isfahan was reported as zero and 1.75%, respectively [27, 37]. Between 1986 and 1991, *T. rubrum* constituted only 4.25% of all agents of dermatophytosis in Iran but it was the main anthropophilic species and the dominant agent of *Tinea unguium* [8]. After 1990s, its occurrence steadily increased and in many areas of the country it recently emerged as the main or the second most prevalent agent of dermatophytosis [18, 19, 21, 30, 55, 57]. In Central and North Europe, the high isolation rate of *T. rubrum* during last decades was noted to be connected with the growing incidence of *Tinea pedis* [5]. Likewise, in surveys conducted after 2000s, the increased frequency of *T. rubrum* in Iran was accompanied with an increased incidence of *Tinea pedis* [21, 25, 30, 46, 79, 121]. As illustrated in Table 3, this species was more frequently isolated from *Tinea pedis* and *Tinea unguium*. An update on dermatophytosis by Seebacher et al. [5] indicated that mycoses by *T. rubrum*, especially *Tinea pedis* and onychomycosis, have increased in central Europe. Generally, consistent with the universal trend, rising incidence of infections by *T. rubrum* is also predicted for Iran and some host and pathogen factors can impute to this rise. These include: (1) the higher capacity of some *T. rubrum* strains to spread [151]; (2) the expanded use of private baths which are the main source for vertical transmission of the species within families [152]; (3) use of occlusive shoes which develops humidity leading to *Tinea pedis/Tinea unguium;* and (4) expansion of social communications leading to greater tendency to the human-to-human than animal-to-human model of infection transmission.

The anthropophilic species *T. schoenleinii* and *T. violaceum* were once the most important agents of scalp and skin lesions in Iran [8, 37, 48, 50, 72], but in this review these two species ranked the second (21%) and the third (14.6%) most frequent agents of scalp infection after *M. canis*. From Table 3, it is figured out that great majority of *T. schoenleinii* infections was as scalp infection. However, in 1970s, it was even the dominant agent of *Tinea unguium* in Iran [76] and in 1980s it ranked the second most common agent of *Tinea corporis* in one study from Isfahan [37]. In East Azerbaijan, between 1985 and 1991, *T. schoenleinii* was demonstrated as the leading cause of dermatophytosis by Mohammadi [50] and Kazemi et al. [48]. The species even was described as the most common cause of dermatophytosis in Hamadan in the late 1990s [27]. In terms of timeframe, most cases of infection by *T. schoenleinii* (88.9.1%) occurred between 1980 and 1999 (Table 2). But, currently the elimination of *T. schoenleinii* and *T. violaceum* is a great variation in the spectrum of pathogenic dermatophytes species in most areas of Iran. Whereas some authors mentioned that *Tinea capitis* by *T. schoenleinii* (favus) is currently limited to some endemic regions like China and Iran [153], apparently no or rare episodes of infection by the species have been traced in the most recent surveys from Iran [21, 34, 109]. Now, it seems that infection by *T. schoenleinii* is only common in some African countries [5].

A similar story was revealed for *T. violaceum* in Iran. The highest rate of isolation for the species was during 1980 to 1999, afterward it decreased markedly in the current decade (Table 3). *Trichophyton violaceum* is a species with endemism confined to some North and East African countries that was imported from these regions to other parts of the world during the twentieth century. The taxon was reported as the most prominent species involved in *Tinea capitis* and *Tinea corporis* in most areas of Africa and currently it is a common cause of scalp infection in Africa and West Asia [148]. Additionally, the theory that *T. violaceum* (along with *T. schoenleinii*) has predilection to infect hair rather than skin [124] was confirmed statistically in this review where 76.3% of infections by this species in Iran were as *Tinea capitis*. The first documented report on human infection with this species in Iran was by Asgari and Sateri [7] from Bandar Abbas. In 1970 they found *T. violaceum* (80%) and *T. soudanense* (15%) as agents of endothrix *Tinea capitis*. Between 1986 and 1991, *T. violaceum* was reported as the second agent of scalp infection in Iran [8]. Currently, the disappearance of *T. violaceum* and *T. schoenleinii* along with reemergence of *T. tonsurans* and decline in the isolation rate of *M. canis* represent the most important alteration in the spectrum of dermatophytes causing *Tinea capitis* in Iran. There is no follow-up data regarding the efficacy of antifungal agents for treatment of dermatophytosis in Iran but, based on multicenter trial studies from Canada, Brazil and South Africa, one more likely explanation for the observed decrease in the isolation rate of *T. violaceum* (and *T. schoenleinii*) compared to the increase of *T. tonsurans* is the greater susceptibility of this species to routinely used antifungals compared to *T. tonsurans* [154]. There are also data supporting that scalp infection due to *T. violaceum* mostly occurs in children, rather than adults [155], and then gradual decrease of infection by the species in Iran can also be interpreted to as improvement in the sanitary standards for children.

As revealed in Tables 2 and 3, the cattle ringworm fungus *T. verrucosum* was one of the most common species causing dermatophytosis in Iran, especially *Tinea corporis* and scalp infection [27, 37, 72]. But, with the advent of sequence-based studies in the second decade of the twenty-first century in Iran, the validity of culture-based reports on the isolation of this zoophilic species from the past onwards was questioned [21, 25, 109]. At present, in Iran, it is known that part of the human isolates which personate the morphological characteristics of *T. verrucosum* are indeed a diverse range of dermatophyte species including *T. mentagrophytes* and *T. verrucosum* [21, 25, 57, 109]. Since 2008 till now, only six sequence-based proven cases of human *T. verrucosum* clinical isolates were detected in Khuzestan ($n = 1$) and Guilan ($n = 5$) of which four isolates were deposited in the Westerdijk Fungal Biodiversity Institute (CBS database, Utrecht, Netherlands) as *T. verrucosum* CBS 130944 to CBS 130947. Then, the true frequency of human dermatophytosis by *T. verrucosum* in Iran is much lower than those reported.

The human-adapted fungus *T. tonsurans* is a ubiquitous dermatophyte known as very contagious and hardly treatable species which can create epidemics especially among athletes in martial arts club [3, 156]. Infection with this taxon was uncommon in the past through the country. Within 1986–1991, it constituted 1.3% of all *Tinea* agents in Iran, and was exclusively isolated from *Tinea capitis* [8]. The considerable emergence of *T. tonsurans* as agent of dermatophytosis in Iran dates back to the late 1980s. This period was, in fact, concurrent with the highest rate of *T. schoenleinii*, *T. violaceum*, and *M. canis* isolation and with the sudden increase in the incidence of *Tinea capitis* and *Tinea corporis*. As shown in Table 2, majority of the *T. tonsurans* isolates (89.1%) were recovered after 2000. In addition, this species was predominantly isolated from cases of *Tinea corporis* (60%) and *Tinea capitis* (27.3%), findings which were statistically significant and in agreement with the global changes in the etiology of *Tinea capitis* towards by *T. tonsurans* [3]. In the last two decades, new presentations of *Tinea corporis* or *Tinea capitis* called *Tinea corporis gladiatorum* and *Tinea capitis gladiatorum* were described among sportsmen with extensive bodily contact, especially judoists and wrestlers, throughout the world [40, 69, 157–159]. Almost all studies pertaining to the mycological aspects of *Tinea gladiatorum* unanimously reported *T. tonsurans* as its most common causative agent. In three reports from Tehran, Qazvin and Sari cities between 2004 and 2008, frequency of this species as agent of *Tinea gladiatorum* ranged from

82.7 to 100% [40, 69, 157]. As a national sport, wrestling is very popular in Iran and many young individuals are involved in it through the country. Given the expansion of combat sports, especially wrestling, further incidence of both *T. tonsurans* and *Tinea corporis gladiatorum/Tinea capitis gladiatorum* in Iran may be expected.

Regarding *T. simii*, only two cases caused by this less-known species were reported from Yasuj and Ahvaz, among which only one was verified by ITS-rDNA sequencing of the isolate [53, 62]. *Trichophyton simii* is a rare dermatophyte pathogen that causes infection in a broad variety of animals and human with majorities of records regarding its isolation from India. Incidence of infection by this fungus, even in endemic area of Indian subcontinent, is little known [160]. Like other uncommon species, application of sequence-based procedures is needed to define the incidence rate of dermatophytosis by this fungus in Iran.

The species that is being identified recently in Iran is *T. benhamiae* (formerly *Trichophyton* species of *Arthroderma benhamiae*). It has been reported as an emerging zoonotic dermatophyte in some parts of Europe, especially Germany and Czech and has Guinea pig and some rodents as animal sources [161, 162]. Some strains of the species share mycological features with *T. mentagrophytes* and generally the definitive species identification relies on ITS-sequencing [161]. The epidemiological aspects of infection by this fungus in Iran is unknown. Until 2013, *T. benhamiae* has not been detected as agent of dermatophytosis in Iran, not as a result of infrequency, but because of misidentifications as other morphologically similar species. In 2013, it was reported for the first time from two cases of *Tinea capitis* and *Tinea faciei* in Tehran [21]. Since that time, the taxon has been isolated from cases of *Tinea corporis, Tinea capitis, Tinea manuum, Tinea pedis*, and *Tinea unguium* in Tehran, Ahvaz, Hamadan, Yasuj, and Shiraz [25, 29, 55, 109]. Among *T. benhamiae* isolates with demographic data in this review, 50% were from *Tinea corporis* and *Tinea faciei* which is consistent with the European trend of infection by the species [161]. Currently, the only explicit confirmation for *T. benhamiae* identification and its differentiation from the closely related species, e.g. *T. erinacei* and *T. eriotrephon* is ITS sequencing [161], thus the true incidence and epidemiologic trend of infection by this species in Iran remain to be determined.

Another recognized species with low frequency in this review was *T. erinacei,* a zoonotic species with hedgehogs as its primary and preferred host. The species causes infection in hedgehogs to a high rate of 20–47%, but it is a very uncommon pathogen for human [163, 164]. The first case of human dermatophytosis by *T. erinacei* was described from New Zealand in 1960 [165]. Since less than 50 cases of human infection by this zoophilic species have been reported worldwide, most of which go back to the twenty-first century and from Asian countries where the hedgehog is popular as pet, e.g. Japan, Korea, Thailand, and Taiwan [164, 166–168]. Surprisingly, Khosravi et al. [8] during their study of dermatophytosis in Iran reported 60 cases of human infection by *T. erinacei,* all of which originated from scalp infection. But, given the absence of molecular verification, validity of those isolations is questionable. To date, the only confirmed case of human dermatophytosis by *T. erinacei* comes from Rezaei-Matehkolaei et al. [21] in a study on dermatophytosis in Tehran. One explanation for such infrequency of infection by the species can be attributed to this fact that hedgehog is clearly not popular as pet in Iran. Similarly to *T. benhamiae*, the thorough identification of this less-known species relies on sequence-based procedures but, at the time of this review, in resource-limited countries such as Iran, the sequencing is relatively costly and difficult to establish in regular clinical diagnostics. Then, the incidence rate of human infection by *T. erinacei* should be considered with caution, but it is estimated to be less than that of *T. benhamiae*.

Trichophyton eriotrephon, which was detected from two cases of dermatophytosis in this review, is a pathogenic species closely related to *T. benhamiae*, *T. erinacei*, and *T. concentricum*. Up to now, there are only two published reports on the clinical isolation of the species worldwide, both of which originated from human dermatophytosis. The first one was in 1925 by Papegaay [169] from a case of *Tinea corporis* in Netherlands. Two other strains identified by ITS sequencing were isolated in Tehran, Iran, during 2008–2010 from two patients with *Tinea capitis* and *Tinea faciei* [21]. Recently, two additional ITS sequences from unpublished clinical isolates from Shiraz, South of Iran, were deposited in the GenBank under the accession numbers KP789415 and KP789416. The new isolates were recovered from cases of *Tinea manuum* and *Tinea pedis*. Considering the very low isolation rate of the species and that all the newly described isolates originated from human dermatophytosis in Iran, the epidemiological particularities of the infection by this mysterious species remain to be discovered. Additionally, similar to *T. benhamiae* and *T. erinacei*, sequencing is mandatory for identification of this species. Therefore, the actual abundance of the species within either Iranian or global dermatophytes population might be greater than what has been discovered so far.

Dermatophytosis due to *Microsporum* species, until 1970s, was uncommon in Iran [76]. *Microsporum canis* was reported as the main species causing *Tinea corporis* in Iran in 1978 [150] and *Tinea capitis* in 1994 [8]. According to Seebacher et al. [5], there was still in 2008 a predominance of *M. canis* among agents of *Tinea capitis* in Mediterranean countries of Europe such as Spain. Based on the new update on dermatophytosis by Coulibaly et al. [126], a number of reports highlighted the reemergence of the species in some Eastern and Northern African countries, especially in *Tinea capitis*. Here, the fungus was also found to be the most common species causing scalp infection in Iran, especially in 1980s. But, its isolation rate has decreased in the new era, most likely because of the decreased occurrence of scalp infection. As inferred from Tables 2 and 3, *Tinea capitis* and *M. canis* exhibited the highest incidence rates during 1980–1989, but they decreased markedly after 2000. Such synchronizations indicate a direct correlation between the presence of *M. canis* and *Tinea capitis*. Another attributable reason for the decline in *M. canis* isolation rate is the increased use of griseofulvin for treatment of *Tinea capitis* after 1980s since some recent investigations concluded that *Microsporum* species are more susceptible than *Trichophyton* species to griseofulvin which was used for treatment of *Tinea capitis* for many years [170]. Today, *M. canis* has been replaced by *T. tonsurans* as etiological agent of most *Tinea* infections including *Tinea capitis/ Tinea corporis*. *Microsporum ferrugineum,* as one member of the former *Arthroderma otae* complex, was found as agent of 12 cases of dermatophytosis in

Iran (Table 2). It has limited endemicity in the Balkans, some parts of Middle East, East Asia and Africa [2, 142]. The species is known for causing *Tinea capitis* but the epidemiological aspects of the infection by this non-indigenous species in Iran are unknown. In this review, 10 of 12 *M. ferrugineum* isolates were recovered from scalp infections but except two cases from Tehran [21] which were identified by ITS-rDNA sequencing, the identity of the other isolates is questionable. Conclusive differentiation of this species from *M. canis* and *M. audouinii* requires sequence-based strategies, e.g. ITS-RFLP or ITS sequencing [26]. Thus, given that in most mycological centers in Iran, the dermatophyte isolates are identified by conventional culture-based methods, the true incidence of this species remains incalculable. Likewise, the accuracy of the 15 reported cases of infection by *M. audouinii* in Iran could not be verified as this species is non-endemic of the country, and both reports on isolation of this fungus were solely based on the macroscopic and microscopic morphology of the colonies on cultures [23, 58].

In the present data set, *Nannizzia gypsea* was found as the dominant geophilic species causing dermatophytosis in Iran. Generally, prevalence of human dermatophytosis by *N. gypsea* compared to anthropophilic and zoophilic taxa is very low. In different reports the incidence of infection by the species ranged from 0.04 to 7% in Europe, 0.6 to 0.87% in the USA, and 0.1 to 6.5% in some Asian countries [171]. To the best of our knowledge, the highest rate (20.9%) was recently reported in one study from Egypt [172]. In Iran, the species contributed in less than 1% [21, 25, 46] to 7.5% [42] of the causative agents of dermatophytosis. *Nannizzia gypsea* is known to most frequently cause *Tinea corporis* but it was also responsible for other clinical conditions, especially *Tinea capitis* [171–173]. Concordance with this matter, 87.4% of all cases of dermatophytosis by *N. gypsea* in Iran had been correlated to *Tinea corporis* and *Tinea capitis* (Table 3).

Nannizzia fulva and *N. incurvata* were reported in only two cases of human dermatophytosis in Iran (1 case each), both of them after 2010 and identified by ITS sequencing [32, 174]. The two taxa are geophilic species morphologically and genetically very close to *N. gypsea* and very seldom engaged in human infections [174–176]. Recently, some sequence-based reports underline that *N. fulva* and *N. incurvata* can be overlooked due to misidentification with *N. gypsea* when identification is solely based on the classical morphological parameters, and thus the true prevalence of *N. fulva* and *N. incurvata* could have been underestimated [32, 174, 176, 177]. There were also some reports from the past to now, regarding the isolation of a few number of non-indigenous species, e.g. *T. concentricum* and *T. soudanense*, and also less common species, e.g. *A. uncinatum*, *L. gallinae*, *N. nana*, *P. cookei*, and *N. persicolor*, from cases of infections in Iran (Table 3). Of note, all studies citing the isolation of the above mentioned species solely used conventional morphological methods. Hence, it is a matter of discussion whether infections by these species are indeed rare or their true incidence underestimated.

4.9 Antifungal Susceptibility Studies on Dermatophytes

Compared to other fungal genera such as *Candida*, *Aspergillus,* and *Malassezia,* and also to the epidemiological aspects of dermatophyte infections, the in vitro antifungal susceptibility patterns of pathogenic dermatophyte species have less routinely been covered, worldwide. Likewise, the area of AFST against dermatophytes has no long history in Iran. We found limited number of studies relevant to this issue, most of which were from the new era and after 2010, and also geographically distributed to some limited provinces. The world's oldest in vitro antifungal susceptibility assay on dermatophytes was documented in 1971 by Shadomy [178]. He evaluated the in vitro activity of clotrimazole against dermatophytes compared with nystatin, griseofulvin, and pyrrolnitrin and found that nystatin and griseofulvin were less active than clotrimazole against *Trichophyton*, *Microsporum*, and *Epidermophyton* species; however pyrrolnitrin and clotrimazole demonstrated approximately similar effects. After that, numerous studies have been focused on the in vitro / in vivo assessment of different classes of antifungal agents for treatment of dermatophytosis, but not in Iran. The first authentic study in Iran was carried out in 2004 by Chadeganipour et al. [22] in Isfahan, center of Iran, by using only griseofulvin. They found for all tested dermatophyte isolates a geometric mean of griseofulvin minimum inhibitory concentration (MIC) between 0.43 and 0.95 μg/mL. On that study, three strains (of *M. canis, T. verrucosum* and *T. mentagrophytes*) with high MIC values for griseofulvin were considered as resistant. However, one of the most comprehensive evaluation on drug susceptibility profile of dermatophytes was conducted by Adimi et al. [179] in 2012. Using CLSI broth microdilution method (M38-A), they analyzed the effects of 10 antifungal agents against a collection of 320 clinical isolates belonging to 15 different species of dermatophytes. Based on the obtained results, itraconazole demonstrated the lowest MIC value while fluconazole had the highest MICs for most tested dermatophytes. Moreover, on that report the most common pathogenic species in Iran, e.g. *T. rubrum, TMTI*SG, *E. floccosum, T. tonsurans* and *M. canis* showed acceptable susceptibility to the principal drugs of choice for dermatophytosis, itraconazole and terbinafine. Also, the only data available so far regarding the AFST of the extremely rare species worldwide *T. eriotrephon* were provided on that study and all antifungals showed very low MICs against the two clinical isolates studied. Similarly, Afshari et al. [147] in Arak city found the lowest and the highest MIC values against tested dermatophytes for itraconazole and fluconazole, respectively. In 2016, the efficacy of six antifungals against a large collection of 316 clinical isolates representative of *TMTI*SG, *T. rubrum, M. canis, E. floccosum, T. tonsurans*, and *T. benhamiae* from Ahvaz, Shiraz, and Yasuj was evaluated by Ansari et al. [29]. They found that terbinafine was the most effective antifungal agent against all examined species followed by itraconazole, griseofulvin, and fluconazole. By contrast, in another study from Mazandaran, Baghi et al. [180] evaluated the antifungal susceptibilities of 100 clinical dermatophyte isolates belonging to *TMTI*SG, *T. rubrum*, *T. tonsurans, E. floccosum*, and *M. canis* species towards 12 antifungal agents. They found that luliconazole, anidulafungin, and caspofungin presented the most potent in vitro activities while fluconazole had the highest MIC values. The activity of the novel triazole, efinaconazole, compared with luliconazole, terbinafine, lanoconazole, itraconazole, and fluconazole against a collection of 120 *T. interdigitale* and *T. rubrum* clinical isolates was the matter of an in vitro assessment by Rezaei-Matehkolaei et al. [109]. Based on the geometric mean (GM) of the obtained MICs they concluded that efinaconazole (GM = 0.007 μg/mL), lanoconazole (GM = 0.002 μg/mL), and luliconazole (GM = 0.0005 μg/mL) are promising candidates for the treatment of onychomycosis due to *T. interdigitale* and *T. rubrum*. The in vitro activity of luliconazole against mentioned species was similar to that reported by Baghi et al. [180].

In 2018, Salehi et al. [181] for the first time in Iran searched for terbinafine resistance among clinical dermatophyte isolates. Out of 97 tested isolates belonging to four species, 5 isolates showed reduced susceptibility to terbinafine (MICs > 32 μg/mL), among which only two isolates of *T. rubrum* and *T. tonsurans* demonstrated Leu393Phe substitution in the squalene epoxidase protein. In the most recent study, Taghipour et al. [122] evaluated the in vitro terbinafine resistance in 140 isolates of *TMTI*SG. They reported five terbinafine resistant isolates, all of which belonged to *T. mentagrophytes* genotype VIII (MICs ≥ 32 μg/mL). Allahdadi et al. [127] determined the antifungal susceptibility profile of dermatophytes originated from the scalp of asymptomatic carriers using four drugs, e.g. griseofulvin, terbinafine, itraconazole, and fluconazole. Fifteen isolates belonging to *T. tonsurans*, *T. interdigitale*, and *T. rubrum* species were highly susceptible to terbinafine (MIC_{90} = 0.064 μg/mL), followed by itraconazole (MIC_{90} = 0.125 μg/mL), while griseofulvin (MIC_{90} = 4 μg/mL) and fluconazole (MIC_{90} = 64 μg/mL) did not show proper in vitro activities. Finally, Ansari et al. [182] evaluated seven antifungal agents against a set of 40 clinical *E. floccosum* isolates from four cities (Tehran, Shiraz, Ahvaz, and Mashhad). The geometric mean across all isolates were, in increasing order, as follows: terbinafine (GM = 0.018 μg/mL), posaconazole (GM = 0.022 μg/mL), itraconazole (GM = 0.034 μg/mL), and voriconazole (GM = 0.045μg/mL), which exhibited low MICs against all tested strains, whereas caspofungin (GM = 0.22 μg/mL), ketoconazole (GM = 0.41 μg/mL), and griseofulvin (GM = 0.62 μg/mL) demonstrated higher MICs. Their study showed that terbinafine and posaconazole were the most potent antifungal drugs against *E. floccosum* strains.

5 Conclusion

During the past six decades, numerous investigations have been directed on dermatophytoses in Iran which markedly improved our knowledge on the epidemiological features of these infections. Within this period, the taxonomy of dermatophytes has widely changed but unfortunately, there have been very few researches that underpinned to the concept of dermatophytes species at the relevant times. Moreover, studies on tracking of dermatophytosis sources, routes of transmission, and risk factors predisposing the infections have been ignored or rare, and therefore research in these areas is still required. In conclusion, this review revealed important changes in the mycological and clinical aspects of dermatophytoses in Iran over time. They remained as dermatological problems among Iranian population though the rate of incidence and dynamic of the infectious agents have changed. *Tinea capitis*, in particular, is declining over the country and its anthropophilic agents *T. schoenleinii* and *T. violaceum* are being eliminated and replaced by another anthropophilic species *T. tonsurans*. On the contrary, *Tinea pedis*, *Tinea unguium*, and *Tinea corporis* are slowly increasing. The anthropophilic species *E. floccosum*, *T. rubrum*, *T. tonsurans* and *T. interdigitale* are currently responsible for the major part of dermatophytosis in Iran. A striking epidemiological feature is the emergence of infection with the less-known zoophilic species *T. benhamiae* with unknown reservoirs, which will be a significant challenge to control. Similar to the global trend, studies in Iran have also shown that terbinafine and itraconazole, among current antifungals, have still the best in vitro efficacy against dermatophytes. Furthermore, the recently FDA-approved azole drugs luliconazole, lanoconazole, and efinaconazole presented promising results compared to old antifungal agents for treatment of dermatophytoses.

References

1. Ameen M. Epidemiology of superficial fungal infections. Clin Dermatol. 2010;28 (2):197–201.
2. Weitzman I, Summerbell RC. The dermatophytes. Clin Microbiol Rev. 1995;8(2):240–59.
3. Hayette M-P, Sacheli R. Dermatophytosis, trends in epidemiology and diagnostic approach. Curr Fungal Infect Rep. 2015;9(3):164–79.
4. De Hoog GS, Dukik K, Monod M, Packeu A, Stubbe D, Hendrickx M, et al. Toward a novel multilocus phylogenetic taxonomy for the dermatophytes. Mycopathologia. 2017;182 (1–2):5–31.
5. Seebacher C, Bouchara J-P, Mignon B. Updates on the epidemiology of dermatophyte infections. Mycopathologia. 2008;166(5–6):335–52.
6. Dogra S, Shaw D, Rudramurthy SM. Antifungal drug susceptibility testing of dermatophytes: laboratory findings to clinical implications. Ind Derm Online J. 2019;10(3):225.
7. Asgari M, Sateri H. Cutaneous mycoses in santhern, Iran (scalp ringworm). Iran J Public Health. 1973:10–27.
8. Khosravi A, Aghamirian M, Mahmoudi M. Dermatophytoses in Iran. Mycoses. 1994;37 (1–2):43–8.
9. Farnagh F, Rafiei M, Farivar Sadri M. The study of etiologic causes of dermatophyte in the location of foot and groin, and the possibility of association of dermatophytosis of these two locations together. Tehran Univ Med J. 2001;1:43–8.
10. Foladvand MA, Naeimi B. Superficial fungal infections of the skin in the fishermen port Bushehr. Iran South Med J. 1999;1.
11. Ghaednia B, Zaini F, Kordbacheh P, Hashemi SJ, Mehrabi MR, Mirbakhsh M. Fungal infections of people exposed to shrimp in Bushehr. Iran South Med J. 1999:1.
12. Asl Rahnamae Akbari N, Adibpour M, Salehpoor A, Noa Khahi A. Investigation of etiological agents of tinea capitis in children under twelve years dermatophytosis suspected infections admitted to the clinic No. 2 Tabriz University of Medical Sciences. Med J Tabriz Univ Med Sci Health Service. 2009;31(2).

13. Kazemi AH, Ghiasi S. Survey of otomycosis in north-western area of Iran (1447–2004). J Mazandaran Uni Med Sci. 2005;15(48):112–9.
14. Kordbacheh P, Moghadami M, Asadi MA. Survey of surface and cutaneous fungal diseases in schools and kindergartens in Mahallat and the suburbs. J School Pub Health Inst Publ Health Res. 1994;23(106):61–71.
15. Moghadami M, Emami M. Investigation causative agents of tinea cruris. J School Pub Health Inst Publ Health Res. 1987.
16. Mosavi A, Amin S, Safizadeh H, Hadizadeh S. Epidemiology of dermatophytosis in patients referred to the medical mycology laboratory of Afzalipoor Faculty of Medicine in Kerman in 2007–2011. J Dermatol Cosmet. 2012;3(2):114–23.
17. Hashemi SJ, Moheb Ali M, Zarehee Z. Epidemiology of tinea capitis in Meshkin Shahr primary schools. J Ardabil Univ Med Sci. 2005;5(3):278–81.
18. Mikaeili A, Mahmod A, Rezaei M, Ebrahimi A. The dermatophytes species frequency in referral patients to medical mycology lab of Kermanshah-2012. Med J Mashhad Univ Med Sci. 2015;57(9):956–61.
19. Zamani S, Sadeghi G, Yazdinia F, Moosa H, Pazooki A, Ghafarinia Z, et al. Epidemiological trends of dermatophytosis in Tehran, Iran: a five-year retrospective study. J Mycol Méd. 2016;26(4):351–8.
20. Sadeghi Tafti H, Falahati M, Kordbacheh P, Mahmoudi M, Safara M, Rashidian S, et al. A survey of the etiological agents of scalp and nail dermatophytosis in Yazd, Iran in 2014–2015. Curr Med Mycol. 2015;1(4):1–6.
21. Rezaei-Matehkolaei A, Makimura K, de Hoog S, Shidfar MR, Zaini F, Eshraghian M, et al. Molecular epidemiology of dermatophytosis in Tehran, Iran, a clinical and microbial survey. J Med Mycol. 2013;51(2):203–7.
22. Chadeganipour M, Nilipour S, Havaei A. *In vitro* evaluation of griseofulvin against clinical isolates of dermatophytes from Isfahan. Mycoses. 2004;47(11–12):503–7.
23. Sadeghi G, Abouei M, Alirezaee M, Tolouei R, Shams-Ghahfarokhi M, Mostafavi E, et al. A 4-year survey of dermatomycoses in Tehran from 2006 to 2009. J Mycol Méd. 2011;21 (4):260–5.
24. Asadi AM, Dehghani R, Sharif Mohammad R. Epidemiologic study of onychomycosis and tinea pedis in Kashan, Iran. Jundishapur J Microbiol. 2009;2009(2):61–4.
25. Abastabar M, Rezaei-Matehkolaei A, Shidfar MR, Kordbacheh P, Mohammadi R, Shokoohi T, et al. A molecular epidemiological survey of clinically important dermatophytes in Iran based on specific RFLP profiles of beta-tubulin gene. Iran J Public Health. 2013;42 (9):1049.
26. Rezaei-Matehkolaei A, Makimura K, de Hoog GS, Shidfar MR, Satoh K, Najafzadeh MJ, et al. Multilocus differentiation of the related dermatophytes *Microsporum canis*, *Microsporum ferrugineum* and *Microsporum audouinii*. J Med Microbiol. 2012;61(1):57–63.
27. Omidynia E, Farshchian M, Sadjjadi M, Zamanian A, Rashidpouraei R. A study of dermatophytoses in Hamadan, the governmentship of West Iran. Mycopathologia. 1996;133 (1):9–13.
28. Aghamirian MR, Ghiasian SA. Onychomycosis in Iran: epidemiology, causative agents and clinical features. Nippon Ishinkin Gakkai Zasshi. 2010;51(1):23–9.
29. Ansari S, Hedayati MT, Zomorodian K, Pakshir K, Badali H, Rafiei A, et al. Molecular characterization and *in vitro* antifungal susceptibility of 316 clinical isolates of dermatophytes in Iran. Mycopathologia. 2016;181(1–2):89–95.
30. Aghamirian MR, Ghiasian SA. Dermatophytoses in outpatients attending the dermatology center of Avicenna Hospital in Qazvin, Iran. Mycoses. 2008;51(2):155–60.
31. Chadeganipour M, Mohammadi R, Shadzi S. A 10-year study of dermatophytoses in Isfahan, Iran. J Clin Lab Anal. 2016;30(2):103–7.
32. Nouripour-Sisakht S, Rezaei-Matehkolaei A, Abastabar M, Najafzadeh MJ, Satoh K, Ahmadi B, et al. *Microsporum fulvum*, an ignored pathogenic dermatophyte: a new clinical isolation from Iran. Mycopathologia. 2013;176(1–2):157–60.

33. Hashemi S, Gerami M, Zibafar E, Daei M, Moazeni M, Nasrollahi A. Onychomycosis in Tehran: mycological study of 504 patients. Mycoses. 2010;53(3):251–5.
34. Salari S, Mousavi SAA, Hadizadeh S, Izadi A. Epidemiology of dermatomycoses in Kerman province, southeast of Iran: a 10-years retrospective study (2004–2014). Microb Pathog. 2017;110:561–7.
35. Sepahvand A, Abdi J, Shirkhani Y, Fallahi S, Tarrahi M, Soleimannejad S. Dermatophytosis in western part of Iran, Khorramabad. Asian J Biol Sci. 2009;2(3):58–65.
36. Gerami Shoar M, Zomorodian K, Emami M, Tarazoei B, Saadat F. Study and identification of the etiological agents of onychomycosis in Tehran, capital of Iran. Iran J Public Health. 2002:100–4.
37. Chadegani M, Momeni A, Shadzi S, Javaheri MA. A study of dermatophytoses in Esfahan (Iran). Mycopathologia. 1987;98(2):101–4.
38. Pakshir K, Hashemi J. Dermatophytosis in Karaj, Iran. Ind J Dermatol. 2006;51(4):262.
39. Bassiri-Jahromi S, Khaksar AA. A etiological agents of tinea capitis in Tehran (Iran). Mycoses. 2006;49(1):65–7.
40. Aghamirian MR, Ghiasian SA. A clinico-epidemiological study on tinea gladiatorum in Iranian wrestlers and mat contamination by dermatophytes. Mycoses. 2011;54(3):248–53.
41. Moshir M, Tabatabai A, Poureslami M, Shamshiri A. Prevalence rate of different types of yeast infection in patients visiting the dermatology clinic of HAZRAT RASUL-E AKRAM hospital Razi. J Med Sci. 2002;8(26):466–7.
42. Rastegar Lari A, Akhlaghi L, Falahati M, Alaghehbandan R. Characteristics of dermatophytoses among children in an area south of Tehran, Iran. Mycoses. 2005;48(1):32–7.
43. Mirzahoseini H, Omidinia E, Shams-Ghahfarokhi M, Sadeghi G, Razzaghi-Abyaneh M. Application of PCR-RFLP to rapid identification of the main pathogenic dermatophytes from clinical specimens. Iran J Public Health. 2009;38(1):18–24.
44. Ghasemi Z, Falahati M, Zaini F, Ghaffarpour G-H, Ahmadi F, Eskandar S-E. Causative agents of tinea unguium in Razi Hospital, Tehran, Iran in 2010 and 2011. J Dermatol Cosmet. 2015;6 (2):79–84.
45. Bassiri-Jahromi S, Sadeghi G, Asghari Paskiaee F. Evaluation of the association of superficial dermatophytosis and athletic activities with special reference to its prevention and control. Int J Dermatol. 2010;49(10):1159–64.
46. Bassiri-Jahromi S, Khaksari AA. Epidemiological survey of dermatophytosis in Tehran, Iran, from 2000 to 2005. Indian J Dermatol Venereol Leprol. 2009;75(2):142–7.
47. Mohamadi P, Masomi H. Prevalence of dermatophytosis in Shahrood. Med J Tabriz Univ Med Sci Health Service. 1995;27:81.
48. Kazemi A, Jamali R, Sinafar H, Noroozi A. Survey dermatophytosis in Shabestar (1991–3). Med J Tabriz Univ Med Sci Health Service. 1994;32.
49. Jamali R, Kazemi A, Lakestani A. Study of Tina capitis agents in Tabriz city. Urmia Med J. 1996;6.
50. Mohamadi P. Study of dermatophytosis in 1800 patients referring to Haft-e-Tir hospital in Tabriz University of Medical Sciences in 1364–1384. Med J Tabriz Univ Med Sci Health Service. 1988;27(9).
51. Razi A, Khoaeyani A, Mohamadi P, Sohrabi A, Nori M. Surface and cutaneous fungal infections in east Urmia villages in 1992–1997. Med J Tabriz Univ Med Sci Health Services. 1999;25.
52. Zaini F, Mahmoudi M, Mehbod A, Kordbacheh P, Safara M. Fungal nail infections in Tehran, Iran. Iran J Publ Health. 2009;38(3):46–53.
53. Zarei Mahmoudabadi A, Yaghoobi R. Tinea corporis due to *Trichophyton simii*—a first case from Iran. Sabouraudia. 2008;46(8):857–9.
54. Sheikhi N, Zaker Bostanabad S, Mirza AS. Comparison and evaluation of laboratory diagnostic method -smear, culture, molecular- in the fungal infection. New Cell Mol Biotechnol J. 2017;7(25):19–26.

55. Farokhipor S, Ghiasian S, Nazeri H, Kord M, Didehdar M. Characterizing the clinical isolates of dermatophytes in Hamadan city, central west of Iran, using PCR-RLFP method. J Mycol Méd. 2018;28(1):101–5.
56. Forghani F, Nasrollahi Omran A, Kouchaki M, Mirzaie A. Frequency of dermatophytosis in wrestling and bodybuilding halls in Challous, 2010. Med Lab J. 2013;7(3):31–7.
57. Fallahi AA, Rezaei-Matehkolaei A, Rezaei S. Epidemiological status of dermatophytosis in Guilan, north of Iran. Curr Med Mycol. 2017;3(1):20.
58. Mahboobi A, Baghestani S, Hamedi Y, Heidari M, Vahdani M. Epidemiology of dermatophytosis in Bandar Abbas, Iran 2003–2004. Bimonthly J Hormozgan Univ Med Sci. 2006;9(4):227–34.
59. Sadri MF, Farnaghi F, Danesh-Pazhooh M, Shokohi A. The frequency of tinea pedis in patients with tinea cruris in Tehran, Iran. Mycoses. 2000;43(1–2):41–4.
60. Falahati M, Akhlaghi L, Lari AR, Alaghehbandan R. Epidemiology of dermatophytoses in an area south of Tehran, Iran. Mycopathologia. 2003;156(4):279–87.
61. Kazemi A, Sadr Karimi A. Survey tinea unguium in northwest of Iran (1996–1999). Med J Mashhad Univ Med Sci. 2006;48(87).
62. Ansari S, Hedayati MT, Nouripour-Sisakht S, Rezaei-Matehkolaei A, Jannesar R, Mohammadi H, et al. A 9-month-old girl from Iran with extensive erythematous plaques due to *Trichophyton simii*, a zoophilic dermatophyte. Mycopathologia. 2016;181(5–6):451–5.
63. Soltanpoor A, Morovati H, Dehghan P, Rejali M. The epidemiology of dermatophytosis among elementary school children in Koohrang city, Iran, in 2014 and 2015. J Isfahan Med School. 2015;33(352):1–10.
64. Moghadami M, Shidfar MR. A study of onychomycosis in Tehran. Med J Islam Repub Iran. 1989;3(3):143–9.
65. Naseri A, Fata A, Najafzadeh MJ, Shokri H. Surveillance of dermatophytosis in northeast of Iran (Mashhad) and review of published studies. Mycopathologia. 2013;176(3–4):247–53.
66. Nikpoor N, Leppard BJ. Ringworm on the scalp due to *Epidermophyton floccosum*. Clin Exp Dermatol J. 1977;2:69–71.
67. Nikookhah F, Azamian A, Mahzounieh M. A study on fungal infection athlete's foot among soccer players in the Esfahan city sport clubs. J Med Sci. 2007;7(5):913–5.
68. Hashemi S, Salami A, Hashemi S. An epidemiological study of human dermatophytosis in Karaj (2001). Arch Razi Inst. 2005;60:45–54.
69. Hedayati MT, Afshar P, Shokohi T, Aghili R. A study on tinea gladiatorum in young wrestlers and dermatophyte contamination of wrestling mats from sari, Iran. Br J Sports Med. 2007;41 (5):332–4.
70. Behzadi Z, Parham S, Fata A, Najafzadeh MJ. The molecular identification of dermatophytes in patients suspected to dermatophytosis in Ghayen (South Khorasan province) in 2015–16: a short report. J Rafsanjan Univ Med Sci. 2018;16(10):983–90.
71. Diba K, Gheibi A, Deilami Z, Yekta Z, Hazrati KH. Use of the molecular and conventional methods for the identification of human dermatophytosis in Urmia lake's peripheral area. J North Khorasan Univ Med Sci. 2012;6(3).
72. Chadeganipour M, Shadzi S, Dehghan P, Movahed M. Prevalence and aetiology of dermatophytoses in Isfahan, Iran. Mycoses. 1997;40(7–8):321–4.
73. Khosravi A, Mansouri P. Onychomycosis in Tehran, Iran: prevailing fungi and treatment with itraconazole. Mycopathologia. 2001;150(1):9–13.
74. Mahmoudabadi AZ. First case of *Microsporum ferrugineum* from Iran. Mycopathologia. 2006;161(5):337–9.
75. Didehdar M, Shokohi T, Khansarinejad B, Sefidgar SAA, Abastabar M, Haghani I, et al. Characterization of clinically important dermatophytes in north of Iran using PCR-RFLP on ITS region. J Mycol Méd. 2016;26(4):345–50.
76. Ardehali M. Dermatophytic agents of tinea unguium in Iran. Int J Dermatol. 1973;12 (5):322–3.

77. Afshar P, Khodavaisy S, Kalhori S, Ghasemi M, Razavyoon T. Onychomycosis in north-east of Iran. Iran J Microbiol. 2014;6(2):98.
78. Afshar P, Vahedi L, Ghasemi M, Mardanshahi A. Epidemiology of tinea capitis in Northeast Iran: a retrospective analysis from 1998 to 2012. Int J Dermatol. 2016;55(6):640–4.
79. Ghojoghi A, Falahati M, Abastabar M, Ghasemi Z, Ansari S, Farahyar S, et al. Molecular identification and epidemiological aspects of dermatophytosis in Tehran, Iran. Res Mol Med. 2015;3(3):11–6.
80. Chadeganipour M, Nilipour S, Ahmadi G. Study of onychomycosis in Isfahan, Iran. Mycoses. 2010;53(2):153–7.
81. Hallaji Z, Sadri M, Dadvar F. Analysis of 58 cases of tinea capitis in Tehran Razi hospital. Acta Med Iran. 2001;39(2):109–12.
82. Mahmoudabadi AZ. A study of dermatophytosis in south west of Iran (Ahwaz). Mycopathologia. 2005;160(1):21–4.
83. Mahmoudabadi AZ, Yaghoobi R, Sadeghi B. A large outbreak of tinea capitis in a primary school. J Infect. 2007;54(6):e247–e8.
84. Rezvani SM, Sefidgar SAA, Hasanjani Roushan MR. Clinical patterns and etiology of dermatophytosis in 200 cases in Babol. North Iran Caspian J Intern Med. 2010;1(1):23–6.
85. Mahmoudabadi AZ, Izadi B. Prevalence of cutaneous mycoses among workers. Turk J Med Sci. 2011;41(2):291–4.
86. Asadi A, Davoodabadi A, Samei A. Prevalence of superficial and cutaneous fungal diseases in recreational center of addicts in Tehran during the year 1376. J Kashan Univ Med Sci Health Services. 1999;3(3):74–80.
87. Asadi MA, Houshyar H, Dehghani R, Ershadi A, Arbabi M, Doroudgar A, et al. Prevalence of skin and superficial mycoses in schools of the city of Kashan and its suburbs. Feyz J Kashan Univ Med Sci Health Services. 2001;4:41–7.
88. Zarrin M, Poosashkan M, Mahmoudabadi A, Mapar M. Prevalence of superficial fungal infection in primary school children in Ahvaz, Iran. Macedonian J Med Sci. 2011;4(1):89–92.
89. Asl Rahnamae Akbari N, Adibpour M, Salehpoor A, Kazemi A. Prevalence of nail fungal infections (onychomycosis). Med J Tabriz Univ Med Sci Health Service. 2004;27(2).
90. Amir Rajab N, Rafie A, Omidian M, Mapar M, Yaghobi R, Rasaei S, et al. Tinea capitis in Ahvaz. Iran J Infect Dis. 2007;37:71–5.
91. Ansarin H, Ghaffarpoor GH, Mehraban F. The survey of frequency and etiology of mycoses of the head and body among primary school students in Varamin. Iran J Med Sci. 2001;8(24).
92. Bassiri-Jahromi S, Khaksar AA. Study of onychomycosis in submitted patients to the Pasteur institute of Iran, 1993–1999. J Fac Med. 2002;60(5):364–70.
93. Bassiri-Jahromi S, Khaksar AA. Surveillance of dermatophytosis and the causative agents among children referred to Pasteur Institute of Iran from 2005 to 2006. Res Med. 2008;32 (4):321–6.
94. Khaksar AA, Basiri-Jahromi S, Amirkhani A, Poorabdolahi A. Survey on the prevalence of tinea capitis in primary school children in coastline of Mazandaran province. Teb va Tazkiye Ministry Health Med Edu. 1996;25:18–22.
95. Zarei MA. A survey of 382 suspected patients with tinea capitis (AHWAZ). Jundishapur Sci Med J. 1997;22:45–52.
96. Zamanian A, Farshchian M, Nazeri H. The prevalence of tinea pedis in Sina hospital staff in Hamadan in 2000. Iran J Dermatol. 2003;6(4).
97. Zaini F, Ghagary A. Survey epidemiology and mycology tinea capitis in schools and kindergartens of the Chabahar port. Iran J Public Health. 1989;18:1–12.
98. Salehpoor A, Adibpour M, Attari M-A-A, Asl Rahnamae Akbari N. Investigating the incidence of thymus epithelium in patients referred to the Mycology Laboratory of Tabriz University of Medical Sciences. Med J Tabriz Univ Med Sci Health Service. 2004;38(62).
99. Shokohi T, Hajheidari Z, Haghani I, Khalilian A, Aghili SR, Miahi S. The study of 101 cases of onychomycosis and associate factors in patients referred to Boali Sina hospital and Toba dermatology outpatient clinics in sari. Mazandaran Univ Med Sci. 2009;19(71):33–43.

100. Alizadeh N, Sadr Eshkevari Sh, Golchei J, Falahi AA. A descriptive study of types of tinea in Gilan. Iran J Dermatol. 2004;7(4).
101. Alizadeh N, Sadr Eshkevari S, Golchaei J, Fallahi AA. Deep dermatophytosis in a patient with rheumatoid arthritis treated with immunosuppressive drugs: a case report. Iran J Dermatol. 2005;6(3):35–8.
102. Kazemi A. Tinea unguium in the north-west of Iran (1996–2004). Rev Iberoam Micol. 2007;24(2):113–7.
103. Bassiri-Jahromi S, Khaksar AA, Amirkhani A. Prevalence of fungal infections among wrestlers in Tehran. Pejouhesh dar Pezeshki (Res Med). 2006;30(3).
104. Rashidian S, Falahati M, Kordbacheh P, Mahmoudi M, Safara M, Sadeghi Tafti H, et al. A study on etiologic agents and clinical manifestations of dermatophytosis in Yazd, Iran. Curr Med Mycol. 2015;1(4):20–5.
105. Khaksar AA, Basiri-Jahromi S, Amirkhani A, Mortazavi R. Study of fungal infections in the physiological and mental disabled in Tehran. Teb va Tazkiye Ministry Health Med Edu. 1997;25.
106. Naseri A, Fata A, Khosravi AR. The first case of *Microsporum persicolor* infection in Iran. Jundishapur J Microbiol. 2012;5(1):362–4.
107. Chadeganipour M, Mohammadi R. Causative agents of onychomycosis: a 7-year study. J Clin Lab Anal. 2016;30(6):1013–20.
108. Yazdanfar A. Tinea capitis in primary school children in Hamedan (west of Iran). Int J Med Med Sci. 2010;2(2):029–33.
109. Rezaei-Matehkolaei A, Rafiei A, Makimura K, Gräser Y, Gharghani M, Sadeghi-Nejad B. Epidemiological aspects of dermatophytosis in Khuzestan, southwestern Iran, an update. Mycopathologia. 2016;181(7–8):547–53.
110. Mohamadi P, Babaei Sh, Rabiei H. Evaluation of dermatophytosis in children of primary school girls in Tabriz. Med J Tabriz Univ Med Sci Health Service. 2002;45.
111. Edalatkhah H, Golforoshan F, Azimi H, Mohammadi P, Razi A. Prevalence of various species of dermatophytes in patients referring to the dermatology clinic of Tabriz haft-e-Tir hospital. J Ardabil Univ Med Sci. 2006;6(1):47–52.
112. Emami M, Asgari M, Alilou M. Evaluation of dermatophytic contamination of garrisons in south of Iran. Iran J Publ Health. 1973;3(4).
113. Asgari M, Alilou M, Sheiban Z. Results of preliminary studies on prevalence and the etiologic agents of tinea capitis in Iran. Pahlavi Med J. 1971;2:422–34.
114. Wayne P. Clinical and Laboratory Standards Institute: Reference method for broth dilution antifungal susceptibility testing of yeasts; approved standard. In: CLSI document M27-A3 and Supplement S, vol. 3; 2008. p. 6–12.
115. Rodriguez-Tudela J, Arendrup M, Arikan S, Barchiesi F, Bille J, Chryssanthou E et al. EUCAST definitive document E. DEF 9.1: Method for the determination of broth dilution minimum inhibitory concentrations of antifungal agents for conidia forming moulds. Def. 2008;9:1–13.
116. Ogasawara Y, Hiruma M, Muto M, Ogawa H. Clinical and mycological study of occult tinea pedis and tinea unguium in dermatological patients from Tokyo. Mycoses. 2003;46 (3–4):114–9.
117. Gupta AK, Versteeg SG, Shear NH. Onychomycosis in the 21st century: an update on diagnosis, epidemiology, and treatment. J Cutan Med Surg. 2017;21(6):525–39.
118. Daniel CR III, Jellinek NJ. The pedal fungus reservoir. Arch Dermatol. 2006;142(10):1344–6.
119. Toukabri N, Dhieb C, El Euch D, Rouissi M, Mokni M, Sadfi-Zouaoui N. Prevalence, etiology, and risk factors of tinea pedis and tinea unguium in Tunisia. Can J Infect Dis Med Microbiol. 2017;2017:6835725.
120. Emami M, Asgari M, Alilou M. Investigation of environmental pollution of dermatophytes in southern garrisons. Iran J Publ Health. 1974;3(4).

121. Hamedifard M, Hashemi SJ, Ghazvini RD, Zareei M, Hosseinpour L, Boroujeni ZB. Mycological study of superficial-cutaneous mycoses in Tehran, Iran. Infect Epidemiol Med. 2017;3(2):60–5.
122. Taghipour S, Pchelin IM, Zarei-Mahmoudabadi A, Ansari S, Katiraei F, Rafiei A, et al. *Trichophyton mentagrophytes* and *T. interdigitale* genotypes are associated with particular geographic areas and clinical manifestations. Mycoses. 2019;62(11):1084–91.
123. Tirado-Sánchez A, Bonifaz A. Tinea capitis: current review of the literature. Curr Fungal Infect Rep. 2018;12(3):120–6.
124. Hay R. Tinea capitis: current status. Mycopathologia. 2017;182(1–2):87–93.
125. Momtaz M. Distribution of skin diseases in Esfahan. Persian Med J. 1974;3:526–32.
126. Coulibaly O, L'Ollivier C, Piarroux R, Ranque S. Epidemiology of human dermatophytoses in Africa. Med Mycol. 2017;56(2):145–61.
127. Allahdadi M, Hajihossein R, Kord M, Rahmati E, Amanloo S, Didehdar M. Molecular characterization and antifungal susceptibility profile of dermatophytes isolated from scalp dermatophyte carriage in primary school children in Arak city, Center of Iran. J Mycol Méd. 2019;29(1):19–23.
128. Edalatkhah R, Samavati A. Prevalence of dermatophytes verities types in patient who referred to skin Clinic of Tabriz Haft-e-Tir hospital skin clinic 1997–8. J Ardabil Univ Med Sci. 2000;3 (7):31–5.
129. Hassanzadeh Rad B, Hashemi SJ, Farasatinasab M, Atighi J. Epidemiological survey of human dermatophytosis due to zoophilic species in Tehran, Iran. Iran J Public Health. 2018;47(12):1930.
130. Shemer A, Babaev M. Fungal infections (onychomycosis, Tinea pedis, Tinea cruris, Tinea capitis, Tinea manuum, Tinea corporis, different *Candida* infections, and pityriasis versicolor) and mycological laboratory analyses. In: Tur E, Maibach HJ, editors. Gender and dermatology. Berlin: Springer; 2018. p. 235–42.
131. Elewski BE, Rich P, Tosti A, Pariser DM, Scher R, Daniel RC, et al. Onchomycosis: an overview. J Drugs Dermatol. 2013;12(7):s96–s103.
132. Elewski BE. Onychomycosis: pathogenesis, diagnosis, and management. Clin Microbiol Rev. 1998;11(3):415–29.
133. Piraccini BM, Alessandrini A. Onychomycosis: a review. J Fungi (Basel). 2015;1(1):30–43.
134. Coloe SV, Baird RW. Dermatophyte infections in Melbourne: trends from 1961/64 to 1995/ 96. Pathology. 1999;31(4):395–7.
135. Motamedi M, Ghasemi Z, Shidfar MR, Hosseinpour L, Khodadadi H, Zomorodian K, et al. Growing incidence of non-dermatophyte onychomycosis in Tehran, Iran. Jundishapur J Microbiol. 2016;9(8):e40543.
136. Fatahinia M, Jafarpour S, Rafiei A, Taghipour S, Makimura K, Rezaei-Matehkolaei A. Mycological aspects of onychomycosis in Khuzestan Province, Iran: a shift from dermatophytes towards yeasts. Curr Med Mycol. 2017;3(4):26–31.
137. Veraldi S, Schianchi R, Benzecry V, Gorani A. Tinea manuum: a report of 18 cases observed in the metropolitan area of Milan and review of the literature. Mycoses. 2019;62(7):604–8.
138. Moshir M, Tabatabaei A, Poureslami M, Shamshiri A. Evaluation of fungal infection in patient referred to dermatology clinic of Rasoul Akram. Razi J Med Sci. 2001;8(26):466–70.
139. Nowicki R. Dermatophytoses in the Gdansk area, Poland: a 12-year survey. Mycoses. 1996;39 (9–10):399–402.
140. Tao-Xiang N, Zhi-Cheng L, Sao-Mao W, Wen-Zhu L. Analysis of dermatomycoses in Lanzhou district of northwestern China. Mycopathologia. 2005;160(4):281–4.
141. Ismail MT, Al-Kafri A. Epidemiological survey of dermatophytosis in Damascus, Syria, from 2008 to 2016. Curr Med Mycol. 2016;2(3):32.
142. Wisuthsarewong W, Chaiprasert A, Viravan S. Outbreak of tinea capitis caused by *Microsporum ferrugineum* in Thailand. Mycopathologia. 1996;135(3):157–61.
143. Sabota J, Brodell R, Rutecki GW, Hoppes WL. Severe tinea barbae due to *Trichophyton verrucosum* infection in dairy farmers. Clin Infect Dis. 1996;23(6):1308–10.

144. Bonifaz A, Ramírez-Tamayo T, Saúl A. Tinea barbae (tinea sycosis): experience with nine cases. J Dermatol. 2003;30(12):898–903.
145. Duarte B, Galhardas C, Cabete J. Adult tinea capitis and tinea barbae in a tertiary Portuguese hospital: a 11-year audit. Mycoses. 2019;62:1079–83.
146. Trotha R, Gräser Y, Platt J, Köster A, König B, König W, et al. Tinea barbae caused by a zoophilic strain of *Trichophyton interdigitale*. Mycoses. 2003;46(1–2):60–3.
147. Afshari MA, Shams-Ghahfarokhi M, Razzaghi-Abyaneh M. Antifungal susceptibility and virulence factors of clinically isolated dermatophytes in Tehran, Iran. Iran J Microbiol. 2016;8(1):36–46.
148. Su H, Packeu A, Ahmed SA, Al-Hatmi AM, Blechert O, İlkit M, et al. Species distinction in the *Trichophyton rubrum* complex. J Clin Microbiol. 2019;57(9):e00352–19.
149. Ilkit M, Durdu M. Tinea pedis: the etiology and global epidemiology of a common fungal infection. Crit Rev Microbiol. 2015;41(3):374–88.
150. Nikpoor N, Buxton M, Leppard B. Fungal diseases in shiraz. Pahlavi Med J. 1978;9(1):27–49.
151. Ghannoum M, Isham N. Fungal nail infections (onychomycosis): a never-ending story? PLoS Pathog. 2014;10(6):e1004105.
152. Nenoff P, Krüger C, Ginter-Hanselmayer G, Tietz HJ. Mycology–an update. Part 1: dermatomycoses: causative agents, epidemiology and pathogenesis. J Dtsch Dermatol Ges. 2014;12(3):188–210.
153. Ilkit M. Favus of the scalp: an overview and update. Mycopathologia. 2010;170(3):143–54.
154. Gupta A, Dlova N, Taborda P, Morar N, Taborda V, Lynde C, et al. Once weekly fluconazole is effective in children in the treatment of tinea capitis: a prospective, multicentre study. Br J Dermatol. 2000;142(5):965–8.
155. Grigoryan KV, Tollefson MM, Olson MA, Newman CC. Pediatric tinea capitis caused by *Trichophyton violaceum* and *Trichophyton soudanense* in Rochester, Minnesota, United States. Int J Dermatol. 2019;58(8):912–5.
156. El-Fari M, Gräser Y, Presber W, Titez HJ. An epidemic of tinea corporis caused by *Trichophyton tonsurans* among children (wrestlers) in Germany. Mycoses. 2000;43 (5–6):191–6.
157. Bassiri-Jahromi S, Khaksar AA. Outbreak of tinea gladiatorum in wrestlers in Tehran (Iran). Ind J Dermatol. 2008;53(3):132–6.
158. Ilkit M, Ali Saracli M, Kurdak H, Turac-Bicer A, Yuksel T, Karakas M, et al. Clonal outbreak of *Trichophyton tonsurans* tinea capitis gladiatorum among wrestlers in Adana, Turkey. Med Mycol. 2010;48(3):480–5.
159. Hiruma J, Ogawa Y, Hiruma M. *Trichophyton tonsurans* infection in Japan: epidemiology, clinical features, diagnosis and infection control. J Dermatol. 2015;42(3):245–9.
160. Beguin H, Goens K, Hendrickx M, Planard C, Stubbe D, Detandt M. Is *Trichophyton simii* endemic to the Indian subcontinent? Med Mycol. 2013;51(4):444–8.
161. Nenoff P, Uhrlaß S, Krüger C, Erhard M, Hipler UC, Seyfarth F, et al. *Trichophyton* species of *Arthroderma benhamiae*–a new infectious agent in dermatology. J Dtsch Dermatol Ges. 2014;12(7):571–81.
162. Hubka V, Peano A, Cmokova A, Guillot J. Common and emerging dermatophytoses in animals: well-known and new threats. In: Seyedmousavi S, de Hoog GS, Guillot J, Verweij PE, editors. Emerging and epizootic fungal infections in animals. Berlin: Springer; 2018. p. 31–79.
163. Philpot CM, Bowen RG. Hazards from hedgehogs: two case reports with a survey of the epidemiology of hedgehog ringworm. Clin Exp Dermatol. 1992;17(3):156–8.
164. Phaitoonwattanakij S, Leeyaphan C, Bunyaratavej S, Chinhiran K. *Trichophyton erinacei* onychomycosis: the first to evidence a proximal subungual onychomycosis pattern. Case Rep Dermatol. 2019;11(2):198–203.
165. Marples MJ, Smith J. The hedgehog as a source of human ringworm. Nature. 1960;188 (4753):867–8.

166. Kim J, Tsuchihashi H, Hiruma M, Kano R, Ikeda S. Tinea corporis due to *Trichophyton erinacei* probably transmitted from a hedgehog. Med Mycol J. 2018;59(4):E77–E9.
167. Rhee DY, Kim MS, Chang SE, Lee MW, Choi JH, Moon KC, et al. A case of tinea manuum caused by *Trichophyton mentagrophytes* var. *erinacei*: the first isolation in Korea. Mycoses. 2009;52(3):287–90.
168. Hsieh C-W, Sun P-L, Wu Y-H. *Trichophyton erinacei* infection from a hedgehog: a case report from Taiwan. Mycopathologia. 2010;170(6):417–21.
169. Papegaay J. About pathogenic fungi on human skin around Amsterdam. Neth J Med. 1925;69:879–90.
170. Gupta AK, Williams JV, Zaman M, Singh J. *In vitro* pharmacodynamic characteristics of griseofulvin against dermatophyte isolates of *Trichophyton tonsurans* from tinea capitis patients. Med Mycol. 2009;47(8):796–801.
171. Dolenc-Voljč M, Gasparič J. Human infections with *Microsporum gypseum* complex (*Nannizzia gypsea*) in Slovenia. Mycopathologia. 2017;182(11–12):1069–75.
172. Shalaby MFM, El-din AN, El-Hamd MA. Isolation, identification, and *in vitro* antifungal susceptibility testing of dermatophytes from clinical samples at Sohag University Hospital in Egypt. Electron Physician. 2016;8(6):2557–67.
173. Romano C, Massai L, Gallo A, Fimiani M. *Microsporum gypseum* infection in the Siena area in 2005–2006. Mycoses. 2009;52(1):67–71.
174. Rezaei-Matehkolaei A, Makimura K, Graser Y, Seyedmousavi S, Abastabar M, Rafiei A, et al. Dermatophytosis due to *Microsporum incurvatum*: notification and identification of a neglected pathogenic species. Mycopathologia. 2016;181(1–2):107–13.
175. Iwasawa MT, Togawa Y, Akita F, Kambe N, Matsue H, Yaguchi T, et al. Kerion celsi due to *Arthroderma incurvatum* infection in a Sri Lankan child: species identification and analysis of area-dependent genetic polymorphism. Med Mycol. 2012;50(7):690–8.
176. Do N-A, Nguyen T-D, Nguyen K-L, Le T-A. Distribution of species of dermatophyte among patients at a dermatology Centre of Nghean province, Vietnam, 2015–2016. Mycopathologia. 2017;182(11–12):1061–7.
177. Rezaei-Matehkolaei A, Jahangiri A, Mahmoudabadi AZ, Najafzadeh MJ, Nouripour-Sisakht-S, Makimura K. Morpho-molecular characterization of soil inhabitant dermatophytes from Ahvaz, southwest of Iran, a high occurrence of *Microsporum fulvum*. Mycopathologia. 2017;182(7–8):691–9.
178. Shadomy S. *In vitro* antifungal activity of clotrimazole (Bay b 5097). Infect Immun. 1971;4 (2):143–8.
179. Adimi P, Hashemi SJ, Mahmoudi M, Mirhendi H, Shidfar MR, Emmami M, et al. *In-vitro* activity of 10 antifungal agents against 320 dermatophyte strains using microdilution method in Tehran. Iran J Pharm Res. 2013;12(3):537–45.
180. Baghi N, Shokohi T, Badali H, Makimura K, Rezaei-Matehkolaei A, Abdollahi M, et al. *In vitro* activity of new azoles luliconazole and lanoconazole compared with ten other antifungal drugs against clinical dermatophyte isolates. Med Mycol. 2016;54(7):757–63.
181. Salehi Z, Shams-Ghahfarokhi M, Razzaghi-Abyaneh M. Antifungal drug susceptibility profile of clinically important dermatophytes and determination of point mutations in terbinafine-resistant isolates. Eur J Clin Microbiol Infect Dis. 2018;37(10):1841–6.
182. Ansari S, Ahmadi B, Norouzi M, Ansari Z, Afsarian MH, Lotfali E, et al. *Epidermophyton floccosum*: nucleotide sequence analysis and antifungal susceptibility testing of 40 clinical isolates. J Med Microbiol. 2019;68(11):1655–63.

Onychomycosis in Adults: A Clinical Perspective from Mexico

Roberto Arenas, Eder Rodrigo Juárez-Durán, and Paola Garcia-Mendoza

1 Introduction

Onychomycosis is an infection of the nails caused by dermatophytes, yeasts, or non-dermatophyte molds (NDM). Onychomycosis is the most frequent nail disorder. It affects adults between 30 and 60 years of age but its frequency is increasing in younger people. It is commonly associated with other diseases such as peripheral vascular insufficiency, diabetes, and immunosuppression [1, 2].

2 Epidemiology

Worldwide prevalence of onychomycosis ranges from 2 to 50%, and it represents about 50% of all nail diseases [3]. In the United States, the prevalence of the disease is estimated to be at least 10% and it is higher in Latin American countries [4]. It is caused by dermatophytes (80%) or yeasts, particularly *Candida* species (20%), but nowadays, the prevalence of onychomycosis due to non-dermatophyte molds is increasing (4–8%), as well as that of mixed onychomycosis [5]. The main risk factors include hallux valgus, occlusive footwear, warm, damp conditions, trauma, peripheral vascular disease, and immunosuppression [6]. Diabetes is one of the most related conditions, 26% of these patients presents onychomycosis, but the type and severity of onychomycosis have not been correlated. Many of these factors may explain the increase in the prevalence of onychomycosis in menopause women or older population. Fungal skin infection and previous onychomycosis have been proposed as risk factors [7].

R. Arenas (✉) · E. R. Juárez-Durán · P. Garcia-Mendoza
Mycology Section, "Dr. Manuel Gea González" General Hospital, México City, Mexico

J.-P. Bouchara et al. (eds.), *Dermatophytes and Dermatophytoses*,
https://doi.org/10.1007/978-3-030-67421-2_14

3 Clinical Features

Toenail onychomycosis is caused mainly by dermatophytes, particularly *Trichophyton rubrum* [8], but *Candida* species cause about 20% of the cases especially in fingernail onychomycosis in women. Onychomycosis can be classified based on the clinical features: distal and lateral subungual onychomycosis, superficial white onychomycosis, subungual proximal white onychomycosis, and total dystrophic onychomycosis; all these forms can be associated with fungal melanonychia. Hay and Baran [9] suggested a clinical classification including endonyx, mixed, and secondary onychomycosis, but new forms can be added because new pathogenic microorganisms have been identified.

Distal and lateral subungual onychomycosis (DLSO) affects hyponychium and side edges, causing subungual hyperkeratosis and subungual thickening with discoloration of the nail plate [2]. Superficial white onychomycosis (SWO) is characterized by superficial white patches of the nail plate. Affected nails become rough and crumbly. Subungual and proximal onychomycosis (SPO) affects the proximal nail plate and it can be accompanied by paronychia. SWO and SPO are present mainly in immunocompromised hosts (AIDS) or in young children [10, 11]. Onycholysis and leukonychia can be observed when *Candida* is the causal agent, paronychia may also be present.

Total dystrophic onychomycosis (TDO) is characterized by a complete affection of the nail plate. It can be due to the progression of any type of onychomycosis, and it is a primary condition in patients with chronic mucocutaneous candidiasis.

Fungal melanonychia is a dark brown to black pigmentation of the nail and can be associated with a dystrophic plate or secondary to subungual hyperkeratosis. Causal agents are dermatophytes, but also NDM, including some dematiaceous filamentous fungi. The main etiologic agents are *T. rubrum, Neoscytalidium dimidiatum, Alternaria* species, and *Aspergillus niger* [12].

4 Diagnosis

Due to the long duration of the antifungal treatment and its potential side effects, confirmation of the clinical diagnosis is mandatory because of the numerous cases of atypical clinical presentations and of noninfectious onyxis. The biological diagnosis mainly relies on mycological investigations, but dermoscopy may be useful, as well as histological examination.

4.1 Dermoscopy

It is a noninvasive diagnostic tool for the assessment of onychomycosis and follow-up [13–15]. The most important dermoscopic findings are jagged proximal edge with spikes (jagged edge, with sharp longitudinal whitish indentations directed to the proximal nail fold), longitudinal striae or "aurora borealis pattern" and distal irregular termination with "ruin appearance" [16].

4.2 Mycological Study

A sample must be obtained by scraping under the nail and performing a direct examination with sodium or potassium hydroxide (KOH or NaOH) 10–40% mixed with dimethylsulfoxide (DMSO). Chlorazol black can be added to this preparation because it stains fungal chitin and hyphae or spores are easily observed with a dark-blue color. When a fluorescence microscope is available, staining with calcofluor white can be performed [4, 17, 18].

On direct examination of scales or small fragments of the nail through the microscope, hyaline and septate hyphae are observed. Fungal masses (dermatophytoma) are also observed at the microscopical examination, evading drug action and decreasing the growth rate [19]. Other morphological aspects can be observed, including hyphae, pseudohyphae, and blastospores for *Candida* yeasts, or tortuous and hyaline or pigmented hyphae for NDM [17].

Culture is usually performed in Sabouraud dextrose agar (SDA) with antibiotics such as chloramphenicol and cycloheximide (Actidione®), incubated at 24–28 °C, but if a non-dermatophyte mold or yeast is suspected, culture must be done only on SDA with chloramphenicol or chloramphenicol plus gentamicin. DTM (Dermatophyte test medium) is a medium with red phenol as an indicator, and when a dermatophyte grows, a color change from yellow to red is observed. The most frequent isolated dermatophytes are *T. rubrum* and *Trichophyton interdigitale*. When yeasts are the causal agents, the most common species are *Candida albicans* and *Candida parapsilosis,* and among NDM, *Scopulariopsis brevicaulis*, *Fusarium* spp., and *Aspergillus* spp. [17].

4.3 Histopathological Features

Studies of the nail clipping using hematoxylin-eosin (H&E), periodic acid-Schiff (PAS), and Gomori stains must be performed in severe, uncommon cases or when the mycological examination is negative [20, 21]. In dermatophyte infection, septate hyphae running parallel to the nail surface can be observed. In DLSO fungal elements are frequently found on the ventral part of the nail plate unlike in WSO, where fungal elements are found in the dorsal surface of the nail plate [22, 23]. In leukonychia, abundant hyphae and arthrospores can be observed, whereas budding spores and pseudohyphae are found in *Candida* onychomycosis. In onychomycosis caused by NDM, truncated spores with vertically thin arising can be observed. In dermatophytoma, a dense mass of hyphae and spores is found [22]. Histopathological examination sometimes is considered the gold standard for diagnosis, but it does not allow the precise identification of the etiological agent.

4.4 Molecular Biology

Molecular techniques are not routine tests. PCR increases sensitivity and specificity [10–12], especially when we need to know the specific etiological agent [24]. They can be useful for the identification of atypical or mycological negative cases. Polymerase chain reaction (PCR) and real-time PCR help identify fungi and species in clinical samples [25, 26].

5 Treatment

Before initiating treatment, a confirmatory test is mandatory, at least for a global diagnosis of onychomycosis. Treatment can be systemic, topical, or combined, but surgical or chemical avulsion and medical devices are also available.

5.1 Systemic Treatment

There are few oral antifungals, and although systemic therapy has proved to be more successful than topical therapy, it is often recommended to combine for better outcomes. Terbinafine and itraconazole remain as the first line and gold standard when choosing oral treatment [27, 28]. FDA has approved the former two systemic drugs for onychomycosis, but other antifungal drugs may also be used off-label.

5.1.1 Terbinafine

It is the most prescribed drug in the US in the last two decades and preferred over itraconazole due to its better outcomes and fewer drug interactions [29].

Terbinafine inhibits squalene epoxidase in the ergosterol biosynthesis pathway, giving it a broad spectrum against dermatophytes, but it is less active on NDMs and *Candida* species [30].

Dosage differs between adults and pediatric patients

Adults [29]	Pediatric dose [31]	
250 mg daily	< 20 kg	62.5 mg daily
	20–40 kg	125 mg daily
6 weeks for fingernails/12 weeks for toenails		

Pulse therapy may be considered in adult patients. It has been reported that terbinafine pulses result in a 54.64% clinical cure [32], with two cycles of 250 mg/day of terbinafine for 4 weeks on and 4 weeks off, or 500 mg daily for 7 days/month, every other month with a treatment duration of 6 months [33, 34].

No pulse therapy has been proposed in children.

Terbinafine treatment has limited side effects. The most common are headache, gastrointestinal symptoms, and rash that are temporary and do not require to stop treatment. Other more severe but less frequent side effects are liver enzyme abnormalities and taste disturbance [29].

5.1.2 Itraconazole

It is an antifungal drug available since the 1980s; its safety and efficacy have been demonstrated and it is also approved by the FDA. It inhibits the cytochrome p450 enzyme 14 alpha-demethylase, blocking the conversion of lanosterol to ergosterol. So it has a broader spectrum and efficacy against dermatophytes, NDMs, and *Candida* species [30].

Dosage also differs between adults and pediatric patients [35]

Adults	Pediatric dose
200 mg daily for 12 weeks	5 mg/kg/day for 12 weeks

Another new option is itraconazole-SUBA (which stands for Super BioAvailability), following the same regimen, but with half of the dose [36] (Fig. 1a, b).

Pulse therapy has been proposed for itraconazole. It has been reported that itraconazole results in a 65.1% of clinical cure in pulse therapy with 400 mg/day, 1 week each month for 6 months or 200 mg/day, 1 week/month for 6 months, depending on the weight (average weight 60 kg) [32, 37]. Pulse therapy is also possible on children, with 5–7 mg/kg/day, 1 week/month during 3–5 months [38, 39].

5.1.3 Fluconazole

This drug has been approved in Europe for treatment of onychomycosis but is used off-label in the U.S. It is approved for candidiasis in children. Fluconazole belongs to the azoles family and, as itraconazole, it inhibits the cytochrome p450 enzyme 14 alpha-demethylase responsible for demethylation of lanosterol, leading to ergosterol depletion. It is not a first-line treatment because of its lower effectiveness on dermatophyte onychomycosis than terbinafine or itraconazole, and of the longer duration of the treatment [40, 41].

Doses vary depending on the literature, from 150 to 300 mg weekly for 12 months, but some authors have suggested a higher dose (450 mg) [42, 43].

5.1.4 Griseofulvin

Besides its progressive discontinuity and unavailability, in comparison with terbinafine and itraconazole, griseofulvin has a lower efficacy [44].

5.1.5 "At-Risk" Population

A baseline transaminase monitoring has been proposed for pediatric patients with onychomycosis, although in healthy children the monitoring during the therapy may be unnecessary [45].

In diabetic patients, the most common adverse event is gastrointestinal pain, which was reported for both terbinafine and itraconazole therapy. Matricciani et al. [46] suggested that both *Tinea pedis* and onychomycosis could be a risk factor to

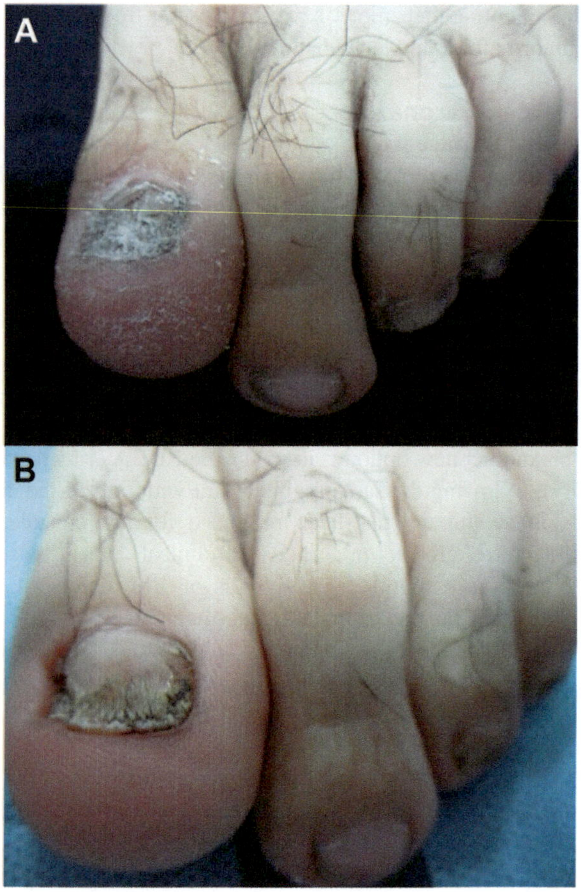

Fig. 1 Total dystrophic onychomycosis due to *Trichophyton rubrum* (**a**) and clinical evolution after 6 weeks of systemic antifungal treatment with itraconazole-SUBA, 50 mg daily (**b**)

develop foot ulcers. Thus, the best available evidence indicates continuous oral terbinafine is as safe as pulse oral itraconazole therapy for treating onychomycosis in people with diabetes. Unlike itraconazole, terbinafine is metabolized by the cytochrome P450 (CYP) 2D6 isoenzyme. This enzyme is not involved in the metabolism of oral hypoglycemic medications, therefore terbinafine can be a safer alternative for people who are unable to take an azole derivative [47].

In HIV infection, onychomycosis has a prevalence of 20–44% and it has been noted a relation between the clinical stage of onychomycosis and a CD4 lymphocyte cell count lower than 450 cells/μl. Also, when starting combined antiretroviral therapy, about 50% of the patients improve clinically without antifungal drugs [48, 49].

Relapse or reinfection is something very common, from 11 to 53% in a 5-year follow-up, and this frequency does not vary with the antifungal drug [50, 51].

5.2 Topical Treatment

It should be used if there is less than 50% involvement of the nail plate or less than three nails involved, as well as in case of superficial onychomycosis or contraindication to oral therapy [52].

Topical antifungals must be used daily for up to 48 weeks, and application should continue until healthy nail growth is complete. Adverse effects with topical antifungals are minimum and usually limited to site-application reactions [53].

FDA has approved three topical therapies for onychomycosis: ciclopirox nail lacquer 8%, efinaconazole solution 10%, and tavaborole solution 5%.

5.2.1 Ciclopirox

Ciclopirox is a hydroxypyridone derivate; it inhibits fungal metal-dependent enzymes by chelating polyvalent cations, like Fe^{3+} which is required for the enzymatic degradation of toxic metabolites.

Ciclopirox has a broad-spectrum activity including dermatophytes, yeasts, some NDMs, and bacteria.

This water-soluble lacquer should be applied after the shower, and the film over the nail can be removed through daily washing [54].

Mycologic cure rates in toenails are 29–36% and complete cure rates 5.5–8.5%. Some adverse effects may occur, strictly local such as burning and periungual erythema [29].

5.2.2 Tavaborole

Tavaborole is a benzoxaborole; it inhibits fungal protein synthesis due to a boron-based structure targeting leucyl-tRNA synthetase, and exhibits antifungal activity against dermatophytes, NDMs, and yeasts [53, 54].

Mycologic cure rates are between 31.1–35.9% and a complete cure is obtained in 9.1% of the cases. Side effects are exfoliation, erythema, and eczema.

5.2.3 Efinaconazole

This triazole derivative inhibits lanosterol 14 alpha-demethylase in the ergosterol biosynthesis pathway. It is active against dermatophytes, NDMs, and *Candida* species. Mycologic cure rates reach 53.4–55.2% and complete cure rate is about 15.2–17.8% [29].

5.3 Off-Label

5.3.1 Amorolfine

Amorolfine, which is a morpholine derivative, inhibits sterol synthesis in the fungal cell wall, leading to the accumulation of not typical spherical sterols in the fungal cell. It is effective against dermatophytes, yeasts, and some molds but not against bacteria. The lacquer is applied once or twice a week on a cleaned nail plate and left for 3–5 min until dry. Among its adverse effects are erythema, burning sensation, nail discoloration, and onycholysis [55]. Although this drug is off-label in the U.S. for treatment of onychomycosis, it has been approved in Europe.

5.3.2 Ciclopirox Plus Hydroxypropyl Chitosan

A combination of ciclopirox with hydroxypropyl chitosan, a water-soluble biopolymer, may also be used. Like amorolfine, this formulation is off-label in the U.S., but it has been approved in Europe. It is applied daily and is eliminated during shower [56].

5.4 Experimental Treatments

Some topical drugs are undergoing clinical investigation, like VT-1161, P-3058, and ME1111.

5.4.1 VT-1161

VT-1161 is a new experimental antifungal drug for the treatment of onychomycosis. It uses a tetrazole ring to target CYP51; this antifungal ability has been tested against *T. rubrum, Trichophyton mentagrophytes,* and *Epidermophyton floccosum*. Unpublished data demonstrate its potential effectiveness in dermatophyte onychomycosis [56].

5.4.2 Reformulations of Terbinafine

Two new formulations of terbinafine are currently being investigated: P-3058 which corresponds to a film-forming solution using the hydroxypropyl chitosan technology, which enhances the penetration of the medication through the nail plate, and Li-P, a terbinafine hydrochloride-loaded liposome film formulation [57].

5.4.3 ME1111

This pyrazole derivative inhibits succinate dehydrogenase of *T. rubrum* and *T. mentagrophytes*. Its effectiveness is similar to ciclopirox, but higher than that of amorolfine, with the advantage to have a low potential for inducing drug resistance [58, 59].

5.4.4 Tazarotene

This retinoid prodrug is converted to its active form, tazarotenic acid, when applied topically and acts by binding to retinoic acid receptors [57].

5.4.5 Acetic Acid-Based Nail Polish

This formulation blocks fungal growth by acidifying the nail, inhibiting acid-sensitive keratinolytic enzymes that are essential for dermatophyte penetration into nails. It is an efficient and safe treatment for mild to moderate cases of onychomycosis.

Comparing the solution based on acetic acid vs. 5% amorolfine nail polish, showed improvement in 96.0% and 79.6% of the cases, respectively, after 180 days of treatment [60].

5.4.6 Tioconazole 28%

Reformulation of the old imidazole drug tioconazole, which exhibits a wide-spectrum antifungal activity, in nanoparticles is currently being investigated. This new formulation helps reducing the minimum concentration required for growth inhibition, nanoparticles acting as reservoirs from where the drug is slowly released over prolonged periods. The advantages are better adherence to treatment and reduction of side effects such as contact dermatitis and local irritation [61].

Antifungals such as amorolfine, efinaconazole, and tavaborole stay in the nail plate at concentrations higher than the minimum inhibitory concentration for dermatophytes more than 28 days after their application [53].

Topical treatment for onychomycosis in children is often used off-label under the same indications as in adults, these therapies are only approved by the FDA for use in adults. It can have a good response due to its easy penetration because of the thin and fast-growing nails of pediatric patients.

5.5 Other Treatments

Lasers perform their fungicidal activity through photothermolysis against fungal chitin, but high temperatures are required; so short pulses minimize pain and reduce complications [62, 63].

It is necessary to consider the number of sessions required, the duration of the treatment, and its cost. Cure rates for laser treatment are lower than those of oral or topical treatments, so they are not recommended as first-line therapy for onychomycosis [29, 64].

6 Remission and Recurrence

To achieve the effectiveness of topical therapy in onychomycosis, patient education is very important. Before starting the treatment, patients should be aware that it requires daily application for at least 6 months for the nails and 12 months for the toenails, depending on the average nail growth rate [54].

Treatment efficacy improves when started early, and if it is given at the same time for *Tinea pedis* [29].

Biofilms have been related to fungal resistance. Therefore, activity against biofilms could help maximize cure rates, combining a higher antifungal dose with surgical or chemical avulsion [65].

The success of topical therapy is defined by the ability to penetrate the nail plate, and this is determined by molecular size, hydrophilicity, the binding of the active drug to keratin and the formulation of the drug.

The clinical cure is difficult when matrix damage and nail dystrophy are associated [65].

Recurrence or reinfection may occur between 10 and 53% according to some authors. To avoid this possibility, it is suggested to continue with topical prophylactic treatment, although the ideal duration of prophylaxis is unknown [29].

References

1. Ribeiro CS, Zaitz C, Framil VM, Ottoboni TS, Tonoli MS, Ribeiro RP. Descriptive study of onychomycosis in a hospital in Sao Paulo. Braz J Microbiol. 2015;46:485–92.
2. Elewski BE. Onychomycosis: pathogenesis, diagnosis, and management. Clin Microbiol Rev. 1998;11(3):415–29.
3. Nakamura RC, Costa MC. Dermatoscopic findings in the most frequent onychopathies: descriptive analysis of 500 cases. Int J Dermatol. 2012;51:483–96.
4. Gold S. Understanding onychomycosis resolving diagnostic dilemmas. Semin Cutan Med Surg. 2016;35(Suppl 3):S48–50.
5. Jayatilake JA, Tilakaratne WM, Panagoda GJ. *Candida* onychomycosis: a mini-review. Mycopathologia. 2009;168(4):165–73.
6. Papini M, Piraccini BM, Difonzo E, Brunoro A. Epidemiology of onychomycosis in Italy: prevalence data and risk factor identification. Mycoses. 2015;58(11):659–64.
7. Ferrari J. Fungal toenail infections. Am Pham Physician. 2015;15(92):132–3.
8. Asz-SIgall D, Tosti A, Arenas R. Tinea unguium: diagnosis and treatment in practice. Mycopathologia. 2017;182(1–2):95–100.
9. Hay RJ, Baran R. Onychomycosis: a proposed revision of the clinical classification. J Am Acad Dermatol. 2011;65(6):1219–27.
10. Baran R, Hay RJ, Tosti A. A new classification of onychomycosis. Br J Dermatol. 1998;139:567–71.
11. Elewsky BE, Hughey C, Sobera JO, Hay R. Fungal Diseases. In: Bolognia J, Jorizzo J, Schaffer J, editors. Dermatology. 3rd ed. New York: Elsevier Saunders; 2012. p. 1251–84.
12. Finch J, Arenas R, Baran R. Fungal melanonychia. J Am Acad Dermatol. 2011;66(5):830–41.
13. Micali G, Lacarrubba F, Massimino D, Schwartz RA. Dermatoscopy: alternative uses in daily clinical practice. J Am Acad Dermatol. 2011;64:1135–46.
14. Piraccini BM, Balestri R, Starance M, Rech G. Nail digital dermoscopy (onychoscopy) in the diagnosis of onychomycosis. J Eur Acad Dermatol Venereol. 2013;27:509–13.
15. Jesús-Silva MA, Fernández Martínez R, Roldán-Marín R, Arenas R. Dermoscopic patterns in patients with a clinical diagnosis of onychomycosis—results of a prospective study including data of potassium hydroxide (KOH) and culture examination. Dermatol Pract Concept. 2015;5 (2):39–44.
16. Crignis G, Valgas N. Dermatoscopy of onychomycosis. Int J Dermatol. 2014;53:80–1.
17. Arenas R. Métodos diagnósticos nas onicomicosis. In: Di Chiacchio N, Belda Junior W, Criado PR, editors. Unhas. Oque há de novo. AtSão Paulo: Atheneu; 2015. p. 11–5.
18. Weinberg JM, Koestenblatt EK, Tutrone WD, Tishler HR, Najarian L. Comparison of diagnostic methods in the evaluation of onychomycosis. J Am Acad Dermatol. 2003;49:193–7.
19. Escalante K, Martínez E, Torres-Guerrero E, Arroyo S, Arenas R. Onychomycosis with dermatophytoma. A comparison among the results of treatments with oral terbinafine, topical 40% urea in monotherapy and combination therapy. Dermatologia Kliniczna. 2013;15 (2):67–70.
20. André J, Sass U, Richert B, Theunis A. Nail pathology. Clin Dermatol. 2013;31:526–39.
21. Baraldi A, Jones SA, Guesné S, Traynor MJ, McAuley WJ, Brown MB, et al. Human nail plate modifications induced by onychomycosis: implications for topical therapy. Pharm Res. 2015;32:1626–33.

22. Stephen S, Tosti A, Rubin AI. Diagnostic applications of nail clippings. Dermatol Clin. 2015;33:289–301.
23. Mayer E, Izhak OB, Bergman R. Histopathological periodic acid–Schiff stains of nail clippings as a second-line diagnostic tool in onychomycosis. Am J Dermatopathol. 2012;34:270–3.
24. Gupta AK, Simpson FC. Diagnosing onychomycosis. Clin Dermatol. 2013;31(5):540–3.
25. Karakousis A, Tan L, Ellis D, Alexiou H, Wormald PJ. An assessment of the efficiency of fungal DNA extraction methods for maximizing the detection of medically important fungi using PCR. J Microbiol Methods. 2006;65(1):38–48.
26. Romanelli AM, Fu J, Herrera ML, Wickes BL. A universal DNA extraction and PCR amplification method for fungal rDNA sequence-based identification. Mycoses. 2014;57(10):612–22.
27. Yin Z, Xu J, Luo D. A meta-analysis comparing long-term recurrences of toenail onychomycosis after successful treatment with terbinafine versus itraconazole. J Dermatolog Treat. 2012;23(6):449–52.
28. Feng X, Xiong X, Ran Y. Efficacy and tolerability of amorolfine 5% nail lacquer in combination with systemic antifungal agents for onychomycosis: a meta-analysis and systematic review. Dermatol Ther. 2017;30(3):1–6.
29. Lipner SR, Scher RK. Onychomycosis: treatment and prevention of recurrence. J Am Acad Dermatol. 2019;80(4):853–67.
30. Jellinek NJ. Understanding Onychomycosis treatment: mechanisms of action and formulation. Semin Cutan Med Surg. 2015;34(3 Suppl):S51–3.
31. Tosti A, Piraccini BM, Iorizzo M. Management of onychomycosis in children. Dermatol Clin. 2003;21:507–9.
32. Ranawaka RR, Nagahawatte A, Gunasekara TA, Weerakoon HS, de Silva SH. Randomized, double-blind, comparative study on efficacy and safety of itraconazole pulse therapy and terbinafine pulse therapy on nondermatophyte mold onychomycosis: a study with 90 patients. J Dermatolog Treat. 2016;27(4):364–72.
33. Gupta AK, Paquet M, Simpson F, Tavakkol A. Terbinafine in the treatment of dermatophyte toenail onychomycosis: a meta-analysis of efficacy for continuous and intermittent regimens. J Eur Acad Dermatol Venereol. 2013;27:267–72.
34. Nenoff P, Krüger C, Paasch U, Ginter-Hanselmayer G. Mycology - an update part 3: dermatomycoses: topical and systemic therapy. J Dtsch Dermatol Ges. 2015;13(5):387–410.
35. Gupta AK, Paquet M. Systemic antifungals to treat onychomycosis in children: a systematic review. Pediatr Dermatol. 2013;30(3):294–302.
36. Abuhelwa AY, Foster DJ, Mudge S, Hayes D, Upton RN. Population pharmacokinetic modeling of itraconazole and hydroxyitraconazole for oral SUBA-itraconazole and sporanox capsule formulations in healthy subjects in fed and fasted states. Antimicrob Agents Chemother. 2015;59(9):5681–96.
37. Zhang L, Xu H, Shi Y, Yu J, Tao Y, Li X. An exploration of the optimum dosage and number of cycles of itraconazole pulse therapy for severe onychomycosis. Mycoses. 2018;61(10):736–42.
38. Huang PH, Paller AS. Itraconazole pulse therapy for dermatophyte onychomycosis in children. Arch Pediatr Adolesc Med. 2000;154:614–8.
39. Heikkilä H, Stubb S. Onychomycosis in children: treatment results of forty-seven patients. Acta Derm Venereol. 2002;82:484–5.
40. Feldstein S, Totri C, Friedlander SF. Antifungal therapy for onychomycosis in children. Clin Dermatol. 2015;33(3):333–9.
41. Solís-Arias MP, García-Romero MT. Onychomycosis in children. A review. Int J Dermatol. 2017;56(2):123–30.
42. Gupta AK, Drummond-Main C, Paquet M. Evidence-based optimal fluconazole dosing regimen for onychomycosis treatment. J Dermatolog Treat. 2013;24(1):75–80.
43. Hryncewicz-Gwóźdź A, Plomer-Niezgoda E, Kalinowska K, Czarnecka A, Maj J, Jagielski T. Efficacy of fluconazole at a 400 mg weekly dose for the treatment of onychomycosis. Acta Derm Venereol. 2015;95(2):251–2.

44. Taheri A, Davis SA, Huang KE, Feldman SR. Onychomycosis treatment in the United States. Cutis. 2015;95(5):E15–21.
45. Patel D, Castelo-Soccio LA, Rubin AI, Streicher JL. Laboratory monitoring during systemic terbinafine therapy for pediatric onychomycosis. JAMA Dermatol. 2017;153(12):1326–7.
46. Matricciani L, Talbot K, Jones S. Safety and efficacy of tinea pedis and onychomycosis treatment in people with diabetes: a systematic review. J Foot Ankle Res. 2011;4:26.
47. Venkatakrishnan K, Von Moltke L, Greenblatt D. Effects of the antifungal agents on oxidative drug metabolism: clinical relevance. Clin Pharmacokinet. 2000;38:111–80.
48. Ruíz-López P, Moreno-Coutiño G, Fernández-Martínez R, Espinoza-Hernández J, Rodríguez-Zulueta P, Reyes-Terán G. Evaluation of improvement of onychomycosis in HIV-infected patients after initiation of combined antiretroviral therapy without antifungal treatment. Mycoses. 2015;58(9):516–21.
49. Moreno-Coutiño G, Arenas R, Reyes-Terán G. Clinical presentation of onychomycosis in HIV/AIDS: a review of 280 Mexican cases. Indian J Dermatol. 2011;56(1):120–1.
50. Sigurgeirsson B, Olafsson JH, Steinsson JB, Paul C, Billstein S, Evans EG. Long-term effectiveness of treatment with terbinafine vs itraconazole in onychomycosis: a 5-year blinded prospective follow-up study. Arch Dermatol. 2002;138(3):353–7.
51. Piraccini BM, Sisti A, Tosti A. Long-term follow-up of toenail onychomycosis caused by dermatophytes after successful treatment with systemic antifungal agents. J Am Acad Dermatol. 2010;62:411–4.
52. Gupta AK, Mays RR, Versteeg SG, Piraccini BM, Takwale A, Shemer A, et al. Global perspectives for the management of onychomycosis. Int J Dermatol. 2018;58(9):1118–29.
53. Gupta AK, Foley KA, Versteeg SG. New antifungal agents and new formulations against dermatophytes. Mycopathologia. 2017;182(1–2):127–41.
54. Lipner SR, Ko D. Optimizing topical therapy for onychomycosis: the importance of patient education. Cutis. 2018;102(6):389–90.
55. Tabara K, Szewczyk AE, Bienias W, Wojciechowska A, Pastuszka M, Oszukowska M, et al. Amorolfine vs. ciclopirox – lacquers for the treatment of onychomycosis. Postepy Dermatol Alergol. 2015;32(1):40–5.
56. Baran R, Tosti A, Hartmane I, Altmeyer P, Hercogova J, Houdelkova V, et al. An innovative water-soluble biopolymer improves efficacy of ciclopirox nail lacquer in the management of onychomycosis. J Eur Acad Dermatol Venereol. 2009;23(7):773–81.
57. Gupta AK, Studholme C. Novel investigational therapies for onychomycosis: an update. Expert Opin Investig Drugs. 2016;25(3):297–305.
58. Kubota-Ishida N, Takei-Masuda N, Kaneda K, Nagira Y, Chikada T, Nomoto M. et al. *In vitro* human onychopharmacokinetic and pharmacodynamic analyses of ME1111, a new topical agent for onychomycosis. Antimicrob Agents Chemother. 2017;62(1):e00779–17.
59. Hui X, Jung EC, Zhu H, Maibach HI. Antifungal ME1111 *in vitro* human onychopharmacokinetics. Drug Dev Ind Pharm. 2017;43(1):22–9.
60. Eertmans F, Doss N, Rossel B, Adriaens E. Daily application of an aqueous, acidifying, peelable nail polish versus weekly amorolfine for topical onychomycosis treatment: a prospective, randomized, blinded trial. Dermatol Ther (Heidelb). 2018;8(3):463–73.
61. Flores FC, Rosso RS, Cruz L, Beck RC, Silva CB. An innovative polysaccharide nano-based nail formulation for improvement of onychomycosis treatment. Eur J Pharm Sci. 2017;100:56–63.
62. Nijenhuis-Rosien L, Kleefstra N, van Dijk PR, Wolfhagen MJHM, Groenier KH, Bilo HJG, et al. Laser therapy for onychomycosis in patients with diabetes at risk for foot ulcers: a randomized, quadruple-blind, sham-controlled trial (LASER-1). J Eur Acad Dermatol Venereol. 2019;33(11):2143–50.
63. Yeung K, Ortner VK, Martinussen T, Paasch U, Haedersdal M. Efficacy of laser treatment for onychomycotic nails: a systematic review and meta-analysis of prospective clinical trials. Lasers Med Sci. 2019;34(8):1513–25.

64. Weber GC, Firouzi P, Baran AM, Bölke E, Schrumpf H, Buhren BA, et al. Treatment of onychomycosis using a 1064-nm diode laser with or without topical antifungal therapy: a single-center, retrospective analysis in 56 patients. Eur J Med Res. 2018;23(1):53.
65. Gupta AK, Versteeg SG, Shear NH, Piguet V, Tosti A, Piraccini BM. A practical guide to curing onychomycosis: how to maximize cure at the patient, organism, treatment, and environmental level. Am J Clin Dermatol. 2019;20(1):123–33.

Tinea capitis in School Children: Current Status

Ifeanyichukwu Emmmanuel Eke and Emeka Innocent Nweze

Abstract

The growing incidence of scalp ringworm infection globally has generated a lot of public health interest. While the disease remains a common superficial infection among young schoolchildren, it has increasingly been reported in older persons. The disease is caused by some keratin-degrading filamentous fungi, called dermatophytes, especially by zoophilic or anthropophilic species. In this chapter, we describe the pathogenesis of dermatophytes, focusing on the ability of the pathogen to attach and invade the epidermal layers of the skin. The chapter also discusses the epidemiologic spread of the disease in different parts of the world, particularly its growing incidence in older people and infants. When not properly diagnosed or promptly treated, ringworm infection can cause complications leading to other conditions such as folliculitis, alopecia, and dermatophytid reactions. The chapter highlights the complications of the infection, as well as current methods used in the diagnosis, treatment, and community management of the infection. Overall, it provides a recent and comprehensive overview of *Tinea capitis* and may serve as a useful resource for educators, clinicians, and students, as well as in helping to formulate adequate health policies for managing *Tinea capitis* infection.

I. E. Eke
Department of Microbiology and Molecular Genetics, Michigan State University, East Lansing, MI, USA

Department of Microbiology, University of Nigeria, Nsukka, Nigeria

E. I. Nweze (✉)
Department of Microbiology, University of Nigeria, Nsukka, Nigeria
e-mail: emeka.nweze@unn.edu.ng

J.-P. Bouchara et al. (eds.), *Dermatophytes and Dermatophytoses*,
https://doi.org/10.1007/978-3-030-67421-2_15

Keywords

Scalp ringworm · Pathogenesis · Epidemiology · Schoolchildren · Diagnosis · Treatment · Community management · *Tinea capitis* · Dermatophytes

1 Introduction

Dermatophytes, which are keratin-degrading filamentous fungi, are the major causative agents of dermatomycosis. They have humans, animals, and the soil as major reservoirs, and are usually transmitted through fomites or direct skin contacts. Due to the characteristic concentric lesions caused by these fungi, dermatophytoses are also known as *Tinea* or ringworm infections [1–3].

Tinea capitis (TC), also known as scalp ringworm, is a dermatophytic infection which affects the scalp. It is a common childhood infection, although an increasing incidence has been reported in aged people and infants [4, 5]. Clinical features of TC range from a scaly scalp to inflammation and alopecia [6]. Apart from the economic toll of the disease, it also diminishes the self-confidence of the patients and affects their self-esteem [7].

In this chapter, we describe recent findings regarding the pathogenesis of dermatophytes, pointing out its relevance to the clinical manifestations of TC. The factors involved in the adhesion of the pathogen to the epidermis, as well as those that enable the pathogen to invade the scalp and elicit a host immune response are examined. The epidemiological distribution of the disease among schoolchildren and its growing incidence among infants and old people are also discussed. From the classic use of the Wood's lamp to molecular diagnostics, the various techniques used in the diagnosis of TC are also thoroughly examined. Current therapeutic strategies for TC are also reviewed, as well as current policies advocated in the community management of the infection. It is hoped that this chapter will provide not just a comprehensive introduction to newcomers in the field, but also stimulate more advanced research to better understand the infection.

2 Pathogenesis and Pathophysiology

During adverse environmental conditions, dermatophytes sporulate into quiescent structures that have low metabolic rates and are physiologically adapted for easy dispersal in the environment. When these structures, which may be microconidia, macroconidia, or arthrospores, reach a favorable environment, they reactivate into metabolically active hyphae. The clinical manifestations of dermatophytosis normally result from the degradation of keratin-rich tissues and the inflammatory response of the infected host [1, 2, 8]. Accordingly, there are four steps which must be recognized when discussing the pathogenesis of dermatophytes: adhesion, germination, invasion, and inflammation.

In the first step (adhesion), the dispersed fungal spores make use of specialized surface molecules in attaching to the epidermal surface of host organisms. Chief among these are adhesins which are usually expressed on the surface of the fungal spores [1, 2]. While the biochemical nature of the fungal adhesins is not well-known, Esquenazi and others [3, 9] suggested their glycoprotein nature and showed the special affinity of *Trichophyton rubrum* and *Trichophyton mentagrophytes* adhesins for sugar residues such as mannose and galactose present on the surface of susceptible host tissues. Morphologically, Kaufman et al. [10] reported the presence of fungal projections known as fibrils adhesins, which are thought to connect dermatophytes to the surface of the stratum corneum layer, preventing the disconnection of the invading fungal pathogen from the epidermis. In addition to adhesin-mediated attachment, extracellular proteases secreted by dermatophytes have also been implicated in the adherence step of fungal infection [8, 10–13]. Though the precise mechanism of secreted protease-mediated adherence is still obscure, two hypotheses are currently used to explain it. First, proteases anchored to the surface of the fungal pathogen serve as ligands for binding to different moieties present on the surface of host cells. Second, the proteases are secreted outside the fungal spores, where they serve as enzymes that modify different ligands on the fungal or host's cell surface, thereby enhancing the adherence of the pathogen [11].

During germination, the adherent spores sense a favorable environment, swell and produce germ tubes that give rise to new active hyphae. The new hyphae are able to penetrate into the stratum corneum, the outermost layer of the epidermis, and are not easily removed by the desquamation events that happen regularly in the skin. Usually, the newly formed fungal hyphae secrete enzymes such as lipases, ceramidases, keratinases, and subtilisins that degrade the cortex layer and keratin fibers of the epidermis [1, 2, 6, 10, 14, 15]. This may lead to a partial hair loss, an important clinical sign of TC. Based on the hair shaft invasion pattern (Table 1), the fungal infection can be ectothrix—where the fungus destroys the hair cuticle and is found on the hair surface; endothrix—where the fungus is found inside the hair shaft, with the hair cuticle remaining intact; and favus—characterized by the formation of yellowish cup-shaped crusts, which contain the fungal hyphae around the hair shafts [1, 6, 17].

In the last step, which usually entails evasion and host's immune response, the metabolic byproducts of the invading fungus diffuse into the inner parts of the epidermis, leading to various clinical responses such as erythema, vesicle formation, or pruritis [14, 18]. Typically, the fungi induce the proliferation and migration of host T-lymphocytes and macrophages to the infected site, causing hypersensitivity reactions. The severe inflammatory appearance of TC, known as kerions, may mimic a bacterial folliculitis and may accompany systemic manifestations such as fever, leukocytosis, papular eruptions, and lymphadenopathy. However, the degree of immune response depends on the dermatophyte species as well the physiopathological status of the host [6, 14, 15]. Generally, acute and highly inflammatory TC results from zoophilic dermatophytes infection, while a chronic and weak inflammatory infection is associated with anthropophilic species [1, 2, 14]. The invading fungus is

Table 1 Invasion patterns of dermatophytes that cause *Tinea capitis*

Hair invasion pattern	Organism			Clinical features
	Zoophilic	Anthropophilic	Geophilic	
Endothrix	*T. mentagrophytes*	*T. rubrum* *T. violaceum* *T. tonsurans* *T. soudanense* *T. yaoundei*[a]		Breakage of hair shaft at scalp level, leaving behind swollen hair stubs that gives rise to a pattern of black dots
Ectothrix	*M. canis* *T. verrucosum* *N. nana*[b]	*M. audouinii*	*N. gypsea*[b]	Usually scaly and inflamed; breakage of hair shafts 2–3 mm or more above scalp level
Favus		*T. schoenleinii*		Hyphae-containing cup-shaped crusts known as scutula; yellow in color and found at the base of the hair shafts; inflammatory

M—Microsporum, N—Nannizzia, T—Trichophyton
[a]The species *Trichophyton yaoundei* is considered as synonym of *Trichophyton violaceum* according to the recent taxonomic revision of dermatophytes from de Hoog et al. [16]
[b]*Nannizzia gypsea* and *Nannizzia nana*: formerly *Microsporum gypseum* and *Microsporum nanum*

retained in the stratum corneum, hardly penetrating beyond the epidermal surface and its appendages [18].

3 Epidemiology

3.1 Occurrence in Schoolchildren

School-aged children, usually those between 3 and 12 years of age, are the major groups affected by TC. This prevalence of TC among prepubertal children can be presumptively explained by the inability of such children to synthesize adequate amount of fungistatic fatty acids in their sebum. Overcrowding in classrooms, sharing of personal belongings, low socioeconomic conditions, infected pet contact, and a family history of dermatophytic infections are factors which also predispose children to TC [7, 19].

Generally, the prevalence rates of pediatric TC infection differ from one region of the world to the other, although male children are more affected than their female colleagues [19–22]. Speculatively, the predominance of the infection among males can be attributed to their short hairs, which allows for the easier attachment and implantation of fungal spores. In contrast to males, females tend to be more conscious of their personal hygiene, wear protective hair covers, and apply antifungal hair products [19, 23, 24]. Furthermore, females tend to reach puberty faster (9–12 years) than boys do (after 12 years), producing fungistatic acidic sebum at an earlier age [25]. However, some studies have also reported TC to be more prevalent

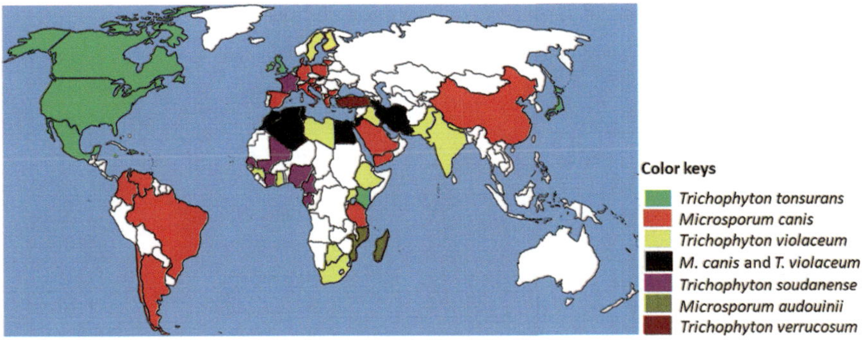

Fig. 1 Predominant etiologic agents of pediatric *Tinea capitis* in different parts of the world

in females than males [26, 27]. Therefore, more studies are probably needed to better understand the gender predilection of pediatric TC in different parts of the world.

Additionally, the etiology of pediatric TC varies from one country or region of the world to another (Fig. 1). *Trichophyton tonsurans* is the major etiologic agent of TC in North America [20, 28–33], while *Microsporum canis* is the prevalent etiologic agent in South America [34–40]. In Europe, the causative agent of pediatric TC varies from *T. tonsurans* [41–45] to *Trichophyton soudanense* [46, 47] and *Trichophyton violaceum* [48, 49], although *M. canis* remains the major agent in most parts of the continent [21, 50–61]. In North Africa, *M. canis* and *T. violaceum* are implicated in most cases of pediatric TC [23, 26, 27, 62–67]. *Trichophyton violaceum* is also a predominant etiologic agent in the southern and eastern parts of Africa [68–72], although species such as *Microsporum audouinii* (formerly known as *Microsporum langeronii*), *M. canis* and *T. tonsurans* have equally been reported as predominant agents in the region [24, 73–76]. *Trichophyton soudanense* is endemic in West Africa and has been implicated as a predominant etiologic agent of TC among schoolchildren in the region [25, 77–82]. In addition to *T. soudanense*, some studies have also reported *T. tonsurans, T. rubrum*, and *T. mentagrophytes* as prevalent species that cause TC in Nigeria [83–85], while others have shown *T. violaceum* to be predominant in Ghana and Guinea [86, 87].

Similar to the pattern observed in northern Africa, *M. canis* and *T. violaceum* were reported as the predominant causative agents of TC in the Middle Eastern countries of Kuwait, Israel, Iraq, Yemen, Saudi Arabia, Jordan, and Iran [22, 88–98]. In some other Asian countries, *Trichophyton verrucosum, M. canis, T. tonsurans*, and *T. violaceum* have also been recognized as predominant species in pediatric TC [99–108].

3.2 Occurrence in Infants

Since the 1950s, TC has been recognized in infants but appears to be underreported [109]. The similarity of TC to other infant scalp pathologies increases its chances of

Table 2 Case reports of infant *Tinea capitis* within the last 20 years (1999–2019)

World region	Age (months)[a]	Sex	Causative agent	References
Europe	0.5	F	*T. rubrum*	[111]
	3	F	*T. rubrum*	[112]
	6	M	*T. tonsurans*	[113]
	24	F	*T. tonsurans*	[113]
	8	F	*M. canis*	[114]
	12	M	*M. canis*	[110]
	20	F	*T. rubrum*	[115]
	28	M	*T. verrucosum*	[116]
North America	0.6	M	*T. tonsurans*	[109]
	0.47	M	*M. canis*	[109]
	12	F	*M. canis*	[109]
	16	F	*T. tonsurans*	[109]
	0.7	F	*M. canis*	[117]
	8	F	*M. canis*	[118]
South America	0.67	F	*T. rubrum*	[119]
	0.93	M	*T. mentagrophytes*	[119]
	1.83	M	*T. mentagrophytes*	[119]
Asia	0.53	M	*M. canis*	[120]
	0.93	M	*T. rubrum*	[121]
	3	M	*N. gypsea*[b]	[122]
	0.83	F	*T. mentagrophytes*	[123]
	9	F	*M. canis*	[124]
Africa	24	F	*M. canis*	[125]
	2	F	*T. verrucosum*	[126]

M—Microsporum, N—Nannizzia, T—Trichophyton

[a]Ages which were originally in days or weeks were converted to months under the assumption that an average of 30 days makes up 1 month

[b]*Nannizzia gypsea*: formerly *Microsporum gypseum*

being misdiagnosed and mistreated [109, 110]. In the last two decades, many cases of infant TC have been highlighted across the globe (Table 2).

This infection may be transmitted to infants through: (1) contacts with asymptomatic children or adults; (2) use of infected objects such as combs, towels, pillows, toys, and brushes; and (3) exposure to carrier pets. It has also been suggested that a humid environment, an indiscriminate use of broad-spectrum antibiotics, and a fragile integumentary system predispose infants to TC [117, 127]. In a bid to highlight the possibility of a mother-to-infant transmission of dermatophytes, Mapelli et al. [111] reported the case of a 15-day-old newborn who was diagnosed with TC and whose mother had *Tinea corporis*. In both cases, mycological and molecular analyses shown that *T. rubrum* was the etiologic agent, highlighting the possibility of transmission during delivery or in the early days of life. Zaraa et al. [127] conducted a 12-year retrospective study of TC in an African hospital and reported that among a total number of 881 cases, 35 were infants with an average age of 20.16 months (3.97%). Males were the most affected group (57.1%), while *M. canis* was the most frequent etiologic agent (62%). Contact with animals was recorded in 28.6% of the cases, while family members presented with TC in only two cases (5.7%). Similarly, Meziou et al. [128] retrospectively reviewed 4003 cases of ringworm diagnosed over an 11-year period in an African hospital and reported a 6.12% prevalence of infant TC. Contact with domestic animals was reported in 2.9% of the cases, while a family history of dermatophytosis was found in 8.7% of the infected children. Mycological examination implicated *T. violaceum* as the major agent (51%), followed by *M. canis* (37%), *T. tonsurans* (8%), and *T. mentagrophytes* (4%).

Even when diagnosed properly, treatment of TC may still pose a dilemma in infants, and this may be due to their fragile immune system, the recalcitrant nature of TC, and the toxicity of the antifungal drugs [110]. Furthermore, there are relatively few treatment regimens for TC in infants, and in several countries, not many systemic drugs are approved for treatment of these infections. Thus, there is a need for proper mycological examination of suspected cases of infant TC in order to guide the treatment options.

3.3 Occurrence in Adults

Until recently, TC was considered to be an exclusive childhood infection, with rare occurrence in post-pubertal grown-ups. Adults usually have a number of features that protect them from TC. These include, but are not limited to, increased production of fungistatic acidic sebum, an increase in hair thickness, sweat, and the presence of a highly developed competitive microflora such as *Malassezia* spp. [129]. *Malassezia*, formerly known as *Pityrosporum*, are lipase-secreting organisms whose activities increase the availability of inhibitory fatty acids in the scalp [14]. Unfortunately, the prevalence of TC in the aging population is gradually rising, and this is attributed to the rising cases of geriatric chronic infections such as diabetes and hypertension, increased use of corticosteroids or immunosuppressants, and regular contact with prepubertal child carriers or pets [1, 5]. Furthermore, the low index suspicion of TC in adults, and the atypical clinical manifestations of the infection, hamper its diagnosis and treatment in adults [130].

Among 185 reports of TC in an Asian hospital, 44.3% were adults older than 20 years of age while the remaining were child patients [5]. Evaluating the cases of adult TC, the authors reported a higher incidence of the infection in the female participants (75.61%), with postmenopausal subjects being the most affected (93.5%). In a related study, Lova-Navarro et al. [130] retrospectively examined 289 cases of TC in southern Spain and found that 11.4% of the patients were adults. 90.9% of the adult patients were women, with most of them over the menopause age (70%). The predominance of TC among older female populations may be due to their reduced blood estrogen levels that concomitantly lead to a decrease in the production of fungi-inhibiting acidic sebum [5, 130].

Similar to pediatric TC, the etiologic agents of adult *Tinea capitis* differ from one country to another. Khosravi et al. [131] examined cases of TC in an Iranian dermatology center and found 20.6% of the cases in adults. *Trichophyton* species were implicated in 84% of the cases, with *T. violaceum* as the predominant species (32%). Similarly, *T. violaceum* was reported to be the predominant agent of adult TC in Egypt (56.9%), followed by *M. audouinii* (19%), *M. canis* (15.5%), and *Trichophyton schoenleinii* (8.6%) [132]. In a broader study, Mebazaa et al. [133] retrospectively examined 1137 cases of TC in Tunisia, with 5.27% of the cases (60/1137) affecting adults. Among them, *T. violaceum* was isolated in 60% of the cases, while *M. canis*, *T. schoenleinii*, and *T. verrucosum* represented 20%, 12% and 3.5% of the cases, respectively. The average age of the patients was 35 years, with females predominating (70%). Some studies have also reported *M. canis* as a predominant cause of adult TC [5, 134]. This pathogen can easily be transferred to humans through contact with pets such as dogs and cats [135]. This is mostly true for the aged population where longer contact with emotional support animals is the case, and where their weak defense systems may be unable to withstand the development of TC.

In addition to hormonal changes which were earlier mentioned as a predisposing factor to TC in aged women, their increased proximity to children and the use of shared hairdressing equipment have been noted by several authors as a predisposing factor to the infection [133, 136].

4 Current Diagnostic Methods

Patients who have alopecia or a mild scalp scaling should be examined immediately for evidence of TC (Fig. 2). The Wood's lamp, an instrument with an ultraviolet light of 320–400 nm spectrum, has classically been used as a preliminary tool in the diagnosis of TC. It is particularly helpful for ectothrix *Microsporum* infections whose members show a yellow-green fluorescence and *T. schoenleinii* which exudes a blue/dull green fluorescence under the lamp. Unfortunately, fluorescence is not a feature of most members of *Trichophyton* genus, hence a Wood's examination is of little significance in the diagnosis of TC caused by these species [17, 137].

In order to make a provisional diagnosis that may be enough to initiate treatment, a close inspection of the scalp and infected hair may be performed with the aid of a dermoscope—"the stethoscope of the dermatologist" [138]. The use of this method, known as trichoscopy, is not completely validated in the diagnosis of TC, and is based on the observation that the visual features of scalp ringworm differ based on the infecting dermatophyte. For instance, "comma-shaped," "corkscrewed," "black dots," and dystrophic hairs have been associated with different etiologic agents of TC [6, 139–145]. Recently, Dhaille et al. [146] highlighted comma-shaped hairs as a common trichoscopic feature of *Trichophyton* TC, while zigzag-shaped hairs were predominant in patients with TC caused by *Microsporum*. Also, in a bid to establish a classification system for the trichoscopic signs of TC, Aqil et al. [147] reported the association of V-shaped hair, crusts, erythema, and follicular pustules with

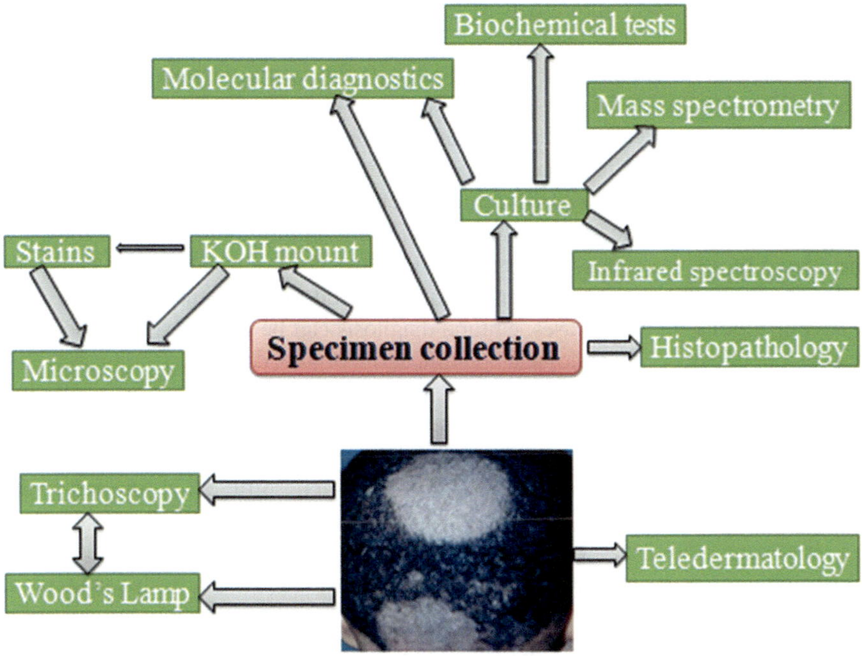

Fig. 2 Techniques used in the diagnosis of *Tinea capitis*

inflammatory TC, while noninflammatory TC was characterized mostly with scales and follicular keratosis. In another study, Lacarrubba et al. [148] used a high-magnification dermoscope and uncovered translucent, easily deformable and horizontal white bands as additional clinical signs of *M. canis* TC. As an innovation, Tang et al. [149] showed the use of a dermoscope attached with a portable Wood's lamp as a reliable tool to diagnose *M. canis* TC. In addition to the horizontal white bands which they observed, the bright green fluorescence of the hair shafts, typical of *Microsporum* infection, increased the confidence of the authors in their findings.

In both cases of Wood's light examination and trichoscopy, the disease and its etiologic agent need to be confirmed with direct microscopic examination and fungal cultures. Here, scalp or hair specimens from the infected area are painlessly collected with forceps, scalpel blade, toothbrush or needle, and transferred to a glass slide for potassium hydroxide (KOH) preparation or to appropriate media for fungal culture. The KOH slide preparation may be briefly heated over a flame or allowed to stay at room temperature for 15–20 min. In either case, the KOH dissolves the host cellular material, exposing the fungal hyphae and spores for rapid identification under the light microscope. Also, dyes such as Congo red, Chicago blue, Parker ink, chlorazole black, or calcofluor have recently been introduced onto KOH mounts, allowing an easier visualization of fungal elements by phase-contrast or fluorescence microscopy [137, 150–152].

On the other hand, culture-dependent diagnosis involves the inoculation of the specimen onto a selective medium and examining for the morphological or microscopic features of colonies that develop after 10–21 days of incubation. In order to control the population of fast-growing saprophytic contaminants, media used in culturing dermatophytes typically contain antibiotics such as gentamicin sulfate, chlortetracycline or chloramphenicol, and antifungals such as cycloheximide. In addition, they may contain pH indicators such as phenol red which changes color based on the metabolic state of the fungi. While culture-dependent diagnosis is hampered by its time-consuming nature and may give rise to a high percentage of false results, it still remains the conventional method for the definitive diagnosis of *Tinea* infections. In a few cases, cultural diagnosis can be taken a step further with the biochemical or physiological identification of the developed colonies. This entails the characterization of the fungi based on its mating type, urease activity, and nutritional requirements. A detailed review of this subject is found in Robert and Pihet [153]. Another challenge with cultural diagnosis is the demand for the presence of an experienced staff capable to recognize the subtle differential features of the colonies and to identify the fungus to the species level [154, 155].

However, with the advent of molecular diagnostics, identification of dermatophytes even to the strain-level has been made easier, faster, and more accurate. Commercially available molecular test kits, specific for dermatophyte identification, are gradually complementing or replacing conventional diagnostic methods [154, 156]. These molecular techniques, usually polymerase chain reaction (PCR)-based, employ the use of species-specific markers such as the internal transcribed spacers (ITS), mitochondrial DNA, and genes such as chitin synthase I, beta-tubulin and topoisomerase II [157]. In a study involving 155 suspected cases of dermatophytosis, nested PCR was used to amplify the chitin synthase gene, detecting a dermatophyte in 83.8% of the cases [158]. KOH microscopy had a positivity rate of 70% while fungal culture identified dermatophytes in 25.8%. In another study, qRT-PCR was shown to have more sensitivity than fungal culture and microscopy, giving rise to 21.2% more dermatophyte-positive results [159]. Also, multiplex PCR has recently been used by Trovato et al. [156] to implicate *T. tonsurans* as the etiologic agent of a case of TC, allowing the timely initiation of antifungal therapy. Sometimes, in order to enhance the sensitivity of dermatophyte detection in infected scalp or hair samples, PCR can be coupled with restriction fragment length polymorphism (PCR-RFLP) or enzyme-linked immunosorbent assay (PCR-ELISA) [68, 160]. A detailed review of the use of molecular biology techniques in the diagnosis of *Tinea* infections can be found in the literature [161, 162].

Finally, it is worthy to note that matrix-assisted laser desorption/ionization-time-of-flight/mass spectrometry (MALDI-TOF/MS) can now be used for the speedy identification of different dermatophyte species. This method identifies pathogens on the basis of the *m/z* ratios of highly abundant protein fragments but is presently restricted to the identification of pathogens which have already been cultured [163–166]. Fourier transform infrared spectroscopy has also been used in the identification of clinically relevant dermatophytes [153].

5 Complications of *Tinea capitis* Infection

The inflammatory subtypes of TC, especially kerions and favus, may lead to permanent scarring and hair loss of the affected scalp region. When TC is misdiagnosed or its treatment delayed, there are also chances of bacterial superinfection and folliculitis [13]. Dermatophytid reactions (id reaction)—a delayed hypersensitivity reaction characterized by eczematous lesions, papules, scaly patches, and pustules around the ears or other parts of the body—may also complicate the *Tinea* infection [6, 14]. It is worthy to note that id reaction can occur before or after initiation of systemic therapy, and its immediate recognition is expedient in order to distinguish it from adverse drug effects. The spread of the fungal pathogen from the scalp to other parts of the skin can also result in *Tinea corporis*. Even after undergoing standard treatment and clearing from the scalp, the infection may reoccur [110], and there is a possibility of a change in the skin color of the affected area [167].

6 Current Therapeutic Strategies

Over the years, a significant improvement in the treatment of TC has been observed. While this is partly due to the development of new drugs, other important innovations include the identification of species-specific response variations to different drugs, and the optimization of factors such as weight-dependent dosages, drug bioavailability, and treatment duration. Pending confirmatory diagnostic results, it might be helpful to initiate treatment on the basis of one or more clinical features of TC. In such cases, treatment should be initiated to reflect the history of the patient, immunologic status, clinical presentation, local epidemiology of TC, and the most likely etiologic agent [137, 167].

Since the 1960s, griseofulvin is seen as the "gold-standard" for treatment of TC [168, 169]. It is a fungistatic drug which interferes with the function of fungal microtubules, arresting cell division and cell wall synthesis [137]. Clinically, griseofulvin is the most effective antifungal agent against ectothrix *Microsporum* infections, particularly *M. canis* and *M. audouinii*, and its absorption and bioavailability are significantly improved with dietary fat intake. While there are scarce reports of in vitro resistance of dermatophytes to griseofulvin, clinical administration of the drug was shown to be less effective against endothrix *Trichophyton* species [137, 167, 170]. Typically, griseofulvin is given for a period of 6–8 weeks in cases of *Microsporum* infection, while 12–18 weeks are recommended for *Trichophyton* infection [137]. The long treatment duration of griseofulvin leads to less compliance in patients and ultimately contributes to treatment failures [168]. While griseofulvin is considered a first-line drug for TC, it is important to point out that newer drugs such as terbinafine and azole antifungals are increasingly being used to replace it [169]. In addition to the long treatment course, the high cost of griseofulvin and its limited cure rate can also be blamed for the decline in its use.

Terbinafine and azole antifungals are considered as cost-effective therapies, which are able to achieve high cure rates under a short treatment course. They act by inhibiting the synthesis of membrane ergosterol, leading to the leakage of the intracellular contents of the fungal pathogen. Terbinafine is usually fungicidal but is more effective against *Trichophyton* than *Microsporum* infections. It has good absorption and retention rates in the body and is usually given for a short period of 4 weeks, although a longer course and a much higher dosage may be required for ectothrix infections [6, 167, 170]. Itraconazole, fluconazole, and in some countries, ketoconazole, are major azole drugs used in the treatment of TC. Depending on their tissue concentration, they may be fungistatic or fungicidal and may match the efficacy rate of griseofulvin, especially for *Trichophyton* infections [167, 171].

When used without a corresponding systemic therapy, topical antifungal drugs are of little advantage in the treatment of TC. While the clinical appearance may initially improve with the administration of topical drugs, dermatophytes at the root of the hair follicle are usually unaffected, and this leads to a high relapse rate of the infection [6, 167]. Nonetheless, novel drug delivery methods such as nanoparticle formulations have been advocated as a strategy to increase the skin penetrating ability of topical antifungals and enhance their therapeutic efficacy [18]. Also, some topical antifungal agents have sporicidal activity and are known to limit the transmission of fungal spores in both infected patients and carriers. Topical antifungal agents mostly come in the form of shampoos, creams or lotions, and they include but are not limited to povidone-iodine, ketoconazole, ciclopirox, selenium sulfide, and zinc pyrithione [6, 137, 168].

Though this is still controversial, oral or topical corticosteroids can be used to reduce itching and general discomfort that comes with inflammatory TC [137, 167]. Also, while some studies have reported the use of wet compresses or surgical means to remove crusts in scalp kerions [172, 173], others have shown that surgical incision does not contribute to the healing process and may even complicate the local treatment of the infection [174–176]. Furthermore, the possibility of a secondary bacterial infection should be properly considered and subsequently treated with requisite topical or oral antibiotics [6].

Finally, with the emergence of antifungal resistance among dermatophytes, it might be helpful to consider their antifungal susceptibility profile prior to the initiation of therapy [64, 177]. Among a number of methods, agar-based disk diffusion method remains a good option for the in vitro susceptibility testing of dermatophytes. It is inexpensive, easy to perform, fast, and does not demand a particular expertise or the use of specialized equipment. Mueller-Hinton medium, usually without a supplementary carbohydrate source or dye, is reported as a typical medium used in the susceptibility testing of dermatophytes [178]. This medium supports the growth of the fungi and allows easy reading of the inhibition zone edges.

7 Community Management

With the advent of the "internet" age and an increase in migration, the world is gradually becoming a global village. In the absence of effective health policies, some infectious diseases which are thought to be geographically restricted can quickly spread from one region to another. As mentioned earlier, TC can spread through direct skin contact, or indirectly through fomites. Thus, a concerted public health effort is needed to ensure its effective management and combat its spread.

It is common knowledge that schoolchildren are the most affected age group, and infected children can quickly transmit the infection to their classmates, resulting in school outbreaks. The community can enact policies to ensure that children with signs of scalp ringworm are immediately diagnosed and placed under appropriate therapy. The training of schoolteachers or school health officials might be necessary to help in the early identification of cases at the school level. While the suspension of the infected child from school is still controversial, it is generally agreed that placing the child under a compulsory systemic and adjuvant topical therapy will highly reduce the risk of transmission. Infected children should also be asked not to share objects such as helmets, scarves, hairbrushes, combs, and toys, as these may become fomites for the transmission of the pathogen. The child may also be prohibited from participating in games where a prolonged close physical contact is the norm rather than the exception, for example, wrestling [167]. While enforcing these school policies, it is expedient to note that TC is a skin disease affecting the aesthetics and self-confidence of the patient. Therefore, the psychological implications of the disease should always be considered when implementing the health policy [7].

Furthermore, a regular investigation of TC among schoolchildren might prove effective in the early identification and management of the infection [73, 85]. This routine epidemiological surveillance is particularly necessary since some studies have suggested a higher risk of infected cases in classrooms where there are many carriers, although it is important to quickly point out that this might be an effect rather than a cause [179]. It might also be expedient to carry out a thorough examination of places such as playgrounds and other areas used by children as playgrounds, especially when one or more children have been reported with the disease [180, 181].

It is no doubt that poverty, overcrowding, and illiteracy are underlying social determinants of TC [6, 7, 182]. In fact, it was previously regarded as a "mark of poverty and destitution" in some countries [6]. Therefore, community actions targeting poverty, education, and housing are important in addressing the infection. Also, governmental and nongovernmental agencies should do more to educate the public on the etiology, spread, and control of the superficial infection [183].

Moreover, the nonlife threatening nature of TC and the cost of seeking healthcare to treat it tend to discourage patients from seeking timely clinical intervention. Such patients may resort to self-medication, tentatively leading to the spread of the infection as well as to a possible increase in the antifungal drug resistance of the pathogens. An essential measure in developing nations and in economically disadvantaged regions of otherwise developed countries is to establish more outdoor subsidized dermatological centers to cater for the health needs of the community. In areas where skin doctors are unavailable, primary healthcare providers should be made aware that the definitive endpoint for TC treatment is mycological clearance, instead of clinical cure. Therefore, a follow-up with adequate mycological screening is needed at the end of the standard treatment duration and should be continued until the documentation of a mycological cure [137].

When a zoophilic dermatophyte such as *M. canis* or *T. verrucosum* is the identified cause of TC in a family or community, the animal source should be immediately identified and properly treated [6, 182]. Proactively, communities should initiate policies that will ensure regular dermatological screening of pets such as dogs, cats, and rabbits, and restrict the uncontrolled movement of stray animals.

8 Recommendations for Future Studies

The virulence factors of dermatophytes are not well-known. More studies, especially at the molecular level, should be done to elucidate the pathogenesis of dermatophytes as well as the resulting clinical signs of the infection. This will help in uncovering loopholes that can be exploited in the rational development of new therapeutic and prophylactic strategies to counter the infection. Also, with the restrictive use of oral therapy in infants, the scientific community should look towards improving the bio-delivery and therapeutic efficacy of topical drugs for TC. More studies are also needed to understand the impact of corticosteroids in the treatment of this disease.

Till date, the high predisposition of prepubescent children and post-menopausal women to TC has not been fully grasped, demanding more attention from the scientific community. While TC is regarded as a pediatric infection which affects the self-confidence of the patients, there are relatively scarce reports on the psycho-social impact of the disease in children. Therefore, pediatric dermatologists and mental health experts are encouraged to take up collaborative projects to understand the psychological impact of pediatric TC. Studies are also needed to investigate the role of climate and other environmental conditions in the epidemiologic spread of TC. Finally, more studies need to be done to enhance the routine use of molecular biology techniques in the diagnosis of these infections and the diagnostic accuracy of trichoscopy should be validated with studies involving large numbers of participants.

Acknowledgement The authors are grateful to Aaron Kren and Shelby Shearer from the Department of Biology, Wayne State University, for their helpful suggestions at the early stages of the manuscript preparation.

References

1. Faway E, de Rouvroit CL, Poumay Y. *In vitro* models of dermatophyte infection to investigate epidermal barrier alterations. Exp Dermatol. 2018;27(8):915–22.
2. Martinez-Rossi NM, Peres NTA, Rossi A. Pathogenesis of dermatophytosis: sensing the host tissue. Mycopathologia. 2017;182(1–2):215–27.
3. Esquenazi D, Alviano CS, de Souza W, Rozental S. The influence of surface carbohydrates during *in vitro* infection of mammalian cells by the dermatophyte *Trichophyton rubrum*. Res Microbiol. 2004;155(3):144–53.
4. Ginter-Hanselmayer G, Weger W, Ilkit M, Smolle J. Epidemiology of Tinea capitis in Europe: current state and changing patterns. Mycoses. 2007;50(2):6–13.
5. Park SK, Park SW, Yun SK, Kim HU, Park J. Tinea capitis in adults: a 18-year retrospective, single-centre study in Korea. Mycoses. 2019;62(7):609–16.
6. Hay RJ. Tinea capitis: current status. Mycopathologia. 2017;182(1):87–93.
7. Nweze EI, Eke IE. Dermatophytes and dermatophytosis in the eastern and southern parts of Africa. Med Mycol. 2018;56(1):13–28.
8. Robati AK, Khalili M, Hazaveh SJH, Bayat M. Assessment of the subtilisin genes in *Trichophyton rubrum* and *Microsporum canis* from dermatophytosis. Comp Clin Pathol. 2018;27(5):1343–7.
9. Esquenazi D, de Souza W, Alviano CS, Rozental S. The role of surface carbohydrates on the interaction of microconidia of *Trichophyton mentagrophytes* with epithelial cells. FEMS Immunol Med Microbiol. 2003;35(2):113–23.
10. Kaufman G, Horwitz BA, Duek L, Ullman Y, Berdicevsky I. Infection stages of the dermatophyte pathogen *Trichophyton*: microscopic characterization and proteolytic enzymes. Med Mycol. 2007;45(2):149–55.
11. Baldo A, Mathy A, Tabart J, Camponova P, Vermout S, Massart L, et al. Secreted subtilisin Sub3 from *Microsporum canis* is required for adherence to but not for invasion of the epidermis. Br J Dermatol. 2010;162(5):990–7.
12. Baldo A, Tabart J, Vermout S, Mathy A, Collard A, Losson B, et al. Secreted subtilisins of *Microsporum canis* are involved in adherence of arthroconidia to feline corneocytes. J Med Microbiol. 2008;57:1152–6.
13. Băguţ ET, Baldo A, Mathy A, Cambier L, Antoine N, Cozma V, et al. Subtilisin Sub3 is involved in adherence of *Microsporum canis* to human and animal epidermis. Vet Microbiol. 2012;160(3–4):413–9.
14. Tainwala R, Sharma YK. Pathogenesis of dermatophytoses. Indian J Dermatol. 2011;56 (3):259–61.
15. Vermout S, Tabart J, Baldo A, Mathy A, Losson B, Mignon B. Pathogenesis of dermatophytosis. Mycopathologia. 2008;166:267–75.
16. de Hoog GS, Dukik K, Monod M, Packeu A, Stubbe D, Hendrickx M, et al. Toward a novel multilocus phylogenetic taxonomy for the dermatophytes. Mycopathologia. 2017;182 (1–2):5–31.
17. Elewski BE. Tinea capitis: a current perspective. J Am Acad Dermatol. 2000;42(1):1–20.
18. Bangia R, Sharma G, Dogra S, Katare OP. Nanotechnological interventions in dermatophytosis: from oral to topical, a fresh perspective. Expert Opin Drug Deliv. 2019;16 (4):377–96.
19. Farag AGA, Hammam MA, Ibrahem RA, Mahfouz RZ, Elnaidany NF, Qutubuddin M, et al. Epidemiology of dermatophyte infections among school children in Menoufia Governorate, Egypt. Mycoses. 2018;61(5):321–5.
20. Mirmirani P, Tucker LY. Epidemiologic trends in pediatric Tinea capitis: a population-based study from Kaiser Permanente Northern California. J Am Acad Dermatol. 2013;69(6):916–21.
21. Prohic A. An epidemiological survey of Tinea capitis in Sarajevo, Bosnia and Herzegovina over a 10-year period. Mycoses. 2008;51(2):161–4.

22. Nawaf AM, Joshi A, Zaki A, Nour-Eldin O, Al-Sheltawy M, El-Adawy I, et al. Tinea capitis among children and adolescents in the Farwaniya region of Kuwait. J Dermatol. 2003;30(12):904–9.
23. Kallel A, Hdider A, Fakhfakh N, Belhadj S, Belhadj-Salah N, Bada N, et al. Tinea capitis: main mycosis child. Epidemiological study on 10 years. J Mycol Med. 2017;27(3):345–50.
24. Moto JN, Maingi JM, Nyamache AK. Prevalence of Tinea capitis in school going children from Mathare, informal settlement in Nairobi, Kenya. BMC Res Notes. 2015;8:274.
25. Fulgence KK, Abibatou K, Vincent D, Henriette V, Etienne AK, Kiki-Barro PC, et al. Tinea capitis in schoolchildren in southern Ivory Coast. Int J Dermatol. 2013;52(4):456–60.
26. Ellabib MS, Agaj M, Khalifa Z, Kavanagh K. *Trichophyton violaceum* is the dominant cause of Tinea capitis in children in Tripoli, Libya: results of a two year survey. Mycopathologia. 2002;153(3):145–7.
27. El Mezouari E, Hocar O, Atarguine H, Akhdari N, Amal S, Moutaj R. Tinea capitis in the military hospital Avicenna (Morocco): review of 8 years (2006-2013). J Mycol Med. 2016;26(1):e1–5.
28. Coloe JR, Diab M, Moennich J, Diab D, Pawaskar M, Balkrishnan R, et al. Tinea capitis among children in the Columbus area, Ohio, USA. Mycoses. 2010;53(2):158–62.
29. East-Innis A, Rainford L, Dunwell P, Barrett-Robinson D, Nicholson AM. The changing pattern of Tinea capitis in Jamaica. West Indian Med J. 2006;55(2):85–8.
30. Marcoux D, Dang J, Auguste H, McCuaig C, Powell J, Hatami A, et al. Emergence of African species of dermatophytes in Tinea capitis: a 17-year experience in a Montreal pediatric hospital. Pediatr Dermatol. 2018;35(3):323–8.
31. Welsh O, Welsh E, Ocampo-Candiani J, Gomez M, Vera-Cabrera L. Dermatophytoses in Monterrey, México. Mycoses. 2006;49(2):119–23.
32. López-Martínez R, Manzano-Gayosso P, Hernández-Hernández F, Bazán-Mora E, Méndez-Tovar LJ. Dynamics of dermatophytosis frequency in Mexico: an analysis of 2084 cases. Med Mycol. 2010;48(3):476–9.
33. Raccurt CP, Dorsainvil D, Boncy M, Boncy J, Auguste G. The emergence of *Trichophyton tonsurans* in Port-au-Prince, Haiti. Med Mycol. 2009;47(2):197–200.
34. Veasey JV, Miguel BAF, Mayor SAS, Zaitz C, Muramatu LH, Serrano JA. Epidemiological profile of Tinea capitis in São Paulo City. An Bras Dermatol. 2017;92(2):283–4.
35. Capote AM, Ferrara G, Panizo MM, García N, Alarcón V, Reviakina V, et al. Superficial mycoses: casuistry of the mycology Department of the Instituto Nacional de Higiene "Rafael Rangel", Caracas, Venezuela (2001-2014). Investig Clin. 2016;57(1):47–58.
36. Cortés GA, Mardones VF, Zemelman DV. Aetiology of childhood alopecia. Rev Chil Pediatr. 2015;86(4):264–9.
37. Alvarado Z, Pereira C. Fungal diseases in children and adolescents in a referral Centre in Bogota, Colombia. Mycoses. 2018;61(8):543–8.
38. Santos PE, Córdoba S, Rodero LL, Carrillo-Muñoz AJ, Lopardo HA. Tinea capitis: two years experience in a paediatric hospital of Buenos Aires, Argentina. Rev Iberoam Micol. 2010;27(2):104–6.
39. Mazza M, Refojo N, Davel G, Lima N, Dias N, Passos da Silva CMF, et al. Epidemiology of dermatophytoses in 31 municipalities of the province of Buenos Aires, Argentina: a 6-year study. Rev Iberoam Micol. 2018;35(2):97–102.
40. Moraes MS, Godoy-Martínez P, Alchorne MM, Boatto HF, Fischman O. Incidence of Tinea capitis in São Paulo, Brazil. Mycopathologia. 2006;162(2):91–5.
41. Timen A, Bovée L, Leentvaar-Kuijpers A, Peerbooms PG, Coutinho RA. Tinea capitis in primary school age children in southeastern Amsterdam: primarily due to *Trichophyton tonsurans*. Ned Tijdschr Geneeskd. 1999;143(1):24–7.
42. Fuller LC, Child FC, Midgley G, Higgins EM. Scalp ringworm in south-East London and an analysis of a cohort of patients from a paediatric dermatology department. Br J Dermatol. 2003;148(5):985–8.

43. Buckley DA, Brockwell M, Leeming JG, Moss C. Scalp ringworm epidemic in Birmingham: a decline? Clin Exp Dermatol. 2000;25(4):349.
44. Hackett BC, O'Connell K, Cafferkey M, O'Donnell BF, Keane FM. Tinea capitis in a paediatric population. Ir Med J. 2006;99(10):294–5.
45. Nasir S, Ralph N, O'Neill C, Cunney R, Lenane P, O'Donnell B. Trends in Tinea capitis in an Irish pediatric population and a comparison of scalp brushings versus scalp scrapings as methods of investigation. Pediatr Dermatol. 2014;31(5):622–3.
46. Gits-Muselli M, Benderdouche M, Hamane S, Mingui A, Feuilhade de Chauvin M, Guigue N, et al. Continuous increase of *Trichophyton tonsurans* as a cause of Tinea capitis in the urban area of Paris, France: a 5-year-long study. Med Mycol. 2017;55(5):476–84.
47. Alshawa K, Lacroix C, Benderdouche M, Mingui A, Derouin F, Feuilhade de Chauvin M. Increasing incidence of *Trichophyton tonsurans* in Paris, France: a 15-year retrospective study. Br J Dermatol. 2012;166(5):1149–50.
48. Heikkilä H, Stubb S. Ringworm of the scalp among immigrants in Finland. Acta Derm Venereol. 2004;84(4):333–4.
49. Drakensjö IT, Chryssanthou E. Epidemiology of dermatophyte infections in Stockholm, Sweden: a retrospective study from 2005-2009. Med Mycol. 2011;49(5):484–8.
50. Calabrò G, Patalano A, Fiammenghi E, Chianese C. Tinea capitis in Campania, Italy: a 9-year retrospective study. G Ital Dermatol Venereol. 2015;150(4):363–7.
51. Asticcioli S, Di Silverio A, Sacco L, Fusi I, Vincenti L, Romero E. Dermatophyte infections in patients attending a tertiary care hospital in northern Italy. New Microbiol. 2008;31(4):543–8.
52. Ziegler W, Lempert S, Goebeler M, Kolb-Mäurer A. [Tinea capitis: pathogen spectrum and epidemiology in temporal change]. J Dtsch Dermatol Ges. 2016;14(8):818–26.
53. Chokoeva AA, Zisova L, Sotiriou E, Miteva-Katrandzhieva T. Tinea capitis: a retrospective epidemiological comparative study. Wien Med Wochenschr. 2017;167(3–4):51–7.
54. del Boz J, Crespo V, Rivas-Ruiz F, de Troya M. A 30-year survey of paediatric Tinea capitis in southern Spain. J Eur Acad Dermatol Venereol. 2011;25(2):170–4.
55. Binder B, Lackner HK, Poessl BD, Propst E, Weger W, Smolle J, et al. Prevalence of Tinea capitis in southeastern Austria between 1985 and 2008: up-to-date picture of the current situation. Mycoses. 2011;54(3):243–7.
56. Kaštelan M, Utješinović-Gudelj V, Prpić-Massari L, Brajac I. Dermatophyte infections in Primorsko-Goranska county, Croatia: a 21-year survey. Acta Dermatovenerol Croat. 2014;22 (3):175–9.
57. Lange M, Nowicki R, Barańska-Rybak W, Bykowska B. Dermatophytosis in children and adolescents in Gdansk, Poland. Mycoses. 2004;47(7):326–9.
58. Saunte DM, Svejgaard EL, Haedersdal M, Frimodt-Møller N, Jensen AM, Arendrup MC. Laboratory-based survey of dermatophyte infections in Denmark over a 10-year period. Acta Derm Venereol. 2008;88(6):614–6.
59. Dolenc-Voljc M. Dermatophyte infections in the Ljubljana region, Slovenia, 1995-2002. Mycoses. 2005;48(3):181–6.
60. Paškevičius A, Švedienė J. Distribution and species composition of causative agents of dermatophytoses in Lithuania. Acta Dermatovenerol Croat. 2013;21(2):99–104.
61. Starova A, Balabanova-Stefanova M, V'lckova-Laskoska M. Dermatophytes in republic of Macedonia. Prilozi. 2010;31(1):317–26.
62. Benmezdad A, Moulahem T, Benyezzar M, Djaballah M, Beldjoudi W, Fendri AH. Tinea capitis in the university hospital of Constantine (Algeria). J Mycol Med. 2012;22(4):354–6.
63. Hamroune Z, Mazouz A, Benelmouffok AB, Kellou D. Evolution of Tinea capitis observed in mycology laboratory of Institute Pasteur of Algeria from 1995 to 2015. J Mycol Med. 2016;26 (4):337–44.
64. Doss RW, El-Rifaie AA, Radi N, El-Sherif AY. Antimicrobial susceptibility of Tinea capitis in children from Egypt. Indian J Dermatol. 2018;63(2):155–9.
65. Zaki SM, Ibrahim N, Aoyama K, Shetaia YM, Abdel-Ghany K, Mikami Y. Dermatophyte infections in Cairo, Egypt. Mycopathologia. 2009;167(3):133–7.

66. Elmaataoui A, Zeroual Z, Lyagoubi M, Aoufi S. Tinea capitis etiology in Ibn Sina Hospital in Rabat (Morocco). J Mycol Med. 2012;22(3):261–4.
67. Saghrouni F, Bougmiza I, Gheith S, Yaakoub A, Gaïed-Meksi S, Fathallah A, et al. Mycological and epidemiological aspects of Tinea capitis in the Sousse region of Tunisia. Ann Dermatol Venereol. 2011;138(8–9):557–63.
68. Wiegand C, Mugisha P, Mulyowa GK, Elsner P, Hipler UC, Gräser Y, et al. Identification of the causative dermatophyte of Tinea capitis in children attending Mbarara Regional Referral Hospital in Uganda by PCR-ELISA and comparison with conventional mycological diagnostic methods. Med Mycol. 2017;55(6):660–8.
69. Teklebirhan G, Bitew A. Prevalence of dermatophytic infection and the spectrum of dermatophytes in patients attending a tertiary hospital in Addis Ababa, Ethiopia. Int J Microbiol. 2015;2015:653419.
70. Bitew A. Dermatophytosis: prevalence of dermatophytes and non-dermatophyte fungi from patients attending Arsho Advanced Medical Laboratory, Addis Ababa, Ethiopia. Dermatol Res Pract. 2018;2018:8164757.
71. Thakur R. Spectrum of dermatophyte infections in Botswana. Clin Cosmet Investig Dermatol. 2015;8:127–33.
72. Morar N, Dlova NC, Gupta AK, Aboobaker J. Tinea capitis in Kwa-Zulu Natal, South Africa. Pediatr Dermatol. 2004;21(4):444–7.
73. Sidat MM, Correia D, Buene TP. Tinea capitis among children at one suburban primary school in the City of Maputo, Mozambique. Rev Soc Bras Med Trop. 2007;40(4):473–5.
74. Carod JF, Ratsitorahina M, Raherimandimby H, Hincky Vitrat V, Ravaolimalala Andrianaja V, Contet-Audonneau N. Outbreak of Tinea capitis and corporis in a primary school in Antananarivo, Madagascar. J Infect Dev Ctries. 2011;5(10):732–6.
75. Komba EV, Mgonda YM. The spectrum of dermatological disorders among primary school children in Dar es Salaam. BMC Public Health. 2010;10:765.
76. Ayaya SO, Kamar KK, Kakai R. Aetiology of Tinea capitis in school children. East Afr Med J. 2001;78(10):531–5.
77. Ayanbimpe GM, Taghir H, Diya A, Wapwera S. Tinea capitis among primary school children in some parts of Central Nigeria. Mycoses. 2008;51(4):336–40.
78. Ngwogu AC, Otokunefor TV. Epidemiology of dermatophytoses in a rural community in Eastern Nigeria and review of literature from Africa. Mycopathologia. 2007;164:149–58.
79. Hogewoning AA, Adegnika AA, Bouwes Bavinck JN, Yazdanbakhsh M, Kremsner PG, van der Raaij-Helmer EM, et al. Prevalence and causative fungal species of Tinea capitis among schoolchildren in Gabon. Mycoses. 2011;54(5):e354–9.
80. Diongue K, Diallo MA, Ndiaye M, Badiane AS, Seck MC, Diop A, et al. Causative agents of superficial mycoses isolated in Dakar, Senegal: retrospective study from 2011 to 2015. J Mycol Med. 2016;26(4):368–76.
81. Kechia FA, Kouoto EA, Nkoa T, Nweze EI, Fokoua DC, Fosso S, et al. Epidemiology of Tinea capitis among school-age children in Meiganga, Cameroon. J Mycol Med. 2014;24 (2):129–34.
82. Coulibaly O, Kone AK, Niaré-Doumbo S, Goïta S, Gaudart J, Djimdé AA, et al. Dermatophytosis among schoolchildren in three eco-climatic zones of Mali. PLoS Negl Trop Dis. 2016;10(4):e0004675.
83. Adesiji YO, Omolade FB, Aderibigbe IA, Ogungbe O, Adefioye OA, Adedokun SA, et al. Prevalence of Tinea capitis among children in Osogbo, Nigeria, and the associated risk factors. Diseases. 2019;7(1). pii:E13.
84. Ayanlowo O, Akinkugbe A, Oladele R, Balogun M. Prevalence of Tinea capitis infection among primary school children in a rural setting in south-West Nigeria. J Public Health Afr. 2014;5(1):349.
85. Nweze EI, Okafor JI. Prevalence of dermatophytic fungal infections in children: a recent study in Anambra State, Nigeria. Mycopathologia. 2005;160:239–43.

86. Cisse M, Diare FS, Kaba A, Magassouba E, Keïta M, Ecra EJ. [Tinea capitis in department of dermatology and venerology in the University hospital of Donka at Conakry, Guinea]. Bull Soc Pathol Exot. 2006;99(1):32–3. In French.
87. Hogewoning AA, Duijvestein M, Boakye D, Amoah AS, Obeng BB, van der Raaij-Helmer EM, et al. Prevalence of symptomatic Tinea capitis and associated causative organisms in the Greater Accra Region, Ghana. Br J Dermatol. 2006;154(4):784–6.
88. Razzaq Adel AA, Sultan AO, Basmiah AM, Aftab A, Nabel N. Prevalence of Tinea capitis in southern Kuwait. Mycoses. 2007;50(4):317–20.
89. Ali-Shtayeh MS, Arda HM, Abu-Ghdeib SI. Epidemiological study of Tinea capitis in schoolchildren in the Nablus area (West Bank). Mycoses. 1998;41(5–6):243–8.
90. Ali-Shtayeh MS, Yaish S, Jamous RM, Arda H, Husein EI. Updating the epidemiology of dermatophyte infections in Palestine with special reference to concomitant dermatophytosis. J Mycol Med. 2015;25(2):116–22.
91. Mashiah J, Kutz A, Ben Ami R, Savion M, Goldberg I, Gan Or T, et al. Tinea capitis outbreak among paediatric refugee population, an evolving healthcare challenge. Mycoses. 2016;59 (9):553–7.
92. Al-Duboon AH, Muhsin TM, Al-Rubaiy KK. Tinea capitis in Basrah. Iraq Mycoses. 1999;42 (4):331–3.
93. Mahmoud AL. A study of dermatophytoses in Sana'a, Yemen Republic. Mycoses. 2002;45 (3–4):105–8.
94. Abanmi A, Bakheshwain S, El Khizzi N, Zouman AR, Hantirah S, Al Harthi F, et al. Characteristics of superficial fungal infections in the Riyadh region of Saudi Arabia. Int J Dermatol. 2008;47(3):229–35.
95. Abu-Elteen KH, Abdul Malek M. Prevalence of dermatophytoses in the Zarqa district of Jordan. Mycopathologia. 1999;45(3):137–42.
96. Abu Shaqra QM, Al Momani W. Cases of Tinea capitis as encountered in a private practice laboratory from Jordan. J Mycol Med. 2011;21(1):24–7.
97. Bassiri Jahromi S, Khaksar AA. Aetiological agents of Tinea capitis in Tehran (Iran). Mycoses. 2006;49(1):65–7.
98. Sadeghi Tafti H, Falahati M, Kordbachch P, Mahmoudi M, Safara M, Rashidian S, et al. A survey of the etiological agents of scalp and nail dermatophytosis in Yazd, Iran in 2014-2015. Curr Med Mycol. 2015;1(4):1–6.
99. Cai W, Lu C, Li X, Zhang J, Zhan P, Xi L, et al. Epidemiology of superficial fungal infections in Guangdong, Southern China: a retrospective study from 2004 to 2014. Mycopathologia. 2016;181(5–6):387–95.
100. Shinoda H, Nishimoto K. Clinical study of 57 cases of infection with *Trichophyton tonsurans* examined at a dermatology clinic in Saga Prefecture, Japan. Nippon Ishinkin Gakkai Zasshi. 2007;48(2):79–84.
101. Bhat YJ, Zeerak S, Kanth F, Yaseen A, Hassan I, Hakak R. Clinicoepidemiological and mycological study of Tinea capitis in the pediatric population of Kashmir Valley: a study from a tertiary care centre. Indian Dermatol Online J. 2017;8(2):100–3.
102. Grover C, Arora P, Manchanda V. Tinea capitis in the pediatric population: a study from North India. Indian J Dermatol Venereol Leprol. 2010;76(5):527–32.
103. Vyas A, Pathan N, Sharma R, Vyas L. A clinicomycological study of cutaneous mycoses in Sawai Man Singh Hospital of Jaipur, North India. Ann Med Health Sci Res. 2013;3(4):593–7.
104. Jahangir M, Hussain I, Khurshid K, Haroon TS. A clinico-etiologic correlation in Tinea capitis. Int J Dermatol. 1999;38(4):275–8.
105. Balci E, Gulgun M, Babacan O, Karaoglu A, Kesik V, Yesilkaya S, et al. Prevalence and risk factors of Tinea capitis and Tinea pedis in school children in Turkey. J Pak Med Assoc. 2014;64(5):514–8.
106. Chadeganipour M, Mohammadi R, Shadzi S. A 10-year study of dermatophytoses in Isfahan, Iran. J Clin Lab Anal. 2016;30(2):103–7.

107. Mikaeili A, Kavoussi H, Hashemian AH, Shabandoost Gheshtemi M, Kavoussi R. Clinico-mycological profile of Tinea capitis and its comparative response to griseofulvin versus terbinafine. Curr Med Mycol. 2019;5(1):15–20.
108. Jha BN, Garg VK, Agrawal S, Khanal B, Agarwalla A. Tinea capitis in Eastern Nepal. Int J Dermatol. 2006;45(2):100–2.
109. Zampella JG, Kwatra SG, Blanck J, Cohen B. Tinea in tots: cases and literature review of oral antifungal treatment of Tinea capitis in children under 2 years of age. J Pediatr. 2017;183:12–8.
110. Mandras N, Roana J, Cervetti O, Panzone M, Tullio V. A case report of Tinea capitis in infant in first year of life. BMC Pediatr. 2019;19(1):65.
111. Mapelli ET, Borghi E, Cerri A, Sciota R, Morace G, Menni S. Tinea corporis due to *Trichophyton rubrum* in a woman and Tinea capitis in her 15-day-old baby: molecular evidence of vertical transmission. Mycopathologia. 2012;173(2–3):135–8.
112. Valari M, Stefanaki C, Karteri G, Zarkadi E, Katsambas A. Tinea capitis due to *Trichophyton rubrum* in a 3-month-old infant. Mycoses. 2006;49(5):439–40.
113. Ravenscroft J, Goodfield MJ, Evans EG. *Trichophyton tonsurans* Tinea capitis and Tinea corporis: treatment and follow-up of four affected family members. Pediatr Dermatol. 2000;17(5):407–9.
114. Koumantaki E, Georgala S, Rallis E, Papadavid E. *Microsporum canis* Tinea capitis in an 8-month-old infant successfully treated with 2 weekly pulses of oral itraconazole. Pediatr Dermatol. 2001;18(1):60–2.
115. Romano C. Case reports. Four paediatric cases of Tinea capitis due to unusual agents. Mycoses. 1999;42(5–6):421–5.
116. Blömer RH, Keilani N, Faber A, Rodeck B, Krüger C, Uhrlaß S, et al. Tinea capitis profunda due to *Trichophyton verrucosum* with cMRSA superinfection in an infant. Hautarzt. 2012;63(8):648–52.
117. Atanasovski M, El Tal AK, Hamzavi F, Mehregan DA. Neonatal dermatophytosis: report of a case and review of the literature. Pediatr Dermatol. 2011;28(2):185–8.
118. Michaels BD, Del Rosso JQ. Tinea capitis in infants: recognition, evaluation, and management suggestions. J Clin Aesthet Dermatol. 2012;5(2):49–59.
119. Larralde M, Gomar B, Boggio P, Abad ME, Pagotto B. Neonatal kerion celsi: report of three cases. Pediatr Dermatol. 2010;27(4):361–3.
120. Mosseri R, Finkelstein Y, Garty BZ. Topical treatment of Tinea capitis in a neonate. Cutis. 2002;69(2):88–90.
121. Chang SE, Kang SK, Choi JH, Sung KJ, Moon KC, Koh JK. Tinea capitis due to *Trichophyton rubrum* in a neonate. Pediatr Dermatol. 2002;19(4):356–8.
122. Ambooken B, Binitha MP, Chandran B. Kerion due to *Microsporum gypseum* in a 1-month-old infant. Int J Trichol. 2013;5(4):214–6.
123. Balasubramanian S, Vindhiya K, Dhanalakshmi K, Ramkumar R. Kerion in a neonate. Indian J Paediatr Dermatol. 2017;18(4):333–4.
124. Kobayashi M, Yamamoto O, Suenaga Y, Asahi M. Kerion celsi in an infant treated with oral terbinafine. J Dermatol. 2001;28(2):108–9.
125. Betlloch-Mas I, Albares-Tendero MP, Soro-Martínez MP, Pérez-Crespo M. Persistent desquamation of the scalp as a manifestation of Tinea capitis in sub-Saharan children. J Immigr Minor Health. 2015;17(5):1588–90.
126. Adefemi SA. Tinea capitis in a 21-day old neonate. SA Fam Pract. 2010;52(5):428–9.
127. Zaraa I, Hawilo A, Trojjet S, El Euch D, Mokni M, Ben Osman A. Letter: Tinea capitis in infants in their first 2 years of life: a 12-year study and a review of the literature. Dermatol Online J. 2012;18(7):16.
128. Meziou TJ, Dammak A, Zaz T, Mseddi M, Boudaya S, Bouzid L, et al. Scalp ringworm Tinea capitis in Tunisian infants. Med Mal Infect. 2011;41(9):486–8.
129. Narang K, Pahwa M, Ramesh V. Tinea capitis in the form of concentric rings in an HIV positive adult on antiretroviral treatment. Indian J Dermatol. 2012;57(4):288–90.

130. Lova-Navarro M, Gómez-Moyano E, Martínez Pilar L, Fernandez-Ballesteros MD, Godoy-Díaz DJ, Vera-Casaño A, et al. Tinea capitis in adults in southern Spain. A 17-year epidemiological study. Rev Iberoam Micol. 2016;33(2):110–3.
131. Khosravi AR, Shokri H, Vahedi G. Factors in etiology and predisposition of adult Tinea capitis and review of published literature. Mycopathologia. 2016;181(5–6):371–8.
132. El-Khalawany M, Shaaban D, Hassan H, Abdalsalam F, Eassa B, Abdel Kader A, et al. A multicenter clinicomycological study evaluating the spectrum of adult Tinea capitis in Egypt. Acta Dermatovenerol Alp Pannonica Adriat. 2013;22(4):77–82.
133. Mebazaa A, Oumari KE, Ghariani N, Mili AF, Belajouza C, Nouira R, et al. Tinea capitis in adults in Tunisia. Int J Dermatol. 2010;49(5):513–6.
134. Cervetti O, Albini P, Arese V, Ibba F, Novarino M, Panzone M. Tinea capitis in adults. Adv Microbiol. 2014;4:12–4.
135. Nweze EI. Dermatophytosis in domesticated animals. Rev Inst Med Trop Sao Paulo. 2011;53 (2):95–9.
136. John AM, Schwartz RA, Janniger CK. The kerion: an angry Tinea capitis. Int J Dermatol. 2018;57(1):3–9.
137. Fuller LC, Barton RC, Mohd Mustapa MF, Proudfoot LE, Punjabi SP, Higgins EM. British Association of Dermatologists' guidelines for the management of Tinea capitis 2014. Br J Dermatol. 2014;171(3):454–63.
138. Elghblawi E. Tinea capitis in children and trichoscopic criteria. Int J Trichol. 2017;9(2):47–9.
139. Amer M, Helmy A, Amer A. Trichoscopy as a useful method to differentiate Tinea capitis from alopecia areata in children at Zagazig University Hospitals. Int J Dermatol. 2017;56 (1):116–20.
140. Brasileiro A, Campos S, Cabete J, Galhardas C, Lencastre A, Serrão V. Trichoscopy as an additional tool for the differential diagnosis of Tinea capitis: a prospective clinical study. Br J Dermatol. 2016;175(1):208–9.
141. Ekiz O, Sen BB, Rifaioğlu EN, Balta I. Trichoscopy in paediatric patients with Tinea capitis: a useful method to differentiate from alopecia areata. J Eur Acad Dermatol Venereol. 2014;28 (9):1255–8.
142. Neri I, Starace M, Patrizi A, Balestri R. Corkscrew hair: a trichoscopy marker of Tinea capitis in an adult white patient. JAMA Dermatol. 2013;149(8):990–1.
143. Mapelli ET, Gualandri L, Cerri A, Menni S. Comma hairs in Tinea capitis: a useful dermatoscopic sign for diagnosis of Tinea capitis. Pediatr Dermatol. 2012;29(2):223–4.
144. Slowinska M, Rudnicka L, Schwartz RA, Kowalska-Oledzka E, Rakowska A, Sicinska J, et al. Comma hairs: a dermatoscopic marker for Tinea capitis: a rapid diagnostic method. J Am Acad Dermatol. 2008;59(5):S77–9.
145. Hughes R, Chiaverini C, Bahadoran P, Lacour JP. Corkscrew hair: a new dermoscopic sign for diagnosis of Tinea capitis in black children. Arch Dermatol. 2011;147(3):355–6.
146. Dhaille F, Dillies AS, Dessirier F, Reygagne P, Diouf M, Baltazard T, et al. A single typical trichoscopic feature is predictive of Tinea capitis: a prospective multicentre study. Br J Dermatol. 2019;181:1046.
147. Aqil N, BayBay H, Moustaide K, Douhi Z, Elloudi S, Mernissi FZ. A prospective study of Tinea capitis in children: making the diagnosis easier with a dermoscope. J Med Case Rep. 2018;12(1):383.
148. Lacarrubba F, Verzì AE, Micali G. Newly described features resulting from high-magnification dermoscopy of Tinea capitis. JAMA Dermatol. 2015;151(3):308–10.
149. Tang J, Ran X, Ran Y. Ultraviolet dermoscopy for the diagnosis of Tinea capitis. J Am Acad Dermatol. 2017;76(2S1):S28–30.
150. Ahdy AM, Sayed-Ahmed MZ, Younis EE, Baraka HN, El-khodery SA. Prevalence and potential risk factors of dermatophytosis in Arabian horses in Egypt. J Equine Vet Sci. 2016;37:71–6.

151. Pihet M, Clément N, Kauffmann-Lacroix C, Nail-Billaud S, Marot A, Pilon F, et al. Diagnosis of dermatophytosis: an evaluation of direct examination using MycetColor® and MycetFluo®. Diagn Microbiol Infect Dis. 2015;83(2):170–4.
152. Tambosis E, Lim C. A comparison of the contrast stains, Chicago blue, chlorazole black, and Parker ink, for the rapid diagnosis of skin and nail infections. Int J Dermatol. 2012;51 (8):935–8.
153. Robert R, Pihet M. Conventional methods for the diagnosis of dermatophytosis. Mycopathologia. 2008;166(5–6):295–306.
154. Pihet M, Le Govic Y. Reappraisal of conventional diagnosis for dermatophytes. Mycopathologia. 2017;82(1–2):169–80.
155. Gümral R, Döğen A, Ilkit MM. Comparison of the contamination rates of culture media used for isolation and identification of dermatophytes. Turk J Med Sci. 2015;45(3):587–92.
156. Trovato L, Oliveri S, Domina M, Patamia I, Scalia G, De Pasquale R. Molecular diagnosis of kerion celsi caused by *Trichophyton tonsurans* in a Italian child. Med Mycol Case Rep. 2019;24:72–4.
157. Mochizuki T, Takeda K, Anzawa K. Molecular markers useful for intraspecies subtyping and strain differentiation of dermatophytes. Mycopathologia. 2017;182(1–2):57–65.
158. Garg J, Tilak R, Garg A, Prakash P, Gulati AK, Nath G. Rapid detection of dermatophytes from skin and hair. BMC Res Notes. 2009;2:60.
159. Ohst T, Kupsch C, Gräser Y. Detection of common dermatophytes in clinical specimens using a simple quantitative real-time TaqMan polymerase chain reaction assay. Br J Dermatol. 2016;174(3):602–9.
160. Beifuss B, Bezold G, Gottlöber P, Borelli C, Wagener J, Schaller M, et al. Direct detection of five common dermatophyte species in clinical samples using a rapid and sensitive 24-h PCR-ELISA technique open to protocol transfer. Mycoses. 2011;54(2):137–45.
161. Verrier J, Monod M. Diagnosis of dermatophytosis using molecular biology. Mycopathologia. 2017;182(1–2):193–202.
162. Gräser Y, Czaika V, Ohst T. Diagnostic PCR of dermatophytes--an overview. J Dtsch Dermatol Ges. 2012;10(10):721–6.
163. Azrad M, Freidus V, Kassem R, Peretz A. Identification of dermatophytes by MALDI-TOF MS technology in the clinical laboratory. Int J Mass Spectr. 2019;440:32–6.
164. L'Ollivier C, Ranque S. MALDI-TOF-based dermatophyte identification. Mycopathologia. 2017;182(1–2):183–92.
165. Tchernev G, Penev PK, Nenoff P, Zisova LG, Cardoso JC, Taneva T, et al. Onychomycosis: modern diagnostic and treatment approaches. Wien Med Wochenschr. 2013;163(1–2):1–12.
166. de Respinis S, Tonolla M, Pranghofer S, Petrini L, Petrini O, Bosshard PP. Identification of dermatophytes by matrix-assisted laser desorption/ionization time-of-flight mass spectrometry. Med Mycol. 2013;51(5):514–21.
167. Bennassar A, Grimalt R. Management of Tinea capitis in childhood. Clin Cosmet Investig Dermatol. 2010;3:89–98.
168. Alkeswani A, Cantrell W, Elewski B. Treatment of Tinea capitis. Skin Appendage Disord. 2019;5:201–10.
169. Kakourou T, Uksal U. Guidelines for the management of Tinea capitis in children. Pediatr Dermatol. 2010;27(3):226–8.
170. Tey HL, Tan AS, Chan YC. Meta-analysis of randomized, controlled trials comparing griseofulvin and terbinafine in the treatment of Tinea capitis. J Am Acad Dermatol. 2011;64 (4):663–70.
171. Shemer A, Plotnik IB, Davidovici B, Grunwald MH, Magun R, Amichai B. Treatment of Tinea capitis – griseofulvin versus fluconazole – a comparative study. J Dtsch Dermatol Ges. 2013;11(8):737–41.
172. Putranti IO, Primanita C. Severe kerion celsi effectively treated with skin debridement and antifungals. J Gen Proced Dermatol Venereol Indones. 2018;2(3):93–7.

173. Patel GA, Schwartz RA. Tinea capitis: still an unsolved problem? Mycoses. 2011;54 (3):183–8.
174. von Laer Tschudin L, Laffitte E, Baudraz-Rosselet F, Dushi G, Hohlfeld J, de Buys Roessingh AS. Tinea capitis: no incision nor excision. J Pediatr Surg. 2007;42(8):E33–6.
175. Thoma-Greber E, Zenker S, Röcken M, Wolff H, Korting HC. Surgical treatment of Tinea capitis in childhood. Mycoses. 2003;46(8):351–4.
176. Gibbon KL, Goldsmith P, Salisbury JA, Bewley AP. Unnecessary surgical treatment of fungal kerions in children. BMJ. 2000;320(7236):696–7.
177. Shalaby MF, El-Din AN, El-Hamd MA. Isolation, identification, and *in vitro* antifungal susceptibility testing of dermatophytes from clinical samples at Sohag University Hospital in Egypt. Electron Physician. 2016;8(6):2557–67.
178. Nweze EI, Mukherjee PK, Ghannoum MA. Agar-based disk diffusion assay for susceptibility testing of dermatophytes. J Clin Microbiol. 2010;48(10):3750–2.
179. Kolivras A, Lateur N, De Maubeuge J, Scheers C, Wiame L, Song M. Tinea capitis in Brussels: epidemiology and new management strategy. Dermatology. 2003;206(4):384–7.
180. Nweze EI. Dermatophytosis among children of Fulani/Hausa herdsmen living in isolated camps in southeastern Nigeria. Rev Iberoam Micol. 2010;27(4):191–4.
181. Nweze EI, Okafor JI. Prevalence of dermatological fungal infections in children: a recent study in Anambra, Nigeria. Mycopathologia. 2005;160:239–43.
182. Nweze EI, Eke IE. Dermatophytosis in northern Africa. Mycoses. 2016;59(3):137–44.
183. Nweze EI. Etiology of dermatophytes amongst children in northeastern Nigeria. Med Mycol. 2001;39:181–4.

Part V

Laboratory and Molecular Diagnosis of Dermatophytosis

The Potential of Molecular Diagnostics in Routine Dermatology

Christiane Kupsch and Yvonne Gräser

Abstract

PCR methods allow the detection of dermatophytes from clinical samples and have the potential of being very sensitive, specific, and fast. However, in order to fully exploit this potential, a number of challenges need to be met. Especially the close phylogenetic relationship of the dermatophytes impedes species-specific detection and beyond that, the implementation of PCR techniques involves some technical and regulatory challenges. Regardless of whether in-house methods are used or a commercial kit is chosen, there are a number of aspects that need to be considered and which will be addressed here.

Keywords

Commercial kit · Differentiation · PCR · Regulatory affairs · Specificity · Sensitivity · Taxonomy

1 Introduction

In addition to conventional dermatophyte diagnostics, i.e. direct microscopy and cultivation with subsequent morphological differentiation, molecular methods are becoming increasingly important. Today, multitudes of MALDI-TOF and PCR techniques are available for the detection and differentiation of pathogenic dermatophytes. Also, companies have long recognized the potential of molecular dermatophyte diagnostics and have launched commercial test systems on the market.

C. Kupsch (✉) · Y. Gräser
National Reference Laboratory for Dermatophytes, Institute for Microbiology, Infectious Diseases and Immunology, Charité—University Medicine Berlin, Berlin, Germany
e-mail: christiane.kupsch@charite.de

J.-P. Bouchara et al. (eds.), *Dermatophytes and Dermatophytoses*,
https://doi.org/10.1007/978-3-030-67421-2_16

Molecular techniques are very sensitive and fast, they lead to results within days. Their specificity, however, is very heterogeneous. For PCR it depends, among other things, on the selected DNA regions and the detection method. In principle, a single sequence polymorphism between two species is sufficient for their differentiation, so that PCR techniques have the potential to specifically identify any species. However, this potential is often not fully exploited.

This chapter gives an overview of the currently available molecular techniques with a focus on the advantages and challenges of using PCR methods for routine diagnostics.

2 Matrix-Assisted Laser Desorption Ionization: Time of Flight/Mass Spectrometry (MALDI-TOF/MS)

MALDI-TOF/MS is a method that is based on the analysis of ionized fragments of biomolecules like peptides, lipids, and saccharides. The analyte is co-crystallized with a laser energy-absorbing matrix. By laser excitation, ionized molecular fragments are dissolved from the mixture and separated according to their mass, resulting in specific mass spectra.

The method has been proven for the identification of bacteria and a number of pathogenic fungi [1]. Some of the manufacturers of MALDI-TOF/MS systems (bioMérieux, Bruker Daltonics, Andromas) provide identification platforms for yeasts such as *Candida* and *Cryptococcus* species and for filamentous fungi such as *Aspergillus*, *Fusarium*, and some dermatophyte species as well [2]. For the detection of dermatophytes, often in-house reference databases are used supplemental to manufacturer libraries [3]. Due to the high degree of phylogenetic similarity, the differentiation of dermatophytes species is challenging. For some closely related species, the differences in mass spectra have been reported to be insufficient or misleading, while for some other species they are too heterogeneous [3]. Complications have been repeatedly described for the discrimination within the species clusters *Trichophyton rubrum/Trichophyton soudanense/Trichophyton violaceum*, *Trichophyton interdigitale/Trichophyton mentagrophytes/Trichophyton quinckeanum/Trichophyton tonsurans*, and *Microsporum audouinii/ Microsporum canis* [4–7].

3 PCR-Based Techniques

Over the last 15 years, a considerable number of PCR techniques have been developed for the detection and differentiation of pathogenic dermatophytes. PCR techniques are fast, leading to diagnosis within 24–48 h. They are very sensitive and, unlike culture diagnostics, they have no problems in detecting pathogens that are growth inhibited by antimycotic therapy. However, the specificity of the various methods varies greatly and strongly depends on the DNA target selected to distinguish between species and the method for the detection of amplified DNA.

3.1 Targets

For the detection and differentiation of pathogenic fungi, the ribosomal genes, such as 18S rDNA, and especially the so-called internal transcribed spacers ITS2 and ITS1, are most frequently used. In contrast to other pathogenic fungi, however, dermatophyte species have a very high degree of phylogenetic similarity. The rather conserved ribosomal genes are not sufficiently variable for the differentiation of this group of fungi. The more variable ITS region is suitable for the differentiation of all known dermatophyte species, even if some differ only by one or two single nucleotide polymorphisms (SNPs).

Besides ITS, the genes encoding chitinase-1 (CHS1), topoisomerase-2 (TOP2), β-tubulin (BTU), or translation elongation factor 1-α (EF-1-α) are used. For the *CHS1* gene, Kano et al. [8, 9] and Cafarchia et al. [10] respectively published a phylogenetic analysis of eight and five different dermatophyte species showing considerable sequence differences, but apart from *T. rubrum* and *T. violaceum,* none of the closely related species groups have been analyzed [9]. Some authors show the suitability of *CHS1* gene for a universal dermatophyte detection in clinical nail [11], skin and hair samples [12, 13]. The *TOP2* gene has been used for dermatophyte detection using RFLP and has shown to be sufficient for genera and in part for species identification, but differentiation of the *T. mentagrophytes* species complex was not possible [14, 15]. The *BTU* gene is more variable than the ITS regions with respect to the genus *Microsporum,* but has some drawbacks compared to the ITS regions for the genus *Trichophyton*. No differences can be found between *Trichophyton benhamiae* and *Trichophyton concentricum*, and only one SNP between *Trichophyton schoenleinii* and *T. quinckeanum* and two between *T. rubrum* and *T. violaceum* [16]. One study shows the use of the *BTU* gene to detect and differentiate ten different dermatophyte species from 603 clinical samples using RFLP analyses [17]. More recently the *EF1-α* gene has been identified as a sufficient identification marker for dermatophytes. It is similarly variable as the ITS regions [18] and even more diverse for the species *T. tonsurans* and *Trichophyton equinum* (14 SNPs vs. only one SNP within the ITS) and *M. canis* and *Microsporum ferrugineum* (ten bases differences vs. two SNPs within the ITS) [19, 20]. *EF1-α* has been used in addition to ITS for characterizing clinical isolates in recent studies [21–23].

4 Methods

Several post-PCR procedures can be used to visualize the amplified DNA to detect the presence of a pathogen. Differences in length and pattern of PCR amplified DNA, with or without subsequent enzymatic digest, can be identified by electrophoretic separation in a gel matrix (e.g. agarose). This applies to PCR fingerprinting, amplified fragment length polymorphism (AFLP), restriction fragment length polymorphism (RFLP), and randomly amplified polymorphic DNA (RAPD). The uses of these techniques have been reviewed in great detail by Cafarchia et al. [24] and therefore only the quite popular RFLP analysis is briefly cited here. RFLP has been shown useful for *Trichophyton* spp. detection directly from nail fragments of patients with suspected onychomycosis using the 28S ribosomal DNA as the target [25, 26]. Successful dermatophyte detection and a distinct pattern of different genera have been shown using the ITS region [27–29], the *BTU* [18, 30] or *TOP2* gene [9, 31] as RFLP target. The differentiation of *Trichophyton* species, especially within the *T. mentagrophytes* species complex is limited, however. It has been shown that the success of RLFP for species differentiation can be enhanced by a combined analysis of the pattern produced by more than one restriction enzyme [18, 31]. Also a combination with high-resolution melting analysis has been described [30]. The advantage of these techniques is that they do not require expensive reagents or equipment apart from the PCR thermal cycler, but a detailed species differentiation cannot be achieved.

Another approach to detect PCR products is to hybridize them to specific probes. In PCR ELISA, microarray or lateral flow techniques, PCR products are labeled during amplification enabling their subsequent detection. The label can be a fluorescent molecule, which emits light of a certain wavelength by laser excitation, or a target for an antibody, which in turn carries a dye that emits light after chemical activation. In either case, a signal is detected only when a probe fits exactly to a PCR product.

Real-time PCR methods allow detection during the PCR reaction without the need for post-PCR steps. The detection is based on fluorescent dyes. Universal DNA detection can be achieved with SYBR-Green and other dyes, which intercalate to double-stranded DNA and dissolve as soon as the DNA melts into single strands. This way, the amplification of the DNA can be monitored in real time. In addition to this universal detection of any amplified DNA, specific detection is achieved by fluorescence-labeled probes. Fluorescence is measured when the probe binds its complementary sequence within a PCR product, while no signal will be detected when the probes target is not present within a sample. Real-time PCR can also be used to quantify DNA when its amplification is compared to dilution series of standards with defined DNA concentrations. Using special kinds of probes, the so-called melting curve analysis can be performed. Directly after PCR amplification, increasing the temperature stepwise causes the probes to separate (melt) from the PCR products and the fluorescence decreases. The melting temperature is individual for each probe and depends on the degree of identity between probe and target.

5 Implementation Aspects of Molecular Methods

5.1 Time and Costs

While culture diagnostics is a well-established procedure in most laboratories at present, the changeover to molecular biological techniques may require some effort. However, the investment strongly depends on which devices for PCR analyses or MALDI-TOF/MS analyses are already in use and to what extent they can be utilized for dermatophyte diagnostics. In addition to the equipment purchase, the costs for consumables for molecular biological methods are higher than those for culture. In PCR, fluorescent labels or kits are expensive consumables.

In contrast, however, the working time required for pathogen identification is much shorter. Classical morphological identification requires (1) direct microscopy of the clinical material to detect the presence of fungal elements, (2) inoculation and growth of a culture, (3) microscopy of the culture, and (4) evaluation of macro- and micromorphological characteristics to identify the species. Morphological characteristics, such as macroconidia, often develop only after several weeks of growth. To classify a sample as negative for dermatophytes, incubation of the clinical material over 4 weeks is necessary [32].

MALDI-TOF analyses also depend on the growth of fungal cultures for a minimum of 3 days [7, 33]. Compared to culture diagnostics this saves time but the costs for material and manpower for the culture are still incurred. Sample preparation and analysis is straightforward. Proteins are usually extracted with acetonitrile from the culture material, mixed with a matrix and dried, which requires time in the range of minutes. The mass spectrometric separation itself is automated and the evaluation is facilitated by software guidance.

PCR analyses do not require culture growth, as they can be performed directly from clinical material. As a first step, DNA is extracted from the clinical material. This requires between 10 min and 12 h incubation time, depending on the method or kit used (see Sect. 5.3). The DNA extracts are then subjected to the PCR reaction (1–2 h). For non-real-time PCRs, a separate detection step is performed (another 1–2 h), followed by a software-assisted or manual evaluation.

In the case of molecular biological techniques, the analysis is completed only days after receipt of a sample, so the communication of the results and the billing can take place promptly instead of a few weeks later.

The time required for molecular methods, in particular for PCR techniques, is also more consistent because it does not depend on the speed or uniqueness of morphological development and can therefore be estimated more accurately. The higher the number of samples the less time is needed per sample because they can be processed in parallel. Common PCR machines allow the simultaneous amplification of 48–96 samples. The higher costs of PCR compared to culture analysis can therefore be offset to some extent by reduced staff costs.

5.2 Application of In-House Methods

In recent years, some diagnostic laboratories have established in-house methods or Laboratory Developed Tests (LDTs) for the identification of dermatophytes. LDTs are developed and operated within a single clinical laboratory facility. Such methods are used both for research purposes and for diagnostics. Approximately 40% of diagnostic laboratories from Germany and other European countries use in-house techniques in the external quality assessment "genome detection of dermatophytes" (RV 492) offered by Instand e.V. in Germany [34].

Challenges in the validation of in-house methods. In order to develop and validate a DNA-based method capable to differentiate dermatophyte species, suitable reference strains need to be used. Sequences of the ITS region of all dermatophyte species are available in publicly accessible databases such as NCBI or the fungal collection of the Westerdijk Institute. However, the entries in these databases are partially incorrect due to the use of outdated and misleading taxonomic terms. The reasons for this lie in extensive taxonomic changes that have been implemented in recent years.

Instead of morphological characteristics alone, researchers have begun to use genetic data to determine the relationship between species. The inclusion of genetic relationships has revolutionized the taxonomy of many species, including the kingdom of fungi. For example, the dual naming of teleomorphs and anamorphs, which represent different manifestations of the same species, was abolished [35]. Also, the dermatophyte taxonomy has changed in the last decade and has led to a simplified species concept of this group of fungi, as shown in Fig. 1. Unfortunately, this simplification has nevertheless led to some difficulties, since the database entries of isolates, some of which are decades old, have not been extensively updated and therefore databases still contain partly obsolete and misleading identifications and designations.

In order to avoid these problems, it is recommended to use only type strains for the development of a diagnostic method. The same applies for the sequence-based identification of clinical isolates. A current overview of type strains can be found in de Hoog et al. [36]. Furthermore, it should be noted that new sequence variants occur from time to time. To avoid misidentifications or non-identifications of such variants the current literature has to be checked continuously in order to classify them correctly. An example is the species *T. mentagrophytes*, for which now more than eight different ITS genotypes are known [21, 37].

Regulatory challenges. In the European Union, with Regulation (EU) 2017/746, a comprehensive law on in vitro diagnostica (IVD) came into force in May 2017, which not only strengthens the requirements for test kits but also covers LDTs and for these, prescribes comprehensive quality management in accordance with the EN ISO 15189 standard as well as a detailed risk management. This regulation is legally binding throughout the EU from 26 May 2022.

In the United States, the Centers for Medicare and Medicaid Services (CMS) are mandated by the Clinical Laboratory Improvement Amendments (CLIA) to ensure the quality of LDTs. According to CLIA regulations, all facilities that perform tests for the diagnosis, prevention, or treatment of disease, impairment or health assessment of human materials must meet certain federal requirements. To obtain a certificate, a laboratory must undergo evaluation every 2 years and meet CLIA quality standards for quality control, proficiency testing, quality assurance, personnel qualification, and sample integrity and record keeping [38]. Since 2010, the Food and Drug Administration (FDA) has been discussing the creation of a regulatory framework for FDA monitoring of in vitro clinical tests (IVCTs), such as test kits and LDTs. At the end of 2018, a draft law VALID (Verifying Accurate, Leading-Edge IVCT Development) Act was published [39].

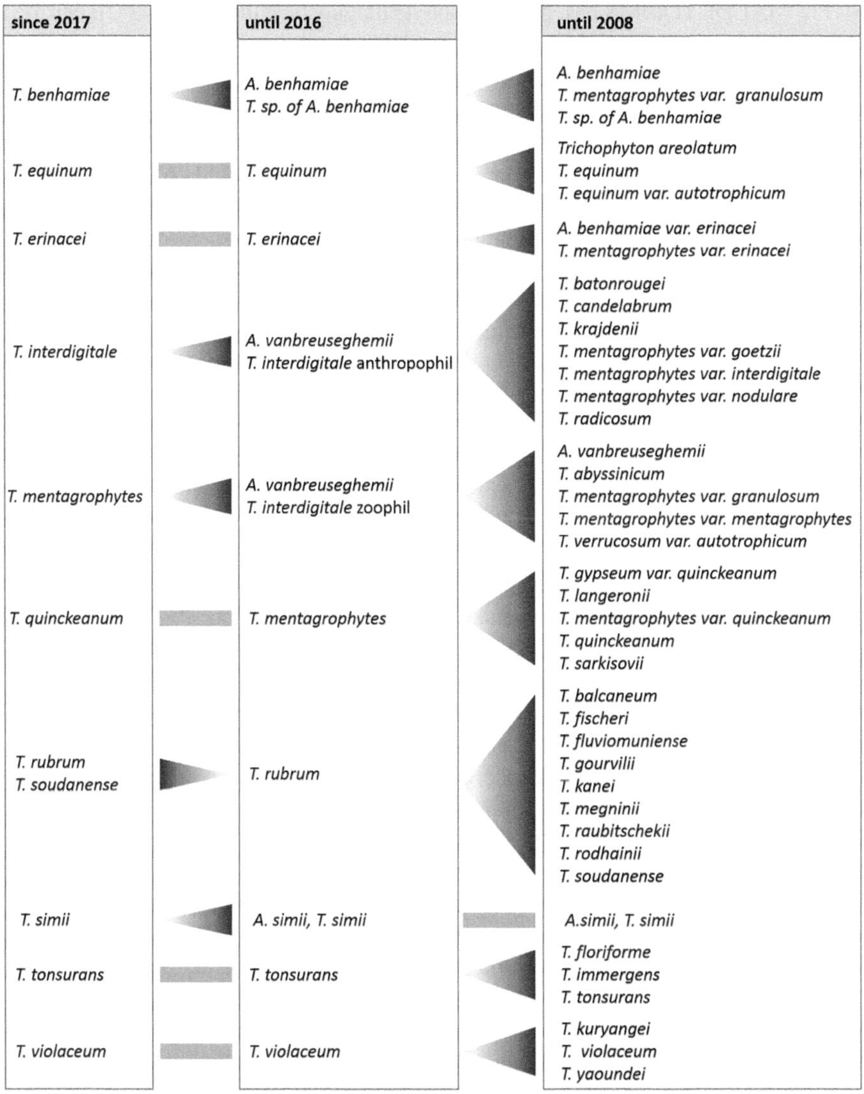

Fig. 1 Todays taxonomy of the genus *Trichophyton* and taxonomic changes since 2008. No changes applied for *T. concentricum*, *T. schoenleinii*, and *T. verrucosum*, which are therefore not listed

5.3 Commercial Kits

The use of commercial kits has the advantage that they meet the required regulatory standards, provide reproducible results and ease the evaluation of the analyses by detailed instructions for use or software support. Thus, the human factor described above plays a subordinate role here.

For MALDI-TOF/MS based dermatophyte diagnostics, there are two CE certified IVD systems and associated databased available for diagnostic use in Europe. The Bruker MALDI Biotyper IVD library contains the seven species *Epidermophyton floccosum, M. canis, Microsporum gypseum* (correct name *Nannizzia gypsea*), *T. interdigitale*, *T. mentagrophytes*, *T. rubrum*, and *T. tonsurans*. The system was evaluated in one study analyzing 115 clinical isolates and achieved correct identifications in 54% [5]. The low percentage is primarily caused by the failure of the method to identify *T. rubrum* at the species level.

The bioMérieux Vitek MS v3.0 system is CE certified and additionally approved by the FDA for diagnostic use in the United States. It allows the identification of the nine species *M. audouinii, M. canis, M. gypseum (= N. gypsea), E. floccosum, T. rubrum, T. interdigitale, T. tonsurans, Trichophyton verrucosum*, and *T. violaceum*. A study analyzing 291 isolates of these species with the Vitek MS v3.0 system achieved correct species identifications for 246 of the strains (84.5%). The major problems were in the identification of *T. violaceum* (41% correct), *T. verrucosum* (58% correct), *T. tonsurans*, and *M. audouinii* (both 91% correct) [40]. Also for the Bruker MALDI Biotyper CA system, an FDA-approved library is available; this however does not contain any dermatophyte species.

For PCR methods, to date seven companies provide CE-marked commercial kits, fulfilling the regulatory requirements for IVD in Europe (Fig. 2) [36–43]. When considering the selection of a PCR-kit it should be noted that there are major differences between the kits not only in terms of the underlying methods and techniques, but also in the scope of pathogen identification and differentiation. The spectrum ranges from universal dermatophyte detection with *T. rubrum* as the only differentiated species [44, 45] or the detection of the various dermatophyte genera with some species discrimination [46–48] to a comprehensive species differentiation [41, 42, 49]. The more recent kits can distinguish *T. tonsurans* of *T. interdigitale* and *T. rubrum* of *T. violaceum*, *T. verrucosum* of *T. benhamiae* as well as *M. canis* of *M. audouinii* [49]. Difficulties still exist in the clear differentiation of very closely related species such as *T. rubrum* and *T. soudanense*, *T. benhamiae* and *Trichophyton erinacei*, *T. tonsurans* and *T. equinum* or *T. interdigitale* of *T. mentagrophytes* (formerly zoophilic strains of *T. interdigitale*) [36]. Figure 2 gives an overview of the currently available kits for dermatophyte diagnostics and the species differentiations they can provide.

The effort required for DNA extraction prior to PCR varies from kit to kit. Some kits contain reagents for DNA extraction; others recommend the use of DNA extraction kits. The most frequently recommended kit is the QIAamp® DNA Mini Kit (Qiagen GmbH, Hilden) with an extended (90 min to 12 h) initial lysis step. The Dermatophyte Kit from Statens Serum Institute and the DermaGenius® 2.0 kit each contain a two-buffer system for DNA extraction that requires only 10 min of incubation.

complexity →

Realtime-PCR	PCR + GE or Realtime-PCR	Realtime-PCR	PCR + GE	Realtime-PCR	PCR + LF	Realtime-PCR	PCR + Microarray
Dermatophytes Bio-Evolution (France)	SSI Dermatophyte (Denmark)	FTD® Dermatophytes (Malta)	Mentype® MycoDerm^{QS} (Germany)	DermaDYN (Israel)	Mentype® MycoDerm^{QS} LF (Germany)	DermaGenius ® 2.0 (Netherlands)	Euroarray Dermatomycosis (Germany)
dermatophyte (universal)	dermatophyte (universal)		*Trichophyton spp.*		dermatophyte (universal)	dermatophyte (universal)	dermatophyte (universal)
		T. tonsurans/ T. equinum		*T. tonsurans/ T. equinum*	*T. tonsurans/ T. equinum*	*T. tonsurans/ T. equinum*	*T. tonsurans* *T. equinum*
						T. interdigitale	*T. interdigitale*
		T. mentagr. SC/ *T. benhamiae* SC	*T. mentagr.* SC	*T. mentagr.* SC/ *T. benhamiae* SC	*T. mentagr.* SC	*T. mentagr. SC*	*T. mentagrophytes*
							T. quinckeanum
							T. schoenleinii
							T. simii
					T. benhamiae SC	*T. benhamiae/ T. erinacei*	*T. benhamiae*
							T. concentricum
							T. erinacei
						T. verrucosum	*T. verrucosum*
	T. rubrum	*T. rubrum*	*T. rubrum*	*T. rubrum*	*T. rubrum*	*T. rubrum*	*T. rubrum*
		T. violaceum		*T. violaceum*	*T. violaceum*	*T. violaceum*	*T. violaceum*
		E. floccosum	*E. floccosum*	*E. floccosum*	*E. floccosum*	*E. floccosum*	*E. floccosum*
		M. canis SC	*M. canis* SC	*M. canis* SC	*M. canis* SC	*M. canis/ M. ferru.*	*M. canis*
						M. audouinii	*M. audouinii*
							M. ferrugineum
			N. gypsea	*N. gypsea*	*N. gypsea*		*N. gypsea*
							N. incurvata
							N. fulva
							N. persicolor

Fig. 2 Growing complexity of PCR kits in dermatophyte species differentiation. *FTD* fast track diagnostics, *GE* gel electrophoresis, *LF* Lateral Flow, *M. ferru. M. ferrugineum*, *T. mentagr. T. mentagrophytes*, *SC* species complex, *SSI* Statens Serum Institute. *T. mentagr.* SC comprises *T. interdigitale*, *T. mentagrophytes*, *T. quinckeanum*, *T. schoenleinii*, *T. simii*; *T. benhamiae* SC comprises *T. benhamiae*, *T. concentricum*, *T. erinacei*, *T. verrucosum*

6 The Impact of PCR Diagnostics on Therapy

6.1 Sensitivity and Speed of PCR

A number of studies have shown that PCR methods are considerably more sensitive than cultures. Especially from nail samples, and also from skin and hair samples, it is often not possible to grow a culture, which can lead to false negative results [25, 50–53]. Patients misdiagnosed with a negative result are not treated with antifungals, causing infections to persist, exuberate, and in addition, this increases the probability of transmission to other people. If the cultivation is successful, it takes about 4 weeks until the result is available. During this period, often a tentative diagnosis is made according to the clinical picture and treatment is started. An incorrect initial diagnosis is particularly critical in patients with severe and painful infections, e.g. *Tinea capitis*, which is easily confused with a bacterial infection, as it delays the healing process [54–56]. As PCR analyses can be performed directly from clinical material, these methods make it possible to clarify within a few days whether or not and which fungal pathogen, if any, is present.

A frequently discussed, seeming disadvantage of PCR is that it also detects dead dermatophyte cells. However, there are no studies available to prove this assumption. On the other hand, a few studies indicate that the results of PCR analyses correlate with the success of therapy. In patients with *Tinea genitalis* caused by *T. mentagrophytes*, it was shown that the results of PCR were negative 2 weeks after the first negative culture result [21]. Another study by Iwanaga et al. [57] conducted on nail samples from onychomycosis patients under therapy showed that the amount of pathogen detected by real-time PCR decreased significantly after 8 and 16 weeks of oral terbinafine therapy. The fact that even after 16 weeks of therapy small amounts of fungal DNA were still detected was explained by the survival of fungal cells in the form of arthrospores [57]. Arthrospores originate from the breakdown of vegetative hyphae and can serve as an enduring form. They are difficult to attack by antimycotics and can lead to a recurrence of the disease at any time [58]. This explanation is quite plausible, because in the study of Iwanaga et al. [57], mainly older patients with an average age of 72 years were examined, in whom the outgrowth of the infected nail plate is clearly slowed compared to younger people, due to a reduced rate of longitudinal nail growth [59]. The survival of dormant fungal cells inside the nail is also supported by follow up studies, which after 18 months show a complete cure in only 76% of elderly patients receiving a 3-month terbinafine therapy [60]. Overall, the available data indicate that PCR methods do not detect dead cells but dormant cells that are missed in the culture due to their reduced division activity. Thus, PCR may be more suitable than culture for therapy control, although this needs to be proven by further studies.

6.2 Specificity of PCR

Species differentiation plays a major role in therapy decisions when it comes to determining (1) the genus or (2) whether an anthropophilic or a zoophilic pathogen is responsible for an infection. For infections caused by species of the genus *Microsporum,* the treatment with griseofulvin has been shown to be more effective than terbinafine, which is commonly used to cure trichophytosis [61]. Species within the genus *Trichophyton* are preferably treated with terbinafine or, as a second choice, itraconazole. Since a few years, however, terbinafine resistant strains of *T. mentagrophytes* and *T. rubrum* originated in India have started to spread worldwide. This means that terbinafine will no longer blindly be the drug of choice in the future. Irrespective of this the species-specific differentiation of anthropophilic (e.g. *T. interdigitale*) and zoophilic (e.g. *Trichophyton (T.) mentagrophytes*) dermatophytes is enormously important for the choice of an efficient therapeutic strategy. In the case of zoophilic pathogens, there is a rather low potential for human-to-human transmission, as these pathogens are not perfectly adapted to the human host. For a successful therapy of such zoonoses, however, it is crucial to identify and treat the animals that are the source of the infection in order to avoid a reinfection of the affected patient and to prevent infection of further persons in contact with the animal in question. Anthropophilic pathogens, on the other hand, are highly contagious. Preventive action, particularly in the form of disinfection, must be taken to avoid transmission to family members and other contact persons. The German guideline even stipulates a quarantine period for affected patients under adequate therapy of 1 weeks to prevent outbreaks. An inaccurate diagnosis, which does not consistently distinguish between anthropophilic and zoophilic pathogens, causes potential reinfections, new infections and possibly unnecessary sick leave and with that not only a higher suffering of the patients but also a multiplication of the costs for the health system.

7 Conclusion

PCR methods provide a valuable tool for clearly distinguishing closely related species that have different host specificities. Their use in routine diagnostics makes therapeutic decisions more effective. At present, however, the full potential of PCR is hardly exploited, but further developments in this field (e.g., regarding the search for more variable DNA regions) are to be expected in the near future.

References

1. Patel R. MALDI-TOF mass spectrometry: transformative proteomics for clinical microbiology. Clin Chem. 2013;59:340–2.
2. Patel R. A moldy application of MALDI: MALDI-TOF mass spectrometry for fungal identification. J Fungi (Basel). 2019;5:4.

3. L'Ollivier C, Ranque S. MALDI-TOF-based dermatophyte identification. Mycopathologia. 2017;182:183–92.
4. Theel ES, Hall L, Mandrekar J, Wengenack NL. Dermatophyte identification using matrix-assisted laser desorption ionization–time of flight mass spectrometry. J Clin Microbiol. 2011;49:4067–71.
5. Karabıçak N, Karatuna O, İlkit M, Akyar I. Evaluation of the Bruker matrix-assisted laser desorption–ionization time-of-flight mass spectrometry (MALDI-TOF MS) system for the identification of clinically important dermatophyte species. Mycopathologia. 2015;180:165–71.
6. Packeu A, Hendrickx M, Beguin H, Martiny D, Vandenberg O, Detandt M. Identification of the *Trichophyton mentagrophytes* complex species using MALDI-TOF mass spectrometry. Med Mycol. 2013;51:580–5.
7. L'Ollivier C, Cassagne C, Normand A-C, Bouchara J-P, Contet-Audonneau N, Hendrickx M, et al. A MALDI-TOF MS procedure for clinical dermatophyte species identification in the routine laboratory. Med Mycol. 2013;51:713–20.
8. Kano R, Nakamura Y, Watari T, Watanabe S, Takahashi H, Tsujimoto H, et al. Phylogenetic analysis of 8 dermatophyte species using chitin synthase 1 gene sequences. Mycoses. 1997;40:411–4.
9. Kano R, Okabayashi K, Nakamura Y, Ooka S, Kashima M, Mizoguchi M, et al. Differences among chitin synthase 1 gene sequences in *Trichophyton rubrum* and *T. violaceum*. Med Mycol. 2000;38:47–50.
10. Cafarchia C, Otranto D, Weigl S, Campbell BE, Parisi A, Cantacessi C, et al. Molecular characterization of selected dermatophytes and their identification by electrophoretic mutation scanning. Electrophoresis. 2009;30:3555–64.
11. Dhib I, Fathallah A, Charfeddine IB, Meksi SG, Said MB, Slama F, et al. Evaluation of chitin synthase (*CHS1*) polymerase chain reaction assay in diagnosis of dermatophyte onychomycosis. J Mycol Med. 2012;22:249–55.
12. Garg J, Tilak R, Garg A, Prakash P, Gulati AK, Nath G. Rapid detection of dermatophytes from skin and hair. BMC Res Notes. 2009;2:60.
13. Kano R, Hirai A, Muramatsu M, Watari T, Hasegawa A. Direct detection of dermatophytes in skin samples based on sequences of the chitin synthase 1 (*CHS1*) gene. J Vet Med Sci. 2003;65:267–70.
14. Kamiya A, Kikuchi A, Tomita Y, Kanbe T. PCR and PCR–RFLP techniques targeting the DNA topoisomerase II gene for rapid clinical diagnosis of the etiologic agent of dermatophytosis. J Dermatol Sci. 2004;34:35–48.
15. Kanbe T, Suzuki Y, Kamiya A, Mochizuki T, Kawasaki M, Fujihiro M, et al. Species-identification of dermatophytes *Trichophyton*, *Microsporum* and *Epidermophyton* by PCR and PCR-RFLP targeting of the DNA topoisomerase II genes. J Dermatol Sci. 2003;33:41–54.
16. Rezaei-Matehkolaei A, Mirhendi H, Makimura K, Hoog D, Sybren G, Satoh K, et al. Nucleotide sequence analysis of beta tubulin gene in a wide range of dermatophytes. Med Mycol. 2014;52:674–88.
17. Abastabar M, Mirhendi H, Rezaei-Matehkolaei A, Shidfar MR, Kordbacheh P, Makimura K. Restriction analysis of β-tubulin gene for differentiation of the common pathogenic dermatophytes. J Clin Lab Anal. 2014;28:91–6.
18. Mirhendi H, Makimura K, Hoog D, Sybren G, Rezaei-Matehkolaei A, Najafzadeh MJ, et al. Translation elongation factor 1-α gene as a potential taxonomic and identification marker in dermatophytes. Med Mycol. 2015;53:215–24.
19. Rezaei-Matehkolaei A, Makimura K, de Hoog GS, Shidfar MR, Satoh K, Najafzadeh MJ, et al. Multilocus differentiation of the related dermatophytes *Microsporum canis*, *Microsporum ferrugineum* and *Microsporum audouinii*. J Med Microbiol. 2012;61:57–63.
20. Rezaei-Matehkolaei A, Makimura K, de Hoog GS, Shidfar MR, Satoh K, Najafzadeh MJ, et al. Discrimination of *Trichophyton tonsurans* and *Trichophyton equinum* by PCR-RFLP and by β-tubulin and translation elongation factor 1-α sequencing. Med Mycol. 2012;50:760–4.

21. Kupsch C, Czaika V-A, Deutsch C, Gräser Y. *Trichophyton mentagrophytes* – a new genotype of zoophilic dermatophyte causes sexually transmitted infections. J Dtsch Dermatol Ges. 2019;17:493–501.
22. Nenoff P, Verma SB, Vasani R, Burmester A, Hipler U-C, Wittig F, et al. The current Indian epidemic of superficial dermatophytosis due to *Trichophyton mentagrophytes* – a molecular study. Mycoses. 2019;62:336–56.
23. Uhrlaß S, Schroedl W, Mehlhorn C, Krüger C, Hubka V, Maier T, et al. Molecular epidemiology of *Trichophyton quinckeanum* – a zoophilic dermatophyte on the rise. J Dtsch Dermatol Ges. 2018;16:21–32.
24. Cafarchia C, Iatta R, Latrofa MS, Gräser Y, Otranto D. Molecular epidemiology, phylogeny and evolution of dermatophytes. Infect Genet Evol. 2013;20:336–51.
25. Bontems O, Hauser PM, Monod M. Evaluation of a polymerase chain reaction-restriction fragment length polymorphism assay for dermatophyte and nondermatophyte identification in onychomycosis. Br J Dermatol. 2009;161:791–6.
26. Monod M, Bontems O, Zaugg C, Léchenne B, Fratti M, Panizzon R. Fast and reliable PCR/sequencing/RFLP assay for identification of fungi in onychomycoses. J Med Microbiol. 2006;55:1211–6.
27. Mirzahoseini H, Omidinia E, Shams-Ghahfarokhi M, Sadeghi G, Razzaghi-Abyaneh M. Application of PCR-RFLP to rapid identification of the main pathogenic dermatophytes from clinical specimens. Iran J Public Health. 2009;38:18–24.
28. Elavarashi E, Kindo AJ, Kalyani J. Optimization of PCR—RFLP directly from the skin and nails in cases of dermatophytosis, targeting the ITS and the 18S ribosomal DNA regions. J Clin Diagn Res. 2013;7:646–51.
29. Jackson CJ, Barton RC, Evans EGV. Species identification and strain differentiation of dermatophyte fungi by analysis of ribosomal-DNA intergenic spacer regions. J Clin Microbiol. 1999;37:931–6.
30. Sun P-L, Lin H-C, Hsiao P-F, Lu J-J. Application of restriction fragment length polymorphism (RFLP) and high resolution melting (HRM) analysis of beta-tubulin gene for species and strains differentiation in *Trichophyton mentagrophytes* species complex. J Dermatol Sci. 2016;83:247–50.
31. Kanbe T. Molecular approaches in the diagnosis of dermatophytosis. Mycopathologia. 2008;166:307–17.
32. Mayser P. Leitlinie Tinea capitis [Internet]. 2019 [cited 2019 Aug 5]. Available from https://www.awmf.org/leitlinien/detail/ll/013-033.html
33. De Respinis S, Tonolla M, Pranghofer S, Petrini L, Petrini O, Bosshard PP. Identification of dermatophytes by matrix-assisted laser desorption/ionization time-of-flight mass spectrometry. Med Mycol. 2013;51:514–21.
34. Kupsch C, Gräser Y. Genomnachweis von Dermatophyten. Hautarzt. 2019;70:627–37.
35. Hawksworth DL, Crous PW, Redhead SA, Reynolds DR, Samson RA, Seifert KA, et al. The Amsterdam declaration on fungal nomenclature. IMA Fungus. 2011;2:105–12.
36. de HGS, Dukik K, Monod M, Packeu A, Stubbe D, Hendrickx M, et al. Toward a novel multilocus phylogenetic taxonomy for the dermatophytes. Mycopathologia. 2017;182:5–31.
37. Nenoff P, Verma SB, Uhrlaß S, Burmester A, Gräser Y. A clarion call for preventing taxonomical errors of dermatophytes using the example of the novel *Trichophyton mentagrophytes* genotype VIII uniformly isolated in the Indian epidemic of superficial dermatophytosis. Mycoses. 2019;62:6–10.
38. Sun F, Bruening W, Uhl S, Ballard R, Tipton K, Schoelles K. Quality, regulation and clinical utility of laboratory-developed molecular tests [Internet]. Rockville: Agency for Healthcare Research and Quality (US); 2010 [cited 2019 Aug 5]. Available from http://www.ncbi.nlm.nih.gov/books/NBK284837/
39. Genzen JR. Regulation of laboratory-developed tests. A clinical laboratory perspective. Am J Clin Pathol. 2019;152:122–31.

40. Rychert J, Slechta ES, Barker AP, Miranda E, Babady NE, Tang Y-W, et al. Multicenter evaluation of the Vitek MS v3.0 system for the identification of filamentous fungi. J Clin Microbiol. 2018;56. pii:e01353-17.
41. Kosanke S, Harder M, Brasch J, Nenoff P, Cavalar M, Uhrlass S, et al. Dermatomycosis-a novel and rapid detection of causative fungal agents with a DNA-based microarray (EuroArray dermatomycosis). J Mol Diagn. 2017;19:978.
42. Wittig F, Uhrlaß S, Gaajetaan G, Dingemans G, Krueger C, Nenoff P. Detection and identification of pathogenic dermatophytes using multiplex real-time PCR assays. Med Mycol. 2018;58 (Suppl 2):S82.
43. Kroschwald L, Weber C, Augustin M, Kurtzer R, Bröse E, Wiedenfeld A, et al. Rapid DNA-probe based screening assay for differential diagnosis of dermatomycoses. Mycoses. 2015;58:63–4.
44. Kondori N, Tehrani PA, Strömbeck L, Faergemann J. Comparison of dermatophyte PCR kit with conventional methods for detection of dermatophytes in skin specimens. Mycopathologia. 2013;176:237–41.
45. Sherman S, Goshen M, Treigerman O, Ben-Zion K, Carp M-J, Maisler N, et al. Evaluation of multiplex real-time PCR for identifying dermatophytes in clinical samples—a multicentre study. Mycoses. 2018;61:119–26.
46. Petinataud D, Berger S, Ferdynus C, Debourgogne A, Contet-Audonneau N, Machouart M. Optimising the diagnostic strategy for onychomycosis from sample collection to FUNGAL identification evaluation of a diagnostic kit for real-time PCR. Mycoses. 2016;59:304–11.
47. Mehlig L, Garve C, Ritschel A, Zeiler A, Brabetz W, Weber C, et al. Clinical evaluation of a novel commercial multiplex-based PCR diagnostic test for differential diagnosis of dermatomycoses. Mycoses. 2014;57:27–34.
48. Kondori N, Abrahamsson A-L, Ataollahy N, Wennerås C. Comparison of a new commercial test, dermatophyte-PCR kit, with conventional methods for rapid detection and identification of *Trichophyton rubrum* in nail specimens. Med Mycol. 2010;48:1005–8.
49. Uhrlaß S, Wittig F, Koch D, Krüger C, Harder M, Gaajetaan G, et al. [Do the new molecular assays-microarray and realtime polymerase chain reaction-for dermatophyte detection keep what they promise?]. Hautarzt. 2019;70:618–26. In German.
50. Wisselink GJ, van Zanten E, Kooistra-Smid AMD. Trapped in keratin; a comparison of dermatophyte detection in nail, skin and hair samples directly from clinical samples using culture and real-time PCR. J Microbiol Methods. 2011;85:62–6.
51. Pankewitz F, Nenoff P, Uhrlaß S, Bezold G, Winter I, Gräser Y. Development of a novel polymerase chain reaction-enzyme-linked immunosorbent assay for the diagnosis of *Trichophyton rubrum* onychomycosis. Br J Dermatol. 2013;168:1236–42.
52. Kupsch C, Ohst T, Pankewitz F, Nenoff P, Uhrlaß S, Winter I, et al. The agony of choice in dermatophyte diagnostics - performance of different molecular tests and culture in the detection of *Trichophyton rubrum* and *Trichophyton interdigitale*. Clin Microbiol Infect. 2016;22:735. e11–7.
53. Bergman A, Heimer D, Kondori N, Enroth H. Fast and specific dermatophyte detection by automated DNA extraction and real-time PCR. Clin Microbiol Infect. 2013;19:E205–11.
54. Bougrine A, Villeneuve-Tang C, Bouffard D, Rouleau D, Chartier S. Kerion of the vulva caused by *Trichophyton mentagrophytes*. J Cutan Med Surg. 2014;18:206–9.
55. Trovato L, Oliveri S, Domina M, Patamia I, Scalia G, De Pasquale R. Molecular diagnosis of kerion celsi caused by *Trichophyton tonsurans* in a Italian child. Med Mycol Case Rep. 2019;24:72–4.
56. Martin ES, Elewski BE. Tinea capitis in adult women masquerading as bacterial pyoderma. J Am Acad Dermatol. 2003;49:177–9.
57. Iwanaga T, Ushigami T, Anzawa K, Mochizuki T. Pathogenic dermatophytes survive in nail lesions during oral terbinafine treatment for tinea unguium. Mycopathologia. 2017;182:673–9.
58. Seebacher C. Action mechanisms of modern antifungal agents and resulting problems in the management of onychomycosis. Mycoses. 2003;46:506–10.

59. Baran R. The nail in the elderly. Clin Dermatol. 2011;29:54–60.
60. Loo DS. Onychomycosis in the elderly. Drugs Aging. 2007;24:293–302.
61. Nenoff P, Krüger C, Paasch U, Ginter-Hanselmayer G. Mycology – an update part 3: dermatomycoses: topical and systemic therapy. J Dtsch Dermatol Ges. 2015;13:387–411.

Laboratory Diagnosis of Dermatophytosis

Sophie Brun and Marc Pihet

Abstract

Dermatophytoses are a very common cause of consultation in dermatology. The diagnosis of these superficial infections, which include a large variety of diseases involving skin, nails, and hair, is usually not clinically obvious and requires mycological analysis, particularly for scalp and nail infections. Direct microscopic examination of clinical samples using clearing agents provides a quick response and should be combined with cultures on specific media. Histological analysis, which is considered the "gold standard" for the diagnosis of onychomycosis, is seldom performed, and as direct examination, it only allows visualization of fungal elements without precise identification of the pathogen. Accurate identification of the causative agent is necessary to initiate an appropriate treatment and also for setting prophylactic measures. However, conventional methods often lack sensitivity, and species identification from cultures may require up to 4 weeks. Thus, the use of fluorochromes may greatly enhance the sensitivity of direct examination. Likewise, the use of culture media containing antifungal deactivators could enhance the sensitivity of cultures. Finally, several PCR-based kits are now commercially available; nevertheless, the biological diagnosis of dermatophytosis still relies on standard mycological procedures, including direct examination and cultures of appropriate clinical specimens, which remain the most informative and the least expensive procedures.

S. Brun
Parasitology-Mycology Department, Avicenne Hospital, AP-HP, Sorbonne Paris Nord University, Bobigny, France

M. Pihet (✉)
Parasitology-Mycology Department, Health Biology Institute, Angers University Hospital, Angers, France

Host-Pathogen Interactions Study Group (EA 3142), University of Angers, Angers, France
e-mail: mapihet@chu-angers.fr

J.-P. Bouchara et al. (eds.), *Dermatophytes and Dermatophytoses*,
https://doi.org/10.1007/978-3-030-67421-2_17

Keywords

Dermatophytes · Dermatophytosis · Mycological diagnosis · Direct examination · Specific culture media

1 Clinical Context

Dermatophytoses are related to the invasion of skin, nails, and hair by dermatophytes. These ubiquitous filamentous fungi produce keratinolytic enzymes and may degrade keratin (hairs, horns, feathers) in the environment, as well as in animals and humans. They are thus responsible in humans for most superficial mycoses affecting the skin (*Tinea corporis*, *Tinea cruris*, *Tinea pedis*, . . .), hair (*Tinea capitis*), beard (*Tinea barbae*), or nails (*Tinea unguium*). Due to a hyperspecialization, some dermatophyte species may affect only human keratin (*Epidermophyton floccosum*, *Microsporum audouinii*, *Trichophyton rubrum*, *Trichophyton tonsurans*, *Trichophyton soudanense*, . . .).

Although usually not painful and strictly superficial, these infections should not be neglected, as they may affect the quality of life of the patients. In addition, *Tinea pedis* and onychomycosis may be risk factors for the development of an acute cellulitis of the leg or erysipela, particularly for patients with diabetes [1]. Likewise, a severe immunodeficiency or a rare genetic susceptibility may allow dissemination into the skin and / or a deep infection.

Dermatophytoses are infections very commonly encountered in private dermatological practice, since they affect up to one-third of the population [2]. According to Moreno and Arenas [3], *Tinea unguium* represents about 50% of etiologies of nail diseases and this could be even greater in the future with the aging population. However, the clinical differentiation of dermatophytosis from other nonmycotic dermatitis may be difficult. Symptoms are usually nonspecific and may also be atypical, particularly for dystrophic nails or recurring skin lesions. In most cases, contribution of the laboratory is required to confirm the diagnosis, particularly when a systemic or a prolonged treatment is needed (as for *Tinea capitis* or *Tinea unguium*), because of its cost and its potential side effects [4]. Indeed, a prolonged antifungal therapy for several weeks or months is often necessary for successful treatment of onychomycosis.

Moreover, precise identification of the fungal isolate may guide the clinician to set up prophylactic measures, such as intrafamily investigation or pets' treatment, faced to the detection of an anthropophilic or a zoophilic dermatophyte species, respectively. This is particularly important for some anthropophilic dermatophyte species because of their potential to cause intrafamily or scholar epidemics [5].

The absence of standardization of sampling and mycological procedures, as well as the lack of commercialized reagents, have hampered for a while the diagnosis of dermatophytosis. Due to the recent availability of PCR-based kits, the place of the conventional diagnosis is now frequently questioned [6]. Sampling methods according to the localization and the type of lesions, the techniques used for direct examination, and the different steps for isolation and identification of the main dermatophyte species encountered in clinical or veterinary practice are discussed in this chapter.

2 Objectives

The biological diagnosis of dermatophytosis allows: (*i*) to confirm the diagnosis of mycosis, first with direct microscopic examination of biological samples; (*ii*) to identify the fungus involved, especially for epidemiological purposes; and (*iii*) to follow up the therapeutic efficacy, particularly in the context of contagious *Tinea capitis*. Conversely, a negative result leads the diagnosis to other skin, nail, or hair disorder (nummular eczema, *pityriasis rosea*, psoriasis, *alopecia areata, ...*).

In medical laboratory, the mycological diagnosis of dermatophytosis should include the following steps:

- Collection, during examination of the patient, of clinical information (epidemiological context, risk factors, evolution of lesions, previous antifungal treatments);
- Examination under UV light of the whole scalp to detect fluorescence in case of *Tinea capitis*;
- Obtention of a high-quality clinical sample;
- Detection of the fungus in tissues by direct microscopic examination of biological samples;
- Isolation of the dermatophyte by cultures on specific media;
- Identification of the fungus at the genus or species level;
- Interpretation of the results, particularly in case of discrepancy between direct examination and cultures, or when molds are detected by cultures.

3 Methods

3.1 Patient Interview

In addition to clinical data, questioning the patient may provide epidemiological information that may be essential for the diagnosis of dermatophytosis. It can help to define the mode of contamination and eventually to identify the dermatophyte involved in the infection, and also to propose prophylactic measures.

Arthroconidia are responsible for the dissemination of dermatophytes. These spores are able to resist in the soil, on the fur of mammals, and on fomites such as clothes, shoes or toys. Contamination occurs through direct (human, animal, or telluric contact) or, more often, through indirect contact via contaminated objects. Typical examples of contamination are contact with an infected animal (young cat and *Microsporum canis*, cattle and *Trichophyton verrucosum*), with the soil (gardening and *Nannizzia gypsea*), with objects (contaminated clippers and *M. audouinii* or *T. tonsurans*), or with hair in children's communities with *Tinea capitis*. Questioning the patient about the practice of particular sports (swimming, judo, wrestling,. . .), some professional contributing factors (e.g. safety footwear), contact with pets or cattle, stays in tropical area, . . . may be helpful to define how the contamination occurred.

The development of dermatophytes on their hosts depend on general and local factors (age, genetic predisposition, underlying diseases such as diabetes or immunodeficiency), and also on immunosuppressive treatments, especially corticosteroids. The growth of the fungus requires a local "permissiveness" of the epidermis, hair follicle, or nail: skin scratch, skin maceration, pre-pubertal absence of sebaceous secretion (*Tinea capitis* in children), and nail micro-traumas. Usually, anthropophilic dermatophytes affect preferably covered parts of the body without inflammatory reaction, whereas zoophilic and geophilic dermatophytes are more often responsible for inflammatory lesions on uncovered parts of the body.

3.2 Specimen Collection

Quality of the sampling notably affects the performances of the different steps of mycological examination. Sampling needs to be adapted to the localization of lesions: skin, nail, or hair fragments are thus collected with sterile equipment in sterile containers. In the case of multiple lesions, specimens from different sites must be collected separately. A sufficient amount of sample should be harvested from the edge of the infected area, which corresponds to the active zone of the lesion, where the fungus is still active. Nails for example should be sampled at the junction between the healthy areas and the affected ones. An optimal sampling should allow collecting enough material to perform microscopic examination, as well as cultures, ideally on two media. Some authors recommend cleaning the lesion site with alcohol before sampling, in order to remove contaminants such as bacteria or conidia from saprophytic fungi [7]. Due to electrostatic attraction, specimens should be collected in sterile glass recipients (Petri dishes) rather than plastic containers. Moreover, to avoid false-negative cultures, samples should be performed before any local or systemic antifungal treatment [8]. An absence of treatment for at least 3 months should be proposed in case of systemic treatment or local application of film-forming solutions, and for 15 days for a topical antifungal treatment [8]. Packaging and transport of samples will be performed in sterile and hermetically closed containers. As dermatophytes are able to survive for a long time in a dry atmosphere and at room temperature in skin, hair, and nail fragments, their shipment is allowed without risk of damage.

Ringworms and intertriginous lesions should be scraped from the edge, at the periphery of lesions (on potential vesicles), using a dermal blunt curette (Brocq's curette) rather than scalpel blades or vaccinostyles. Swabbing with a swab previously moistened with sterile saline may be useful after scraping, to harvest the remaining material on the skin. Upon examination under UV light (Wood's lamp) of an inguino-crural lesion, less frequently an axillary or interdigito-plantar lesion, a red fluorescence ("brick" or "coral") is suggestive of erythrasma (due to *Corynebacterium minutissimum*).

The sampling technique greatly influences the culture yield for nail lesions [9], and methods differ according to the type of onychomycosis. Indeed, in subungual onychomycosis, viable fungal hyphae may be encountered only near the nail bed, particularly at the edge of the lesion. The free edge of the nail will be cut and discarded because of its frequent contamination by environmental molds. All the subungual material will be collected using a blunt curette rather than a vaccinostyle, until the limit of detachment of the nail plate. Conversely, for white superficial onychomycosis (leukonychia), scratching the white areas at the surface of the nail plate allows collecting infected material. Microscopic examination of nail samples requires very thin nail fragments that may be obtained by conventional scraping, but some authors recommend the use of a microdrill to obtain nail dust [9]. In vivo confocal microscopy has also been proposed for direct examination by Piérard [10]. This noninvasive method allows the visualization of fungal hyphae in situ using a microscope objective directly applied on the nail surface, but its cost has hampered its use in routine laboratories [11].

In case of *Tinea capitis*, hair roots and crusts are scraped with a dermal blunt curette, then a swab is applied to harvest the remaining material on the scalp. Apart from very inflammatory, oozing, or suppurating lesions for which the sampling with a curette would be painful, the exclusive use of swabbing should be avoided. In the above-mentioned cases, at least two humidified sterile swabs should be applied, one for direct microscopic examination and the other for cultures. In folliculitis and trichophytic *Tinea capitis*, hair is broken usually very shortly (black dots in the Petri dish), whereas in microsporic *Tinea capitis*, hair is broken at 2–3 mm of the scalp. Hair fragments will be collected by scraping, or picked from the infected area using tweezers with slight traction (infected hair is coming very easily). Sampling of hair should be preceded by examination of the whole scalp under UV light (Wood's lamp). The metabolic pathways used by some dermatophytes species (*M. canis*, *M. audouinii* and *Trichophyton schoenleinii*) to degrade keratin lead to the presence, in colonized tissues, of spontaneously fluorescent compounds under UV light. Thus, the use of Wood's lamp, through a bright green fluorescence, may attest of a microsporic or a favic *Tinea capitis* and guide the sampling procedure. Dermatophytosis cannot be ruled out in the absence of fluorescence, as no fluorescence is usually observed in microid, megasporic, or endothrix *Tinea capitis*. Additionally, asymptomatic carriers may be detected by rubbing the whole scalp or hair with two humidified sterile swabs in human, or using a sterile piece of carpet or a tooth-brush for pets [12]. Otherwise, associated lesions should also be sampled.

According to the European Onychomycosis Observatory, most general physicians and about 60% of dermatologists do not perform sampling in case of a suspected dermatophytosis [13]. This may explain the inefficiency of some antifungal treatments and may be detrimental for the patients taking into account the high frequency of onychomycosis due to non-dermatophytic molds, which are largely more difficult to eradicate. Before beginning treatment, it is therefore highly recommended to take nail or hair samples for mycological analysis, which should be performed in a specialized laboratory with an experienced staff and according to rigorous procedures [8].

3.3 Direct Examination

Direct microscopic examination is an efficient screening technique, as it allows a quick response, thus allowing clinicians to prescribe an antifungal treatment pending culture results [2]. Isolation and identification of the fungus may indeed take several weeks. The detection of hyphae by direct examination of the clinical samples confirms the diagnosis of mycosis (Fig. 1). This simple technique is achievable in most laboratories of microbiology and in less than 1 h. Examination is usually done at $\times 10$ or $\times 20$ magnification for branching hyphae, with confirmation using $\times 40$ objective. It may be performed with skin scales, hair, or nails fragments, or with serous fluids from inflammatory lesions (paronychia, kerion, or sycosis).

Microscopic examination may also allow the detection of different morphological aspects which may help to specify the type of fungus involved, especially in onychomycosis (regular hyphae, sometimes dissociated in arthroconidia for dermatophytes; irregular and poorly colored or vesicular hyphae in a mold infection; blastoconidia associated or not to pseudohyphae in yeast infections). In case of *Tinea*

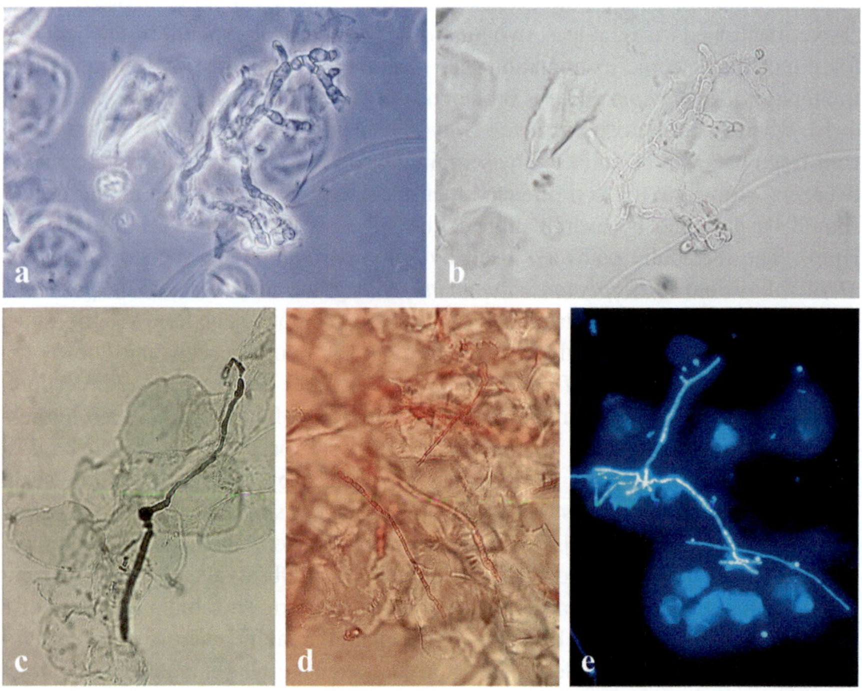

Fig. 1 Direct examination of nail fragments showing septate hyphae. Phase-contrast microscopy (**a**) increases visibility and contrast compared to observation under white light illumination (**b**). Fungal elements are more easily visualized using stains such as chlorazol black E (**c**) or Congo red (**d**), but their detection is greatly facilitated using fluorochromes such as calcofluor white (**e**) (**e**, courtesy of Pr. Raymond Robert, Angers, France)

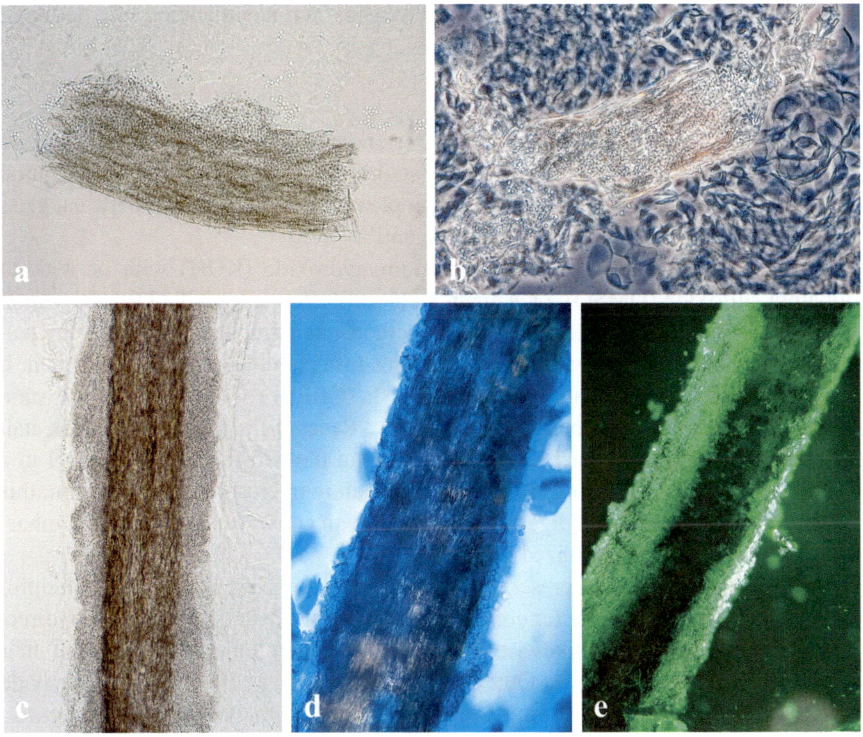

Fig. 2 Direct examination of hair fragments showing arthroconidia inside (**a** and **b**—endothrix) or outside hair shaft (**c**, **d**, and **e**—ectothrix) using white light illumination (**a**, **c**), phase-contrast microscopy (**b**), lactic blue (**d**) or Blankophor (**e**)

capitis or *Tinea barbae*, the determination of the hair invasion pattern (endothrix, ectothrix, or favic) can steer towards the dermatophyte involved (Fig. 2).

Nevertheless, the quality of the specimen and the skill of the observer determine the success of mycological examination. Indeed, up to 30% of false-negative results have been reported in routine practice [14]. Sensitivity of direct examination appears to be lower when samples are uneasy to be collected, particularly in case of proximal onychomycosis or total onychodystrophy. Conversely, too superficial or too distal sampling in onychomycosis may also highlight conidia from saprophytic molds, leading to false-positive results [15]. In some cases, cultures remain sterile or the absence of sporulation of the isolated fungus hampers its identification. Diagnosis of dermatophytosis will be formally maintained, regardless of the subsequent outcome of cultures. Thus, about one-third of the diagnoses of dermatophytosis (nearly half of the diagnoses of dermatophytic onychomycosis of the feet) are established on direct microscopic examination exclusively.

The clinician will be quickly informed of the positivity of direct microscopic examination, as a preliminary response ("presence of fungal hyphae, probable dermatophytosis"). A complement of response will then be addressed with the culture results (negativity or isolation of a fungus, then identification of a dermatophyte, or a mold on nail samples).

3.3.1 Clearing Agents

Before direct microscopic examination, samples need to be treated with a dissociating agent (combined or not with stain), which facilitates the visualization of fungal elements by keratin digestion. Samples are usually mounted between glass slide and coverslip in a drop of clearing agent.

Most laboratories use 10–30% potassium hydroxide (KOH) with or without dimethyl sulfoxide (DMSO), or Amann's chloral-lactophenol (CL) [16]. Digestion by KOH, which is one of the cheapest methods, is particularly suitable for nail fragments and other thick material, but requires fast examination of the specimen. It may be supplemented with chlorazole black E (CBE) or Congo red, which stain fungal elements in black or red, respectively. Recently, a method for thick nail samples was proposed, which involves dissolving the nail fragments in KOH in a microtube during several hours to several days before microscopic examination: this technique enables complete digestion of keratin with conservation of fungal hyphae, which are more easily visualized [17].

The use of CL allows to preserve the fungal structures and is particularly helpful to study the hair invasion (endothrix, ectothrix, or favic type) or when direct examination is differed. Likewise, it is recommended for conservation of hair samples for teaching purposes. Other dissociating agents as 10% sodium hydroxide (NaOH), or detergents as sodium dodecyl sulfate (SDS), have also been proposed.

Furthermore, phase-contrast microscopy facilitates the observation of unstained specimens, since it provides a clear difference in brightness between the fungal elements and the background, resulting in greater sensitivity (Figs. 1a, b and 2b).

3.3.2 Stains and Fluorochromes

Use of stains or fluorochromes may allow easier visualization of fungal hyphae or conidia, particularly for nail specimens, and thus can increase sensitivity of direct examination (Figs. 1c–e and 2e) [18, 19]. Some vegetable fibers may also be stained, but in most cases they are easily distinguished from fungal hyphae by their irregular shape and larger size, and the absence of septa.

Various stains, which can be associated with clearing agents, have been proposed [20, 21]. Cotton blue C4B (Bacti-lab inc., R.A.L. or Bio-Rad, associated with lactic acid and phenol) and Blue-Black Ink permanent (Parker Quink®), or CBE (Sigma-Aldrich) stain the fungal elements in deep blue or black, respectively [20]. However, these stains are not specific and some artifacts ("mosaic-fungus") may be seen. Nevertheless, in the study performed by Lilly et al. [16], KOH-CBE was as sensitive (94.3% vs. 98.8%, NS) as histological analysis using periodic acid-Schiff (PAS), which stains polysaccharides such as glycosaminoglycans. Congo red, which has been proposed by Slifkin and Cumbie [22] for direct examination, is commercially available in a kit containing SDS for the digestion of keratinized structures (Mycetcolor®, Biosynex). Fungal elements are red-colored on a pink or light orange background with this stain, which binds to polysaccharides of the fungal cell wall

Table 1 Comparison of the performances of various stains for direct microscopic examination in suspected cases of dermatophytoses [19]

	Direct examination (number of positive samples)			
Clinical samples (total number)	CL	KOH + CBE	MycetColor®	MycetFluo®
Nails ($n = 51$)	14	13	13	25
Glabrous skin ($n = 38$)	16	16	16	20
Hair ($n = 13$)	13	13	13	13
Total ($n = 102$)	43	42	42	58

CL chloral-lactophenol, *KOH* potassium hydroxide, *CBE* chlorazol black E

such as β-D-glucans, and therefore allows to exclude some artifacts [19]. Moreover, it offers a better safety profile than chlorazole black. Chicago Sky Blue 6B (CSB stain®) is another stain associating rapid detection and low cost, and allowing better contrast [23].

When a fluorescence microscope with appropriate filter sets (400–440 nm blue filter) is available, the use of fluorochromes is highly recommended such as calcofluor white (CW), Blankophor P Flüssig® (Bayer) or Uvitex 2B [24]. These distilbene derivatives bind to chitin, a polymer of *N*-acetyl-D-glucosamine, which is one of the main polysaccharides of the fungal cell wall. They greatly facilitate the detection of fungal hyphae and spores, and thus improve the sensitivity and reduce the reading time. Calcofluor white or Blankophor (0.1%, w/v) may be used in KOH 10–30% (1:1, v/v) or in SDS (as in MycetFluo® from Biosynex, which is a CW-based kit), allowing immediate observation of the preparation. Fungal elements in CW- or Blankophor-stained specimens appear blue when using a fluorescence microscope equipped with a 330–380 nm excitation filter and an emission filter of > 420 nm, or brightly fluoresce in green with a fluorescein filter system. In a study performed on 240 nail samples, Abdelrahman et al. [25] reported a higher specificity and sensitivity using CW versus KOH alone (88% and 72%, respectively; $p = 0.0116$). In another study performed on 102 samples (skin, hair, and nail specimens) from patients with suspected dermatophytosis, MycetFluo® allowed a higher number of positive samples to be detected compared to other methods (Table 1) [19]. Indeed, fungal elements, which leap out on a dark field, are easily identified even by a nonexperienced staff, particularly in thick specimens such as nail samples (Fig. 1). This was also highlighted by Ovrén et al. [26], who showed that the reading time of direct examination was considerably reduced using Blankophor staining, particularly with an unexperienced staff. Therefore, the use of fluorochromes associated with clearing agents should be recommended as it enhances the sensitivity of direct examination and reduces the reading time, especially for large series of samples.

3.3.3 Histological Examination

In onychomycosis and deep mycosis, histological examination of biopsy remains the "gold standard" for direct examination. Achten's technique with Gomori's methenamine silver (GMS) and PAS stain seems to be the more sensitive method for diagnosis of onychomycosis [27]. GMS has recently been proposed by some authors as a confirmatory test following screening by KOH [28]. Histological examination of nail plate may allow confirming the diagnosis of fungal infection with direct in situ visualization of fungal elements: hyphae appear red (PAS) or gray-black (GMS) within the host tissue [29]. False-negative results may thus be excluded. However, results may depend on the quantity of sampling and on the clinical presentation of onychomycosis. Indeed, false-negative results may be observed, mainly in early nail infections [30]. In addition, histological analysis does not allow species identification. Some morphological traits of hyphae may suggest a mold rather than a dermatophyte. In particular, some mold species such as *Scopulariopsis brevicaulis* may produce thin perforating hyphae that induce micro-fractures of nails [31]. A recent immunochromatographic test using monoclonal antibodies showed promising results for detection of dermatophytes on alcohol- and formalin-fixed samples, but it is restricted to the *Trichophyton* genus [32]. Accurate identification of the pathogen cannot be achieved through histology alone, and mycological cultures remain necessary [2]. Moreover, biopsies are relatively invasive procedures; therefore they are seldom performed, and more simple methods have been proposed.

The Hotchkiss and MacManus's staining technique, adapted from PAS of pathologists, stands for a good compromise between direct examination and histology [33]. The stain reveals the presence of the fungus (purplish-red "fuchsia" stained) in nails and allows to define its morphology. It also permits a long preservation of the preparations. By the analysis of toenail clippings from 108 patients, Blake et al. [34] found that PAS stain had the highest sensitivity, compared to direct examination using CW and fungal cultures. Other authors also proposed cyanoacrylate surface skin scraping (CSSS), a strip biopsy where a fragment of horny layer is removed together with an adhesive, as a simple diagnostic tool for direct examination [35]. CSSS showed a negative predictive value of 92% and a positive predictive value of 52% compared to culture, but it can only be used for skin lesions.

3.4 Cultures

Pending improvement and generalization of molecular methods, fungal culture is still needed for definitive identification. It is a valuable complement to direct examination and histological analysis, which do not provide identification at the species level. Indeed, precise identification is useful, since prophylactic measures may vary depending on the dermatophyte species. Moreover, in case of onychomycosis or *Tinea capitis*, which both require prolonged therapy, confirmation of the diagnosis can help to improve medication adherence [2]. However, cultures often lack sensitivity, they are time-consuming and identification of filamentous fungi at the species level using morphological characteristics requires an experienced staff. The increasing use of MALDI-TOF mass spectrometry in microbiology laboratories should facilitate the identification of dermatophyte species from cultures, particularly in case of atypical isolates [36].

3.4.1 Isolation of Dermatophytes

Samples are usually inoculated on Sabouraud's agar supplemented with antibiotic (s) (chloramphenicol and/or gentamicin) and cycloheximide (Actidione®), and eventually with vitamins. Indeed, the use of a selective medium is essential to avoid contamination by bacteria, yeasts, or saprophytic fungi that can be present on skin or hair surface. Cycloheximide inhibits or limits the growth of most contaminating molds. However, as some cycloheximide-susceptible fungi, including molds like *Neoscytalidium dimidiatum* and some pathogenic yeasts, may be responsible for nail, sole, or palm infections, Sabouraud's agar without cycloheximide should be used in parallel to selective medium. Several Sabouraud's media with or without cycloheximide are commercially available as agar plates or slants (Becton-Dickinson, bioMérieux, Bio-Rad, . . .). Agar plates offer a larger surface than slants for fungal growth. However, the use of agar slants is often preferred in case of prolonged incubation in order to avoid drying of the culture medium.

If the Sabouraud's agar denomination remains the same, the composition and the pH of the culture medium may vary from one manufacturer to another. This may result in variations of the performances, as already demonstrated for yeasts and opportunistic molds [37]. Alternatively, commercially available potato dextrose (EO Labs) or potato flake (Remel) agar containing cycloheximide and chloramphenicol can be used for primary isolation instead of Sabouraud's agar [38].

Cultures are usually incubated at 25 ± 2 °C, but higher incubation temperatures (30–32 °C) can be used if *T. verrucosum* infection is suspected. Cultures should be monitored once or twice a week. The delay for optimal growth varies according to the species. Colonies of *E. floccosum* or *N. gypsea* will be identified in 4–5 days, but a longer time (up to 2 weeks) may be necessary for other dermatophytic species. So, an incubation time of 3–4 weeks is usually recommended before considering cultures as negative. However, through a study including 5459 dermatophyte isolates collected over a 4-year period, Rezusta et al. [39] showed that a 17-days incubation time was enough to recover 98.8% of isolates and could make the technique more effective. Nevertheless, the authors do not evoke the slow growing species such as *T. verrucosum* or the pseudodermatophyte *Arachnomyces nodosetosus* (formerly *Onychocola canadensis*), which show typical morphological traits only after 3–4 weeks.

Plates or agar slants with the Dermatophyte Test Medium (DTM, Becton-Dickinson) initially proposed by Taplin [40], have been reported for presumptive identification of dermatophytes. Due to the alkaline by-products generated during growth of dermatophytes, the color of this medium changes to deep red. However, several false-positive as well as some false-negative results have been reported, and modifications were proposed to overcome the limitations of the former [41, 42]. Several in-office culture systems based on the use of DTM are currently available for veterinary (Dermatophytes kit, Biovet) or medical practice (InTray™ DM, BioMed Diagnostics), and are also commercialized in dual-compartment plates with DTM on one side and Sabouraud's dextrose agar (Sab-Duet®, Bacti-Lab) or rapid sporulating medium (RSM) (Derm-Duet®, Bacti-Lab) on the other side. However, the usefulness of such devices in mycology laboratories is controversial. For some authors, DTM culture seems to be an effective and convenient method to confirm the diagnosis of onychomycosis and initiate the treatment without further identification [43]. Conversely, Scherer et al. [44] found that only 50% of positive DTM cultures correlated with a positive culture for dermatophytes, and that the association of direct examination using a fluorochrome and cultures in a mycology laboratory was superior to in-office DTM cultures for the diagnosis of onychomycosis.

3.4.2 Identification Culture Media

Sabouraud isolation medium, which contains dextrose, promotes mycelial growth, but sporulation may be lacking; so when identification of dermatophyte isolates is not achievable directly from primary cultures, subcultures may be performed on specific media that stimulate both conidiation and the production of pigments. The morphology of the dermatophytes indeed greatly differs according to the culture medium. Some media allow differentiation of the species (Christensen's urea agar medium, rice-agar-Tween, BCP-casein), while others help to identify specific dermatophytes (brain-heart-agar, peptone 3% agar, ...) [21]. Several media are commonly used, but to our knowledge, only potato dextrose agar (Difco), brain-heart infusion (Sigma), malt chloramphenicol agar (bioMérieux), and Borelli's lactrimel agar (Biosynex) media are commercially available. In the study performed by Gumral et al. [45], Borelli's lactrimel medium demonstrated the lowest rate of contamination by molds.

3.5 Identification of Dermatophytes

In clinical laboratories, identification of dermatophytes classically relies on morphological traits of the fungus from macroscopic and microscopic examination of the cultures, thus requiring skill and expertise. In addition, complementary tests for atypical isolates, as search for urease activity on commercially available urea-indole broth or Christensen's urea agar medium, or the in vitro hair perforation test on sterile light hair [46], are of limited interest, because of the taxonomic changes in this group of filamentous fungi and of the development of molecular tools. Dermatophytes species were initially classified in three genera *Epidermophyton*, *Microsporum*, and *Trichophyton*, but recent phylogenetic studies based on multi-loci sequencing resulted in a major revision of their taxonomy [47]. Nine genera are now differentiated: the *Arthroderma* and *Nannizzia* genera, which initially included the sexual forms of dermatophytes, are today considered as distinct entities from the *Microsporum* and *Trichophyton* genera, and new genera were described such as *Lophophyton* and *Paraphyton*. For more details on the present taxonomy of dermatophytes, refer to the corresponding chapter in this book.

Observation of teleomorph states of dermatophytes, which are heterothallic species, may be useful for the identification, but mating tests (i.e. sexual reproduction using complementary (+ and −) isolates of the same species) are limited to specialized laboratories. Likewise, nutritional requirements for some vitamins or amino acids, as well as assimilation of sorbitol or other carbon or nitrogen sources, have been reported in the identification process, but they can be studied only in specialized laboratories [48].

Other potential tools have also been proposed for species identification of dermatophytes, such as Fourier transform infrared spectroscopy (FTIR-S) [49] or determination of the volatile organic compound profiles by gas chromatography-mass spectrometry [50]. Nevertheless, the potential interest of these methods has been hampered by the generalization in microbiology laboratories of MALDI-TOF mass spectrometry [51]. This technology, which is less costly than DNA sequencing, is useful for the identification of dermatophytes from cultures, but preliminary experiments also showed its ability to identify *T. rubrum* directly from nail samples [52]. CE-marked commercial databases are poorly developed for filamentous fungi, leading to the development of "home" reference spectra libraries for more accurate species identification. The contribution of mass spectrometry to the diagnosis of dermatophytosis will be developed in a specific chapter.

3.6 Alternative Methods to the Conventional Diagnosis

As false-negative results are frequently reported for direct examination, methods have been proposed to overcome the lack of sensitivity of microscopy, while providing a faster response than culture to the clinician.

Molecular tools displayed the highest sensitivity compared to other methods in many studies [53]. Several kits are now commercially available, allowing detection of fungal DNA directly from clinical samples in less than 48 h. Some of these kits enable the identification of dermatophytes species, molds, or yeasts responsible for cutaneous mycoses, whereas others (FTD® Dermatophytes, Fast-track Diagnostics) focus exclusively on the most commonly encountered species (*T. rubrum, T. interdigitale, T. tonsurans, M. canis, M. audouinii*, ...). Other kits target only *T. rubrum* DNA (Dermatophyte PCR kit®, SSI Diagnostica) or provide a "pan-dermatophyte" DNA detection (Bio-Evolution).

In addition, several molecular approaches have been developed for identification at the species level of dermatophytes from cultures or biological samples, such as PCR-RFLP, ITS sequencing, or oligonucleotide hybridization, but these tests are still restricted to a few reference laboratories. For further details, see the specific chapter in this book.

Recently, an immunochromatographic test (Dermatophyte Test Strip—Diafactory *Tinea unguium*, Biosynex; or FungiCheck *Tinea unguium*, HFL Laboratories), which allows the fast detection of an antigen common to different dermatophyte species from nail samples, was evaluated by Tsunemi et al. [54]. In this multicenter study, it provided a very good positive concordance rate with direct microscopy and polymerase chain reaction (PCR). This is a reliable, convenient, and quick method to test for *Tinea unguium*; nevertheless, this screening test requires confirmation by a specialist and its rather high cost has to be considered in daily practice.

4 Interpretation of Results

4.1 Discrepancies Between Direct Examination and Cultures

Despite the presence of fungal hyphae on direct microscopic examination, primary cultures may remain negative for dermatophytes (Table 2). This negativity does not invalidate the diagnosis nor attests a clinical recovery when control samples are performed, as long as the direct examination remains positive. The discrepancies between direct examination and cultures may be explained by an insufficient amount of collected material, the presence of nonviable fungal hyphae in old lesions, or by an antifungal treatment initiated before sampling. False-negative results of cultures may also be related to a too short incubation time, a nonsuitable temperature, or the presence of contaminating molds which can prevent the development of the pathogen. Antifungal drugs remain for a long time within the horny layer of the epidermis and the nail tablet, and may thus inhibit the growth of the pathogen in primary cultures. A medium containing lecithin and polysorbate 80 was developed, which minimizes the carryover effect of antifungals [70]. Compared to Sabouraud's dextrose agar, this medium called CDSAM (Combined Deactivators-Supplemented Agar Medium) revealed, in a study encompassing 44 patients with *Tinea pedis*, a greater efficiency for the recovery of dermatophytes from patients who had already undergone an antifungal treatment [71].

Table 2 Discrepancies between direct examination and primary cultures for the detection of dermatophytes [48]

Reference	Number of samples	Clearing agent and stain	Percentage of positivity	
			Direct examination	Cultures
Arca et al. [55]	52	KOH 20%	77	23
Kardjeva et al. [56]	195	KOH 30%	93	22
Arabatzis et al. [57]	92	KOH	43	33
Garg et al. [58]	152	KOH 20%	63.4	25
Gupta et al. [59]	62	KOH 20%	77.4	22.6
Bontems et al. [60]	4363	CW	53	13
Kondori et al. [61]	177	Blankophor	46	34
Litz and Cavagnolo [62]	550	KOH 10%	40	22
Beifuss et al. [63]	204	CBE	100	59.8
Li et al. [64]	104	KOH 15%	50.9	27.8
Luk et al. [65]	120	KOH 10%	29.2	10
Chandran et al. [66]	107	KOH 30%	100	35.5
Dhib et al. [67]	201	KOH 30%	71.1	65.6
Mehlig et al. [68]	253	CBE	34.4	31.6
Wlodek et al. [69]	38,591	Not specified	59.2	35.5

KOH potassium hydroxide, *CW* calcofluor white, *CBE* chlorazol black E

4.2 Detection of Molds on Primary Cultures of Nail Specimens

Difficulties in interpreting culture results frequently occur when only molds like *Fusarium*, *Aspergillus*, or *Scopulariopsis* species are recovered from nail fragments. Species identification has to be compared to the results of direct microscopic examination, usually showing irregular and vesicular hyphae for molds. The abundance of fungal elements should also be specified. In some cases, typical conidia from *S. brevicaulis* or brown hyphae from dematiaceous fungi such as *N. dimidiatum*, may be visualized [72].

The clinical sample may contain dead dermatophyte material, and the molds present may colonize the skin or nail surface. Indeed, up to 20% of clinical specimens from patients with *Tinea unguium* contain contaminating molds associated with a dermatophyte [73]. In addition, the presence of some rapidly growing molds may hinder the recovery of dermatophytes. However, although uncommon, real mixed infections may occur. Isolation of the same fungal species in pure culture from successive specimens (ideally in most seeding points on the agar plate or slant) will allow establishing the pathogenicity of the mold [74]. It is unlikely to recover the same mold as a saprophyte in another sampling, but this seems to be widely misunderstood by practitioners [38].

5 Conclusion

Treatment of dermatophytosis such as *Tinea capitis* or onychomycosis often requires several months of antifungal therapy. Therefore, an accurate diagnosis is essential before starting treatment. Epidemiological and clinical data are important, because they can help identifying the dermatophyte, especially for *Tinea corporis* and *Tinea capitis*. Moreover, the isolation and identification of a dermatophyte allow specifying the mode of contamination and thus enable to predict the complementary measures (intrafamilial investigation or in a collectivity, treatment of pets, . . .). Conventional mycological analysis may display a low sensitivity due to the quality of clinical sample, to the technique used or to a previous antifungal treatment, and the slow growth of dermatophytes in vitro may delay the final result. Identification at the species level has been impeded for a while by the lack of commercially available specific culture media and the low discriminating power of historical tests. However, progress in this field are expected from the development of fungal databases for dermatophyte identification using MALDI-TOF mass spectrometry. Finally, molecular tools using PCR-based kits are now commercially available but they target only one or a few dermatophyte species. Therefore, despite an excellent sensitivity and specificity in many studies and a faster response using PCR, routine diagnosis of these superficial mycoses still relies on conventional methods. Microscopic visualization of fungal elements in clinical samples, which is greatly facilitated by the use of fluorochromes, remains essential, allowing a quick response to clinicians in order to initiate antifungal treatment. Indeed, a positive direct examination confirms by itself the diagnosis of mycosis, whatever the subsequent result of the cultures.

Progress, however, are needed in culture media to increase sensitivity of fungal cultures that yet remain an indispensable complement to direct examination. Nevertheless, regardless of the techniques used, the quality of the results primarily depends on the quality of sampling and on the expertise of the mycologist.

Acknowledgements The authors thank Pr. Raymond Robert (Angers, France) for providing microphotographs of direct examination of nail scales using CW (MycetFluo®).

References

1. Bristow IR, Spruce MC. Fungal foot infection, cellulitis and diabetes: a review. Diabet Med. 2009;26(5):548–51. https://doi.org/10.1111/j.1464-5491.2009.02722.x.
2. Feuilhade de Chauvin M. Mycological analysis in dermatology. Ann Dermatol Venereol. 2018;145(10):623–32. https://doi.org/10.1016/j.annder.2018.05.006.
3. Moreno G, Arenas R. Other fungi causing onychomycosis. Clin Dermatol. 2010;28(2):160–3. https://doi.org/10.1016/j.clindermatol.2009.12.009.
4. Hainer BL. Dermatophyte infections. Am Fam Physician. 2003;67(1):101–8.
5. Vingataramin Y, Akhoundi M, Bruel C, Izri A, Brun S. Epidemiological and molecular characterization of a *Trichophyton tonsurans* epidemic in schools of a city in the northern suburbs of Paris, France. Clin Microbiol Infect. 2019;25(4):529–30. https://doi.org/10.1016/j.cmi.2018.11.026.
6. Petinataud D, Berger S, Ferdynus C, Debourgogne A, Contet-Audonneau N, Machouart M. Optimising the diagnostic strategy for onychomycosis from sample collection to FUNGAL identification evaluation of a diagnostic kit for real-time PCR. Mycoses. 2016;59(5):304–11. https://doi.org/10.1111/myc.12471.
7. Elewski BE. Onychomycosis: pathogenesis, diagnosis, and management. Clin Microbiol Rev. 1998;11(3):415–29.
8. Groupe de travail de la Société Française de Dermatologie, Recommandations pour la pratique clinique. Onychomycoses : modalités de diagnostic et de prise en charge. Ann Dermatol Venereol. 2007;134:5S7–16.
9. Qureshi HS, Ormsby HA, Kapadia N. Effects of modified sample collection technique on fungal culture yield: nail clipping/scraping versus microdrill. J Pak Med Assoc. 2004;54(6):301–5.
10. Piérard GE. *In vivo* confocal microscopy: a new paradigm in dermatology. Dermatology. 1993;186(1):4–5.
11. Rothmund G, Sattler EC, Kaestle R, Fischer C, Haas CJ, Starz H, et al. Confocal laser scanning microscopy as a new valuable tool in the diagnosis of onychomycosis - comparison of six diagnostic methods. Mycoses. 2013;56(1):47–55. https://doi.org/10.1111/j.1439-0507.2012.02198.x.
12. Moriello KA. Diagnostic techniques for dermatophytosis. Clin Tech Small Anim Pract. 2001;16(4):219–24. https://doi.org/10.1053/svms.2001.27597.
13. Effendy I, Lecha M, Feuilhade de Chauvin M, Di Chiacchio N, Baran R, European Onychomycosis Observatory. Epidemiology and clinical classification of onychomycosis. J Eur Acad Dermatol Venereol. 2005;19(Suppl 1):8–12. https://doi.org/10.1111/j.1468-3083.2005.01281.x.
14. Summerbell RC, Cooper E, Bunn U, Jamieson F, Gupta AK. Onychomycosis: a critical study of techniques and criteria for confirming the etiologic significance of nondermatophytes. Med Mycol. 2005;43(1):39–59.
15. Gianni C, Cerri A, Crosti C. Unusual clinical features of fingernail infection by *Fusarium oxysporum*. Mycoses. 1997;40:455–9.
16. Lilly KK, Koshnick RL, Grill JP, Khalil ZM, Nelson DB, Warshaw EM. Cost-effectiveness of diagnostic tests for toenail onychomycosis: a repeated-measure, single-blinded, cross-sectional

evaluation of 7 diagnostic tests. J Am Acad Dermatol. 2006;55(4):620–6. https://doi.org/10.1016/j.jaad.2006.03.033.
17. Ishida H, Noriki S. Novel detection technique of dermatophytes: tube method. J Dermatol. 2019; https://doi.org/10.1111/1346-8138.15037. [Epub ahead of print].
18. Elewski BE. Diagnostic techniques for confirming onychomycosis. J Am Acad Dermatol. 1996;35(3 Pt 2):S6–9.
19. Pihet M, Clement N, Kauffmann-Lacroix C, Nail-Billaud S, Marot A, Pilon F, et al. Diagnosis of dermatophytosis: an evaluation of direct examination using MycetColor® and MycetFluo®. Diagn Microbiol Infect Dis. 2015;83(2):170–4. https://doi.org/10.1016/j.diagmicrobio.2015.06.022.
20. Panasiti V, Borroni RG, Devirgiliis V, Rossi M, Fabbrizio L, Masciangelo R, et al. Comparison of diagnostic methods in the diagnosis of dermatomycoses and onychomycoses. Mycoses. 2006;49(1):26–9. https://doi.org/10.1111/j.1439-0507.2005.01185.x.
21. Robert R, Pihet M. Conventional methods for the diagnosis of dermatophytosis. Mycopathologia. 2008;166(5–6):295–306. https://doi.org/10.1007/s11046-008-9106-3.
22. Slifkin M, Cumbie R. Congo red as a fluorochrome for the rapid detection of fungi. J Clin Microbiol. 1988;26(5):827–30.
23. Prakash R, Prashanth HV, Ragunatha S, Kapoor M, Anitha TK, Krishnamurthy V. Comparative study of efficacy, rapidity of detection, and cost-effectiveness of potassium hydroxide, calcofluor white, and Chicago sky blue stains in the diagnosis of dermatophytoses. Int J Dermatol. 2016;55(4):e172–5. https://doi.org/10.1111/ijd.13037.
24. Hamer EC, Moore CB, Denning DW. Comparison of two fluorescent whiteners, Calcofluor and Blankophor, for the detection of fungal elements in clinical specimens in the diagnostic laboratory. Clin Microbiol Infect. 2006;12(2):181–4. https://doi.org/10.1111/j.1469-0691.2005.01321.x.
25. Abdelrahman T, Letscher-Bru V, Waller J, Noacco G, Candolfi E. Dermatomycoses : comparaison des performances du Calcofluor et de la potasse à 30% pour l'examen direct des squames et des ongles. J Mycol Med. 2006;16:87–91.
26. Ovren E, Berglund L, Nordlind K, Rollman O. Dermatophytosis: fluorostaining enhances speed and sensitivity in direct microscopy of skin, nail and hair specimens from dermatology outpatients. Mycoses. 2016;59(7):436–41. https://doi.org/10.1111/myc.12491.
27. Velasquez-Agudelo V, Cardona-Arias JA. Meta-analysis of the utility of culture, biopsy, and direct KOH examination for the diagnosis of onychomycosis. BMC Infect Dis. 2017;17(1):166. https://doi.org/10.1186/s12879-017-2258-3.
28. Karaman BFO, Acikalin A, Unal I, Aksungur VL. Diagnostic values of KOH examination, histological examination, and culture for onychomycosis: a latent class analysis. Int J Dermatol. 2019;58(3):319–24. https://doi.org/10.1111/ijd.14255.
29. Piérard GE, Arrese JE, Pierre S, Bertrand C, Corcuff P, Leveque JL, et al. Microscopic diagnosis of onychomycoses. Ann Dermatol Venereol. 1994;121(1):25–9.
30. Suarez SM, Silvers DN, Scher RK, Pearlstein HH, Auerbach R. Histologic evaluation of nail clippings for diagnosing onychomycosis. Arch Dermatol. 1991;127(10):1517–9.
31. English MP. Nails and fungi. Br J Dermatol. 1976;94(6):697–701.
32. Noriki S, Ishida H. Production of an anti-dermatophyte monoclonal antibody and its application: immunochromatographic detection of dermatophytes. Med Mycol. 2016;54(8):808–15. https://doi.org/10.1093/mmy/myw037.
33. Lachaud L, Sasso M, Rispail P, Bourgeois N. Biological diagnosis of onychomycoses. Direct examination after simplified technique of PAS staining. J Mycol Méd. 2014;24(4):279–86. https://doi.org/10.1016/j.mycmed.2014.10.007.
34. Blake N, Zhu J, Hernandez G, Juliano PJ. A retrospective review of diagnostic testing for onychomycosis of the foot. J Am Podiatr Med Assoc. 2015;105(6):503–8. https://doi.org/10.7547/14-063.1.
35. Lousbergh D, Buntinx F, Piérard GE. Diagnosing dermatomycosis in general practice. Fam Pract. 1999;16(6):611–5.

36. L'Ollivier C, Ranque S. MALDI-TOF-based dermatophyte identification. Mycopathologia. 2017;182(1–2):183–92. https://doi.org/10.1007/s11046-016-0080-x.
37. Brun S, Bouchara JP, Bocquel A, Basile AM, Contet-Audonneau N, Chabasse D. Evaluation of five commercial Sabouraud gentamicin-chloramphenicol agar media. Eur J Clin Microbiol Infect Dis. 2001;20(10):718–23.
38. Weitzman I, Summerbell RC. The dermatophytes. Clin Microbiol Rev. 1995;8(2):240–59.
39. Rezusta A, de la Fuente S, Gilaberte Y, Vidal-Garcia M, Alcala L, Lopez-Calleja A, et al. Evaluation of incubation time for dermatophytes cultures. Mycoses. 2016;59(7):416–8. https://doi.org/10.1111/myc.12484.
40. Taplin D, Zaias N, Rebell G, Blank H. Isolation and recognition of dermatophytes on a new medium (DTM). Arch Dermatol. 1969;99(2):203–9.
41. Salkin IF, Padhye AA, Kemna ME. A new medium for the presumptive identification of dermatophytes. J Clin Microbiol. 1997;35(10):2660–2.
42. Gromadzki S, Ramani R, Chaturvedi V. Evaluation of new medium for identification of dermatophytes and primary dimorphic pathogens. J Clin Microbiol. 2003;41(1):467–8.
43. Pariser D, Opper C. An in-office diagnostic procedure to detect dermatophytes in a nationwide study of onychomycosis patients. Manage Care. 2002;11(3):43–8, 50.
44. Scherer WP, Kinmon K. Dermatophyte test medium culture versus mycology laboratory analysis for suspected onychomycosis. A study of 100 cases in a geriatric population. J Am Podiatr Med Assoc. 2000;90(9):450–9.
45. Gumral R, Dogen A, Ilkit MM. Comparison of the contamination rates of culture media used for isolation and identification of dermatophytes. Turk J Med Sci. 2015;45(3):587–92.
46. Shadomy HJ, Philpot CM. Utilization of standard laboratory methods in the laboratory diagnosis of problem dermatophytes. Am J Clin Pathol. 1980;74(2):197–201.
47. de Hoog GS, Dukik K, Monod M, Packeu A, Stubbe D, Hendrickx M, et al. Toward a novel multilocus phylogenetic taxonomy for the dermatophytes. Mycopathologia. 2017;182 (1–2):5–31. https://doi.org/10.1007/s11046-016-0073-9.
48. Pihet M, Le Govic Y. Reappraisal of conventional diagnosis for dermatophytes. Mycopathologia. 2017;182(1–2):169–80. https://doi.org/10.1007/s11046-016-0071-y.
49. Bastert J, Korting HC. Diagnostic methods for species-specific differentiation of various dermatophytes. Possibilities and limitations. Dtsch Med Wochenschr. 1998;123(16):499–503. https://doi.org/10.1055/s-2007-1024001.
50. Verscheure M, Gofflot S, Beguin H, Marlier M, Belot JL, Nolard N, et al. Evaluation of volatile metabolites as taxonomic tool for identification of dermatophytes. Mycoses. 2002;45(Suppl 2):67.
51. Cassagne C, Normand AC, L'Ollivier C, Ranque S, Piarroux R. Performance of MALDI-TOF MS platforms for fungal identification. Mycoses. 2016;59(11):678–90. https://doi.org/10.1111/myc.12506.
52. Hollemeyer K, Jager S, Altmeyer W, Heinzle E. Proteolytic peptide patterns as indicators for fungal infections and nonfungal affections of human nails measured by matrix-assisted laser desorption/ionization time-of-flight mass spectrometry. Anal Biochem. 2005;338(2):326–31. https://doi.org/10.1016/j.ab.2004.12.019.
53. Verrier J, Monod M. Diagnosis of dermatophytosis using molecular biology. Mycopathologia. 2017;182(1–2):193–202. https://doi.org/10.1007/s11046-016-0038-z.
54. Tsunemi Y, Hiruma M. Clinical study of Dermatophyte Test Strip, an immunochromatographic method, to detect tinea unguium dermatophytes. J Dermatol. 2016;43(12):1417–23. https://doi.org/10.1111/1346-8138.13348.
55. Arca E, Saracli MA, Akar A, Yildiran ST, Kurumlu Z, Gur AR. Polymerase chain reaction in the diagnosis of onychomycosis. Eur J Dermatol. 2004;14(1):52–5.
56. Kardjeva V, Summerbell R, Kantardjiev T, Devliotou-Panagiotidou D, Sotiriou E, Graser Y. Forty-eight-hour diagnosis of onychomycosis with subtyping of *Trichophyton rubrum* strains. J Clin Microbiol. 2006;44(4):1419–27. https://doi.org/10.1128/JCM.44.4.1419-1427.2006.

57. Arabatzis M, Bruijnesteijn van Coppenraet LE, Kuijper EJ, de Hoog GS, Lavrijsen AP, Templeton K, et al. Diagnosis of common dermatophyte infections by a novel multiplex real-time polymerase chain reaction detection/identification scheme. Br J Dermatol. 2007;157(4):681–9. https://doi.org/10.1111/j.1365-2133.2007.08100.x.
58. Garg J, Tilak R, Singh S, Gulati AK, Garg A, Prakash P, et al. Evaluation of pan-dermatophyte nested PCR in diagnosis of onychomycosis. J Clin Microbiol. 2007;45(10):3443–5. https://doi.org/10.1128/JCM.02367-06.
59. Gupta AK, Zaman M, Singh J. Fast and sensitive detection of *Trichophyton rubrum* DNA from the nail samples of patients with onychomycosis by a double-round polymerase chain reaction-based assay. Br J Dermatol. 2007;157(4):698–703. https://doi.org/10.1111/j.1365-2133.2007.08110.x.
60. Bontems O, Hauser PM, Monod M. Evaluation of a polymerase chain reaction-restriction fragment length polymorphism assay for dermatophyte and nondermatophyte identification in onychomycosis. Br J Dermatol. 2009;161(4):791–6. https://doi.org/10.1111/j.1365-2133.2009.09291.x.
61. Kondori N, Abrahamsson AL, Ataollahy N, Wenneras C. Comparison of a new commercial test, Dermatophyte-PCR kit, with conventional methods for rapid detection and identification of *Trichophyton rubrum* in nail specimens. Med Mycol. 2010;48(7):1005–8. https://doi.org/10.3109/13693781003743130.
62. Litz CE, Cavagnolo RZ. Polymerase chain reaction in the diagnosis of onychomycosis: a large, single-institute study. Br J Dermatol. 2010;163(3):511–4. https://doi.org/10.1111/j.1365-2133.2010.09852.x.
63. Beifuss B, Bezold G, Gottlober P, Borelli C, Wagener J, Schaller M, et al. Direct detection of five common dermatophyte species in clinical samples using a rapid and sensitive 24-h PCR-ELISA technique open to protocol transfer. Mycoses. 2011;54(2):137–45. https://doi.org/10.1111/j.1439-0507.2009.01771.x.
64. Li HC, Bouchara JP, Hsu MM, Barton R, Chang TC. Identification of dermatophytes by an oligonucleotide array. J Clin Microbiol. 2007;45(10):3160–6. https://doi.org/10.1128/JCM.00829-07.
65. Luk NM, Hui M, Cheng TS, Tang LS, Ho KM. Evaluation of PCR for the diagnosis of dermatophytes in nail specimens from patients with suspected onychomycosis. Clin Exp Dermatol. 2012;37(3):230–4. https://doi.org/10.1111/j.1365-2230.2011.04258.x.
66. Chandran NS, Pan JY, Pramono ZA, Tan HH, Seow CS. Complementary role of a polymerase chain reaction test in the diagnosis of onychomycosis. Australas J Dermatol. 2013;54(2):105–8. https://doi.org/10.1111/ajd.12027.
67. Dhib I, Fathallah A, Yaacoub A, Hadj Slama F, Said MB, Zemni R. Multiplex PCR assay for the detection of common dermatophyte nail infections. Mycoses. 2014;57(1):19–26. https://doi.org/10.1111/myc.12096.
68. Mehlig L, Garve C, Ritschel A, Zeiler A, Brabetz W, Weber C, et al. Clinical evaluation of a novel commercial multiplex-based PCR diagnostic test for differential diagnosis of dermatomycoses. Mycoses. 2014;57(1):27–34. https://doi.org/10.1111/myc.12097.
69. Wlodek C, Trickey A, de Berker D, Johnson EM. Trends in laboratory-diagnosed onychomycosis between 2006 - 2014 in the south west of England. Br J Dermatol. 2017;176(1):237–40. https://doi.org/10.1111/bjd.14804.
70. Nakashima T, Nozawa A, Ito T, Majima T, Yamaguchi H. Development of a new medium useful for the recovery of dermatophytes from clinical specimens by minimizing the carryover effect of antifungal agents. Microbiol Immunol. 2002;46(2):83–8.
71. Adachi M, Watanabe S. Evaluation of combined deactivators-supplemented agar medium (CDSAM) for recovery of dermatophytes from patients with tinea pedis. Med Mycol. 2007;45(4):347–9. https://doi.org/10.1080/13693780601186531.
72. Gugnani HC, Oyeka CA. Foot infections due to *Hendersonula toruloidea* and *Scytalidium hyalinum* in coal miners. J Med Vet Mycol. 1989;27(3):167–79.

73. Summerbell RC. Epidemiology and ecology of onychomycosis. Dermatology. 1997;194(Suppl 1):32–6.
74. Meireles TE, Rocha MF, Brilhante RS, Cordeiro Rde A, Sidrim JJ. Successive mycological nail tests for onychomycosis: a strategy to improve diagnosis efficiency. Braz J Infect Dis. 2008;12 (4):333–7.

MALDI-TOF-Based Identification of Dermatophytes

Anne-Cécile Normand, Alicia Moreno-Sabater, Arnaud Fekkar, and Renaud Piarroux

Abstract

Dermatophytes and other keratinophilic fungi are common pathogens. Yet, their correct morphological identification remains a challenge for the mycologist. In the past few years, matrix-assisted laser desorption/ionization–time-of-flight mass spectrometry (MALDI-TOF MS) has become a widely used technique for pathogen identification in microbiology laboratories. The application of mass spectrometry for the identification of dermatophytes faces specific difficulties and benefits from a limited number of studies. In this chapter, we review the general principles of mass spectrometry and detail the technical aspects, such as sample preparation, type of targets, and scope of the different matrixes available. Various systems are currently available that use different basic spectrum processing and variable identification algorithms. We describe these different systems and the

A.-C. Normand
AP-HP, Groupe Hospitalier La Pitié-Salpêtrière, Service de Parasitologie Mycologie, Paris, France

A. Moreno-Sabater
AP-HP, Groupe Hospitalier Saint-Antoine, Service de Parasitologie Mycologie, Paris, France

Sorbonne Université, Inserm, CNRS, Centre d'Immunologie et des Maladies Infectieuses, Cimi-Paris, Paris, France

A. Fekkar
AP-HP, Groupe Hospitalier La Pitié-Salpêtrière, Service de Parasitologie Mycologie, Paris, France

Sorbonne Université, Inserm, CNRS, Centre d'Immunologie et des Maladies Infectieuses, Cimi-Paris, Paris, France

R. Piarroux (✉)
AP-HP, Groupe Hospitalier La Pitié-Salpêtrière, Service de Parasitologie Mycologie, Paris, France

Sorbonne Université, Inserm, Institut Pierre Louis d'Epidemiologie et de Santé Publique, Paris, France
e-mail: renaud.piarroux@aphp.fr

J.-P. Bouchara et al. (eds.), *Dermatophytes and Dermatophytoses*,
https://doi.org/10.1007/978-3-030-67421-2_18

different databases available to date and their respective performance for the identification of dermatophytes and propose a review of the literature.

Keywords

Mass spectrometry · Devices · Methods · Algorithms · Database · Assessment

1 Introduction: Why Use Mass Spectrometry to Make an Accurate Identification?

Dermatophytosis is a common cutaneous infection exhibiting a large range of clinical presentations [1]. The main species implicated in human pathology are *Trichophyton rubrum*, *Trichophyton interdigitale*, *Trichophyton mentagrophytes*, *Trichophyton tonsurans*, *Trichophyton violaceum*, *Trichophyton soudanense*, *Epidermophyton floccosum*, *Microsporum canis*, and *Microsporum audouinii*, but physicians can encounter many other species. Even if identification methods have to deal with numerous species, mycological diagnosis is fundamental to identify the etiological agent or to guide clinicians to a differential diagnosis with similar skin disorders [2]. It could also be useful to guide the treatment which may depend on the incriminated dermatophyte species. Thus, for *Tinea capitis* recent reports have revealed that griseofulvin has a higher cure rate in *Microsporum* spp. infections than in *T. tonsurans* infections [3], and inversely, terbinafine shows a higher cure rate in children infected with *Trichophyton* spp., than in children infected with *Microsporum* spp. [4, 5]. Dermatophyte identification is also valuable for identifying the source of contamination. Thus, about a dozen possible species, from the three different origins, anthropophilic, zoophilic, or geophilic, can be implicated in *Tinea corporis* infections, needing the implementation of different prophylactic measures to prevent their spread.

In addition to its contribution to the therapeutic patient management, accurate identification of dermatophytes is also the first step of epidemiological surveys. Indeed, the geographical distribution of dermatophyte species is multifarious and differs between continents [6, 7]. In the last 20 years, we have witnessed changes in the dermatophyte epidemiology intimately linked to globalization. These changes included the increased prevalence of a known dermatophytes species in certain areas such as *T. tonsurans* [8], but they were also related to the import of exotic dermatophytoses after international travel, which has forced mycologists to alert the medical community about the risk of dealing with uncommon dermatophytic species [9, 10] such as *Trichophyton concentricum*. Epidemiology is also being impacted by emerging zoophilic species, such as *Trichophyton benhamiae,* or *Trichophyton quinckeanum,* most of them transmitted by new pets [11–15]. Although less frequent than anthropophilic or zoophilic species, new species of geophilic dermatophytes may be responsible for emerging diseases in the future [16]. Another important but less studied factor that can impact dermatophyte epidemiology is the increased number of long-term immunosuppressed patients

[17]. Altogether, these findings probe the need for a correct identification of dermatophytes for patient management as well as for the survey of epidemiological trends influenced by modern trends in population movements, social habits, or medical treatments.

Classical morphological identification of dermatophytes is realized after culture of infected skin, nail, or hair samples [18]. Identification is performed by macroscopic examination of colony morphology and microscopic features. Morphological identification of dermatophytes is particularly difficult as they can show variations in the color and shape of colonies but also be devoid of microscopic features such as macroconidia or microconidia, hindering their identification. While morphological differences inside a same genus and species may induce a misdiagnosis, they are now well described and can be overcome by the expertise of a well-trained mycologist. Nevertheless, identification becomes challenging for those species which are morphologically related and when complete anamnesis of the patient or contacts with a certain animal or pets does not make it possible to orientate towards a particular species either. In these cases, sequencing of the internal transcribed spacer (ITS) regions 1 and 2 of ribosomal DNA (rDNA) or the beta tubulin gene may allow to overcome the limitation of the morphological methods [19, 20]. However, despite the clear advantage of fungal identification by molecular biology over the morphological diagnosis, major improvements still need to be done. Identification by sequence comparison is often problematic using the publicly available NCBI database since strains can be misidentified by sequence depositors [21]. Furthermore, taxonomic classification is not yet clearly established. In the last years, taxonomic changes have been proposed depending of the fungal genes used for sequencing [22]. Following this new classification, the number of genera has increased, enhancing the difficulties for their identification. Even if studies have confirmed this new taxonomy [23], phylogeny for the genera *Epidermophyton* and *Nannizzia* was contradictory to the ITS phylogeny, casting doubt on the status of these genera and on the capacity of the molecular methods to become the gold standard for dermatophyte identification. Furthermore, PCR methods have shown a great progress for dermatophyte identification in epidemiological studies, but their benefits for clinical routine analysis of dermatological samples have to be put in balance with the price of molecular biology reagents, and especially the time spent conducting laboratory manipulations. In this context, matrix-assisted laser desorption/ionization–time-of-flight mass spectrometry (MALDI-TOF MS) can represent an alternative, which will be presented here.

2 Principle of MALDI-TOF MS Identification

The identification of infectious agents, whether prokaryotic or eukaryotic, has undergone a revolution in recent years with the advent of MALDI-TOF mass spectrometry, a diagnostic approach invented in 1987 by Koichi Tanaka who further earned the Nobel Prize in 2002 [24]. Koichi Tanaka showed that the protein molecules, when adequately placed in a solid or viscous phase, can be ionized using soft laser desorption, demonstrating the applicability of laser technology to biological macromolecules. MALDI-TOF MS stands for Matrix Assisted Laser Desorption Ionization Time of Flight Mass spectrometry. Its principle consists in preparing a biological sample, covering it with a so-called matrix, introducing it into a mass spectrometer, ionizing the molecules and measuring the time that each molecule takes to cover the distance between the origin and the detector [25]. The result is a spectrum that can be compared to a database of spectra, whether it is a "homemade" or marketed database.

To perform MALDI-TOF MS identification, a biological sample is deposited on a plate, also called a target, then dried and covered by an organic compound that is able to absorb energy. The role of this compound, called "matrix," is to facilitate the ionization process. Proteins, mainly ribosomal proteins, of the biological sample co-crystallize with the matrix that covers them, forming complexes. Then, the target is manually introduced into the machine through an airlock chamber in which the vacuum has been made. There the protein–matrix complexes are submitted to hundreds of laser shots whose wavelength is located on the post-UV light (337 nm). Due to the energy absorbed by the complexes, a soft ionization takes place and a gaseous cloud is formed, without fragmentation or decomposition of the molecules as it was the case with old mass spectrometry techniques. The complexes formed by the proteins of the sample and the matrix are energetically ablated from the target. The matrix molecules are temporarily excited and ionized by the laser shots. However, as they are very unstable, they soon transfer one positive ion (H+) to the nearby proteins. Thus, every protein of the biological sample carries one (and generally only one) positive charge in the gaseous cloud at the time they are reaching the flight tube. Charged molecules are then accelerated in an electric field that applies a constant kinetic energy on the ions. As the kinetic energy is equal to half the product of the mass and the squared velocity, the constant kinetic energy allows the ions with the lowest mass to travel the fastest in the tube, leading to the progressive sorting of the proteins according to their increasing mass. The calculation of the time spent by the groups of ionized molecules to travel through the time of flight tube enables analytic software to calculate the mass of the proteins impacted on the detector located at the end of the tube, following the equation: $m/z = 2eE(t/d)^2$ (where E is the accelerating voltage, e is the elementary charge, t is the time of flight, d is the length of the tube, m is the mass of the ionized molecule and z is its charge). The separation between the molecules can be enhanced by a reflection mode in which the detector at the end of the tube is replaced by an ion reflector that can be assimilated to a mirror. In such case the detector is located in the midst of the tube, in opposite direction of the initial flight path. The various steps described here are illustrated in Fig. 1.

The data produced by the detector—here a series of intensity values reflecting the successive arrival of charged molecules over time—are then analyzed by a computer which summarizes them in a succession of intensity peaks during the course of time, each intensity peak translating the presence of molecules with increasing m/z. The obtained profile, which can be considered as a bar code, is further compared to other profiles corresponding to a set of known references. Various algorithms have been

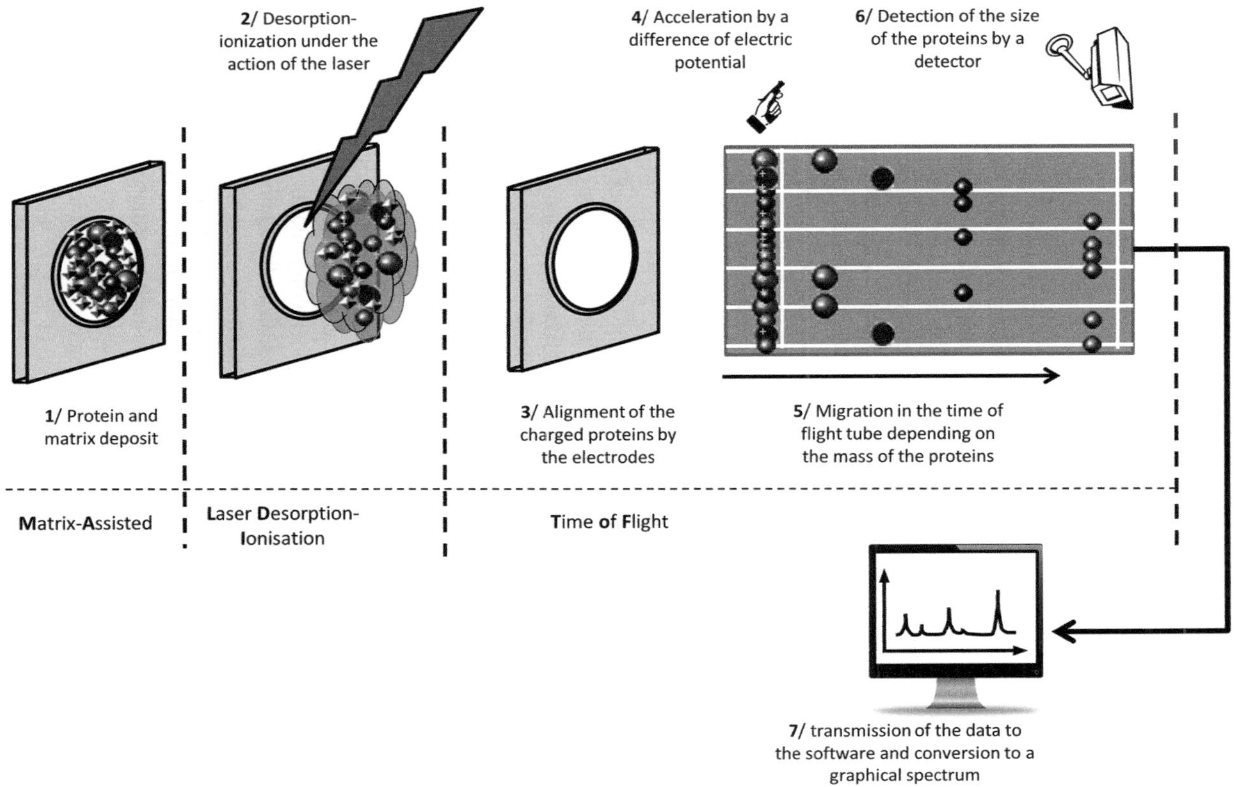

Fig. 1 General principle of operation of a MALDI-TOF type mass spectrometer

designed to perform this comparison and to quantify the level of similarity between the spectrum to be identified and the different reference spectra. The best score obtained indicates the most similar reference spectrum and therefore the most likely identification. Many factors at each stage have a significant impact on the reliability of the identification process.

3 MALDI-TOF MS Devices

There are currently four suppliers that share the market of MALDI-TOF MS in the world: Bruker with the Microflex machine, bioMérieux with the Vitek MS machine, Shimadzu with the Axima Performance machine, and Asta with the IDSys LT machine. The Vitek MS machine and the Axima Performance machine are both manufactured by Shimadzu. In every case, the spectral acquisition process uses similar analytical methods based on ionized molecule separation, signal detection, and amplification [26]. Differences between the systems can be found in sample preparation, spectra preprocessing, reference databases, and individual spectra comparison against the reference spectra.

Regarding the brand of the device, studies report the use of different mass spectrometers with dermatophytes: AXIMA (by Shimadzu) [27, 28], Bruker [29–31], Vitek (bioMérieux) [32], or Andromas [33] (this latter being no more available). To date, the newly developed Korean IDSys LT (ASTA) has not been assessed in scientific publications for dermatophytes, but only for yeast identification [34]. On dermatophytes, the different brands produce similar results; the crucial point being not the mass spectrometer itself but the exhaustiveness of the database used with regard to the dozens of dermatophyte species likely to be found in pathology.

4 Culture and Sample Preparation

With few exceptions that do not concern the subject treated here, mass spectrometry requires a culture step. Regarding the identification of dermatophytes, the most commonly used medium is Sabouraud chloramphenicol-gentamicin, with or without cycloheximide. A new medium, ID-Fungi plates (Conidia, Quincieux, France), has recently been marketed [35]. This culture medium is covered with a membrane intended to facilitate the recovery of the fungal colony for identification by mass spectrometry. Aside from the practical side, the culture medium type has little or no impact on the performance of MALDI-TOF identification. Majority of laboratories use the Sabouraud dextrose agar (supplemented with antibiotics). Some publications report the use of other media, which do not compromise the identification. Alternatively to the Sabouraud culture medium, Erhard et al. [27] tested the use of potato dextrose agar, malt extract agar, and Kimmig agar. Although the mass spectra varied, this did not substantially affect the identification results obtained. As for the study published by Theel et al. [36], six different media were tested, namely chocolate agar, inhibitory mold agar, modified dermatophyte test medium, Mycosel agar, nutrient agar, and Sabouraud agar. Although the results are not detailed, the authors indicate that they did not observe any significant differences depending on the culture medium used.

Different methods of preparation of biological samples have been proposed, depending on the type of microorganisms that were investigated. The easiest method is the direct deposit method that is mostly used for bacteria and that consists on depositing a smear of the colony on the target and covering it with one drop of matrix. Another extraction method, used for yeasts and for Gram-positive bacteria, consists on adding a drop of formic acid on the microbial smear on the target prior to cover it with the matrix. An extraction method for complex organisms such as mycobacteria, *Nocardia*, and filamentous fungi has also been developed. For those organisms, a first inactivation step in ethanol is done. To do so, a piece of mycelium is scraped from the culture plate and suspended in a tube containing a solution of EtOH 75% v/v (900 μL absolute ethanol in 1200 μL final volume). After inactivation, the supernatant is eliminated by centrifugation and gentle suction. Then the biological material is mixed into an equal volume of formic acid 70% and acetonitrile 100% (the volume used for this step depends on the size of the pellet and can easily be adjusted, from 10 to 50 μL of both reagents). After a last centrifugation, 1 μL of the supernatant is deposited on the target and covered with the matrix. This method is recommended by various suppliers and has been used in many studies [29, 30, 32, 36–38]. Other extraction techniques have been tested, which can be found in the medical literature [39–41]. Dermatophytes, like other fungi, have a cell wall mainly composed of polysaccharides. Since MALDI-TOF identification uses a protein spectrum, it is essential to extract the proteins properly, which requires using a method that degrades the fungal cell wall. Studies by Erhard et al. [27], Nenoff et al. [28] and Riat et al. [42] used a direct extraction approach after deposition on the target with a solution containing the matrix (in this case 2,5-dihydroxy benzoic acid) in an aqueous solution of acetonitrile acidified with trifluoroacetic acid. Other approaches were considered, including formic acid extraction with or without acetonitrile. Packeu et al. [29] compared two extraction protocols for the identification of 176 clinical isolates of dermatophytes (mainly *T. rubrum* and *T. interdigitale*) using MALDI BioTyper 3.0 software and a Microflex device (Brüker Daltonics, Germany). The authors used direct deposition of fungal material on the target with or without prior extraction by formic acid. By this approach there was only 33% correct identification without formic acid and 40% with formic acid. However, it should be noted that the reference spectra used to develop the database came from a complete extraction (formic acid + acetonitrile). The authors therefore proposed a complete extraction by formic acid and acetonitrile, which was associated with an increased growth time. As a result of these changes, they obtained far better identification performances, from 80% for 7-day-old cultures to 100% for 14-day-old cultures. Although the authors changed several parameters, it is likely that extraction plays an important role in improving these results. In this study, the use of mass spectrometry revealed an incorrect identification rate by morphological-based methods of 7%. Yet, another study reported an excellent rate of identification (97.8%) regardless of the incubation time [30].

More recently, da Cunha et al. [31] compared a previously published fast method versus a complete extraction. The fast method based on a direct transfer of mycelium onto the ground steel target overlaid with 1 μL of alpha-cyano-4-hydroxycinnamic acid matrix was found to be highly relevant for other molds [42]. Using this method, 67% of isolates of various species were correctly identified using a homemade database. However, the correct identification rate increased to 98% when the authors used the full extraction protocol described by L'Ollivier et al. [30] that uses ethanol, formic acid, and then acetonitrile. A direct deposit therefore may be an option in first intention provided that the sample is free from any agar, as this later can negatively impact the identification. It has to be followed by a complete extraction in case of failure.

5 MALDI-TOF MS Targets

Up to date, there are three types of targets that can be used for MALDI-TOF MS identification of biological samples for Bruker Microflex users and one type of target slides for bioMérieux Vitek MS users. For Bruker equipment, polished-steel targets are the most widely used targets around the world. These targets are designed for any type of sample deposit, from direct deposit to complete chemical extraction. Recently developed to enhance the quality of spectra obtained by direct deposits are the Biotargets, which are one-use targets, with a unique barcode. With this type of plate harboring an identification barcode, it is impossible for the user to use the same position twice (spot). This avoids the problems associated with insufficient washing of the plate, as the user can be sure that each position is only used once. Although they were designed for direct deposits, Biotargets proved to give also good identification spectra with the other methods of extraction on yeasts [43]. Another type of target used for microorganisms identification with Bruker equipment is the ground-steel target. This particular target is of use for direct deposits, but its ground surface renders the other extractions difficult to perform as it had been shown for yeasts [43, 44]. The targets used for Vitek MS are disposable slides that possess a unique barcode, and dedicated positions for the calibration standard [45].

6 MALDI-TOF MS Matrixes

The matrix composition also has significant impact on the quality of the spectra. The choice of the matrix to use for MALDI-TOF MS identification of biological samples depends on the type of compounds that are targeted. Most matrices are acidic, easily excited, leading to a loss of proton for the benefit of the ionization of the ribosomal protein. All available matrices have in common a conjugated pi system under the form of a benzene ring with attached functional groups. Five matrices are commonly used in mass spectrometry, and among them, three are commonly used in microbial identification: 3,5-dimethoxy-4-hydroxycinnamic acid also called sinapinic acid (SA), α-cyano-4-hydroxycinnamic acid (HCCA or CHCA), and 2,5-dihydroxybenzoic acid also called gentianic acid (DHB). HCCA and DHB are the most widely used for fingerprinting, as they both allow the detection of peptides smaller than 10,000 Daltons. HCCA is considered as more efficient than DHB in ionization, leading to a better signal intensity. DHB produces less matrix clusters, leading to less background signal and is thus interesting to use for smaller peptides and for high concentrations of peptides. In MALDI-TOF MS microbial identification, HCCA matrix is favored because it allows the detection of small amounts of peptides.

7 MALDI-TOF MS Spectra Treatment and Identification Algorithms

The different MALDI-TOF systems available operate following different algorithms, either for the treatment of the raw spectra or for the identification processes, leading to differences in the identification scoring.

The Maldi biotyper software, associated to the Bruker devices, preprocesses the raw spectrum and converts it into a mass list and the corresponding intensities. The mass list is then compared to the mass list of each reference spectra in the database. A scoring is performed, called a pattern matching, using three components. First, the Rel Score (percentage of correspondence of the reference spectrum) is calculated. To do so, the number of common peaks between the query spectrum and the reference spectrum is calculated. The Rel Score is the result of the division of the number of common peaks by the total number of peaks of the reference. Second, the Rel P-Num is calculated (percentage of correspondence of the query spectrum). Here, the number of common peaks is divided by the total number of peaks of the query spectrum. Last, the I-Corr is calculated. I-Corr means correlation of the Intensities, i.e. the sum of the ratios of the intensities. For each of the common peaks, the less intense value between the query spectrum and the reference spectrum is divided by the more intense of the two values and all ratios are added. Hence, the maximum score of the Rel Score is 1, the maximum score of the Rel P-Num is 1 and the maximum score of the I-Corr is 1. All three components are then added to obtain a score that ranges from 0 to 3. Bruker considers (for bacteria; no adaptations were officially developed for fungus, and even less for dermatophytes) that a score > 2 corresponds to a highly probable species identification, a score between 1.7 and 1.999 corresponds to a probable genus identification, and a score < 1.7 indicates a nonreliable identification [46].

bioMérieux proposes an identification platform called Vitek MS. For each query spectrum, the masses are divided into 1300 predefined intervals called bins, and ranging from 3000 to 17,000 Daltons. In each bin, only the peak with the maximum intensity is kept for comparison to the reference spectrum, and the other values in the bin are discarded. To each of the reference spectra, a species specific weight is attributed, using an algorithm called "Advances Spectra Classifier," using common peaks between several reference spectra for the same species. Then the bin scores of the query spectrum are multiplied by the weight of the weighted-bins of each species included as reference. A confidence value is then calculated for each species available. According to the manufacturer, values between 60.0 and 99.9 indicate a reliable discrimination of species [47].

The Axima/Saramis software used by Shimadzu proceed to an extraction of the peak list that comprises the mass and the intensity of each peak. The Saramis system is decomposed into two phases. During the first line identification, the query spectrum is compared to a SuperSpectra database, and if the first line does not allow a proper identification, a second test is performed. SuperSpectra are built by computing 10–20 different isolates of one particular species and the conserved mass signals are summarized in a consensus spectrum per species. Then, each mass of the consensus spectrum is compared to all of the other consensus spectra of the database and specific peaks are given a positive weight while nonspecific peaks are excluded, making random matches improbable. Since SuperSpectra contain only species-specific peaks, some isolates of a given species cannot be identified with this first-line of identification. Then, all references contained in the database are individually compared to the query spectrum as a second-line of identification.

Using the Andromas platform, the raw spectrum is also converted into a mass peak list, but only peaks with a relative intensity above 0.1 are kept for comparison. The score is calculated using the number of common peaks between the query spectrum and the reference spectrum, by dividing the number of common peaks by the total number of peaks of the reference spectrum, and it is expressed as a percentage of similarity between the sample spectrum and a database reference spectrum. The percentages are classified into three categories: a "good identification" is obtained when the score is greater or equal to 65%, and if the difference of scores with the second best hit is > 10%. In theory, all good identifications are given at the species level. If the score is comprised between 60 and 64.99%, it needs to be confirmed. Finally, if the score is lower than 60%, or if the difference between the first two proposed species is less than 10%, it is categorized into the no-identification category [48].

8 MALDI-TOF MS Databases

At the end of 2007, Erhard et al. [27] published a study assessing the accuracy of mass spectrometry for the identification of dermatophytes causing onychomycosis and *Tinea pedis*. Isolates belonging to 17 different species were used to build a homemade database (using the SARAMIS software package from AnagnosTec) in which data from individual species were summarized in consensus spectra or superspectra. These consensus spectra contained only the mass signals that were present in at least 80% of the individual mass spectra. Then the authors used 20 clinical isolates of dermatophytes characterized by their morphological features and ITS1, ITS2, and 5.8S rDNA sequencing to assess the robustness of the approach. The isolates tested belonged to the following species: *T. rubrum* ($n = 13$), *T. interdigitale* ($n = 4$), and one isolate each of *T. tonsurans*, *M. canis*, and *T. benhamiae*. All isolates belonged to species that were used to build the database and they were all successfully identified by MALDI-TOF approach with a high level of confidence, showing the effectiveness of this approach.

Since this pioneering study, commercial databases have been set up for MALDI-TOF MS identification purpose. In contrast to bacteria and yeasts, they are still poorly developed for identification of filamentous fungi, especially for identification of dermatophytes. The database marketed by bioMérieux comprises 96 species of filamentous fungi among which 18 are dermatophyte species (*E. floccosum, M. audouinii, M. canis, Nannizzia fulva, Nannizzia gypsea, Nannizzia persicolor, Nannizzia praecox, T. benhamiae, Trichophyton equinum, Trichophyton erinacei, T. interdigitale, T. mentagrophytes, T. rubrum, Trichophyton schoenleinii, Trichophyton terrestre, T. tonsurans, Trichophyton verrucosum*, and *T. violaceum*). The filamentous fungi database commercialized by Bruker totalizes 159 species among which 22 are non-dermatophytic keratinophilic fungi and 20 are dermatophyte species (*Arthroderma amazonicum, Arthroderma eboreum, Arthroderma flavescens, Arthroderma gloriae, Arthroderma lenticulare, Chrysosporium keratinophilum, Chrysosporium shanxiense, E. floccosum, M. canis, N. fulva, N. gypsea, Nannizzia incurvata, N. persicolor, N. praecox, Paraphyton cookei, T. benhamiae, T. equinum, T. erinacei, T. interdigitale, T. rubrum, T. tonsurans*, and *T. violaceum*). The Vitek database contains a low number of dermatophyte species, but the main human pathogenic species are represented. The Bruker database takes into account the new taxonomy, but the absence of *M. audouinii* is to be noted. Moreover, Bruker published a note in the release of their new database stating that the distinction between *T. equinum, T. interdigitale, T. mentagrophytes*, and *T. tonsurans*, and between *T. rubrum* and *T. violaceum* is difficult, explaining the high amount of misidentification in those two complexes when using this database. Another released note for the latest version of the fungal database is that a former *T. rubrum* reference is now a *T. interdigitale* and that a former reference of *M. equinum* is now a *Penicillium* species. By themselves those two notes explain some of the discrepancies that were described in articles on fungal identification using the Bruker database [30].

The lack of comprehensiveness of the libraries proposed by mass spectral providers led many teams to build "in-house" databases or to supplement commercial databases with homemade reference spectra. Theel et al. [36] compared the Bruker MALDI Biotyper library version 3.0 with this same database enhanced with 20 homemade reference spectra. The improved database allowed identification at the genus level in 93% of cases versus 37.4% for the Biotyper database and identification at the species level for 59.6% of isolates versus 20.5% for Biotyper. More recently, in a study published by Da Cunha et al. [31], the comparison between the Bruker Filamentous fungi library v 1.0 database and a homemade database also showed tremendous progress when adding new references: the homemade database allowed to reach up to 98% of good identification (cutoff LogScore above 1.7) against only 37% with the manufacturer database. It should be noted that even when the species were present in the database, many isolates were poorly identified (e.g. 40% of *T. rubrum* isolates).

Identification of dermatophytes is a good example illustrating the difficulties to create a suitable MALDI-TOF database. This is particularly complicated as taxonomic difficulties are still encountered with other methods, further impacting the mass spectrometric identification. In 2017, De Hoog et al. [22] proposed a new taxonomy for dermatophytes. In this article, the authors kept cautious about the delimitation between some species such as *T. equinum* and *T. tonsurans, T. rubrum* and *T. violaceum*, or *M. audouinii, M. canis* and *Microsporum ferrugineum*. Hence, the phenotypic diversity is probably larger than the genomic and protein diversity. The same tendency observed with DNA sequences probably occurs with protein profiles as closely related species are prone to have highly similar mass spectrometric profiles. However, the distinction between related species by mass spectrometry is distorted by the fact that MS-references-names lean on DNA identification. As DNA identification of dermatophyte species is blurred between some species, the capacity of mass spectrometry to distinguish between some closely related species depends on the accuracy of the identification of the references which varies according to changes in the taxonomy.

In addition, whereas DNA sequencing of fungal colonies does not change with the aging of the culture, it is not the case with protein expression, and a MS profile obtained under special culture conditions may differ from the profile obtained for the same fungus cultivated under standard growth conditions. Consequently, the profile obtained may vary according to the age of the cultures. Likewise, changing the growth temperature, or the culture medium may impact the identification results obtained with mass spectrometry as exposed in Fig. 2 for one isolate of *T. tonsurans* grown at 30 °C on three different culture media, for 4–20 days.

Finally, even if every MALDI-TOF user is free to add references in their database, they must respect the recommendations of the manufacturers when using a commercial database if their purpose is to obtain a certification for their laboratory. In the certification objectives, the validation of an in-house reference database and of local identification algorithms requires the validation of the process by one or several scientific publications or at least by a thorough local validation of the different identification parameters that are modified.

In 2017, an online application has been developed that contains an upgradable database built from complete formic acid/acetonitrile extraction obtained from solid Sabouraud agar cultures. The objective was to allow matching spectra from cultures as young as possible. One advantage of this application is that every user around the word can have access through internet to a large database built in for identification purposes. Moreover, as online users from many countries share the same identification approach with the same database, this opens the door to large clinical and epidemiological studies as already done for other molds genera [49]. The tool can easily be upgraded with new references, inappropriate references can be deleted, and names of the reference strains can be changed in regard to the current taxonomy. The MSI identification algorithm has proven to be more efficient than the commercial algorithm by Bruker, even when references are built with the same strains [50]. In 2019, following a multicentric study on pathogenic *Aspergillus* species, the online identification system has been upgraded with references of cryptic *Aspergillus*

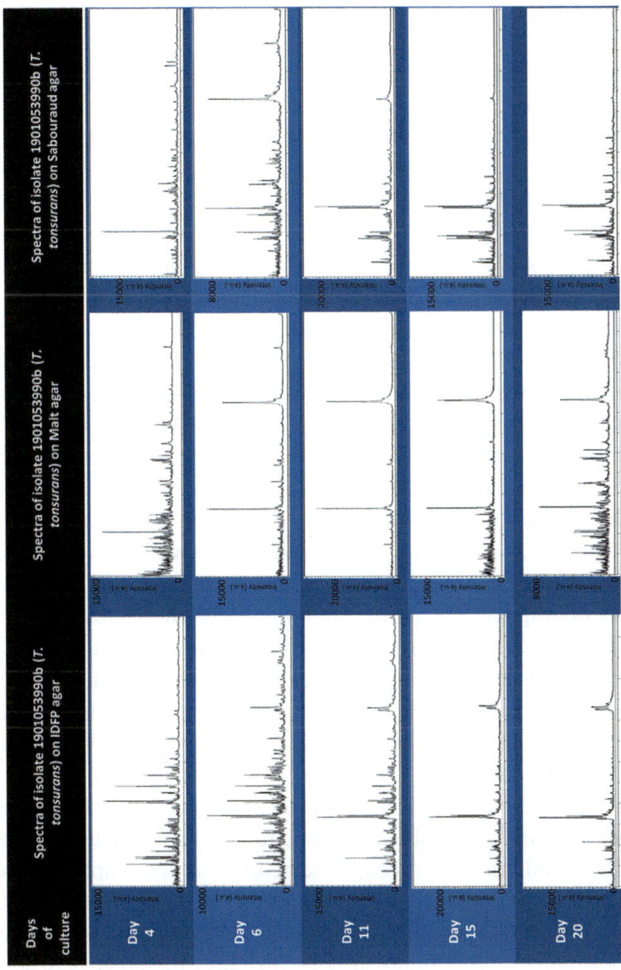

Fig. 2 Variations of mass spectra as a function of the growth medium and the age of the culture

species, allowing the application to discriminate between closely related species. The same upgrading is ongoing regarding dermatophytes, in order to provide the users a larger panel of reference spectra, using several culture media and consistent with physiological changes of the fungus with maturation and aging of the colonies.

9 Global Results of Dermatophyte Identification by MALDI-TOF MS, Limits of the Method

A search in the NCBI Pubmed database with the occurrences "dermatophytes AND MALDI-TOF" gave only 35 items in August 2019. This particularly low result shows how far we still have to go in this area. The intrinsic difficulties of identification, linked not only to the pleomorphism (phenomenon of disappearance of characteristic elements) of isolates but also to the variability of phenotypic characteristics or evolutionary taxonomy, explain the limited number of studies devoted to this theme. The main results of these different studies are summarized in Table 1.

Most studies used homemade databases to enhance their correct identification rates. Several studies report that it is appropriate to lower the identification threshold recommended by the manufacturer, thus leading to a gain in sensitivity in identification without loss of specificity. On the contrary, the use of database held by the mass spectrometer manufacturers leads to poor identification levels. In the most recently available publication on the topic by Tartor et al. [54], the authors reported the use of Biotyper version 3.1, and the reference database version 3.4.119.0 from Bruker Daltonics yielded good identification for *M. canis* but was unable to correctly identify *T. verrucosum*, *T. violaceum*, and *T. mentagrophytes*.

This overview of bibliographic data and our medical practice in a routine laboratory allow us to highlight some particular points regarding the identification of dermatophytes by MALDI-TOF mass spectrometry. First, although MALDI-TOF mass spectrometry represents an advance over the morphological identification of dermatophytes, progress is still needed at all steps of the process from culture to final analysis of the digital data. Starting from the culture, the recovery of a fungal fragment is sometimes complicated, some colonies being strongly rooted in the culture medium. The recent marketing of specific media covered with a membrane will facilitate this step and allow obtaining spectra not contaminated by the protein peaks of the culture medium [35]. As explained earlier in this chapter, an extraction based on acid associated with acetonitrile provides better results than a rapid method in which the fungus extract is placed directly on the plate and covered by the matrix. This additional extraction step significantly complicates the process of identifying dermatophytes by mass spectrometry and makes the automation of the process more complicated. In the future, solutions will have to be proposed to improve this step and to make it as reproducible as possible. Once the sample processed, MALDI-TOF identification of dermatophytes faces the same pitfalls as the identification of other filamentous fungi. In most cases, the results are satisfactory, provided that the species is present in the database, the difficulty being to have access to the most exhaustive databases possible in terms of number of represented species. However,

Table 1 Currently available publications reporting the use of MALDI-TOF mass spectrometry for the identification of dermatophytes

				Database			Evaluation				
Reference	Year[a]	Device	Sample preparation	Type	Number of strains used	Number of species	Number of isolates tested	Number of species tested	Gold standard for identification	Identification accuracy	Comments
Erhard et al. [27]	2007	SARAMIS	Direct extraction on plate	Homemade	Unknown	17	20	5	Morphological criteria and ITS sequencing	100%	Studied isolates were mainly *T. rubrum* and *T. interdigitale*
Theel et al. [36]	2011	Microflex LT (Bruker)	Complete extraction [b]	Bruker Biotyper Library version 3.0	Unknown	Unknown	171	7	28S rRNA sequencing	37.4% (genus) 20.5% (species)	Lowering the recommended cut-off values allowed better genus and species identification. Approximately 20% of isolates required reanalysis/second extraction
				Homemade supplement to Bruker Biotyper Library version 3.0	20 supplemental strains	9				93% (genus) 59.6% (species)	
Alshawa et al. [33]	2012	Andromas	70% formic acid extraction	Homemade	50	12	381	10	Conventional culture criteria	91.6%	This study included 21 *Neoscytalidium* isolates for which 85.7% correct identification was obtained
Nenoff et al. [28]	2012	Axima	Direct extraction in matrix solution	Homemade	285	21	285	21	Conventional diagnosis and, for 164 isolates, ITS sequencing	78.2–99.3%	The buildup of the database and its subsequent evaluation used the same isolates. Moderate agreement with conventional diagnosis (78.2%), but good agreement (99.3%) with PCR techniques

(continued)

Table 1 (continued)

				Database			Evaluation				
Reference	Year[a]	Device	Sample preparation	Type	Number of strains used	Number of species	Number of isolates tested	Number of species tested	Gold standard for identification	Identification accuracy	Comments
De Respinis et al. [51]	2012	Axima	25% formic acid extraction + lipid elimination	Homemade	108	18	141	9	Morphological criteria and ITS sequencing	95.8%	90.8% of correct identification by morphological criteria
Packeu et al. [52]	2013	Microflex LT (Bruker)	Complete extraction [b]	Homemade	17	6	54	6	Molecular and morphological methods from a culture collection center	89%	Correlation of spectra with phylogenetic data. Isolates used for evaluation included those which were used to create the reference spectra
L'Ollivier et al. [30]	2013	Ultraflex (Bruker)	Complete extraction [b]	Homemade	48	17	134	11	Conventional diagnosis and ITS sequencing	97.8%	Accurate identification using small colonies obtained after 3–6 days of culture. Correct identification for 4 of 6 (67%) *M. canis* isolates. Authors tested the effect of cycloheximide in the culture media
Packeu et al. [29]	2014	MicroFlex (Bruker)	Direct extraction on plate by 70% formic acid/	Homemade	195	58	168	8	Microscopy and sequencing in case of discordance	Up to 100%	Correct identification varied from 40% (direct deposit after 3 days culture) to 80%

			complete extraction [b]								(complete extraction after 7 days incubation) and 100% (complete extraction, 14 days incubation)
Calderaro et al. [37]	2014	Microflex LT (Bruker)	Complete extraction [b]	Homemade supplement to Bruker Database version 3.1.2.0	24	13	64	11	Conventional and molecular procedures	Up to 100%	No accurate identification within the *T. mentagrophytes* complex
Karabiçak et al. [38]	2015	Autoflex III (Bruker)	Complete extraction [b]	Biotyper Library version 3.1.2.0	Unknown	Unknown	126	11	Phenotypic identification; 11 references strains	70.6% (genus) 51.6% (species)	Eight of the ten species used to build up the supplemental database were absent from the Biotyper Library. Results obtained by lowering the spectral score cut-off recommended by the manufacturer.
				Homemade supplement to Biotyper Library	10 supplemental reference strains	10				96.8% (genus) 89.7% (species)	
da Cunha et al. [31]	2018	Microflex LT (Bruker)	Direct transfer on plate/ complete extraction	Homemade	60	14	149 clinical isolates	13	Phenotypic and molecular identification (rRNA or ITS)	95.3%	85% correct identification by direct transfer (95% for *T. rubrum*)
				Homemade			1104 spectra from 276 strains	5 + 3 belonging to the *T. rubrum* complex		97%	Authors pointed out the difficulties related to ITS-based classification: discordant results were found according to the

(continued)

Table 1 (continued)

Reference	Year[a]	Device	Sample preparation	Database			Evaluation				Comments
				Type	Number of strains used	Number of species	Number of isolates tested	Number of species tested	Gold standard for identification	Identification accuracy	
											database used (ISHAM or NCBI)
				Bruker filamentous fungi version 1.0	51 spectra	15	60	14		31.7%	Performance of Bruker database was assessed using the strains used for supplemental homemade database
Hedayati et al. [53]	2019	Microflex (Bruker)	Complete extraction from subculture on brain-heart infusion liquid medium	Biotyper version 3.0 and the database version 3.1.66	Unknown	Unknown	93	5	RFLP[c] of the ITS region and calmodulin gene sequencing	83.9% (genus) 78.5% (species)	Most of the isolates were *T. interdigitale* and *T. rubrum*. Species identification was improved by lowering the spectral score cut-off.

[a]Year the publication was available online in Pubmed
[b]Fungal sample is mixed with ethanol (75% in pure water). After centrifugation the pellet is resuspended in 70% formic acid; acetonitrile is then added
[c]Restriction fragment length polymorphism

as for other filamentous species, the presence of a single reference per species is not sufficient. Indeed it is necessary to multiply the reference spectra by including several strains per species and several subcultures per strain in order to take into account the phenotypic variations within a species and among the subcultures of the same isolate [55]. In particular, databases will require improvements to account for variations in spectra observed on isolates of different ages. In the near future, progress can also be expected from the use of better algorithms to differentiate closely related species. Thus, it is desirable to develop research on protein profiles in order to make the algorithms better at identifying specific protein peaks. Another approach would be to use artificial intelligence solutions based on deep neural networks to differentiate the mass spectra and to attribute them to the right fungal species. Our team started working on these approaches and some encouraging results are already obtained.

10 Conclusion

When realizing dermatophyte identification, a good morphologist is prone to realize a correct diagnostic, but she/he frequently faces difficulties such as the lack of fructification, pigmentation problems, and variation in the composition of the culture media due to a change of suppliers implying a modification in the morphological aspect of the colonies. However, no one will dare contradict senior mycologists who are renowned for their experience in dermatophyte identification. Conversely, as a new technique that claims to be applicable worldwide, mycologists have extensively challenged mass spectrometry. This challenge is more profitable than deleterious because it pushes to seek more and more successful solutions. We can thus anticipate that mass spectrometry and computer analyses of spectra will progress further in the coming years, each advance being rapidly generalizable. Classical morphological identification can also progress as every mycologist can acquire experience, but this progress is much less generalizable. The know-how of one does not transfer so easily to another. Viewed as a competition, this changing balance between classical morphological and modern numerical approaches can provoke frustrations among some mycologists. However, it can also be an opportunity, giving the biologist more time and latitude to focus on the clinical significance of the obtained identification. This is all the more important since once the results are more precise and more reproducible, it will be easier to link an identified species to a particular clinical and / or epidemiological situation and thus to improve the service given to the clinician and to the patient.

References

1. Bouchara JP, Mignon B, Chaturvedi V. Dermatophytes and dermatophytoses: a thematic overview of state of the art, and the directions for future research and developments. Mycopathologia. 2017;182(1–2):1–4. https://doi.org/10.1007/s11046-017-0114-z.

2. Libon F, Nikkels-Tassoudji N, Dezfoulian B, Arrese JE, Nikkels AF. Non-dermatophyte dermatoses mimicking dermatophytoses in humans. Mycopathologia. 2017;182(1–2):101–11. https://doi.org/10.1007/s11046-016-0059-7.
3. Abdel-Rahman SM, Wright KJ, Navarre HC. Griseofulvin only modestly diminishes persistence of *Trichophyton tonsurans* on the scalp of carriers. J Pediatr Pharmacol Ther. 2009;14 (2):94–9. https://doi.org/10.5863/1551-6776-14.2.94.
4. Baudraz-Rosselet F, Monod M, Jaccoud S, Frenk E. Efficacy of terbinafine treatment of tinea capitis in children varies according to the dermatophyte species. Br J Dermatol. 1996;135 (6):1011–2. https://doi.org/10.1046/j.1365-2133.1996.d01-1117.x.
5. Mock M, Monod M, Baudraz-Rosselet F, Panizzon RG. Tinea capitis dermatophytes: susceptibility to antifungal drugs tested *in vitro* and *in vivo*. Dermatology (Basel). 1998;197(4):361–7. https://doi.org/10.1159/000018032.
6. Coulibaly O, L'Ollivier C, Piarroux R, Ranque S. Epidemiology of human dermatophytoses in Africa. Med Mycol. 2018;56(2):145–61. https://doi.org/10.1093/mmy/myx048.
7. Seebacher C, Bouchara JP, Mignon B. Updates on the epidemiology of dermatophyte infections. Mycopathologia. 2008;166(5–6):335–52. https://doi.org/10.1007/s11046-008-9100-9.
8. Gits-Muselli M, Benderdouche M, Hamane S, Mingui A, Feuilhade de Chauvin M, Guigue N, et al. Continuous increase of *Trichophyton tonsurans* as a cause of tinea capitis in the urban area of Paris, France: a 5-year-long study. Med Mycol. 2017;55(5):476–84. https://doi.org/10.1093/mmy/myw107.
9. Pihet M, Bourgeois H, Mazière J-Y, Berlioz-Arthaud A, Bouchara JP, Chabasse D. Isolation of *Trichophyton concentricum* from chronic cutaneous lesions in patients from the Solomon Islands. Trans R Soc Trop Med Hyg. 2008;102(4):389–93. https://doi.org/10.1016/j.trstmh.2008.01.002.
10. Esposto MC, Lazzarini C, Prigitano A, Olivi A, Monti M, Tortorano AM. *Trichophyton concentricum* in skin lesions in children from the Salomon Islands. G Ital Dermatol Venereol. 2015;150(5):491–4.
11. Monod M, Fratti M, Mignon B, Baudraz-Rosselet F. [Dermatophytes transmitted by pets and cattle]. Rev Med Suisse. 2014;10(424):749–53.
12. Touitou R, Bachmeyer C, Hennequin C, Moreno-Sabater A. Image gallery: a new source of zoophilic dermatophyte infection. Br J Dermatol. 2018;179(1):e4. https://doi.org/10.1111/bjd.16685.
13. Bloch M, Cavignaux R, Debourgogne A, Dorin J, Machouart M, Contet-Audonneau N. Du cochon d'Inde à l'Homme: épidémie de teigne à *Trichophyton mentagrophytes* var. *porcellae* dans les animaleries nancéiennes. J Mycol Med. 2016;26(3):227–32. https://doi.org/10.1016/j.mycmed.2016.05.005.
14. Nenoff P, Uhrlaß S, Krüger C, Erhard M, Hipler UC, Seifarth F, et al. *Trichophyton* species of *Arthroderma benhamiae* - a new infectious agent in dermatology. J Dtsch Dermatol Ges. 2014;12(7):571–81. https://doi.org/10.1111/ddg.12390.
15. Uhrlaß S, Schroedl W, Mehlhorn C, Krüger C, Hubka V, Maier T, et al. Molecular epidemiology of *Trichophyton quinckeanum* – a zoophilic dermatophyte on the rise. J Dtsch Dermatol Ges. 2018;16(1):21–32. https://doi.org/10.1111/ddg.13408.
16. Hayette M-P, Sacheli R. Unusual species of dermatophytes: rarely identified or new? Mycopathologia. 2017;182(1–2):203–13. https://doi.org/10.1007/s11046-016-0066-8.
17. Lipner SR, Scher RK. Onychomycosis: clinical overview and diagnosis. J Am Acad Dermatol. 2019;80(4):835–51. https://doi.org/10.1016/j.jaad.2018.03.062.
18. Pihet M, Le Govic Y. Reappraisal of conventional diagnosis for dermatophytes. Mycopathologia. 2017;182(1–2):169–80. https://doi.org/10.1007/s11046-016-0071-y.
19. Verrier J, Monod M. Diagnosis of dermatophytosis using molecular biology. Mycopathologia. 2017;182(1–2):193–202. https://doi.org/10.1007/s11046-016-0038-z.

20. Sabou M, Denis J, Boulanger N, Forouzanfar F, Glatz I, Lipsker D, et al. Molecular identification of *Trichophyton benhamiae* in Strasbourg, France: a 9-year retrospective study. Med Mycol. 2018;56(6):723–34. https://doi.org/10.1093/mmy/myx100.
21. Normand AC, Packeu A, Cassagne C, Hendrickx M, Ranque S, Piarroux R. Nucleotide sequence database comparison for routine dermatophyte identification by internal transcribed spacer 2 genetic region DNA barcoding. J Clin Microbiol. 2018;56(5):e00046-18. https://doi.org/10.1128/JCM.00046-18.
22. de Hoog GS, Dukik K, Monod M, Packeu A, Stubbe D, Hendrickx M, et al. Toward a novel multilocus phylogenetic taxonomy for the dermatophytes. Mycopathologia. 2017;182(1–2):5–31. https://doi.org/10.1007/s11046-016-0073-9.
23. Baert F, Stubbe D, D'hooge E, Packeu A, Hendrickx M. Updating the taxonomy of dermatophytes of the BCCM/IHEM collection according to the new standard: a phylogenetic approach. Mycopathologia. 2019; https://doi.org/10.1007/s11046-019-00338-7.
24. Tanaka K. The origin of macromolecule ionization by laser irradiation (Nobel lecture). Angew Chem Int Ed Engl. 2003;42(33):3860–70. https://doi.org/10.1002/anie.200300585.
25. Ahmed FE. Utility of mass spectrometry for proteome analysis: part I. Conceptual and experimental approaches. Expert Rev Proteomics. 2008;5(6):841–64. https://doi.org/10.1586/14789450.5.6.841.
26. Posteraro B, De Carolis E, Vella A, Sanguinetti M. MALDI-TOF mass spectrometry in the clinical mycology laboratory: identification of fungi and beyond. Expert Rev Proteomics. 2013;10(2):151–64. https://doi.org/10.1586/epr.13.8.
27. Erhard M, Hipler U-C, Burmester A, Brakhage AA, Wöstemeyer J. Identification of dermatophyte species causing onychomycosis and tinea pedis by MALDI-TOF mass spectrometry. Exp Dermatol. 2008;17(4):356–61. https://doi.org/10.1111/j.1600-0625.2007.00649.x.
28. Nenoff P, Erhard M, Simon JC, Muylowa GK, Herrmann J, Rataj W, et al. MALDI-TOF mass spectrometry - a rapid method for the identification of dermatophyte species. Med Mycol. 2013;51(1):17–24. https://doi.org/10.3109/13693786.2012.685186.
29. Packeu A, Bel AD, l'Ollivier C, Ranque S, Detandt M, Hendrickx M. Fast and accurate identification of dermatophytes by matrix-assisted laser desorption ionization–time of flight mass spectrometry: validation in the clinical laboratory. J Clin Microbiol. 2014;52(9):3440–3. https://doi.org/10.1128/JCM.01428-14.
30. L'Ollivier C, Cassagne C, Normand A-C, Bouchara JP, Contet-Audonneau N, Hendrickx M, et al. A MALDI-TOF MS procedure for clinical dermatophyte species identification in the routine laboratory. Med Mycol. 2013;51(7):713–20. https://doi.org/10.3109/13693786.2013.781691.
31. da Cunha KC, Riat A, Normand A-C, Bosshard PP, de Almeida MTG, Piarroux R, et al. Fast identification of dermatophytes by MALDI-TOF/MS using direct transfer of fungal cells on ground steel target plates. Mycoses. 2018;61(9):691–7. https://doi.org/10.1111/myc.12793.
32. De Respinis S, Monnin V, Girard V, Welker M, Arsac M, Cellière B, et al. Matrix-assisted laser desorption ionization-time of flight (MALDI-TOF) mass spectrometry using the Vitek MS system for rapid and accurate identification of dermatophytes on solid cultures. J Clin Microbiol. 2014;52(12):4286–92. https://doi.org/10.1128/JCM.02199-14.
33. Alshawa K, Beretti J-L, Lacroix C, Feuilhade de Chauvin M, Dauphin B, Quesne G, et al. Successful identification of clinical dermatophyte and *Neoscytalidium* species by matrix-assisted laser desorption ionization-time of flight mass spectrometry. J Clin Microbiol. 2012;50(7):2277–81. https://doi.org/10.1128/JCM.06634-11.
34. Lee H, Park J-H, Oh J, Cho S, Koo J, Park IC, et al. Evaluation of a new matrix-assisted laser desorption/ionization time-of-flight mass spectrometry system for the identification of yeast isolation. J Clin Lab Anal. 2019;33(2):e22685. https://doi.org/10.1002/jcla.22685.
35. ID Fungi Plates. Conidia. http://conidia.fr/en/id-fungi-plates/. Accessed 28 Oct 2019.
36. Theel ES, Hall L, Mandrekar J, Wengenack NL. Dermatophyte identification using matrix-assisted laser desorption ionization-time of flight mass spectrometry. J Clin Microbiol. 2011;49(12):4067–71. https://doi.org/10.1128/JCM.01280-11.

37. Calderaro A, Motta F, Montecchini S, Gorrini C, Piccolo G, Piergianni M, et al. Identification of dermatophyte species after implementation of the in-house MALDI-TOF MS database. Int J Mol Sci. 2014;15(9):16012–24. https://doi.org/10.3390/ijms150916012.
38. Karabıçak N, Karatuna O, İlkit M, Akyar I. Evaluation of the Bruker matrix-assisted laser desorption–ionization time-of-flight mass spectrometry (MALDI-TOF MS) system for the identification of clinically important dermatophyte species. Mycopathologia. 2015;180 (3):165–71. https://doi.org/10.1007/s11046-015-9898-x.
39. Cassagne C, Cella A-L, Suchon P, Normand A-C, Ranque S, Piarroux R. Evaluation of four pretreatment procedures for MALDI-TOF MS yeast identification in the routine clinical laboratory. Med Mycol. 2013;51(4):371–7. https://doi.org/10.3109/13693786.2012.720720.
40. Lau AF, Drake SK, Calhoun LB, Henderson CM, Zelazny AM. Development of a clinically comprehensive database and a simple procedure for identification of molds from solid media by matrix-assisted laser desorption ionization-time of flight mass spectrometry. J Clin Microbiol. 2013;51(3):828–34. https://doi.org/10.1128/JCM.02852-12.
41. Hettick JM, Green BJ, Buskirk AD, Kashon ML, Slaven JE, Janotka E, et al. Discrimination of *Aspergillus* isolates at the species and strain level by matrix-assisted laser desorption/ionization time-of-flight mass spectrometry fingerprinting. Anal Biochem. 2008;380(2):276–81. https://doi.org/10.1016/j.ab.2008.05.051.
42. Riat A, Hinrikson H, Barras V, Fernandez J, Schrenzel J. Confident identification of filamentous fungi by matrix-assisted laser desorption/ionization time-of-flight mass spectrometry without subculture-based sample preparation. Int J Infect Dis. 2015;35:43–5. https://doi.org/10.1016/j.ijid.2015.04.013.
43. Normand A-C, Gabriel F, Riat A, Cassagne C, Bourgeois N, Huguenin A, et al. Optimization of MALDI-ToF mass spectrometry for yeast identification: a multicenter study. Med Mycol. 2019;58:639. https://doi.org/10.1093/mmy/myz098.
44. Riat A, Rentenaar RJ, van Drongelen AM, Barras V, Bertens LC, Vlek AL, et al. Ground steel target plates in combination with direct transfer of clinical *Candida* isolates improves frequencies of species-level identification by matrix-assisted laser desorption ionization-time of flight mass spectrometry in comparison with polished steel target plates. J Clin Microbiol. 2015;53(6):1993–5. https://doi.org/10.1128/JCM.03632-14.
45. VITEK; MS-DS SLIDE | Mass Spec Reagents & Consumables | Mass Spectrometry Reagents & Consumables | MASS SPECTROMETRY | ID/AST Auto | Bacteriology | bioMerieux Clinical USA. https://biomerieuxdirect.com/clinical/Bacteriology/ID-AST-Auto/MASS-SPECTROMETRY/Mass-Spectrometry-Reagents-%26-Consumables/Mass-Spec-Reagents-%26-Consumables/VITEK%26reg%3B-MS-DS-SLIDE/p/410893. Accessed 28 Oct 2019.
46. Cassagne C, Normand A-C, L'Ollivier C, Ranque S, Piarroux R. Performance of MALDI-TOF MS platforms for fungal identification. Mycoses. 2016;59(11):678–90. https://doi.org/10.1111/myc.12506.
47. Dupont D, Normand A-C, Persat F, Hendrickx M, Piarroux R, Wallon M. Comparison of matrix-assisted laser desorption ionization time of flight mass spectrometry (MALDI-TOF MS) systems for the identification of moulds in the routine microbiology laboratory. Clin Microbiol Infect. 2019;25(7):892–7. https://doi.org/10.1016/j.cmi.2018.10.013.
48. Bille E, Dauphin B, Leto J, Bougnoux ME, Beretti JL, Lotz A, et al. MALDI-TOF MS Andromas strategy for the routine identification of bacteria, mycobacteria, yeasts, *Aspergillus* spp. and positive blood cultures. Clin Microbiol Infect. 2012;18(11):1117–25. https://doi.org/10.1111/j.1469-0691.2011.03688.x.
49. Imbert S, Normand AC, Gabriel F, Cassaing S, Bonnal C, Costa D, et al. Multi-centric evaluation of the online MSI platform for the identification of cryptic and rare species of *Aspergillus* by MALDI-TOF. Med Mycol. 2019;57(8):962–8. https://doi.org/10.1093/mmy/myz004.
50. Normand A-C, Cassagne C, Gautier M, Becker P, Ranque S, Hendrickx M, et al. Decision criteria for MALDI-TOF MS-based identification of filamentous fungi using commercial and

in-house reference databases. BMC Microbiol. 2017;17(1):25. https://doi.org/10.1186/s12866-017-0937-2.
51. de Respinis S, Tonolla M, Pranghofer S, Petrini L, Petrini O, Bosshard PP. Identification of dermatophytes by matrix-assisted laser desorption/ionization time-of-flight mass spectrometry. Med Mycol. 2013;51(5):514–21. https://doi.org/10.3109/13693786.2012.746476.
52. Packeu A, Hendrickx M, Beguin H, Martiny D, Vandenberg O, Detandt M. Identification of the *Trichophyton mentagrophytes* complex species using MALDI-TOF mass spectrometry. Med Mycol. 2013;51(6):580–5. https://doi.org/10.3109/13693786.2013.770605.
53. Hedayati MT, Ansari S, Ahmadi B, Taghizadeh Armaki M, Shokohi T, Abastabar M, et al. Identification of clinical dermatophyte isolates obtained from Iran by matrix-assisted laser desorption/ionization time-of-flight mass spectrometry. Curr Med Mycol. 2019;5(2):22–6. https://doi.org/10.18502/cmm.5.2.1157.
54. Tartor YH, Abo Hashem ME, Enany S. Towards a rapid identification and a novel proteomic analysis for dermatophytes from human and animal dermatophytosis. Mycoses. 2019;62(12):1116–26. https://doi.org/10.1111/myc.12998.
55. Normand A-C, Cassagne C, Ranque S, L'Ollivier C, Fourquet P, Roesems S, et al. Assessment of various parameters to improve MALDI-TOF MS reference spectra libraries constructed for the routine identification of filamentous fungi. BMC Microbiol. 2013;13:76. https://doi.org/10.1186/1471-2180-13-76.

Part VI

Antifungal Resistance of Dermatophytes

Antifungal Susceptibility Testing of Dermatophytes

Nathan P. Wiederhold

Abstract

In order to help guide treatment, antifungal susceptibility testing is often performed in clinical microbiology laboratories. The results of these in vitro assays are used by clinicians to change therapy should resistance be detected, continue with current regimens with susceptible results, and as an aide in determining possible reasons for treatment failure. Antifungal susceptibility testing can be performed against dermatophytes, although fewer assay formats are recommended for these types of fungi compared to yeasts and molds. In addition, clinical breakpoints and epidemiologic cut-off values are not currently available for dermatophytes to help with the interpretation of these results. However, recently studies have demonstrated that reduced in vitro susceptibility to terbinafine, an allylamine commonly used in the treatment of dermatophyte infections, may be associated with clinical failures.

Keywords

Antifungal susceptibility testing · Azoles · Allylamines · Resistance · Clinical breakpoint · Epidemiologic cut-off value · Broth microdilution · Dermatophytes

1 Introduction

Antifungal susceptibility testing is performed in order to provide clinicians information that helps guide therapy in patients with fungal infections. As with antibiotic susceptibility testing, antifungal susceptibility testing performed in clinical

N. P. Wiederhold (✉)
Fungus Testing Laboratory, Department of Pathology and Laboratory Medicine, University of Texas Health Science Center at San Antonio, San Antonio, TX, USA
e-mail: wiederholdn@uthscsa.edu

J.-P. Bouchara et al. (eds.), *Dermatophytes and Dermatophytoses*,
https://doi.org/10.1007/978-3-030-67421-2_19

Table 1 Examples of variables that can influence in vitro susceptibility testing results for dermatophytes

Variable	Example
Medium used for subcultures (for adequate production of conidia)	Potato dextrose/potato flake agar, Sabouraud dextrose agar, oatmeal agar
Fungal morphology tested	Microconidia/macroconidia, hyphae, arthroconidia
Starting inoculum size	0.5×10^4 to 5×10^5 conidia/mL
Broth	RPMI, Sabouraud dextrose, yeast-peptone dextrose
Incubation temperature	28–35 °C
Incubation period prior to endpoint reading	72–168 h
Endpoint used	Visual growth inhibition, optical density, colorimetric

microbiology laboratories are in vitro phenotypic assays that assess the ability of a particular drug to inhibit the growth of a specific organism over a range of concentrations. The endpoint that is read, the minimum inhibitory concentration (MIC), is the lowest concentration of a drug that inhibits the organism, and this value can be used to determine if an isolate is susceptible or resistant to the specific drugs being tested. Although these types of assays appear to be relatively straightforward, there are a number of variables that can influence the results (Table 1). In addition, it is important to remember that the MIC value that is obtained is only one component of many that should be taken into consideration when decisions are made regarding an individual's antifungal regimen, and that other factors, including those related to the host and drug, may be more important in determining clinical outcomes. Susceptibility testing is also performed as part of surveillance studies to detect changes in susceptibility and resistance patterns over time, and in the pre-clinical development of investigational agents as a means of assessing the spectrum of activity and in vitro potency of therapeutic candidates. In this chapter, a general overview of antifungal susceptibility testing will be provided, including strengths and weaknesses and its role helping guide treatment decisions. Descriptions of the various antifungal susceptibility assays that are utilized in clinical microbiology laboratories and how useful these are for testing against dermatophytes will also be discussed, along with recent evidence showing associations between in vitro terbinafine susceptibility results, mechanisms of resistance, and clinical outcomes.

2 Standardized Methods for Antifungal Susceptibility Testing of Dermatophytes

In the United States and Europe, the two primary organizations that have established standard methods for antifungal susceptibility testing are the Clinical and Laboratory Standards Institute (CLSI) and the European Union Committee on Antimicrobial Susceptibility Testing (EUCAST), respectively. Both organizations have established

Table 2 Differences between CLSI and EUCAST broth microdilution procedures for antifungal susceptibility testing against filamentous fungi

Parameter	CLSI	EUCAST
Microtiter plate	96-well round bottom	96-well flat bottom
RPMI glucose concentration	0.2%	2%
Test inoculum size	0.4–5 × 10^4 conidia/mL	1–2.5 × 10^5 conidia/mL
Incubation period (dermatophytes)	96 h	Not specified for dermatophytes, but incubation longer than 72 h not recommended for other filamentous fungi
Endpoint for dermatophytes	80% inhibition of growth (100% for amphotericin B)	100% inhibition of growth

broth microdilution methodologies for testing against yeasts and filamentous fungi, which have undergone rigorous evaluations for interlaboratory reproducibility [1, 2]. For the broth microdilution methods, 96-well plates are used that contain doubling dilutions of antifungals into which fungi are inoculated. In addition, CLSI also describes methods for broth macrodilution that some labs may employ for certain species and for safety reasons [1]. Although there have been efforts to harmonize the methods between these two groups, important differences still remain, including microtiter plate type (round bottom vs. flat bottom wells), inoculum size, glucose concentration, and specifically for dermatophytes the endpoints used to define the MIC [1, 2]. Differences between the CLSI and EUCAST methods, as they pertain to filamentous fungi, including dermatophytes, are detailed in Table 2. Despite these differences, the MIC results obtained by each method are relatively comparable, including against dermatophytes [3, 4].

Starting Material—In order to perform antifungal susceptibility testing against filamentous fungi, including dermatophytes, the correct starting inoculum size is needed, and both CLSI and EUCAST recommend that conidia or spores be used rather than hyphae. Studies have demonstrated that higher inocula can negatively influence the in vitro activity of certain antifungals, leading to elevated MICs for some drugs (e.g., griseofulvin and fluconazole) but not others (itraconazole and terbinafine) [5]. For most dermatophyte species, including *Trichophyton mentagrophytes, Trichophyton tonsurans, Epidermophyton floccosum*, and *Microsporum canis*, sufficient production of conidia can be achieved by culturing the isolates on potato dextrose or potato flake agar for a period of ~4 to 7 days prior to harvesting for susceptibility testing [6]. However, for *Trichophyton rubrum* a high percentage of isolates will fail to sporulate under these conditions. In order to overcome this, isolates of *T. rubrum* may be subcultured on oatmeal cereal agar, which has been shown to lead to a higher percentage of isolates that produce conidia [6]. However, even with the use of this agar type ~15% of *T. rubrum* isolates may fail to sporulate. In order to overcome this limitation, which can also occur with other molds, a method of inoculum preparation that contains a mycelial mixture of hyphae and conidia, has been proposed [7]. This fragmented mycelium method, which involves a brief homogenization step after collection of the hyphae/conidia mixture from a subculture with sufficient growth, has been shown to result in comparable MIC values for the EUCAST and CLSI broth microdilution methods, including against dermatophytes, when compared to the use of conidia alone for the starting inoculum material [4, 7]. However, the use of fragmented mycelia is not currently part of either the EUCAST or CLSI methods. In vivo, dermatophytes may produce arthroconidia, which are produced by fragmentation of hyphae and are considered to be the primary cause of infection by this type of fungi [8, 9]. While in vitro susceptibility studies primarily use microconidia or hyphae, the susceptibility of arthroconidia to clinically available antifungals may be reduced, but this may be both antifungal and species-specific [9, 10]. In addition, the in vitro production of dermatophyte arthroconidia is not a standardized practice in clinical microbiology laboratories.

Growth Medium—Different types of media have been evaluated for their ability to support the growth of dermatophytes in broth dilution assays. These include Sabouraud dextrose, yeast-nitrogen base with 0.5% glucose, and antibiotic medium #3 broths, and the chemically defined medium Roswell Park Memorial Institute (RPMI) 1640 (pH adjusted to 7.0 and buffered with 0.165 M 3-(*N*-morpholino) propanesulfonic acid [MOPS], without bicarbonate, and with glutamine and phenol red). In a study that evaluated various conditions for the optimization of susceptibility testing against dermatophytes, it was reported that both RPMI and Sabouraud dextrose supported optimal growth [5]. However, both CLSI and EUCAST recommend the use of RPMI as the growth medium for broth dilution susceptibility testing of fungi, including yeasts, molds, and dermatophytes, since this is a chemically defined medium in contrast to other media, which may be subject to lot-to-lot variation. The recommended glucose concentration in RPMI is different between the CLSI (0.2% w/v) and EUCAST (2% w/v) methods. Recently, some have questioned the use of RPMI for susceptibility testing of dermatophytes since the nutrient content of this medium does not represent that available in skin and nails. Human nail is made up of different types of keratin, which are proteins containing cross-linked disulfide bonds, with relatively low amounts of carbohydrates. In contrast, the RPMI broth used in susceptibility testing contains up to 2% glucose and few proteins. Some researchers have begun to employ media containing differences sources of keratin. One group has reported that fungi cultured in a medium of keratin derived from human nail powder (sodium phosphate buffer supplemented with powdered human nail material) had a metabolic profile similar to that observed in RPMI [11]. They also reported that the MICs of terbinafine and ciclopirox were increased between 2- and 16-fold in this keratin-based medium compared to RPMI but that these antifungals were still inhibitory at clinically relevant concentration. This may be clinically relevant as the human keratin medium may more accurately mimic conditions encountered in onychomycosis. However, additional studies to determine the reproducibility of these results and correlations with clinical outcomes are needed.

Incubation Time and Temperature—Compared to most yeasts and many other filamentous fungi that cause disease in humans, growth of dermatophytes is relatively slow. Because of this, a longer duration of incubation may be needed in order to observe sufficient growth within the growth control well used in broth susceptibility assays. CLSI recommends an incubation period of 4 days (96 h) for testing against dermatophytes [1], although some studies that have used these methods have read results after a shorter period of incubation (i.e., 72 h) as long as there was sufficient growth of the growth control at the chosen timepoint [5, 8]. EUCAST does not make a specific recommendation for the incubation period for dermatophytes, although studies based on this method have used periods of 2–9 days for incubation [4]. During this period, the plates or tubes should be incubated without agitation at ~35 °C (range 34–37 °C per EUCAST). It is recognized that a lower temperature (~30 °C) may be better suited for some fungi, and studies have used lower temperatures for susceptibility testing of dermatophytes [5, 12].

Endpoints—As previously described the endpoint used for susceptibility testing is called the minimum inhibitory concentration (MIC). For broth dilution testing, this is the lowest concentration of the antimicrobial that inhibits growth of the organism being tested. Both CLSI and EUCAST recommend a visual reading of this endpoint for filamentous fungi, including dermatophytes [1, 2]. Per the EUCAST standard, the MIC for antifungals other than echinocandins against filamentous fungi is a 100% inhibition of growth, also defined as a lack of visual growth [2]. CLSI also uses this endpoint for amphotericin B and the azoles against filamentous fungi [13]. However, for dermatophytes an 80% inhibition of growth (reduction of 80% growth in comparison to the growth control) is the endpoint recommended for the azoles, terbinafine, and griseofulvin, while 100% inhibition is the endpoint for amphotericin B. It is well recognized that endpoint determination can be a substantial source of interlaboratory variability, especially for the azoles and allylamines where a small amount of turbidity may persist after a pronounced reduction in growth at lower drug concentrations (i.e., trailing) [8]. With the CLSI method, the 80% inhibition of growth has been reported to result in less variability, especially for fungistatic drugs, such as the azoles (e.g., fluconazole, isavuconazole, itraconazole, ketoconazole, posaconazole, voriconazole) and allylamines (terbinafine), where a trailing endpoint may be observed [5, 8].

Reproducibility and Quality Control—Clinical antifungal susceptibility testing results should be reproducible in order for the assays to be trusted by clinicians. However, given the numerous variables that can influence these in vitro results (i.e., inoculum size, growth medium, incubation temperature, endpoints used), the acceptable error of these assays is a 2-dilution (or four-fold MIC) difference. As most clinical microbiology laboratories do not routinely perform repeat testing to assess for reproducibility, quality control or reference isolates should be included in the assay runs as a means to assess for the quality of the assay runs and if the results are acceptable to be released to the ordering clinicians and other health care providers caring for the patients. CLSI has set reference MIC ranges for several antifungals against two well-described dermatophytes isolates [14], which were evaluated in multicenter studies [15, 16]: (1) *T. mentagrophytes* ATCC MYA-4439 (MRL 1957), for which reference MIC ranges are available for ciclopirox, griseofulvin, itraconazole, posaconazole, voriconazole, and terbinafine; and (2) *T. rubrum* ATCC MYA-4438 (MRL 666), for which reference MIC ranges are available for ciclopirox, fluconazole, and voriconazole, but not for itraconazole or terbinafine.

3 Interpretation of Susceptibility Results

Susceptibility testing provides a measure of the in vitro activity of a particular antimicrobial or panel of antimicrobial drugs against a particular microbe. However, the MIC values alone are not enough to establish interpretations of susceptible or resistant that clinicians are accustomed to and use to make decisions regarding therapeutic management. At or below these thresholds, isolates are considered susceptible to the antimicrobials tested; above the thresholds the isolates are considered resistant. Clinical breakpoints are the MIC thresholds that are used to interpret these in vitro results. It is important to remember that clinical breakpoints are not naturally occurring values but are instead set by committees, which consider different information and data sets in order to establish these thresholds. Information that is considered includes: (1) MIC distributions for a particular antimicrobial agent and microbial species, which often come from multiple laboratories and different studies; (2) pharmacokinetic parameters of the particular antimicrobial agent and if antimicrobial concentrations at or above the MIC values can be clinically achieved; (3) pharmacokinetic/pharmacodynamic parameters that best predict in vivo efficacy from in vitro and in vivo models of infections; (4) results from clinical trials demonstrating associations/correlations between MIC values and clinical responses in the study patients [17, 18]. Once set, clinical breakpoints are subject to review and revisions based on new information. Both CLSI and EUCAST have set clinical breakpoints for some of the azoles and echinocandins against more prevalent *Candida* species [19–21]. EUCAST has also set breakpoints for some azoles against *Aspergillus* species [2, 20], and CLSI has proposed breakpoints for voriconazole against *A. fumigatus*. However, neither group has set breakpoints for any antifungal against dermatophytes, and no breakpoints exist for terbinafine, other squalene epoxidase inhibitors, or griseofulvin against any fungal species.

It is also important to note that a susceptible MIC value does not necessarily predict clinical success, nor does a resistant MIC value predict clinical failure. Several other factors are key contributors to clinical response. Host factors are very important in determining clinical outcomes, and include immune status, the extent and severity of the underlying disease, and other comorbidities. In addition, for dermatophyte infections, the presence of arthroconidia at the site of infection may influence outcomes, as these forms are less susceptible to antifungals than other forms of these fungi [9, 10, 22]. The pharmacokinetics of the drug, including the concentration and overall exposure that is achieved at the site of the infection is also an important factor. For example, in dermatophyte infections the concentrations of the antifungals achieved in the stratum corneum and nails, the sites of infections, are probably more important than that achieved within the bloodstream (serum or plasma) [22]. Terbinafine levels within the bloodstream of humans are low, while concentrations within the corneum stratum and nails are high, and overall exposure, as measured by the area under the concentration curve (AUC), is high due to an extended half-life in these tissues [23]. Thus, in humans terbinafine is primarily used for the treatment of dermatophyte infections within the skin and nails. In contrast to the prolonged exposure observed with terbinafine within the stratum corneum [24], griseofulvin concentrations build up rapidly within this tissue but also fall rapidly once discontinued [25]. Thus, a prolonged treatment time is required in order to achieve clinical response with this antifungal, and relapses may occur once discontinued. Other factors that may influence clinical outcomes include drug-drug interactions, where antifungal drug concentrations may be lowered below effective concentrations if patients receive concomitant medications that induce metabolism, or toxicity may be observed if coadministered with drugs that inhibit their metabolism and levels exceed those associated with safe use. Patient adherence to the prescribed medications is also important as missed doses may result in subtherapeutic concentrations and potentially the development of resistance [17, 26]. Patient adherence can also be affected by the adverse effects/toxicities that they are experiencing. Thus, as previously noted, results of antifungal susceptibility testing are only one variable that can influence clinical outcomes. However, chances of a poor outcome are higher when in vitro resistance is detected. The ability of susceptibility testing to predict clinical response has been summed up as the "90/60 rule." In general terms, the 90/60 rule comes from observations that infections caused by susceptible isolates respond to appropriate therapy ~90% of the time, while those caused by resistant isolates may still respond to therapy ~60% of the time [17, 26]. Examples of this "rule" can be found for both antibacterial and antifungal susceptibility testing, including testing against dermatophytes. In a recent study that included 30 patients with *Tinea corporis* or *Tinea cruris* due to *Trichophyton interdigitale* in which both terbinafine MICs and clinical histories were available, 83.3% of patients whose infection was caused by an isolate with a terbinafine MICs < 1 μg/mL responded to therapy, while 66.7% of those with infecting isolates that had higher terbinafine MICs (≥ 1 μg/mL) failed to respond to treatment with this allylamine [27]. Interestingly, some patients whose isolates had elevated terbinafine MICs responded to higher doses and/or longer durations of treatment, highlighting the importance of achieving adequate exposures of antifungals at the sites of infection. Based on these results and those previously published, some have suggested that *Trichophyton* isolates with terbinafine MICs < 1 μg/mL may be considered susceptible to this antifungal while those with higher MIC values may be considered resistant [27, 28].

The observation that elevated terbinafine MICs are associated with clinical failure, is supported by other data, including studies that have evaluated mechanisms of resistance to the squalene epoxidase inhibitors. Terbinafine, and related compounds such as naftifine, butenafine, tolnaftate, and tolciclate, are selective inhibitors of squalene epoxidase (also known as squalene monooxygenase), an enzyme that converts squalene to 2,3-oxidosqualene within the fungal cell membrane, and is upstream of lanosterol 14α-demethylase in the ergosterol biosynthesis pathway [29]. Inhibition of squalene epoxidase leads to an accumulation of squalene, which can be toxic to fungi. Studies that have used clinical dermatophytes isolates cultured from patients who have failed terbinafine therapy have reported nonsynonymous point mutations within the gene that encodes squalene epoxidase that lead to amino acid changes within this enzyme, including those as codons Leu393, Phe397, Phe415, and His440 [28–30]. In six consecutive *T. rubrum* isolates collected from a single patient who had failed therapy with terbinafine (phenotypic MIC > 4 μg/mL vs. < 0.001 μg/mL against isolates from patients who had responded), inhibition of squalene epoxidase enzyme activity by this allylamine was also markedly reduced (IC50 30 μmol/L) compared to that against normal enzyme (19 nmol/L) [28]. In addition, cross-resistance was observed against other squalene epoxidase inhibitors, but not against azoles, griseofulvin, or amphotericin B [31]. Others have also reported associations between terbinafine exposures, clinical failures, and point mutations in the squalene epoxidase gene leading to amino acid changes in this enzyme and reduced in vitro susceptibility to this antifungal [27, 30, 32–35]. However, the exact clinical breakpoint that should be used for terbinafine resistance in dermatophytes remains unknown, and further clinical validation using a larger number of geographically diverse isolates is needed.

4 Epidemiologic Cut-Off Values

When clinical breakpoints are not available for certain drugs or species of fungi, epidemiological cut-off values (ECVs or ECOFFs) may be used to provide some guidance to clinicians. These values are thresholds derived from MIC distributions of wild-type isolates for a specific drug, and in general, the ECV will cover 95–97.5% of the isolates within the wild-type MIC distribution, although there are various statistical means used to calculate the ECV. The purpose of the ECV, which should be both antifungal- and species-specific, is to help identify isolates less likely to respond to treatment because of the development of acquired resistance [36, 37]. ECVs are also useful in tracking the emergence of reduced susceptibility/resistance, and are frequently used by CLSI and EUCAST in making clinical breakpoint decisions. However, it should be remembered that ECVs are not the same as clinical breakpoints, which also consider the pharmacokinetics of a drug, pharmacokinetic/pharmacodynamic relationships, and clinical outcomes associated with in vitro susceptibility results. Although CLSI and EUCAST have published ECVs for several antifungals against different species of yeasts and molds, no ECV values are currently available for any drugs against any dermatophyte species.

5 Other Phenotypic Assays for Antifungal Susceptibility Testing

In addition to the CLSI and EUCAST broth dilution methods, commercially available antifungal susceptibility assays are available for use. These assays are often used by clinical microbiology laboratories for susceptibility testing against yeasts due to their ease of use and commercial availability. In contrast, laboratories that use the CLSI or EUCAST methods either must make the antifungal trays themselves, which requires trained personnel, additional time, and that appropriate quality assurance measures be in place, making them technically challenging and costly, or have the plates custom made by a manufacturer, for which options may be limited. Types of commercially available antifungal susceptibility tests include colorimetric assays, disc diffusion and gradient diffusions assays, and those performed by automated instruments. While these assays have proven useful for antifungal susceptibility testing of yeasts, variability has been reported when used for filamentous fungi, and few data are available for dermatophytes.

Sensititre YeastOne Colorimetric Assay—One of the most popular commercially available assays used in clinical microbiology testing is the YeastOne Sensititre assay (Thermo Scientific, formerly TREK Diagnostics). This assay uses a 96-well cell culture plate, broth microdilution format similar to that used by CLSI and EUCAST methods, but also includes a blue colorimetric dye, resazurin (alamarBlue), for endpoint detection. Metabolically active cells convert the resazurin to resorufin, causing a color change from blue to red, and for yeasts it is the lowest antifungal concentration where this color change begins to occur that is used for the MIC endpoint. Although several studies had demonstrated very good agreement between the YeastOne assay and broth microdilution reference methods [38–45], similar results have not been observed for dermatophytes. In one study that included 49 dermatophyte isolates and compared YeastOne MIC results to those obtained by the CLSI M38 method for four antifungals (amphotericin B, fluconazole, itraconazole, and ketoconazole), agreement between the two assays ($\leq$ 2 dilution difference in MIC values) was variable depending upon the drug, ranging from 67.3% (fluconazole) to 87.7% (itraconazole) [46]. Variability was also noted between dermatophyte species for each antifungal, and MICs were generally lower with the YeastOne assay, possibly due to the use of the color change as the endpoint rather than growth inhibition as used by the CLSI method. Worse agreement between the two methods was reported in a second study that compared these two assays for fluconazole, itraconazole, and voriconazole against 46 dermatophytes isolates, including those of *T. mentagrophytes*, *T. rubrum*, and *Nannizzia gypsea* (formerly *Microsporum gypseum*) [47]. In this study, agreement ranged between 0 and 83.3%, and as before, this varied by both antifungal and dermatophyte species with lower MICs observed with the colorimetric assay. In both studies, the authors concluded that the YeastOne colorimetric assay cannot be recommended for antifungal susceptibility testing of dermatophytes.

Etest and Disk Diffusion—Another commercially available assay that is used by clinical microbiology laboratories for antifungal susceptibility testing is the Etest (bioMerieux, formerly AB Biodisk). This assay employs a plastic strip that contains a concentration gradient of a particular antifungal agent, which is placed onto the surface of an agar plate that has been inoculated with a fungal isolate. The antifungal agent then diffuses into the agar, and the MIC is read after incubation as the concentration at which the elliptical zone of inhibition crosses the strip. Good essential agreement ($>$ 90% within $\leq$ 2 dilutions) has been reported between the

Etest method and the CLSI and EUCAST broth microdilution reference methods for yeasts [38, 39]. However, studies comparing this assay to the reference methods for dermatophyte species have shown mixed results. One multicenter study that compared Etest to the CLSI broth microdilution method against 30 dermatophyte isolates reported the best agreement between the two assays for amphotericin B (97%), followed by itraconazole (80%) and ketoconazole (77%), and the worst with fluconazole (27%) [12]. Interlaboratory variability was also noted between the different dermatophyte species. Others that have evaluated the agreement between Etest and reference methods have reported less promising results. In a single center study that included 46 dermatophyte isolates, 30 of which were *T. mentagrophytes* and eight each of *T. rubrum* and *N. gypsea*, agreement between the Etest and CLSI M38 methods was 19.5% for itraconazole, 45.6% for fluconazole, and 52.1% for voriconazole [48]. Interestingly, the incubation temperatures used in these two studies were different (28 °C for the first and 35 °C in the second). Similar to the Etest results, studies have reported varying results for disk diffusion antifungal susceptibility testing of dermatophytes compared to that of the CLSI broth microdilution method. One study that used RPMI as the growth medium for both the disk diffusion (agar) and broth microdilution assays reported very low agreement between the two methods for fluconazole, itraconazole, and voriconazole against 46 dermatophyte isolates, with the disk diffusion zone diameters of voriconazole actually increasing in relation to higher MIC values as measured by broth microdilution [48]. In contrast, another study that evaluated disk diffusion testing against dermatophytes and used Mueller-Hinton agar as the growth medium, reported good agreement between terbinafine zone diameters and broth microdilution MICs against 19 dermatophyte isolates [49]. This study included five *T. rubrum* isolates with elevated terbinafine MICs ($\geq$ 4 μg/mL) for which no zone diameters were observed. Thus, although easier to setup and perform than broth microdilution, Etest and disk diffusion assays require additional testing and optimization before they can be recommended for antifungal susceptibility testing against dermatophytes.

6 Conclusions

Antifungal susceptibility testing is frequently performed by clinical microbiology laboratories as a means of detecting resistance in order to predict potential clinical failures or as a means of determining why a patient may not be responding to a particular treatment regimen. Numerous variables may affect the results of these in vitro assays, and methods have been standardized by CLSI and EUCAST in order to limit variability observed between laboratories. Although commercially available antifungal susceptibility assays are available, mixed results and marked variability have been reported when evaluated against dermatophytes, and these are not currently recommended for testing of dermatophytes in clinical microbiology laboratories. Finally, clinical breakpoints and ECVs have not been established for antifungals against dermatophytes, and this may limit the clinical utility of susceptibility testing against these fungi. However, our understanding of resistance mechanisms and in vitro resistance and their relationship to clinical failure with terbinafine and other squalene epoxidase inhibitors is increasing, which may allow for clinical breakpoints to be set in the future.

References

1. CLSI. Reference method for broth dilution antifungal susceptibility testing of filamentous fungi, third edition. Document M38Ed3. Wayne: Clinical and Laboratory Standards Institute; 2017.
2. Arendrup MC, Meletiadis J, Mouton JW, Lagrou K, Hamal P, Guinea J. EUCAST definitive document e.Def 9.3.1: method for the determination of broth dilution minimum inhibitory concentrations of antifungal agents for conidia forming moulds. Basel: EUCAST; 2017.
3. Dabas Y, Xess I, Singh G, Pandey M, Meena S. Molecular identification and antifungal susceptibility patterns of clinical dermatophytes following CLSI and EUCAST guidelines. J Fungi (Basel). 2017;3(2)
4. Risslegger B, Lass-Florl C, Blum G, Lackner M. Evaluation of a modified EUCAST fragmented-mycelium inoculum method for *in vitro* susceptibility testing of dermatophytes and the activity of novel antifungal agents. Antimicrob Agents Chemother. 2015;59 (6):3675–82.
5. Norris HA, Elewski BE, Ghannoum MA. Optimal growth conditions for the determination of the antifungal susceptibility of three species of dermatophytes with the use of a microdilution method. J Am Acad Dermatol. 1999;40(6 Pt 2):S9–13.
6. Jessup CJ, Warner J, Isham N, Hasan I, Ghannoum MA. Antifungal susceptibility testing of dermatophytes: establishing a medium for inducing conidial growth and evaluation of susceptibility of clinical isolates. J Clin Microbiol. 2000;38(1):341–4.
7. Schmalreck A, Willinger B, Czaika V, Fegeler W, Becker K, Blum G, et al. Susceptibility screening of hyphae-forming fungi with a new, easy, and fast inoculum preparation method. Mycopathologia. 2012;174(5–6):467–74.
8. Thatai P, Sapra B. Critical review on retrospective and prospective changes in antifungal susceptibility testing for dermatophytes. Mycoses. 2016;59(10):615–27.
9. Coelho LM, Aquino-Ferreira R, Maffei CM, Martinez-Rossi NM. *In vitro* antifungal drug susceptibilities of dermatophytes microconidia and arthroconidia. J Antimicrob Chemother. 2008;62(4):758–61.
10. Arrese JE, Pierard-Franchimont C, Pierard GE. A plea to bridge the gap between antifungals and the management of onychomycosis. Am J Clin Dermatol. 2001;2(5):281–4.
11. Mercer DK, Stewart CS, Miller L, Robertson J, Duncan VMS, O'Neil DA. Improved methods for assessing therapeutic potential of antifungal agents against dermatophytes and their application in the development of NP213, a novel onychomycosis therapy candidate. Antimicrob Agents Chemother. 2019;63(5)
12. Fernandez-Torres B, Carrillo-Munoz A, Ortoneda M, Pujol I, Pastor FJ, Guarro J. Interlaboratory evaluation of the etest for antifungal susceptibility testing of dermatophytes. Med Mycol. 2003;41(2):125–30.
13. CLSI. Reference method for broth dilution antifungal susceptibility testing of yeasts; approved standard - fourth edition. Document M27Ed4. Wayne: Clinical and Laboratory Standards Institute; 2017.
14. CLSI. Performance standards for antifungal susceptibility testing of filamentous fungi, first edition. Document M61. Wayne: Clinical and Laboratory Standards Institute; 2017.
15. Ghannoum MA, Arthington-Skaggs B, Chaturvedi V, Espinel-Ingroff A, Pfaller MA, Rennie R, et al. Interlaboratory study of quality control isolates for a broth microdilution method (modified CLSI M38-A) for testing susceptibilities of dermatophytes to antifungals. J Clin Microbiol. 2006;44(12):4353–6.

16. Ghannoum MA, Chaturvedi V, Espinel-Ingroff A, Pfaller MA, Rinaldi MG, Lee-Yang W, et al. Intra- and interlaboratory study of a method for testing the antifungal susceptibilities of dermatophytes. J Clin Microbiol. 2004;42(7):2977–9.
17. Rex JH, Pfaller MA. Has antifungal susceptibility testing come of age? Clin Infect Dis. 2002;35 (8):982–9.
18. Rex JH, Pfaller MA, Galgiani JN, Bartlett MS, Espinel-Ingroff A, Ghannoum MA, et al. Development of interpretive breakpoints for antifungal susceptibility testing: conceptual framework and analysis of *in vitro-in vivo* correlation data for fluconazole, itraconazole, and *Candida* infections. Subcommittee on antifungal susceptibility testing of the National Committee for Clinical Laboratory Standards. Clin Infect Dis. 1997;24(2):235–47.
19. Arendrup MC, Cuenca-Estrella M, Lass-Florl C, Hope W. EUCAST-AFST. EUCAST technical note on the EUCAST definitive document EDef 7.2: method for the determination of broth dilution minimum inhibitory concentrations of antifungal agents for yeasts EEef 7.2 (EUCAST-AFST). Clin Microbiol Infect. 2012;18(7):E246–7.
20. EUCAST. EUCAST antifungal agents breakpoint tables for interpretation of MICs 2018. Available from http://www.eucast.org/astoffungi/clinicalbreakpointsforantifungals/.
21. CLSI. Performance standards for antifungal susceptibility testing of yeasts, first edition. Document M60. Wayne: Clinical and Laboratory Standards Institute; 2017.
22. Khurana A, Sardana K, Chowdhary A. Antifungal resistance in dermatophytes: recent trends and therapeutic implications. Fungal Genet Biol. 2019;132:103255.
23. Faergemann J, Zehender H, Denouel J, Millerioux L. Levels of terbinafine in plasma, stratum corneum, dermis-epidermis (without stratum corneum), sebum, hair and nails during and after 250 mg terbinafine orally once per day for four weeks. Acta Derm Venereol. 1993;73(4):305–9.
24. Prescribing information, Lamisil oral tablet. East Hanover: Novartis Pharmaceuticals, Inc.; 2005.
25. Epstein WL, Shah VP, Riegelman S. Griseofulvin levels in stratum corneum. Study after oral administration in man. Arch Dermatol. 1972;106(3):344–8.
26. Pfaller MA, Rex JH, Rinaldi MG. Antifungal susceptibility testing: technical advances and potential clinical applications. Clin Infect Dis. 1997;24(5):776–84.
27. Khurana A, Masih A, Chowdhary A, Sardana K, Borker S, Gupta A, et al. Correlation of *in vitro* susceptibility based on mics and squalene epoxidase mutations with clinical response to terbinafine in patients with *Tinea corporis/cruris*. Antimicrob Agents Chemother. 2018;62 (12):e01038.
28. Favre B, Ghannoum MA, Ryder NS. Biochemical characterization of terbinafine-resistant *Trichophyton rubrum* isolates. Med Mycol. 2004;42(6):525–9.
29. Salehi Z, Shams-Ghahfarokhi M, Razzaghi-Abyaneh M. Antifungal drug susceptibility profile of clinically important dermatophytes and determination of point mutations in terbinafine-resistant isolates. Eur J Clin Microbiol Infect Dis. 2018;37(10):1841–6.
30. Yamada T, Maeda M, Alshahni MM, Tanaka R, Yaguchi T, Bontems O, et al. Terbinafine resistance of *Trichophyton* clinical isolates caused by specific point mutations in the squalene epoxidase gene. Antimicrob Agents Chemother. 2017;61(7):e00115-17.
31. Mukherjee PK, Leidich SD, Isham N, Leitner I, Ryder NS, Ghannoum MA. Clinical *Trichophyton rubrum* strain exhibiting primary resistance to terbinafine. Antimicrob Agents Chemother. 2003;47(1):82–6.
32. Singh A, Masih A, Khurana A, Singh PK, Gupta M, Hagen F, et al. High terbinafine resistance in *Trichophyton interdigitale* isolates in Delhi, India harbouring mutations in the squalene epoxidase gene. Mycoses. 2018;61(7):477–84.
33. Rudramurthy SM, Shankarnarayan SA, Dogra S, Shaw D, Mushtaq K, Paul RA, et al. Mutation in the squalene epoxidase gene of *Trichophyton interdigitale* and *Trichophyton rubrum* associated with allylamine resistance. Antimicrob Agents Chemother. 2018;62(5):e02522
34. Schosler L, Andersen LK, Arendrup MC, Sommerlund M. Recurrent terbinafine resistant *Trichophyton rubrum* infection in a child with congenital ichthyosis. Pediatr Dermatol. 2018;35(2):259–60.

35. Digby SS, Hald M, Arendrup MC, Hjort SV, Kofoed K. Darier disease complicated by terbinafine-resistant *Trichophyton rubrum*: a case report. Acta Derm Venereol. 2017;97 (1):139–40.
36. Turnidge J, Paterson DL. Setting and revising antibacterial susceptibility breakpoints. Clin Microbiol Rev. 2007;20(3):391–408.
37. Pfaller MA, Diekema DJ. Progress in antifungal susceptibility testing of *Candida* spp. by use of Clinical and Laboratory Standards Institute broth microdilution methods, 2010 to 2012. J Clin Microbiol. 2012;50(9):2846–56.
38. Alexander BD, Byrne TC, Smith KL, Hanson KE, Anstrom KJ, Perfect JR, et al. Comparative evaluation of etest and sensititre yeastone panels against the clinical and laboratory standards institute M27-A2 reference broth microdilution method for testing *Candida* susceptibility to seven antifungal agents. J Clin Microbiol. 2007;45(3):698–706.
39. Cuenca-Estrella M, Gomez-Lopez A, Alastruey-Izquierdo A, Bernal-Martinez L, Cuesta I, Buitrago MJ, et al. Comparison of the Vitek 2 antifungal susceptibility system with the Clinical and Laboratory Standards Institute (CLSI) and European Committee on Antimicrobial Susceptibility Testing (EUCAST) broth microdilution reference methods and with the Sensititre YeastOne and Etest techniques for *in vitro* detection of antifungal resistance in yeast isolates. J Clin Microbiol. 2010;48(5):1782–6.
40. Espinel-Ingroff A, Pfaller M, Messer SA, Knapp CC, Holliday N, Killian SB. Multicenter comparison of the Sensititre YeastOne colorimetric antifungal panel with the NCCLS M27-A2 reference method for testing new antifungal agents against clinical isolates of *Candida* spp. J Clin Microbiol. 2004;42(2):718–21.
41. Espinel-Ingroff A, Pfaller M, Messer SA, Knapp CC, Killian S, Norris HA, et al. Multicenter comparison of the sensititre YeastOne colorimetric antifungal panel with the National Committee for Clinical Laboratory Standards M27-a reference method for testing clinical isolates of common and emerging *Candida* spp., *Cryptococcus* spp., and other yeasts and yeast-like organisms. J Clin Microbiol. 1999;37(3):591–5.
42. Pfaller MA, Chaturvedi V, Diekema DJ, Ghannoum MA, Holliday NM, Killian SB, et al. Comparison of the Sensititre YeastOne colorimetric antifungal panel with CLSI microdilution for antifungal susceptibility testing of the echinocandins against *Candida* spp., using new clinical breakpoints and epidemiological cutoff values. Diagn Microbiol Infect Dis. 2012;73 (4):365–8.
43. Pfaller MA, Chaturvedi V, Diekema DJ, Ghannoum MA, Holliday NM, Killian SB, et al. Clinical evaluation of the Sensititre YeastOne colorimetric antifungal panel for antifungal susceptibility testing of the echinocandins anidulafungin, caspofungin, and micafungin. J Clin Microbiol. 2008;46(7):2155–9.
44. Pfaller MA, Espinel-Ingroff A, Jones RN. Clinical evaluation of the Sensititre YeastOne colorimetric antifungal plate for antifungal susceptibility testing of the new triazoles voriconazole, posaconazole, and ravuconazole. J Clin Microbiol. 2004;42(10):4577–80.
45. Pfaller MA, Jones RN, Microbiology Resource Committee College of American Pathologist. Performance accuracy of antibacterial and antifungal susceptibility test methods: report from the College of American Pathologists microbiology surveys program (2001-2003). Arch Pathol Lab Med. 2006;130(6):767–78.
46. Pujol I, Capilla J, Fernandez-Torres B, Ortoneda M, Guarro J. Use of the Sensititre colorimetric microdilution panel for antifungal susceptibility testing of dermatophytes. J Clin Microbiol. 2002;40(7):2618–21.
47. Castro C, Serrano MC, Valverde A, Peman J, Almeida C, Martin-Mazuelos E. Comparison of the Sensititre YeastOne colorimetric antifungal panel with the modified Clinical and Laboratory Standards Institute broth microdilution (M38-A) method for antifungal susceptibility testing of dermatophytes. Chemotherapy. 2008;54(6):427–30.
48. Mendez CC, Serrano MC, Valverde A, Peman J, Almeida C, Martin-Mazuelos E. Comparison of E-test, disk diffusion and a modified CLSI broth microdilution (M38-A) method for *in vitro*

testing of itraconazole, fluconazole and voriconazole against dermatophytes. Med Mycol. 2008;46(2):119–23.
49. Nweze EI, Mukherjee PK, Ghannoum MA. Agar-based disk diffusion assay for susceptibility testing of dermatophytes. J Clin Microbiol. 2010;48(10):3750–2.

Terbinafine and Itraconazole Resistance in Dermatophytes

Michel Monod, Marc Feuermann, and Tsuyoshi Yamada

Abstract

Terbinafine-resistant dermatophytes in patients were exceptional before the second decade of this century. However, acquired resistance to commonly used antifungal compounds has recently emerged in several countries. Resistance towards terbinafine is generated by missense mutations in the squalene epoxidase enzyme targeted by the drug, while recorded resistance towards azoles is due to the overexpression of genes encoding multidrug transporters of the ABC family. At present, approximately 1% of *Trichophyton rubrum* isolates from *Tinea pedis* and onychomycosis in Switzerland are resistant to terbinafine. Terbinafine-resistant *T. rubrum* was also isolated from extended *Tinea corporis* in patients more susceptible to fungal infections and requiring continuous treatment. Repeated topical and systemic treatments with terbinafine have likely contributed to the development of terbinafine resistance in patients. The prevalence of *T. rubrum* resistant to terbinafine in Europe contrasts with that of resistant *Trichophyton mentagrophytes* isolates in India (30–70%). If drugs on the open market and overmedication can partly explain this alarming situation, several indications also allow the suspicion of an origin linked to environmental problems.

M. Monod (✉)
Department of Dermatology, Centre Hospitalier Universitaire Vaudois (CHUV), Lausanne, Switzerland
e-mail: Michel.Monod@chuv.ch

M. Feuermann
Swiss-Prot Group, SIB Swiss Institute of Bioinformatics, Geneva, Switzerland

T. Yamada
Teikyo University Institute of Medical Mycology, Tokyo, Japan

Asia International Institute of Infectious Disease Control, Teikyo University, Tokyo, Japan

J.-P. Bouchara et al. (eds.), *Dermatophytes and Dermatophytoses*,
https://doi.org/10.1007/978-3-030-67421-2_20

Keywords

Trichophyton · *Microsporum* · Terbinafine · Itraconazole · Amphotericin B · Antifungal drug resistance · ABC transporters · MFS transporters

1 Introduction

Drug resistance in dermatophytes is becoming a real problem in dermatology. On one hand, the insensitivity of a fungus to an antifungal agent may be intrinsic to the species. On the other hand, acquired resistance may occur after prolonged antifungal treatments or because of the presence of antifungal drugs in the environment. With a low mutation frequency in a yeast population or mycelium, resistant strains can be selected from the fungus that was previously susceptible after selective pressure by continuous exposure to an antifungal agent.

Drug resistance acquired in human pathogenic fungi was first studied in *Candida albicans* and then in other yeasts and filamentous fungi, in particular *Aspergillus fumigatus*. Most resistance mechanisms are due either to missense mutations in genes encoding drug targets or to their overexpression [1–6], or to the overexpression of genes encoding multidrug transporters [6–10]. A combination of these three mechanisms results in an additive effect [2, 11]. The objective of this chapter is to briefly review the different resistance mechanisms of dermatophytes and to examine their importance in public health. Identifying the mechanism of resistance can be useful in selecting the appropriate treatment.

2 Antifungal Drugs Against Dermatophytes

The antifungal drugs currently available for the treatment of dermatophytosis are terbinafine, azole compounds, such as itraconazole and fluconazole, and griseofulvin. Terbinafine and azole compounds target the biosynthetic pathway of ergosterol [12]. A defect in the synthesis of ergosterol leads to an alteration in the structure of the plasma membrane and its function [12, 13]. Terbinafine prevents the conversion of squalene epoxide to lanosterol, the early step of ergosterol biosynthesis from squalene, by inhibiting squalene epoxidase (SQLE) [14–16]. Inhibiting SQLE results in an intracellular accumulation of squalene, which is toxic to the fungal cells [13]. Azole compounds act downstream of the SQLE reaction by inhibiting the lanosterol 14-α-demethylase (encoded by *ERG11* in *C. albicans* and *CYP51* in filamentous fungi), which prevents the demethylation of lanosterol in the ergosterol biosynthetic pathway [13]. Inhibition of the lanosterol 14-α-demethylase leads to the accumulation of sterol precursors in the fungal cell and constitutes a roadblock in the synthesis of ergosterol.

Griseofulvin inhibits fungal mitosis by disrupting the mitotic spindle through interacting with polymerised microtubules [17]. Griseofulvin was isolated from *Penicillium griseofulvum* in 1939 [18], and it was the first systemic antifungal agent to be used in both human and animals in the 1960s following the successful oral therapy of experimental dermatophytosis in Guinea pigs [19]. Griseofulvin is only administered orally, but it has been shown to be carcinogenic and affects fertility and organogenesis in rodents at high doses [20, 21]. However, griseofulvin remains the treatment of choice in many cases of *Tinea capitis* [22, 23].

3 Antifungal Susceptibility Testing

Methods for testing the susceptibility of filamentous fungi are similar to those used in bacteriology or for yeasts [24]. The minimum inhibitory concentration (MIC) of an antifungal agent remains its lowest concentration that substantially inhibits the growth of an organism. Microdilution, checkerboard, disc diffusion and Etest assays can be performed in the same way for filamentous fungi as for bacteria and yeasts, provided that spores are produced in sufficient quantities to be used as colony-forming units (CFU). Therefore, the absence or low sporulation of *Trichophyton rubrum* has always been a major complication and a limiting factor for performing routine antifungal susceptibility testing. Different media [e.g. Sabouraud's dextrose agar (SDA) diluted 1/10, potato dextrose agar (PDA) and oatmeal agar (OTA)], on which *T. rubrum* is supposed to sporulate easily, are proposed in the literature and in guidelines from the Clinical and Laboratory Standards Institute (CLSI) [24, 25]. However, variability between strains remains present in these media regarding quantitative sporulation. Large quantities of *T. rubrum* spores, even with poorly sporulating strains, can be obtained with high CO_2 tensions and PDA growth medium [26], and the availability of large quantities of spores allows the preparation of standardised inocula to perform assays. The effect of high CO_2 tension on *T. rubrum* sporulation was described by Chin and Knight in 1957 [27].

Strict application of the protocols established by the CLSI for broth dilution antifungal susceptibility testing [24] is also difficult for *T. rubrum*, and other dermatophyte species, due to the use of RPMI 1640 medium. Dermatophytes grow poorly in this synthetic medium in contrast to yeasts and *Aspergillus* spp., and were even reported to not grow in RPMI 1640 medium [26]. Therefore, in our opinion, it is more advisable to use SDA medium for performing reproducible antifungal tests.

4 Importance of the Identification of the Infectious Agent Before Immediately Searching Resistance

In *Tinea capitis*, the therapeutic response varies according to the dermatophyte species involved. The cure rate with terbinafine is good in cases of infection by the anthropophilic species *Trichophyton violaceum* and *Trichophyton soudanense*, but not in cases of infection by the anthropophilic *Microsporum audouinii* and zoophilic species, such as *Microsporum canis*, *Trichophyton benhamiae* and *Trichophyton mentagrophytes* [28]. The clinical efficacy of terbinafine is not related to the *in vitro* susceptibility of the species. Griseofulvin is used in many cases of

Tinea capitis, which are insensitive to terbinafine and azoles [22, 23]. In most cases of *Tinea corporis*, dermatophyte identification is not so important, as these mycoses respond well to topical standard treatments, regardless of the dermatophyte species.

Insensitivity of onychomycosis to standard treatments using terbinafine and azoles is generally due to non-dermatophyte fungi (NDF), which are especially difficult to cure [29]. In these cases, *Fusarium* sp., *Acremonium* sp. and *Aspergillus* sp. are often identified as the sole infectious agents by PCR using extracted DNA from the nail as a target. The topical application of an amphotericin B solution was efficient in curing NDF onychomycosis [30].

In conclusion, correct species identification is important in cases of superficial fungal infections, especially hair and nail infections, to prescribe adequate treatments and before suspecting acquired fungal resistance.

5 Resistance of *T. rubrum* and *Trichophyton interdigitale* to Terbinafine

Two cases of terbinafine resistance were first reported in the 2000s in *T. rubrum* isolated from onychomycoses [31–33] (Table 1). Single-point mutations were revealed in the gene encoding SQLE, targeted by terbinafine. In both cases, the mutation generated a missense substitution in the same region of the enzyme (L393F in case 1 and F397L in case 2). These two cases were apparently exceptional when they were discovered, but since then, terbinafine-resistant *T. rubrum* isolates have been reported in several countries in Europe and Asia (Table 1). Terbinafine-resistant *T. rubrum* isolates can be easily identified routinely on Sabouraud agar medium containing 2 μg/ml terbinafine (Fig. 1) as well as with a broth microdilution method [40]. Cases of resistant *Trichophyton interdigitale* were also identified in Switzerland, Denmark and Japan [34, 35, 41]. The DNA sequence of the gene encoding SQLE in *T. rubrum* terbinafine-resistant isolates revealed single-point mutations, which led to amino acid substitutions at one of four amino acid positions (L393, F397, F415 and H440) within SQLE. In one case, SQLE was found to contain two amino acid substitutions (I121M and V237I), which were located in the pocket where terbinafine binds [35]. The high prevalence of the mutations L393F and F397L identified in the two first cases of terbinafine resistance in *T. rubrum* is particularly striking (Table 1).

The substituted amino acids in position 393, 397, 415 and 440, which were detected in terbinafine-resistant strains of *Trichophyton*, were introduced by transformation into a terbinafine-sensitive strain of *T. mentagrophytes* (formerly *Arthroderma vanbreuseghemii*), and terbinafine-resistant phenotypes were generated [34]. The terbinafine MIC of *T. mentagrophytes* transformants was comparable to that of the corresponding clinical isolates. Therefore, terbinafine resistance in *Trichophyton* clinical isolates could be reliably attributed to the amino acid substitution detected in the SQLE protein.

Resistant *T. rubrum* isolates with F397V and F415S mutations in SQLE grow slowly compared to sensitive isolates or isolates with L393F or F397L mutations

Table 1 Recorded and published cases of resistance of *T. rubrum* and other *Trichophyton* species to terbinafine (2003–2020)

Species	Type of infection	Mutations	Country	References
T. rubrum (*n* = 1)	*Tinea unguium*	L393F	USA	Mukherjee et al. [31] Osborne et al. [32]
T. rubrum (*n* = 1)	Not given	F397L	Switzerland	Osborne et al. [33]
T. rubrum (*n* = 16)	*Tinea pedis* and *Tinea unguium*	L393F (*n* = 4) L393S (*n* = 2) F397L (*n* = 4) F397I (*n* = 1) F397V (*n* = 1) F415I (*n* = 1) F415S (*n* = 1) F415V (*n* = 1) H440Y (*n* = 1)	Switzerland	Yamada et al. [34]
T. rubrum (*n* = 12)	*Tinea pedis* (*n* = 3) *Tinea unguium* (*n* = 3) *Tinea corporis* (*n* = 4) *Tinea corporis* in a patient with Darier disease (*n* = 1)[a] *Tinea corporis* in a patient with congenital ichthyosis (*n* = 1)[b]	L393F (*n* = 1) L393S (*n* = 2) F397L (*n* = 6) F415S (*n* = 1) H440Y (*n* = 1) I121M/V237I	Denmark	Saunte et al. [35]
T. rubrum (*n* = 2)	Not given (patients with only *Tinea unguium* excluded)	F397L	India	Rudramurthy et al. [36]
T. rubrum (*n* = 1)	Not given	L393F	Iran	Salehi et al. [37]
T. rubrum (*n* = 1)	*Tinea pedis*	L393F	Japan	Suzuki et al. [38]
T. rubrum (*n* = 1)	*Tinea unguium*	F397L	Japan	Noguchi et al. [39]
T. rubrum (*n* = 8)	Mostly *Tinea corporis*	F397L	India	Ebert et al. [40]
T. interdigitale (*n* = 1)	*Tinea unguium*	F397L	Switzerland	Yamada et al. [34]
T. interdigitale (*n* = 1)	*Tinea pedis*	Possible overexpression of genes encoding	Japan	Hiruma et al. [41]

(continued)

Table 1 (continued)

Species	Type of infection	Mutations	Country	References
		ABC-transporter proteins		
T. interdigitale ($n = 2$)	*Tinea pedis* ($n = 1$) *Tinea pedis, corporis* and *cruris* ($n = 1$)	L393F F397L	Denmark	Saunte et al. [35]
T. mentagrophytes ($n = 20$) (under *T. interdigitale*)	*Tinea corporis* *Tinea cruris*	L393F ($n = 8$) F397L ($n = 12$)	India	Singh et al. [42]
T. mentagrophytes (under *T. interdigitale*)	*Tinea corporis* *Tinea cruris*	F397L ($n = 4$)	India	Rudramurthy et al. [36]
T. mentagrophytes ($n = 1$)	*Tinea corporis*	F397L ($n = 1$)	Bahrain	Süß et al. [43]
T. mentagrophytes ($n = 199$)	Mainly *Tinea corporis* and *Tinea cruris*	L393F ($n = 6$) L393S ($n = 7$) S395P ($n = 1$) F397L ($n = 180$) Q408L ($n = 2$) H440Y ($n = 2$) S443P ($n = 1$)	India	Ebert et al. [40]
T. mentagrophytes ($n = 1$)	*Tinea corporis*	Q408L ($n = 1$)	Switzerland	Hsieh et al. [44]
T. mentagrophytes ($n = 5$)	*Tinea corporis*	L393S ($n = 1$) F397L ($n = 4$)	Iran	Taghipour et al. [45]
T. mentagrophytes ($n = 74$)	Mainly *Tinea corporis*	F397L ($n = 74$)	India	Shaw et al. [46]
T. tonsurans ($n = 1$)	Not indicated	L393F	Iran	Salehi et al. [37]

[a]Case report by Digby et al. [47]
[b]Case report by Schøsler et al. [48]

[34]. The *T. mentagrophytes* transformants in which the SQLE gene had been substituted by an allele with the F397V mutation also exhibited a slow-growth phenotype. These results suggest that mutations reduce the affinity of terbinafine for SQLE and, at the same time, reduce the specific activity of the enzyme causing growth retardation. Similarly, in *C. albicans*, a G464S mutation in the sterol 14-α-demethylase encoded by *ERG11* reduced both the inhibitory effect of fluconazole and the enzyme catalytic activity [49]. Therefore, a slow-growing dermatophyte isolate may reflect resistance.

Repeated systemic treatments with terbinafine have likely contributed to the development of terbinafine resistance in patients. Currently, *T. rubrum* isolates resistant to terbinafine are recovered from *Tinea pedis* and onychomycosis in Switzerland with a frequency of approximately 1% (26 cases among 2056 tested isolates) [34; unpublished results]. *Trichophyton rubrum* resistant to terbinafine was

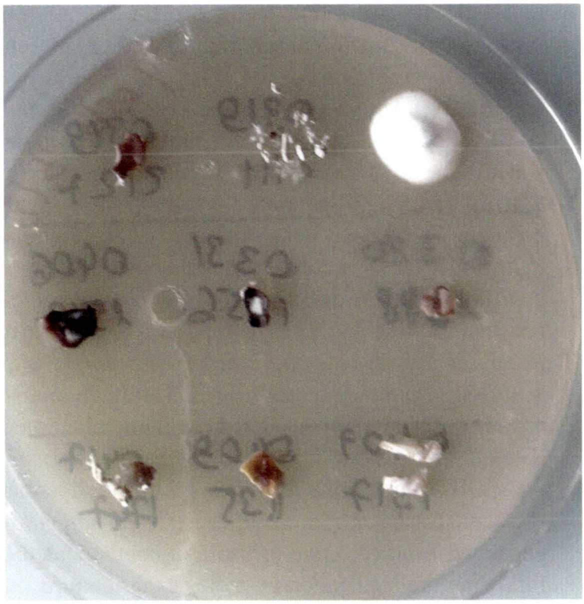

Fig. 1 Terbinafine-resistant *Trichophyton rubrum* isolate. Nine isolates of *T. rubrum* were tested on Sabouraud agar medium containing 2 μg/ml terbinafine. The MIC of terbinafine had previously been estimated at 1 μg/ml

also isolated from extended chronic *Tinea corporis* in patients more susceptible to fungal infection, for instance, in patients suffering from skin diseases with keratinisation disorders, such as congenital ichthyosis and Darier disease [47, 48]. Resistance was acquired after increased and continuous exposure to terbinafine to combat the first infection.

6 Resistance of *T. mentagrophytes* to Terbinafine

A particular genotype of *T. mentagrophytes* (*T. mentagrophytes* type VIII) endemic in India is the cause of an epidemic of *Tinea cruris* and *Tinea corporis* resistant to terbinafine [50]. *Trichophyton* terbinafine-resistant clinical isolates have a reported prevalence of 17–71% depending on the studies [36, 40, 42]. The DNA sequence of the gene encoding SQLE revealed single-point mutations identical to those previously recorded for *T. rubrum*, leading to amino acid substitutions at one of the four previously mentioned amino acid positions (L393, F397, F415 and H440) within the SQLE protein like in *T. rubrum*. The amino acid substitution F397L in the SQLE of resistant *T. mentagrophytes* was highly prevalent, till 91% in a recent multicentre study [40]. Three novel substitutions, S395P, Q408L and S443P, were discovered in resistant strains (Table 1). The mutation Q408L was also reported in a case report in Geneva with disseminated *Tinea corporis* [44].

Terbinafine resistance in *T. mentagrophytes* is an emerging problem in India. The high prevalence of resistant isolates is particularly striking. If drugs on the open market and overmedication partly explain this alarming situation, several factors give rise to the suspicion of an origin linked to environmental problems:

1. *High concentrations of antifungal compounds in the environment*: In India, high concentrations of antibiotic and antifungal compounds were recorded in wastewater in the environment of bulk drug manufacturing industries [51]. Dissemination of extended-spectrum beta-lactamase- and carbapenemase-producing pathogens were associated with high concentrations of antibiotics in sewers of industrial areas. Voriconazole and fluconazole concentrations were also measured up to 2500 and 237 μg/L, respectively. However, the presence of fungi potentially resistant to these antifungal compounds was not investigated.
2. *Reservoir of T. mentagrophytes in soil*: Several studies have reported *T. mentagrophytes* isolates from soil in India and Nepal [52, 53]. *Trichophyton mentagrophytes* is considered as zoophilic species, causing inflammatory dermatophytosis but there is no evidence for transmission of *T. mentagrophytes* type VIII from animals to humans. It is possible that *T. mentagrophytes* isolates resistant to azoles, like azole-resistant *A. fumigatus* isolates (see below), were generated from selective pressure in contaminated soil, either by wastewater or following the application of azoles in agriculture.
3. *The example of emerging azole-resistant strains of A. fumigatus in hospitals*: It is widely accepted that patients with azole-resistant aspergillosis had been colonised by azole-resistant environmental isolates before developing an infection [54, 55]. Azole-resistant *A. fumigatus* has been isolated in hospitals with a high prevalence (5–15%) in several countries. Although resistant strains may have emerged in response to long-term drug therapy, many *A. fumigatus*-resistant clinical isolates are from the environment.

Now, it is imperative to identify the reservoir of *T. mentagrophytes*-type VIII as a possible source of infection. Microbiological analyses and measurements of the level of antifungal agents in soil and wastewater in the patient's environment would be useful to identify the origin of resistant strains, in addition to overmedication. The possible transmission from the environment to humans, and from man to man, should also be studied to control this epidemic of dermatophytosis resistant to conventional treatments.

7 Azole Resistance in *T. rubrum*

Itraconazole is widely used for the treatment of dermatophyte infections [22, 23]. A terbinafine-resistant *T. rubrum* strain, isolated in Switzerland and called TIMMM20092, also showed reduced sensitivity to azole compounds (Fig. 2). This strain had been isolated from a patient with *Tinea pedis*, which was insensitive to systemic and topical terbinafine treatments and itraconazole. A mutation of the *CYP51A* gene encoding lanosterol 14-α-demethylase, which is targeted by azoles, was not detected in this strain. In contrast, the resistance of TIMM20092 to azoles

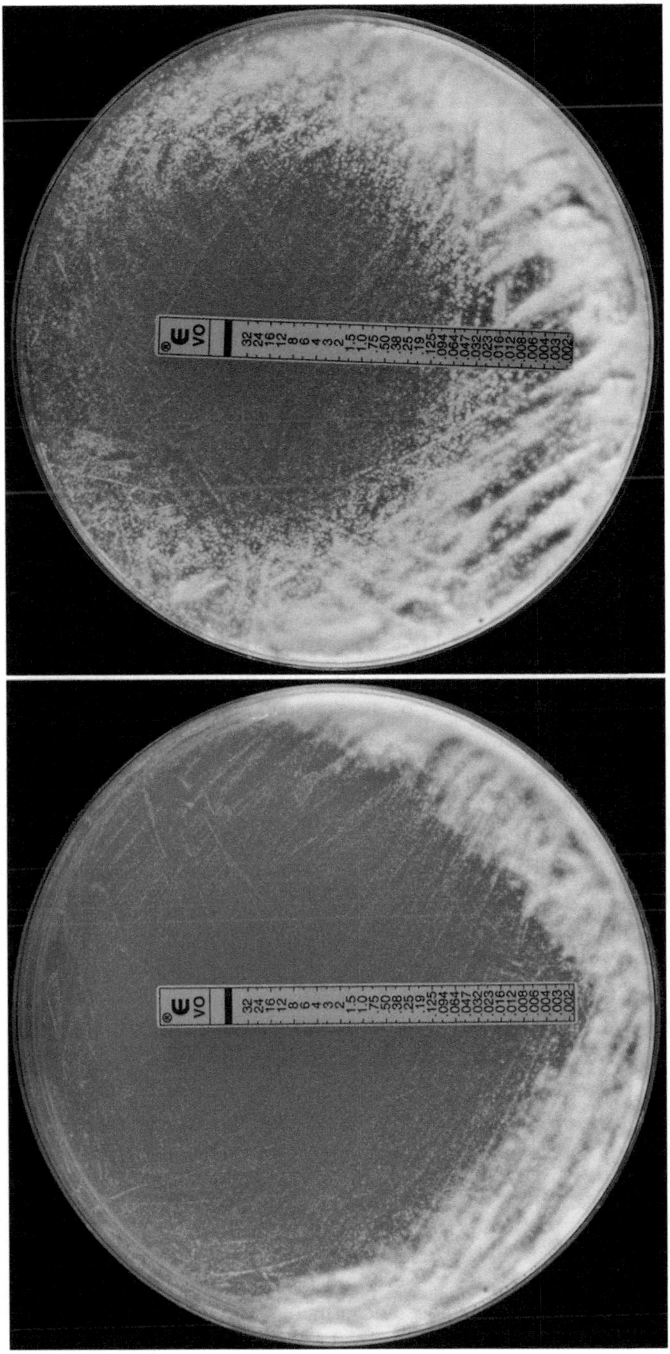

Fig. 2 *Trichophyton rubrum* strain showing reduced sensitivity to azole compounds (right) in comparison to a sensitive strain (left). Voriconazole Etest strips were placed on fungal lawns (10^6 CFU) prepared on Sabouraud agar plates

Table 2 Susceptibility and resistance to itraconazole (ITC), voriconazole (VRC), fluconazole (FLC), ketoconazole (KTC), miconazole (MIC) and cycloheximide (CYH) of *S. cerevisiae* transformed with different plasmids encoding *T. rubrum* ABC and MFS transporters[a]

Locus name	Name	Type	VRC	ITC	KTC	FLC	MIC	CYH	Comments[b]
TERG_02508	*TruMDR1*	ABC	+	-	-	+	+	-	Close to *A. fumigatus* atrI
TERG_08613	*TruMDR2*	ABC	-	+	-	-	-	-	Close to *A. fumigatus* mdr1
TERG_02186	*TruMDR3*	ABC	+	+	+	+	+	-	Close to *A. fumigatus* abcC/cdr1B/atrE
TERG_04952	*TruMDR5*	ABC	-	+	-	-	-	-	Close to *A. fumigatus* abcE
TERG_01623	*TruMFS1*	MFS	+	+	+	+	+	+	Close to *A. fumigatus* aflT
TERG_08336	*TruMFS2*	MFS	+	-	-	+	-	-	Close to *A. fumigatus* mdrA

The constructs showing resistance in *S. cerevisiae* are marked with a + sign and a grey shading
[a]Data from Monod et al. [56]
[b]Closest homologs from *A. fumigatus* are indicated

was inhibited by milbemycin oxime, leading to the hypothesis that the resistance was mediated by the overexpression of genes encoding efflux pumps.

Among 39 ATP-binding cassette (ABC) transporters and 170 major facilitator superfamily (MFS) transporters identified *in silico* within the *T. rubrum* genome, four ABC transporters (TruMDR1, TruMDR2, TruMDR3 and TruMDR5) and two MFS transporters (TruMFS1 and TruMFS2) were able to operate as azole efflux pumps when their encoding genes were overexpressed in *Saccharomyces cerevisiae* [56]. TruMDR3 and TruMFS1 were able to act as an efflux pump towards fluconazole, itraconazole, ketoconazole, miconazole and voriconazole (Table 2). TruMFS1 was also able to transport cycloheximide suggesting that it acts as a pleiotropic drug transporter with a broad substrate spectrum (Table 2). In contrast, TruMDR1 and TruMFS2 only transported fluconazole and voriconazole, two azoles of similar molecular structure, whereas TruMDR1 and TruMDR5 were specific for itraconazole (Table 2).

The DNA sequences of the genes *MDR1* and *MDR2* (for multidrug resistance 1 and 2, respectively) were firstly reported by Cervelatti et al. [57] and Fachin et al. [58]. The overexpression of the genes encoding TruMDR2 and TruMDR3 in *T. rubrum* TIMM20092 was revealed by qRT-PCR analysis. Transcription of *TruMDR1*, *TruMDR2*, *TruMDR3* and *TruMFS2* was also up-regulated upon exposure to itraconazole and voriconazole in TIMM20092, as well as in azole-sensitive strains as previously observed with MDR1 and MDR2 in dermatophyte species [59]. TruMDR3 is closely related to *A. fumigatus* cdr1B [10] and to *C. albicans* CDR1 and CDR2 [7, 8]. Milbemycin oxime inhibited TruMDR3 as it inhibited CDR1 and CDR2 in *C. albicans* [60–62]. In contrast, this inhibitor showed no effect on TruMDR1, TruMDR2 and TruMDR5.

Gain-of-function mutations in transcription factors, mediating the expression of genes encoding efflux pumps, have been described in *Candida* spp. and *A. fumigatus*

[63, 64]. The overexpression of genes encoding TruMDR2 and TruMDR3 leads to the assumption that there was a gain-of-function mutation in a transcription factor that remains to be identified in dermatophytes.

While resistance to azoles remains exceptional in *T. rubrum*, so far, 15% of *T. mentagrophytes* type VIII isolates were less sensitive to itraconazole and voriconazole among 297 strains in the recent multicentre study by Ebert *at al.* [40]. To date, we have no data on the possible resistance mechanisms in these isolates.

8 Conclusion

Today, we have to be aware that acquired resistance of dermatophytes to antifungal drugs exists and can no longer be considered exceptional. However, the identification of the infectious agent remains very important in cases of nail and hair infections for the prescription of appropriate treatment because of the possible intrinsic resistance. Standard treatments with terbinafine and azoles are inefficient in cases of non-dermatophyte mold onychomycosis. In addition, some dermatophyte species (*M. audouinii* and zoophilic species such as *M. canis*, *T. benhamiae* and *T. mentagrophytes*) do not respond to terbinafine and azole treatments in many cases of *Tinea capitis* and require the use of griseofulvin.

In cases of confirmed *Trichophyton* infections that are insensitive to terbinafine, it is essential to test the sensitivity of the fungus to this antifungal compound. We now routinely test all isolates of *T. rubrum* and *T. interdigitale* in our laboratory for possible resistance to terbinafine by depositing a small block of agar with the growing fungus on Sabouraud agar medium containing 0.02 μg/ml terbinafine (two times the MIC). We then examine if detected resistance in *Trichophyton* clinical isolates had to be imputed to an amino acid substitution in the SQLE protein. *Trichophyton* resistance to terbinafine treatment is an emerging problem, and a switch to azole-based treatment may be necessary to cure such cases of onychomycosis.

References

1. Sanglard D, Ischer F, Koymans L, Bille J. Amino acid substitutions in the cytochrome P-450 lanosterol 14alpha-demethylase (CYP51A1) from azole-resistant *Candida albicans* clinical isolates contribute to resistance to azole antifungal agents. Antimicrob Agents Chemother. 1998;42:241–53.
2. Perea S, López-Ribot JL, Kirkpatrick WR, McAtee RK, Santillán RA, Martínez M, et al. Prevalence of molecular mechanisms of resistance to azole antifungal agents in *Candida albicans* strains displaying high-level fluconazole resistance isolated from human immunodeficiency virus-infected patients. Antimicrob Agents Chemother. 2001;45:2676–84.
3. Mellado E, Garcia-Effron G, Alcazar-Fuoli L, Cuenca-Estrella M, Rodriguez-Tudela JL. Substitutions at methionine 220 in the 14alpha-sterol demethylase (Cyp51A) of *Aspergillus fumigatus* are responsible for resistance *in vitro* to azole antifungal drugs. Antimicrob Agents Chemother. 2004;48:2747–50.

4. Chen J, Li H, Li R, Bu D, Wan Z. Mutations in the *cyp51A* gene and susceptibility to itraconazole in *Aspergillus fumigatus* serially isolated from a patient with lung aspergilloma. J Antimicrob Chemother. 2005;55:31–7.
5. Mellado E, Garcia-Effron G, Alcázar-Fuoli L, Melchers WJ, Verweij PE, Cuenca-Estrella M, et al. A new *Aspergillus fumigatus* resistance mechanism conferring *in vitro* cross-resistance to azole antifungals involves a combination of cyp51A alterations. Antimicrob Agents Chemother. 2007;51:1897–904.
6. Hagiwara D, Watanabe A, Kamei K, Goldman GH. Epidemiological and genomic landscape of azole resistance mechanisms in *Aspergillus* fungi. Front Microbiol. 2016;7:1382.
7. Sanglard D, Kuchler K, Ische F, Pagani JL, Monod M, Bille J. Mechanisms of resistance to azole antifungal agents in *Candida albicans* isolates from AIDS patients involve specific multidrug transporters. Antimicrob Agents Chemother. 1995;39:2378–86.
8. Sanglard D, Ischer F, Monod M, Bille J. Cloning of *Candida albicans* genes conferring resistance to azole antifungal agents: characterization of CDR2, a new multidrug ABC transporter gene. Microbiology. 1997;143(Pt 2):405–16.
9. Sanglard D, Ischer F, Bille J. Role of ATP-binding-cassette transporter genes in high-frequency acquisition of resistance to azole antifungals in *Candida glabrata*. Antimicrob Agents Chemother. 2001;45:1174–83.
10. Fraczek MG, Bromley M, Buied A, Moore CB, Rajendran R, Rautemaa R, et al. The cdr1B efflux transporter is associated with non-cyp51a-mediated itraconazole resistance in *Aspergillus fumigatus*. J Antimicrob Chemother. 2013;68:1486–96.
11. Sanglard D, Ischer F, Calabrese D, Micheli M, Bille J. Multiple resistance mechanisms to azole antifungals in yeast clinical isolates. Drug Resist Update. 1998;1:255–65.
12. Leyden J. Pharmacokinetics and pharmacology of terbinafine and itraconazole. J Am Acad Dermatol. 1998;38(5 Pt 3):S42–7.
13. Odds FC, Brown AJ, Gow NA. Antifungal agents: mechanisms of action. Trends Microbiol. 2003;11:272–9.
14. Ryder NS. Specific inhibition of fungal sterol biosynthesis by SF 86-327, a new allylamine antimycotic agent. Antimicrob Agents Chemother. 1985;27:252–6.
15. Ryder NS, Mieth H. Allylamine antifungal drugs. Curr Top Med Mycol. 1991;4:158–88.
16. Favre B, Ryder NS. Characterization of squalene epoxidase activity from the dermatophyte *Trichophyton rubrum* and its inhibition by terbinafine and other antimycotic agents. Antimicrob Agents Chemother. 1996;40:443–7.
17. Gull K, Trinci AP. Griseofulvin inhibits fungal mitosis. Nature. 1973;244(5414):292–4.
18. Oxford AE, Raistrick H, Simonart P. Studies in the biochemistry of micro-organisms: Griseofulvin, C(17)H(17)O(6)Cl, a metabolic product of *Penicillium griseo-fulvum* Dierckx. Biochem J. 1939;33:240–8.
19. Gentles JC. Experimental ringworm in guinea pigs: oral treatment with griseofulvin. Nature. 1958;182(4633):476–7.
20. Barich LL, Nakai T, Schwarz J, Barich DJ. Tumour-promoting effect of excessively large doses of oral griseofulvin on tumours induced in mice by methylcholanthrene. Nature. 1960;187:335–6.
21. Knasmüller S, Parzefall W, Helma C, Kassie F, Ecker S, Schulte-Hermann R. Toxic effects of griseofulvin: disease models, mechanisms, and risk assessment. Crit Rev Toxicol. 1997;27:495–537. Review. Erratum in: Crit Rev Toxicol. 1998;28:102.
22. Hay RJ. Tinea capitis: current status. Mycopathologia. 2017;182:87–93.
23. Hay RJ. Therapy of skin, hair and nail fungal infections. J Fungi (Basel). 2018;4(3):99.
24. M38-A protocols. National Committee for Clinical Laboratory Standards. Reference method for broth dilution antifungal susceptibility testing of filamentous fungi. Approved standard M38-A2; 2nd ed. Wayne: Clinical and Laboratory Standards Institute; 2008.
25. Jessup CJ, Wamer J, Isham N, Hasan I, Ghannoum MA. Antifungal susceptibility testing of dermatophytes: establishing a medium for inducing conidial growth and evaluation of susceptibility of clinical isolates. J Clin Microbiol. 2000;38:341–4.

26. Laurent A, Monod M. Production of *Trichophyton rubrum* microspores in large quantities and its application to evaluate amorolfine/azole compound interactions *in vitro*. Mycoses. 2017;60:581–6.
27. Chin B, Knight S. Growth of *Trichophyton mentagrophytes* and *Trichophyton rubrum* in increased carbon dioxide tensions. J Gen Microbiol. 1957;16:642–6.
28. Baudraz-Rosselet F, Monod M, Jaccoud S, Frenk E. Efficacy of terbinafine treatment of tinea capitis in children varies according to the dermatophyte species. Br J Dermatol. 1996;135:1011–2.
29. Baudraz-Rosselet F, Ruffieux C, Lurati M, Bontems O, Monod M. Onychomycosis insensitive to systemic terbinafine and azole treatments reveals non-dermatophyte moulds as infectious agents. Dermatology. 2010;220:164–8.
30. Lurati M, Baudraz-Rosselet F, Vernez M, Spring P, Bontems O, Fratti M, et al. Efficacious treatment of non-dermatophyte mould onychomycosis with topical amphotericin B. Dermatology. 2011;223:289–92.
31. Mukherjee PK, Leidich SD, Isham N, Leitner I, Ryder NS, Ghannoum MA. Clinical *Trichophyton rubrum* strain exhibiting primary resistance to terbinafine. Antimicrob Agents Chemother. 2003;47:82–6.
32. Osborne CS, Leitner I, Favre B, Ryder NS. Amino acid substitution in *Trichophyton rubrum* squalene epoxidase associated with resistance to terbinafine. Antimicrob Agents Chemother. 2005;49:2840–4.
33. Osborne CS, Leitner I, Hofbauer B, Fielding CA, Favre B, Ryder NS. Biological, biochemical, and molecular characterization of a new clinical *Trichophyton rubrum* isolate resistant to terbinafine. Antimicrob Agents Chemother. 2006;50:2234–6.
34. Yamada T, Maeda M, Alshahni MM, Tanaka R, Yaguchi T, Bontems O, et al. Terbinafine resistance of *Trichophyton* clinical isolates caused by specific point mutations in the squalene epoxidase gene. Antimicrob Agents Chemother. 2017;61(7):e00115–17.
35. Saunte DML, Hare RK, Jørgensen KM, Jørgensen R, Deleuran M, Zachariae CO, et al. Emerging terbinafine resistance in *Trichophyton*: Clinical characteristics, squalene epoxidase gene mutations, and a reliable EUCAST method for detection. Antimicrob Agents Chemother. 2019;63(10):e01126–19.
36. Rudramurthy SM, Shankarnarayan SA, Dogra S, Shaw D, Mushtaq K, Paul RA, et al. Mutation in the squalene epoxidase gene of *Trichophyton interdigitale* and *Trichophyton rubrum* associated with allylamine resistance. Antimicrob Agents Chemother. 2018;62(5):e02522–17.
37. Salehi Z, Shams-Ghahfarokhi M, Razzaghi-Abyaneh M. Antifungal drug susceptibility profile of clinically important dermatophytes and determination of point mutations in terbinafine-resistant isolates. Eur J Clin Microbiol Infect Dis. 2018;37:1841–6.
38. Suzuki S, Mano Y, Furuya N, Fujitani K. Discovery of terbinafine low susceptibility *Trichophyton rubrum* strain in Japan. Biocontrol Sci. 2018;23:151–4.
39. Noguchi H, Matsumoto T, Hiruma M, Kimura U, Kano R, Yaguchi T, et al. Tinea unguium caused by terbinafine-resistant *Trichophyton rubrum* successfully treated with fosravuconazole. J Dermatol. 2019;46(12):e446–7.
40. Ebert A, Monod M, Salamin K, Burmester A, Uhrlaß S, Wiegand C, et al. Alarming India wide phenomenon of antifungal resistance in dermatophytes: A multicentre study. Mycoses. 2020;63:717–28.
41. Hiruma J, Kitagawa H, Noguchi H, Kano R, Hiruma M, Kamata H, et al. Terbinafine-resistant strain of *Trichophyton interdigitale* strain isolated from a tinea pedis patient. J Dermatol. 2019;46:351–3.
42. Singh A, Masih A, Khurana A, Singh PK, Gupta M, Hagen F, et al. High terbinafine resistance in *Trichophyton interdigitale* isolates in Delhi, India harbouring mutations in the squalene epoxidase gene. Mycoses. 2018;61:477–84.
43. Süß A, Uhrlaß S, Ludes A, Verma SB, Monod M, Krüger C, et al. [Extensive tinea corporis due to a terbinafine-resistant *Trichophyton mentagrophytes* isolate of the Indian genotype in a young infant from Bahrain in Germany]. Hautarzt. 2019;70(11):888–96.

44. Hsieh A, Quenan S, Riat A, Toutous-Trellu L, Fontao L. A new mutation in the SQLE gene of *Trichophyton mentagrophytes* associated to terbinafine resistance in a couple with disseminated tinea corporis. J Mycol Med. 2019;29:352–5.
45. Taghipour S, Shamsizadeh F, Pchelin IM, Rezaei-Matehhkolaei A, Zarei Mahmoudabadi A, Valadan R, et al. Emergence of terbinafine resistant *Trichophyton mentagrophytes* in Iran, harboring mutations in the squalene epoxidase (*SQLE*) gene. Infect Drug Resist. 2020;13:845–50.
46. Shaw D, Singh S, Dogra S, Jayaraman J, Bhat R, Panda S, et al. MIC and upper limit of wild-type distribution for 13 antifungal agents against a *Trichophyton mentagrophytes-Trichophyton interdigitale* complex of Indian origin. Antimicrob Agents Chemother. 2020;64(4):e01964–19.
47. Digby SS, Hald M, Arendrup MC, Hjort SV, Kofoed K. Darier disease complicated by terbinafine-resistant *Trichophyton rubrum*: a case report. Acta Derm Venereol. 2017;97:139–40.
48. Schøsler L, Andersen LK, Arendrup MC, Sommerlund M. Recurrent terbinafine resistant *Trichophyton rubrum* infection in a child with congenital ichthyosis. Pediatr Dermatol. 2018;35:259–60.
49. Kelly SL, Lamb DC, Loeffler J, Einsele H, Kelly DE. The G464S amino acid substitution in *Candida albicans* sterol 14alpha-demethylase causes fluconazole resistance in the clinic through reduced affinity. Biochem Biophys Res Commun. 1999;262:174–9.
50. Nenoff P, Verma SB, Vasani R, Burmester A, Hipler UC, Wittig F, et al. The current Indian epidemic of superficial dermatophytosis due to *Trichophyton mentagrophytes*-a molecular study. Mycoses. 2019;62:336–56.
51. Lübbert C, Baars C, Dayakar A, Lippmann N, Rodloff AC, Kinzig M, et al. Environmental pollution with antimicrobial agents from bulk drug manufacturing industries in Hyderabad, South India, is associated with dissemination of extended-spectrum beta-lactamase and carbapenemase-producing pathogens. Infection. 2017;45:479–91.
52. Chatterjee A, Chattopadhyay D, Chatterjee D, Sengupta DN. Isolation of dermatophytes from rural and urban soil samples in premises of infected and non-infected animals. Int J Zoonoses. 1983;10:22–7.
53. Gugnani HC, Paliwal-Joshi A, Rahman H, Padhye AA, Singh TS, Das TK, et al. Occurrence of pathogenic fungi in soil of burrows of rats and of other sites in bamboo plantations in India and Nepal. Mycoses. 2007;50:507–11.
54. Snelders E, Huis In't Veld RA, Rijs AJ, Kema GH, Melchers WJ, Verweij PE. Possible environmental origin of resistance of *Aspergillus fumigatus* to medical triazoles. Appl Environ Microbiol. 2009;75:4053–7.
55. Verweij PE, Snelders E, Kema GH, Mellado E, Melchers WJ. Azole resistance in *Aspergillus fumigatus*: a side-effect of environmental fungicide use? Lancet Infect Dis. 2008;9:789–95.
56. Monod M, Feuermann M, Salamin K, Fratti M, Makino M, Mahdi Alshahni M, et al. *Trichophyton rubrum* azole resistance mediated by a new ABC transporter, TruMDR3. Antimicrob Agents Chemother. 2019;63(11). pii: AAC.00863-19.
57. Cervelatti EP, Fachin AL, Ferreira-Nozawa MS, Martinez-Rossi NM. Molecular cloning and characterization of a novel ABC transporter gene in the human pathogen *Trichophyton rubrum*. Med Mycol. 2006;44:141–7.
58. Fachin AL, Ferreira-Nozawa MS, Maccheroni W Jr, Martinez-Rossi NM. Role of the ABC transporter TruMDR2 in terbinafine, 4-nitroquinoline N-oxide and ethidium bromide susceptibility in *Trichophyton rubrum*. J Med Microbiol. 2006;55(Pt 8):1093–9.
59. Martins MP, Franceschini ACC, Jacob TR, Rossi A, Martinez-Rossi NM. Compensatory expression of multidrug-resistance genes encoding ABC transporters in dermatophytes. J Med Microbiol. 2016;65:605–10.
60. Holmes AR, Lin YH, Niimi K, Lamping E, Keniya M, Niimi M, et al. ABC transporter Cdr1p contributes more than Cdr2p does to fluconazole efflux in fluconazole-resistant *Candida albicans* clinical isolates. Antimicrob Agents Chemother. 2008;52:3851–62.

61. Niimi K, Harding DR, Holmes AR, Lamping E, Niimi M, Tyndall JD, et al. Specific interactions between the *Candida albicans* ABC transporter Cdr1p ectodomain and a D-octapeptide derivative inhibitor. Mol Microbiol. 2012;85:747–67.
62. Silva LV, Sanguinetti M, Vandeputte P, Torelli R, Rochat B, Sanglard D. Milbemycins: more than efflux inhibitors for fungal pathogens. Antimicrob Agents Chemother. 2013;57:873–86.
63. Coste A, Turner V, Ischer F, Morschhäuser J, Forche A, Selmecki A, et al. A mutation in Tac1p, a transcription factor regulating CDR1 and CDR2, is coupled with loss of heterozygosity at chromosome 5 to mediate antifungal resistance in *Candida albicans*. Genetics. 2006;172:2139–56.
64. Camps SM, Dutilh BE, Arendrup MC, Rijs AJ, Snelders E, Huynen MA, et al. Discovery of a HapE mutation that causes azole resistance in *Aspergillus fumigatus* through whole genome sequencing and sexual crossing. PLoS One. 2012;7:e50034.

Part VII

Treatment of Dermatophytoses

New Antifungal Agents and New Formulations Against Dermatophytes

Aditya K. Gupta, Maanasa Venkataraman, and Emma M. Quinlan

Abstract

There are a plethora of antifungal agents available for the treatment of onychomycosis. Systemic antifungal drugs are generally recommended for severe onychomycosis, while topical antifungal agents and devices are recommended for mild to moderate onychomycosis. Oral terbinafine is the gold standard treatment of onychomycosis, with high cure rates and proven treatment success. Novel therapies and newer antifungal formulations are under development. New topical formulations focus on enhanced drug penetration into the nail plate through multiroute administration to eliminate the infection from the nail bed. Effective management of onychomycosis is vital to combat increasing treatment failures, antifungal resistance and incidences of non-dermatophyte mold onychomycosis and mixed infections. Patients should be educated about nail hygiene, proper sanitization of shoes and socks, wearing proper footwear in communal showers and public pools, keeping feet dry, and prophylaxis post-treatment. Important steps in maximizing treatment success are combination therapy and inclusion of agents targeting fungal biofilms.

Keywords

Onychomycosis · Antifungal · *Tinea unguium* · Terbinafine · Itraconazole

A. K. Gupta (✉)
Division of Dermatology, Department of Medicine, University of Toronto, Toronto, ON, Canada

Mediprobe Research Inc., London, ON, Canada
e-mail: agupta@execulink.com

M. Venkataraman · E. M. Quinlan
Mediprobe Research Inc., London, ON, Canada

J.-P. Bouchara et al. (eds.), *Dermatophytes and Dermatophytoses*,
https://doi.org/10.1007/978-3-030-67421-2_21

1 Introduction

Onychomycosis is a fungal nail infection caused by dermatophytes, yeasts, and non-dermatophyte molds (NDM). Dermatophyte infections account for 90% of onychomycosis around the world. The general characteristics of onychomycosis are nail discoloration, subungual hyperkeratosis, onycholysis, and sometimes paronychia. The unpleasant appearance of the nail can have a negative psychological impact in patients, which may cause insecurity and affect their day-to-day activities and social life. Onychomycosis should be treated to prevent further secondary infections, to restore normal, healthy nails, and to enhance the quality of life of the patient [1]. Treatment aims to eradicate the fungal pathogen (mycological cure) and if possible return to a normal, healthy appearing nail (clinical cure).

The choice of treatment depends on several factors: causative organism, severity of infection (number of digits and area of the affected nail), comorbidities such as psoriasis, diabetes, HIV, cardiovascular disease, and the presence of *Tinea pedis*. Oral and topical antifungal drugs are currently used to treat onychomycosis. Presently, systemic drugs such as terbinafine, itraconazole (FDA-approved in the USA) and fluconazole (off-label in North America) are used in cases of severe onychomycosis, while topical drugs such as efinaconazole, tavaborole, ciclopirox (FDA-approved), and amorolfine (off-label in North America) are used in treating mild to moderate onychomycosis.

With new research directions, more treatment options are becoming available to patients. The general trend appears to be toward novel topical treatments, as well as reformulations of existing antifungal drugs. Additionally, new systemic treatments are under investigation. Devices and other physical modalities are being investigated as monotherapies and combination therapies. There are various characteristics of fungi that must be studied in order to effectively manage onychomycosis, namely their lifecycle, their ability to form biofilms, and their resistance to antifungals.

2 Systemic Treatments

Griseofulvin and ketoconazole were among the first oral antifungal drugs used in the treatment of onychomycosis. Griseofulvin was discovered in 1939 from *Penicillium griseofulvum*. It is effective only against dermatophytes. Its mechanism of action is by inhibiting fungal nucleic acid synthesis and mitosis at metaphase; it may also affect the morphology of fungal cells. Griseofulvin therapy requires a longer dosing period despite its low affinity to keratin which allows more free-drug to be present in the nail. It is not suitable for long-term therapy due to its poor absorption in the gut and low cure rates [2].

Ketoconazole was patented in 1977 and approved in the USA in 1981 for the treatment of onychomycosis [2]. It is effective against dermatophytes, *Candida* species, and shows low activity against NDMs. As an imidazole, it inhibits 14α demethylase that converts lanosterol to ergosterol, which affects fungal cell membrane synthesis. It is no longer recommended for onychomycosis treatment due to its severe side effects such as hepatitis (1:2000–1:15,000 patients) and inhibition of certain cytochrome P450 pathways. New systemic antifungal drugs were discovered and later approved for the treatment of onychomycosis. Terbinafine and itraconazole were introduced and were preferred by both patients and physicians due to their high efficacy and comparatively better adverse effects (AE) profile. Typically, systemic antifungals are recommended for severe infections due to safety issues and potential drug–drug interactions [3–6]. The fungal cell membrane biosynthesis and the site of action of antifungal agents are depicted in Fig. 1.

3 Terbinafine

Terbinafine was discovered in 1974. It was first available in Europe in 1991, Canada in 1993, and the USA in 1996 for dermatophyte onychomycosis of toenails and/or fingernails. The chemical name of terbinafine is (E)-*N*,6,6-trimethyl-*N*-(naphthalen-1-ylmethyl)hept-2-en-4-yn-1-amine. It is prescribed as Lamisil (terbinafine hydrochloride) tablets for oral consumption. Currently, the patent is assigned to Dr. Anton Stutz and Novartis AG [2, 7–9]. In 2007, the FDA approved terbinafine granules for the treatment of *Tinea capitis* in children above the age of 4 years [10].

3.1 Pharmacology

It is an allylamine which inhibits the enzyme squalene epoxidase, thus affecting ergosterol biosynthesis and therefore preventing the formation of the fungal cell membrane [11]. Terbinafine has both fungistatic and fungicidal effects. Inhibition of the enzyme squalene epoxidase leads to depletion of ergosterol, an essential structural aspect of the fungal cell membrane, which in turn inhibits the growth of fungi. Additionally, accumulation of squalene leads to deposition of lipid vesicles on the fungal cell membrane which disrupts the membrane permeability, thus killing the fungal cells. This antifungal agent is effective against dermatophytes and shows some activity against NDMs, but is less effective against *Candida* species [12].

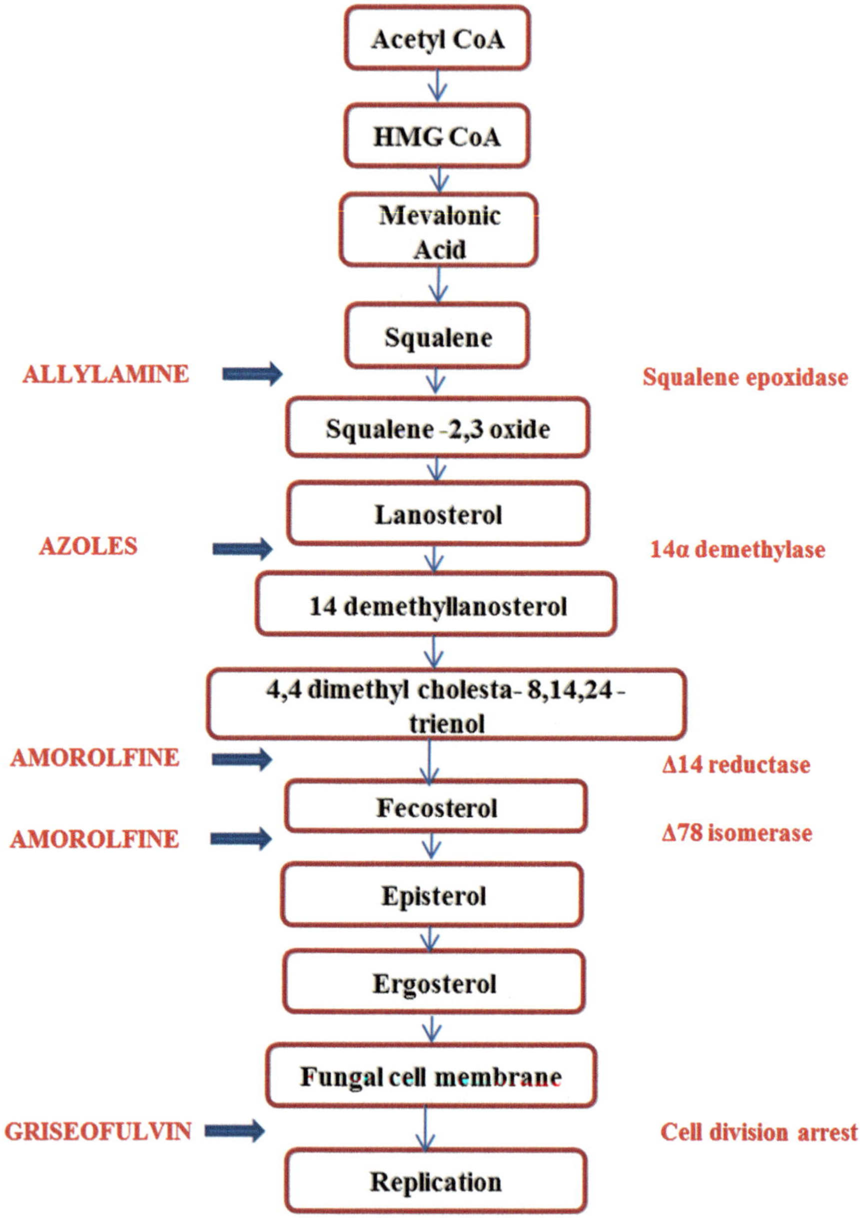

Fig. 1 Fungal cell membrane biosynthesis and the site of action of antifungal agents

3.2 Pharmacokinetics

The bioavailability of terbinafine is high and is not affected whether taken with or without food; terbinafine gets metabolized by the liver and its bioavailability decreases to 40% [2]. Terbinafine is detected early in the nail and reaches 0.45 μg/g within a week, equivalent to 10–100 times the minimum inhibitory concentration (MIC) for dermatophytes. When terbinafine is administered orally for 6–12 weeks, it can reach concentrations up to 0.52–1.01 μg/g and is detectable in the nails for 30–36 weeks after completion of therapy [13]. It gets absorbed and circulated quickly, reaching 1.3 μg/mL in plasma with just a single 250 mg dose. Systemic exposure reaches a steady state in 10–14 days [8]. Terbinafine is reported to have long half-lives ranging from 24 to 156 days [14]. It is lipophilic and has strong affinity toward plasma proteins.

3.3 Dosage

Treatment regimen is usually 250 mg daily for 6 weeks for fingernail and 12 weeks for toenail infections. It has high cure rates, with a mycological cure rate of 70% and a complete cure rate (mycological cure plus 0% clinical involvement) of 38% at week 48 (follow-up after 36 weeks of treatment) for toenails [9].

3.4 Adverse Effects and Drug Interactions

Some common side effects of terbinafine include gastrointestinal problems such as nausea, abdominal pain, etc., headaches, rashes, and altered sense of taste (0.6–2.8%); however, these are generally not serious enough to discontinue treatment [15]. Adverse side effects associated with the liver such as hepatitis are observed in 1:50,000–1:120,000 patients, and liver enzyme abnormalities such as elevated serum aminotransferase levels (two times the upper limit of normal) have been observed in 1–6% of the patients on terbinafine therapy [9, 16–18]. Hence, liver function tests (LFT) should be performed prior to treating with terbinafine [19]. Regular monitoring of liver function during treatment duration is advised (by author AKG).

Terbinafine is considered an FDA pregnancy category B drug. It is advisable to avoid terbinafine medication during pregnancy and breastfeeding, since the presence of terbinafine is detected in the breast milk of lactating women. Terbinafine is an inhibitor of CYP2D6 enzyme and will affect other drugs that are substrates of CYP2D6, such as antidepressants, antiarrhythmic class 1C, etc., leading to drug–drug interactions [9].

3.5 Clinical Studies

Randomized, double-blind trials showed the superior efficacy of oral terbinafine therapy over other systemic drugs such as itraconazole, fluconazole, and griseofulvin in toenail dermatophyte onychomycosis. It was observed that the tolerability profile of oral terbinafine was higher compared to griseofulvin and similar to itraconazole and fluconazole [20]. An independent comparative study performed with continuous terbinafine (250 mg/day for 12 weeks) and pulsed itraconazole regimen (400 mg/day for 1 week on and 3 weeks off, 3 pulses) showed that the mycological cure rates were not statistically different when treated with continuous terbinafine or pulsed itraconazole regimen. However, clinical cure was favorable for patients over 55 years or female patients treated with terbinafine [21]. Terbinafine is used off-label to treat pediatric onychomycosis and found to be safe and effective [22].

4 Itraconazole

Itraconazole was synthesized in 1980 with the chemical formula (±)-1-[(RS)-sec-butyl]-4-[p-[4-[p-[[(2R,4S)-2-(2,4-dichlorophenyl)-2-(1H-1,2,4-triazol-1ylmethyl)-1,3-dioxolan-4-yl]methoxy]phenyl]-1-piperazinyl]phenyl]-Δ2–1,2,4-triazolin-5-one [23]. The patent was initially obtained by Janssen Pharmaceutica (Dr. Paul Janssen) for the discovery of the backbone of itraconazole [24]. Itraconazole was initially approved for systemic mycoses in the USA in 1992 and for onychomycosis 3 years later (1995). The pulsed regimen was approved in the USA in 1997 for fingernail onychomycosis. Itraconazole was approved in Canada in 1993 and in Australia in 1998 for the treatment of onychomycosis. It is marketed as Sporanox capsules, which contain itraconazole as the active compound [2].

4.1 Pharmacology

Itraconazole is a triazole which inhibits 14α-demethylase that converts lanosterol to ergosterol which in turn inhibits fungal cell wall biosynthesis. It has broader antifungal activity than terbinafine, against dermatophytes, NDMs (*Aspergillus*, *Fusarium*, etc.), and *Candida* spp. [25], which may be an effective consideration in treating mixed infections [26].

4.2 Pharmacokinetics

The bioavailability of itraconazole is 55% which may be affected in a fasting state due to reduced gastric acidity [2]. The drug is detectable in toenails within 2 weeks of starting therapy; it reaches a mean maximum concentration of 0.8 μg/g in nails at the end of continuous regimen (200 mg/day for 12 weeks). The daily dosage of 200 mg itraconazole could lead up to twice the concentration in the subungual area of the nail than the distal region in 10 days [27, 28]. Itraconazole may be found in toenails up to 11 months from the time of initialization when treated with 3 pulses of itraconazole [2, 29]. The plasma concentration reaches up to 0.3 μg/mL after administering a single dose of 200 mg itraconazole and reaches a steady state in 15 days. The half-life of a single dose of itraconazole is 17 h, while repeated administration is reported to have half-lives of 34–42 h. It is highly lipophilic and has a strong affinity toward plasma proteins, especially albumin [23].

4.3 Dosage

Itraconazole is most commonly administered as a continuous regimen of 200 mg/day for 12 weeks, or as a pulsed regimen of 200 mg twice daily (or 400 mg/day) for 1 week, followed by 3 weeks of no medication (2 pulses for fingernail and 3 pulses for toenail) [30]. The mycological cure rate is reported as 54%, whereas the complete cure rate (mycological cure plus visibly clear nail with minimal deformities) is 14% for the continuous regimen [23].

4.4 Adverse Effects and Drug Interactions

Side effects reported are similar to those caused by terbinafine, such as gastrointestinal problems including nausea, headaches, and upper respiratory tract infections. More severe side effects have been known to occur, including hypertriglyceridemia, elevated transaminases (0.5–5%; two times upper limit of normal), and hepatic injury—about 1 in 500,000 patients develop hepatitis [17, 31–34]. The pulse regimen has an AE profile comparable to that of continuous itraconazole therapy [35, 36]. Itraconazole has the most contraindications, especially for patients who exhibit congestive heart failure. Itraconazole intervenes with CYP3A4, which can lead to potential drug–drug interactions. It is a category C pregnancy drug and should be avoided during pregnancy and at least 2 months before planning a pregnancy. Itraconazole is also excreted into breast milk, so its use should be delayed until breastfeeding is completed [23].

4.5 Clinical Studies

Itraconazole therapy (off-label) is reported to be safe and effective for the treatment of pediatric onychomycosis [22, 37, 38]. Treatment studies on dermatophyte toenail onychomycosis patients with diabetes mellitus revealed that both pulsed itraconazole and continuous terbinafine worked effectively and are safe to use [39].

5 Fluconazole

Fluconazole was discovered in 1981 by Ken Richardson and a group of scientists at Pfizer Central Research in Sandwich, Kent, UK [40]. The chemical formula of fluconazole is 2,4-difluoro-α,α^{1}-bis(1H-1,2,4-triazol-1-ylmethyl) benzyl alcohol [41]. It is prescribed as Diflucan tablets which have fluconazole as the active compound. It has been approved in over 27 countries and used as an off-label treatment for onychomycosis in the USA and Canada. Fluconazole was first approved in UK in 1988 for the treatment of superficial mycoses in humans and in Finland and China in 1993 for the treatment of onychomycosis [2].

5.1 Pharmacology

As a triazole, fluconazole targets the ergosterol pathway like any other azole by inhibiting 14α-demethylase. Fluconazole is effective against dermatophytes, *Candida* species, and NDMs [42, 43]. The data supporting its use are limited, especially in the case of NDMs and superficial candidiasis in patients with AIDS or other immunocompromised individuals [26].

5.2 Pharmacokinetics

The bioavailability of fluconazole exceeds 90%. Due to the high bioavailability, a low concentration of fluconazole such as 50 mg is sufficient to be detected in nail clippings on the second day of therapy. With the continuous treatment regimen of 150 mg weekly for 12 months, the peak concentration observed in the nail is 8.54 μg/g. It could be detected in the nail plate up to 6 months after the end of treatment with the standard dose [2]. The half-life of fluconazole is much shorter than other systemic antifungals and is reported to be 27–36 h in healthy individuals [44, 45]. Thus, the fluconazole treatment period is longer compared to the other two oral antifungal drugs. It does not bind to plasma proteins (11–12%), is hydrophilic, and found evenly distributed throughout tissue fluids [2, 41].

5.3 Dosage

The dosing regimen of fluconazole ranges from 150 to 400 mg once a week for 6–9 months for fingernail and a longer duration of 9–18 months for toenail onychomycosis. Longer duration of fluconazole treatment (more than 6 months) at 150 mg once weekly may result in higher cure rates for toenail and fingernail dermatophyte onychomycosis compared to treatment periods of 6 months or less [44]. It is proven to be mycologically and clinically effective in treating toenail distal subungual onychomycosis caused by dermatophytes. Although it is superior to topical treatments, fluconazole is not as effective as terbinafine and itraconazole against dermatophytes, with mycological cure rates of 47–62% and clinical cure rates of 28–36% [42, 46, 47].

5.4 Adverse Effects and Drug Interactions

The absorption of fluconazole is not dependent on gastric pH, and it can cause hepatic side effects like liver enzyme abnormalities (mild to moderate elevations in 5% and severe elevations in ~1%), liver injury and failure (up to 10%) [48–51]. Immunocompromised patients or patients with AIDS are at greater risk of developing hepatic diseases [49]. It has a favorable AE profile and regularly reported effects include nausea, headache, abdominal pain, diarrhea, etc. Fluconazole inhibits CYP2C9 and thus cannot be used with some medications. Additionally, it is a category D pregnancy drug: unfavorable effects on human fetuses have been reported [52]. The drug is also excreted into breast milk and should not be used until after pregnancy and breastfeeding.

5.5 Clinical Studies

In a multicenter, double-blind, randomized, placebo-controlled study, patients diagnosed with distal subungual dermatophyte onychomycosis were treated with different fluconazole dosing regimens of 150, 300, or 450 mg weekly for a maximum of 12 months. The mycological and clinical cures at the end of treatment were reported to be 47–62% and 28–36% respectively [42]. Fluconazole 3–6 mg/kg dose once a week for 12–16 weeks for fingernails and 18–26 weeks for toenails is the recommended regimen (off-label) for treating onychomycosis in children [53].

6 Other Off-Label Oral Treatments

In addition to fluconazole, other off-label oral treatments include pulsed terbinafine and booster therapy (with terbinafine or itraconazole). A pulsed regimen of 250 mg terbinafine/day for 4 weeks and then 4 weeks off without any treatment (2 pulse cycles) had comparable mycological and clinical cure rates to continuous daily 250 mg for 12 weeks regimen from a meta-analysis of published literature [54]. A recent meta-analysis concluded that 250 mg continuous terbinafine regimen and 400 mg pulsed itraconazole regimen (1 pulse = 400 mg daily for 1 week and 3 weeks off) were superior to fluconazole and topical agents such as ciclopirox 8% nail lacquer, efinaconazole 10% topical solution, and tavaborole 5% solution [46].

In a non-randomized, open study, terbinafine-pulsed regimen was proposed as an alternative, economical treatment option for dermatophyte onychomycosis. The standard terbinafine-pulsed regimen considered for the treatment of onychomycosis was 500 mg/day for 1 week and 3 weeks off (2 or 3 pulses for fingernail and 3 or 4 pulses for toenail). This study compared the efficacy of the standard continuous regimen (250 mg/day for 3 months), standard pulse regimen mentioned above, and altered pulse regimen (500 mg/day for 1 week a month every 3 months). Total cure was the primary outcome, which includes clear, uninfected nail and negative fungal culture. The efficacy and AE profile of both the pulsed terbinafine regimens were similar to the conventional continuous regimen. The longer duration of treatment with pulsed regimen every 3 months might reduce the overall development of side effects and be more economical for the patient compared to the continuous regimen. However, this method might not be suitable for the elderly due to poor compliance, discontinuation, or not following up for subsequent pulses [55].

Terbinafine and itraconazole are recommended antifungal agents for booster or supplemental therapy [56]. The idea of booster therapy surfaced when certain conditions or patients were difficult to treat. This was found to be effective in patients who had thick nail plate > 2 mm, slow-growing nails, $> 75\%$ nail plate involvement, lateral involvement of onychomycosis, and immunosuppression. Additional 4 weeks of 250 mg/day terbinafine or pulsed itraconazole 6–9 months after initiation of the antifungal therapy is administered if there is $< 50\%$ improvement from baseline, to improve cure rates and reduce the risk of recurrence [57, 58].

7 Topical Treatments

Although systemic antifungals are effective in treating onychomycosis and achieving higher cure rates, the focus is shifting towards topical drugs that have been recently developed and approved for the treatment of onychomycosis. Physicians and patients may prefer topical treatment methods over oral drugs due to the adverse side effects associated with the prolonged use of systemic antifungals. The dosage and duration of oral therapies seem to concern patients who want easier and more pragmatic treatment methods. Several studies have shown the tolerability and efficacy of topical drugs when used as monotherapy in treating onychomycosis. Until recently, ciclopirox nail lacquer (8% solution) was the only FDA-approved topical drug for treating mild to moderate toenail onychomycosis. Amorolfine 5% nail lacquer is approved only in Europe and is not available in North America. Due to poor nail penetration of existing drugs, difficulty in application due to the requirement of debridement of the nails at regular intervals (ciclopirox nail lacquer), and low cure rates, there is a need for newer antifungal agents. Topical solutions tend to be more effective than topical nail lacquers as they have multiple routes of entry such as through the transungual (dorsal to ventral surface of nail plate into the nail bed and nail matrix) and the subungual space (hyponychium or damaged onychodermal band and lateral nail folds). Lacquers require nail filing before application and have limited penetrating abilities [59, 60]. Efinaconazole 10% solution and tavaborole 5% solution are twenty-first century topical drugs that were approved recently by the FDA for toenail dermatophyte onychomycosis. They seem to have higher or equivalent cure rates to ciclopirox nail lacquer.

8 Why Is It Hard to Treat a Nail Infection?

The two main important characteristics of topical treatment include the structure and properties of drug molecules, and the vehicle in which it is formulated and delivered. These, along with the site or route of administration, determine treatment success. The nail is a unique organ with a set of physiochemical properties that affect the efficacy of a topical drug. To understand the working mechanism of topical drugs, please refer the nail anatomy which is explained in detail in a previous chapter.

The nail plate is made of keratin and consists of several layers of cornified cells which are attached to the underlying nail bed. Affinity toward keratin determines the bioavailability of the drug in the nail unit. Drug molecules that have low affinity toward keratin are available in higher quantities in free form and have higher rate of release from bound keratin, thus increasing the concentration of active compound of the drug in the nail bed and nail matrix.

Molecular size of the drug and porosity of the nail are key factors that impact drug penetration into the nail. Topical drugs need to be smaller in size than the nail pores to easily pass through the nail. Nail porosity can be enhanced by hydration of the nail plate which causes swelling and increased entry of the drugs into the nail plate. For mild to moderate cases of distal and/or lateral subungual onychomycosis (DLSO) and superficial white onychomycosis (SWO) topical therapy is recommended [61–63].

9 Ciclopirox

Ciclopirox belongs to the class of drugs called hydroxypyridinones. The chemical name is 6-cyclohexyl-1-hydroxy-4-methyl-2(1*H*)-pyridinone. Ciclopirox 8% nail lacquer formulation was approved by the FDA in 1999 for fingernail and toenail onychomycosis. It is marketed as Penlac which has ciclopirox as the active compound. Currently, the patent is held by Medicis Pharmaceutical Corp [59]. The initial patent for Ciclopirox nail lacquer (8% solution) was held by Karl Kramer and Manfred Bohm (Germany).

9.1 Pharmacology

Ciclopirox has a unique mechanism of action which does not affect the ergosterol pathway; instead, it chelates multivalent metal agents such as Fe^{3+} and Al^{3+}, inhibiting metal-dependent enzymes from clearing peroxides from the fungal cells. It also affects membrane transfer by interrupting sodium and potassium ion channels. It may also affect nutrient uptake in fungal cells, thereby affecting macromolecule synthesis. It has a broad spectrum of activity against dermatophytes, *Candida* species, and NDMs. It also has antibacterial activity and anti-inflammatory properties [4, 59, 64].

9.2 Pharmacokinetics

Ciclopirox nail lacquer (8% solution) penetrates the nail plate just after a single topical application. The dorsal nail plate (superficial layer) has the highest concentration, 7812 μg/g, and the subsequent lower layers accumulate slower than the top layer (34 μg/g) [65]. The systemic exposure of ciclopirox observed throughout the treatment period is 10–24.6 ng/mL [59].

9.3 Dosage

The treatment regimen consists of daily application for 48 weeks with regular debridement of the affected nail. The mycological cure rates at the end of treatment period (week 48) from two double-blind, placebo-controlled studies are 29% and 36%, almost complete cure rates (mycological cure and $\leq$ 10% target nail involvement) are 6.5% and 12%, and complete cure rates (mycological cure plus clear nail) are 5.5% and 8.5% [59]. Ciclopirox nail lacquer forms a film on the nail plate after initial application. The solvent evaporates, thus increasing the concentration of ciclopirox from 8% up to 34.8%. This accumulation of ciclopirox exceeds the MIC observed in vitro against fungal pathogens [59, 64, 65].

9.4 Adverse Effects and Drug Interactions

The adverse reactions observed in patients treated with ciclopirox are mild and include erythema in the surrounding skin, burning, and tingling sensation upon application. It has a favorable and localized AE profile, minimal concomitant drug–drug interactions, and found to be safe [66].

9.5 Clinical Studies

Ciclopirox nail lacquer (8% solution) has been recommended for the treatment of onychomycosis caused by dermatophytes, NDMs, and yeasts [67]. It is used as an adjunct along with an oral therapy or chemical avulsion method with urea or bifonazole. It is an alternative for patients for whom oral therapy is not effective [3, 5, 68–70]. Ciclopirox is approved for the treatment of onychomycosis in children aged 12 years and older [59].

10 Amorolfine

Amorolfine belongs to the class of drugs called morpholines which was introduced in the early 1980s. It was first approved in Europe for toenail dermatophyte onychomycosis in 1991. The patent was assigned to Galderma UK Limited. It is not available in North America for onychomycosis. The market name is Loceryl 5% nail lacquer which has amorolfine hydrochloride as the active ingredient. Curanail 5% nail lacquer is a duplicate product of Loceryl nail lacquer, but is used to treat mild onychomycosis. It was approved in UK on April 7, 2006 [60].

10.1 Pharmacology

Amorolfine is recommended for the treatment of onychomycosis without matrix or lunula involvement and only when fewer than two nails are affected. It targets the ergosterol pathway by inhibiting 14Δ reductase and 78Δ isomerase enzymes which leads to alteration of membrane permeability and ultimately fungal cell death. Amorolfine also affects chitin synthesis by inhibiting chitin synthase. Additionally, it inhibits NADH oxidase and succinate cytochrome c reductase. It has a broad spectrum of fungistatic and fungicidal activity against dermatophytes, yeasts, and non-dermatophyte molds, but is not effective against *Aspergillus* and *Fusarium* species [4].

10.2 Pharmacokinetics

Like ciclopirox, amorolfine concentration in the nail reaches an initial peak after a single application of the lacquer. It can be detected in the nail for 2 weeks after ceasing treatment. Systemic absorption is low and could not be detected in blood [64]. There is no indication of drug accumulation in the body following prolonged use [60].

10.3 Dosage

The treatment regimen consists of once or twice weekly application for 6 months. The mycological cure rate is 60%, and complete cure (mycological cure plus complete cure or < 10% involvement of target nail) is 38% when applied daily for 6 months and followed up after 3 months [64]. Another multicenter, randomized study investigated the efficacy of once-weekly application of amorolfine 5% lacquer for 6 months. The mycological and complete cure rates at 9 months (mycological cure plus clinical cure or < 10% target nail involvement) were 71% and 46%, respectively [64]. Amorolfine lacquer forms a film on the nail plate which then releases the active drug, which travels through the nail plate, accumulating at the site of infection [60]. The concentration of the active drug on the surface of the nail plate increases from 5% to 27% as the solvent evaporates [64].

10.4 Adverse Effects and Drug Interactions

Very mild adverse effects are observed with the application of amorolfine. These include pain and burning sensation, nail discoloration, erythema, etc. No potential drug–drug interactions have been observed with the use of amorolfine [71].

10.5 Clinical Studies

A systematic review and meta-analysis showed that amorolfine had higher efficacy in treating onychomycosis when used in combination with other systemic antifungal drugs as opposed to monotherapy [72]. A multicenter, randomized, open study with three groups—amorolfine 5% lacquer once a week for 15 months with oral terbinafine 250 mg daily for 6 weeks initially vs. amorolfine 5% lacquer once a week for 15 months and 250 mg terbinafine daily for 12 weeks vs. control (250 mg terbinafine daily for 12 weeks)—showed higher efficacy and cost-effectiveness of combination therapy [73]. Several studies suggest that amorolfine 5% nail lacquer is a suitable off-label topical therapy for pediatric onychomycosis [74]. It could be used as a monotherapy to treat onychomycosis without matrix involvement and in combination with oral antifungals to manage severe onychomycosis with nail matrix involvement [64, 75].

11 Efinaconazole

Efinaconazole 10% is the first topical drug to be approved from the azole family of drugs. It was approved by the FDA for toenail onychomycosis in 2014 and by Health Canada in 2013 for mild to moderate onychomycosis. It is marketed under the name Jublia and is assigned to Valeant Pharmaceuticals International Inc. (now Bausch Health in North America). The chemical name of efinaconazole is (2R,3R)-2-

(2,4-Difluorophenyl)-3-(4-methylene-1-piperidinyl)-1-(1H-1,2,4-triazol-1-yl)-2-butanol [76].

11.1 Pharmacology

The mechanism of action of efinaconazole is similar to other azoles: it inhibits lanosterol 14α-demethylase which affects the ergosterol pathway, which in turn inhibits fungal cell membrane formation. It has broad-spectrum activity compared to other topical drugs such as amorolfine and systemic drugs such as terbinafine. Efinaconazole is effective against dermatophytes, and shows high activity against *Candida* and NDMs. It is easy to use compared to the other topical lacquers since it does not require regular debridement of the affected nails [76].

11.2 Pharmacokinetics

It is detected in blood at low levels compared to systemic drugs when administered topically. It reaches a mean plasma concentration of 0.67 ng/mL after 28 days from the start of therapy. H3 is the major metabolite of efinaconazole. Efinaconazole reaches a half-life of 29.9 h in plasma after daily application in 10 toenails for 7 days [76]. The plasma protein binding of efinaconazole is not clinically relevant due to its low systemic exposure.

11.3 Dosage

The treatment regimen includes once daily topical application for 48 weeks on the nail plate, lateral nail grooves, infected nail bed, and hyponychium using a brush applicator. It could be detected in plasma 2 weeks after the end of treatment. The mycological cure rates at week 52 (4 weeks follow-up after 48-week treatment regimen) for two double-blind, randomized trials were 55.2% and 53.4%, almost complete cure rates (mycological cure and $\leq$ 5% target nail involvement) were 26.4% and 23.4%, and complete cure rates (mycological cure plus 0% clinical involvement) were 17.8% and 15.2% [76].

The efficacy of efinaconazole is observed to be comparable to amorolfine but higher than ciclopirox. This higher efficacy may be attributed to the higher drug penetration into the nail to reach the nail bed. The main properties of efinaconazole that aid in penetration are low surface tension, poor solubility in water, and low keratin affinity. This allows the drug to penetrate and spread across in the nail bed since it is administered transungually and subungually (under the distal free edge of the nail) [77]. Efinaconazole has low affinity to keratin which allows the drug to be available in free form at a higher concentration as opposed to other antifungal agents whose activity is inhibited due to keratin binding. The severity of the infection and the digits involved may not affect the efficacy of efinaconazole [68, 70, 78–80].

11.4 Adverse Effects and Drug Interactions

Efinaconazole has a favorable AE profile. There might be mild events which include ingrown toenail, redness, itching, stinging, burning, pain, etc. It is an FDA pregnancy category C drug and should be avoided during pregnancy and breastfeeding. There are no known drug interactions when applied topically [76].

11.5 Clinical Studies

Efinaconazole 10% solution is reported to be effective in treating onychomycosis patients with diabetes. It is well tolerated and safe to use [81].

A phase IV, multicenter, open-label study showed that efinaconazole 10% topical solution could be used to treat mild to severe distal lateral subungual onychomycosis affecting $\geq$ 20% target nail in children. The safety, pharmacokinetics, and efficacy of Jublia were studied in the enrolled children population (6–16 years of age). The treatment regimen consisted of daily application for 48 weeks with a follow-up after 4 weeks. The efficacy achieved in treating pediatric onychomycosis was higher compared to the adult population due to faster-growing nails and elimination of infection, good response to treatment, etc. The mycological and complete cure rates were higher at week 52 (65% and 40%, respectively). The systemic exposure was observed to be low, and the drug was deemed safe to use for pediatric onychomycosis [80]. Efinaconazole is approved by the FDA for the treatment of onychomycosis in children aged 6 years and older [76].

12 Tavaborole

Tavaborole, also known as AN2960, is another novel twenty-first-century antifungal drug that was approved for adult toenail dermatophyte onychomycosis by the FDA in 2014. It belongs to a new class of drugs called oxaboroles. They are boron-containing antifungal agents. The chemical name of tavaborole is 5-fluoro-1,3-dihydro-1-hydroxy-2,1-benzoxaborole. It is marketed as Kerydin (tavaborole 5%) invented by Stephen Baker and team, and is a trademark of Anacor Pharmaceuticals, Inc. [82, 83].

12.1 Pharmacology

Tavaborole has a unique mechanism of action where it inhibits protein synthesis in fungi by inhibiting aminoacyl-transfer ribonucleic acid (tRNA) synthetase. Tavaborole has a broad spectrum of activity and is effective against dermatophytes (especially against *Trichophyton rubrum* and *Trichophyton mentagrophytes*), while exhibiting low activity against yeasts and molds [3, 68, 84, 85].

12.2 Pharmacokinetics

The systemic exposure of tavaborole is low as it could not be detected in plasma even after several days of dosing. It has higher antifungal activity compared to ciclopirox since it can penetrate the nail plate and retain activity in the presence of keratin. It is detected in the nail up to 3 months after the end of treatment in concentrations higher than the MIC for dermatophytes [82, 84].

12.3 Dosage

The dosing regimen is similar to other topical drugs, once daily for 48 weeks. The mycological cure rates at week 52 (4 weeks follow-up after a 48-week regimen) for two multicenter, double-blind, vehicle-controlled studies were 31.1% and 35.9%, almost complete cure rates (mycological cure and $\leq$ 10% target nail involvement) were 15.3% and 17.9%, and complete cure rates (mycological cure plus 0% clinical involvement) were 6.5% and 9.1% [86]. The higher efficacy of tavaborole is due to its high solubility in water, low molecular weight, and antifungal activity in the presence of keratin. The alcohol-based solution spreads easily along the nail folds and into the nail bed [82]. The mycological cure rates of the oral and topical antifungal agents currently in use are depicted in Fig. 2 [9, 23, 41, 59, 64, 76, 82].

12.4 Adverse Effects and Drug Interactions

The adverse events associated with tavaborole application are mild and localized. It has a favorable AE profile which is limited to erythema, exfoliation, dermatitis, etc. It is an FDA pregnancy category C drug and should be avoided during pregnancy and breastfeeding. Tavaborole has not been shown to have any potent drug–drug interactions since it does not interact with cytochrome P450 enzymes in vitro. Tavaborole is highly specific and has 1000-fold higher affinity to fungal tRNA synthetase compared to its human counterpart [87].

12.5 Clinical Studies

A phase IV, open-label, single-arm study demonstrated the safety, pharmacokinetics, and efficacy of tavaborole 5% in treating pediatric toenail onychomycosis. The patient population had moderate to severe distal subungual onychomycosis ($\geq$ 20% target toenail involvement) and treated with 5% tavaborole topical solution once daily for 48 weeks. The complete cure rate was 8.5% at week 52 and the almost

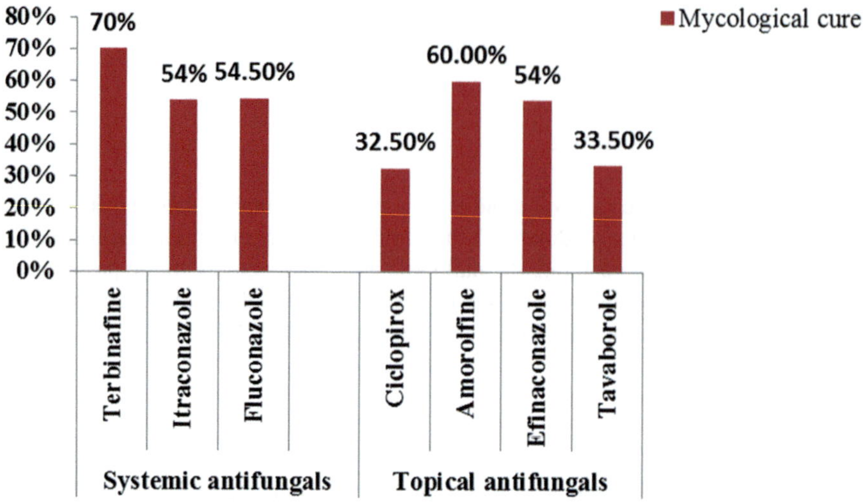

Fig. 2 Comparison of mycological cure rate of systemic and topical antifungal agents currently in use for the treatment of onychomycosis. *54.50% is the average of 47% in study 1; 62% in study 2. +32.50% is the average of 29% in study 1; 36% in study 2. ^54% is the average of 53.4% in study 1; 55.3% in study 2. ⫞33.5% is the average of 31.1% in study 1; 35.9% in study 2

Table 1 The AE profile of systemic antifungal agents

Systemic antifungal agents	Terbinafine	Itraconazole	Fluconazole
Gastrointestinal tract problems	Nausea, abdominal pain, diarrhea, etc.	Nausea, abdominal pain, diarrhea, etc.	Nausea, abdominal pain, vomiting, diarrhea, etc.
Neurological effects	Headache	Headache, dizziness	Headache
Others	Loss of taste and smell	Hypertriglyceridemia, upper respiratory tract infection	Rash
Hepatitis	1:50,000–1:120,000	1:500,000	0–31.6:10,000 (acute liver failure)

complete cure rate (up to < 5% of affected nail plate and negative mycology) was 14.9%. It was well tolerated with mild adverse events [88]. Tavaborole is approved by the FDA for the treatment of onychomycosis in children aged 6 years and older [82].

The AE profiles of systemic and topical antifungal agents are summarized in Tables 1 and 2, respectively [9, 16, 17, 23, 31, 41, 48, 59, 60, 76, 82, 89, 90].

Table 2 The AE profile of topical antifungal agents

Topical antifungal agents	Ciclopirox	Amorolfine	Efinaconazole	Tavaborole
Localized AE profile	Erythema, ingrown toenail, irritation, burning, etc.	Nail injury, pruritus, erythema, contact dermatitis	Contact dermatitis, pain, ingrown toenail, etc.	Erythema, contact dermatitis, exfoliation, etc.

13 Device-Based and Other Physical Treatments

The use of physical modalities has been a topic of exploration in recent years as a means of counteracting the poor penetration of topical antifungals, long treatment durations, adverse events, and low-to-moderate efficacies of conventional therapies. Medical devices are approved by the FDA due to similarity to predicate devices, but are not required to show efficacy in phase III trials [91–93]. They include lasers, photodynamic therapy (PDT) and iontophoresis. Devices can be used as monotherapies or in combination with existing antifungal agents.

14 Laser Systems

Lasers employ different mechanisms, such as photochemical, photothermal, and photomechanical, which may affect the human tissue in several ways. Dermal lasers rely on selective photothermolysis to heat fungal mycelium, and the presence of chitin in the fungal cells causes specific heat accumulation that results in fungicidal effects [92, 94]. These lasers could be long-pulsed, short-pulsed, or Q-switched. As of January 2012, five laser systems were approved by the FDA for "temporary increase of clear nail." All of them are neodymium-yttrium lasers of wavelength 1064 nm [95]. They are PinPointe FootLaser (PinPointe USA, Inc.; Chico, CA), Cutera GenesisPlus Laser System (Cutera, Inc.; Brisbane, CA), CoolTouch VARIA laser (CoolTouch, Inc.; Roseville, CA), JOULE ClearSense (Sciton, Inc.), and Light Age Q-Clear laser (Light Age, Inc.; Somerset, NJ). These are the first device-based treatments approved for onychomycosis. The first four approved laser systems are short-pulsed (μs) and the fifth laser system is Q-switched (ns) [96].

The traditional, long-pulsed lasers produce 5–30 ms pulses with 0.7 ms thermal relaxation time which heats the tissue, requiring constant cooling to avoid pain and necrosis of the surrounding, uninfected tissue. The short-pulsed lasers produce 300–650 μs pulses which are shorter than the thermal relaxation time of the fungal pathogen, which in turn allows heat to accumulate in the fungal cell wall and ultimately kills the fungi while enabling better dissipation of heat in the surrounding tissue [97]. Pulse duration and pulse repetition rate are two key factors that increase the thermal load on fungal pathogens while preserving the surrounding normal tissue. The duration between pulses should be long enough for the heat to dissipate to the surrounding tissue but short enough for the heat to accumulate in the fungal cell wall. The CoolTouch VARIA laser operates in the presence of a cryogen cooling system to constantly cool the surrounding tissue which helps keep the patient comfortable during treatment.

Lasers are approved by the FDA for improving cosmetic appearance by increasing the amount of clear nail [92, 93, 97]. However, the effects of lasers as a monotherapy for onychomycosis are inconsistently reported and cure rates tend to be lower than those of oral and topical antifungals. Additionally, there is limited evidence that they eradicate fungal pathogens because there is a lack of reporting in the literature with few randomized controlled trials [98–100]. Two studies (non-randomized, uncontrolled, retrospective trials) report an average mycological cure rate of 11% ($N = 18$) [101, 102]. Lasers may be better for addressing the approved cosmetic indications rather than as a medical antifungal treatment since the clinical improvement was found in only 36% of patients receiving laser therapy (five studies, $N = 69$) [101–106]. The treatment guidance provided by FDA for laser systems is different from the guidance provided for other antifungal drugs. The primary endpoint for laser treatment is aesthetic/cosmetic since the outcome is limited to temporary clearance of the infected nail and not necessarily the organism itself (mycological endpoint). The inclusion criteria and efficacy outcomes for lasers differ from drugs, and onychomycosis endpoints like mycological, clinical, and complete cure are rarely reported due to the lack of randomized controlled trials [81–83]. This causes difficulties in making meaningful comparisons between drugs and devices.

Fractional CO_2 lasers fractionate the nail and laser energy is split into thousand tiny beams to treat the infected fraction of the nail. It is often associated with another physical therapy such as photodynamic therapy or with topical drugs to enhance their penetration and efficacy. The number of sessions and passes per session depends on the severity of onychomycosis and the number of digits involved. Several studies have been performed to evaluate the safety and efficacy of fractional CO_2 lasers combined with a topical agent and it has been shown as a possible alternative for systemic therapy [107–109]. For example, a complete cure rate of 50% has been reported in a study that used fractional carbon dioxide lasers in conjunction with a topical antifungal cream [107]. An increase in efficacy has also been observed in combinations like terbinafine coupled with Nd:YAG laser treatment, but no specific combination has been approved by the FDA or Health Canada [110].

15 Photodynamic Therapy (PDT)

Photodynamic therapy causes fungal apoptosis by irradiating specific wavelengths of light upon the fungal cells that are pretreated with photosensitizers like 5-aminolevulinic acid (ALA), methyl aminolevulinate (MAL), and methylene blue (MB) [97]. The excited photosensitizers produce highly reactive oxygen species that lead to fungal apoptosis. It appears to be effective in treating patients, as negative culture and/or microscopy has been observed in 67% of patients involved in photodynamic therapy studies in recent years (six studies, $N = 58$) [111–116]. However, many of these studies were open label, and with a limited number of subjects.

16 Iontophoresis

In preliminary studies of iontophoresis, it has been demonstrated that it can deliver topical antifungal drugs to the site of infections using a small electrical current that is applied to the biological membranes which then transports hydrophobic or charged molecules to the site of infection [117]. For example, this allows for the release of topical terbinafine cream at fungicidal concentrations to eradicate dermatophyte strains [117]. In intact cadaver toe models, iontophoresis delivers higher concentrations of terbinafine (7–13 folds higher) compared with passive delivery [118].

17 Other Physical Alterations

Physical alterations aid antifungals to penetrate the nail, an otherwise difficult-to-cross compact keratin structure, to the site of infection [119]. For example, nail drilling produces passages through the nail. The portion of clear nail improves in patients when nail drilling (horizontal lines 2 mm apart) is combined with oral and topical formulations of terbinafine (250 mg and 1%, respectively) when compared to topical terbinafine alone ($P = 0.038$) [119]. Precise laser poration can also be used to create 100 μm diameter pores with minimal damage to surrounding tissues. The complete poration of the nail increases drug permeation by 2–3 orders of magnitude compared to untreated nails, ultimately increasing the efficacy of topical treatments [120]. The nail can be either partially or completely removed through surgery or chemical aid to allow for topical therapies to pass through to the infection [26, 67]. Chemical nail removal requires urea or bifonazole, but infectious organisms can remain if they are found beneath the nail fold [121]. Thus, topical agents are applied after chemical avulsion or surgical removal to eliminate fungal pathogens from the nail bed to prevent the newly growing nail from becoming infected. The complete cure rates with these compounds are 20.8% and 27.7%, respectively [122–124].

18 New Antifungal Agents and Treatment Strategies

It has been an era of healthcare innovation. Antifungal agents with higher efficacy, favorable AE profile, and minimal drug–drug interactions have been approved [79]. Given the armamentarium of treatment options for onychomycosis, cure rates are not high combined with adverse effects. Treatment and management of onychomycosis have been difficult for both patients and physicians. Antifungal agents and treatment strategies should be chosen in a way that they do not affect routine activities, are effective, and are safe to use. However, the relapse and reinfection rates are high (10–53%) and around 20–25% of people do not respond to therapy and have poor prognosis [79]. There is a pressing demand for new antifungal agents and therapies new formulations to enhance delivery and drug release (penetration enhancers), novel molecular targets in the fungal pathogens, and remodeling existing drugs.

19 Novel Oral Therapies and Formulations

19.1 Meltrex Technology

A novel formulation of 200 mg itraconazole tablet has been developed using Meltrex technology (Hot melt extrusion technology). This technology is effective in designing a "solid solution" which has the equivalent dosage of two 100 mg tablets. This aids in sustained drug release and may improve the bioavailability of the active drug. The technique uses heat and pressure to form a homogenous mixture of vehicle polymer and the active drug. In a randomized, phase III study of one itraconazole 200 mg formulation tablet vs. two 100 mg tablets, it was shown that the itraconazole 200 mg novel formulation had similar efficacy in achieving mycological, clinical, and complete cures as the 100 mg tablets. This novel formulation improved patient compliance and overall outcome [125–127].

19.2 VT-1161

VT-1161 is a novel, oral antifungal belonging to the tetrazole group of drugs. It is developed by Mycovia Pharmaceuticals for the treatment of recurrent vulvovaginal candidiasis and onychomycosis. It is highly selective in inhibiting fungal cytochrome P45051 or CYP51. A randomized, phase IIb trial to study the efficacy and safety of VT-1161 (300 or 600 mg once a week for 10 or 22 weeks following a loading period of 2 weeks with daily dosing of 300 or 600 mg) with placebo in treating onychomycosis showed the mycological cure rate to be 61–72% in the VT-1161 treatment groups vs. ~12% in the placebo group and complete cure (mycological cure plus 0% target nail involvement) to be 32–40% in the VT-1161 treatment group vs. 0% in the placebo group at week 48 [68, 128, 129].

19.3 Ravuconazole and Derivative Drug Products

Ravuconazole is a triazole drug developed by Bristol-Myers Squibb as an oral formulation. It is structurally similar to fluconazole. The mechanism of action is similar to other azoles in inhibiting 14α-demethylase which affects the ergosterol pathway. The drug is in phase II study. A regimen of ravuconazole 200 mg/day for

12 weeks was proposed as a potential, favorable treatment option for onychomycosis in a randomized, double-blind phase I/II study by Gupta et al. [130].

Several prodrugs and drug products of ravuconazole are in advanced stages of clinical trials for the treatment of onychomycosis. Fosravuconazole L-lysine ethanolate (F-RVCZ) is a novel triazole antifungal drug which was approved in Japan in 2018 for the treatment of onychomycosis. It is a prodrug of ravuconazole. It is marketed under the name Nailin by Sato Pharmaceutical Co, Ltd., Tokyo, Japan [131]. A randomized, multicenter, double-blind, phase III trial in Japan showed higher efficacy of fosravuconazole (100 mg daily for 12 weeks) with complete cure of 59.4% and mycological cure of 82% compared to placebo at week 48. This drug is yet to be approved in the USA for the treatment of onychomycosis [132].

19.4 Other Oral Drugs in Development

Posaconazole is approved by the FDA for invasive *Candida* and *Aspergillus* infections. It is an oral antifungal in development for the treatment of onychomycosis. A phase IIb multicenter trial assessed the safety and efficacy of four different dosage regimens of posaconazole for adult toenail onychomycosis. At week 48, all four groups had higher complete cure rates (79% mycological cure and 46% clinical cure for 400 mg/day for 24 weeks dosage regimen) compared to placebo. It was well tolerated with mild adverse effects [133]. A similar phase II, randomized, double-blind, dose-ranging study (100–400 mg/week for 24 or 26 weeks vs. placebo) with albaconazole (novel triazole) showed that it was effective in treating distal subungual onychomycosis with a dose-dependent mycological cure of 34–71% and complete cure (mycological cure plus 100% clear nail) of 12–33% at week 52 [134].

20 Novel Topical Formulations

20.1 HPCH Technology

P-3051 is a new ciclopirox 8% nail hydrolacquer which has been developed based on hydroxyl-propyl chitosan (HPCH) technology in Europe. The HPCH technology was developed by Polichem SA (Lugano, Switzerland). The active compound remains the same in this new formulation (ciclopirox 8%). The P-3051 drug is a water-soluble, film-forming biopolymer, unlike commercial ciclopirox nail lacquer which forms a water-insoluble lacquer on a vinyl resin. In a multicenter, randomized, placebo-controlled trial in Europe, the novel P-3051 drug was studied for its efficacy and safety. This study compared the new hydrolacquer with the existing commercial ciclopirox 8% nail lacquer (Penlac) for a treatment period of 48 weeks (daily application), followed by a washout period of 4 weeks, and then a follow-up period of 8 weeks. The complete cure (primary endpoint) was 5.7% at week 52 for the P-3051 group vs. 3.2% in the Penlac group. This increased to 12.7% after follow-up in the P-3051 group vs. 5.8% in the Penlac group. Thus, the P-3051 drug appears to be more effective in treating onychomycosis compared to the commercial ciclopirox 8% nail lacquer Penlac [68, 135].

In vitro drug permeation experiments with HPCH lacquers showed strong affinity to keratin membrane, resulting in efficient active drug release from the biopolymer [136]. In another randomized, evaluator-blinded clinical trial, high mycological and complete cure rates (100% vs. 81.7% respectively) were observed at week 48 with P-3051 compared to amorolfine nail lacquer (35% vs. 11.7% respectively) [68].

Similarly, P-3058 is a topical formulation of terbinafine hydrochloride 10% solution using HPCH technology by Polichem S.A. A multicenter, randomized phase III trial assessed the safety and efficacy of P-3058 in the treatment of mild to moderate distal lateral subungual onychomycosis of toenail when administered once weekly for 48 weeks with a follow-up period of 12 weeks with results yet to be released [137].

20.2 Penetration Enhancers

The WO2019088005A1 patent describes a recent, novel chemical penetration enhancer which has an antifungal drug such as luliconazole or efinaconazole along with a chemical such as ethyl acetate which enhances the drug penetration and drug delivery by hydrating the nail or breaking keratin bonds. Similarly, CN109303815A patent is a novel plant-based formulation developed in 2018 against onychomycosis which has plant extracts such as turpentine, petrolatum, azone, salicylic acid among other compounds [138].

20.3 ME1111

ME1111 is a novel topical formulation that is being developed for the treatment of onychomycosis. The chemical formula of ME1111 is 2-(3, 5-dimethyl-1H-pyrazol-1-yl)-5-methylphenol. It was discovered by Meiji Seika Pharma Co, Ltd., Tokyo, Japan. It has a small molecular weight of 202.25 Da and has been seen to more effectively penetrate the nail plate than ciclopirox. It has a unique mechanism of action by inhibiting the succinate dehydrogenase enzyme of the electron transport chain which in turn has an effect on ATP production [68, 139].

20.4 Topical Terbinafine Formulations

A randomized, multicenter, phase III trial in North America with MOB-015, a topical formulation of terbinafine 10% solution by Moberg Pharma, has met their primary endpoint of complete cure (negative fungal culture and clear nail) of 4.5% at week 52 and secondary endpoint of mycological cure (absence of fungal pathogen) 70%, which is higher than that achieved with existing topical antifungals, in

December 2019. There are no safety issues or severe adverse events observed [140, 141].

TDT-067 is another novel, carrier-based formulation of terbinafine in a transfersome at 15 mg/mL. This transfersome vehicle was developed by Targeted Delivery Technologies Ltd. It is a vesicle made up of lipid aggregates which activate the hydrophilic pathway in the nail, which in turn ensure high levels of drug penetration into the target site. TDT-067 is a liquid spray which is used on the nail and surrounding skin. Clinical trials are ongoing which will shed light on the efficacy and safety of TDT-067 in the treatment of onychomycosis [98, 142].

20.5 Polyurethane (PU) Polymer

A novel formulation of excipients combining two topical antifungal drugs terbinafine and ciclopirox in a 10% polyurethane (PU) polymer was reported to be superior in drug release, permeation, and antifungal activity in vitro [68].

20.6 Nitric Oxide in Treating Onychomycosis

Nitric oxide (NO), a free radical present in plants and animals is proven to exercise fungistatic activity through different mechanisms, including upregulation of macrophages, DNA damage, and enzyme inactivation. Antifungal resistance against NO has not been reported to date, making it a potential treatment option for onychomycosis. A unique formulation of nitric oxide, NO nanoparticles, was tested in animal models with *T. mentagrophytes* dermatophytosis where it showed high mycological and clinical cure rates. It was also found to be effective against *Candida* spp. and NDM infections. Hydrogel-based NO nanoparticles were synthesized and combined with a topical antifungal agent such as efinaconazole 10% solution to study their synergy and ability to overcome the limitation of poor penetration of topical agents. The NO nanoparticles aid in tissue penetration of efinaconazole due to its small size and achieve sustained release of NO which improves the antifungal properties [143]. The released NO reacts with thiol groups (–SH) in the keratin fibers to form nitrosothiols (–SNO) which exert antifungal properties against *T. rubrum* and *T. mentagrophytes*. Similarly, acidified nitrite cream for 16 weeks is a potential topical treatment shown to be effective in treating *Tinea pedis* [144, 145]. A recent study by Zhao et al. [146] has shed light on the role of endogenous NO in fungal physiological and metabolic pathways, including growth, germination, infection, resistance, etc., in molds. The authors have proposed a possible treatment method by targeting the NO production in these molds [146]. It remains unclear due to the conflicting studies in literature. Further research is recommended to understand the role of endogenous NO in fungal cells and the potential of NO as a topical treatment for onychomycosis.

21 Novel Physical Treatment Methods

21.1 Non-thermal Plasma

A small pilot clinical study showed that non-thermal plasma could be used to treat toenail onychomycosis. Non-thermal plasma was created by using dielectric insulators where atmospheric molecules were ionized and converted into ions, electrons, ozone, hydroxyl radicals, and nitric oxide. Mycological cure was achieved in 15.4% of the treated patients. A previous in vitro study with cold atmospheric plasma treatment inhibited the growth of *T. rubrum*. Further research and large-scale clinical studies are required to assess the long-term safety and efficacy of this treatment method [147].

21.2 Laser Poration

Laser poration uses a femtosecond-pulsed laser to irradiate a highly focused, high-energy beam guided by hollow-core photonic crystal fibers. This high-energy beam creates pores of 100 μm in diameter in the nail plate which then enhances the permeation of topical treatments. Another study combined laser poration with nanocapsule formulation of tioconazole for the treatment of onychomycosis. These nanocapsules hold the active drug in a stable concentration and aids in sustained drug release [68, 148].

22 Management of Onychomycosis and Preventative Measures

22.1 FDA Guidelines for the Treatment of Onychomycosis with Oral and Topical Antifungals

The FDA defines the treatment of onychomycosis as reduction or elimination of the fungal organism and growth of full clear fingernail after 6 months and toenail after 12 months after the start of treatment [91]. The elimination of the fungal pathogen is known as mycological cure which is determined by a negative microscopic stain and a negative fungal culture. The growth of a normal, healthy nail is known as clinical cure. Complete cure is defined as mycological cure plus 0% target nail involvement or 100% clear nail [91]. Mycological cure is seen as the primary outcome of most treatment methods. The observed mycological, almost complete cure/effective treatment, and complete cure rates of oral and topical antifungal drugs approved by the FDA for the treatment of onychomycosis are listed in Table 3 [9, 23, 59, 76, 82].

Table 3 Treatment success with FDA-approved antifungal agents for toenail dermatophyte onychomycosis

Antifungal drugs for toenail dermatophyte onychomycosis	Mycological cure (%)	Almost complete cure/ Effective treatment (%)	Complete cure (%)
Systemic drugs			
Terbinafine	70	59	38
Itraconazole	54	35	14
Topical drugs			
Ciclopirox	29–36	6.5–12	5.5–8.5
Efinaconazole	53.4–55.3	31–46	15.2–18.8
Tavaborole	31.1–35.9	26.1–27.5	6.5–9.1

Mycological cure—negative KOH and negative fungal culture
Complete cure—0% clinical involvement or clear nail plus negative mycological cure
Almost complete cure/effective treatment—$\leq$ 5–10% target nail involvement or $>$ 5 mm new unaffected nail growth plus negative mycological cure

22.2 What Factors Affect Cure?

There are several factors that affect cure including age, severity, and duration of the infection, treatment method, environmental factors such as antifungal resistance, biofilms, etc. There is limited evidence that the HLA-DR8 allele (MHC) found in patients with onychomycosis might cause poor response to therapy or lead to prolonged duration of the infection. Special populations such as the elderly and children, and patients with comorbidities such as diabetes mellitus, vascular disease, and immunocompromised individuals (e.g., AIDS, hepatitis) are less likely to achieve complete cure.

22.3 Relapse and Reinfection

Relapse and reinfection are observed in 10–53% of patients with onychomycosis [149, 150]. Relapse is the recurrence of the same infection due to incomplete cure or poor response to therapy while reinfection is acquiring a new infection after complete cure or successful treatment with therapy. There are multiple factors that contribute to relapse or reinfection. Clinically normal nail is difficult to restore even if the infecting organism is eradicated in severe, dystrophic onychomycosis due to the extensive damage to the nail bed and matrix. Other factors include comorbidity, predisposition, belonging to a special population (i.e., elderly, diabetic, AIDS, etc.), severity of the infection, family history, occupation, and poor personal hygiene [3, 68, 151–154].

In addition, the nail exhibits some immune privilege: there is no protective adaptive immune response generated by the body against onychomycosis. Thus, reinfection and recurrence occur even with the help of therapeutic agents [155]. Relapse has been found more frequently in patients treated with itraconazole compared to those treated with terbinafine, and relapses generally occur within

2.5 years of cure [156–159]. On the other hand, there is very little information available about relapse rates in topical treatment use [160]. It is becoming increasingly difficult to distinguish between relapse and reinfection due to coexisting *Tinea pedis* infection in 59% of the cases, creating a reservoir of fungal pathogens surrounding the nail [161].

22.4 Antifungal Resistance

One of the emerging problems with the treatment of onychomycosis is antifungal resistance. Fungal pathogens develop resistance due to the increased use of broad-spectrum antifungal drugs. There are several reports across the globe of developing dermatophyte resistance to the most commonly used antifungal agent terbinafine. Missense mutations in the squalene epoxidase gene have been found in *T. rubrum* and *T. mentagrophytes* in Switzerland and India. Due to increasing immigration and travel, terbinafine resistance is spreading globally. As a result, these pathogens show resistance to terbinafine, requiring increased dosage due to an increase in the MIC [149, 162]. This is referred to as clinical resistance where higher-than-safe dosage is required to inhibit the pathogen [163]. To combat this resistance, in vitro susceptibility testing is recommended to identify the antifungal agents that are not effective, i.e., resistant. Increasing azole resistance is observed in *Candida* species which overexpress and upregulate MDR genes coding for efflux pumps that expel azoles out of the cells [149, 162]. *Candida dubliniensis* shows in vitro resistance to fluconazole by multiple mechanisms which include exoenzyme production and phenotype switching [164]. *Fusarium* species show antifungal resistance to a variety of agents, highlighting the difficulty in treating non-dermatophyte mold onychomycosis [165].

22.5 Biofilms

Increasing research shows the potential of fungi to form biofilms which might be another major reason for widespread antifungal resistance. Fungi are conventionally thought to be planktonic, free-floating species in the environment. Recent findings suggest otherwise: fungi are capable of forming biofilms and adapting themselves to their environment. Fungal biofilms offer various advantages such as virulence factors (adherence, filamentation, hydrophobicity, etc.) against the host immune system, community-based differential gene expression, and increased resistance to antifungal agents through the secretion of extracellular matrix (ECM). This ECM provides structural protection to biofilms and prevents any external physical disruption while reducing penetration of the antifungal drugs to the site of infection. There are several dermatological conditions such as acne, atopic dermatitis, etc., that have associated fungal biofilms growing on epithelial surfaces.

An in vitro experiment showed that *T. rubrum* and *T. mentagrophytes* formed biofilms 72 h after inoculation. The previously mentioned in vitro susceptibility testing should be done against fungal biofilms and not on free, suspended cells to devise efficient treatment strategies. Multiple targets have been identified to disrupt and disintegrate biofilms including glycoproteins on the cell surface which aid in ECM synthesis and attachment. Antifungals such as amphotericin B and echinocandins seem to be effective against biofilms. Agents targeting biofilms should be considered in the treatment regimen of onychomycosis to yield better success [166–168]. An in vitro study showed the efficacy of Nd:YAG laser in eradicating *Candida albicans* and *Fusarium oxysporum* biofilms grown on healthy human nail fragments [149, 169]. Additionally, mechanical disruption of the nail may be considered through lasers, drilling, or other devices to physically disrupt biofilms [167, 170, 171].

22.6 Mixed Infections

Increasing prevalence of non-dermatophyte mold (NDM) onychomycosis globally has created new research questions addressing the existing diagnostic methods, treatment options, and management of onychomycosis. Clinical studies demonstrate the possibility of NDMs as the reason for treatment failure and recurrence. Accurate diagnosis and identification of the causative organism are important in determining the treatment method and duration as it can have a direct impact on cure. Mixed infections with an NDM and a co-infecting dermatophyte such as *T. rubrum* may be difficult to treat with the existing oral or topical monotherapy because of their complex clinical presentation and requirement of longer treatment duration [172].

22.7 Preventative Measures

In order to prevent recurrence, patients should be examined for the presence of *Tinea pedis* and treated along with onychomycosis [1, 79, 173]. Topical terbinafine and naftifine hydrochloride may be used for treating *Tinea pedis* as both agents can be found in the stratum corneum several weeks after treatment [152]. Topical azoles, ciclopirox cream, and other effective topical agents are also considerations. Treatment of *Tinea pedis* will not only decrease the fungal reservoir in the skin, but improve the efficacy of topical onychomycosis treatments when co-administered [155]. When using an oral antifungal agent it is not necessary to use a topical antifungal agent, but combination therapy can improve treatment outcomes for difficult-to-treat cases. Patients are advised to employ a number of strategies to reduce the chances of relapse or recurrence. Many are effective for those at higher risk of developing onychomycosis, such as elderly people, diabetic patients, HIV-positive, and dialysis patients. For example, wearing appropriate footwear in public showers and at pools, keeping feet cool and dry, trimming toenails regularly, maintaining proper nail hygiene, proper sanitization of socks at high temperatures (>60 °C), and using ozone gas or UV light to sanitize shoes [149, 161, 174].

22.8 Combination Therapy

Studies with combination therapy are being increasingly reported [161, 162]. The mycological cure rate increased to 88.2% when oral terbinafine was used in combination with topical 8% ciclopirox nail lacquer compared to 64.7% when just oral terbinafine was used [175]. Amorolfine 5% nail lacquer in combination with continuous itraconazole (200 mg once daily for 6 weeks) was studied in a randomized clinical study in Spain and was found to be safe and efficient in treating patients with severe onychomycosis [176]. Fractional CO_2 lasers seem to enhance the absorption of topical drugs: a laser session every 2 weeks combined with 1% terbinafine cream was observed to be safe and efficient for the treatment of onychomycosis [108]. Physical, chemical, or mechanical treatment methods are often combined with topical drugs to enhance its absorption by the nail plate and improve its efficacy [177].

22.9 Prophylaxis

Using prophylactic treatment can also help prevent or delay recurrence. For example, efinaconazole can be applied twice weekly post-treatment for 3 or more years [1, 152, 178]. This medication remains in the nail for a long time and provides effective antifungal activity for 2 weeks post-application [152, 179]. Tavaborole can be used similarly as it is retained in the nail for 14 days post-application [180]. Amorolfine nail lacquer is another antifungal agent for prophylactic treatment. Several studies demonstrate the prophylactic effect of amorolfine in the management of onychomycosis. Amorolfine when used in half the frequency of its usual regimen after successfully treating a dermatophyte onychomycosis with an oral therapy seems effective in achieving complete cure and prevents recurrence [75].

23 Conclusion

Onychomycosis is a chronic infection which is difficult to treat due to its multifactorial nature. The current global scenario of onychomycosis persuades us to rethink the existing conventional diagnostic methods and reform treatment strategies. Accurate confirmatory diagnosis is necessary to identify the right etiological agent and to initiate treatment. Novel permeation enhancers, new therapeutic agents, combinational therapy, prophylaxis post-treatment, addition of agents targeting fungal biofilms, disinfection of fungal reservoirs, patient education and lifestyle modifications, are some of the considerations for effective management of onychomycosis.

References

1. Gupta AK, Versteeg SG, Shear NH. Onychomycosis in the 21st century: an update on diagnosis, epidemiology, and treatment. J Cutan Med Surg. 2017;21(6):525–39.
2. Wolverton SE, Wu JJ. Comprehensive dermatologic drug therapy e-book. Amsterdam: Elsevier Health Sciences; 2019. 186p.
3. Gupta AK, Uro M, Cooper EA. Onychomycosis therapy: past, present, future. J Drugs Dermatol. 2010;9(9):1109–13.
4. Baran R, editor. Baran & Dawber's diseases of the nails and their management. 5th ed. Hoboken, NJ: Wiley-Blackwell; 2019.
5. Gupta AK, Sauder DN, Shear NH. Antifungal agents: an overview. Part I. J Am Acad Dermatol. 1994;30(5 Pt 1):677–98; quiz 698–700.
6. Ketoconazole. In: LiverTox: clinical and research information on drug-induced liver injury [Internet]. Bethesda, MD: National Institute of Diabetes and Digestive and Kidney Diseases; 2012. http://www.ncbi.nlm.nih.gov/books/NBK547869/. Cited 27 Jan 2020.
7. Anton Stutz inventions, patents and patent applications – Justia patents search [Internet]. https://patents.justia.com/inventor/anton-stutz. Cited 18 Feb 2020.
8. Novartis Pharmaceuticals Corporation. Product monograph Lamisil (terbinafine hydrochloride) [Internet]. 2011. http://webprod3.hc-sc.gc.ca/dpd-bdpp/info.do?lang=eng&code=13990. Cited 21 Aug 2012.
9. Novartis. LAMISIL (terbinafine hydrochloride) Tablets, 250 mg [Internet]. Drugs@FDA. http://www.accessdata.fda.gov/drugsatfda_docs/label/2012/020539s021lbl.pdf. Cited 15 Mar 2013.
10. Terbinafine Granules Approved by FDA for Tinea Capitis [Internet]. https://www.jwatch.org/pa200711210000001/2007/11/21/terbinafine-granules-approved-fda-tinea-capitis. Cited 14 Feb 2020.
11. Ryder NS. Terbinafine: mode of action and properties of the squalene epoxidase inhibition. Br J Dermatol. 1992;126(Suppl 39):2–7.
12. Gupta AK, Sauder DN, Shear NH. Antifungal agents: an overview. Part II. J Am Acad Dermatol. 1994;30(6):911–33; quiz 934–6.
13. Schatz F, Bräutigam M, Dobrowolski E, Effendy I, Haberl H, Mensing H, et al. Nail incorporation kinetics of terbinafine in onychomycosis patients. Clin Exp Dermatol. 1995;20(5):377–83.
14. Debruyne D, Coquerel A. Pharmacokinetics of antifungal agents in onychomycoses. Clin Pharmacokinet. 2001;40(6):441–72.
15. Doty RL, Bromley SM. Effects of drugs on olfaction and taste. Otolaryngol Clin North Am. 2004;37(6):1229–54.
16. Terbinafine. In: LiverTox: clinical and research information on drug-induced liver injury [Internet]. Bethesda, MD: National Institute of Diabetes and Digestive and Kidney Diseases; 2012. http://www.ncbi.nlm.nih.gov/books/NBK548617/. Cited 16 Jan 2020.
17. Gupta AK, Shear NH. A risk-benefit assessment of the newer oral antifungal agents used to treat onychomycosis. Drug Saf. 2000;22(1):33–52.
18. Wang Y, Lipner SR. Retrospective analysis of adverse events with systemic onychomycosis medications reported to the United States Food and Drug Administration. J Dermatolog Treat. 2020:1–5.
19. Gupta AK, Ryder JE, Skinner AR. Treatment of onychomycosis: pros and cons of antifungal agents. J Cutan Med Surg. 2004;8(1):25–30.
20. Darkes MJM, Scott LJ, Goa KL. Terbinafine. Am J Clin Dermatol. 2003;4(1):39–65.
21. Cohen AD, Medvesovsky E, Shalev R, Biton A, Chetov T, Naimer S, et al. An independent comparison of terbinafine and itraconazole in the treatment of toenail onychomycosis. J Dermatolog Treat. 2003;14(4):237–42.
22. Gupta AK, Adamiak A, Cooper EA. The efficacy and safety of terbinafine in children. J Eur Acad Dermatol Venereol. 2003;17(6):627–40.

23. Janssen Pharma. SPORANOX® (itraconazole) capsules [Internet]. Drugs@FDA. http://www.accessdata.fda.gov/drugsatfda_docs/label/2012/020083s048s049s050lbl.pdf.
24. Hector R. The many lives of itraconazole. Expert Opin Ther Pat. 2003;13(2):141–8.
25. Korting HC, Schöllmann C. The significance of itraconazole for treatment of fungal infections of skin, nails and mucous membranes. J Dtsch Dermatol Ges. 2009;7(1):11–9.
26. Gupta AK, Drummond-Main C, Cooper EA, Brintnell W, Piraccini BM, Tosti A. Systematic review of nondermatophyte mold onychomycosis: diagnosis, clinical types, epidemiology, and treatment. J Am Acad Dermatol. 2012;66(3):494–502.
27. De Doncker P, Decroix J, Piérard GE, Roelant D, Woestenborghs R, Jacqmin P, et al. Antifungal pulse therapy for onychomycosis. A pharmacokinetic and pharmacodynamic investigation of monthly cycles of 1-week pulse therapy with itraconazole. Arch Dermatol. 1996;132(1):34–41.
28. Willemsen M, De Doncker P, Willems J, Woestenborghs R, Van de Velde V, Heykants J, et al. Posttreatment itraconazole levels in the nail: New implications for treatment in onychomycosis. J Am Acad Dermatol. 1992;26(5, Part 1):731–5.
29. De Doncker P. Pharmacokinetics of orally administered antifungals in onychomycosis. Int J Dermatol. 1999;38(Suppl 2):20–7.
30. Sigurgeirsson B, Billstein S, Rantanen T, Ruzicka T, di Fonzo E, Vermeer BJ, et al. L.I.ON. Study: efficacy and tolerability of continuous terbinafine (Lamisil) compared to intermittent itraconazole in the treatment of toenail onychomycosis. Lamisil vs. Itraconazole in Onychomycosis. Br J Dermatol. 1999;141(Suppl 56):5–14.
31. Itraconazole. In: LiverTox: clinical and research information on drug-induced liver injury [Internet]. Bethesda, MD: National Institute of Diabetes and Digestive and Kidney Diseases; 2012. http://www.ncbi.nlm.nih.gov/books/NBK548273/. Cited Jan 16 2020.
32. Elewski B, Tavakkol A. Safety and tolerability of oral antifungal agents in the treatment of fungal nail disease: a proven reality. Ther Clin Risk Manag. 2005;1(4):299–306.
33. Lavrijsen AP, Balmus KJ, Nugteren-Huying WM, Roldaan AC, van't Wout JW, Stricker BH. Hepatic injury associated with itraconazole. Lancet. 1992;340(8813):251–2.
34. Van Peer A, Woestenborghs R, Heykants J, Gasparini R, Gauwenbergh G. The effects of food and dose on the oral systemic availability of itraconazole in healthy subjects. Eur J Clin Pharmacol. 1989;36(4):423–6.
35. Gupta AK, Stec N, Bamimore MA, Foley KA, Shear NH, Piguet V. The efficacy and safety of pulse *vs.* continuous therapy for dermatophyte toenail onychomycosis. J Eur Acad Dermatol Venereol. 2020;34(3):580–8.
36. Havu V, Brandt H, Heikkilä H, Hollmen A, Oksman R, Rantanen T, et al. A double-blind, randomized study comparing itraconazole pulse therapy with continuous dosing for the treatment of toe-nail onychomycosis. Br J Dermatol. 1997;136(2):230–4.
37. Huang P-H, Paller AS. Itraconazole pulse therapy for dermatophyte onychomycosis in children. Arch Pediatr Adolesc Med. 2000;154(6):614–8.
38. Gupta AK, Nolting S, de Prost Y, Delescluse J, Degreef H, Theissen U, et al. The use of itraconazole to treat cutaneous fungal infections in children. Dermatology (Basel). 1999;199 (3):248–52.
39. Gupta AK, Gover MD, Lynde CW. Pulse itraconazole *vs.* continuous terbinafine for the treatment of dermatophyte toenail onychomycosis in patients with diabetes mellitus. J Eur Acad Dermatol Venereol. 2006;20(10):1188–93.
40. Fluconazole – an overview | ScienceDirect Topics [Internet]. https://www.sciencedirect.com/topics/pharmacology-toxicology-and-pharmaceutical-science/fluconazole. Cited 18 Feb 2020.
41. Pfizer. DIFLUCAN® (Fluconazole tablets) (Fluconazole for oral suspension) [Internet]. 2014. https://www.accessdata.fda.gov/drugsatfda_docs/label/2014/019949s060,020090s044lbl.pdf.
42. Scher RK, Breneman D, Rich P, Savin RC, Feingold DS, Konnikov N, et al. Once-weekly fluconazole (150, 300, or 450 mg) in the treatment of distal subungual onychomycosis of the toenail. J Am Acad Dermatol. 1998;38(6 Pt 2):S77–86.

43. Bueno JG, Martinez C, Zapata B, Sanclemente G, Gallego M, Mesa AC. *In vitro* activity of fluconazole, itraconazole, voriconazole and terbinafine against fungi causing onychomycosis. Clin Exp Dermatol. 2010;35(6):658–63.
44. Gupta AK, Drummond-Main C, Paquet M. Evidence-based optimal fluconazole dosing regimen for onychomycosis treatment. J Dermatolog Treat. 2013;24(1):75–80.
45. Brown SJ. Efficacy of fluconazole for the treatment of onychomycosis. Ann Pharmacother. 2009;43(10):1684–91.
46. Gupta AK, Daigle D, Foley KA. Network meta-analysis of onychomycosis treatments. Skin Appendage Disord. 2015;1:74–81.
47. Gupta AK, Cooper EA. Update in antifungal therapy of dermatophytosis. Mycopathologia. 2008;166(5–6):353–67.
48. Fluconazole. In: LiverTox: clinical and research information on drug-induced liver injury [Internet]. Bethesda, MD: National Institute of Diabetes and Digestive and Kidney Diseases; 2012. http://www.ncbi.nlm.nih.gov/books/NBK548300/. Cited 16 Jan 2020.
49. Spernovasilis N, Kofteridis DP. Pre-existing liver disease and toxicity of antifungals. J Fungi (Basel). 2018;4(4):133.
50. Brammer KW, Farrow PR, Faulkner JK. Pharmacokinetics and tissue penetration of fluconazole in humans. Rev Infect Dis. 1990;12(Suppl 3):S318–26.
51. Muñoz P, Moreno S, Berenguer J, Bernaldo de Quirós JC, Bouza E. Fluconazole-related hepatotoxicity in patients with acquired immunodeficiency syndrome. Arch Intern Med. 1991;151(5):1020–1.
52. Stevens DA. The new generation of antifungal drugs. Eur J Clin Microbiol Infect Dis. 1988;7(6):732–5.
53. Sethi A, Antaya R. Systemic antifungal therapy for cutaneous infections in children. Pediatr Infect Dis J. 2006;25(7):643–4.
54. Gupta AK, Paquet M, Simpson F, Tavakkol A. Terbinafine in the treatment of dermatophyte toenail onychomycosis: a meta-analysis of efficacy for continuous and intermittent regimens. J Eur Acad Dermatol Venereol. 2013;27(3):267–72.
55. Sprenger AB, Purim KSM, Sprenger F, Queiroz-Telles F. A week of oral terbinafine pulse regimen every three months to treat all dermatophyte onychomycosis. J Fungi (Basel). 2019;5(3):82.
56. Gupta AK, Baran R, Summerbell R. Onychomycosis: strategies to improve efficacy and reduce recurrence. J Eur Acad Dermatol Venereol. 2002;16(6):579–86.
57. Gupta AK, Konnikov N, Lynde CW. Single-blind, randomized, prospective study on terbinafine and itraconazole for treatment of dermatophyte toenail onychomycosis in the elderly. J Am Acad Dermatol. 2001;44(3):479–84.
58. Gupta AK, Del Rosso JQ. An evaluation of intermittent therapies used to treat onychomycosis and other dermatomycoses with the oral antifungal agents. Int J Dermatol. 2000;39(6):401–11.
59. Aventis Pharma. PENLAC® Nail Lacquer (ciclopirox) topical solution, 8% [Internet]. https://www.accessdata.fda.gov/drugsatfda_docs/label/2004/21022s004lbl.pdf. 31 Jan 2004.
60. Curanail 5% Nail Lacquer (Amorolfine Hydrochloride) PL 10590/0049 [Internet]. MHRA. http://www.mhra.gov.uk/home/groups/l-unit1/documents/websiteresources/con2023787.pdf. Cited 8 Mar 2016.
61. Del Rosso JQ. The role of topical antifungal therapy for onychomycosis and the emergence of newer agents. J Clin Aesthet Dermatol. 2014;7(7):10–8.
62. Angelo T, Borgheti-Cardoso LN, Gelfuso GM, Taveira SF, Gratieri T. Chemical and physical strategies in onychomycosis topical treatment: a review. Med Mycol. 2017;55(5):461–75.
63. Pollak RA, Jo Siu WJ, Tatsumi Y, Pillai R. Efinaconazole topical solution, 10%: factors contributing to onychomycosis success. J Fungi (Basel). 2015;1(2):107–14.
64. Tabara K, Szewczyk AE, Bienias W, Wojciechowska A, Pastuszka M, Oszukowska M, et al. Amorolfine vs. ciclopirox – lacquers for the treatment of onychomycosis. Postepy Dermatol Alergol. 2015;32(1):40–5.

65. Bohn M, Kraemer KT. Dermatopharmacology of ciclopirox nail lacquer topical solution 8% in the treatment of onychomycosis. J Am Acad Dermatol. 2000;43(4 Suppl):S57–69.
66. Gupta AK, Fleckman P, Baran R. Ciclopirox nail lacquer topical solution 8% in the treatment of toenail onychomycosis. J Am Acad Dermatol. 2000;43(4 Suppl):S70–80.
67. Tosti A, Piraccini BM, Lorenzi S. Onychomycosis caused by nondermatophytic molds: clinical features and response to treatment of 59 cases. J Am Acad Dermatol. 2000;42(2 Pt 1):217–24.
68. Gupta AK, Stec N. Recent advances in therapies for onychomycosis and its management. F1000Res. 2019;8
69. Seebacher C. Action mechanisms of modern antifungal agents and resulting problems in the management of onychomycosis. Mycoses. 2003;46(11–12):506–10.
70. Lipner S, Scher RK. Onychomycosis: current and future therapies. Cutis. 2014;93(2):60–3.
71. Haria M, Bryson HM. Amorolfine. A review of its pharmacological properties and therapeutic potential in the treatment of onychomycosis and other superficial fungal infections. Drugs. 1995;49(1):103–20.
72. Feng X, Xiong X, Ran Y. Efficacy and tolerability of amorolfine 5% nail lacquer in combination with systemic antifungal agents for onychomycosis: a meta-analysis and systematic review. Dermatol Ther. 2017;30(3) https://doi.org/10.1111/dth.12457.
73. Baran R, Feuilhade M, Combernale P, Datry A, Goettmann S, Pietrini P, et al. A randomized trial of amorolfine 5% solution nail lacquer combined with oral terbinafine compared with terbinafine alone in the treatment of dermatophytic toenail onychomycoses affecting the matrix region. Br J Dermatol. 2000;142(6):1177–83.
74. Eichenfield LF, Friedlander SF. Pediatric onychomycosis: the emerging role of topical therapy. J Drugs Dermatol. 2016;16(2):105–9.
75. Sigurgeirsson B, Olafsson JH, Steinsson JT, Kerrouche N, Sidou F. Efficacy of amorolfine nail lacquer for the prophylaxis of onychomycosis over 3 years. J Eur Acad Dermatol Venereol. 2010;24(8):910–5.
76. Valeant Canada LP. PrJUBLIA™ Efinaconazole topical solution, 10% w/w. 2013.
77. Gupta AK, Simpson FC. Efinaconazole: a new topical treatment for onychomycosis. Skin Therapy Lett. 2014;19(1):1–4.
78. Gupta AK, Mays RR, Versteeg SG, Shear NH, Friedlander SF. Onychomycosis in children: safety and efficacy of antifungal agents. Pediatr Dermatol. 2018;35(5):552–9.
79. Lipner SR, Scher RK. Efinaconazole in the treatment of onychomycosis. Infect Drug Resist. 2015;8:163–72.
80. Eichenfield LF, Elewski B, Sugarman J, Rosen T, Gupta A, Pillai R, et al. Safety, pharmacokinetics, and efficacy of Efinaconazole 10% topical solution for the treatment of onychomycosis in pediatric patients. J Drugs Dermatol. 2020;19(9):867–72.
81. Gupta AK, Studholme C. Update on Efinaconazole 10% topical solution for the treatment of onychomycosis. Skin Therapy Lett. 2016;21(6):7–11.
82. Anacor Pharmaceuticals Inc. KERYDIN™ (tavaborole) topical solution, 5% [Internet]. US Food and Drug Administration; 2014. http://www.accessdata.fda.gov/scripts/cder/drugsatfda/index.cfm?fuseaction=Search.DrugDetails. Cited 14 July 2014
83. Generic Kerydin Availability – Drugs.com [Internet]. https://www.drugs.com/availability/generic-kerydin.html. Cited 18 Feb 2020.
84. Abastabar M, Haghani I, Shokohi T, Hedayati MT, Aghili SR, Jedi A, et al. Low *in vitro* antifungal activity of Tavaborole against yeasts and moulds from onychomycosis. Antimicrob Agents Chemother. 2018;62(12):e1632–18.
85. Gupta AK, Foley KA, Versteeg SG. New antifungal agents and new formulations against dermatophytes. Mycopathologia. 2017;182(1–2):127–41.
86. Gupta AK, Daigle D. Tavaborole (AN-2690) for the treatment of onychomycosis of the toenail in adults. Expert Rev Anti-Infect Ther. 2014;12(7):735–42.

87. Aly R, Gupta AK, Winter T, Zane LT, Vlahovic T. Tavaborole in difficult-to-treat onychomycosis cases: a post-hoc assessment of phase III subjects. J Drugs Dermatol. 2017;16(10):1016–21.
88. Rich P, Spellman M, Purohit V, Zang C, Crook TJ. Tavaborole 5% topical solution for the treatment of toenail onychomycosis in pediatric patients: results from a phase 4 open-label study. J Drugs Dermatol. 2019;18(2):190–5.
89. Gayam V, Khalid M, Dahal S, Garlapati P, Gill A. Hyperacute liver injury following intravenous fluconazole: a rare case of dose-independent hepatotoxicity. J Family Med Prim Care. 2018;7(2):451.
90. Lawson C-A, Karlowsky JA. Fluconazole-induced hepatotoxicity: review of published case reports. Can J Hosp Pharm. 1998;51(2):3.
91. US Food and Drug Administration. Medical devices and clinical trial design for the treatment or improvement in the appearance of fungally-infected nails – draft guidance for industry and FDA staff [Internet]. 2015. http://www.fda.gov/downloads/MedicalDevices/DeviceRegulationandGuidance/GuidanceDocuments/UCM431312.pdf. Cited 22 May 2015.
92. Gupta AK, Foley KA, Versteeg SG. Lasers for onychomycosis. J Cutan Med Surg. 2017;21(2):114–6.
93. Hay R. Therapy of skin, hair and nail fungal infections. J Fungi (Basel). 2018;4(3):99.
94. Gupta AK, Simpson FC, Heller DF. The future of lasers in onychomycosis. J Dermatolog Treat. 2016;27(2):167–72.
95. Gupta AK, Simpson F. Newly approved laser systems for onychomycosis. J Am Podiatr Med Assoc. 2012;102(5):428–30.
96. Vlahovic TC. Onychomycosis: evaluation, treatment options, managing recurrence, and patient outcomes. Clin Podiatr Med Surg. 2016;33(3):305–18.
97. Gupta AK, Simpson FC. New pharmacotherapy for the treatment of onychomycosis: an update. Expert Opin Pharmacother. 2015;16(2):227–36.
98. Nenoff P, Grunewald S, Paasch U. Laser therapy of onychomycosis. J Dtsch Dermatol Ges. 2014;12(1):33–8.
99. Nair AB, Vaka SRK, Murthy SN. Transungual delivery of terbinafine by iontophoresis in onychomycotic nails. Drug Dev Ind Pharm. 2011;37(10):1253–8.
100. Nematollahi AR, Badiee P, Nournia E. The efficacy of ultraviolet irradiation on *Trichophyton* species isolated from nails. Jundishapur J Microbiol. 2015;8(6):e18158.
101. Carney C, Cantrell W, Warner J, Elewski B. Treatment of onychomycosis using a submillisecond 1064-nm neodymium:yttrium-aluminum-garnet laser. J Am Acad Dermatol. 2013;69(4):578–82.
102. Karsai S, Jäger M, Oesterhelt A, Weiss C, Schneider SW, Jünger M, et al. Treating onychomycosis with the short-pulsed 1064-nm-Nd:YAG laser: results of a prospective randomized controlled trial. J Eur Acad Dermatol Venereol. 2017;31(1):175–80.
103. Hees H, Jäger MW, Raulin C. Treatment of onychomycosis using the 1 064 nm Nd:YAG laser: a clinical pilot study. J Dtsch Dermatol Ges. 2014;12(4):322–9.
104. Noguchi H, Miyata K, Sugita T, Hiruma M, Hiruma M. Treatment of onychomycosis using a 1064 nm Nd:YAG laser. Med Mycol J. 2013;54(4):333–9.
105. Kimura U, Takeuchi K, Kinoshita A, Takamori K, Hiruma M, Suga Y. Treating onychomycoses of the toenail: clinical efficacy of the sub-millisecond 1,064 nm Nd: YAG laser using a 5 mm spot diameter. J Drugs Dermatol. 2012;11(4):496–504.
106. Renner R, Grüsser K, Sticherling M. 1,064-nm diode laser therapy of onychomycosis: results of a prospective open treatment of 82 toenails. Dermatology (Basel). 2015;230(2):128–34.
107. Lim E-H, Kim H, Park Y-O, Lee Y, Seo Y-J, Kim C-D, et al. Toenail onychomycosis treated with a fractional carbon-dioxide laser and topical antifungal cream. J Am Acad Dermatol. 2014;70(5):918–23.
108. Shi J, Li J, Huang H, Permatasari F, Liu J, Xu Y, et al. The efficacy of fractional carbon dioxide (CO_2) laser combined with terbinafine hydrochloride 1% cream for the treatment of onychomycosis. J Cosmet Laser Ther. 2017;19(6):353–9.

109. de Oliveira GB, Antonio JR, Antonio CR, Tomé FA. The association of fractional CO_2 laser 10.600nm and photodynamic therapy in the treatment of onychomycosis. An Bras Dermatol. 2015;90(4):468–71.
110. Xu Y, Miao X, Zhou B, Luo D. Combined oral terbinafine and long-pulsed 1,064-nm Nd: YAG Laser treatment is more effective for onychomycosis than either treatment alone. Dermatol Surg. 2014;40(11):1201–7.
111. Watanabe D, Kawamura C, Masuda Y, Akita Y, Tamada Y, Matsumoto Y. Successful treatment of toenail onychomycosis with photodynamic therapy. Arch Dermatol. 2008;144 (1):19–21.
112. Piraccini BM, Rech G, Tosti A. Photodynamic therapy of onychomycosis caused by *Trichophyton rubrum*. J Am Acad Dermatol. 2008;59(5 Suppl):S75–6.
113. Gilaberte Y, Aspiroz C, Martes MP, Alcalde V, Espinel-Ingroff A, Rezusta A. Treatment of refractory fingernail onychomycosis caused by nondermatophyte molds with methylaminolevulinate photodynamic therapy. J Am Acad Dermatol. 2011;65(3):669–71.
114. Sotiriou E, Koussidou-Eremonti T, Chaidemenos G, Apalla Z, Ioannides D. Photodynamic therapy for distal and lateral subungual toenail onychomycosis caused by *Trichophyton rubrum*: preliminary results of a single-centre open trial. Acta Derm Venereol. 2010;90 (2):216–7.
115. Souza LWF, Souza SVT, Botelho AC d C. Distal and lateral toenail onychomycosis caused by *Trichophyton rubrum*: treatment with photodynamic therapy based on methylene blue dye. An Bras Dermatol. 2014;89(1):184–6.
116. Aspiroz C, Fortuño-Cebamanos B, Rezusta A, Gilaberte Y. Photodynamic therapy with methyl-aminolevulinate can be useful in the management of Scytalidium infections. Actas Dermosifiliogr. 2013;104(8):725–7.
117. Nair AB, Kim HD, Chakraborty B, Singh J, Zaman M, Gupta A, et al. Ungual and trans-ungual iontophoretic delivery of terbinafine for the treatment of onychomycosis. J Pharm Sci. 2009;98 (11):4130–40.
118. Nair AB, Kim HD, Davis SP, Etheredge R, Barsness M, Friden PM, et al. An *ex vivo* toe model used to assess applicators for the iontophoretic ungual delivery of terbinafine. Pharm Res. 2009;26(9):2194–201.
119. Shemer A, Gupta AK, Amichai B, Farhi R, Baran R, Daniel CR, et al. An open comparative study of nail drilling as adjunctive treatment for toenail onychomycosis. J Dermatolog Treat. 2016;27(5):480–3.
120. Vanstone S, Cordery SF, Stone JM, Gordeev SN, Guy RH. Precise laser poration to control drug delivery into and through human nail. J Control Release. 2017;268:72–7.
121. Lecha M, Effendy I, Feuilhade de Chauvin M, Di Chiacchio N, Baran R, Taskforce on Onychomycosis Education. Treatment options—development of consensus guidelines. J Eur Acad Dermatol Venereol. 2005;19(Suppl 1):25–33.
122. Lahfa M, Bulai-Livideanu C, Baran R, Ortonne JP, Richert B, Tosti A, et al. Efficacy, safety and tolerability of an optimized avulsion technique with Onyster® (40% urea ointment with plastic dressing) ointment compared to bifonazole-urea ointment for removal of the clinically infected nail in toenail onychomycosis: a randomized evaluator-blinded controlled study. Dermatology (Basel). 2013;226(1):5–12.
123. Gupta AK, Lynch LE. Management of onychomycosis: examining the role of monotherapy and dual, triple, or quadruple therapies. Cutis. 2004;74(1 Suppl):5–9.
124. Gupta AK, Paquet M, Simpson FC. Therapies for the treatment of onychomycosis. Clin Dermatol. 2013;31(5):544–54.
125. Repka MA, Majumdar S, Kumar Battu S, Srirangam R, Upadhye SB. Applications of hot-melt extrusion for drug delivery. Expert Opin Drug Deliv. 2008;5(12):1357–76.
126. Thiry J, Lebrun P, Vinassa C, Adam M, Netchacovitch L, Ziemons E, et al. Continuous production of itraconazole-based solid dispersions by hot melt extrusion: preformulation, optimization and design space determination. Int J Pharm. 2016;515(1–2):114–24.

127. Maddin S, Quiring J, Bulger L. Randomized, placebo-controlled, phase 3 study of itraconazole for the treatment of onychomycosis. J Drugs Dermatol. 2013;12(7):758–63.
128. Elewski B, Kempers S, Bhatia N, Blauvelt A, Curelop S, Brand S, et al. Efficacy and safety of VT-1161 in a randomized, double-blind, placebo-controlled study of four oral VT-1161 regimens in the treatment of patients with moderate-to-severe Distal-Lateral Subungual Onychomycosis (DLSO). In: Fourth international summit for nail diseases, 2017, Athens, Greece.
129. Gupta AK, Stec N. Emerging drugs for the treatment of onychomycosis. Expert Opin Emerg Drugs. 2019;24(4):213–20.
130. Gupta AK, Leonadri C, Pierce P, Conetta B. A phase I/II randomized, double-blind, placebo-controlled, dose-ranging study evaluating the efficacy, safety and pharmacokinetics of ravuconazole in the treatment of onychomycosis. J Eur Acad Dermatol Venereol. 2005;19 (4):437–43.
131. Oral Antifungal Agent Nailin® Capsules 100mg Approved In JAPAN | News Release:2018 | Eisai Co., Ltd. [Internet]. https://www.eisai.com/news/2018/news201806.html. Cited 28 Jan 2020.
132. Watanabe S, Tsubouchi I, Okubo A. Efficacy and safety of fosravuconazole L-lysine ethanolate, a novel oral triazole antifungal agent, for the treatment of onychomycosis: a multicenter, double-blind, randomized phase III study. J Dermatol. 2018;45(10):1151–9.
133. Elewski B, Pollak R, Ashton S, Rich P, Schlessinger J, Tavakkol A. A randomized, placebo- and active-controlled, parallel-group, multicentre, investigator-blinded study of four treatment regimens of posaconazole in adults with toenail onychomycosis. Br J Dermatol. 2012;166 (2):389–98.
134. Sigurgeirsson B, van Rossem K, Malahias S, Raterink K. A phase II, randomized, double-blind, placebo-controlled, parallel group, dose-ranging study to investigate the efficacy and safety of 4 dose regimens of oral albaconazole in patients with distal subungual onychomycosis. J Am Acad Dermatol. 2013;69(3):416–25.
135. Baran R, Tosti A, Hartmane I, Altmeyer P, Hercogova J, Koudelkova V, et al. An innovative water-soluble biopolymer improves efficacy of ciclopirox nail lacquer in the management of onychomycosis. J Eur Acad Dermatol Venereol. 2009;23(7):773–81.
136. Monti D, Saccomani L, Chetoni P, Burgalassi S, Saettone MF, Mailland F. *In vitro* transungual permeation of ciclopirox from a hydroxypropyl chitosan-based, water-soluble nail lacquer. Drug Dev Ind Pharm. 2005;31(1):11–7.
137. Study to evaluate the efficacy and safety of P-3058 nail solution in the treatment of onychomycosis – full text view – ClinicalTrials.gov [Internet]. https://clinicaltrials.gov/ct2/show/NCT03094468. Cited 28 Jan 2020.
138. Kreutz T, de Matos SP, Koester LS. Recent patents containing permeation enhancers for nail delivery. Recent Pat Drug Deliv Formul. 2019;13(3):203–18.
139. Takahata S, Kubota N, Takei-Masuda N, Yamada T, Maeda M, Alshahni MM, et al. Mechanism of action of ME1111, a novel antifungal agent for topical treatment of onychomycosis. Antimicrob Agents Chemother. 2016;60(2):873–80.
140. Moberg Pharma AB. Moberg Pharma meets primary endpoint for MOB-015 in a Phase 3 study for the treatment of onychomycosis [Internet]. News Powered by Cision. 2019. https://news.cision.com/moberg-pharma/r/moberg-pharma-meets-primary-endpoint-for-mob-015-in-a-phase-3-study-for-the-treatment-of-onychomycos,c2984987. Cited 7 Jan 2020.
141. Gupta AK, Surprenant MS, Kempers SE, Pariser DM, Rensfeldt K, Tavakkol A. Efficacy and safety of topical terbinafine 10% solution (MOB-015) in the treatment of mild to moderate distal subungual onychomycosis: A randomized, multicenter, double-blind, vehicle-controlled phase 3 study. J Am Acad Dermatol. 2020 Jun 22:S0190-9622(20)31153-1. https://doi.org/10.1016/j.jaad.2020.06.055. Epub ahead of print.
142. Ghannoum M, Isham N, Herbert J, Henry W, Yurdakul S. Activity of TDT 067 (terbinafine in Transfersome) against agents of onychomycosis, as determined by minimum inhibitory and fungicidal concentrations. J Clin Microbiol. 2011;49(5):1716–20.

143. Costa-Orlandi CB, Mordorski B, Baltazar LM, Mendes-Giannini MJS, Friedman JM, Nosanchuk JD, et al. Nitric oxide releasing nanoparticles as a strategy to improve current onychomycosis treatments. J Drugs Dermatol. 2018;17(7):717–20.
144. Finnen MJ, Hennessy A, McLean S, Bisset Y, Mitchell R, Megson IL, et al. Topical application of acidified nitrite to the nail renders it antifungal and causes nitrosation of cysteine groups in the nail plate. Br J Dermatol. 2007;157(3):494–500.
145. Weller R, Ormerod AD, Hobson RP, Benjamin NJ. A randomized trial of acidified nitrite cream in the treatment of tinea pedis. J Am Acad Dermatol. 1998;38(4):559–63.
146. Zhao Y, Lim J, Xu J, Yu J-H, Zheng W. Nitric oxide as a developmental and metabolic signal in filamentous fungi. Mol Microbiol. 2020;113(5):872–82.
147. Lipner SR, Friedman G, Scher RK. Pilot study to evaluate a plasma device for the treatment of onychomycosis. Clin Exp Dermatol. 2017;42(3):295–8.
148. Flores FC, Chiu WS, Beck RCR, da Silva CB, Delgado-Charro MB. Enhancement of tioconazole ungual delivery: combining nanocapsule formulation and nail poration approaches. Int J Pharm. 2018;535(1–2):237–44.
149. Gupta AK, Versteeg SG, Shear NH, Piguet V, Tosti A, Piraccini BM. A practical guide to curing nychomycosis: how to maximize cure at the patient, organism, treatment, and environmental level. Am J Clin Dermatol. 2019;20(1):123–33.
150. How common is relapse or reinfection in onychomycosis? [Internet]. https://www.medscape.com/answers/1105828-22690/how-common-is-relapse-or-reinfection-in-onychomycosis. Cited 28 Jan 2020.
151. Scher RK, Baran R. Onychomycosis in clinical practice: factors contributing to recurrence. Br J Dermatol. 2003;149(Suppl 65):5–9.
152. Gupta AK, Elewski BE, Rosen T, Caldwell B, Pariser DM, Kircik LH, et al. Onychomycosis: strategies to minimize recurrence. J Drugs Dermatol. 2016;15(3):279–82.
153. Tosti A, Elewski BE. Onychomycosis: practical approaches to minimize relapse and recurrence. Skin Appendage Disord. 2016;2(1–2):83–7.
154. Lipner SR, Scher RK. Onychomycosis: treatment and prevention of recurrence. J Am Acad Dermatol. 2019;80(4):853–67.
155. Gupta AK, Studholme C. Novel investigational therapies for onychomycosis: an update. Expert Opin Investig Drugs. 2016;25(3):297–305.
156. Yin Z, Xu J, Luo D. A meta-analysis comparing long-term recurrences of toenail onychomycosis after successful treatment with terbinafine versus itraconazole. J Dermatolog Treat. 2012;23(6):449–52.
157. Gupta AK, Cooper EA, Paquet M. Recurrences of dermatophyte toenail onychomycosis during long-term follow-up after successful treatments with mono- and combined therapy of terbinafine and itraconazole. J Cutan Med Surg. 2013;17(3):201–6.
158. Tosti A, Piraccini BM, Stinchi C, Colombo MD. Relapses of onychomycosis after successful treatment with systemic antifungals: a three-year follow-up. Dermatology (Basel). 1998;197 (2):162–6.
159. Sigurgeirsson B, Olafsson JH, Steinsson JB, Paul C, Billstein S, Evans EGV. Long-term effectiveness of treatment with terbinafine vs itraconazole in onychomycosis: a 5-year blinded prospective follow-up study. Arch Dermatol. 2002;138(3):353–7.
160. Rosen T, Gold LS. Antifungal drugs for onychomycosis: efficacy, safety, and mechanisms of action. Semin Cutan Med Surg. 2016;35(3, Suppl 3):51–5.
161. Gupta AK, Baran R. Ciclopirox nail lacquer solution 8% in the 21st century. J Am Acad Dermatol. 2000;43(4 Suppl):S96–102.
162. Monod M, Méhul B. Recent findings in onychomycosis and their application for appropriate treatment. J Fungi (Basel). 2019;5(1):20.
163. Pfaller MA. Antifungal drug resistance: mechanisms, epidemiology, and consequences for treatment. Am J Med. 2012;125(1 Suppl):S3–13.
164. Jabra-Rizk MA, Falkler WA, Meiller TF. Fungal biofilms and drug resistance. Emerg Infect Dis. 2004;10(1):14–9.

165. Gianni C, Cerri A, Crosti C. Non-dermatophytic onychomycosis. An underestimated entity? A study of 51 cases. Mycoses. 2000;43(1–2):29–33.
166. Pierce CG, Thomas DP, López-Ribot JL. Effect of tunicamycin on *Candida albicans* biofilm formation and maintenance. J Antimicrob Chemother. 2009;63(3):473–9.
167. Gupta AK, Daigle D, Carviel JL. The role of biofilms in onychomycosis. J Am Acad Dermatol. 2016;74(6):1241–6.
168. Nusbaum AG, Kirsner RS, Charles CA. Biofilms in dermatology. Skin Therapy Lett. 2012;17 (7):1–5.
169. Vila TVM, Rozental S, de Sá Guimarães CMD. A new model of *in vitro* fungal biofilms formed on human nail fragments allows reliable testing of laser and light therapies against onychomycosis. Lasers Med Sci. 2015;30(3):1031–9.
170. Andes D, Nett J, Oschel P, Albrecht R, Marchillo K, Pitula A. Development and characterization of an *in vivo* central venous catheter *Candida albicans* biofilm model. Infect Immun. 2004;72(10):6023–31.
171. Song JW, Shin JH, Kee SJ, Kim SH, Shin MG, Suh SP, et al. Expression of *CgCDR1*, *CgCDR2*, and *CgERG11* in *Candida glabrata* biofilms formed by bloodstream isolates. Med Mycol. 2009;47(5):545–8.
172. Gupta AK, Taborda VBA, Taborda PRO, Shemer A, Nakrieko K-A. High prevalence of mixed infections in global onychomycosis. In: 2020 American Academy of Dermatology annual meeting, March 20, Denver, Colorado.
173. Ogasawara Y, Hiruma M, Muto M, Ogawa H. Clinical and mycological study of occult tinea pedis and tinea unguium in dermatological patients from Tokyo. Mycoses. 2003;46 (3–4):114–9.
174. Tosti A, Hay R, Arenas-Guzmán R. Patients at risk of onychomycosis—risk factor identification and active prevention. J Eur Acad Dermatol Venereol. 2005;19(Suppl 1):13–6.
175. Avner S, Nir N, Henri T. Combination of oral terbinafine and topical ciclopirox compared to oral terbinafine for the treatment of onychomycosis. J Dermatolog Treat. 2005;16 (5–6):327–30.
176. Lecha M. Amorolfine and itraconazole combination for severe toenail onychomycosis; results of an open randomized trial in Spain. Br J Dermatol. 2001;145(Suppl 60):21–6.
177. Cohen PR, Scher RK. Topical and surgical treatment of onychomycosis. J Am Acad Dermatol. 1994;31(3, Part 2):S74–7.
178. Gupta AK, Mays RR, Versteeg SG, Piraccini BM, Takwale A, Shemer A, et al. Global perspectives for the management of onychomycosis. Int J Dermatol. 2019;58(10):1118–29.
179. Sakamoto M, Sugimoto N, Kawabata H, Yamakawa E, Kodera N, Pillai R, et al. Transungual delivery of efinaconazole: its deposition in the nail of onychomycosis patients and *in vitro* fungicidal activity in human nails. J Drugs Dermatol. 2014;13(11):1388–92.
180. Hold KM, Sanders V, Bu W, Baker SJ, Plattner JJ, Maples KR, et al. Nail penetration and nail concentration of AN2690, a novel broad-spectrum antifungal agent in development for the topical treatment of onychomycosis [Internet]. Palo Alto, CA: Anacor Pharmaceuticals Inc.. http://www.anacor.com/news/posters/aaps_2006_final.pdf.

Are Natural Products an Alternative Therapy for Dermatophytosis?

Mónica Zuzarte, Graciliana Lopes, Eugénia Pinto, and Lígia Salgueiro

Abstract

Dermatophytosis is an important health concern with an increasing negative impact due to unsuccessful therapeutic approaches. Natural products have emerged as promising alternative/complementary agents, due to long traditional uses and increasing scientific recognition. Current knowledge on the antidermatophytic activity of natural products, mainly plant extracts with an up-

M. Zuzarte
Faculty of Medicine, University of Coimbra, Coimbra Institute for Clinical and Biomedical Research (iCBR), Health Sciences Campus, Azinhaga de Santa Comba, Coimbra, Portugal

University of Coimbra, Center for Innovative Biomedicine and Biotechnology (CIBB), Health Sciences Campus, Azinhaga de Santa Comba, Coimbra, Portugal

Clinical Academic Centre of Coimbra (CACC), Coimbra, Portugal

G. Lopes
CIIMAR, Interdisciplinary Centre of Marine and Environmental Research of the University of Porto, Terminal de Cruzeiros do Porto de Leixões, Matosinhos, Portugal

Department of Biology, FCUP, Faculty of Sciences, University of Porto, Rua do Campo Alegre, Porto, Portugal

E. Pinto (✉)
CIIMAR, Interdisciplinary Centre of Marine and Environmental Research of the University of Porto, Terminal de Cruzeiros do Porto de Leixões, Matosinhos, Portugal

Microbiology Service, Biological Sciences Department, Faculty of Pharmacy, University of Porto, Porto, Portugal
e-mail: epinto@ff.up.pt

L. Salgueiro
Chemical Engineering Department of FCTUC, Chemical Process Engineering and Forest Products Research Centre (CIEPQPF), Rua Sílvio Lima, Coimbra, Portugal

Faculty of Pharmacy, University of Coimbra, Health Sciences Campus, Azinhaga de Santa Comba, Coimbra, Portugal

J.-P. Bouchara et al. (eds.), *Dermatophytes and Dermatophytoses*,
https://doi.org/10.1007/978-3-030-67421-2_22

to-date status on their activity, mechanism of action, developed pharmaceutical formulations, and effectiveness in animal models of infection and patients with dermatophytosis is presented. Although the information available on this matter points out an undeniable potential of these compounds, their usage by the pharmaceutical industry and medical community remains a challenge.

Keywords

Dermatophytosis therapy · Dermatophytes · Natural products · Non-volatile natural extracts · Volatile natural extracts

1 Introduction

Dermatophytosis is a serious public health concern that affects ca. 20–25% of the world population [1] and has relevant implications in the healthcare costs, especially in poor developing countries, where mycoses appear endemically [2]. These infections are generally non-invasive, but in immunocompromised patients they may progress to life-threatening disseminated infections. Moreover, pediatric and geriatric populations seem to be more susceptible and an increased number of infections has been observed in recent years [3, 4].

Although conventional antifungals are available, including azoles, allylamines, and morpholine derivatives, several side effects contribute to therapy failure. Beyond that, dermatophytosis is often refractory to therapy, with frequent relapses and resistance [5], thus stressing out the need of new effective therapeutic strategies.

Natural products have traditionally been used during centuries for their antifungal properties, including antidermatophytic effects [6] with several studies pointing out the huge potential of plant extracts and their isolated compounds, as stated in different reviews [7, 8]. Over the years, several methodologies including in vitro and in vivo approaches have been developed to evaluate and/or validate the antidermatophytic activity of these extracts/compounds. In addition, several cytotoxicity tests have also been used to assess their safety profile and different pharmaceutical formulations developed to improve local applications. Nevertheless, clinical trials are still lacking, thus hampering the use of these natural products by the pharmaceutical industries.

Despite the documented effectiveness of natural products, several limitations continue to discourage their use in the clinic. One of the main limitations is the difficulty to ensure the access and adequate supply of these products by the pharmaceutical industry [9]. Indeed, natural products are generally complex and difficult to synthesize, being found in low quantities in natural sources and with frequent chemical variations, that can compromise their bioactivity [10].

The present manuscript provides an updated review on the current knowledge on the use of natural products for treatment of dermatophytosis, with a major focus on plant extracts and their main compounds. First, a brief summary on dermatophytes and dermatophytosis is provided followed by a systematized revision on the antidermatophytic effect of both non-volatile and volatile extracts and isolated metabolites with reference to putative mechanisms of action. A final section addresses in vivo validations in animal models of infection and patients diagnosed with dermatophytosis as well as pharmaceutical formulations that have been developed over the years.

2 Dermatophytes and Dermatophytosis

Dermatophytes are classified according to their main habitat or host as geophilic (found in decomposing keratinized structures), zoophilic (present in animals), or anthropophilic (occurring in humans) [11]. These aerobic fungi invade and infect keratinized layers of the skin, hair, and nails, thus causing dermatophytosis [12].

These infections may vary from mild to very intense, depending on several factors such as site of infection, virulence of the infecting strain, host reaction to the metabolic products produced by the fungi, as well as environmental factors [13, 14]. Dermatophytosis, also called ringworm infection, is caused by fungi from the genera *Trichophyton*, *Microsporum*, *Nannizzia*, and *Epidermophyton*. The main symptoms of these infections include skin scaling, gray loops in the skin, hair loss and loosening, and nail deformities [15].

Diagnosis of dermatophytosis is quite difficult, with some infections presenting symptoms similar to other clinical conditions [16]. For an accurate diagnosis, mycological analyses, biochemical, and physiological tests and/or molecular approaches may be required [17]. Currently available treatments include both oral and topical formulations, the most common being antifungal azoles (e.g., clotrimazole, miconazole, econazole, oxiconazole, tioconazole) and allylamines (e.g., terbinafine and naftifine). Topical formulations tend to cause mild skin reactions at the site of application [18] and oral antifungals are responsible for major side effects including hepatotoxicity, neurotoxicity, nephrotoxicity, hematological reactions, and rare skin problems like Stevens–Johnson syndrome [19]. In addition, drug interactions have been reported, which may compromise antifungals effectiveness [20]. Essentially, dermatophytosis remains challenging to eradicate due to frequent recurrence, antifungal resistance, long treatment required, and side effects of most antifungal drugs. To increase efficacy, some approaches combine oral and topical formulations or antifungals with different mechanisms of action. More recently, the combination of conventional antifungals with natural products, such as essential oils, has also shown promising results [21].

3 Natural Products with Antidermatophytic Activity

Plants represent a huge source of bioactive molecules due to their almost limitless ability to synthesize compounds of different functional groups, including phenolic compounds, alkaloids, and terpenoids. Indeed, plants and their extracts have long been used in developing countries to treat several disorders and, in industrialized healthcare systems, have become increasingly recognized mainly as complementary and/or alternative medicines [22]. The huge potential of plant extracts is undeniable, with several scientific studies validating many of the traditional uses ascribed to them. Nevertheless, it is difficult to attribute the biological effect to a single compound or compound class, since the final biological activity is frequently the result of synergisms and/or antagonisms between the different compounds that make up the extract [23]. Despite this, current research on medicinal plants still under-appreciates the value of complex mixtures in comparison to single isolated constituents [24].

Regarding fungal infections, and in particular dermatophytosis, natural products have emerged as promising therapeutic candidates. The majority of these studies focused on plant extracts but, more recently, marine organisms have also emerged as good candidates for the development of new and innovative drugs [25]. Overall, these extracts can inhibit the growth of pathogens and some have fungicidal effects. A considerable number of studies have assessed the antidermatophytic activity of plant extracts with only some studies exploring more in-depth the potential of active isolated compounds [22]. For this purpose, several in vitro susceptibility tests have been applied, the most common being the agar-based disk diffusion assay, the broth dilution method, and the vapor phase test [26]. Antifungal susceptibility testing of dermatophytes is not easy to perform due to their slow growth rates and different phenotypic presentations, depending on the strain and testing conditions. It is also difficult to compare the results obtained by different research groups, since the minimal inhibitory and minimal fungicidal concentrations (MIC and MFC) can vary according to the selected test and the culture conditions [27]. Moreover, different approaches have also been used to disclose possible mechanisms of action, namely addressing molecular targets of conventional antifungals. These include enzymes and other molecules involved in fungal cell wall synthesis, cell membrane synthesis, DNA synthesis, and mitosis [28]. Finally, in vivo models have been developed to validate these effects. In the following section, the antidermatophytic potential of natural extracts, mainly plant extracts, separated into non-volatile and volatile extracts, and some isolated compounds is presented. In addition, putative mechanisms of action underlying some of the reported effects are also mentioned.

4 Non-volatile Natural Extracts and Compounds

Antifungal screening assays are very common with associated phytochemical analysis that point out possible active compounds. Recently, a review highlighted the most effective medicinal plants for dermatophytosis in traditional medicine as *Azadirachta indica*, *Capparis spinosa*, *Anagallis arvensis*, *Juglans regia*, *Inula viscosa*, *Phagnalon rupestre*, *Plumbago europaea*, *Ruscus aculeatus*, *Ruta chalepensis*, *Salvia fruticosa*, *Artemisia judaica*, *Ballota undulate*, *Cleome amblyocarpa*, *Peganum harmala*, *Teucrium polium*, *Aegle marmelos*, *Artemisia sieberi*, *Cuminum cyminum*, *Foeniculum vulgare*, *Heracleum persicum*, *Mentha spicata*, *Nigella sativa*, and *Rosmarinus officinalis*. In some of these plants, the presence of phenolic compounds, including flavonoids, tannins, and anthocyanins, was associated with their antifungal effect [29]. Several studies tend to access the antifungal effect of non-volatile extracts. Regarding dermatophytes, the alcoholic, hydro-alcoholic, chloroform, and hexane extracts are, in general, more active than the aqueous ones [7, 30]. For example, Svetaz et al. [31] screened 327 extracts, obtained from plants from different Latin American countries and used in traditional medicine, against the dermatophytes *Trichophyton rubrum*, *Trichophyton mentagrophytes*, *Epidermophyton floccosum*, and *Nannizzia gypsea* (formerly *Microsporum gypseum*) and showed that 40% of the ethanolic extracts were active against the tested fungi. Furthermore, the methanol, ethyl acetate, and acetone extracts of the marine sponge *Sigmadocia carnosa* were tested against the same species and the methanol extract, characterized by the presence of terpenoids, steroids, alkaloids, saponins, and glycosides, was the most active [32]. More recently, 48 extracts, including hexane, chloroform, ethyl acetate, ethanol, methanol, and water extracts, obtained from plants and algae, were assessed against *T. rubrum*, *Trichophyton interdigitale,* and *Trichophyton tonsurans*. Overall, the plant extracts exhibited stronger antidermatophytic activity than the algae ones. The ethanol and methanol extracts from the leaves of *Rhapis excelsa*, and the methanol extracts from the leaves of *Syzygium myrtifolium* were highly active. It is important to underline that in the latter the water extracts were also very effective [33].

Regarding isolated compounds, phenolic derivatives, alkaloids, and saponins are the most promising antidermatophytic agents. Therefore, a more detailed review on extracts with high contents of these compounds as well as active isolated compounds is discussed below and presented in Table 1. For a better understanding, Table 1 only includes studies that test the antidermatophytic effects of a maximum of three compounds, with reference to the plant source, dermatophyte species/strains tested, and antifungal activity, by indicating the respective MIC values, when available. More exhaustive studies, evaluating more than three compounds, are discussed in the manuscript, with most attention being given to studies or reviews performed in the last years. In addition, chemical structures of bioactive compounds, one from each chemical class of phenolic derivatives, as well as an example of an alkaloid and a saponin, are schematized in Fig. 1.

4.1 Phenolic Compounds

Phenolic compounds are a large class of plant secondary metabolites, characterized by the presence of, at least, one hydroxyl group associated with an aromatic ring. These secondary metabolites are the most abundant in plants, where they normally occur as esters or glycosides [61]. In plants, these compounds can act as antioxidants, structural polymers, attractants, UV screens, signal compounds, and in the defense response. Two basic biosynthetic pathways are generally involved: the shikimic acid pathway, responsible for the synthesis of the majority of polyphenols in plants, and the malonic acid pathway, with great importance in fungi and bacteria [61]. These compounds range from simple structures with one aromatic ring to highly complex polymeric compounds such as lignins and tannins. Several classifications have been proposed for the division of phenolic compounds, those

Table 1 Antidermatophytic compounds isolated from bioactive non-volatile natural extracts

Compound class	Bioactive compound	Source (extract)	Tested species and antifungal effect (MIC)	Antifungal assay	Ref.
Coumarins	Aurapten (7-geranyloxycoumarin)	*Baccharis darwinii* (petroleum ether extract)	*N. gypsea*, *T. rubrum,* and *T. mentagrophytes* (> 250 μg/mL)	Microbroth dilution	[34]
	Daphnoretin	*Loeselia mexicana* (ethanolic extract)	*T. mentagrophytes* and *T. rubrum* (25 μg/mL)	Modified agar dilution	[35]
	5′-Hydroxy aurapten (anisocoumarin H)	*Baccharis darwinii* (petroleum ether extract)	*N. gypsea*, *T. rubrum* and *T. mentagrophytes* (62.5 μg/mL)	Microbroth dilution	[34]
	4-Methoxy-1-methyl-3-(2′*S*-hydroxy-3′-ene butyl)-2-quinolone	*Skimmia laureola* (ethanolic extract)	*M. canis* (200 μg/mL)	Agar tube dilution	[36]
	5′-Oxoaurapten (diversinin)	*Baccharis darwinii* (petroleum ether extract)	*N. gypsea*, *T. rubrum* and *T. mentagrophytes* (15.6 μg/mL)	Microbroth dilution	[34]
	Scopoletin	*Loeselia mexicana* (ethanolic extract)	*T. mentagrophytes* and *T. rubrum* (12.5 μg/mL)	Modified agar dilution	[35]
	Ulopterol	*Skimmia laureola* (ethanolic extract)	*M. canis* (200 μg/mL)	Agar dilution	[36]
	Umbelliferone	*Loeselia mexicana* (ethanolic extract)	*T. mentagrophytes* and *T. rubrum* (25 μg/mL)	Modified agar dilution	[35]
Quinones	Damnacanthal	*Saprosma fragrans* (ethanolic extract)	*T. mentagrophytes* (1.56 μg/mL)	n.i.	[37]
	3,4-Dihydroxy-1-methoxy anthraquinone-2-carboxaldehyde	*Saprosma fragrans* (ethanolic extract)	*T. mentagrophytes* (12.5 μg/mL)	n.i.	[37]
	Embelin	*Oxalis erythrorhiza* (dichloromethane extract)	*E. floccosum*, *M. canis*, *T. mentagrophytes* and *T. rubrum* (50 μg/mL); *N. gypsea* (100 μg/mL)	Agar dilution	[38]

	Lawsone (naphthoquinone)	Commercial	*N. gypsea* (62.5 μg/mL); *T. rubrum* (125 μg/mL); *T. mentagrophytes* (250 μg/mL)	Modified agar dilution	[39]
	Lawsone methyl ether (naphthoquinone)	Synthesized	*N. gypsea* (3.9 μg/mL); *T. mentagrophytes* and *T. rubrum* (7.8 μg/mL)	Modified agar dilution	[39]
	Methylene-3,3′-bilawsone (naphthoquinone)	Synthesized	*N. gypsea* (500 μg/mL); *T. mentagrophytes* and *T. rubrum* (> 500 μg/mL)	Modified agar dilution	[39]
	Thymoquinone (naphthoquinone)	Commercial	*E. floccosum* and *M. canis* (125 μg/mL); *T. interdigitale*, *T. mentagrophytes* and *T. rubrum* (250 μg/mL)	Agar diffusion	[40]
			M. canis (62 μg/mL); *N. gypsea* and *T. mentagrophytes* (125 μg/mL)	Disk diffusion and macrobroth dilution	[41]
Flavonoids	Chrysin	Commercial	*N. gypsea*, *T. mentagrophytes,* and *T. rubrum* (250 μg/mL)	Disk diffusion and macrobroth dilution	[42]
	Derrubon 5-methyl ether	*Derris eriocarpa* (methanolic extract)	*N. gypsea* and *T. mentagrophytes* (25 μg/mL); *T. terrestre* (50 μg/mL)	Microbroth dilution	[43]
	Flavone glycoside	*Vitex negundo* (ethanolic extract)	*T. mentagrophytes* (6.25 μg/mL)	Disk diffusion	[44]
	7-Hydroxy-4′,6-dimethoxy-isoflavone	*Myroxylon peruiferum* (ethanolic extract)	*T. rubrum* (156–1250 μg/mL, depending on the strain)	Microbroth dilution	[45]
	4′-Hydroxy-5,7-dimethoxy-6-(3-methyl-2-butenyl)-isoflavone	*Derris eriocarpa* (methanolic extract)	*N. gypsea* (12.5 μg/mL); *T. mentagrophytes* (25 μg/mL); *T. terrestre* (100 μg/mL)	Microbroth dilution	[43]
	Myricetin-3-O-β-allopyranoside	*Plinia cauliflora* (ethanolic extract)	*T. rubrum* (125 μg/mL); *M. canis* (250 μg/mL)	Microbroth dilution	[46]

(continued)

Table 1 (continued)

Compound class	Bioactive compound	Source (extract)	Tested species and antifungal effect (MIC)	Antifungal assay	Ref.
	Myricetin-3-O-β-galactopyranoside	*Plinia cauliflora* (ethanolic extract)	*T. rubrum* (125 μg/mL); *M. canis* (500 μg/mL)	Microbroth dilution	[46]
	Pinocembrin	Urban propolis (ethanolic extract)	*T. mentagrophytes* and *T. rubrum* (31.2 μg/mL); *N. gypsea* (125 μg/mL)	Microbroth dilution	[42]
	Orientin	*Piper solmsianum* var. *solmsianum* (methanolic extract)	*M. canis* (7 μg/mL); *T. mentagrophytes* and *T. rubrum* (8 μg/mL); *E. floccosum* and *N. gypsea* (9 μg/mL)	Agar dilution	[47]
	1-(3′,4′,5′-Trimethoxyphenyl)-2-methoxy-2-(4″-methoxyphenyl)-ethane-1-ol	*Derris eriocarpa* (methanolic extract)	*N. gypsea* and *T. terrestre* (100 μg/mL); *T. mentagrophytes* (> 100 μg/mL)	Microbroth dilution	[43]
	Mixture of quercetin-3-O-β-glucopyranoside and quercetin-3-O-β-galactopyranoside	*Plinia cauliflora* (ethanolic extract)	*M. canis* and *T. rubrum* (500 μg/mL)	Microbroth dilution	[46]
Lignans	Conocarpan	*Piper fulvescens* (dichloromethane extract)	*T. mentagrophytes* (8 μg/mL); *N. gypsea* (16 μg/mL)	Broth dilution	[48]
	Eupomatenoid 3	*Piper regnellii* (hydroalcoholic extract)	*T. rubrum* (50 μg/mL)	Microbroth dilution and spore germination inhibition	[49]
	Eupomatenoid 5	*Piper fulvescens* (dichloromethane extract)	*N. gypsea* and *T. mentagrophytes* (no inhibitory effects at concentrations < 256 μg/mL)	Broth dilution	[48]
	Eupomatenoid 5	*Piper regnellii* (hydroalcoholic extract)	*T. rubrum* (6.2 μg/mL)	Microbroth dilution and spore germination inhibition	[49]

	Eupomatenoid 6	*Piper fulvescens* (dichloromethane extract)	*N. gypsea* (0.5 μg/mL); *T. mentagrophytes* (1 μg/mL)	Broth dilution	[48]
	Honokiol	*Magnolia obovate* (ethanolic extract)	*E. floccosum*, *N. gypsea* ,and *T. mentagrophytes* (25 μg/mL)	Diameter of growth inhibitory zone and macrobroth dilution	[50]
	Magnolol	*Magnolia obovate* (ethanolic extract)	*E. floccosum* (25 μg/mL); *N. gypsea* and *T. mentagrophytes* (50 μg/mL)	Diameter of growth inhibitory zone and macrobroth dilution	[50]
	3′-Methyl-nordihydroguaiaretic acid	Urban propolis (methanolic extract)	*T. rubrum* (31.2 μg/mL); *N. gypsea* and *T. mentagrophytes* (62.5 μg/mL)	Microbroth dilution	[42]
	Nordihydroguaiaretic acid	Urban propolis (methanolic extract)	*N. gypsea*, *T. mentagrophytes,* and *T. rubrum* (62.5 μg/mL)	Microbroth dilution	[42]
Tannins	Dieckol	*Ecklonia cava* (marine brown alga) (methanolic extract)	*T. rubrum* (200 μM)	Microbroth dilution and viability fluorescence microscopy	[51]
	Punicalagin	*Punica granatum* (hydroalcoholic extract)	*T. rubrum* (62.5 μg/mL)	Microbroth dilution; disk diffusion; hyphal growth inhibition by fluorescence microscopy and conidial germination inhibition	[52]
Alkaloids	6,7-Dehydro-19β-hydroxyschizozygine	*Schizozygia coffaeoides* (dichloromethane extract)	*N. gypsea* (1.95 μg/mL); *T. interdigitale* (< 3.9 μg/mL); *T. tonsurans* (3.9 μg/mL); *E. floccosum* and *T. mentagrophytes* (> 1.95 μg/mL)	Microbroth dilution	[53]
	Glabradine	*Stephania glabra* (ethanolic extract)	*M. canis* and *N. gypsea* (25 μg/mL); *T. rubrum* (50 μg/mL)	Disk diffusion	[54]

(continued)

Table 1 (continued)

Compound class	Bioactive compound	Source (extract)	Tested species and antifungal effect (MIC)	Antifungal assay	Ref.
	Isoschizogaline	*Schizozygia coffaeoides* (dichloromethane extract)	*E. floccosum*, *N. gypsea*, *T. interdigitale*, *T. mentagrophytes*, and *T. tonsurans* (> 500 μg/mL)	Microbroth dilution	[53]
	Schizozygine	*Schizozygia coffaeoides* (dichloromethane extract)	*E. floccosum*, *N. gypsea*, *T. interdigitale*, *T. mentagrophytes*, and *T. tonsurans* (> 500 μg/mL)	Microbroth dilution	[53]
Saponins	Bivittoside-D	*Bohadschia vitiensis* (methanolic extract)	*T. mentagrophytes* (0.78 μg/mL)	Microbroth dilution	[55]
	CAY-1	*Capsicum frutescens* (n.i.)	*M. canis*, *T. mentagrophytes*, *T. rubrum*, and *T. tonsurans* (10–20 μg/mL)	Microbroth dilution and inhibition of metabolic activity (XTT)	[56]
	Flindersine	*Toddalia asiatica* (ethyl acetate extract)	*E. floccosum*, *T. mentagrophytes*, and *T. rubrum* (62.5 μg/mL)	Disk diffusion and microbroth dilution	[57]
	(25R)-Spirost-7-en-2α,3β,5α-triol-3-O-[α-L-rhamnopyranosyl (1→2)-[β-D-galactopyranosyl (1→3)] β-D-glucopyranoside	*Agapanthus africanus* (ethanolic extract)	*T. mentagrophytes* (15.6 μg/mL)	Inhibition of fungal growth and in vivo assays	[58]
	Sakurasosaponin	*Jacquinia flammea* (methanolic extract)	*N. gypsea* and *T. rubrum* (31.2 μg/mL); *T. mentagrophytes* (62.5 μg/mL)	Microbroth dilution	[59]
	Spirostanic saponin SC-2	*Solanum chrysotrichum* (n.i.)	*N. gypsea*, *T. mentagrophytes*, and *T. rubrum* (n.a.)	Ultrastructural changes	[60]

E. Epidermophyton, *M. Microsporum*, *N. Nannizzia*, *T. Trichophyton*, *n.i.* not identified, *n.a.* not available

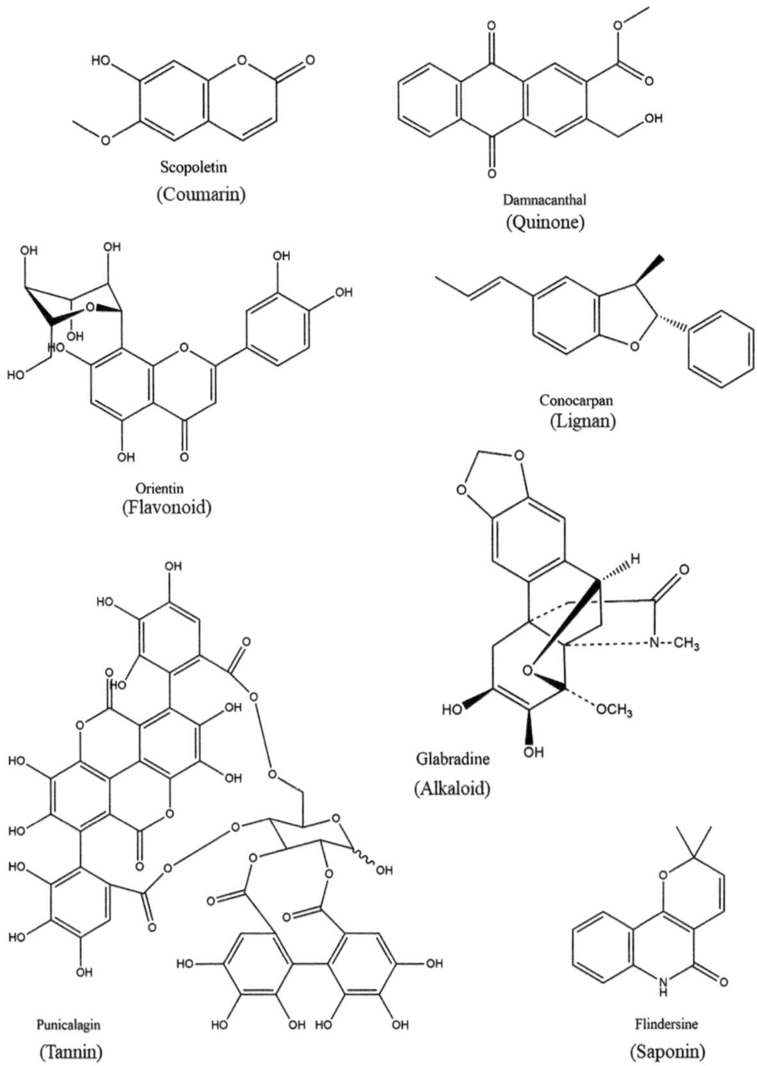

Fig. 1 Examples of non-volatile compounds, from different chemical classes, with antidermatophytic potential

based on structural criteria being the most adopted. Various bioactivities have also been associated with these compounds such as anti-aging, anti-inflammatory, anti-oxidant, antimicrobial, and anti-proliferative [62]. Those with the most promising antidermatophytic activity isolated from natural sources include coumarins, quinones, flavonoids, lignans, and tannins, and will be considered in more detail in the following sections.

4.1.1 Coumarins

The name coumarin derives from the French term *coumarou* that refers to *Dipteryx odorata* (*Coumarouna odorata*) seeds, one of the sources from which coumarins were first isolated in 1820. Coumarins can be classified as simple coumarins, furocoumarins, dihydrofurocoumarins, pyranocoumarins, phenylcoumarins, and biscoumarins [63]. Derivates of benzo-α-pyrone such as coumarin, aesculetin, umbelliferone, and scopoletin are common in plants both in the free state or as glycosides [64]. Plants from the Apiaceae family are the major source of coumarins, these compounds being present in the integument of seeds, fruits, flowers, roots, leaves, and stems. These compounds are involved in disease and pest resistance, as well as UV-tolerance. Usually, coumarins have a sweet aroma that resembles new-mown hay, making them very attractive compounds in perfumery. In the pharmaceutical industry, they are extensively used as precursor molecules in the synthesis of several synthetic anticoagulants [65]. Moreover, coumarins have other biological effects, including antitumor, anti-inflammatory, antioxidant, anti-HIV, and antimicrobial activities, and are used in the clinic for the treatment of rheumatoid arthritis [66]. In what concerns the antidermatophytic potential of these compounds, some studies are pointed out in Table 1, while others are discussed below.

Céspedes et al. [67] carried out an exhaustive study using several coumarins extracted from the aerial parts of *Tagetes lucida*, such as 7,8-dihydroxycoumarin, umbelliferone, scoparone, esculetin, 6-hydroxy-7-methoxycoumarin, herniarin, and scopoletin. Also, synthesized derivatives, namely 6,7-diacetoxycoumarin, 6-methoxy-7-acetylcoumarin, and 6-acetoxy-7-methoxycoumarin, were tested as well as commercially available compounds like 8-methoxypsoralen, 8-acetyl-7-hydroxycoumarin, 7,8-dihydroxy-6-methoxycoumarin, 6,7-dimethoxy-4-methylcoumarin, 5,7-dihydroxy-4-methylcoumarin, 4-hydroxycoumarin, 4-hydroxy-6,7-dimethylcoumarin, naringenin, glycoside-7-rhamnonaringin, and rutin. Overall, the dimethoxy compounds 6,7-dimethoxy-4-methylcoumarin and scoparone showed a strong effect against *T. mentagrophytes* with a total inhibition of the fungal growth at 125 μg/mL [67]. Several coumarins were also isolated from *Pterocaulon* species and assessed, alone or combined, for their antidermatophytic effect together with the hexane extracts from which they were obtained. Only the mixture of prenyletin and prenyletin-methyl-ether showed activity against *N. gypsea*, *T. mentagrophytes*, and *T. rubrum*, although lower than the hexane extract of *Pterocaulon balansae* (MIC = 31.25–625 μg/mL), thus suggesting a synergistic effect between the compounds of the hexane extracts and/or the presence of active minor compounds [68]. Coumarins complexed with metals seem to exert a higher antifungal effect when compared to uncomplexed coumarins. Indeed, synthesized coumarin-derived compounds and their transition metal complexes (cobalt, copper, nickel, and zinc) were screened against several dermatophyte strains, and metal complexes were more active. This effect is probably due to an increased lipophilic nature of the central metal atom that allowed a more efficient permeation through the fungus [69]. Importantly, the use of coumarin derivatives in antifungal therapy, which was patented in 2007, is based, in part, on the discovery that some glycosidic coumarin compounds (e.g., esculin) are effective in the treatment of onychomycosis and other dermatophytic infections. This effect is due to the ability of dermatophytes to metabolize glycosidic groups attached to coumarin compounds, converting the compounds into the active coumarin core, which is effective against the fungi [70]. Accordingly, it has been demonstrated that β-glucosidases produced by dermatophytes can hydrolyze, for example, esculin to esculetin, with concomitant antifungal activity. This is quite interesting since esculin has a good aqueous solubility, allowing to be formulated and delivered topically as an inactive prodrug in a water-based gel or cream. In this way, glycosylated coumarins may be delivered as inactive prodrugs and then converted into active antifungals in situ [71].

4.1.2 Quinones

Quinones are oxygen-containing compounds with a C_6–C_4 skeleton, in which a dione is conjugated with an aromatic nucleus. Quinones are classified according to their aromatic carbon skeleton into benzoquinones, naphthoquinones, anthraquinones, and phenanthrenequinones [72], naphthoquinones being the most common. Many African plants from the Clusiaceae, Bignoniaceae, Aloeaceae, and Rubiaceae families are known sources of these compounds [73]. Overall, quinones play an important role in plant defense and are very relevant for energy production, being involved in several oxidative processes such as photosynthesis [74]. Quinones, particularly naphthoquinones, are well known for their antibacterial, antifungal, and antitumoral activities [73]; however, their medicinal use remains limited in part because of their toxicity. Indeed, these compounds are known to be highly toxic because of several mechanisms of action that include arylation of the thiol groups of proteins, intercalation into DNA, and generation of free radicals [72].

Studies on the antidermatophytic activity of quinones, mainly naphthoquinones, are shown in Table 1 and highlighted below. These compounds are indeed very effective in inhibiting fungal growth, presenting low MIC values that in some cases are similar to that of antifungal drugs or even lower. For example, the naphthoquinone plumbagin, found in the extract of the stem bark of *Diospyros crassiflora*, which exhibited a huge antifungal activity, was pointed out as a promising antidermatophytic agent when compared to ketoconazole [75]. Also, lapachol and β-lapachone showed a low MIC value when compared to ketoconazole, thus emphasizing their antidermatophytic potential [76]. Furthermore, 2-benzenesulphinyl-(1,4)-naphthoquinone and 14 derivatives were assessed against *T. tonsurans*, with the synthesized molecules leading to an inhibition higher than 75%. Worth noticing, six derivatives were more effective than amphotericin B [77]. The vapor activity of quinone derivatives was also tested, with thymoquinone showing the most potent vapor activity, followed by menadione and ubiquinone. These results foresee interesting applications, for example, in shoes and head coverings or in the disinfection of rooms [78]. Naphthodianthrones such as hypericin, hypericin tetrasulphonic acid, and fagopyrin on *Microsporum canis* and *T. rubrum* showed distinct inhibitory effects that were correlated with structural differences between the compounds [79]. Similarly, an antifungal assay with 27 simple and heterocycle-fused prenyl-1,4-naphthoquinones and 1,4-anthraquinones against several human pathogenic strains, including *Trichophyton* species, emphasized that the most promising compounds were those with one or two chlorine atoms attached to the quinone ring [80].

4.1.3 Flavonoids

Flavonoids are compounds with a 15-carbon core (C_6–C_3–C_6). These compounds are formed by two aromatic rings (A and B) linked by a three-carbon chain that may also form a ring (C). The presence or absence of this third ring, as well as the position of ring B, allows flavonoids division into the following groups: flavanones, flavanonols, flavones, isoflavones, flavonols, flavanols, anthocyanidins and deoxyanthocyanidins, chalcones, dihydrochalcons, and aurones [81]. Flavonoids occur in several plant parts and are involved in plant protection against microorganisms, serve as UV filters and are also inducers of nitrogen-fixing bacteria in plants from the Fabaceae family. Flavonoids are well-known antioxidants, also recognized as anti-inflammatory, antiallergic, hepatoprotective, antithrombotic, antiviral, and anticarcinogenic agents [82]. Several antidermatophytic flavonoids have also been described, as pointed out in Table 1 and detailed next.

For example, epigallocatechin 3-O-gallate was tested for its antifungal effect on 35 clinical isolates of dermatophytes. All isolates were susceptible to this compound, with MIC values ranging from 0.5 to 16 μg/mL [83]. The flavonoids quercetin, *trans*-chalcone, ellagic acid, luteolin, galangin, and genistein were also assessed for their antifungal activity against *T. rubrum* and effects on fatty acid synthase (FAS) and ergosterol synthesis, two promising antifungal targets. Quercetin and *trans*-chalcone showed the most promising antifungal activity among the flavonoids tested. Although quercetin was a more effective FAS inhibitor, *trans*-chalcone showed lower MIC values and it simultaneously inhibited the synthesis of fatty acids and ergosterol [84]. Moreover, the flavonoids isolated from the twigs of *Dorstenia barteri*, namely isobavachalcone, 4-hydroxylonchocarpin, kanzonol C, and amentoflavone were tested against several strains of bacteria and fungi, including the dermatophytes. All compounds were quite effective with MIC values ranging from 1.2 to 39.1 μg/mL [85]. Hydroxychavicol isolated from *Piper betle* showed an effective inhibitory activity on 27 strains (MTCC/ATCC or clinical isolates) of the dermatophytes *E. floccosum*, *M. canis*, *N. gypsea*, *T. mentagrophytes*, and *T. rubrum* [86].

Approaches aiming to define structure/activity relations have also been performed. For example, 17 synthetic chalcones and analogs, based on the flavonoid xanthoxyline, were tested for their antifungal effects. The data show that neither the presence of a "xanthoxyline-like" substitution pattern nor a 2′-OH substituent on ring A are sufficient for the antifungal activity. Contrarily, the chalcone 3-(2-chlorophenyl)-1-(2′-hydroxy-4′,6′-dimethoxyphenyl)prop-2-en-1-one, with a Cl atom in the *ortho* position of ring B, was very effective against several isolates of *T. rubrum* [87]. In addition, it has been suggested, mainly in bacteria, that flavonoids can affect cell wall permeability and porins and are also able to complex with extracellular proteins [88].

4.1.4 Lignans

Lignans are dimer-based natural phenolic compounds derived from the oxidative coupling of phenylpropanoids [89]. According to their structure, lignans can be classified into different groups: proper lignans, neolignans, sesquilignans, dineolignans, norlignans, and hybrid lignans, with the first two being derived from the same phenylpropane (C_6–C_3) unit that makes up the lignin polymer [90]. In lignans, the phenylpropanoid units are bound by the C8–C8′ position of the aliphatic chain, while in neolignans the binding of the two phenylpropanoid units occurs at any position other than the C8 and C8′ [91]. Lignans can be found in different parts of vascular plants, wound resins being the major source of these compounds. Lignans tend to occur in low concentrations but in some cases such as flax and sesame seeds, they can reach values higher than 300 mg/100 g [90]. Examples of lignans found in foods include secoisolariciresinol (e.g., cashews), matairesinol (e.g., flax), lariciresinol (e.g., buckwheat), medioresinol (e.g., sesame seeds), pinoresinol (e.g., asparagus), syringaresinol (e.g., barley), and sesamin and sesamolin (sesame seeds) [90]. Lignans play relevant roles in plant structure by strengthening cell walls and are also involved in water transport and plant protection against pathogens [92]. The bioactive potential of these compounds is also known due to their anticancer, antioxidant, antimicrobial, anti-inflammatory, and immuno-suppressive activities [93]. Although the antifungal activity of lignans has been known for a long time, only a few studies focused on dermatophytes (Table 1). The crude extracts from the genus *Piper* have shown promising antidermatophytic effects that encouraged the isolation and identification of some bioactive compounds [48, 49]. For example, in *Piper regnelli*, the active chloroform fraction obtained from hydroalcoholic crude extract showed a similar activity to the isolated neolignan and eupomatenoid-5 against *T. rubrum* (Table 1). In another study three neolignans, conocarpan, euromatenoid 5, and euromatenoid 6, from the dichloromethane extracts of *Piper fulvescens* leaves (Table 1) were analyzed and a structure–activity relation was pointed out by suggesting that the absence of a methoxy group at the 3′ position of the phenyl-propenyl-benzofuran structure is relevant for the antidermatophytic activity, whereas the saturation of the 2–3 bond plays a role in selectivity [48]. Also, a guided fractionation of urban propolis from Argentina was performed, with eight fractions showing a high antifungal activity against dermatophytes (MIC = 16.0–62.5 μg/mL). From these, the lignans 3′-methyl-nordihydroguaiaretic acid and nordihydroguaiaretic acid were isolated and showed strong activities against *N. gypsea*, *T. mentagrophytes*, and *T. rubrum*, as shown in Table 1 [42, 94]. Regarding combined antidermatophytic effects of lignans or neolignans with antifungal drugs, only one report addressed the combined effect of nyasol isolated from *Anemarrhena asphodeloides* with several antifungals [95]. Regarding the mode of action of lignans, it is known that these compounds are able to inactivate microbial adhesins, enzymes, and other proteins, through non-specific forces (hydrogen bonding, covalent bonding, and hydrophobic interactions). They are also able to inhibit cell wall polymer synthesis or assembly, thus justifying their antimicrobial effects [96].

4.1.5 Tannins

The name *tannin* comes from the tanning process, in which tannins can be used due to their ability to bind and precipitate proteins. Tannins are a group of phenolic compounds that present a great structural diversity, being divided into three groups: hydrolyzable tannins, that consist of a central glucose molecule linked to molecules of gallic acid (gallotannins) or hexahydroxydiphenic acid (ellagitannins); condensed tannins, also called proanthocyanidins, that are polymeric flavans, not readily hydrolyzable; and complex tannins that consist of a 3-flavonol unit linked (C–C) to gallotannins or ellagitannins [81]. Many tannins are glycosides that can be found in different beverages, foods, and medicinal plants. Tannins also occur in brown seaweeds, where they consist of phlorotannins that result from the polymerization of phloroglucinol units [61, 97]. Overall, tannins are involved in defense mechanisms, namely plant protection against infections, herbivores, and oxidative stress. They are also used as protein precipitating agents in wine and beer production [61].

Tannins have long been recognized for their anti-inflammatory and antiseptic effects, being applied in diarrhea and tumors of the stomach or duodenum [61]. Although the biological activities of several extracts rich in tannins have been extensively explored, studies concerning the antidermatophytic potential of isolated compounds are quite limited, as already pointed out in a recent review [8]. Table 1 shows two studies on the antidermatophytic effect of isolated tannins and more extensive studies are addressed below.

Twenty-eight medicinal plants from the south of India were screened for their antifungal activity against *T. mentagrophytes*, *T. rubrum*, and *Trichophyton soudanense*, with the aqueous extract of *Terminalia chebula* showing inhibitory effects on the three dermatophytes, plausibly because of the presence of tannins. Nevertheless, chebulinic acid, a known tannin of *T. chebula,* was ineffective [98]. Regarding seaweed phlorotannins, purified extracts from *Cystoseira nodicaulis, C. usneoides,* and *Fucus spiralis* were screened for their antifungal activity, with *E. floccosum* and *T. rubrum* being the most susceptible among dermatophytes [99]. Despite the antidermatophytic potential of these compounds, their chemical synthesis is difficult, which constitutes a drawback for advancing to therapy [100].

The antifungal properties of tannins may be related to their ability to inactivate microbial adhesins, enzymes, and transport proteins, as well as their capacity to form complexes with polysaccharides [101]. Furthermore, phlorotannins seem to exert their antifungal effect by disturbing both the chitin levels in the cell wall and the respiratory chain function [102].

4.2 Alkaloids

Alkaloids are a highly diverse group of compounds with a nitrogen atom in a heterocyclic ring, organized into different classes according to their carbon skeletal structure [103]. There are two broad divisions of alkaloids: non-heterocyclic or atypical alkaloids, sometimes called "protoalkaloids" and heterocyclic or typical alkaloids, divided into 14 groups according to their ring structure [104]. These compounds are predominantly found in some plant families such as Solanaceae, Papaveraceae, Ranunculaceae, and Amaryllidaceae [105]. Alkaloids play a defensive role against herbivores and pathogens [103]. The biological effects of these metabolites are rather diverse, including antimalarial, anticancer, cholinomimetic, vasodilatory, antiarrhythmic, analgesic, antibacterial, and antihyperglycemic activities [106]. Several studies have also reported the antifungal effects of alkaloids. Regarding dermatophytes, relevant studies are mentioned in Table 1 or discussed next.

The ethanolic extract of the bark of the *Phellodendron amurense* trees, with a total alkaloid content of 7.58 mg/mL, showed an antifungal activity against *T. mentagrophytes* similar to that of clotrimazole [107]. These effects were also validated in rabbits, as discussed in the section In vivo *studies and pharmaceutical formulations*. The acetone extract of *Citrus reticulata* wood, together with six isolated compounds including a new acridone, citruscridone, was tested against *N. gypsea*. Only citruscridone and valencic acid were able to slightly inhibit the fungal growth [108]. In addition, the methanolic extract from *Glaucium oxylobum* showed good activity against *E. floccosum*, *M. canis*, *N. gypsea,* and *T. mentagrophytes*, the alkaloids dicentrine, glaucine, protopine, and α-allocryptopine being identified as the active compounds of the extract [109]. Also, the indole alkaloids erchinines A and B showed a significant antifungal effect against *T. rubrum*, comparable to that of griseofulvin [110]. Three bisbenzylisoquinoline alkaloids, namely cycleanine, cocsoline, and *N*-desmethylcycleanine, were extracted from *Albertisia villosa*, a subtropical medicinal plant widely used in traditional African medicines. Cycleanine was the most abundant (85%) and accounted for all the activities of the crude extract against *M. canis* [111]. The antifungal potential of alkaloids has been related to their ability to modify cell membrane permeability, impair mitochondrial function, produce oxidative stress, and modulate heme enzymes in the fungi [112].

4.3 Saponins

The name "saponin" derives from the Latin word "*sapo*" that means soap. Indeed, these compounds have foaming properties and behave like soap in water solutions [113]. Saponins are glycosylated compounds and are classified into steroidal or triterpenic saponins, according to the nature of their aglycone skeleton. In this way, steroidal saponins are steroidal aglycones consisting of a C_{27} spirostane skeleton and generally comprising a six-ring structure whereas triterpenic saponins are triterpenic aglycones, consisting of a C_{30} skeleton and comprising a pentacyclic structure [91, 114].

Steroidal saponins accumulate in crop plants including yam, alliums, asparagus, fenugreek, yucca, and ginseng. Triterpenoid saponins are mostly found in legumes such as soybeans, beans, and peas. These compounds are relevant in plant defense against pathogens, pests, and herbivores and are also involved in plant growth and development [115]. Regarding their biological effects, saponins are potent membrane permeabilizing agents with immunostimulatory, hypocholesterolemic, anticarcinogenic, anti-inflammatory, antimicrobial, antiprotozoan, molluscicidal, and antioxidant properties. Some saponins impair gut protein digestion and uptake of vitamins and minerals [115]. In what concerns the antifungal potential of saponins, most studies focused on yeasts but some reported promising antidermatophytic compounds, as shown in Table 1 and discussed next.

For example, steroidal and oleanane-type triterpenoid saponins from *Paullinia pinnata* were assessed against *E. floccosum*, *Trichophyton equinum*, *Microsporum audouinii*, *N. gypsea*, and *T. mentagrophytes*. The (3*β*,16*α*-hydroxy)-3-O-(2-′-acetamido-2′-deoxy-*β*-D-glucopyranosyl) echinocystic acid exhibited the best antidermatophytic activity (MIC = 6.25–25 μg/mL), the activity being in some cases comparable to that of griseofulvin [116]. Also, oleanane-type and lupine-type triterpenoid saponins, together with 12 analogs isolated from the extract of *Sapindus mukorossi* pulps were tested against *T. rubrum*, and the oleanolic acids 3-O-*β*-D-xylopyranosyl-(1→3)-*α*-L-rhamnopyranosyl-(1→2)-*α*-L-arabinopyranoside and 3-O-*α*-L-arabinopyranosyl-(1→3)-*α*-L-rhamnopyranosyl-(1→2)-*α*-L-arabinopyranoside showed promising inhibitory effects with an MIC80 value of 8 μg/mL (defined by the author as the minimum concentration required to inhibit 80% of fungal growth) [117]. Spirostanol saponins isolated from the leaves of *Solanum hispidum* showed antimycotic activity, with 6*α*-O-[*β*-D-xylopyranosyl-(1→3)-*β*-D-quinovopyranosyl]-(25*S*)-5*α*-spirostan-3*β*-ol being the most active against both *T. mentagrophytes* and *T. rubrum* [118]. Furthermore, the fractionation of the crude dichloromethane-methanol extract of *Polyscias fulva* led to the separation of 10 saponins that showed activity against several dermatophyte species, including *E. floccosum,* as well as several *Microsporum* (*M. audouinii, M. canis,* and *M. ferrugineum*), *Nannizzia* (*N. gypsea*), and *Trichophyton* (*T. equinum, T. mentagrophytes*, *T. rubrum*, *T. terrestre*, and *T. violaceum*) species [119]. A mixture of spirostanol saponins isolated from *Allium ursinum* was active at a concentration of 400 μg/mL against *M. canis* and *T. mentagrophytes* [120]. In a more recent assay, 21 new triterpenoid saponins (glinusopposides A–U) were isolated from *Glinus oppositifolius*, with glinusopposides B, Q, T, and U showing strong inhibitory effects against *N. gypsea* and *T. rubrum* [121].

Structure–activity relationships were also investigated by testing 19 saponins and three parent triterpenoids against *N. gypsea*, *T. interdigitale*, and *T. tonsurans*. Overall, aglycones were less effective than glycosides, monodesmosidic glycosides of medicagenic acid being the most active compounds, especially the 3-O-*β*-D-glucopyranoside [122]. In an earlier study, kalopanaxsaponins isolated from *Kalopanax pictus* were ineffective against *E. floccosum*, *N. gypsea,* and *T. mentagrophytes*. The authors suggested that the 28-O-glycoside instead of the 17-free carboxyl remove the activity, rather than the polarity of the molecule, due to an increment of sugar moiety [123]. In another study on synthetic saponins and cardiac glycosides, it was shown that both hemolytic and antifungal activities of diosgenyl (DIO)-glycosides (steroid saponins) are generally parallel to each other while almost all hemolytic methyl glycyrrhetinate (GAM)-glycosides (triterpenoid saponins) exhibit no antifungal effects. It seems that DIO-glycosides are able to penetrate into fungal cell walls, thus combining with ergosterol in the cell membranes [124].

The antifungal properties of saponins can be attributed to their ability to complex with sterols in fungal membranes, thereby causing pore formation and loss of membrane integrity. Aggregation of the saponin-sterol complexes in the membrane may be mediated by interactions with the sugar residues of the saponin, which can explain the loss of biological activity of aglycones [125].

5 Volatile Natural Extracts: Essential Oils (EOs), Terpenes, and Phenylpropanoids

The International Standard Organization on Essential Oils [126] and the European Pharmacopoeia [127] define an essential oil (EO) as the product obtained from plant raw material by hydrodistillation, steam distillation or dry distillation, or by a suitable mechanical process (for *Citrus* fruits). Essential oils are complex mixtures of volatile compounds, usually with a strong odor, rarely colored, soluble in organic solvents but insoluble in water. Monoterpenes ($C_{10}H_{16}$) and sesquiterpenes ($C_{15}H_{24}$) are usually the main group of compounds found in EOs. In some cases, phenylpropanoids such as anetol, apiol, and myristicin are also important compounds [128, 129]. In addition, some fatty acids and their esters or, more rarely, nitrogen and sulfur derivatives can also occur [128]. Terpenic compounds are synthesized via the mevalonate or deoxyxylulose phosphate biosynthetic pathway while phenylpropanoids are formed through the shikimate pathway [130, 131]. Essential oils are produced and accumulated in specialized secretory structures (secretory cells, secretory cavities, secretory ducts, glandular trichomes, epidermal cells, or osmophores) that minimize the risk of autotoxicity and can be found in all plant organs [132]. These secondary metabolites play a relevant role as signaling agents. Indeed, EOs are involved in the attraction of pollinators and seed dispersers, in plant protection against microorganisms, insects, and herbivores as well as in allelopathic interactions [129, 133]. Essential oil quality strongly depends upon several extrinsic (ecological and environmental aspects) and intrinsic (sexual, seasonal, ontogenetic, and genetic variations) factors [134, 135]. Genetic variations may result in the expression of different metabolic pathways and, consequently, quantitative and qualitative variations in oil composition may occur. Therefore, when significant differences are found, an intraspecific category (chemotype) is defined. Commercialized EOs require a standardized chemical profile according to analytical guidelines such as the European Pharmacopoeia and International Standard Organization (ISO), that allow quality control of both commercialized EOs and plants from which they are obtained [136]. Hyphenation of gas chromatography (GC) with mass spectrometry is the analytical technique mostly used for accurate compound identification [128]. More than 3000 EOs are known, but only about 150 have commercial relevance in global markets [137] with industrial-scale productions being achieved mainly for orange (*Citrus sinensis*), cornmint (*Mentha arvensis*), eucalyptus cineole-type (*Eucalyptus globulus*), citronella (*Cymbopogon winterianus*), peppermint (*Mentha* x *piperita*), lemon (*Citrus limon*), eucalyptus citronellal type (*Corymbia citriodora*), clove leaf (*Syzygium aromaticum*), cedarwood (*Juniperus virginiana*), aromatic litsea (*Litsea cubeba*), and lemongrass (*Cymbopogon citratus*) [136].

Aromatic plants and their EOs have been used for centuries by different cultures and for distinct purposes [138]. Nowadays these volatile extracts are valued in several industries, being applied in food products and beverages, perfumes, and cosmetics, and in supplements, nutraceuticals, and pharmaceuticals. It is of major relevance that many aromatic species are included in the Generally Recognized as Safe (GRAS) list fully approved by the US Food and Drug Administration (FDA) and Environmental Protection Agency (EPA, USA) for addition to food and beverages, thus highlighting their industrial relevance. Many of the referred applications rely on EOs organoleptic properties and/or bioactive potential. Indeed, several effects have been described for these compounds including fungicidal, bactericidal, anti-inflammatory, antioxidant, anticarcinogenic, and insecticidal. Importantly, EOs lipophilicity and small molecular size enable a direct passage through cell membranes (passive diffusion), thus avoiding limitations by the cell transport machinery [139]. These features make EOs very interesting for the development of innovative products, mainly in the medical field, since different routes of administration (topical, *per-os,* and inhalatory) can be considered.

One of the most studied biological effects of EOs and their isolated compounds is the antifungal effect, that has gained a renewed interest due to the increasing number of drug-resistant fungal isolates. For example, several studies have demonstrated the effectiveness of these compounds against dermatophytes. Table 2 summarizes the most relevant studies on the antidermatophytic activity of EOs performed over the last years, with reference to the main compounds present in the oil, order of the most sensitive species/strains (based on MIC values when available), and methodology used to assess the antifungal effect. Studies addressing more than three species or revisions were not included in the table and are referred below.

In 2016, Chaftar et al. [179] published the results of a large comparative study on the antifungal activity of 19 EOs (14 commercial and five isolated from plants collected in different Tunisian regions). Two clinical isolates were assessed, namely *T. mentagrophytes* and *T. rubrum*. Overall, the EO of *Cymbopogon citratus* was the most active followed by *Thymus vulgaris*, *Mentha* x *piperita*, *Citrus sinensis*, *Origanum vulgare*, *Cymbopogon martinii* var. *motia*, *Syzygium aromaticum* and *Pinus laricio.* Houël et al. [168] analyzed the antifungal effect of seven EOs collected in the French Guiana, namely *Achetaria guianensis*, *Cymbopogon citratus*, *Mikania micrantha*, *Otacanthus azureus*, *Piper hispidum*, *Protium heptaphyllum*, and *Vouacapoua americana.* The most active EOs were those of *C. citratus*, *O. azureus,* and *P. heptaphyllum* against *M. canis*, *N. gypsea,* and *T. mentagrophytes*. Interestingly, the MIC values recorded for *C. citratus* and *O. azureus* EOs against *T. mentagrophytes* (8 μg/mL) were similar to that of fluconazole. In another study, 20 commercial EOs were analyzed (*Boswellia sacra*, *Citrus bergamia*, *C. limon*, *C. medica*, *Cinnamomum zeylanicum*, *Eucalyptus globulus*, *Foeniculum vulgare*, *Helichrysum italicum*, *Illicium verum*, *Litsea cubeba*, *Mentha spicata*, *Myrtus communis*, *Ocimum basilicum*, *Origanum majorana*,

Table 2 Antidermatophytic activity of essential oils

Plant source (site of collection)	Main compounds	Tested species and antifungal effect (MIC)	Antifungal assay	Ref.
Angelica major (Portugal)	α-Pinene (21.8%) and *cis-β*-ocimene (30.4%)	*E. floccosum, M. canis, T. mentagrophytes,* and *T. rubrum* (0.32 μL/mL); *N. gypsea* and *T. interdigitale* (0.64 μL/mL); *T. verrucosum* (1.25 μL/mL)	Macrobroth dilution	[140]
Apium nodiflorum (Portugal and Sardinia)	*Sample 1*: Myristicin (29.1%), dillapiol (22.5%) and limonene (16.7%) *Sample 2*: Dillapiol (70.8%) and limonene (14.4%)	*Sample 1: M. canis* (0.04 μL/mL); *N. gypsea* (0.08 μL/mL); *E. floccosum, T. mentagrophytes, T. interdigitale,* and *T. rubrum* (0.16 μL/mL); and *T. verrucosum* (0.32 μL/mL) *Sample 2: M. canis* (0.04 μL/mL); *E. floccosum, N. gypsea* and *T. rubrum* (0.08 μL/mL); *T. mentagrophytes* and *T. interdigitale* (0.16 μL/mL); *T. verrucosum* (0.32 μL/mL)	Macrobroth dilution	[141]
Artemisia judaica (Jordan)	Piperitone (30.4%), camphor (16.1%) and ethylcinnamate (11.0%)	*E. floccosum, M. canis, N. gypsea, T. mentagrophytes, T. rubrum, T. interdigitale,* and *T. verrucosum* (0.64 μL/mL)	Macrobroth dilution	[142]
Azorella cryptantha (Argentina)	*Sample 1*: α-Pinene (21.9%) and α-thujene (12.5%) *Sample 2*: n.i. (22.7%) and α-pinene (9.6%)	*Sample 1: T. mentagrophytes* and *T. rubrum* (500 μg/mL); *N. gypsea* (1000 μg/mL) *Sample 2: N. gypsea, T. mentagrophytes,* and *T. rubrum* (125 μg/mL)	Microbroth dilution	[143]
Chenopodium ambrosioides (India)	*m*-Cymene (43.9%) and myrtenol (13.3%)	*T. rubrum* (350 ppm); *N. gypsea* (700 ppm)	Food poisoning technique	[144]
Commiphora molmol (Iran)	Furanoeudesma 1,3-diene (50.0%) and menthofuran (13.5%)	*M. canis* and *T. rubrum* (25 μg/mL); *N. gypsea* and *T. verrucosum* (50 μg/mL); *T. mentagrophytes* (100 μg/mL)	Food poisoning technique and microbroth dilution	[145]
Cuminum cyminum (India)	Cumin aldehyde (61.7%) and γ-terpinen-7-al (12.3%)	*N. gypsea, T. mentagrophytes, T. rubrum,* and *T. tonsurans* (n.a)	Growth inhibition	[146]

(continued)

Table 2 (continued)

Plant source (site of collection)	Main compounds	Tested species and antifungal effect (MIC)	Antifungal assay	Ref.
Curcuma longa (n.i)	Terpinolene (26.4%)	*E. floccosum*, *N. gypsea*, *T. mentagrophytes*, *T. rubrum*, and *T. violaceum* (n.a)	Food poisoning technique and mycelial growth inhibition	[147]
Cymbopogon citratus (Portugal)	Geranial (45.7%) and neral (32.5%)	*T. mentagrophytes* (200–400 μg/mL); *T. rubrum* (200–> 400 μg/mL, depending on the strain); *T. interdigitale* (> 400 μg/mL)	Microbroth dilution and effect on conidial germination and morphology	[147]
Cymbopogon martini (India)	*trans*-Geraniol (60.9%) and *β*-elemene (12.3%)	*T. rubrum* (150 ppm); *N. gypsea* (200 ppm)	Food poisoning technique	[144]
Daucus carota subsp. *carota* (Portugal)	Geranyl acetate (29.0%), *α*-pinene (27.2%) and 11 *α*H-himachal-4-en-1*β*-ol (9.2%)	*E. floccosum* and *T. rubrum* (0.32 μL/mL); *M. canis*, *N. gypsea*, *T. mentagrophytes*, *T. interdigitale,* and *T. verrucosum* (0.64 μL/mL)	Macrobroth dilution	[148]
Eremurus persicus (Iran)	Limonene (16.3%) and geranylgeraniol (15.2%)	*T. schoenleinii* (0.02–0.06%); *M. canis* (0.03–0.17%); *N. gypsea* (0.04–0.09%); *T. rubrum* (0.07–0.12%)	Microbroth dilution	[149]
Eucalyptus citriodora (Corymbia citriodora) (Algeria)	Citronellal (69.8%), citronellol (10.6%) and isopulegol (4.7%)	*M. canis* and *T. rubrum* (0.6 μL/mL); *T. mentagrophytes* (1.25 μL/mL); *N. gypsea* (5 μL/mL)	Disk diffusion; disk volatilization and agar dilution	[150]
Eucalyptus smithii (commercial)	1,8-Cineole (72.2%)	*T. rubrum* (62.5 μg/mL); *T. mentagrophytes* (125–> 1000 μg/mL, depending on the strain); *M. canis* and *N. gypsea* (500 μg/mL)	Microbroth dilution and scanning electron microscopy	[151]
Eugenia caryophyllus (n.i.)	Eugenol (71.8%), eugenyl acetate (18.4%) and *trans-β*-caryophyllene (7.5%)	*M. canis* (62.5–125 μg/mL); *N. gypsea* (125 μg/mL)	Microbroth dilution; inhibition of mycelial radial growth and inhibition of conidia germination	[152]
Foeniculum vulgare (Portugal)	*trans*-Anethole (47.0%), *α*-phellandrene (11.0%), *α*-pinene (10.1%) and fenchone (10.8%)	*M. canis* (0.64–1.25 μL/mL); *E. floccosum*, *T. rubrum* and *T. verrucosum* (1.25 μL/mL); *N. gypsea*, *T. mentagrophytes,* and *T. interdigitale* (2.5 μL/mL)	Macrobroth dilution	[153]

Foeniculum vulgare (China)	*trans*-Anethole (63.3%) and pinene (11.1%)	*T. tonsurans* (25 µg/mL); *N. gypsea* and *T. rubrum* (50 µg/mL); *T. mentagrophytes* (100 µg/mL)	Microbroth dilution; effect on mycelial growth and weight; spore germination; ergosterol content; ultrastructure and succinate dehydrogenase, malate dehydrogenase and ATPase activities	[154]
Hirtellina lobelii (Lebanon)	*α*-Bisabolol (34.5%), fokienol (12.0%) and *T*-muurolol (6.8%)	*T. tonsurans* (8 µg/mL); *T. soudanense* and *T. violaceum* (16 µg/mL); *T. mentagrophytes* (32 µg/mL); *T. rubrum* (64 µg/mL)	Microbroth dilution and checkerboard	[155]
Homalomena aromatica (India)	Linalool (62.5%)	*N. gypsea*, *T. mentagrophytes*, and *T. rubrum* (n.a)	Agar well diffusion and minimal inhibitory concentration	[156]
Juniperus communis subsp. *alpina* (Portugal)	Sabinene (26.2%), terpinen-4-ol (13.0%), *α*-pinene (12.9%) and limonene (10.4%)	*M. canis* and *T. rubrum* (0.32 µL/mL); *E. floccosum*, *N. gypsea*, and *T. mentagrophytes* (0.64 µL/mL); *T. interdigitale* (1.25 µL/mL); *T. verrucosum* (2.5 µL/mL)	Macrobroth dilution	[157]
Lavandula luisieri (Portugal)	Oxygenated necrodane derivative (36.0%) and 1,8-cineol (8.9%)	*T. mentagrophytes* (200–400 µg/mL); *T. rubrum* (200–> 400 µg/mL, depending on the strain); *T. interdigitale* (> 400 µg/mL)	Microbroth dilution and effect on conidial germination and morphology	[147]
Lavandula multifida (Portugal)	Carvacrol (42.8%) and *cis-β*-ocimene (27.4%)	*E. floccosum*, *M. canis*, *N. gypsea*, *T. mentagrophytes*, *T. interdigitale*, *T. rubrum*, and *T. verrucosum* (0.16 µL/mL)	Macrobroth dilution	[158]
Lavandula viridis (Portugal)	1,8-Cineole (34.5%), camphor (13.4%), *α*-pinene (9.0%) and linalool (7.9%)	*E. floccosum*, *M. canis*, *T. rubrum*, and *T. verrucosum* (0.32 µL/mL); *T. mentagrophytes* and *T. interdigitale* (0.32–0.64 µL/mL); *N. gypsea* (0.64 µL/mL)	Macrobroth dilution	[159]
Lippia alba (Brazil)	Linalool (66.0%)	*T. rubrum* (39 µg/mL); *E. floccosum* (156 µg/mL); *N. gypsea* (312 µg/mL)	Microbroth dilution; effect on peptidase/keratinase activities; BSA and keratin cleavage by sodium dodecyl sulfate polyacrylamide gel electrophoresis	[160]

(continued)

Table 2 (continued)

Plant source (site of collection)	Main compounds	Tested species and antifungal effect (MIC)	Antifungal assay	Ref.
Lonicera japonica (Republic of Korea)	*trans*-Nerolidol (16.3%) and caryophyllene oxide (11.2%)	*M. canis* (62.5–250 μg/mL, depending on the strain); *T. rubrum* (125–500 μg/mL, depending on the strain); *T. mentagrophytes* (500 μg/mL)	Food poisoning technique; agar dilution; spore germination and growth kinetics	[161]
Matricaria recutita (Iran)	Chamazulene (61.3%) and isopropyl hexadecanoate (12.7%)	*M. canis*, *N. gypsea*, *T. mentagrophytes*, *T. rubrum* and *T. tonsurans* (n.a.)	Disk diffusion and inhibition of fungal growth	[162]
Metasequoia glyptostroboides (Republic of Korea)	α-Pinene (29.5%)	*T. mentagrophytes* (62.5 μg/mL); *M. canis* (62.5–500 μg/mL, depending on the strain); *T. rubrum* (125–500 μg/mL, depending on the strain)	Disk diffusion; minimum inhibitory concentration; spore germination and growth kinetics (for *M. canis*)	[163]
Myrtus nivellei (Sahara)	1,8-Cineole (37.5%) and limonene (25.0%)	*E. floccosum*, *M. canis*, and *T. rubrum* (0.64 μL/mL); *N. gypsea*, *T. mentagrophytes*, *T. interdigitale*, and *T. verrucosum* (1.25 μL/mL)	Macrobroth dilution	[164]
Nandina domestica (Republic of Korea)	1-Indolizino carbazole (19.7%), 2-pentanone (16.4%) and mono phenol (12.1%)	*M. canis* and *T. mentagrophytes* (62.5–125 μg/mL, depending on the strain); *T. rubrum* (62.5–500 μg/mL, depending on the strain)	Disk diffusion and minimum inhibitory concentration	[165]
Oenanthe crocata (Portugal)	*trans-β*-Ocimene (31.3%), sabinene (29.0%) and *cis-β*-ocimene (12.3%)	*E. floccosum*, *M. canis*, *N. gypsea*, and *T. rubrum* (0.08 μL/mL); *T. mentagrophytes* and *T. interdigitale* (0.16 μL/mL); *T. verrucosum* (0.64–1.25 μL/mL)	Macrobroth dilution	[166]
Origanum majorana (Mexico)	*Colorless oil*: *allo*-Ocimene (28.5%), terpinen-4-ol (23.1%), thymol (16.3%) and *trans*-hydrate sabinene (15.3%) *Yellow oil:* Terpinen-4-ol (27.7%), thymol (24.6%) and *trans*-hydrate sabinene (10.5%)	*E. floccosum*, *M. canis*, *N. gypsea*, *T. mentagrophytes*, *T. rubrum*, and *T. tonsurans* (n.a.)	Disk diffusion	[167]

Otacanthus azureus (French Guiana)	*β*-Copaen-4-*α*-ol (23.3%) and *α*-humulene (10.6%)	*T. rubrum* (4 μg/mL); *N. gypsea* (31 μg/mL); *T. mentagrophytes* and *T. tonsurans* (62 μg/mL); *M. canis* (> 500 μg/mL)	Microbroth dilution and checkerboard	[168]
Rosmarinus officinalis (Mexico)	*Colorless oil*: Terpinen-4-ol (45.9%) and 1,8-cineole (23.8%) *Yellow oil*: Terpinen-4-ol (87.6%)	*E. floccosum, M. canis, N. gypsea, T. mentagrophytes, T. rubrum,* and *T. tonsurans* (n.a.)	Disk diffusion	[167]
Salvia multicaulis (Lebanon)	*Sample 1: trans*-Nerolidol (12.3%) *Sample 2: α*-Pinene (16.3%) and eucalyptol (16.3%)	*T. violaceum* (8 μg/mL); *T. soudanense* (32 μg/mL); *T. mentagrophytes, T. rubrum,* and *T. tonsurans* (64 μg/mL)	Microbroth dilution	[169]
Schinus lentiscifolius (Brazil)	*γ*-Eudesmol (12.8%) and elemol (10.5%)	*M. canis* (0.004–0.08 μg/mL, depending on the strain); *T. rubrum* (0.008 μg/mL); *N. gypsea* and *T. mentagrophytes* (0.016 μg/mL)	Microbroth dilution; time-kill assay (for *T. rubrum*); sorbitol protection assay; ergosterol effect; electron microscopy and checkerboard	[170]
Syzygium aromaticum (Commercial)	Eugenol (85.3%)	*M. canis* (0.08–0.16 μL/mL); *E. floccosum, N. gypsea, T. mentagrophytes,* and *T. rubrum* (0.16 μL/mL)	Macrobroth dilution	[171]
Thapsia villosa (Portugal)	Limonene (57.5%) and methyleugenol (35.9%)	*E. floccosum, M. canis, T. mentagrophytes,* and *T. rubrum* (0.64 μL/mL); *N. gypsea, T. interdigitale,* and *T. verrucosum* (1.25 μL/mL)	Macrobroth dilution	[172]
Thymus camphoratus (Portugal)	1,8-Cineole (15.5%), *α*-pinene (12.7%) and borneol (8.5%)	*E. floccosum, M. canis, N. gypsea, T. mentagrophytes,* and *T. rubrum* (570 μg/mL); *T. interdigitale* and *T. verrucosum* (1110 μg/mL)	Macrobroth dilution	[173]
Thymus carnosus (Portugal)	Borneol (29.0%), camphene (19.5%) and *α*-pinene (8.9%)	*T. rubrum* (570 μg/mL); *E. floccosum, M. canis, T. mentagrophytes,* and *T. interdigitale* (1110 μg/mL); *N. gypsea* and *T. verrucosum* (2230 μg/mL)	Macrobroth dilution	[173]
Thymus x viciosoi (Portugal)	Carvacrol (30.0%), thymol (18.0%) and *p*-cymene (19.0%)	*T. mentagrophytes* (0.08 μL/mL); *E. floccosum, M. canis, N. gypsea,* and *T. rubrum* (0.16 μL/mL)	Macrobroth dilution	[174]

(continued)

Table 2 (continued)

Plant source (site of collection)	Main compounds	Tested species and antifungal effect (MIC)	Antifungal assay	Ref.
Thymus vulgaris (Mexico)	*o*-Cymene (37.1%), *γ*-terpinene (20.0%) and thymol (17.0%)	*E. floccosum, M. canis, N. gypsea, T. mentagrophytes, T. rubrum,* and *T. tonsurans* (n. a)	Disk diffusion	[167]
Vitex agnus-castus (Italy)	*Leaves:* Bicyclogermacrene (16.2%) and *(E)-β*-farnesene (10.1%) *Flowers:* Bicyclogermacrene (9.7%) *Fruits: (E)-β*-Farnesene (14.4%) and bicyclogermacrene (14.0%)	*E. floccosum, M. canis* and *T. rubrum* (0.64 μL/mL); *N. gypsea* and *T. mentagrophytes* (1.25 μL/mL)	Macrobroth dilution	[175]
Vitex rivularis (Cameroon)	*Sample 1*: Germacrene D (12.6%) *Sample 2*: Germacrene D (20.6%)	*T. rubrum* (0.16 μL/mL); *E. floccosum, T. mentagrophytes,* and *N. gypsea* (0.32 μL/mL); *M. canis* (0.64 μL/mL)	Macrobroth dilution	[176]
Zataria multiflora (Commercial)	*Sample 1*: Thymol (25.8%) and carvacrol (34.4%) *Sample 2*: Thymol (38.0%) and carvacrol (30.6%) *Sample 3*: Thymol (41.2%) and carvacrol methyl ether (28.3%)	*Sample 1: N. gypsea* and *T. mentagrophytes* (0.03 μL/mL); *T. rubrum* (0.06 μL/mL); *M. canis* and *T. schoenleinii* (0.125 μL/mL) *Sample 2: T. mentagrophytes* and *T. rubrum* (0.03 μL/mL); *N. gypsea* (0.06 μL/mL); *T. schoenleinii* (0.125 μL/mL); *M. canis* (0.25 μL/mL) *Sample 3: N. gypsea, T. mentagrophytes* and *T. rubrum* (0.03 μL/mL); *T. schoenleinii* (0.06 μL/mL); *M. canis* (0.125 μL/mL)	Elastase production; microbroth dilution and inhibitory effects on mycelium growth	[177]
Ziziphora tenuior (Jordan)	Pulegone (46.8%), *p*-menth-3-en-8-ol (12.5%), isomenthone (6.6%) and 8-hydroxymenthone (6.2%)	*E. floccosum, T. rubrum* and *T. verrucosum* (0.64 μL/mL); *M. canis* (0.64–1.25 μL/mL); *N. gypsea, T. mentagrophytes,* and *T. interdigitale* (1.25 μL/mL)	Macrobroth dilution	[178]

E. Epidermophyton, *M. Microsporum*, *N. Nannizzia*, *T. Trichophyton*, *n.i.* not identified, *n.a.* not available

O. vulgare, *Pelargonium graveolens*, *R officinalis*, *Santalum album*, *Satureja montana*, and *Thymus serpyllum*) against clinical animal isolates of *M. canis*, *N. gypsea*, *Trichophyton erinacei*, *T. mentagrophytes*, and *T. terrestre*. The EOs of *T. serpyllum* (rich in thymol), *O. vulgare* (with high amounts of carvacrol), and *L. cubeba* (with high amounts of neral) were the most effective [180]. Furthermore, Ethiopian EOs were analyzed by Nasir et al. [181] with *Thymus schimperi* and *Cinnamomum zeylanicum* being more active against clinical isolates of *Microsporum* spp. and *Trichophyton* spp. In this study, no chemical analysis of the volatile extracts was conducted, thus disabling the identification of putative active compounds. In the following year, Sadhasivam et al. [182] studied another group of commercial EOs including *Boswellia serrata*, *Carum carvi*, *Coriandrum sativum*, *Cyperus scariosus*, *Elettaria cardamom*, *Ocimum sanctum*, *Piper nigrum*, *Syzygium aromaticum*, and *S. cumini*, with *B. serrata* EO showing the lowest MIC value, followed by *S. aromaticum* oil with maximum potency against *Trichophyton* spp. Then, the antimicrobial activity of the EOs of six conifers harvested in Lebanon, namely *Abies cilicica*, *Cedrus libani*, *Cupressus sempervirens*, *C. macrocarpa*, *Juniperus excelsa*, and *J. oxycedrus*, was also compared, but clinical isolates of *T. mentagrophytes*, *T. rubrum*, *T. soudanense*, *T. violaceum*, and *T. tonsurans* were sensitive to most of the tested EOs [183].

In a review published in 2011 on the antidermatophytic activity of EOs, Zuzarte et al. [26] highlighted the in vitro effectiveness of several EOs and suggested that the antifungal effect of EOs is not due to a single mechanism but results from the effects of different compounds on several cell targets. This could have a major health impact since fungal resistance is unlikely to occur [184]. Later, Flores et al. [185] gathered the information on the use of EOs for the treatment of onychomycosis and pointed out the following effective species: *Allium* spp., *Melaleuca* spp., *Thymus* spp., *Eucalyptus* spp., *Origanum* spp., and *Lavandula* spp. In the same year, Shokri [186] reviewed the potential of *Nigella sativa* EO against a considerable number of dermatophyte clinical isolates and showed a weak to moderate inhibitory effect of this extract. Nevertheless, the major compound present in the EO, thymoquinone, was pointed out as an active antidermatophytic compound, as referred previously in the non-volatile compound's section. More recently, Lopes et al. [8] summarized the existing knowledge on the antidermatophytic activity of natural products, with a special focus on their mechanism of action, available pharmaceutical formulations and effect on animal models of infection and on human dermatophytosis. In addition, the establishment of chemical composition/antifungal activity relationships was attempted, although this task is uneasy since the antifungal activity of the oil may be due to synergistic effects between different compounds, rather than to a single major compound. However, it seems evident that EOs rich in phenolic compounds, such as terpenic phenols (e.g., thymol and carvacrol) or phenylpropanoids (e.g., eugenol, cinnamaldehyde, and benzyl benzoate) are more active, thus suggesting that the presence of a phenol function in the molecules increases their antimicrobial effects [187]. In addition, strong antidermatophytic activities have also been positively correlated with the presence of other oxygenated monoterpenes. Indeed, Lima et al. [188] tested several EOs obtained from medicinal plants in the Central Andes of Argentina and showed that those rich in oxygenated monoterpenes, namely *Lippia integrifolia* and *Satureja parvifolia,* were more active against *N. gypsea, T. mentagrophytes,* and *T. rubrum.* Also, the presence of necrodane derivatives and 1,8-cineole in *Lavandula luisieri* EO and geranial and neral in *Cymbopogon citratus* EO enhanced their antidermatophytic effect [147].

Importantly, EO quality strongly depends on the origin of the plants and on the intraspecific variability that can compromise the antifungal potential. For example, Marongiu et al. [189] showed that two chemotypes of *Crithmum maritimum* with different concentrations of dillapiole presented distinct antifungal effects, the sample with higher amounts of this compound being the most active; different chemotypes for *Lavandula pedunculata* according to the site of collection were investigated by Zuzarte et al. [190] and the one with high amounts of camphor revealed to be the most effective against dermatophytes. Likewise, Maxia et al. [191] investigated the antifungal potential of the volatile extracts of *Daucus carota* subsp. *carota* growing wild in the Mediterranean (Sardinia) and Atlantic coasts (Portugal); EOs from blooming umbels and ripe umbels with mature seeds were obtained with major differences in the main compounds between Sardinian and Portuguese samples (β-bisabolene and 11-α-(H)-himachal-4-en-1-β-ol vs. geranyl acetate and α-pinene, respectively), and the Sardinian EO obtained from flowering umbels being the most effective. More recently, Mahboubi et al. [177] analyzed three commercial EOs of *Zataria multiflora.* Two of these oils belonged to the thymol/carvacrol/*p*-cymene chemotype, while the third belonged to high thymol/low carvacrol/*p*-cymene chemotype, and the later presented greater antidermatophytic activity. Also, EOs of *Pistacia terebinthus* growing in Tunisia or in the Sardinia Island greatly differed, and the Tunisian oils exhibited a higher antifungal activity correlated with elevated amounts of α-pinene [192]. Contrarily, in a study on EOs from *Artemisia sieberi* collected at different harvesting times and at distinct locations in Iran, no significant differences were observed in their antidermatophytic activities, strains of *M. canis, N. gypsea,* and *T. rubrum* being the most sensitive [193].

Although the combination of EOs with conventional antifungal drugs is recommended to improve the outcome, studies regarding their combined effect on dermatophyte strains remain scarce. Dias et al. [147] showed a synergistic effect between *Lavandula luisieri* EO and terbinafine against a terbinafine-resistant *T. rubrum* strain. In addition, the checkerboard method revealed synergistic interactions between *Otacanthus azureus* EO and azole antifungals against *T. mentagrophytes* [168] as well as between *Schinus lentiscifolius* EO and terbinafine [170]. More recently, Khoury et al. [155] reported a synergistic effect of *Hirtellina lobelii* EO with fluconazole and griseofulvin against several *Trichophyton* species, namely *T. mentagrophytes, T. rubrum, T. soudanense,* and *T. violaceum.* These positive effects could enable to reduce the dose of antifungal drugs, thus decreasing the adverse side effects associated with conventional antifungals. Another interesting approach regarding the antidermatophytic activity of EOs assessed the photoactivation of *Citrus aurantifolia* EO to treat common cutaneous fungal infections, including infections caused by *T. rubrum.* Interestingly, natural light exposure enhanced the lethal effect of the EO, probably because of an increase in free radical production. This effect should be considered in clinical settings in which toenail infections are treated with topical application of EO while being exposed to sunlight [194].

Furthermore, aromatic extracts obtained by different extractive methods have also been assessed for their antidermatophytic potential. For example, an aromatic extract of *Nigella sativa* obtained with a soxhlet apparatus showed inhibitory effects against *M. canis*, *N. gypsea*, and *T. mentagrophytes*, being more active than the non-volatile extracts [41]. Also, volatile extracts from *Vitex agnus-castus* obtained by supercritical CO_2 extraction were tested against several dermatophytes; nevertheless, in the majority of the cases, the EO (obtained by hydrodistillation) was more effective [175].

The antidermatophytic activity of individual compounds found in EOs has also been investigated, as summarized in Table 3. In addition, examples of the chemical structures of active volatile compounds from different chemical classes are presented in Fig. 2. Overall, several reports have pointed out the effect of isolated monoterpenes but studies on sesquiterpenes are scarce. An extensive study on isolated compounds, performed by Nardoni et al. [180], evaluated the antifungal effect of 17 compounds against isolates of *M. canis*, *N. gypsea*, *T. mentagrophytes*, *T. erinacei,* and *T. terrestre*. Neral was the most effective compound followed by thymol, carvacrol, eugenol, geranial, geraniol, and fenchone that showed moderate activity against most of the tested fungal species. Contrarily, limonene, *p*-cymene, α-pinene, and γ-terpinene were ineffective. In addition, the antidermatophytic effect of geraniol, nerol, citral, neral, and geranial was assessed against four dermatophyte species [197]. The *trans* isomers showed higher antifungal potential than the *cis*, with *T. rubrum* being the most sensitive to the monoterpenes. Overall, oxygenated monoterpenes and in particular phenolic terpenes, namely thymol and carvacrol, present a higher antifungal capacity in comparison to hydrocarbons. In what concerns sesquiterpenes, synthetic (3R,6E)-2,3-dihydrofarnesol and its enantiomer (3S,6E)-2,3-dihydrofarnesol, as well as (2E,6E)-farnesol were able to inhibit the growth of *E. floccosum*, *M. canis*, *T. mentagrophytes,* and *T. rubrum*, with the last two compounds showing a pronounced effect similar to, or even stronger than, that of fluconazole. Interestingly, these compounds have been described as volatiles released by microorganisms, namely *Candida albicans*, although their relevance *in situ* is unknown [198]. Regarding phenylpropanoids, eugenol and its derivatives were effective against dermatophytes, 4-allyl-2-methoxy-5-nitrophenol being the most effective derivative [199].

The antifungal activity of volatile extracts does not seem to be due to a single mechanism of action, but results from the effect of different compounds on several cell targets. Suggested modes of action include lysis of cell walls and membranes affecting fungus permeability with release of intracellular constituents, impairment in ergosterol biosynthesis, elevated vacuolization, as well as the disintegration of cell organelles such as mitochondria, as pointed out by Zuzarte et al. [26].

Table 3 Main antidermatophytic compounds present in bioactive essential oils

Compound class	Bioactive compound	Tested species and antifungal effect (MIC)	Antifungal assay	Ref.
Monoterpene hydrocarbon	α-Pinene	*T. rubrum* (70 μg/mL); *E. floccosum*, *M. canis*, and *N. gypsea* (140 μg/mL); *T. mentagrophytes* and *T. interdigitale* (280 μg/mL); *T. verrucosum* (1110 μg/mL)	Macrobroth dilution	[173]
	p-Cymene	*T. rubrum* (1.25 μL/mL); *M. canis* (2.5 μL/mL); *E. floccosum* and *T. mentagrophytes* (5 μL/mL); *N. gypsea* (10 μL/mL)	Macrobroth dilution	[174]
	(R)-(+)-Limonene	*E. floccosum*, *M. canis* and *T. rubrum* (0.08 μL/mL); *N. gypsea* (0.08–0.16 μL/mL); *T. mentagrophytes*, *T. interdigitale*, and *T. verrucosum* (0.16 μL/mL)	Macrobroth dilution	[172]
	β-Ocimene	*M. canis* and *T. rubrum* (0.32–0.64 μL/mL); *E. floccosum*, *N. gypsea*, *T. mentagrophytes*, and *T. interdigitale* (0.64 μL/mL); *T. verrucosum* (0.64–1.25 μL/mL)	Macrobroth dilution	[140]
	Sabinene	*T. rubrum* (512 μg/mL)	Microbroth dilution	[183]
		T. rubrum (0.16 μL/mL); *E. floccosum*, *M. canis*, *T. mentagrophytes*, and *T. interdigitale* (0.32 μL/mL); *N. gypsea* (0.32–0.64 μL/mL); *T. verrucosum* (0.64 μL/mL)	Macrobroth dilution	[166]
	α-Terpinene	*T. rubrum* (512 μg/mL)	Microbroth dilution	[183]
	γ-Terpinene	*T. rubrum* (512 μg/mL)	Microbroth dilution	[183]
Oxygenated monoterpene	Borneol	*E. floccosum*, *M. canis*, *N. gypsea*, *T. mentagrophytes*, *T. interdigitale*, *T. rubrum*, and *T. verrucosum* (2300 μg/mL)	Macrobroth dilution	[173]
	Camphor	*E. floccosum*, *M. canis*, *N. gypsea*, *T. mentagrophytes*, and *T. rubrum* (> 20 μL/mL)	Macrobroth dilution	[159]
		T. rubrum (512 μg/mL)	Microbroth dilution	[183]
	Citronellal	*T. rubrum* (512 μg/mL)	Microbroth dilution	[183]
	Citronellol	*T. rubrum* (8–1024 μg/mL, depending on the strain)	Microbroth dilution; effect on dry mycelium weight; conidia germination and morphogenesis; in vitro infectivity; release of intracellular material; sorbitol assay and sterol quantitation	[195]
		T. rubrum (256 μg/mL)	Microbroth dilution	[183]

	Fenchone	*E. floccosum* and *T. rubrum* (1.25–2.5 μL/mL); *N. gypsea* (2.5 μL/mL); *M. canis* and *T. mentagrophytes* (2.5–5 μL/mL)	Macrobroth dilution	
	Geraniol	*T. rubrum* (16–32 μg/mL, depending on the strain)	Microbroth dilution; effect on dry mycelium weight; conidia germination and morphogenesis; in vitro infectivity; release of intercellular material; sorbitol assay and sterol quantitation	[195]
	Linalool	*T. mentagrophytes* and *T. rubrum* (1.25 μL/mL); *E. floccosum*, *N. gypsea*, and *T. verrucosum* (1.25–2.5 μL/mL); *M. canis* and *T. interdigitale* (2.5 μL/mL)	Macrobroth dilution	[159]
	Terpinen-4-ol	*T. rubrum* (64 μg/mL)	Microbroth dilution	[183]
Oxygenated sesquiterpene	*α*- Bisabolol	*E. floccosum*, *M. canis*, *N. gypsea*, *T. mentagrophytes*, *T. rubrum*, *T. tonsurans*, and *T. violaceum* (n.a.)	Inhibition of growth and spore germination; ultrastructure alterations (TEM)	[196]
		T. tonsurans (8 μg/mL); *T. soudanense* (16 μg/mL); *T. mentagrophytes*, *T. rubrum*, and *T. violaceum* (32 μg/mL)	Microbroth dilution	[155]
Phenolic terpene	Carvacrol	*M. canis*, *N. gypsea* and *T. mentagrophytes* (0.04 μL/mL); *E. floccosum*, *T. interdigitale*, and *T. rubrum* (0.08 μL/mL); *T. verrucosum* (0.16 μL/mL)	Macrobroth dilution	[158]
	Thymol	*T. mentagrophytes*, *T. rubrum*, and *N. gypsea* (62.5 μg/mL)	Microbroth dilution	[143]
		N. gypsea, *T. mentagrophytes* and *T. rubrum* (62.5 μg/mL)	Microbroth dilution	[188]
		M. canis (0.08 μL/mL); *E. floccosum*, *N. gypsea*, *T. mentagrophytes*, and *T. rubrum* (0.16 μL/mL)	Macrobroth dilution	[174]
Phenylpropanoid	Eugenol	*M. canis* (0.08 μL/mL); *E. floccosum*, *N. gypsea*, *T. mentagrophytes*, and *T. rubrum* (0.16 μL/mL)	Macrobroth dilution	[171]
		N. gypsea (62.5 μg/mL); *M. canis* (125–250 μg/mL)	Microbroth dilution; inhibition of mycelial radial growth and conidia germination	[152]
	Isoeugenol	*N. gypsea* (250–500 μg/mL); *M. canis* (500 μg/mL)	Microbroth dilution; inhibition of mycelial radial growth and conidia germination	[152]
	Methyleugenol	*E. floccosum*, *M. canis*, *N. gypsea*, *T. mentagrophytes*, *T. interdigitale*, *T. rubrum*, and *T. verrucosum* (0.32 μL/mL)	Macrobroth dilution	[172]
	Methylisoeugenol	*M. canis* (125–250 μg/mL); *N. gypsea* (250–500 μg/mL)	Microbroth dilution; inhibition of mycelial radial growth and conidia germination	[152]

E. Epidermophyton, *M. Microsporum*, *N. Nannizzia*, *T. Trichophyton*, *n.a.* not available

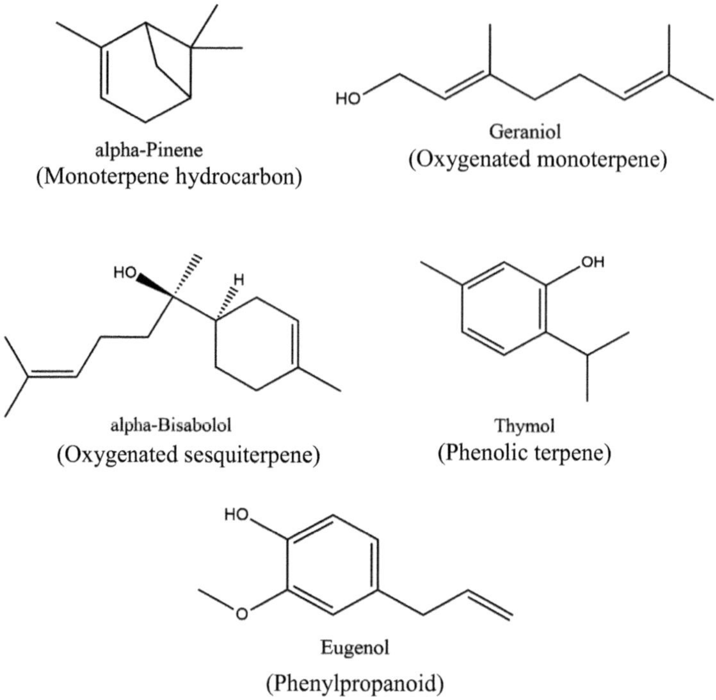

Fig. 2 Examples of volatile compounds, from different chemical classes, with antidermatophytic potential

6 In Vivo Studies and Pharmaceutical Formulations

Despite the promising results obtained in vitro in the numerous studies that have been performed, in vivo validations are required to assess how the whole organism will respond to the use of natural products. In this aim, different animal models have been tested, mainly guinea pigs. Also, specific formulations (e.g., ointments, creams, oils, nail lacquers) have been developed according to the nature, chemical characteristics, and stability of the therapeutic compounds as well as the site of application. Next, several studies are presented reporting the in vivo validation of pharmaceutical formulations using either animal models or, in a few cases, patients in clinical trials.

A flavonoid isolated from the most active extract of *T. chebula* stems was tested on infected mice to treat dermatophytosis. For this purpose, an ointment was prepared with the isolated apigenin (5 mg/g). A complete recovery from the infection was recorded within 12 days of treatment, as observed with terbinafine [200]. On rats, an ointment prepared from the alkaloidal fraction of *Enantia chlorantha* stem bark was tested against dermatophytic infections. The formulation was effective against several dermatophytes and led to a higher reduction of the fungal load at the end of the experiment, in comparison to both the placebo and tioconazole (1% cream) [201].

Regarding studies with rabbits, the ethanolic extract of the bark of *P. amurense* tree, rich in alkaloids, showed a partial antifungal efficacy against an infection induced by *T. mentagrophytes* [107]. Also, the combination of berberine and palmatine significantly elicited the antifungal effect of the first compound in rabbits infected with *M. canis*. Indeed, the combined treatment was more effective than single administration of each compound or clotrimazole [202].

Several studies have been performed on infected guinea pigs. For example, Ayatollahi and Kazemi [203] assessed the therapeutic effects of *M. communis* and *C. zeylanicum* methanolic extracts on dermatophytosis induced by *M. canis* and *T. mentagrophytes*. Both extracts were able to completely treat the infection in shorter periods of time than clotrimazole. The crude extract obtained with a mixture of dichloromethane-methanol from *P. fulva* stem barks was also tested on a *T. mentagrophytes* infection. Thirteen days of daily treatment permitted to cure the infection without irritating effects. These observations validate the traditional uses ascribed to this species. Indeed, the stem bark of this tree is used locally against dermatoses [204]. Furthermore, palm kernel oil also proved to be a good vehicle for the extract and thus, the oil-extract of *P. fulva* seems to be a good candidate for antidermatophytic phytomedicine [204]. Lee et al. [205] tested two volatile compounds, eugenol and nerolidol, isolated from Japanese cypress oil, in a similar model of infection by *N. gypsea*. The compounds were incorporated in vaseline petroleum jelly and applied topically on skin lesions for 3 weeks. The formulation with nerolidol was more active and improvement in the lesions was seen since the first week of application. It is important to underline that both compounds showed a lower degree of hyperkeratosis and inflammatory cell infiltration than econazole used as a positive control. Ointments of *C. martini* and *Chenopodium ambrosioides* EOs were also able to eliminate induced ringworm in 7–21 days [144]. In vivo validations were performed using a mixture of three essential oils, i.e., *O. vulgare* (5%), *R. officinalis* (5%), and *T. serpyllum* (2%) in sweet almond oil. This formulation was topically applied to seven cats with ringworm. Although four cats recovered, no untreated control group was included in the study [206]. This is quite questionable since dermatophytosis may be sometimes self-healing.

Larger animals have also been used to validate the antidermatophytic effect of natural extracts. Infected goats were treated with *Lawsonia inermis* (henna) paste, aqueous and ethanolic extracts, or clotrimazole. The henna paste was the most effective, with the disappearance of the lesions and complete hair repair after 30 days of treatment [207]. A liposomal gel loaded with *Eucalyptus camaldulensis* EO was also effective against *M. canis*, *N. gypsea*, *T. rubrum*, and *T. verrucosum* [208]. Interestingly, the oil entrapped in liposomes remained stable for an extended period, which highlights the relevance of this kind of formulation since the enhanced stability of the active compound may enable a higher therapeutic outcome. Furthermore, horses affected with equine ringworm were treated with a mixture of 25% tea tree oil (*Melaleuca alternifolia*) in sweet almond oil while the control group received enilconazole 2% solution. All treated animals showed complete healing [209], thus stressing out the effectiveness of the plant extract.

A few clinical trials have been conducted in humans. Romero et al. [210] assessed the efficacy of *Ageratina pichinchensis* extract (maceration with a mixture of hexane and ethyl acetate) in the topical treatment of nails in patients with mild to moderate onychomycosis. In this trial, two concentrations (12.6 and 16.8%) were assessed for both tolerability and therapeutic efficacy. Both concentrations of the extract were incorporated in a cosmetic nail lacquer. Following 6 months of treatment, 100% of tolerability was observed in both cases, the therapeutic efficacy being 79% in the group treated with the higher concentration. Hospitalized patients aged from 3 months to 58 years and diagnosed with *Tinea corporis, Tinea capitis, Tinea manuum,* or *Tinea pedis* received a flavonoid-rich extract incorporated in an ointment for 1–5 weeks. At the end of treatment, a complete cure was observed in 64% of patients while 24% showed significant improvement and only 12% registered a little improvement [211]. A fungitoxic cream formulation (extracts of *J. regia* hulls, pulverized roots of *Nardostachys jatamansi* or *Vetiveria zizanioides* or *Catharanthus roseus*, polyols, fixed oil, non-ionic emulsifiers, thickening agent plasticizer, and base) was patented by Bindra et al. [212]. The formulation showed good adherence to the nail surface and dried rapidly. The fungitoxic effects were observed after 4 days of application, with a complete recovery at the end of 190 days of treatment. Importantly, it also prevented the recurrence of the infection. An ethanolic propolis extract was also used as a therapeutic option for patients with onychomycosis. Topical treatments were applied with a 6-month follow-up period. The extract was able to penetrate the nail without cytotoxicity, and permitted a complete mycological and clinical recovery in 56% of the patients [213].

Volatile compounds have also been used in clinical trials, mainly *M. alternifolia* and *Eucalyptus* spp. EOs. Indeed, *M. alternifolia* (tea tree) oil applied twice daily with a formulation consisting of 100% tea tree oil was effective in the treatment of onychomycosis, showing an efficacy similar to that of clotrimazole [214]. Also, a double-blind placebo-controlled study was carried out to examine the clinical efficacy and tolerability of a formulation containing 5% of *M. alternifolia* oil incorporated in a cream to manage toenail onychomycosis in a cohort. After 16 weeks, 80% of the patients using this formulation were cured, with no relapse observed during 1 year of follow-up [215]. In another large randomized, double-blind study, both 25% and 50% tea tree oil solutions were effective in treating *Tinea pedis*, but to less extend than standard topical treatments [216]. The antifungal potential attributed to *Eucalyptus* species has also been confirmed in a pilot study that evaluated the efficacy of an over-the-counter ointment in an outpatient clinic setting. The patients diagnosed with onychomycosis showed positive cultures for both *Candida parapsilosis* and *T. mentagrophytes*. The formulation rich in *Eucalyptus* oil was able to improve the infection in 83% of the patients. However, for a total recovery 48 weeks were needed [217]. *Curcuma longa* EO has also been formulated in an ointment (1% *w/v*) and tested topically on patients diagnosed with *Tinea corporis*. At the end of the treatment, 75% of patients had completely recovered [218].

Although several studies demonstrated the therapeutic potential of natural extracts and isolated compounds in animal models of dermatophytosis, there are still few studies devoted to the evaluation of their toxicity in vivo. Further studies are needed to determine potential toxic effects, toward animals or humans. For instance, Freile et al. [219] evaluated the toxicity of aqueous extracts of a plant with antidermatophytic activity in vivo, using fish and fish embryo-larval stages as models, and demonstrated the low toxicity of the extracts in comparison to ketoconazole. Jagetia et al. [220] evaluated the acute toxicity of a leaf extract of *A. marmelos*, a plant with antidermatophytic activity [221], in Swiss albino male mice, through intraperitoneal injection of the extract. The authors verified that the extract was non-toxic up to a single dose of 1750 mg/kg. Despite several in vitro evaluations of toxicity using human and non-human cell models [222], as well as the therapeutic effectiveness studies, the lack of data regarding toxicity in animal models is certainly a setback in the advance of natural compounds for commercialization. Nevertheless, some commercial formulations mainly with EOs are currently available for the treatment of onychomycosis. For example, Tineacide® (Blaine Labs, California, USA) comprises tea tree and lavender oils in its formulation while "Fungus Stop" (Zane Hellas, Thessaloniki, Greece) used for nail treatment has carvacrol as its main constituent. Interestingly, Vick's VapoRub® (Procter and Gamble, EUA), designed for respiratory problems, is also used in popular medicine to treat onychomycosis [185].

7 Concluding Remarks and Future Perspectives

Dermatophytosis is one of the most common skin diseases worldwide, with tendency to increase among elderly and children. Current therapeutics may be unsuccessful, explaining why natural products are emerging as effective alternatives/complements. Indeed, these extracts are able to act on multiple cell targets, an important feature when considering microorganisms that are intrinsically or became resistant to conventional therapies. Several studies have demonstrated the effectiveness of these natural extracts and their isolated compounds, in particular those of plant origin. Nevertheless, more reliable studies are still required. It is also important to emphasize that standard protocols should be adopted to ensure comparisons between studies. Another aspect to bear in mind is that most assays are conducted on fungal isolates, not reflecting the host's complex microbial environment. Furthermore, pharmacokinetic and pharmacodynamic approaches are lacking and should be considered to evaluate the effectiveness of a possible treatment, as well as toxicological assessments. An additional relevant aspect to consider is the absorption, metabolism, and bioavailability of these compounds in humans, which remains controversial. Regarding onychomycoses, more studies are needed to prove the penetration of active compounds into the nails and their in vivo efficacy.

References

1. Havlickova B, Czaika VA, Friedrich M. Epidemiological trends in skin mycoses worldwide. Mycoses. 2008;51:2–15.
2. Ghannoum MA, Rice LB. Antifungal agents: mode of action, mechanisms of resistance, and correlation of these mechanisms with bacterial resistance. Clin Microbiol Rev. 1999;12 (4):501–17.
3. Mukherjee PK, Leidich SD, Isham N, Leitner I, Ryder NS, Ghannoum MA. Clinical *Trichophyton rubrum* strain exhibiting primary resistance to terbinafine. Antimicrob Agents Chemother. 2003;47(1):82–6.
4. Monod M. Secreted proteases from dermatophytes. Mycopathologia. 2008;166(5–6):285–94.
5. Gupta AK, Cooper EA. Update in antifungal therapy of dermatophytosis. Mycopathologia. 2008;166(5–6):353–67.
6. Ríos JL, Recio MC. Medicinal plants and antimicrobial activity. J Ethnopharmacol. 2005;100 (1–2):80–4.
7. Soares LA, de Sardi J, Gullo FP, de Pitangui N, Scorzoni L, Leite FS, et al. Anti dermatophytic therapy—prospects for the discovery of new drugs from natural products. Braz J Microbiol. 2013;44(4):1035–41.
8. Lopes G, Pinto E, Salgueiro L. Natural products: an alternative to conventional therapy for dermatophytosis? Mycopathologia. 2017;182(1–2):143–67.
9. Cheuka PM, Mayoka G, Mutai P, Chibale K. The role of natural products in drug discovery and development against neglected tropical diseases. Molecules. 2017;22(1):E58.
10. Yun B-W, Yan Z, Amir R, Hong S, Jin Y-W, Lee E-K, et al. Plant natural products: history, limitations and the potential of cambial meristematic cells. Biotechnol Genet Eng Rev. 2012;28(1):47–60.
11. Dahdah MJ, Scher RK. Dermatophytes. Curr Fungal Infect Rep. 2008;2(2):81–6.
12. Piggott C, Friedlander S. Dermatophytes and other superficial fungi. In: Long S, editor. Principles and practice of pediatric infectious diseases. 4th ed. Amsterdam: Elsevier; 2013. p. 1246–50.
13. Weitzman I, Summerbell RC. The dermatophytes. Clin Microbiol Rev. 1995;8(2):240–59.
14. Romani L. Immunity to fungal infections. Nat Rev Immunol. 2011;11(4):275–88.
15. Akinboro A, Olasode O, Onayemi O, Oguntola A, Ajibola A. Prediction of dermatophyte culture by clinical features: saving time and cost in resource-poor settings. Ibnosina J Med Biomed Sci. 2013;5(4):189–95.
16. Ziemer M, Seyfarth F, Elsner P, Hipler U-C. Atypical manifestations of tinea corporis. Mycoses. 2007;50(Suppl 2):31–5.
17. Robert R, Pihet M. Conventional methods for the diagnosis of dermatophytosis. Mycopathologia. 2008;166(5–6):295–306.
18. Gupta AK, Chow M, Daniel CR, Aly R. Treatments of tinea pedis. Dermatol Clin. 2003;21 (3):431–62.
19. Castellsague J, García-Rodríguez L-A, Duque A, Pérez S. Risk of serious skin disorders among users of oral antifungals: a population-based study. BMC Dermatol. 2002;2:14.
20. Gupta AK, Versteeg SG, Shear NH. Common drug-drug interactions in antifungal treatments for superficial fungal infections. Expert Opin Drug Metab Toxicol. 2018;14(4):387–98.
21. Hemaiswarya S, Kruthiventi AK, Doble M. Synergism between natural products and antibiotics against infectious diseases. Phytomedicine. 2008;15(8):639–52.
22. Arif T, Bhosale JD, Kumar N, Mandal TK, Bendre RS, Lavekar GS, et al. Natural products—antifungal agents derived from plants. J Asian Nat Prod Res. 2009;11(7):621–38.
23. Caesar LK, Cech NB. Synergy and antagonism in natural product extracts: when 1 + 1 does not equal 2. Nat Prod Rep. 2019;36(6):869–88.
24. Yarnell E. Synergy in herbal medicines: Part 1. J Restor Med. 2015;4(1):60–73.
25. Cheung RCF, Wong JH, Pan WL, Chan YS, Yin CM, Dan XL, et al. Antifungal and antiviral products of marine organisms. Appl Microbiol Biotechnol. 2014;98(8):3475–94.

26. Zuzarte M, Gonçalves MJ, Canhoto J, Salgueiro L. Antidermatophytic activity of essential oils. Sci Against Microb Pathog. 2011:1167–78.
27. Hadacek F, Greger H. Testing of antifungal natural products: methodologies, comparability of results and assay choice. Phytochem Anal. 2000;11(3):137–47.
28. Golan D, Tahsjian A, Armstrong E, Armstrong A. Principles of pharmacology. The pathophysiologic basics of drug therapy. 3rd ed. Philadelphia, PA: Wolters Kluwer; 2012. p. 620.
29. Sepahvand A, Eliasy H, Mohammadi M, Safarzadeh A, Azarbaijani K, Shahsavari S, et al. A review of the most effective medicinal plants for dermatophytosis in traditional medicine. Biomed Res Ther. 2018;5(6):2378–88.
30. Javed I, Mishra R, Aashiq HA. Antidermatophytic activity of angiospermic plants: a review. Asian J Pharm Clin Res. 2015;8(2):75–80.
31. Svetaz L, Zuljan F, Derita M, Petenatti E, Tamayo G, Cáceres A, et al. Value of the ethnomedical information for the discovery of plants with antifungal properties. A survey among seven Latin American countries. J Ethnopharmacol. 2010;127(1):137–58.
32. Dhayanithi NB, Kumar TTA, Kalaiselvam M, Balasubramanian T, Sivakumar N. Antidermatophytic activity of marine sponge, *Sigmadocia carnosa* (Dendy) on clinically isolated fungi. Asian Pac J Trop Biomed. 2012;2(8):635–9.
33. Sit NW, Chan YS, Lai SC, Lim LN, Looi GT, Tay PL, et al. *In vitro* antidermatophytic activity and cytotoxicity of extracts derived from medicinal plants and marine algae. J Mycol Med. 2018;28(3):561–7.
34. Kurdelas RR, Lima B, Tapia A, Feresin GE, Sierra MG, Rodríguez MV, et al. Antifungal activity of extracts and prenylated coumarins isolated from *Baccharis darwinii* Hook and Arn. (Asteraceae). Molecules. 2010;15(7):4898–907.
35. Navarro-García VM, Rojas G, Avilés M, Fuentes M, Zepeda G. *In vitro* antifungal activity of coumarin extracted from *Loeselia mexicana* Brand. Mycoses. 2011;54(5):e569–71.
36. Ahmad K, Sultana N. Studies on bioassay directed antifungal activity of medicinal plants *Calotropis procera*, *Skimmia laureola*, *Peltophorum pterocarpum* and two pure natural compounds ulopterol and 4-methoxy-1-methyl-3-(2'S-hydroxy-3'-ene butyl)-2-quinolone. J Chem Soc Pak. 2003;25(4):328–30.
37. Singh DN, Verma N, Raghuwanshi S, Shukla PK, Kulshreshtha DK. Antifungal anthraquinones from *Saprosma fragrans*. Bioorg Med Chem Lett. 2006;16(17):4512–4.
38. Feresin GE, Tapia A, Sortino M, Zacchino S, De Arias AR, Inchausti A, et al. Bioactive alkyl phenols and embelin from *Oxalis erythrorhiza*. J Ethnopharmacol. 2003;88(2–3):241–7.
39. Sakunphueak A, Panichayupakaranant P. Comparison of antimicrobial activities of naphthoquinones from *Impatiens balsamina*. Nat Prod Res. 2012;26(12):1119–24.
40. Aljabre SHM, Randhawa MA, Akhtar N, Alakloby OM, Alqurashi AM, Aldossary A. Antidermatophyte activity of ether extract of *Nigella sativa* and its active principle, thymoquinone. J Ethnopharmacol. 2005;101(1–3):116–9.
41. Mahmoudvand H, Sepahvand A, Jahanbakhsh S, Ezatpour B, Ayatollahi Mousavi SA. Evaluation of antifungal activities of the essential oil and various extracts of *Nigella sativa* and its main component, thymoquinone against pathogenic dermatophyte strains. J Mycol Med. 2014;24(4):e155–61.
42. Agüero MB, Svetaz L, Baroni V, Lima B, Luna L, Zacchino S, et al. Urban propolis from San Juan province (Argentina): ethnopharmacological uses and antifungal activity against *Candida* and dermatophytes. Ind Crop Prod. 2014;57:166–73.
43. Zhang H-X, Lunga P-K, Li Z-J, Dai Q, Du Z-Z. Flavonoids and stilbenoids from *Derris eriocarpa*. Fitoterapia. 2014;95:147–53.
44. Sathiamoorthy B, Gupta P, Kumar M, Chaturvedi AK, Shukla PK, Maurya R. New antifungal flavonoid glycoside from *Vitex negundo*. Bioorg Med Chem Lett. 2007;17(1):239–42.
45. Pereira R, Pereira AL, Ferreira MM, Fontenelle ROS, Saker-Sampaio S, Santos HS, et al. Evaluation of the antimicrobial and antioxidant activity of 7-hydroxy-4', 6-dimethoxy-isoflavone and essential oil from *Myroxylon peruiferum* L.f. An Acad Bras Cienc. 2019;91(2): e20180204.

46. Souza-Moreira TM, Severi JA, Rodrigues ER, de Paula MI, Freitas JA, Vilegas W, et al. Flavonoids from *Plinia cauliflora* (Mart.) Kausel (Myrtaceae) with antifungal activity. Nat Prod Res. 2019;33(17):2579–82.
47. De Campos MP, Cechinel Filho V, Da Silva RZ, Yunes RA, Zacchino S, Juarez S, et al. Evaluation of antifungal activity of *Piper solmsianum* C. DC. var. *solmsianum* (Piperaceae). Biol Pharm Bull. 2005;28(8):1527–30.
48. Freixa B, Vila R, Ferro EA, Adzet T, Cañigueral S. Antifungal principles from *Piper fulvescens*. Planta Med. 2001;67(9):873–5.
49. Koroishi AM, Foss SR, Cortez DAG, Ueda-Nakamura T, Nakamura CV, Dias Filho BP. *In vitro* antifungal activity of extracts and neolignans from *Piper regnellii* against dermatophytes. J Ethnopharmacol. 2008;117(2):270–7.
50. Bang KH, Kim YK, Min BS, Na MK, Rhee YH, Lee JP, et al. Antifungal activity of magnolol and honokiol. Arch Pharm Res. 2000;23(1):46–9.
51. Lee MH, Lee KB, Oh SM, Lee BH, Chee HY. Antifungal activities of dieckol isolated from the marine brown alga *Ecklonia cava* against *Trichophyton rubrum*. J Appl Biol Chem. 2010;53(4):504–7.
52. Foss SR, Nakamura CV, Ueda-Nakamura T, Cortez DAG, Endo EH, Dias Filho BP. Antifungal activity of pomegranate peel extract and isolated compound punicalagin against dermatophytes. Ann Clin Microbiol Antimicrob. 2014;13(1):32.
53. Kariba RM, Houghton PJ, Yenesew A. Antimicrobial activities of a new schizozygane indoline alkaloid from *Schizozygia coffaeoides* and the revised structure of isoschizogaline. J Nat Prod. 2002;65:566–9.
54. Semwal DK, Rawat U. Antimicrobial hasubanalactam alkaloid from *Stephania glabra*. Planta Med. 2009;75(4):378–80.
55. Lakshmi V, Srivastava S, Mishra SK, Shukla PK. Antifungal activity of bivittoside-D from *Bohadschia vitiensis* (Semper). Nat Prod Res. 2012;26(10):913–8.
56. Stergiopoulou T, De Lucca AJ, Meletiadis J, Sein T, Boue SM, Schaufele R, et al. *In vitro* activity of CAY-1, a saponin from *Capsicum frutescens*, against *Microsporum* and *Trichophyton* species. Med Mycol. 2008;46(8):805–10.
57. Duraipandiyan V, Ignacimuthu S. Antibacterial and antifungal activity of flindersine isolated from the traditional medicinal plant, *Toddalia asiatica* (L.) Lam. J Ethnopharmacol. 2009;123 (3):494–8.
58. Singh DN, Verma N, Raghuwanshi S, Shukla PK, Kulshreshtha DK. Antifungal activity of *Agapanthus africanus* extractives. Fitoterapia. 2008;79(4):298–300.
59. García-Sosa K, Sánchez-Medina A, Álvarez SL, Zacchino S, Veitch NC, Sima-Polanco P, et al. Antifungal activity of sakurasosaponin from the root extract of *Jacquinia flammea*. Nat Prod Res. 2011;25(12):1185–9.
60. López-Villegas EO, Herrera-Arellano A, de Los Angeles Martínez-Rivera M, Alvarez L, Cano-Nepauseno M, Marquina S, et al. Ultrastructural changes on clinical isolates of *Trichophyton rubrum*, *Trichophyton mentagrophytes*, and *Microsporum gypseum* caused by *Solanum chrysotrichum* saponin SC-2. Planta Med. 2009;75(14):1517–20.
61. Vermerris W, Nicholson R. Families of phenolic compounds and means of classification. In: Vermerris W, Nicholson R, editors. Phenolic compound biochemistry. Dordrecht: Springer; 2008. p. 1–34.
62. Lin D, Xiao M, Zhao J, Li Z, Xing B, Li X, et al. An overview of plant phenolic compounds and their importance in human nutrition and management of type 2 diabetes. Molecules. 2016;21(10):E1374.
63. Venugopala KN, Rashmi V, Odhav B. Review on natural coumarin lead compounds for their pharmacological activity. Biomed Res Int. 2013;2013:963248.
64. Borges F, Roleira F, Milhazes N, Santana L, Uriarte E. Simple coumarins and analogues in medicinal chemistry: occurrence, synthesis and biological activity. Curr Med Chem. 2005;12 (8):887–916.

65. Matos MJ, Santana L, Uriarte E, Abreu OA, Molina E, Yordi EG. Coumarins—an important class of phytochemicals. In: Rao V, editor. Phytochemicals—isolation, characterisation and role in human health. Rijeka: InTech; 2015. p. 113–40.
66. Xu L, Zhao X-Y, Wu Y-L, Zhang W. The study on biological and pharmacological activity of coumarins. In: Asia-Pacific energy equipment engineering research conference. Amsterdam: Atlantis Press; 2015.
67. Céspedes CL, Avila JG, Martínez A, Serrato B, Calderón-Mugica JC, Salgado-Garciglia R. Antifungal and antibacterial activities of Mexican tarragon (*Tagetes lucida*). J Agric Food Chem. 2006;54(10):3521–7.
68. Stein AC, Álvarez S, Avancini C, Zacchino S, von Poser G. Antifungal activity of some coumarins obtained from species of *Pterocaulon* (Asteraceae). J Ethnopharmacol. 2006;107 (1):95–8.
69. Rehman SU, Chohan ZH, Gulnaz F, Supuran CT. *In vitro* antibacterial, antifungal and cytotoxic activities of some coumarins and their metal complexes. J Enzyme Inhib Med Chem. 2005;20(4):333–40.
70. Stewart CS. Use of coumarin derivatives in antifungal therapy. PCT/GB2007/002062; 2007.
71. Mercer DK, Robertson J, Wright K, Miller L, Smith S, Stewart CS, et al. A prodrug approach to the use of coumarins as potential therapeutics for superficial mycoses. PLoS One. 2013;8 (11):e80760.
72. Futuro DO, Ferreira PG, Nicoletti CD, Borba-Santos LP, Da Silva FC, Rozental S, et al. The antifungal activity of naphthoquinones: an integrative review. An Acad Bras Cienc. 2018;90 (1):1187–214.
73. Eyong KO, Kuete V, Efferth T. Quinones and benzophenones from the medicinal plants of Africa. In: Kuete V, editor. Medicinal plant research in Africa: pharmacology and chemistry. Amsterdam: Elsevier; 2013. p. 351–91.
74. Nowicka B, Kruk J. Occurrence, biosynthesis and function of isoprenoid quinones. Biochim Biophys Acta. 2010;1797(9):1587–605.
75. Dzoyem JP, Tangmouo JG, Lontsi D, Etoa FX, Lohoue PJ. *In vitro* antifungal activity of extract and plumbagin from the stem bark of *Diospyros crassiflora* Hiern (Ebenaceae). Phytother Res. 2007;21(7):671–4.
76. Guiraud P, Steiman R, Campos-Takaki G-M, Seigle-Murandi F, de Buochberg M. Comparison of antibacterial and antifungal activities of lapachol and β-lapachone. Planta Med. 1994;60(04):373–4.
77. Wittebolle V, Lemriss S, La Morella G, Errante J, Boiron P, Barret R, et al. Antifungal effects of aminosulphoxide and disulphide derivatives. Mycoses. 2006;49(3):169–75.
78. Inouye S, Uchida K, Takizawa T, Yamaguchi H, Abe S. Evaluation of the effect of terpenoid quinones on *Trichophyton mentagrophytes* by solution and vapor contact. J Infect Chemother. 2006;12(2):100–4.
79. Sytar O, Švedienė J, Ložienė K, Paškevičius A, Kosyan A, Taran N. Antifungal properties of hypericin, hypericin tetrasulphonic acid and fagopyrin on pathogenic fungi and spoilage yeasts. Pharm Biol. 2016;54(12):3121–5.
80. Castro MÁ, Gamito AM, Tangarife-Castaño V, Zapata B, Miguel Del Corral JM, Mesa-Arango AC, et al. Synthesis and antifungal activity of terpenyl-1,4-naphthoquinone and 1,4-anthracenedione derivatives. Eur J Med Chem. 2013;67:19–27.
81. Vermerris M, Nicholson R. Phenolic compound biochemistry. Dordrecht: Springer; 2006.
82. Corradini E, Foglia P, Giansanti P, Gubbiotti R, Samperi R, Laganà A. Flavonoids: chemical properties and analytical methodologies of identification and quantitation in foods and plants. Nat Prod Res. 2011;25(5):469–95.
83. Park BJ, Taguchi H, Kamei K, Matsuzawa T, Hyon SH, Park JC. *In vitro* antifungal activity of epigallocatechin 3-O-gallate against clinical isolates of dermatophytes. Yonsei Med J. 2011;52 (3):535–8.
84. Bitencourt TA, Komoto TT, Massaroto BG, Miranda CES, Beleboni RO, Marins M, et al. *Trans*-chalcone and quercetin down-regulate fatty acid synthase gene expression and reduce

ergosterol content in the human pathogenic dermatophyte *Trichophyton rubrum*. BMC Complement Altern Med. 2013;13:229.
85. Mbaveng AT, Ngameni B, Kuete V, Simo IK, Ambassa P, Roy R, et al. Antimicrobial activity of the crude extracts and five flavonoids from the twigs of *Dorstenia barteri* (Moraceae). J Ethnopharmacol. 2008;116(3):483–9.
86. Ali I, Khan FG, Suri KA, Gupta BD, Satti NK, Dutt P, et al. *In vitro* antifungal activity of hydroxychavicol isolated from *Piper betle* L. Ann Clin Microbiol Antimicrob. 2010;9 (1):7–15.
87. Boeck P, Leal PC, Yunes RA, Cechinel Filho V, López S, Sortino M, et al. Antifungal activity and studies on mode of action of novel xanthoxyline-derived chalcones. Arch Pharm (Weinheim). 2005;338(2–3):87–95.
88. Alvarez MA, Debattista NB, Pappano NB. Antimicrobial activity and synergism of some substituted flavonoids. Folia Microbiol (Praha). 2008;53(1):23–8.
89. Zálešák F, Bon DJ-YD, Pospíšil J. Lignans and neolignans: plant secondary metabolites as a reservoir of biologically active substances. Pharmacol Res. 2019;146:104284.
90. Calvo-Flores FG, Dobado JA, Isac-García J, Martín-Martínez FJ. Structure and chemical properties of lignans. In: Calvo-Flores FG, Dobado JA, Isac-García J, Martín-Martínez FJ, editors. Lignin and lignans as renewable raw materials. New York: Wiley; 2015. p. 313–68.
91. Bruneton J. Pharmacognosy, phytochemistry, medicinal plants. 2nd ed. Paris: Lavoisier; 2008. p. 1136.
92. Shehzad A, Islam SU, Al-Suhaimi EA, Lee Y-S. Pleiotropic effects of bioactive phytochemicals (polyphenols and terpenes). In: Vattem D, Maitin V, editors. Functional foods, nutraceuticals and natural products concepts and applications. Pennsylvania: DEStech; 2016. p. 66.
93. Saleem M, Hyoung JK, Ali MS, Yong SL. An update on bioactive plant lignans. Nat Prod Rep. 2005;22(6):696–716.
94. Agüero MB, Svetaz L, Sánchez M, Luna L, Lima B, López ML, et al. Argentinean Andean propolis associated with the medicinal plant *Larrea nitida* Cav. (Zygophyllaceae). HPLC-MS and GC-MS characterization and antifungal activity. Food Chem Toxicol. 2011;49(9):1970–8.
95. Iida Y, Oh KB, Saito M, Matsuoka H, Kurata H. *In vitro* synergism between nyasol, an active compound isolated from *Anemarrhena asphodeloides*, and azole agents against *Candida albicans*. Planta Med. 2000;66(5):435–8.
96. Zacchino S, Rodríguez G, Santecchia C, Pezzenati G, Giannini F, Enriz R. *In vitro* studies on mode of action of antifungal 8.O.4'-neolignans occurring in certain species of *Virola* and related genera of Myristicaceae. J Ethnopharmacol. 1998;62(1):35–41.
97. Lopes G. Seaweeds from the Portuguese coast: chemistry, antimicrobial and antiinflammatory capacity. PhD Thesis. University of Porto; 2014.
98. Vonshak A, Barazani O, Sathiyamoorthy P, Shalev R, Vardy D, Golan-Goldhirsh A. Screening south Indian medicinal plants for antifungal activity against cutaneous pathogens. Phytother Res. 2003;17(9):1123–5.
99. Lopes G, Sousa C, Silva LR, Pinto E, Andrade PB, Bernardo J, et al. Can phlorotannins purified extracts constitute a novel pharmacological alternative for microbial infections with associated inflammatory conditions? PLoS One. 2012;7(2):e31145.
100. Koivikko R, Loponen J, Pihlaja K, Jormalainen V. High-performance liquid chromatographic analysis of phlorotannins from the brown alga *Fucus vesiculosus*. Phytochem Anal. 2007;18 (4):326–32.
101. Scalbert A. Antimicrobial properties of tannins. Phytochemistry. 1991;30(12):3875–83.
102. Lopes G, Pinto E, Andrade PB, Valentão P. Antifungal activity of phlorotannins against dermatophytes and yeasts: approaches to the mechanism of action and influence on *Candida albicans* virulence factor. PLoS One. 2013;8(8):e72203.
103. Ziegler J, Facchini PJ. Alkaloid biosynthesis: metabolism and trafficking. Annu Rev Plant Biol. 2008;59(1):735–69.

104. Evans W. Trease and Evans' pharmacognosy. 16th ed. Philadelphia: Saunders; 2009. p. 616.
105. Wangchuk P. Plant alkaloids: classification, isolation and drug development. In: Swamy MK, Patra JK, Rudramurthy GR, editors. Medicinal plants: chemistry, pharmacology and therapeutic applications. Boca Raton: Taylor and Francis, CRC; 2019. p. 131–7.
106. Singh N, Sharma B. Toxicological effects of berberine and sanguinarine. Front Mol Biosci. 2018;5:21.
107. Xiao CW, Ji QA, Rajput ZI, Wei Q, Liu Y, Bao GL. Antifungal efficacy of *Phellodendron amurense* ethanol extract against *Trichophyton mentagrophytes* in rabbits. Pakistan Vet J. 2014;34(2):219–23.
108. Phetkul U, Wanlaso N, Mahabusarakam W, Phongpaichit S, Carroll AR. New acridone from the wood of *Citrus reticulata* Blanco. Nat Prod Res. 2013;27(20):1922–6.
109. Morteza-Semnani K, Amin G, Shidfar MR, Hadizadeh H, Shafiee A. Antifungal activity of the methanolic extract and alkaloids of *Glaucium oxylobum*. Fitoterapia. 2003;74(5):493–6.
110. Yu HF, Qin XJ, Ding CF, Wei X, Yang J, Luo JR, et al. Nepenthe-like indole alkaloids with antimicrobial activity from *Ervatamia chinensis*. Org Lett. 2018;20(13):4116–20.
111. Lohombo-Ekomba M-L, Okusa P, Penge O, Kabongo C, Choudhary MI, Kasende O. Antibacterial, antifungal, antiplasmodial, and cytotoxic activities of *Albertisia villosa*. J Ethnopharmacol. 2004;93(2–3):331–5.
112. Khan H, Mubarak MS, Amin S. Antifungal potential of alkaloids as an emerging therapeutic target. Curr Drug Targets. 2017;18(16):1825–35.
113. Biswas T, Dwivedi UN. Plant triterpenoid saponins: biosynthesis, *in vitro* production, and pharmacological relevance. Protoplasma. 2019;256(6):1463–86.
114. Vincken J-P, Heng L, de Groot A, Gruppen H. Saponins, classification and occurrence in the plant kingdom. Phytochemistry. 2007;68(3):275–97.
115. Moses T, Papadopoulou KK, Osbourn A. Metabolic and functional diversity of saponins, biosynthetic intermediates and semi-synthetic derivatives. Crit Rev Biochem Mol Biol. 2014;49(6):439–62.
116. Lunga PK, Qin XJ, Yang XW, Kuiate JR, Du ZZ, Gatsing D. Antimicrobial steroidal saponin and oleanane-type triterpenoid saponins from *Paullinia pinnata*. BMC Complement Altern Med. 2014;14(1):369.
117. Hu Q, Chen YY, Jiao QY, Khan A, Li F, Han DF, et al. Triterpenoid saponins from the pulp of *Sapindus mukorossi* and their antifungal activities. Phytochemistry. 2018;147:1–8.
118. González M, Zamilpa A, Marquina S, Navarro V, Alvarez L. Antimycotic spirostanol saponins from *Solanum hispidum* leaves and their structure-activity relationships. J Nat Prod. 2004;67(6):938–41.
119. Njateng GSS, Du Z, Gatsing D, Nanfack Donfack AR, Feussi Talla M, Kamdem Wabo H, et al. Antifungal properties of a new terpernoid saponin and other compounds from the stem bark of *Polyscias fulva* Hiern (Araliaceae). BMC Complement Altern Med. 2015;15:25.
120. Sobolewska D, Janeczko Z, Kisiel W, Podolak I, Galanty A, Trojanowska D. Steroidal glycosides from the underground parts of *Allium ursinum* L. and their cytostatic and antimicrobial activity. Acta Pol Pharm Drug Res. 2006;63(3):219–23.
121. Zhang D, Fu Y, Yang J, Li XN, San MM, Oo TN, et al. Triterpenoids and their glycosides from *Glinus oppositifolius* with antifungal activities against *Microsporum gypseum* and *Trichophyton rubrum*. Molecules. 2019;24(12):2206–8.
122. Houghton P, Patel N, Jurzysta M, Biely Z, Cheung C. Antidermatophyte activity of medicago extracts and contained saponins and their structure-activity relationships. Phytother Res. 2006;20(12):1061–6.
123. Kim DW, Bang KH, Rhee YH, Lee KT, Park HJ. Growth inhibitory activities of kalopanaxsaponins A and I against human pathogenic fungi. Arch Pharm Res. 1998;21 (6):688–91.
124. Takechi M, Doi K, Wakayama Y. Biological activities of synthetic saponins and cardiac glycosides. Phytother Res. 2003;17(1):83–5.

125. Avis TJ. Antifungal compounds that target fungal membranes: applications in plant disease control. Can J Plant Pathol. 2007;29(4):323–9.
126. ISO 9235. Aromatic natural raw materials—vocabulary. 2013.
127. Council of Europe. European pharmacopoeia. 5th ed. Strasbourg: Council of Europe; 2004.
128. Baser K, Demirci F. Chemistry of essential oils. In: Berger R, editor. Flavours and fragrances—chemistry, bioprocessing and sustainability. Berlin: Springer; 2007. p. 43–86.
129. Bakkali F, Averbeck S, Averbeck D, Idaomar M. Biological effects of essential oils-a review. Food Chem Toxicol. 2008;46(2):446–75.
130. Dewick P. The mevalonate and deoxyxylulose phosphate pathways: terpenoids and steroids. In: Medicinal natural products: a biosynthetic approach. 2nd ed. London: Wiley; 2002. p. 167–285.
131. Dewick P. The shikimate pathway: aromatic amino acids and phenylpropanoids. In: Dewick PM, editor. Medicinal natural products: a biosynthetic approach. 2nd ed. London: Wiley; 2002. p. 121–64.
132. Svoboda K, Svoboda T. In: Syred PM, editor. Secretory structures of aromatic and medicinal plants: a review and atlas of micrographs. Knighton: Microscopix; 2000. p. 60.
133. Theis N, Lerdau M. The evolution of function in plant secondary metabolites. Int J Plant Sci. 2003;164(S3):S93–102.
134. Figueiredo AC, Barroso JG, Pedro LG, Scheffer JJC. Factors affecting secondary metabolite production in plants: volatile components and essential oils. Flavour Fragr J. 2008;23 (4):213–26.
135. Taiz L, Zeiger E. Plant physiology. 5th ed. Sunderland: Sinauer Associates; 2010. p. 782.
136. Zuzarte M, Salgueiro L. Essential oils chemistry. In: Sousa DP, editor. Bioactive essential oils and cancer. Cham: Springer International; 2015. p. 19–61.
137. Barbieri C, Borsotto P. Essential oils: market and legislation. In: El-Shemy H, editor. Potential of essential oils. Rijeka: InTech; 2018. p. 107–27.
138. Elshafie HS, Camele I. An overview of the biological effects of some Mediterranean essential oils on human health. Biomed Res Int. 2017;2017:9268468.
139. Baser K. Handbook of essential oils: science, technology, and applications. New York: CRC; 2010. p. 1112.
140. Cavaleiro C, Salgueiro L, Gonçalves MJ, Hrimpeng K, Pinto J, Pinto E. Antifungal activity of the essential oil of *Angelica major* against *Candida*, *Cryptococcus*, *Aspergillus* and dermatophyte species. J Nat Med. 2015;69(2):241–8.
141. Maxia A, Falconieri D, Piras A, Porcedda S, Marongiu B, Frau MA, et al. Chemical composition and antifungal activity of essential oils and supercritical CO_2 extracts of *Apium nodiflorum* (L.) Lag. Mycopathologia. 2012;174(1):61–7.
142. Abu-Darwish MS, Cabral C, Gonçalves MJ, Cavaleiro C, Cruz MT, Zulfiqar A, et al. Chemical composition and biological activities of *Artemisia judaica* essential oil from southern desert of Jordan. J Ethnopharmacol. 2016;191:161–8.
143. López S, Lima B, Aragón L, Espinar LA, Tapia A, Zacchino S, et al. Essential oil of *Azorella cryptantha* collected in two different locations from San Juan Province, Argentina: chemical variability and anti-insect and antimicrobial activities. Chem Biodivers. 2012;9(8):1452–64.
144. Prasad CS, Shukla R, Kumar A, Dubey NK. *In vitro* and *in vivo* antifungal activity of essential oils of *Cymbopogon martini* and *Chenopodium ambrosioides* and their synergism against dermatophytes. Mycoses. 2010;53(2):123–9.
145. Mahboubi M, Mohammad Taghizadeh Kashani L. The anti-dermatophyte activity of *Commiphora molmol*. Pharm Biol. 2016;54(4):720–5.
146. Romagnoli C, Andreotti E, Maietti S, Mahendra R, Mares D. Antifungal activity of essential oil from fruits of Indian *Cuminum cyminum*. Pharm Biol. 2010;48(7):834–8.
147. Dias N, Dias MC, Cavaleiro C, Sousa MC, Lima N, Machado M. Oxygenated monoterpenes-rich volatile oils as potential antifungal agents for dermatophytes. Nat Prod Res. 2017;31 (4):460–4.

148. Alves-Silva JM, Zuzarte M, Gonçalves MJ, Cavaleiro C, Cruz MT, Cardoso SM, et al. New claims for wild carrot (*Daucus carota* subsp. carota) essential oil. Evid Based Complement Altern Med. 2016; 9045196.
149. Salehi B, Ayatollahi SA, Segura-Carretero A, Kobarfard F, Contreras MM, Faizi M, et al. Bioactive chemical compounds in *Eremurus persicus* (Joub. and Spach) Boiss. essential oil and their health implications. Cell Mol Biol. 2017;63(9):1–7.
150. Tolba H, Moghrani H, Benelmouffok A, Kellou D, Maachi R. Essential oil of Algerian *Eucalyptus citriodora*: chemical composition, antifungal activity. J Mycol Med. 2015;25(4): e128–33.
151. Baptista EB, Zimmermann-Franco DC, Lataliza AAB, Raposo NRB. Chemical composition and antifungal activity of essential oil from *Eucalyptus smithii* against dermatophytes. Rev Soc Bras Med Trop. 2015;48(6):746–52.
152. Pinto SML, Sandoval LVH, Vargas LY. *In vitro* susceptibility of *Microsporum* spp. and mammalian cells to *Eugenia caryophyllus* essential oil, eugenol and semisynthetic derivatives. Mycoses. 2019;62(1):41–50.
153. Cabral C, Miranda M, Gonçalves MJ, Cavaleiro C, Cruz MT, Salgueiro L. Assessment of safe bioactive doses of *Foeniculum vulgare* Mill. essential oil from Portugal. Nat Prod Res. 2017;31(22):2654–9.
154. Zeng H, Chen X, Liang J. *In vitro* antifungal activity and mechanism of essential oil from fennel (*Foeniculum vulgare* L.) on dermatophyte species. J Med Microbiol. 2015;64 (1):93–103.
155. Khoury M, El Beyrouthy M, Ouaini N, Eparvier V, Stien D. *Hirtellina lobelii* DC. essential oil, its constituents, its combination with antimicrobial drugs and its mode of action. Fitoterapia. 2019;133:130–6.
156. Policegoudra RS, Goswami S, Aradhya SM, Chatterjee S, Datta S, Sivaswamy R, et al. Bioactive constituents of *Homalomena aromatica* essential oil and its antifungal activity against dermatophytes and yeasts. J Mycol Med. 2012;22(1):83–7.
157. Cabral C, Francisco V, Cavaleiro C, Gonçalves MJ, Cruz MT, Sales F, et al. Essential oil of *Juniperus communis* subsp. *alpina* (Suter) Čelak needles: chemical composition, antifungal activity and cytotoxicity. Phytother Res. 2012;26(9):1352–7.
158. Zuzarte M, Vale-Silva L, Gonçalves MJ, Cavaleiro C, Vaz S, Canhoto J, et al. Antifungal activity of phenolic-rich *Lavandula multifida* L. essential oil. Eur J Clin Microbiol Infect Dis. 2012;31(7):1359–66.
159. Zuzarte M, Gonçalves M, Cavaleiro C, Canhoto J, Vale-Silva L, Silva M, et al. Chemical composition and antifungal activity of the essential oils of *Lavandula viridis* l'Hér. J Med Microbiol. 2011;60(5):612–8.
160. Costa DCMH, Vermelho AB, Almeida CA, De Souza Dias EP, Cedrola SML, De Fátima A-BM, et al. Inhibitory effect of linalool-rich essential oil from *Lippia alba* on the peptidase and keratinase activities of dermatophytes. J Enzyme Inhib Med Chem. 2014;29(1):12–7.
161. Rahman A, Al-Reza SM, Siddiqui SA, Chang T, Kang SC. Antifungal potential of essential oil and ethanol extracts of *Lonicera japonica* Thunb. against dermatophytes. EXCLI J. 2014;13:427–36.
162. Jamalian A, Shams-Ghahfarokhi M, Jaimand K, Pashootan N, Amani A, Razzaghi-Abyaneh M. Chemical composition and antifungal activity of *Matricaria recutita* flower essential oil against medically important dermatophytes and soil-borne pathogens. J Mycol Med. 2012;22 (4):308–15.
163. Bajpai VK, Yoon JI, Chul KS. Antioxidant and antidermatophytic activities of essential oil and extracts of *Metasequoia glyptostroboides* Miki ex Hu. Food Chem Toxicol. 2009;47 (6):1355–61.
164. Bouzabata A, Bazzali O, Cabral C, Gonçalves MJ, Cruz MT, Bighelli A, et al. New compounds, chemical composition, antifungal activity and cytotoxicity of the essential oil from *Myrtus nivellei* Batt. and Trab.; an endemic species of Central Sahara. J Ethnopharmacol. 2013;149(3):613–20.

165. Bajpai VK, Yoon JI, Kang SC. Antifungal potential of essential oil and various organic extracts of *Nandina domestica* Thunb. against skin infectious fungal pathogens. Appl Microbiol Biotechnol. 2009;83(6):1127–33.
166. Valente J, Zuzarte M, Gonçalves MJ, Lopes MC, Cavaleiro C, Salgueiro L, et al. Antifungal, antioxidant and anti-inflammatory activities of *Oenanthe crocata* L. essential oil. Food Chem Toxicol. 2013;62:349–54.
167. Guerra-Boone L, Alvarez-Román R, Salazar-Aranda R, Torres-Cirio A, Rivas-Galindo VM, de Torres NW, et al. Antimicrobial and antioxidant activities and chemical characterization of essential oils of *Thymus vulgaris*, *Rosmarinus officinalis*, and *Origanum majorana* from northeastern México. Pak J Pharm Sci. 2015;28(1):363–9.
168. Houël E, Rodrigues AMS, Jahn-Oyac A, Bessière JM, Eparvier V, Deharo E, et al. *In vitro* antidermatophytic activity of *Otacanthus azureus* (Linden) Ronse essential oil alone and in combination with azoles. J Appl Microbiol. 2014;116(2):288–94.
169. Fahed L, Stien D, Ouaini N, Eparvier V, El Beyrouthy M. Chemical diversity and antimicrobial activity of *Salvia multicaulis* Vahl essential oils. Chem Biodivers. 2016;13(5):591–5.
170. Danielli LJ, Pippi B, Duarte JA, Maciel AJ, Lopes W, Machado MM, et al. Antifungal mechanism of action of *Schinus lentiscifolius* Marchand essential oil and its synergistic effect *in vitro* with terbinafine and ciclopirox against dermatophytes. J Pharm Pharmacol. 2018;70 (9):1216–27.
171. Pinto E, Vale-Silva L, Cavaleiro C, Salgueiro L. Antifungal activity of the clove essential oil from *Syzygium aromaticum* on *Candida*, *Aspergillus* and dermatophyte species. J Med Microbiol. 2009;58(11):1454–62.
172. Pinto E, Gonçalves M-J, Cavaleiro C, Salgueiro L. Antifungal activity of *Thapsia villosa* essential oil against *Candida*, *Cryptococcus*, *Malassezia*, *Aspergillus* and dermatophyte species. Molecules. 2017;22(10):E1595–605.
173. Alves M, Gonçalves MJ, Zuzarte M, Alves-Silva JM, Cavaleiro C, Cruz MT, et al. Unveiling the antifungal potential of two Iberian thyme essential oils: effect on *C. albicans* germ tube and preformed biofilms. Front Pharmacol. 2019;10:446.
174. Vale-Silva LA, Gonçalves MJ, Cavaleiro C, Salgueiro L, Pinto E. Antifungal activity of the essential oil of *Thymus* x *viciosoi* against *Candida*, *Cryptococcus*, *Aspergillus* and dermatophyte species. Planta Med. 2010;76(9):882–8.
175. Marongiu B, Piras A, Porcedda S, Falconieri D, Goncalves MJ, Salgueiro L, et al. Extraction, separation and isolation of volatiles from *Vitex agnus-castus* L. (Verbenaceae) wild species of Sardinia, Italy, by supercritical CO_2. Nat Prod Res. 2010;24(6):569–79.
176. Cabral C, Gonçalves MJ, Cavaleiro C, Sales F, Boyom F, Salgueiro L. Composition and antifungal activity of the essential oil from cameroonian *Vitex rivularis* Gurke. Nat Prod Res. 2009;23(16):1478–84.
177. Mahboubi M, HeidaryTabar R, Mahdizadeh E. The anti-dermatophyte activity of *Zataria multiflora* essential oils. J Mycol Med. 2017;27(2):232–7.
178. Abu-Darwish MS, Cabral C, Gonçalves MJ, Cavaleiro C, Cruz MT, Paoli M, et al. *Ziziphora tenuior* L. essential oil from Dana Biosphere Reserve (Southern Jordan); chemical characterization and assessment of biological activities. J Ethnopharmacol. 2016;194:963–70.
179. Chaftar N, Girardot M, Labanowski J, Ghrairi T, Hani K, Frère J, et al. Comparative evaluation of the antimicrobial activity of 19 essential oils. Adv Exp Med Biol. 2016;901:1–15.
180. Nardoni S, Giovanelli S, Pistelli L, Mugnaini L, Profili G, Pisseri F, et al. *In vitro* activity of twenty commercially available, plant-derived essential oils against selected dermatophyte species. Nat Prod Commun. 2015;10(8):1473–8.
181. Nasir M, Tafess K, Abate D. Antimicrobial potential of the Ethiopian *Thymus schimperi* essential oil in comparison with others against certain fungal and bacterial species. BMC Complement Altern Med. 2015;15(1):260.
182. Sadhasivam S, Palanivel S, Ghosh S. Synergistic antimicrobial activity of *Boswellia serrata* Roxb. ex Colebr. (Burseraceae) essential oil with various azoles against pathogens associated with skin, scalp and nail infections. Lett Appl Microbiol. 2016;63(6):495–501.

183. Fahed L, Khoury M, Stien D, Ouaini N, Eparvier V, El Beyrouthy M. Essential oils composition and antimicrobial activity of six conifers harvested in Lebanon. Chem Biodivers. 2017;14 (2):e1600235.
184. Palmeira-de-Oliveira A, Salgueiro L, Palmeira-de-Oliveira R, Martinez-de-Oliveira J, Pina-Vaz C, Queiroz J, et al. Anti-*Candida* activity of essential oils. Mini reviews. Med Chem. 2009;9(11):1292–305.
185. Flores FC, Beck RCR, da Silva C. Essential oils for treatment for onychomycosis: a mini-review. Mycopathologia. 2016;181(1–2):9–15.
186. Shokri H. A review on the inhibitory potential of *Nigella sativa* against pathogenic and toxigenic fungi. Avicenna J Phytomed. 2016;6(1):21–33.
187. Tullio V, Nostro A, Mandras N, Dugo P, Banche G, Cannatelli MA, et al. Antifungal activity of essential oils against filamentous fungi determined by broth microdilution and vapour contact methods. J Appl Microbiol. 2007;102(6):1544–50.
188. Lima B, López S, Luna L, Aguero MB, Aragón L, Tapia A, et al. Essential oils of medicinal plants from the Central Andes of Argentina: chemical composition, antifungal, antibacterial and insect-repellent activities. Chem Biodivers. 2011;8:924–36.
189. Marongiu B, Maxia A, Piras A, Porcedda S, Tuveri E, Gonçalves MJ, et al. Isolation of *Crithmum maritimum* L. volatile oil by supercritical carbon dioxide extraction and biological assays. Nat Prod Res. 2007;21(13):1145–50.
190. Zuzarte M, Gonçalves MJ, Cavaleiro C, Dinis AM, Canhoto JM, Salgueiro LR. Chemical composition and antifungal activity of the essential oils of *Lavandula pedunculata* (Miller) Cav. Chem Biodivers. 2009;6(8):1283–92.
191. Maxia A, Marongiu B, Piras A, Porcedda S, Tuveri E, Gonçalves MJ, et al. Chemical characterization and biological activity of essential oils from *Daucus carota* L. subsp. *carota* growing wild on the Mediterranean coast and on the Atlantic coast. Fitoterapia. 2009;80 (1):57–61.
192. Piras A, Marzouki H, Maxia A, Marengo A, Porcedda S, Falconieri D, et al. Chemical characterisation and biological activity of leaf essential oils obtained from *Pistacia terebinthus* growing wild in Tunisia and Sardinia Island. Nat Prod Res. 2017;31(22):2684–9.
193. Mahboubi M, Kazempour N. The antifungal activity of *Artemisia sieberi* essential oil from different localities of Iran against dermatophyte fungi. J Mycol Med. 2015;25(2):e65–71.
194. Fekrazad R, Poorsattar Bejeh Mir A, Ghasemi Barghi V, Shams-Ghahfarokhi M. Eradication of *C. albicans* and *T. rubrum* with photoactivated indocyanine green, *Citrus aurantifolia* essential oil and fluconazole. Photodiagnosis Photodyn Ther. 2015;12(2):289–97.
195. De Oliveira PF, Mendes JM, Lima IO, De Lira Mota KS, De Oliveira WA, De Oliveira Lima E. Antifungal activity of geraniol and citronellol, two monoterpenes alcohols, against *Trichophyton rubrum* involves inhibition of ergosterol biosynthesis. Pharm Biol. 2015;53 (2):228–34.
196. Romagnoli C, Baldisserotto A, Malisardi G, Vicentini CB, Mares D, Andreotti E, et al. A multi-target approach toward the development of novel candidates for antidermatophytic activity: ultrastructural evidence on α-bisabolol-treated *Microsporum gypseum*. Molecules. 2015;20(7):11765–76.
197. Miron D, Battisti F, Silva FK, Lana AD, Pippi B, Casanova B, et al. Antifungal activity and mechanism of action of monoterpenes against dermatophytes and yeasts. Rev Bras Farm. 2014;24(6):660–7.
198. Brasch J, Horter F, Fritsch D, Beck-Jendroschek V, Tröger A, Francke W. Acyclic sesquiterpenes released by *Candida albicans* inhibit growth of dermatophytes. Med Mycol. 2014;52(1):46–55.
199. Carrasco H, Raimondi M, Svetaz L, Di Liberto M, Rodriguez MV, Espinoza L, et al. Antifungal activity of eugenol analogues. Influence of different substituents and studies on mechanism of action. Molecules. 2012;17(1):1002–24.
200. Singh G, Kumar P, Joshi SC. Treatment of dermatophytosis by a new antifungal agent "apigenin.". Mycoses. 2014;57(8):497–506.

201. Nyong EE, Odeniyi MA, Moody JO. *In vitro* and *in vivo* antimicrobial evaluation of alkaloidal extracts of *Enantia chlorantha* stem bark and their formulated ointments. Acta Pol Pharm Drug Res. 2015;72(1):147–52.
202. Xiao CW, Ji QA, Wei Q, Liu Y, Bao GL. Antifungal activity of berberine hydrochloride and palmatine hydrochloride against *Microsporum canis*—induced dermatitis in rabbits and underlying mechanism. BMC Complement Altern Med. 2015;15(1):177.
203. Ayatollahi MSA, Kazemi A. *In vitro* and *in vivo* antidermatophytic activities of some Iranian medicinal plants. Med Mycol. 2015;53(8):852–9.
204. Njateng GSS, Gatsing D, Mouokeu RS, Lunga PK, Kuiate JR. *In vitro* and *in vivo* antidermatophytic activity of the dichloromethane-methanol (1:1 v/v) extract from the stem bark of *Polyscias fulva* Hiern (Araliaceae). BMC Complement Altern Med. 2013;13:95.
205. Lee S-J, Han J-I, Lee G-S, Park M-J, Choi I-G, Na K-J, et al. Antifungal effect of eugenol and nerolidol against *Microsporum gypseum* in a guinea pig model. Biol Pharm Bull. 2007;30 (1):184–8.
206. Mugnaini L, Nardoni S, Pinto L, Pistelli L, Leonardi M, Pisseri F, et al. *In vitro* and *in vivo* antifungal activity of some essential oils against feline isolates of *Microsporum canis*. J Mycol Med. 2012;22(2):179–84.
207. Wagini NH. *In vitro* and *in vivo* anti dermatophytes activity of *Lawsonia Inermis* L. (Henna) leaves against ringworm and its etiological agents. Am J Clin Exp Med. 2014;2(3):51–8.
208. Moghimipour E, Aghel N, Mahmoudabadi AZ, Ramezani Z, Handali S. Preparation and characterization of liposomes containing essential oil of *Eucalyptus camaldulensis* Leaf. Jundishapur J Nat Pharm Prod. 2012;7(3):117–22.
209. Pisseri F, Bertoli A, Nardoni S, Pinto L, Pistelli L, Guidi G, et al. Antifungal activity of tea tree oil from *Melaleuca alternifolia* against *Trichophyton equinum*: an *in vivo* assay. Phytomedicine. 2009;16(11):1056–8.
210. Romero-Cerecero O, Román-Ramos R, Zamilpa A, Jiménez-Ferrer JE, Rojas-Bribiesca G, Tortoriello J. Clinical trial to compare the effectiveness of two concentrations of the *Ageratina pichinchensis* extract in the topical treatment of onychomycosis. J Ethnopharmacol. 2009;126 (1):74–8.
211. Bhadauria S, Kumar P. Broad spectrum antidermatophytic drug for the control of tinea infection in human beings. Mycoses. 2012;55(4):339–43.
212. Bindra R, Singh A, Shawl A, Kumar S. Antifungal herbal formulation for treatment of human nails fungus and process thereof. US Patent No 6,296,838, 2001.
213. Veiga FF, Gadelha MC, da Silva MRT, Costa MI, Kischkel B, de Castro-Hoshino LV, et al. Propolis extract for onychomycosis topical treatment: from bench to clinic. Front Microbiol. 2018;9:779.
214. Buck DS, Nidorf DM, Addino JG. Comparison of two topical preparations for the treatment of onychomycosis: *Melaleuca alternifolia* (tea tree) oil and clotrimazole. J Fam Pract. 1994;38 (6):601–5.
215. Syed TA, Qureshi ZA, Ali SM, Ahmad S, Ahmad SA. Treatment of toenail onychomycosis with 2% butenafine and 5% *Melaleuca alternifolia* (tea tree) oil in cream. Trop Med Int Health. 1999;4(4):284–7.
216. Satchell AC, Saurajen A, Bell C, Barnetson RSC. Treatment of interdigital tinea pedis with 25% and 50% tea tree oil solution: a randomized, placebo-controlled, blinded study. Australas J Dermatol. 2002;43(3):175–8.
217. Derby R, Rohal P, Jackson C, Beutler A, Olsen C. Novel treatment of onychomycosis using over-the-counter mentholated ointment: a clinical case series. J Am Board Fam Med. 2011;24 (1):69–74.
218. Pandey KP, Mishra RK, Kamran A, Mishra P, Bajaj AK, Dikshit A. Studies on antidermatophytic activity of waste leaves of *Curcuma longa* L. Physiol Mol Biol Plants. 2010;16(2):177–85.

219. Freile M, Giannini F, Sortino M, Zamora M, Juárez A, Zacchino S, et al. Antifungal activity of aqueous extracts and of berberine isolated from *Berberis heterophylla.* Acta Farm Bonaer. 2006;25(1):83–8.
220. Jagetia GC, Venkatesh P, Baliga MS. Evaluation of the radioprotective effect of bael leaf (*Aegle marmelos*) extract in mice. Int J Radiat Biol. 2004;80(4):281–90.
221. Balakumar S, Rajan S, Thirunalasundari T, Jeeva S. Antifungal activity of *Aegle marmelos* (L.) Correa (Rutaceae) leaf extract on dermatophytes. Asian Pac J Trop Biomed. 2011;1 (4):309–12.
222. Li W, Zjou J, Xu Y. Study of the *in vitro* cytotoxicity testing of medical devices. Biomed Rep. 2015;3(5):617–20.

GPSR Compliance

The European Union's (EU) General Product Safety Regulation (GPSR) is a set of rules that requires consumer products to be safe and our obligations to ensure this.

If you have any concerns about our products, you can contact us on ProductSafety@springernature.com

In case Publisher is established outside the EU, the EU authorized representative is:

Springer Nature Customer Service Center GmbH
Europaplatz 3
69115 Heidelberg, Germany

Batch number: 10683281

Printed by Printforce, the Netherlands